D1753130

Encyclopedia of Electrochemistry

Edited by A.J. Bard and M. Stratmann

Volume 1
Thermodynamics and
Electrified Interfaces

Volume Edited by E. Gileadi and M. Urbakh

Encyclopedia of Electrochemistry

Editors-in-Chief: Allen J. Bard, Martin Stratmann

Volume 1: Thermodynamics and Electrified Interfaces (Editors: Eliezer Gileadi, Michael Urbakh)
Volume 2: Interfacial Kinetics and Mass Transport (Editor: Ernesto Julio Calvo)
Volume 3: Instrumentation and Electroanalytical Chemistry (Editor: Pat Unwin)
Volume 4: Corrosion and Oxide Films (Editors: Martin Stratmann, Gerald S. Frankel)
Volume 5: Electrochemical Engineering (Editor: Digby D. Macdonald)
Volume 6: Semiconductor Electrodes and Photoelectrochemistry (Editor: Stuart Licht)
Volume 7: Inorganic Electrochemistry (Editors: William E. Geiger, Chris Pickett)
Volume 8: Organic Electrochemistry (Editor: Hans-J. Schäfer)
Volume 9: Bioelectrochemistry (Editor: George S. Wilson)
Volume 10: Modified Electrodes (Editors: Israel Rubinstein, Masamichi Fujihira)
Volume 11: Index

Encyclopedia of Electrochemistry

Edited by A.J. Bard and M. Stratmann

Volume 1
Thermodynamics and
Electrified Interfaces

Volume Edited by E. Gileadi and M. Urbakh

WILEY-VCH

Editors:

Prof. Dr. Allen J. Bard
Department of Chemistry
University of Texas
Austin, TX 78712
USA

Prof. Dr. Martin Stratmann
Max-Planck-Institut
für Eisenforschung
Max-Planck-Str. 1
40237 Düsseldorf
Germany

Prof. Eliezer Gileadi
Prof. Michael Urbakh
School of Chemistry
Tel Aviv University
Tel Aviv 69978
Israel

COVER: Courtesy of Prof. Dr. D. M. Kolb, University of Ulm, Germany

■ This book was careful produced nevertheless, authors, editors, and publisher do not warrant the information contained therein to be free of errors. Readers are advised to keep in mind that statements, data illustrations, procedural details or other items may inadvertently be inaccurate.

Library of Congress Card No: applied for
British Library Cataloguing-in-Publication Data. A catalogue record for this book is available from the British Library.

Die Deutsche Bibliothek-CIP-Cataloguing-in-Publication Data. A catalogue record for this publication is available from Die Deutsche Bibliothek

© WILEY-VCH Verlag GmbH & Co.KGaA, Weinheim, 2002
All rights reserved (including those of translation into other languages). No part of this book may be reproduced in any form – nor transmitted or translated into machine language without written permission from the publishers. Registered names, trademark, etc. used in this book, even when not specifically marked as such are not to be considered unprotected by law.

Printed in the Federal Republic of Germany
Printed on acid-free paper.

Composition: Laserwords Private Ltd, Chennai, India
Printing: betz-druck GmbH, Darmstadt
Bookbinding: J. Schäffer GmbH + Co.KG, Grünstadt
ISBN 3-527-30393-6

Preface

Electrochemistry plays a dominant role in a vast number of research and applied areas. Electrochemical reactions can overcome kinetic limitations, even at very low temperatures, by application of a potential and they are chemically and stereochemically specific, leading to chemical synthesis applications.

Electrochemical analyses are based on the high sensitivity of the reactions. The excellent spatial and temporal control is of importance in the emerging field of nanotechnology. Finally electrochemical reactions are known for a wide range of materials like metals, semiconductors, polymers, and biological systems leading to a large role in a number of rather diverse areas, such as preparative chemistry, analytical chemistry, energy storage, energy conversion, biochemistry, solid state chemistry, materials science, and microelectronics.

While studies of the transport of charged species and thermodynamic considerations dominated early studies, kinetic aspects of electrochemistry have become more important in electrochemical research with an increased understanding of the chemical and electronic structure of the solid/solution interface (accelerated by the application of numerous in situ and ex situ spectroscopic techniques, combined with electrochemical experiments). More recently the introduction of in situ scanning probe techniques has allowed one to follow electrochemical reactions on an atomic or molecular scale.

However, the most important development over the last few decades has been the spread of electrochemical concepts into very different areas of research and development. Based on research by electrochemists, electrochemistry is now used in many fundamental fields, such as the study of new organic and inorganic compounds and biological systems. In more applied areas, it is used to shape materials from the macroscopic to the microscopic scale, to analyze for chemical impurities, to understand and prevent the corrosion of materials, to probe the functioning of living cells, and to convert chemical energy into electricity. Scientists and engineers working in diverse areas need to locate and use electrochemical information.

It is the aim of this encyclopedia to fill this need by providing an up-to-date electrochemical source for engineers and scientists, as well as for students needing a starting point in their search for reliable information. The *Encyclopedia of Electrochemistry* is organized into 10 volumes plus 1 index volume, which concentrate on the major areas of electrochemistry, with each volume containing some 20 articles by experts. These are intended not only for electrochemists, but also for those in other fields who use electrochemistry. The *Encyclopedia of Electrochemistry* also includes the latest reviews of recent achievements in fundamental and applied electrochemistry. The extensive cross-referencing between volumes allows users to deepen their knowledge to the extent defined by their needs.

We hope that the *Encyclopedia of Electrochemistry* facilitates the broad use of electrochemistry in academia and industry but also provides a basis for further progress in this highly interesting and fast moving area of science.

December 2001

Allen J. Bard
Martin Stratmann

About the Editors

Allen J. Bard

Born December 18, 1933, Professor Bard received his early education in the public schools of New York City and attended The City College of New York (B.Sc., summa cum laude, 1955). He did his graduate work at Harvard University with J. J. Lingane (MA, 1956; PhD, 1958) in electroanalytical chemistry. In 1958 he joined the faculty of The University of Texas at Austin where he currently holds the Norman Hackerman/Welch Regents' Chair in Chemistry. His research interests have been in the application of electrochemical methods to the study of chemical problems and include investigations in electro-organic chemistry, photoelectrochemistry, electrogenerated chemiluminescence, and electroanalytical chemistry. He has published three books (*Electrochemical Methods*, with Larry Faulkner, *Integrated Chemical Systems*, and *Chemical Equilibrium*) and over 600 papers and chapters while editing the series *Electroanalytical Chemistry* (21 volumes) and the *Encyclopedia of the Electrochemistry of the Elements* (16 volumes) plus co-editing the monograph, *Standard Potentials in Aqueous Solution*. He is currently editor-in-chief of the *Journal of the American Chemical Society*. The ISI listing of the "50 most cited chemists from 1981–1997" ranks Professor Bard at number 13 (taken from a total of 627,871 chemists surveyed).

Martin Stratmann

Born 20 April 1954, Professor Stratmann studied chemistry at the Ruhr-Universität Bochum and received his diploma in 1980. He finished his PhD in 1982 at the Max Planck Institut für Eisenforschung in Düsseldorf with H.J. Engell on electrochemical studies of phase transformations in rust layers and spent his postdoctoral education with Ernest Yeager at the Case Western Reserve University. His professorship in physical chemistry followed in 1992 at the University of Düsseldorf with electrochemical studies on metal surfaces covered with ultrathin electrolyte layers. In 1994 he took over the Chair in Corrosion Science and Surface Engineering at the University of Erlangen and since 2000 has been a scientific member of the Max Planck Gesellschaft and director at the Max Planck Institut für Eisenforschung in Düsseldorf, heading a department of interface chemistry and surface engineering. Further he is a faculty member of the departments of Materials Science and Chemistry at the Ruhr-Universität Bochum. His research interests concentrate on corrosion related electrochemistry, in particular with emphasis on microscopic aspects and in situ spectroscopy, electrochemistry at buried metal/polymer interfaces – an area where he pioneered novel electrochemical techniques – atmospheric corrosion, adhesion and surface chemistry of reactive metal substrates. He has published more than 150 papers and is co-editor of *Steel Research and Materials and Corrosion*.

Eliezer Gileadi

Born in Hungary in 1932, Professor Gileadi gained his MSc at the Hebrew University, Jerusalem, and his PhD at the University of Ottawa, Canada in 1963. He was Senior Res. Assoc. at the University of Philadelphia from 1963–1966, before joining Tel-Aviv University, where he is now professor. He has been Director of the Gordon Centre for Energy Studies, a member of the National Council for High Education and on the Editorial Board of the *Journal of Electroanalytical Chemistry*. Professor Gileadi has been invited to visit or consult a wide number of institutions, including the Israel Aviation Industry, the universities of NY, Virginia, Belgrade, Ottawa, and Osaka, Brookhaven Nat. Labs, Lockheed Missiles and Space, Case Western Reserve University, Los Alamos National Laboratories, The John Hopkins University, CNRS, the TU-Dresden, and RMIT in Melbourne. He has written 135 scientific papers in journals, five chapters in books, one monograph and two textbooks, and is a Fellow of the American Electrochemical Society and of the American Association for the Advancement of Science.

Michael Urbakh Professor Urbakh was born in Moscow on 18 March 1951 and gained his MSc from Moscow State University in 1973, and his PhD from Moscow Physical University 5 years later. The title of his thesis was "The Influence of Adsorption on Surface Electronic Properties". Following a period as a research fellow at the Academy of Sciences of the USSR, in Moscow, Michael Urbakh joined Tel Aviv University in 1990, becoming a Full Professor in 1997. He is the holder of many academic and professional awards and has over 20 publications to his name. He was a visiting professor in both Germany and France and has been on several organizing committees. His major research interests are theoretical electrochemistry; nanotribology; molecular motors; dynamical properties of liquids in spatial restrictions, frictional properties of nanoscale liquid films; electrodynamics of metal surfaces with account of microscopic effects, and the interpretation of data on electrochemical optical spectroscopy of metals.

Contents

1 Electrode Potentials *1*
  Oleg A. Petrii, Galina A. Tsirlina

2 Electrochemical Double Layers *27*

2.1 Electrochemical Interfaces: At the Border Line *33*
  Alexei A. Kornyshev, Eckhard Spohr, Michael A. Vorotyntsev

2.2 Electrical Double Layers: Theory and Simulations *133*
  Wolfgang Schmickler

2.3 Electrochemical Double Layers: Liquid–Liquid Interfaces *162*
  Alexander G. Volkov, Vladislav S. Markin

2.4 Electrical Double Layers. Double Layers at Single-crystal and Polycrystalline Electrodes *188*
  Enn Lust

2.5 Analyzing Electric Double Layers with the Atomic Force Microscope *225*
  Hans-Jürgen Butt

2.6 Comparison Between Liquid State and Solid State Electrochemistry *253*
  Ilan Riess

2.7 Polyelectrolytes in Solution and at Surfaces *282*
  Roland R. Netz, David Andelman

3 Specific Adsorption *323*

3.1 Introduction *327*
  Boris B. Damaskin, Oleg A. Petrii

3.2 State of Art: Present Knowledge and Understanding *349*
  G. Horányi

3.3 Phase Transitions in Two-dimensional Adlayers at Electrode Surfaces: Thermodynamics, Kinetics, and Structural Aspects *383*
  Thomas Wandlowski

4 Underpotential Deposition 469

4.1 Atomically Controlled Electrochemical Deposition and Dissolution of Noble Metals 471
Shen Ye, Kohei Uosaki

4.2 Electrodeposition of Compound Semiconductors by Electrochemical Atomic Layer Epitaxy (EC-ALE) 513
John L. Stickney, Travis L. Wade, Billy H. Flowers, Raman Vaidyanathan, Uwe Happek

4.3 Electrocatalysis on Surfaces Modified by Metal Monolayers Deposited at Underpotentials 561
Radoslav Adžić

Volume Index 601

1
Electrode potentials

Oleg A. Petrii, Galina A. Tsirlina
Department of Electrochemistry, Moscow State University, Moscow, Russia

1.1	Introductory Remarks	3
1.2	External and Internal Potentials	4
1.3	Electromotive Force (emf) and Gibbs Energy of Reaction	5
1.4	Diffusion Potential Drop	9
1.5	Classification of Electrodes	10
1.6	Membrane Electrodes	12
1.7	Standard Potentials	12
1.8	Specific Features of Certain Reference Electrodes	13
1.9	Values of Redox Potentials	15
1.10	Potential Windows	16
1.11	Experimental Techniques	16
1.12	Effects of Ion Pairing	17
1.13	Absolute Electrode Potential	18
1.14	The Role of Electric Double Layers	20
1.15	Conclusion	23
	References	23

1.1
Introductory Remarks

The term "electrode potential" combines two basic notions: "electrode" and "potential". The electrical potential φ, as known from the physical definition, represents the electrical energy (work term), which is necessary for transferring a unit test charge from infinity in vacuum into the phase under consideration. We call this charge as "probe" (sometimes also "test" or "imaginary") in order to emphasize that it is affected only by the external field and does not interact with the medium via non-Coulombic forces, and, moreover, is sufficiently small to induce any charge redistribution inside the phase. In actual practice, the charges exist only in a combination with certain species (elementary particles, particularly, electrons, and ions). Hence, the value of φ appears to be beyond the reach of experimental determination, a fact that poses the problems concerned with interpretation of the electrode potential and brings to existence numerous potential scales. A number of discussions of various levels can be found in the literature [1–20].

A system that contains two (or more) contacting phases, which includes at least one electronic and one ionic conductor, is usually considered as an electrode in electrochemistry. Electrodes can be based on metals, alloys, any type of semiconductors (namely, oxides and salts with electronic or mixed conduction, electron conducting polymers, various covalent compounds of metals), and also on a variety of composite materials. The corresponding ionic conductors are usually electrolyte solutions or melts, solid electrolytes (particularly, amorphous and polymeric materials), supercritical fluids, and various quasiliquid systems. In certain cases, the terms "electrode" and "electrode potential" are applied also to semipermeable membranes – the systems separating two solutions of different composition (although these "membrane electrodes" do not exactly satisfy the definition given above).

Real electrodes sometimes represent extremely complicated systems, which can include several interfaces, each locating a certain potential drop. These interfaces, together with the phases in contact, take part in the equilibria established in the electrochemical system. This is why the electrode potential is one of the most important notions of electrochemical thermodynamics (to say nothing about its role in electrochemical kinetics), which describes the equilibrium phenomena in the systems containing charged components. The term "component" as applied to the charged species does not mean that the ions are considered as the components amenable to Gibbs' phase rule.

Consideration of equilibria in the systems of this sort requires the application of the notion of electrochemical potential introduced by Guggengheim [21] on the basis of the relationship for the electrochemical Gibbs free energy $\bar{G}$:

$$d\bar{G} = -S\,dT + V\,dp + \sum_i \mu_i\,dN_i + F \sum_i z_i \varphi\,dN_i \quad (1)$$

where S is the entropy, V is the volume of the system, T and p are its temperature and pressure, respectively, μ_i is the chemical potential of the ith species, N_i is the number of moles of the ith component, z_i is the charge of the ith species taking into consideration its sign, and φ is the electrical potential in the location point of the ith species, which is also called the *internal* (or *inner*) potential of the corresponding phase [22–26].

According to Eq. (1), the electrochemical potentials of the components, $\bar{\mu}_i$, are characteristic values for the ith species in any phase α under consideration:

$$\bar{\mu}_i^\alpha = \mu_i^\alpha + z_i F \varphi^\alpha = \left(\frac{\partial \bar{G}}{\partial N_i}\right)_{p,T,N_{j\neq i}} \quad (2)$$

Only these values or the differences of electrochemical potentials referred to phases α and β

$$\bar{\mu}_i^\beta - \bar{\mu}_i^\alpha = (\mu_i^\beta - \mu_i^\alpha) + z_i F(\varphi^\beta - \varphi^\alpha) \quad (3)$$

are experimentally available.

1.2
External and Internal Potentials

The electrical state of any phase α can be characterized by its internal potential, which is a sum of the external (or outer) potential ψ^α induced by free electrostatic charges of the phase and the surface potential χ^α [6, 23–26]:

$$\varphi^\alpha = \psi^\alpha + \chi^\alpha \quad (4)$$

When the free electrostatic charge in phase α turns to zero, $\psi^\alpha = 0$ and $\varphi^\alpha = \chi^\alpha$. The surface potential of a liquid phase is dictated by a certain interfacial orientation of solvent dipoles and other molecules with inherent and induced dipole moments, and also of ions and surface-active solute molecules. For solid phases, it is associated with the electronic gas, which expands beyond the lattice (and also causes the formation of a dipolar layer); other reasons are also possible.

A conclusion of fundamental importance, which follows from Eq. (3), states that the electrical potential drop can be measured only between the points, which find themselves in the phases of one and the same chemical composition. This conclusion was first formulated by Gibbs. Indeed, in this case, $\mu_i^\beta = \mu_i^\alpha$ and $\Delta_\alpha^\beta \varphi = \varphi^\beta - \varphi^\alpha = \dfrac{\bar{\mu}_i^\beta - \bar{\mu}_i^\alpha}{z_i F}$. Otherwise, when the points belong to two different phases, the experimental determination of the potential drop becomes impossible. Furthermore, these arguments imply that the difference between the internal potentials at the interface, $\Delta_\alpha^\beta \varphi$, cannot be measured. The quantity $\Delta_\alpha^\beta \varphi$ is called *Galvani potential* and determines the electrostatic component of the work term corresponding to the transfer across the interface, whereas the "chemical" component is determined by the difference between the chemical potentials. In the literature, the term Galvani potential is also applied to the separate value of internal potential. However, it is not significant, because of the relative nature of this value, which is discussed below.

The electrochemical equilibrium requires the equality of electrochemical potentials for all components in the contacting phases. From this condition of electrochemical equilibrium, the dependence of Galvani potential on the activities of potential-determining ions can be derived, which represents the Nernst equation [27] for a separate Galvani potential. That is, for an interface formed by a metal (M) and solution (S) containing the ions of this metal M^{z+},

$$\Delta_S^M \varphi = \varphi^M - \varphi^S = \text{const} + \frac{RT}{zF} \ln a_{M^{z+}} \quad (5)$$

where φ^M and φ^S are the internal potentials of metal and solution, respectively, and $a_{M^{z+}}$ is the activity of metal ions in solution.

The impossibility to measure a separate Galvani potential rules out the possibility of establishing a concentration corresponding to $\Delta_S^M \varphi = 0$.

A formal consideration of Eq. (5) predicts the infinitely large limiting value of the Galvani potential ($\Delta_S^M \varphi \to -\infty$) when the activity of metal ions approaches zero (a hypothetical solution containing no ions of this sort, or pure solvent). However, this is not the case, because the limit of thermodynamic stability of the solvent will be exceeded. For an ideal thermodynamically stable (hypothetical) solution, this uncertainty can be clarified, if the electronic equilibria between the metal and the solvent are taken into account, which requires the consideration of solvated electrons ($\bar{e}_s$) [7–11]. The latter are real species present in any solution [27]; in aqueous and certain other media, their equilibrium concentration is extremely low, which, however, does not prevent us from formally using the thermodynamic consideration of the electrode potential in the framework of the concept of electronic equilibria. However, in real systems the exchange current density for M^{z+}/M pair appears to be so small in the limit under consideration that equilibrium cannot be maintained, and Eq. (5) no longer applies.

The equilibrium with participation of solvated electrons can be expressed by a scheme

$$M \text{ (solid)} \Longleftrightarrow M \text{ (in solution)}$$
$$\Longleftrightarrow M^{z+} + z\bar{e}_s \quad (6)$$

The dissolved metal atoms can be considered as an electrolyte, which dissociates producing M^{z+} cations and the simplest anions, $\bar{e}_s$. Actually, the condition $a_{M^{z+}} = 0$ cannot be achieved, because a finite value for the "solubility product" of metal exists, and, simultaneously, the condition of electroneutrality is valid. By using the condition of the equality of electrochemical potentials of electrons in the solution and in the metal phases, we obtain

$$\Delta_S^M \varphi = \frac{\mu_{\bar{e}}^M - \mu_{\bar{e}}^S}{F} = \text{const} - \frac{RT}{F} \ln a_{\bar{e}} \quad (7)$$

where $\mu_{\bar{e}}^M$ and $\mu_{\bar{e}}^S$ are the chemical potentials of electrons in metal and solution, respectively.

1.3
Electromotive Force (emf) and Gibbs Energy of Reaction

To study the electrical properties of the M|S interface presented schematically in Fig. 1,

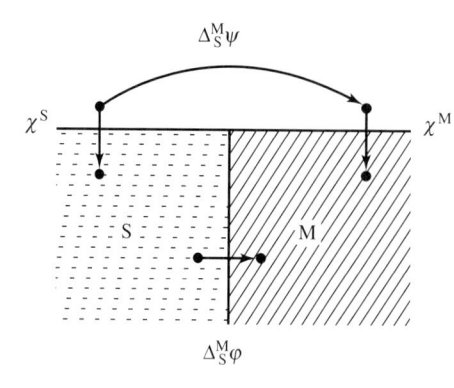

Fig. 1 A scheme of potential drops at metal|solution interface: $\Delta_S^M \varphi$ – metal/solution Galvani potential; χ^S and χ^M the solution/vacuum and metal/vacuum surface potentials, respectively; $\Delta_S^M \psi$ – metal/solution Volta potential.

it is necessary to construct a correctly connected circuit (Fig. 2) – a galvanic cell satisfying the condition of identical metal contacts at both terminals. For the system under discussion, it is a circuit of the following type:

$$M_1|S|M|M_1' \tag{8}$$

in which the electronic conductors M_1' and M_1 consist of the same material, differing only in their electric states (all the interfaces are assumed to be in equilibrium). It is assumed that both terminals of voltmeter used for measurement of the potential consist of the same metal. It is evident that the measuring device with terminals also made of M_1 records the total difference of electric potentials, or emf,

$$E = \Delta_{M_1}^{M_1'} \varphi = \Delta_M^{M_1'} \varphi + \Delta_{M_1}^S \varphi + \Delta_S^M \varphi \tag{9}$$

which corresponds to the route of probe-charge transfer $1 \Rightarrow 2 \Rightarrow 3$ and consists of three Galvani potentials. On the other hand, when going from M_1 to M_1' via the route marked by points $1' \Rightarrow 2' \Rightarrow 3' \Rightarrow 4'$ (Fig. 2), we obtain

$$E = \Delta_{M_1}^{M_1'} \varphi = \Delta_S^M \psi + \Delta_{M_1}^S \psi + \Delta_M^{M_1'} \psi \tag{10}$$

Hence, the potential discussed can be expressed by three values, which are called *Volta potentials* (in some cases, contact potentials). They represent the potential differences between the points just outside the phases M and S in vacuum, and are measurable quantities. The distances from

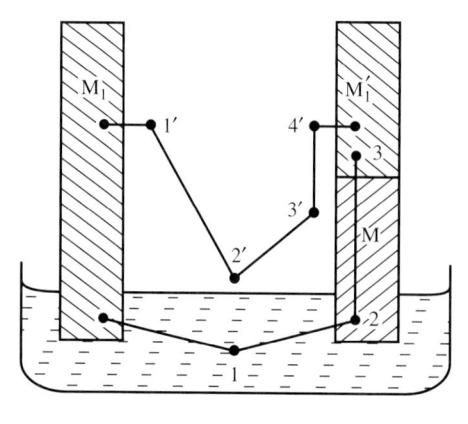

Fig. 2 Illustration of two possible routes of probe-charge transfer in a correctly connected circuit.

these points to the interface should be sufficiently large, as compared with the characteristic distance of molecular and image forces (up to 10^{-4} cm), and, at the same time, not too high to prevent the weakening of interaction with the charges inside the phases. Volta potential is a measurable quantity, because the points between which it is measured are located in the same phase and exclusively in the field of long-range forces. This means that the Volta potential does not depend on the charge of the probe.

For the metal/metal boundary, these potential differences ($\Delta_M^{M_1'} \psi$) can be defined by the difference of work functions [22]. The latter can be obtained from the photoelectron emission (or thermoelectron emission) and also directly by using a circuit

$$M_1|M|vacuum|M_1$$

where M_1 is the reference metal. The main experimental problems are the surface pretreatment (purification) and the elimination of potential drop between the samples of M_1 and M in ultrahigh vacuum.

Volta potential for the interface metal (mercury)/solution can be found by using the following cell reference electrode|solution|inert gas|Hg|solution|reference electrode when the solution flows to the system through the internal walls of a vertical tube, where mercury flows out via a capillary placed axially in a vertical tube and is broken into drops. These mercury drops carry away the free charges, thus eliminating the potential difference in inert gas between Hg and the solution. If a plate of solid metal is used instead of liquid metal, it is necessary to eliminate the potential drop mentioned above. A similar technique can be used for the solution/solution interface. There are also other techniques of Volta potential determination [1–6, 28].

The interface $M_1'|M$ (circuit 1.8) which, in the simplest case, represents a boundary of two metals, is easy to construct. However, the $M_1|S$ contact, which includes an additional electrochemical interface, is rarely feasible, being usually unstable and dependent on the nature of M_1. Generally, in place of M_1, a special electrochemical system, the so-called reference electrode should be included into cell 1.8. The electrode potential can be determined as the emf E of a correctly connected electrical circuit formed by the electrified interface under discussion and a reference electrode. According to this definition, the potential of any reference electrode is assumed to be zero.

Thus, electrochemists deal with the values of potentials, which are, actually, the differences of potentials. The nature of the emf of a circuit for potential measurement depends on the type of the electrodes. If we ignore a less frequent situation of ideally polarizable electrodes, for the majority of systems, two half-reactions take place simultaneously in the cell. For ideally polarizable electrodes, their contribution to emf can be considered, in the first approximation, as the potential drop inside a capacitor formed by the electric double layer. Hence, this drop depends on the electrode free charge (see Sects. 1.3 and 1.5):

$$Ox_1 + n_1\bar{e} \Longleftrightarrow Red_1$$

(on the electrode under study), (11)

$$Ox_2 + n_2\bar{e} \Longleftrightarrow Red_2$$

(on the reference electrode). (12)

Taken together, Eqs. (11) and (12) correspond to the chemical reaction:

$$\nu_1 Red_1 + \nu_2 Ox_2 \Longleftrightarrow \nu_2 Red_2 + \nu_1 Ox_1 \quad (13)$$

the relationship for stoichiometric coefficients being

$$n_1 v_1 = n_2 v_2 = n \qquad (14)$$

where n is the number of electrons transferred across the electrochemical cell in a single act of reaction (13).

From Eq. (3), the emf of the cell, which corresponds to reaction (13), can be expressed as follows:

$$E = \frac{1}{nF}(v_1 \mu_{Red_1} + v_2 \mu_{Ox_2} - v_2 \mu_{Red_2} - v_1 \mu_{Ox_1}) \qquad (15)$$

The right-hand term in parentheses equals the Gibbs energy of reaction (13) with the opposite sign. Hence, the emf of an electrochemical cell (a cell with eliminated diffusion and thermoelectric potential drops is implied, see later) corresponds to the reaction Gibbs free energy:

$$E = -\frac{\Delta G}{nF} \qquad (16)$$

This statement is true also for more complicated reactions with the participation of N reactants and products, and can be generalized in the form of the Nernst equation:

$$E = E^0 - \frac{RT}{nF} \sum_{i=1}^{N} v_i \ln a_i \qquad (17)$$

where the stoichiometric numbers $v_i < 0$ for the reactants and $v_i > 0$ for the products, a_i are their activities, n is the total number of electrons transferred in the coupled electrochemical reactions, and E^0 is the standard potential, which is expressed by the standard Gibbs free energy ΔG^0 as

$$E^0 = -\frac{\Delta G^0}{nF} \qquad (18)$$

Under equilibrium conditions, the mean activities or the activities of neutral molecules (salts, acids) in Eq. (17) can be substituted for partial activities of ions, which cannot be measured experimentally.

The applications of various types of electrochemical cells to chemical thermodynamics are considered in Sect. 2.

Experimentally, the emf can be measured either by compensating the circuit voltage (classical technique which became rare nowadays) or by using a voltmeter of very high internal resistance. The accuracy of emf determination of about 1 µV can be achieved in precise measurements, whereas common devices provide the accuracy of about 1 mV. The potential unit named *Volt*, which is used in the modern literature (particularly, below), is the so-called absolute (abs) Volt; it differs slightly from the international (int) Volt value. The ratio abs/int is 1.00033. To determine the sign of emf, a conventional rule is adopted, which states that the left electrode should be considered as the reference one.

For carrying out the experimental measurements of electrode potentials, a system chosen as the reference electrode should be easy to fabricate, and also stable and reproducible. This means that any pair of reference electrodes of the same type fabricated in any laboratory should demonstrate stable zero potential difference within the limits of experimental error. Additionally, the potential differences between two reference electrodes of different type should remain constant for a long time.

Another point is that the transfer of electricity (although of very low quantity) occurs in the course of emf measurements. Thus, the reference electrode should comply with the requirement of nonpolarizability: when the currents (usually in the nanoampere range) flow across the system, the potential of the reference electrode should remain constant. One of the most important features that determines this requirement is the exchange current density i_0, which expresses the rates of direct and reverse processes under equilibrium conditions. This quantity determines the rate of establishment of equilibrium (the state of the electric double layer is established very rapidly as compared with the total equilibrium).

1.4 Diffusion Potential Drop

The use of reference electrodes frequently poses the problem of an additional potential drop between the electrolytes of the electrode under study and of the reference one. For liquid electrolytes, this drop arises at the solution/solution interface (liquid junction). The symbol : conventionally denotes an interface of two solutions with a diffusion potential drop in between; if this drop is eliminated (see later), then the symbol ∷ is used. In this case, the equilibrium is not exact because of the existence of a diffusion potential drop $\Delta\varphi_{\text{diff}}$. The latter has the meaning of Galvani potential and cannot be measured; however, it can be estimated by adopting a model approach to the concentration distribution of ions in the interfacial region, the models of Planck [29] and Henderson [30, 31] being the most conventional. The general expression for $\Delta\varphi_{\text{diff}}$ at the interface of liquid solutions (1) and (2) is as follows:

$$\Delta\varphi_{\text{diff}} = -\frac{RT}{F}\int_{(1)}^{(2)} \sum \frac{t_i}{z_i} d\ln a_i \quad (19)$$

where t_i is the transport number of the ith ion, that is, the portion of current transferred by this ion through the solution. In the first approximation, $\Delta\varphi_{\text{diff}}$ can be estimated by substituting corresponding concentrations for the partial activities.

In Planck's model for a sharp boundary, for a 1,1 electrolyte,

$$\Delta\varphi_{\text{diff}} = \frac{RT}{F}\ln\xi \quad (20)$$

where function ξ can be found from a transcendent equation in which "+" and "−" denote the sets, which consist of all cations and of all anions, respectively, and λ denotes the limiting conductivity of the corresponding ion:

$$\frac{\xi\sum_{+}\lambda_+^{(2)}c_+^{(2)} - \sum_{+}\lambda_+^{(1)}c_+^{(1)}}{\sum_{-}\lambda_-^{(2)}c_-^{(2)} - \xi\sum_{-}\lambda_-^{(1)}c_-^{(1)}}$$

$$= \frac{\ln\left(\frac{\sum_i c_i^{(2)}}{\sum_i c_i^{(1)}}\right) - \ln\xi}{\ln\left(\frac{\sum_i c_i^{(2)}}{\sum_i c_i^{(1)}}\right) + \ln\xi}$$

$$\times \frac{\xi\sum_i c_i^{(2)} - \sum_i c_i^{(1)}}{\sum_i c_i^{(2)} - \xi\sum_i c_i^{(1)}} \quad (21)$$

The Henderson equation that has gained wider acceptance can be written as follows for concentrations c having the units of normality:

$$\Delta\varphi_{\text{diff}} = \frac{RT}{F} \frac{\sum_i \left(\frac{\lambda_i}{z_i}\right)(c_i^{(2)} - c_i^{(1)})}{\sum_i \lambda_i(c_i^{(2)} - c_i^{(1)})}$$

$$\times \ln \frac{\sum_i \lambda_i c_i^{(1)}}{\sum_i \lambda_i c_i^{(2)}} \quad (22)$$

The solution is given for the case of a smeared out boundary and linear spatial distributions of concentrations.

Generally, Eqs. (20)–(22) yield similar results; however, for junctions with a pronounced difference in ion mobilities (such as HCl : LiCl), the deviation can reach about 10 mV. A specific feature of the Planck equation is the existence of two solutions, the first being close to that of Henderson, and the second one being independent of the solution concentration and of no physical meaning [32].

In practice, in place of model calculations and corresponding corrections, the elimination of the diffusion potential is conventionally applied. This is achieved by introducing the so-called *salt bridges* filled with concentrated solutions of salts, which contain anions and cations of close transport numbers. A widely known example is saturated KCl (4.2 M); in aqueous solutions, potassium and ammonium nitrates are also suitable. However, the requirement of equal transport numbers is less important as compared with that of high concentration of electrolyte solution, which fills the bridge [33, 34]. A suitable version of the salt bridge can be chosen for any type of cells, when taking into account the kind of studies and the features of chosen electrodes.

In melts, an additional problem of thermo-emf arises, and the emf correction can be calculated from thermoelectric coefficients of phases in contact [35].

1.5
Classification of Electrodes

No universally adopted general classification of electrodes exists; however, when dealing with thermodynamic aspects of the electrode potential notion, we dwell on the electrode classification based on the nature of species participating in electrochemical equilibria.

Electrodes of the first kind contain electronic conductors as the reduced forms, and ions (particularly, complex ions) as the oxidized forms. The equilibrium can be established with respect to cations and anions; in the absence of ligands, the cations are more typical. The examples are: $Cu|Cu^{2+}$ ... or ... $Au|[Au(CN)_2]^-$ ....

This group can be supplemented also by amalgam electrodes (or other liquid electrodes) and electrodes fabricated from nonstoichiometric solids capable of changing their composition reversibly (intercalation compounds based on carbons, oxides, sulfides, and multicomponent salts, particularly, Li-intercalating electrodes of batteries).

Electrodes of the second kind contain a layer of a poorly soluble compounds (salt, oxide, hydroxide), which is in contact with a solution of the same anion. The equilibrium is always established with respect to anions. The most typical examples are based on poorly soluble compounds of mercury and silver (Table 1).

The redox polymers, both of organic and inorganic origin (such as polyvinylpyridine modified by redox-active complexes of metals; Prussian blue and related materials), can be considered as a version of electrodes of the second kind; however, the equilibrium is usually established with respect to cations. Electron conducting polymers (polyanyline, polypyrrol, and so forth) also pertain, in the first approximation, to the electrodes of the second kind, which maintain equilibrium with respect to anion. Ion exchange polymer films on electrode surfaces form a subgroup of membrane electrodes.

Rather exotic electrodes of the third kind (with simultaneous equilibria with

Tab. 1 Conventional reference electrodes [36]

Reference electrode	Potential versus SHE (aqueous systems, recommended values for 25 °C) [V]	Analogs	Media
Calomel electrodes		Mercurous bromide, iodide, iodate, acetate, oxalate electrodes	Aqueous and mixed (with alcohols or dioxane)
Saturated (SCE)	0.241(2)		
Normal (NCE)	0.280(1)		
decinormal	0.333(7)		
Silver chloride electrode (saturated KCl)	0.197(6)	Silver cyanide, oxide, bromate, iodate, perchlorate	Aqueous, mixed, abs. alcoholic
		Nitrate	Aprotic
Mercury/mercurous sulfate electrode	0.6151(5)	Ag/Ag$_2$SO$_4$, Pb/Pb$_2$SO$_4$	Aqueous, mixed
Mercury/mercuric oxide electrode	0.098		Aqueous, mixed
Quinhydrone electrode		Chloranil, 1,4-naphtoquinhydrone	Any with sufficient solubility of components
0.01 M HCl	0.586(8)		
0.1 M HCl	0.641(4)		

Note: NCE: Normal calomel electrodes; SCE: saturated calomel electrode.

respect to anions and cations) were also proposed for measuring certain special characteristics, such as the solubility product of poorly soluble salts. The rare examples are Ag|AgCl|PbCl$_2$|Pb(NO$_3$)$_2$... or M|MOx|CaOx|Ca(NO$_3$)$_2$... where M = Hg or Pb, and Ox is the oxalate anion.

When the metal does not take part in the equilibrium and both components of the redox pair find themselves in solution, the system is called a *redox electrode*, for example, a quinhydrone electrodes (Table 1) or electrodes on the basis of transition metal complexes:

Au|[Fe(CN)$_6$]$^{3-}$, [Fe(CN)$_6$]$^{4-}$ ...

or Pt|[Co(EDTA)]$^-$, [Co(EDTA)]$^{2-}$ ...

When equilibrium is established between ions in solution and the gas phase, we have gaseous electrodes; the intermediates of this equilibrium are usually adatoms: Pt, H$_2$|H$^+$..., or Pt, Cl$_2$|Cl$^-$... Palladium, hydrogen-sorbing alloys, and intermetallic compounds pertain to gaseous electrodes. However, at the same time, these systems are example of intercalation processes operating under equilibrium conditions, which brings them close to electrodes of the first kind.

The various kinds of electrodes considered above can be unified, if we take into account the concept of an electronic equilibrium at metal/solution interface, that is, bear in mind that a certain activity of $\bar{e}_s$ in solution, which corresponds to any equilibrium potential, exists.

1.6
Membrane Electrodes

A special comment should be given of membrane electrodes. The potential drop across a membrane consists of the diffusion potential (which can be calculated in a similar manner as that for completely permeable systems) and two so-called Donnan potential drops, named after F. G. Donnan who was the first to consider these systems on the basis of Gibbs thermodynamics [37]; more detailed consideration was given latter in Ref. [38]. The Donnan potential can be expressed via the activities of ions capable of permeating through the membrane, and its value is independent of the activities of the other ions. This fact forms the basis for the widespread experimental technique of measuring the activities in the systems with eliminated diffusion potentials (ion-selective electrodes [39, 40]). The system best known among the latter is the pH-sensitive glass electrode, for which the modes of operation, the selectivity, and microscopic aspects were studied intensively [41]. Sensors of this sort are known for the vast majority of inorganic cations and many organic ions. The selectivity of membranes can be enhanced by constructing enzyme-containing membranes.

The following brief classification of membrane electrodes can be used [42]: inert membranes (cellulose, some sorts of porous glass); ion exchange membranes, in which charged groups are bound with the membrane matrix; solid membranes reversible with respect to certain ions (as for glass electrode); biomembranes. The simplest model of a membrane electrode is the liquid/liquid interface, which is prepared by contacting two immiscible solutions and usually considered together with liquid electrodes in terms of soft electrochemical interface (see Sect. 3.4).

1.7
Standard Potentials

Standard potential values are usually those of ideal unimolal solutions at a pressure of 1 atm (ignoring the deviations of fugacity and activity from pressure and concentration, respectively). A pressure of 1 bar $= 10^5$ Pa was recommended as the standard value to be used in place of 1 atm $= 101\,325$ Pa (the difference corresponds to a 0.34-mV shift of potential). If a component of the gas phase participates in the equilibrium, its partial pressure is taken as the standard value; if not, the standard pressure should be that of the inert gas over the solution or melt. In a certain case, a standard potential can be established in a system with nonunity activities, if the combination of the latter substituted in the Nernst equation equals unity. For any solid component of redox systems, the chemical potential does not change in the course of the reaction, and it remains in its standard state. In contrast to the common thermodynamic definition of the standard state, the temperature is ignored, because the potential of the standard hydrogen (protium) electrode is taken to be zero at any temperature in aqueous and protic media. The zero temperature coefficient of the SHE corresponds to the conventional assumption of the zero standard entropy of H^+ ions. This extrathermodynamic assumption induces the impossibility of comparing the values referred to the hydrogen electrodes, in different solvents. Of the systems regarded as reference electrodes, platinum hydrogen electrodes exhibit i_0 values ranked among the greatest known (10^{-3}–10^{-2} A cm^{-2}, at least, in aqueous acidic solutions). This type of electrodes is used as the reference for tabulating universal potential values [43–47].

The universal definition of the standard potential E^0 of a redox couple Red/Ox is as follows: the standard potential is the value of emf of an electrochemical cell, in which diffusion potential and thermoemf are eliminated. This cell consists of an electrode, on which the Red/Ox equilibria establish under standard conditions, and a SHE.

1.8
Specific Features of Certain Reference Electrodes

An important type of reference electrodes is presented by the so-called *reversible hydrogen electrodes* (RHE) in the same solution, which makes it possible to avoid the liquid junction. These electrodes are preferentially used in electrochemical experiments when investigating the systems with H^+ and H_2 involved in the process under study (adsorption and electrocatalysis on hydrogen-adsorbing surfaces, such as platinum group metals). The RHE can be produced for a wide range of pH; certain special problems associated with neutral solutions can be solved by using buffering agents. In media containing organic components, the possible catalytic hydrogenation of the latter poses an additional limitation for hydrogen electrodes.

In general, electrodes of the second kind are more convenient as reference electrodes, because they do not require a source of gaseous hydrogen. Dynamic hydrogen reference electrodes, which represent a wire of platinum metal (or another metal with low hydrogen overvoltage) cathodically polarized up to the hydrogen evolution, give a possibility of avoiding the use of gaseous hydrogen. Their stable potential values are determined by the existence of the current–potential relationship, which is possible for the electrode processes with high degree of stability. A special type of stable hydrogen electrode is based on Pd hydride (β-phase). Another advantage of the electrodes of the second kind is their applicability in a wider range of temperatures, and also suitability for a wide range of pressures. Operating at high temperatures presents a real challenge in finding a suitable reference electrode (although there are few examples stable up to 250–300 °C). One should never use any reference electrode containing mercury at elevated temperatures. For other reference electrodes (e.g., silver/silver chloride), stability and reproducibility should be tested carefully before it is used at temperatures above 100 °C.

For aqueous hydrogen electrodes, the potentiometric data are now available in both sub- and supercritical regions, up to 723 K and 275 bar [48]. However, their use as the high-temperature reference systems involves numerous complications.

Among reliable systems for high-temperature measurements, gaseous (first of all, oxygen) electrodes based on solid electrolytes of the yttria-stabilized zirconia (YSZ) type can be recommended, which are always applied in the electrochemical cells with solid electrolytes. Gaseous electrodes based on Pt or carbon are also widely used in high-temperature melts, especially those based on oxides (oxygen reference electrode) or chlorides (chlorine reference electrode).

There are lots of systems, especially for electroplating and electrosynthesis, in which electrodes of the first kind can be used, without any liquid junctions (the example is liquid Al in AlF_3-containing melts). More universal systems for melts of various kinds are: a chlorine electrode in equimolar NaCl + KCl melt and Ag/Ag$^+$ electrodes with the range of Ag$^+$ concentrations (0.01–10 mM) corresponding to usual solubility values. Reference electrodes of the second kind can hardly be used in melts because of the high solubility of the majority of inorganic solids.

When studying nonaqueous systems by means of galvanic cells with aqueous or mixed reference electrodes, we cannot avoid liquid/liquid junctions and estimate the corresponding potential drop from any realistic model. In protic nonaqueous media (alcohols, dioxane, acetone, etc.), a hydrogen electrode can be used; it is also suitable for some aqueous/aprotic mixtures. However, the i_0 values for the hydrogen reaction are much lower as compared with purely aqueous solutions. When studies are carried out in nonaqueous media, in order to avoid liquid/liquid junction preference should be given to the reference electrodes in the same solvent as the electrode of interest.

In aprotic (as well as in protic and mixed) media, the two reference systems of choice are ferrocene/ferrocenium and bis(biphenil)chromium(I/0). The pentamethylcyclopentadienyl analog of the former was recently shown to yield higher performance [49, 50]. Among other typical electrodes, Ag/AgNO$_3$ should be mentioned. We can also mention special reference systems suitable for certain solvents, such as amalgam electrodes based on Tl$^-$ and Zn in liquid ammonia and hydrazine, and also Hg/Hg$_2$F$_2$ electrode in pure HF.

An exhaustive consideration of the specific features of various reference electrodes (fabrication, reproducibility and stability, modes of applicability, effects of impurities, necessary corrections) can be found in Refs. [36, 51–54]. Nowadays, certain new findings in this field are possible because of the novel approaches to immobilization of redox centers on the electrode surfaces.

The attempts to interrelate the potential scales in aqueous and nonaqueous solutions have been undertaken and are still in progress. Such a relationship could have been found if the free energy of transfer was known at least for one type of charged species common to the solvents. It is evident that ways of solving this problem can be based on the assumptions beyond the scopes of thermodynamics. Thus, it was mentioned in Ref. [18–20] that the free energy of solvation of Rb$^+$ ion is low and approximately the same in different solvents because of its great size. However, any direct application of rubidium electrodes is hampered by their corrosion activity as well as by the fact that their potential lies in the region of electroreduction of a number of solvents. Another proposition concerns the fact that the free energies of transfer for certain cations and neutral molecules of the redox pairs A/A$^+$ are the same in different solvents, which is also caused by their great sizes [51]. Ferrocen/ferrocenium-like systems were taken as the A/A$^+$ pairs. It is interesting that these scales – rubidium and A/A$^+$ – adequately correlate with one another. Different extrathermodynamic assumptions are compared in Ref. [54].

Special reference electrodes are introduced when considering the thermodynamics of surface phenomena on electrodes [55]. That is, when we use a "virtual" reference electrode potential, which is referred to the electrode reversible with respect to cation or anion in the same solution, changes by a value of $(RT/F) \ln a_\pm$ ($a_\pm$ is a mean activity coefficient), then the thermodynamic relationships can be generalized for a solution of an arbitrary concentration without introducing any notions on the activity of a separate ion (Sect. 5).

1.9
Values of Redox Potentials

The values of redox potentials were tabulated in numerous collections [43–47], the latest collections being critically selected. The potentials of redox systems with participation of radicals and species in excited electronic states are discussed in Refs. [56, 57]. There is no need to measure the potentials for all redox pairs (and, correspondingly, ΔG for all known reactions). If we obtain the partial values for a number of ions and compounds, the characteristic values for any other reactions can be computed. The idea of calculation is based on the fact that the emf value only depends on the initial and final states of the system, being independent of the existence of any intermediate states. This fact is of great importance for systems, for which it is impossible, or extremely difficult, to prepare a reversible electrode (redox couples containing oxygen or active metals). From considerations of the equilibria with participation of a solvated electron, the E^0 value that determines the value of the constant in Eq. (7) was estimated: − 2.87 V (SHE) [56].

It should be mentioned that no standard electrode can be experimentally achieved, because the standard conditions are hypothetical (no system is ideal at a unimolal concentration and the atmospheric pressure). Hence, the majority of tabulated values were recalculated from the data for more dilute solutions, usually after the preliminary extrapolation to the zero ionic strength. When dealing with the reverse procedure (recalculation of equilibrium potentials for dilute solutions from tabulated standard values), we should take into account that in real systems the time of equilibrium establishment increases with decreasing concentration, because of the decrease in rates of half-cell reactions. Hence, the possibility to prepare a reversible electrode is limited to a certain concentration value.

A famous collection of calculated potentials was presented by Pourbaix [46] for aqueous solutions of simplest compounds of elements with oxygen and hydrogen. Although this collection ignores the non-stoichiometry phenomena, which are of great importance for many oxides and hydroxides, it remains helpful for making rough estimates and also predicting the corrosion peculiarities. It should be noted that these predictions cannot be observed in some systems under pronounced kinetic limitations.

The shifts of reversible potentials of oxide electrodes with composition were first considered in Ref. [12] using the example of hydrated MnO_{2-x}, which demonstrates a wide region of inhomogeneity. Later, the considerations of this sort were carried out also for other systems [58].

There are lots of empirical and semiempirical correlations of redox potentials with molecular characteristics of substances, especially for the sequences of related compounds, which can be used for semi-quantitative estimates of potentials of novel systems. The most advanced approaches known from coordination chemistry take into account σ-donating and π-accepting abilities of the ligands, which can be expressed, for example, in terms of Hammett or Taft, and also other effective parameters [59]. For the sequences of complexes with different central ions and related ligands, the steric factors (namely, chelating ability) are known to significantly affect the redox potentials.

A separate field deals with correlating the redox potentials with spectroscopic parameters, such as the energy of the first allowed d–d transition in a complex, the energy of the metal-to-ligand charge-transfer band (i.e., the separation between the HOMO (highest occupied molecular orbital) on the metal and LUMO (lowest unoccupied molecular orbital) on the ligand), and nuclear magnetic resonance (NMR) chemical shifts. The problem of extending these correlations over a wider range of reactions, including those irreversible (for which the kinetics makes a substantial contribution to the formal potential value, the latter being evidently nonthermodynamic) was considered a long time ago [60]. Advanced spectroscopic techniques are widely used for solving the reverse problem of determining the redox potentials of irreversible couples [61].

1.10
Potential Windows

The feasibility of determining the redox potential depends strongly on the solvent, the electrolyte, and the electrode material: all of them should remain inert in potential ranges as wide as possible. In this connection, the term "potential window" is usually used, which characterizes the width of the interval between the potentials of cathodic and anodic background processes. The narrow potential window of water (about 1.3–1.4 V on platinum group metals, and up to 2 V on mercury-like metals, which do not catalyze the hydrogen evolution reaction) stimulates the potentiometric studies in nonaqueous solvents. Aprotic solvents usually demonstrate windows of about 3–3.5 V, if the optimal supporting electrolyte is chosen. An extremely wide window is known, for example, for liquid SO_2, which is of highest interest for measuring extremely positive redox potentials (up to +6.0 V (SCE)) [62]. Low-temperature haloaluminate melts are highly promising systems [63]. Finally, a number of unique mixed solvents with extremely wide windows were found in recent studies of lithium batteries [64].

Among electrode materials, the widest windows are known for transition metal oxides, borides, nitrides, and some specially fabricated carbon-based materials. It should be mentioned that, if the nature of electrode material can affect the formal potential value by changing the mechanism and kinetic parameters, the solvent frequently has a pronounced effect on the equilibrium potential, because of the solvation contribution to free Gibbs energy.

1.11
Experimental Techniques

Direct potentiometry (emf measurements) requires the potential of indicator electrode to be determined by the potential of the redox pair under study. Sometimes this is impossible, because of the low exchange current densities and/or concentrations (particularly, at low solubilities or limited stabilities of the components), when extraneous redox pairs contribute significantly to the measured potential. The sensitivity of direct potentiometry can be enhanced by using preconcentration procedures, particularly in polymeric films on the electrode surfaces [65]. However, it is common to use voltammetry, polarography, and other related techniques in place of potentiometry. The formal potential (determined as the potential between anodic and cathodic current peaks on voltammograms) does not generally coincide with the equilibrium redox potential, the accuracy depending on the reaction mechanism [66–68]. For reversible and simplest quasireversible electrode reactions, the exact determination of standard potentials from the data of stationary voltammetry and polarography is possible. Only these dynamic techniques are suitable for the studies of equilibria with participation of long-living radicals and other excited species, which can be introduced into solution only by in situ electrochemical generating.

If voltammetric and related techniques are used, the ohmic drop should be either compensated (now this is usually done by the software or hardware of electrochemical devices [69]), or reduced by using, for example, a Luggin (Luggin-Gaber) capillary (see in Ref. [12]). Another important technical detail is that the components of reference redox systems (such as ferrocene/ferrocenium) are frequently added immediately into the working compartment when voltammetry-like techniques are applied.

Applications of potentiometry are rather widespread, and its efficiency is high enough when operating with relative potential values. A mention should be made, first of all, of the determination of basic thermodynamic quantities, such as the equilibrium constants for coordination chemistry and the activity coefficients. The latter task covers a wide field of pH-metric applications, and also the analytical techniques based on ion-selective electrodes.

1.12
Effects of Ion Pairing

An important factor affecting redox potentials is also ion pairing, which is conventionally taken into account in terms of activity coefficients. For redox potentials of coordination compounds, the following equation is known [70]:

$$E = E_0 + \frac{RT}{F} \ln \frac{K_{\text{red}}}{K_{\text{ox}}} \qquad (23)$$

where K_{red} and K_{ox} denote the equilibrium constants of a complex (particularly, ion pair) formation. In as much as outer sphere ionic association is highly sensitive to the ionic charge, K_{red} appears to be higher than K_{ox}, and the potential shifts to more positive values with the ionic strength. That is, for usual concentrations of hexacyanoferrates in aqueous solutions, the shift induced by the association with K^+ ions reaches 0.1 V and higher at usual concentrations.

In electroanalysis, Eq. (23) forms the basis of highly sensitive techniques of potentiometric titration. The practical applications are presented by numerous potentiometric sensors, particularly biosensors [71–74]. The potential measurements in microheterogeneous media (microemulsions; anionic, cationic and nonionic micelles) were worked out recently [75]. A separate field is the potentiometric studies of liquid/liquid junctions directed to the determination of ionic transport numbers.

1.13
Absolute Electrode Potential

Although it has long been known that only relative electrode potentials can be measured experimentally, numerous attempts were undertaken to determine the value of potential of an isolated electrode without referring it to any reference system ("absolute electrode potential"). Exhaustive and more particular considerations of the problem of "absolute potential" can be found in Refs. [1–20, 76–86]. These attempts were concentrated around the determination of a separate Galvani potential, $\Delta_S^M \varphi$ (which was named initially as the *absolute electrode potential*) and also included the search for a reference electrode, for which the maximum possible work of any imaginary electrode process equals zero. Sometimes, the problem was formulated as a search for the hypothetical reference state determined as reckoned from the ground state of electron in vacuum (a "physical" scale of energy with the opposite sign). In this connection, the requirement for the reference electrode under discussion was formulated in the absence of any additional electrochemical interfaces.

Finally, these studies have transformed into a reasonable separation of the measured emf for a cell, which consists of an electrode under consideration and a standard reference electrode, in order to determine two electrode potentials referred to individual interfaces (or to the separate half-reactions) by using only the experimentally measurable values.

To illustrate the technique of Galvani potential calculation, we return to the M|S interface. According to scheme in Fig. 1,

$$\Delta_S^M \varphi = \Delta_S^M \psi - \chi^S + \chi^M \quad (24)$$

and the problem reduces to the determination of χ^S and χ^M, because the Volta potential $\Delta_S^M \psi$ can be measured. The estimates of χ^{H_2O} for the water/air interface were made by considering the adsorption effects that inorganic acids, $HClO_4$, and HBr [87, 88] and aliphatic compounds [89] exert on the Volta potential of water/air interface, and also by measuring the temperature coefficient of χ^{H_2O} [90]. In this connection, the measured real hydration energies of ions, $\Delta G_{\text{hydr}}^{(\text{real})}$, were compared with the calculated chemical hydration energies, $\Delta G_{\text{hydr}}^{(\text{chem})}$. The real (total or free) energies were considered as the energy changes, which accompany the transfer of ions from air to solution. This value can be divided into the chemical hydration energy, which is caused by the interaction of an ion with surrounding water molecules, and the change of electric energy that equals $z_i F \chi^{H_2O}$ (where z_i is the charge of the ion of the corresponding sign):

$$\Delta G_{\text{hydr}}^{(\text{real})} = \Delta G_{\text{hydr}}^{(\text{chem})} + z_i F \chi^{H_2O} \quad (25)$$

On the other hand, the real solvation energy can be written as

$$\Delta G_{\text{hydr}}^{(\text{real})} = -\Delta G_{\text{subl}} - \Delta G_{\text{ion}}$$
$$+ z_i F \Delta_{H_2O}^M \psi - z_i F W_e \quad (26)$$

where ΔG_{subl} is the free energy of metal sublimation (atomization), ΔG_{ion} is the free energy of ionization of a metal atom, W_e is the electron work function of M, and $\Delta_{H_2O}^M \psi$ is the Volta potential of the interface corresponding to the equilibrium potential in a solution, in which the concentration of metal ions is unity. The most reliable values of $\Delta G_{\text{hydr}}^{(\text{real})}$ were reported in Ref. [91].

The value of $\Delta G_{\text{hydr}}^{(\text{chem})}$ cannot be measured directly; however, it can be estimated on the basis of information on the liquid structure of water and the structure of its molecule. Having $\Delta G_{\text{hydr}}^{(\text{chem})}$ and $\Delta G_{\text{hydr}}^{(\text{real})}$, we can find χ^{H_2O} from Eq. (24). The estimates of this value given in the literature for water/air interface disagree with one another significantly (from -0.48 up to $+0.29$ V). A critical comparative consideration of various studies leads to a conclusion about a low positive value of χ^{H_2O}, namely, 0.13 V [1–9]. This value corresponds to the infinite dilution and can be changed by changing the solution concentration.

The estimation of χ^M is an even more complicated problem. Figure 3 schematically illustrates a "metal box" corresponding to the jellium model. According to this figure,

$$W_e = V - \varepsilon_F = e_0 \chi^M + V_{\text{ex}} - \varepsilon_F \quad (27)$$

the notations are given in the figure caption. In the framework of the jellium model, we can calculate χ^M by two techniques: directly from the distribution of the electronic gas outside the metal, and also by calculating V_{ex} and ε_F with the subsequent use of Eq. (26). Unfortunately, the jellium model has a serious limitation; the estimates of χ^M will be improved in future as the theory of electronic structure of metals and their surfaces becomes more advanced.

As was correctly reasoned in Ref. [10, 11], the scale of "absolute potential"

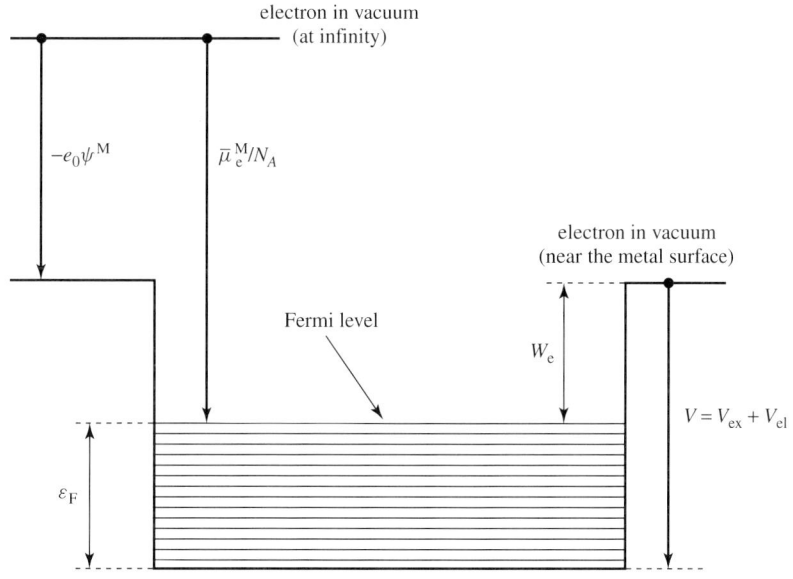

Fig. 3 A scheme of electron energy levels for a model of "metal box" (according to C. Kittel, *Elementary Solid State Physics*, Wiley, New York, 1962): $\bar{\mu}_e^M$ – electrochemical potential of electron in metal, N_A – Avogadro number, ε_F – Fermi energy, ψ^M – outer potential of metal, W_e – electron work function, V – electron potential energy in metal ($V = V_{\text{ex}} + V_{\text{el}}$, V_{ex} – energy of electron exchange, which corresponds to the electron interaction with positively charged jellium, $V_{\text{el}} = e_0 \chi^M$ – the surface component of electron potential energy).

constructed in accordance with Ref. [7] and recommended in Ref. [9] is, in fact, equivalent to the E_K scale first proposed in Refs. [76–81], which should be named after Kanevskii:

$$E_K = -\frac{[\Delta G_{subl} + \Delta G_{ion} + \Delta G_{hydr}^{(real)}]}{z_i F} \quad (28)$$

By introducing this scale, we can separate the total emf of the cell into two quantities, for which the contributions ΔG_{subl}, ΔG_{ion}, and $\Delta G_{hydr}^{(real)}$ can be experimentally determined. For a hydrogen electrode,

$$(E_K)_{H^+,H_2} = -\frac{\left[\frac{1}{2}\Delta G_{diss} + \Delta G_{ion} + \Delta G_{hydr}^{(real)}\right]}{F} \quad (29)$$

which gives the possibility to recommend the value -4.44 ± 0.02 V for the "absolute potential" of hydrogen electrode (at 298.15 K). For other temperatures, the calculations of this sort are limited by the lack of information on the temperature coefficient of work function. It should be mentioned that the uncertainty of absolute potential calculation is substantially higher than the accuracy of direct potential measurements with respect to a reference electrode. The useful comments to Kanevskii's scale can be found in Refs. [16–20, 84]. In the "absolute" scale, the "absolute potential" of any aqueous electrode at 298.15 K can be determined as

$$E_K = (E_K)_{H^+,H_2} + E \approx -4.44 + E \quad (30)$$

where E is the potential referred to the SHE. Two scales are combined for comparison in Fig. 4.

The "absolute potentials" of the hydrogen electrode in a number of nonaqueous solvents are reported in Ref. [9].

The value of E_K can be expressed via metal/vacuum electron work function and Volta potential [10, 11]:

$$E_K = W_e + \Delta_S^M \psi \quad (31)$$

The majority of electrochemical problems can be solved without separating the emf into "absolute potentials". However, it should be mentioned that the problem concerning the structure of the interfacial potential drop becomes the topical problem for the modern studies of electrified interfaces on the microscopic level, particularly, in attempts of testing the electrified interfaces by probe techniques [92–94]. The absolute scales are also of interest for electrochemistry of semiconductors in the context of calibrating the energy levels of materials. This problem is related to another general problem of physical chemistry – the determination of activity coefficient of an individual charged species.

1.14
The Role of Electric Double Layers

Later, a more detailed consideration of the M|S interface is given, which takes into account the formation of electric double layer. Figure 1 shows the potential drops at the metal/solution interface. It follows from this figure that

$$\Delta_S^M \psi = \chi^S + \Delta_S^M \varphi - \chi^M \quad (32)$$

The value of the Galvani potential $\Delta_S^M \varphi$ can be represented as the sum of the potential drop inside the ionic double layer $\Delta \varphi$ and the surface potentials of solution and metal, which have changed as a result

1.14 The Role of Electric Double Layers

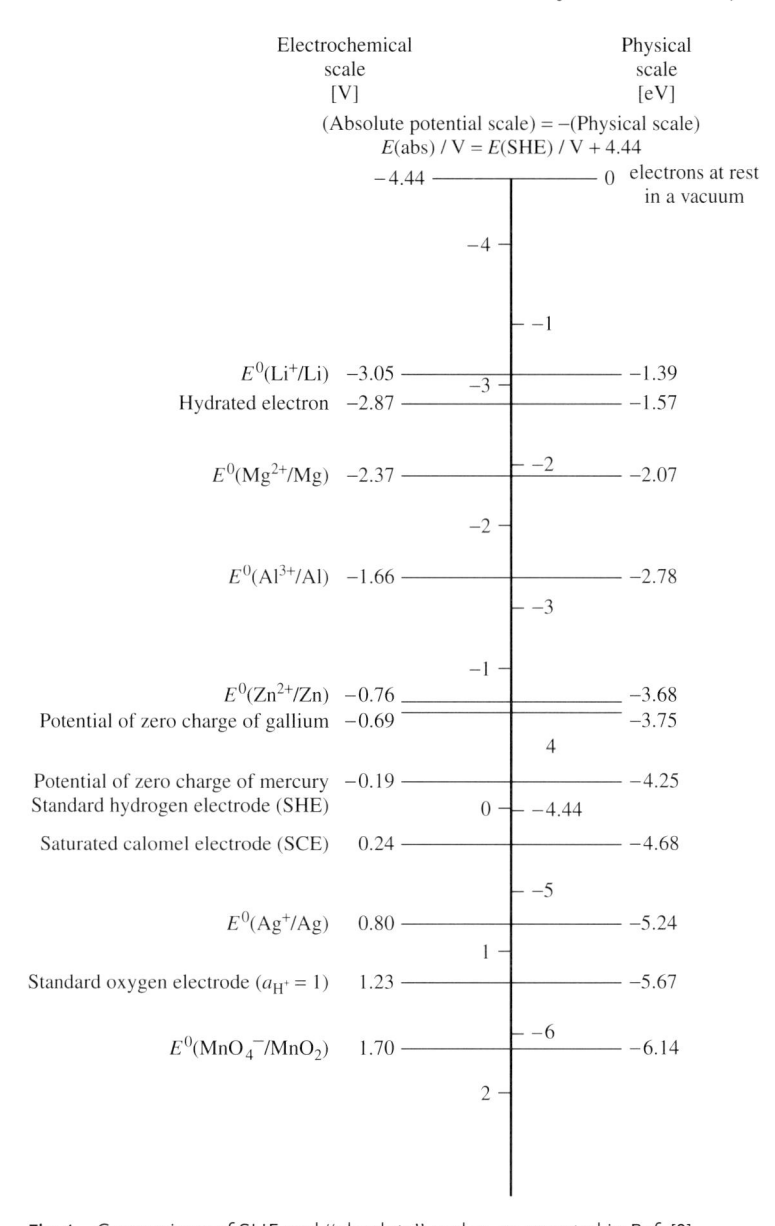

	Electrochemical scale [V]		Physical scale [eV]

(Absolute potential scale) = −(Physical scale)
$E(\text{abs}) / \text{V} = E(\text{SHE}) / \text{V} + 4.44$

	Electrochemical [V]		Physical [eV]	
	−4.44		0	electrons at rest in a vacuum
$E^0(\text{Li}^+/\text{Li})$	−3.05		−1.39	
Hydrated electron	−2.87		−1.57	
$E^0(\text{Mg}^{2+}/\text{Mg})$	−2.37		−2.07	
$E^0(\text{Al}^{3+}/\text{Al})$	−1.66		−2.78	
$E^0(\text{Zn}^{2+}/\text{Zn})$	−0.76		−3.68	
Potential of zero charge of gallium	−0.69		−3.75	
Potential of zero charge of mercury	−0.19		−4.25	
Standard hydrogen electrode (SHE)	0		−4.44	
Saturated calomel electrode (SCE)	0.24		−4.68	
$E^0(\text{Ag}^+/\text{Ag})$	0.80		−5.24	
Standard oxygen electrode ($a_{\text{H}^+}=1$)	1.23		−5.67	
$E^0(\text{MnO}_4^-/\text{MnO}_2)$	1.70		−6.14	

Fig. 4 Comparison of SHE and "absolute" scales, as reported in Ref. [9].

of mutual contact:

$$\Delta_S^M \varphi = \Delta\varphi - \chi^{S(M)} + \chi^{M(S)} \qquad (33)$$

where $\chi^{S(M)}$ and $\chi^{M(S)}$ are the surface potentials of solution and metal at their common interface, respectively. Hence,

$$\Delta_S^M \psi = \chi^S - \chi^{S(M)} + \Delta\varphi + \chi^{M(S)} - \chi^M$$
$$= \Delta\varphi + \delta\chi^M - \delta\chi^S \qquad (34)$$

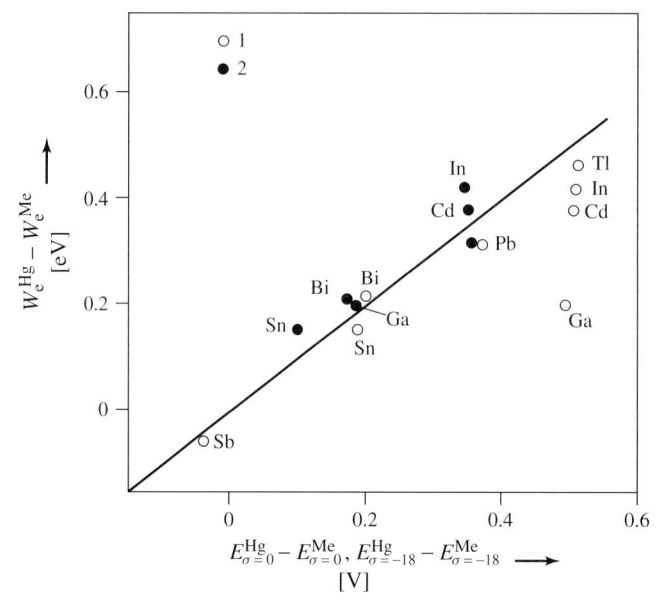

Fig. 5 Correlation of work function differences and the differences of potentials at constant electrode charge σ for $\sigma = 0$ (1) and -18 µC cm^{-2} (2), as reported in Ref. [5].

By writing such relationships for two interfaces, $M_1|S$ and $M_2|S$ and substituting corresponding $\Delta_S^{M_1}\psi$ and $\Delta_S^{M_2}\psi$ into Eq. (10), we obtain

$$E = \Delta_{M_1}^M \psi + (\Delta\varphi_1 - \Delta\varphi_2) \\ + (\delta\chi_1^M - \delta\chi_2^M + \delta\chi_2^S - \delta\chi_1^S) \quad (35)$$

If the metals M_1 and M_2 interact very weakly with the solvent molecules, we can assume that, in the first approximation, the last right-hand term in parentheses is zero, and

$$E \cong \Delta_{M_1}^M \psi + (\Delta\varphi_1 - \Delta\varphi_2) \quad (36)$$

When both metals, M_1 and M_2, are zero charged (their excess surface charge density $\sigma_M = 0$), and the ionic double layers are absent on their surfaces,

$$\Delta E_{\sigma_M=0} \approx \Delta_{M_1}^{M_2}\psi \quad (37)$$

Eq. (36) and (37) obtained by Frumkin [1–5] can be classified as the solution of the famous Volta problem of the nature of emf of electrochemical circuit. Equation (37) demonstrates that the difference of potentials of zero charge (pzc) of two metals is approximately equal to their Volta potential. In as much as the Volta potential $\Delta_{M_1}^{M_2}\psi$ is equal to the difference of work functions of the metals ($W_e^{M_2}$ and $W_e^{M_1}$), the verification of Eq. (37) can be carried out by comparing the differences both in pzc and in work functions (Fig. 5). The data of Fig. 5 confirm Eq. (37) and, simultaneously, demonstrate the possibility of substantial deviations from this relationship for the systems with pronounced solvent/metal interaction (as, for example, for the Ga/water interface).

In the context of concept of zero charge potential [1–5], a specific scale of electrode potentials should be mentioned, a reduced scale, in which the potential of electrode referred to its pzc for given solution composition. It was used actively in the studies of the electric double layer as the so-called *Grahame rational scale* (see Sect. 5).

1.15
Conclusion

As mentioned earlier, the role played by the electrode potential in electrochemistry and surface science is not limited to thermodynamic aspects. The electrode potential is the factor that governs (directly or via the electrode charge) the surface reconstruction and phase transitions in adsorbed layers. It is also the key factor, which determines the Gibbs free energy of reaction and, correspondingly, the Frank–Condon barrier. In electrochemical kinetics, the characteristic value is not the potential itself, but the overvoltage, which is the deviation of the potential from its reversible value in the same solution when the current flows across the electrochemical interface. The potential plays an important role in electrosynthesis, because the controlled potential electrolysis can be performed with high selectivity, in the absence of foreign reagents (oxidizers or reducers). The concept of membrane potential is one of the basic concepts in biophysics. All these wide fields of science demonstrate the interplay with electrochemical thermodynamics and provide the basis for further interpenetration of chemistry, biology, and physics, which was pioneered two centuries ago by the studies on the problem of electrode potential.

References

1. A. N. Frumkin, *Ergeb. Exact. Naturwiss.* **1928**, *7*, 235–273.
2. A. Frumkin, A. Gorodetzkaja, *Z. Phys. Chem.* **1928**, *136*, 215–227.
3. A. N. Frumkin, *Phys. Z. Sowjetunion* **1933**, *4*, 239–261.
4. A. N. Frumkin, *J. Chem. Phys.* **1939**, *7*, 552–553.
5. A. N. Frumkin, *Potentsialy nulevogo zaryada (in Russian) (Potentials of Zero Charge)*, Nauka, Moscow, 1979.
6. R. Parsons in *Modern Aspects of Electrochemistry* (Eds.: J. O'M. Bockris), Butterworths Sci. Publ., London, 1954, pp. 47–102, Vol. 1.
7. S. Trasatti, *J. Electroanal. Chem.* **1975**, *66*, 155–161.
8. S. Trasatti in *Comprehensive Treatise of Electrochemistry* (Eds.: J. O'M. Bockris, B. E. Conway, E. Yeger), Plenum Press, New York, London, 1980, pp. 45–81, Vol. 1.
9. S. Trasatti, *Pure Appl. Chem.* **1985**, *58*, 955–966.
10. A. Frumkin, B. Damaskin, *J. Electroanal. Chem.* **1975**, *66*, 150–154.
11. A. N. Frumkin, B. B. Damaskin, *Dokl. AN SSSR* **1975**, *221*, 395–398.
12. K. J. Vetter, *Electrochemische Kinetik*, Springer-Verlag, Berlin, 1961.
13. R. G. Compton, G. H. W. Sanders, *Electrode Potentials*, Oxford University Press, New York, 1996.
14. J. R. Runo, D. G. Peters, *J. Chem. Educ.* **1993**, *70*, 708–713.
15. M. I. Temkin, *Bull. Acad. Sci. l'URSS* **1946**, *N2*, 235–244.
16. B. V. Ershler, *Usp. Khim.* **1952**, *21*, 237–249.
17. B. V. Ershler, *Zh. Fiz. Khim.* **1954**, *28*, 957–960.
18. V. A. Pleskov, *Usp. Khim.* **1947**, *16*, 254–278.
19. V. A. Pleskov, *Zh. Fiz. Khim.* **1949**, *23*, 104.
20. V. A. Pleskov, *Zh. Fiz. Khim.* **1950**, *24*, 379.
21. E. A. Guggenheim, *J. Phys. Chem.* **1929**, *33*, 842–849.
22. W. Schottky, H. Rothe in *Handbuch der Experimentalphysik*. (Eds.: W. Wien, F. Harms, H. Lenz), Acad. Verlag. m.b.h., Leipzig, 1928, pp. 145–150, Vol. 13 (2.Teil).
23. E. Lange, K. P. Miscenko, *Z. Phys. Chem.* **1930**, *149A*, 1–41.
24. E. Lange in *Handbuch der Experimentalphysik* (Eds.: W. Wien, F. Harms, H. Lenz), Acad. Verlag. m.b.h., Leipzig, 1933, pp. 263–322, Vol. 12 (2.Teil).
25. E. Lange, *Z. Elektrochem.* **1951**, *55*, 76–92.
26. E. Lange, *Z. Elektrochem.* **1952**, *56*, 94–106.
27. W. Nernst, *Z. Phys. Chem.* **1889**, *4*, 129–181.

28. P. P. Edwards in *Advances in Inorganic Chemistry and Radiochemistry* (Eds.: H. J. Emeleus, A. G. Sharpe), Academic Press, New York, 1982, pp. 135–185, Vol. 25.
29. M. Plank, *Ann. Physik.* **1890**, *40*, 561–576.
30. P. Henderson, *Z. Phys. Chem.* **1907** *59*, 118–127.
31. P. Henderson, **1908**, *63*, 325–345.
32. B. B. Damaskin, G. A. Tsirlina, M. I. Borzenko, *Russ. J. Electrochem.* **1998**, *34*, 199–203.
33. J. B. Chloupek, V. Z. Danes, B. A. Danesova, *Collect. Czech. Chem. Commun.* **1933**, *5*, 469–478.
34. J. B. Chloupek, V. Z. Danes, B. A. Danesova, *Collect. Czech. Chem. Commun.* **1933**, *5*, 527–534.
35. N. Q. Minh, L. Redey, *Molten Salt Techniques*, Plenum Press, New York, 1987, Vol. 3.
36. D. J. G. Ives, G. J. Janz, (Eds.), *Reference Electrodes. Theory and Practice* Academic Press, New York, London, 1961.
37. F. G. Donnan, *Z. Elektrochem.* **1911**, *17*, 572–581.
38. F. G. Donnan, E. A. Guggenheim, *Z. Phys. Chem.* **1932**, *162A*, 346–360.
39. W. E. Van der Linden in *Comprehensive Analytical Chemistry*. (Ed.: G. Svehla), Elsevier, Amsterdam, 1981, Vol. 1.
40. J. Vesely, D. Weiss, K. Stulik, *Analysis with Ion-Selective Electrodes*, Ellis Horwood, Chichester, 1978.
41. G. Eisenman, *Biophys. J.* **1962**, *2*, 259–323.
42. J. S. Newman, *Electrochemical Systems*, 2nd ed., Prentice-Hall, Englewood Cliffs, 1991.
43. R. Parsons, *Redox Potentials in Aqueous Solutions: A Selective and Critical Source Book*, Marcel Dekker, New York, 1985, pp. 1–34.
44. A. J. Bard, R. Parsons, J. Jordan, *Standard Potentials in Aqueous Solution*, Marcel Dekker, New York, 1985.
45. M. S. Antelman, F. J. Harris, (Eds.), *The Encyclopedia of Chemical Electrode Potentials*, Plenum Press, New York, London, 1982.
46. M. Pourbaix, *Atlas d'Equilibres Electrochimiques*, Gauthier-Villars, Paris, 1963.
47. S. G. Bratsch, *J. Phys. Chem. Ref. Data* **1989**, *18*, 1–21.
48. K. Eklund, S. N. Lvov, D. D. Macdonald, *J. Electroanal. Chem.* **1997**, *437*, 99–110.
49. J. T. Hupp, *Inorg. Chem.* **1990**, *29*, 5010–5120.
50. J. K. Bashkin, P. J. Kinlen, *Inorg. Chem.* **1990**, *29*, 4507–4509.
51. H. Strehlow, *The Chemistry of Non-Aqueous Solvents*, Academic Press, 1967, pp. 129–171, Vol. 1.
52. J. N. Butler in *Advances in Electrochemistry and Electrochemical Engineering* (Ed.: P. Delahay), Intersci. Publi., New York, 1970, pp. 77–175, Vol. 7.
53. G. Gritzner, J. Kuta, *Pure Appl. Chem.* **1984**, *56*, 461–466.
54. Y. Marcus, *Pure Appl. Chem.* **1986**, *58*, 1721–1736.
55. A. N. Frumkin, *Zh. Fiz. Khim.* **1956**, *30*, 2066–2069.
56. D. M. Stanbury in *Advances in Inorganic Chemistry* (Ed.: A.G. Sykes), Academic Press, INC, New York, 1989, pp. 69–138, Vol. 33.
57. D. D. Wayner, V. D. Parker, *Acc. Chem. Res.* **1993**, *26*, 287–294.
58. G. A. Tsirlina, O. V. Safonova, O. A. Petrii, *Electrochim. Acta* **1997**, *42*, 2943–2946.
59. J. Chatt, *Coord. Chem. Rev.* **1982**, *43*, 337–347.
60. A. A. Vlcek, *Prog. Inorg. Chem.* **1963**, *5*, 211–384.
61. F. Sanchez, P. Perez-Tejeda, F. Perez et al., *J. Chem. Soc., Dalton Trans.* **1999**, *N 17*, 3035–3039.
62. E. Garcia, A. Bard, *J. Electrochem. Soc.* **1990**, *137*, 2752–2759.
63. C. L. Hussey in *Chemistry of Nonaqueous Solutions* (Eds.: G. Mamantov, A. I. Popov), VCH, New York, 1994, pp. 227–275.
64. K. Xu, N. D. Day, C. A. Angell, *J. Electrochem. Soc.* **1996**, *143*, L209–L211.
65. J. Huang, M. S. Wrighton, *Anal. Chem.* **1993**, *65*, 2740–2746.
66. A. J. Bard, L. R. Faulkner, *Electrochemical Methods. Fundamentals and Applications*, Wiley, New York, 2001.
67. D. K. Gosser, *Cyclic Voltammetry: Simulation and Analysis of Reaction Mechanisms*, VCH Publ., Deerfield Beach, 1993.
68. J. F. Cassidy, *Compr. Anal. Chem.* **1992**, *27*, 1–69.
69. R. M. Souto, *Electroanalysis* **1994**, *6*, 531–542.
70. J. Heyrovsky, J. Kuta, *Zaklady Polarografie*, Nakl. Ceskoslov. Academ. Ved, Praha, 1962.
71. J. Janata, *Chem. Rev.* **1990**, *90*, 691–703.
72. J. F. Coetzee, B. K. Deshmukh, C.-C. Liao, *Chem. Rev.* **1990**, *90*, 827–835.
73. M. Alvarez-Icaza, U. Bilitowski, *Anal. Chem.* **1993**, *65*, 525A–533A.
74. J. Wang, *Anal. Chem.* **1995**, *67*, 487R–492R.

75. S. A. Myers, R. A. Mackey, R. A. Brajter-Toth, *Anal. Chem.* **1993**, *65*, 3447–3453.
76. E. A. Kanevskii, *Zh. Fiz. Khim.* **1948**, *22*, 1397–1404.
77. E. A. Kanevskii, *Zh. Fiz. Khim.* **1950**, *24*, 1511–1514.
78. E. A. Kanevskii, *Zh. Fiz. Khim.* **1951**, *25*, 854–862.
79. E. A. Kanevskii, *Zh. Fiz. Khim.* **1952**, *26*, 633–641.
80. E. A. Kanevskii, *Zh. Fiz. Khim.* **1953**, *27*, 296–309.
81. E. A. Kanevskii, *Dokl. Akad. Nauk SSSR* **1981**, *257*, 926–929.
82. V. A. Pleskov, B. V. Ershler, *Zh. Fiz. Khim.* **1949**, *23*, 101–103.
83. B. V. Ershler, V. A. Pleskov, *Zh. Fiz. Khim.* **1951**, *25*, 1258–1260.
84. V. M. Novakovskii, *Zh. Fiz. Khim.* **1956**, *30*, 2820–2822.
85. H. Reiss, A. Heller, *J. Phys. Chem.* **1985**, *89*, 4207–4213.
86. Yu. V. Pleskov, *J. Phys. Chem.* **1987**, *91*, 1691–1692.
87. A. Frumkin, *Z. Phys. Chem.* **1924**, *111*, 190–210.
88. A. Frumkin, *Z. Phys. Chem.* **1924**, *109*, 34–48.
89. A. N. Frumkin, Z. A. Iofa, M. A. Gerovich, *Zh. Fiz. Khim.* **1956**, *30*, 1455–1468.
90. J. Randles, D. Schiffrin, *J. Electroanal. Chem.* **1965**, *10*, 480–484.
91. J. Randles, *Trans. Faraday Soc.* **1956**, *52*, 1573–1581.
92. A. C. Hillier, S. Kim, A. J. Bard, *J. Phys. Chem.* **1996**, *100*, 18808–18817.
93. R. C. Thomas, P. Tangyunyong, J. E. Houston et al., *J. Phys. Chem.* **1994**, *98*, 4493, 4494.
94. Z.-X. Xie, D. M. Kolb, *J. Electroanal. Chem.* **2000**, *481*, 177–182.

2
Electrochemical Double Layers

2.1	Electrochemical Interfaces: At the Border Line	33
	Alexei A. Kornyshev	33
	Eckhard Spohr	33
	Michael A. Vorotyntsev	33
2.1.1	Introduction	33
2.1.2	Basic Concepts	34
2.1.3	Thermodynamics of Electrified Interfaces	39
2.1.4	Classical Models of Electrified Interfaces	41
2.1.4.1	Helmholtz Model	41
2.1.4.2	Gouy–Chapman–Stern–Grahame Model	42
2.1.5	Diffuse and Compact Layer Properties	44
2.1.5.1	Grahame Model in Comparison with Capacitance Data for Liquid Electrodes	44
2.1.5.2	Compact Layer Properties	51
2.1.5.3	Manifestations of the Diffuse Layer in Kinetics of Electrode Reactions	53
2.1.5.4	Electrokinetic Phenomena	56
2.1.6	Solid Electrodes: Effects of Polycrystallinity and Surface Roughness .	58
2.1.6.1	"Uniform" Model of Solid Electrodes. Surface Roughness	58
2.1.6.2	Crystallographic Inhomogeneity Effects for Solid Electrode Surfaces	61
2.1.6.3	EDL Structure for NonUniform Electrode Surfaces	64
2.1.6.4	Surface Roughness Effect on the Diffuse-Layer Capacitance	66
2.1.7	Semi-phenomenological Nonlocal Theory of the Double Layer	68
2.1.7.1	Basic Equations	68
2.1.8	"Dipolar" Models of the Compact Layer	70
2.1.8.1	Two-State Model	70
2.1.8.2	Multistate Models	71
2.1.9	The Role of Metal Electrons	72
2.1.9.1	Surface Electronic Profile and its Response to Charging	72
2.1.9.2	Response to Charging of the Electrochemical Interface and the Compact-Layer Capacitance: Main Qualitative Effects	75
2.1.9.3	Optical Response and the Interfacial Structure	79

2.1.10	Effects of the Molecular Nature of Electrolyte Solutions	82
2.1.11	Ionic Adsorption	93
2.1.11.1	Qualitative Aspects	93
2.1.11.2	Ion Adsorption Models	96
2.1.11.3	Image Potential for Ions in the Diffuse or Compact Layer	99
2.1.11.4	Interaction Between Adsorbed Ions	102
2.1.11.5	Statistical Mechanics of Adsorbed Ion Ensemble	104
2.1.11.6	Ion Adsorption Isotherm	106
2.1.12	Electric Field-Induced Modifications of the Surface Structure	111
2.1.12.1	Electrochemical Missing Row Reconstruction: Facts and Speculations	112
2.1.12.2	Driving Force for Reconstruction	113
2.1.12.3	Ising Model. Order–Disorder Transitions	114
2.1.12.4	Roughening Transitions	116
2.1.13	Soft Electrochemical Interfaces	117
2.1.13.1	Capillary Waves at ITIES. Field Effect on Capillary Waves and Capillary Waves Effect on Capacitance	118
2.1.13.2	Ion Penetration into an "Unfriendly Medium"	122
2.1.14	From Basic Concepts to Surface Electrochemistry	124
	References	125
2.2	**Electrical Double Layers: Theory and Simulations**	**133**
	Wolfgang Schmickler	*133*
2.2.1	Introduction	133
2.2.2	Gouy–Chapman Theory and its Extensions	133
2.2.3	Theoretical Methods and Principles of Computer Simulations	136
2.2.3.1	Integral Equations	137
2.2.3.2	Computer Simulations	138
2.2.4	Electronic Structure Calculations	139
2.2.5	Models for the Solution	140
2.2.5.1	Field-theoretical Approach to the Double Layer	140
2.2.5.2	Ensembles of Hard Spheres	141
2.2.6	The Role of the Metal	143
2.2.6.1	Qualitative Considerations	143
2.2.6.2	The Jellium Model	144
2.2.6.3	Calculations for Metal Clusters	146
2.2.7	The Metal–Solution Interface	147
2.2.7.1	Jellium in Contact with Model Solutions	147
2.2.7.2	Molecular Dynamics Simulations	150
2.2.7.3	Car–Parinello-type Simulations	153
2.2.8	Rough Electrodes	153
2.2.9	Liquid–Liquid Interfaces	154
2.2.9.1	Verwey–Niessen Theory	155
2.2.9.2	Capillary Waves	156
2.2.9.3	Lattice Gas Model	156

2.2.10	Conclusion	158
	Acknowledgments	159
	References	159
2.3	**Electrochemical Double Layers: Liquid–Liquid Interfaces**	**162**
	Alexander G. Volkov	162
	Vladislav S. Markin	162
2.3.1	Polarizable and Nonpolarizable Liquid Interfaces	162
2.3.2	Verwey–Niessen Model	163
2.3.3	The Electrocapillary Equation	167
2.3.4	Zero-Charge Potentials	169
2.3.5	Specific Adsorption at Liquid–Liquid Interfaces	173
2.3.6	Adsorption Isotherm	176
2.3.7	Image Forces	179
2.3.8	Modified Poisson–Boltzmann (MPB) Model	183
	References	187
2.4	**Electrical Double Layers. Double Layers at Single-crystal and Polycrystalline Electrodes**	**188**
	Enn Lust	188
2.4.1	General Aspects	188
2.4.1.1	The Potential of Zero Charge and Electric Double-layer Structure	188
2.4.1.2	Differential Capacitance of Electrode–Electrolyte Solution Interface	188
2.4.1.3	Electrical Double-layer Structure at Solid Electrodes	189
2.4.1.4	Surface Roughness of Solid Electrodes	191
2.4.1.5	The Applicability of the Gouy–Chapman–Stern–Grahame Model to Solid Electrodes	191
2.4.1.5.1	Shape of C, E Curves	191
2.4.1.5.2	Parsons and Zobel Plot Method	192
2.4.1.5.3	Surface Roughness and Shape of Inner-layer Capacitance Curves	196
2.4.1.5.4	Polycrystalline Electrode Models	196
2.4.1.5.5	Debye Length–dependent Surface Roughness Model	199
2.4.1.5.6	Electric Double Layer and Fractal Structure of Surface	201
2.4.2	Experimental Aspects. Short Review of Data	202
2.4.2.1	Silver	202
2.4.2.2	Gold	204
2.4.2.3	Copper	204
2.4.2.4	Lead	205
2.4.2.5	Tin	205
2.4.2.6	Zinc	206
2.4.2.7	Cadmium	206
2.4.2.8	Bismuth	208
2.4.2.9	Antimony	209
2.4.2.10	Iron	210

2 Electrochemical Double Layers

2.4.2.11	Nickel	211
2.4.2.12	Pt-group Metals	211
2.4.3	General Correlations	214
2.4.3.1	Zero-charge Potential and Reticular Density of Planes	214
2.4.3.2	Potential of Zero Charge, Work Function, Interfacial Parameter, and Inner-layer Capacitance	214
2.4.3.3	Data of Quantum-chemical Calculations	220
2.4.3.4	Adsorption of Organic Compounds and Hydrophilicity of Electrode	221
	References	222
2.5	**Analyzing Electric Double Layers with the Atomic Force Microscope**	**225**
	Hans-Jürrgen Butt	*225*
2.5.1	Introduction	225
2.5.2	Surface Forces in Aqueous Electrolyte	226
2.5.2.1	DLVO Forces	226
2.5.2.2	Early Surface Force Measurements	227
2.5.2.3	The Surface Forces Apparatus	228
2.5.2.4	The Osmotic Stress Method	229
2.5.2.5	The Bimorph Surface Force Apparatus	229
2.5.3	Force Measurements with the AFM	230
2.5.3.1	The Practice of Force Measurements	230
2.5.3.2	The Cantilever	232
2.5.3.3	The Probe: Tip or Colloidal Particle	233
2.5.3.4	An Instability: The Jump-in	234
2.5.3.5	AFM-related Techniques	235
2.5.4	Theory of Measuring Surface Forces with the AFM in Aqueous Electrolyte	236
2.5.4.1	Derjaguins Approximation	236
2.5.4.2	The Electrostatic Double-layer Force	237
2.5.4.3	Other Surface Forces in Aqueous Electrolyte	238
2.5.5	AFM Force Measurements of the Electric Double Layer	240
2.5.5.1	Between Solid Surfaces	240
2.5.5.2	High Surface Potentials	241
2.5.5.3	Between Deformable Surfaces	242
2.5.5.4	Imaging Charge Distributions	244
2.5.6	Conclusion and Outlook	248
	References	248
2.6	**Comparison Between Liquid State and Solid State Electrochemistry**	**253**
	Ilan Riess	*253*
	Abstract	*253*
2.6.1	Introduction	253

2.6.2	Electrochemical Cell Geometry	254
2.6.2.1	Cells Based on LEs	254
2.6.2.2	Cells Based on SEs	255
2.6.2.3	Comparing the Two Groups of Cells	256
2.6.3	Charge Transport in Electrolytes	257
2.6.3.1	Charge Transport in LEs	257
2.6.3.2	Charge Transport in SEs	258
2.6.3.3	Summarizing the Differences Between Transport in LEs and SEs	261
2.6.4	Theoretical Current Voltage Relations for Ideal ECs (No Electrode Polarization)	261
2.6.4.1	General Current Density Equations	261
2.6.4.2	Half Cells	262
2.6.4.3	Supporting Electrolyte	262
2.6.4.4	Boundary Conditions	263
2.6.4.5	Relations Between Open-Circuit Voltage and the Nernst Voltage	263
2.6.4.6	Chemical Diffusion Coefficient	265
2.6.4.7	$I-V$ Relations	266
2.6.5	Methods for Characterizing ECs	266
2.6.5.1	Methods for Characterizing LEs and SEs/MIECs	266
2.6.5.2	Methods for Characterizing Elements in the EC	268
2.6.6	Electrode Processes and Overpotentials	268
2.6.6.1	General	268
2.6.6.2	Inner (Galvani) Electrical Potential	269
2.6.6.2.1	Inner Electrical Potential in LSE	269
2.6.6.2.2	Inner Electrical Potential in SSE	270
2.6.6.3	$I-V$ Relations for Charge-Transfer Processes in LSE	271
2.6.6.3.1	The Butler–Volmer Equation for an Elementary Step	271
2.6.6.3.2	The Butler–Volmer Equation for Multistep Reactions	273
2.6.6.4	$I-V$ Relations for Charge-Transfer Processes in SSE	274
2.6.6.4.1	General	274
2.6.6.4.2	Ion Transfer	274
2.6.6.4.3	Electron Transfer	275
2.6.6.4.4	Further Discussion of Ion and Electron Transfer	276
2.6.6.4.5	Relating the Overpotential to a Difference in a Chemical or Electrochemical Potential	277
2.6.6.5	Conclusions for Charge-Transfer Processes	278
2.6.6.6	Diffusion	278
2.6.7	Galvani Potential Distribution versus Electrochemical Potential Distribution	278
2.6.8	Heterogeneous Catalysis in SSE	279
2.6.9	Summary	279
	References	279

2.7	**Polyelectrolytes in Solution and at Surfaces**	**282**
	Roland R. Netz	*282*
	David Andelman	*282*
	Abstract	282
2.7.1	Introduction	282
2.7.2	Neutral Polymers in Solution	283
2.7.2.1	Flexible Chain Statistics	283
2.7.2.2	Semiflexible Chain Statistics	285
2.7.3	Properties of Polyelectrolytes in Solution	287
2.7.3.1	Isolated Polyelectrolyte Chains	288
2.7.3.1.1	Manning Condensation	293
2.7.3.1.2	Self-avoidance and Polyelectrolyte Chain Conformations	294
2.7.3.2	Dilute Polyelectrolyte Solutions	295
2.7.3.3	Semidilute Polyelectrolyte Solution	296
2.7.4	Adsorption of a Single Polyelectrolyte Chain	299
2.7.4.1	Adsorption on Curved Substrates	302
2.7.5	Adsorption from Semidilute Solutions	304
2.7.5.1	Mean-field Theory and its Profile Equations	304
2.7.5.2	Numerical Profiles: Constant ψ_s	306
2.7.5.3	Scaling Results	307
2.7.5.3.1	Low-salt Regime $D \ll \kappa^{-1}$ and $\psi_s =$ constant	308
2.7.5.3.2	High-salt Regime $D \gg \kappa^{-1}$ and $\psi_s =$ constant	309
2.7.5.4	Overcompensation of Surface Charges: Constant σ	313
2.7.5.4.1	Low Salt Limit: $D \ll \kappa^{-1}$	313
2.7.5.4.2	High Salt Limit: $D \gg \kappa^{-1}$	314
2.7.5.5	Final Remarks on Adsorption from Semidilute Solutions	314
2.7.6	Polyelectrolyte Brushes	315
2.7.7	Conclusion	318
	Acknowledgment	319
	References	320

2.1
Electrochemical Interfaces: At the Border Line

Alexei A. Kornyshev
Research Center Jülich, Jülich, Germany and Imperial College of Science, Technology and Medicine, London, UK

Eckhard Spohr
Research Center Jülich, Jülich, Germany

Michael A. Vorotyntsev
Université de Bourgogne, Dijon, France

2.1.1
Introduction

Most of the events in electrochemistry take place at an interface, and that is why *interfacial electrochemistry* constitutes the major part of electrochemical science. Relevant interfaces here are the metal–liquid electrolyte (LE), metal–solid electrolyte (SE), semiconductor–electrolyte, and the interface between two immiscible electrolyte solutions (ITIES). These interfaces are chargeable, that is, when the external potential is applied, charge separation of positive and negative charges on the two sides of the contact occurs. Such an interface can accumulate energy and be characterized by electric capacitance, within the range of *ideal polarizability* beyond which Faraday processes turn on.

The interfaces dielectric–electrolyte are less important in classical electrochemistry, but are central for bioelectrochemistry and colloid science. These are, e.g., ionic crystal–electrolyte, lipid-bilayer–electrolyte, and protein–electrolyte interfaces. Such interfaces are not chargeable, but they may contain charges attached to surfaces. All these interfaces, chargeable or containing fixed charges, zwitter-ions or dipoles, are called *electrified interfaces*.

Properties of the interface play a crucial role in electrochemical energy conversion, electrolysis, electrocatalysis, and electrochemical devices. On the other hand, the chargeable interface can and has been widely used to probe various surface properties. Indeed, due to Debye screening in electrolyte, the electric field at the electrochemical interface is localized in a narrow interfacial region and the interface can be easily charged to tens of $\mu C\,cm^{-2}$ at quite modest potentials. Physicists called electrochemistry *a surface science with a joystick*. The potentiostat is the joystick, which allows to vary the electric field at the interface from zero to $10\,V\,Å^{-1}$. Measuring the given signal dependence on the potential modulation makes it possible to distinguish the terms related to the interface, since the electric field in equilibrium is zero in the bulk. That joystick was not available in surface science, and thus physicists started to immerse solid surfaces into the electrolyte environment in order to probe the interface response to charging.

Distribution of potential near a charged interface is crucial in electrode kinetics, since the potential drop plays the role of a variable driving force of reaction. The reactions take place at the interface and with participation of adsorbed reactants. However, not only reactants but other species as well can adsorb at the interface. Competitive adsorption of these species, ions or dipoles can play, depending on the situation, a catalyzing or inhibiting role on the reaction rates.

On the whole, understanding the structure of the interface and its response to charging is most essential for electrochemistry and its implications for colloid science and biophysics.

The electrochemical interface is composed of molecules (solvent, adsorbed molecular species) and ions (of electrolyte), which can be partially discharged when chemisorbed, electrons and skeleton ions in the case of metal electrodes, electrons and holes in the case of semiconductor electrodes, mobile conducting and immobile skeleton ions in SEs. Molecules and ions are classical objects but electrons, holes with small effective mass, and protons are quantum objects. Interaction between molecules and surfaces is quantum-mechanical in nature in the case of chemisorption. Thus, microscopic description of the interface requires a combination of quantum and classical methods. One can benefit, however, from simple or more involved phenomenological descriptions of the interface.

In this chapter we will focus on chargeable interfaces: metal–LE interface and ITIES. Although similar in concepts and methods, the extensive research area of semiconductor/electrolyte and metal/SE interfaces will not be considered here. We will start with phenomenology, and then move, where possible, to a microscopic theory and simulations. Without a solid phenomenological basis one may risk calculating things from the scratch many times, whereas the calculation of one parameter of the phenomenological theory might be all that is actually needed from a microscopic model! On the other hand, the phenomenology often receives its justification and validity criteria from microscopic theories.

2.1.2
Basic Concepts

Electrified interfaces represent the principal object of "interfacial electrochemistry". Various aspects of this area are described in monographs and textbooks on this subject [1–11]. A brief outline of this information is given in the following text.

Electrified interfaces represent a specific quasi two-dimensional object in the vicinity of the geometrical boundary of two phases containing, in general, mobile electronic and/or ionic charges, and, correspondingly, an electric-field distribution generated by these charges. Practically at least one of these media in contact is conducting, while the second one may be either a conductor of another kind (metal–electrolyte interface, semiconductor–electrolyte interface), or of the same kind (interface of two immiscible electrolytes), or an insulator. As important exceptions, one can point out pure solvent/air or pure solvent/insulator (e.g. silica-water) boundaries where a significant potential drop may be created by the formation of a dipole layer, without the generation of a space-charge region.

The simplest example is given by the free metal surface. Owing to the Pauli principle the metal electrons at the highest occupation levels possess a significant kinetic energy so that their density extends outside the ionic skeleton of the electrode. It results in the formation of a structure characteristic for numerous electrified interfaces; two oppositely charged spatial regions (with zero overall charge), this time a positively charged region inside the ion skeleton due to the depletion of electrons and a negatively charged region formed by "the electronic tail". Because of this "nanocondenser", the bulk metal has got a high positive potential of several Volts with respect to the vacuum. This manifests itself in the value of the electron work function the work needed to withdraw an electron from the metal across its uncharged surface.

A more complicated example is provided by the contact of two electronic conductors, metals or semiconductors. At equilibrium the electrochemical potentials of electrons (i.e. their Fermi energies) must be equal in both bulk phases. The Galvani potential difference between them is determined by the bulk properties of the media in contact, i.e. by the difference between the chemical potentials of electrons:

$$\Delta\varphi^{m1/m2} = \frac{\mu_e^{m1} - \mu_e^{m2}}{e} \quad (1)$$

where e is the elementary charge, since the right-hand-side terms are constant, one cannot change the interfacial Galvani potential in equilibrium conditions without changing the chemical composition of the phases. The difference between the bulk potentials required by Eq. (1) has to be established by the formation of an "electrical double layer" (EDL) owing to the electron exchange between the surface layers of the phases. Finally, one of the surface layers acquires an extra electronic charge, while the adjacent layer of another phase contains an excessive positive charge formed by the uncompensated ionic skeleton. The thickness of each charged layer depends on the screening properties of the corresponding medium (their Thomas–Fermi or Debye screening lengths) so that the whole space-charge region is very thin for a contact of two metals (within 0.1 nm), while it may be much more extended in the case of semiconductors, depending on their bulk properties.

In all cases such interfaces are *nonpolarizable*, since the only way to change the interfacial potential difference is to modify the *bulk* properties of the phase(s), for example, the degree of doping or stoichiometry for semiconductors.

Analogous ionic systems are represented by a contact of two electrolyte solutions in immiscible liquids, which contain a common ion. Owing to its interfacial exchange the electrochemical potential of this ion is to be constant throughout the system, and it leads again to a relation for the interfacial potential difference similar to Eq. (1) (called *Donnan equation* this time), that is, the interface is nonpolarizable without a change of the *bulk media* properties.

The same process of interfacial ion exchange determines the potential difference for other ion conductors, like membranes or SEs. The existence of this potential difference again leads to the formation of the EDL, but this time it is formed by excessive ionic charges in the surface layers of both solutions. Since this EDL is created by the ion transfer across the interface, without charge transport across the bulk media, the Donnan potential drop corresponding to the composition of the phases is established very rapidly.

Similar to the contact of two electronic conductors, the imposition of a *nonequilibrium* potential difference between two ion conductors (having a common ion) from the external circuit results in a dc current passage. Subsequent processes depend on the type of electrode reactions at the reversible electrodes immersed into the ion-conducting phases. If both reactions generate or consume the same exchangeable ion, even a long-term realization of the process does not change the bulk composition of the phases. In the case of the electrode reactions generating or consuming different species, the longtime application of this potential difference will lead to *electrolysis*, that is, to the gradual change of the composition of the bulk phases. The latter is accompanied by the change of the equilibrium values of all interfacial potential drops (between ion conductors and at both reversible electrodes), with the approach to a different equilibrium potential difference.

The contact of solutions that possess more than one common ion, in particular all contacts of solutions with identical or miscible solvents, exhibits a more complicated phenomenon. At the initial stage, all exchangeable ions are transferred across the boundary generating a long-living nonequilibrium structure, "liquid junction (LJ)", which retains its "diffusion potential difference" for an extended time, despite the absence of the thermodynamic equilibrium (different compositions) between the interphase and two bulk phases. If the solutions in contact have the same solvent, this potential drop is usually relatively small, mostly within a few dozen mV or less (but it may sometimes reach a much larger value for dilute solutions even in the same solvent). Elaborated procedures for its calculation as a function of the solutions' composition exist [12]. For nonidentical solvents, this potential difference may be much larger, being primarily dependent on the solvent properties.

The principal object of electrochemical interest is given by another type of electrified interface, contacts of an electronic (liquid or solid metal, semiconductor) and an ionic (liquid solution, SEs, membranes, etc) conductor. For numerous contacts of this kind, one can ensure such ionic composition of the latter that there is practically no dc current across the interface within a certain interval of the externally applied potential. Within this potential interval the system is close to the model of an *ideally polarizable interface*; the change of the potential is accompanied by the relaxation current across the external circuit and the bulk media that vanishes after a certain period. For sufficiently small potential changes, dE, the ratio of the integrated relaxation current, dQ, to dE is independent of the amplitude and it determines the principal electrochemical characteristics of the interface, its differential capacitance per unit surface area, C:

$$C = S^{-1}\frac{dQ}{dE} \qquad (2)$$

S being the surface area. Its value can be measured for each potential within the ideal polarizability interval, and it varies generally as a function of this potential as well as of the nature of the electrode and the solution composition. The EDL for such systems contains an excess (positive or negative) electronic charge at the electrode surface and the ionic countercharge at the solution side. The possibility to freely change the interfacial potential difference implies a qualitative analogy to a condenser whose separated charge is directly related to the above-mentioned quantity, dQ.

This simple picture gives a reasonable idealization of the interface in the case of when the electronic and ionic charges of the EDL are separated by a layer of the solvent molecules, as it is in the case of a "surface-inactive electrolyte" discussed in the following text. The specific adsorption of the ionic-solution components, that is, their significant accumulation in the immediate contact with the metal surface, is usually accompanied by the strong redistribution of the electronic charge near each adsorbed species. As a result they may possess a much lower charge, compared to the one in the bulk solution (it is termed as adsorption with a *partial* or *complete* charge transfer). Under these conditions the charge dQ supplied via the external circuit is not equal to the change of the "free electronic charge" distributed along the surface of the metal. Such "perfectly polarizable electrodes" require a more elaborated description; see Chapter 3.1 of this book [13].

At approaching the limits of the ideal-polarizability interval, the rate of faradaic processes at the interface increases rapidly (discharge of the solvent or solute components), with a dc current passing through the system.

The existence of this "ideal-polarizability" range of potentials is of great practical importance, since it allows one to study an individual faradaic process of the added electroactive species in the presence of the *background electrolyte*, whose presence provides important advantages (diminished Ohmic potential drop, rapid charging of the EDL, etc).

A specific type of the electrified interface is given by the contact of an insulating phase with electrolyte solutions. Since the charged species cannot cross the insulator, the EDL formation originates from the ionization process of surface groups (most frequently, the proton-based dissociation or association) or/and the adsorption of ionic components of the solution. As a result the EDL includes this "fixed" charge at the surface and the counterions in the solution part. For a further discussion, see Ref. [14].

Such charged interfaces are typical for various ionic or polar solids, for ion exchange, lipid or biological membranes, or for any insulating surface in the presence of amphiphilic solute species. Even though one cannot polarize this interface from the external source, the potential difference and the separated charge can change as a function of the solution composition, in particular its pH and ionic strength.

In many cases the solvent molecules and solutes penetrate into the surface layer of the insulator, thus forming a "gel layer" composed of the components of both phases. Its thickness is sometimes significant. Then the charge distribution across such *interphase* is quite different from the simple "condenser" idealization, in particular the overall potential difference between the bulk phases is composed of two contributions for the insulator–gel layer and gel layer–solution interfaces, with a plateau between them, if the gel layer is sufficiently thick [15].

This two-step profile of the potential across the interphase is also characteristic for certain types of *modified electrodes* in which the metal surface is coated with a film, whose thickness significantly exceeds the atomic scale. Such systems represent a much more complicated type of electrified interfaces, since the distribution of charged species depends crucially on the specific properties of the film. In most cases of sufficiently thick films, the profile of the Galvani potential across the interphase possesses a plateau inside the bulk film separating two potential drops at its interfaces with the electrode and the solution [16].

Traditionally, special attention in electrochemistry has been paid to the metal–solution interfaces. The interfacial potential drop and the corresponding electrode charge within the interval of the ideal polarizability can be regulated by the imposed external voltage. The change of this voltage requires a supply of extra electron and ion charges to both sides of the interface across the corresponding bulk media. This charging process is mostly limited by the transport in the solution (as well as inside the electrode in certain cases), since the relaxation time inside the interfacial region itself (Debye–Falkenhagen time determined by the double-layer thickness and the ion-diffusion coefficient, L_D^2/D) is extremely short. Generally the current distribution along the electrode surface is nonuniform due to geometrical reasons (shape and positions of electrodes), hydrodynamic structure, electrode surface properties, and so on. If the effect of these factors is sufficiently small, the charging process can be described by a simple RC circuit composed of the Ohmic resistance of the solution and the interfacial capacitance (multiplied by the surface area, S) defined in Eq. (2). This result is widely used for the measurement of this important interfacial characteristic, with the use of chronoamperometry or electrochemical impedance.

For liquid metals (mercury, gallium) or their alloys, one can measure another interfacial quantity, interfacial tension equal to the specific energy of the interface formation, γ, at different values of the electrode potential, E [1, 17]. At equilibrium the Lippmann equation relates it to the electrode-charge density σ, that is, to the charge Q in Eq. (2) per unit surface area:

$$\sigma = \frac{Q}{S} = -\frac{d\gamma}{dE} \quad (3)$$

This equation enables one to completely determine the charge–potential relation. Moreover, the knowledge of the interfacial tension–potential curves for various electrolyte concentrations enables one to find all other equilibrium characteristics of the interface, without any additional measurements, see the following text.

However, this electrocapillary approach to the interfacial properties is actually less popular than the one based on capacitance curves, because the latter has been more developed instrumentally and can also be used in the case of solid electrodes. Furthermore, it does not require a graphical/numerical differentiation of the experimental curve, which is related to an increase of the measurement errors.

Within the latter method, the charge–potential relation is obtained by the numerical integration of capacitance data for various potentials, according to Eq. (2). The difficulty of this treatment is to find the integration constant, that is, the value of the charge for some potential must be known.

For liquid electrodes, one can directly measure the potential of zero charge (p.z.c.), $E_{\sigma=0}$, with the use of an open-circuit streaming electrode. Then the charge Q for any potential E is obtained by integration of the capacitance curve, $C(E)$, from $E_{\sigma=0}$ to E. For solid electrodes, the experimental p.z.c. determination represents a serious problem since only indirect methods are available [18, 19]. Moreover, the very definition of this important characteristics is ambiguous in this case because of the crystallographic inhomogeneity of the solid surfaces (except for perfect single-crystal faces), a time-variation of their properties and the adsorption of the solution components with the partial charge transfer, see [13, 20] for a further discussion. For systems in which these complications are assumed to be not especially important, the most popular method of the p.z.c. determination is to measure capacitance curves for a series of dilute concentrations of an electrolyte, which is not adsorbed specifically. If the capacitance minimum is observed at about the same potential and the minimum depth increases monotonously with diluting the solution, then this potential of the minimum is identified to $E_{\sigma=0}$ (for symmetrical electrolytes). A more thorough analysis includes the treatment of the whole capacitance curves in accordance with the Grahame theory, see the following text. As soon as the p.z.c. value has been determined, one can calculate the charge-potential curves similar to the case of liquid electrodes. The transition to the charge density, σ, requires the knowledge of the real surface area, A. For liquid electrodes, the latter can be identified with the geometrical area, while its determination is much more complicated in the case of a solid surface [21, 22], see Sect. 2.1.6.

2.1.3
Thermodynamics of Electrified Interfaces

In the aforementioned discussion of Eqs. (2) and (3), the solution composition was considered to be fixed. A comparison of the data for different electrolyte concentrations opens a way to extract extensive additional information on the interfacial properties on the basis of the Gibbs thermodynamics. To exploit this possibility one has to deal with **electrode potentials** (instead of the Galvani potential). This quantity is defined as the voltage between the working electrode and the reference electrode immersed to the same solution and reversible with respect to one of its ionic components, for example, $E_\pm$ for such cation- or anion-reversible electrodes in the case of a binary electrolyte solution, $M^{z_i}A^{-z_i}$. Then the electrocapillary equation gives a relation of the above-mentioned quantities to the chemical potential of the salt, $\mu = \mu_+ + \mu_-$, and the Gibbs adsorptions of cations or anions, $\Gamma_\pm$:

$$d\gamma = -\sigma\, dE_+ - \Gamma_-\, d\mu$$
$$\equiv -\sigma\, dE_- - \Gamma_+\, d\mu \quad (4)$$

Hence,

$$\Gamma_+ = -\left(\frac{\partial \gamma}{\partial \mu}\right)_{E_-}, \quad \Gamma_- = -\left(\frac{\partial \gamma}{\partial \mu}\right)_{E_+} \quad (5)$$

It means that the knowledge of the interfacial tension as a function of any electrode potential and the chemical potential of the electrolyte, that is, of its concentration, enables one to determine all other interfacial quantities by differentiation (Eqs. (3) and (5)).

It is often more useful to choose another electric variable, electrode charge (density) σ instead of electrode potentials. Then, one needs to make a transition to the interfacial pressures (dependent on the aforementioned choice of the reference electrode):

$$\xi_\pm = \gamma + \sigma E_\pm \quad (6)$$

for which

$$d\xi_+ = E_+\, d\sigma - \Gamma_-\, d\mu,$$
$$d\xi_- = E_-\, d\sigma - \Gamma_+\, d\mu \quad (7)$$

One by one, the coefficients in these relations may be represented as functions of the same variables, σ and μ:

$$dE_+ = C^{-1}\, d\sigma - t_-^{m/s}\frac{d\mu}{z_i e},$$
$$dE_- = C^{-1}\, d\sigma + t_+^{m/s}\frac{d\mu}{z_i e} \quad (8a)$$

$$d\Gamma_+ = -t_+^{m/s}\frac{d\sigma}{z_i e} + C_\mu \frac{d\mu}{z_i^2 e^2},$$

$$d\Gamma_- = -t_-^{m/s}\frac{d\sigma}{z_i e} + C_\mu \frac{d\mu}{z_i^2 e^2} \quad (8b)$$

Because of the general relations:

$$E_+ - E_- \equiv -\frac{\mu}{z_i e}, \quad \sigma \equiv z_i e(\Gamma_- - \Gamma_+) \quad (9)$$

the differentials (8a, 8b) contain only three independent characteristics of the interface, each of them providing important information on its structure [23–25]. C is the interfacial capacitance per unit surface area, see Eq. (2); $t_+^{m/s} = 1 - t_-^{m/s}$ are the "interfacial numbers of cations and anions": they show which fraction of the electrode charge variation, $d\sigma$, is compensated by the corresponding ion (at a constant solution composition)

$$t_+^{m/s} = \frac{\partial \sigma_+}{\partial \sigma}, \quad t_-^{m/s} = \frac{\partial \sigma_-}{\partial \sigma} \quad (10)$$

$\sigma_\pm = \pm z_i e \Gamma_\pm$ are the partial ion charges inside the EDL; alternatively, one can use partial cation and anion capacitances related to the contributions of these ions into the change of the whole ion countercharge:

$$C_+^{m/s} \equiv t_+^{m/s} C = \frac{\partial \sigma_+}{\partial E},$$

$$C_-^{m/s} \equiv t_-^{m/s} C = \frac{\partial \sigma_-}{\partial E},$$

$$(\mu = \text{constant}, E = E_+ \text{ or } E_-) \quad (11)$$

$C_\mu = z_i^2 e^2 \partial \Gamma_\pm / \partial \mu$ at constant electrode charge, see the discussion in the following text.

One should emphasize that all these quantities represent **equilibrium** characteristics of the interface, and the aforementioned identities allow one to calculate the others if the charge vs. potential dependence has been found for different electrolyte concentrations [24]. However, this procedure based on integration of $t_+^{m/s} \equiv z_i e \partial E_+/\partial \mu$ over the charge to determine the Gibbs adsorptions of ions, $\Gamma_\pm$, requires to know an "integration constant", that is, the values of $\Gamma_\pm$ as a function of the electrolyte concentration for some electrode charge, for example, $\sigma = 0$. The slope of the latter dependence is directly related to the aforementioned parameter C_μ.

Such information (supplementary to capacitance data) may be obtained from electrocapillary experiments. According to Eq. (3), the interfacial tension, γ, passes the maximum at the p.z.c. as a function of the potential (at a fixed concentration). In dilute solutions this maximum value of γ for the Hg–water interface is almost independent of the concentration of the "surface-inactive electrolyte" (see following text). The data for more concentrated solutions recalculated with the use of Eq. (5) show a linear dependence on the bulk concentration, c:

$$\Gamma_\pm = -z_H c \quad (12)$$

where the factor, z_H, is within 0.2–0.4 nm [26, 27].

For the interpretation of this finding, one must recollect that the Gibbs adsorption represents an integrated difference of the actual concentration profile for this species across the interface, and the "unperturbed" one in which the concentration is equal to the bulk value up to the Gibbs plane and zero beyond it. The position of the Gibbs adsorption plane is chosen here from the condition of the zero adsorption of the solvent.

Since the electrolyte concentration in the above-mentioned experiments was not too high (less than or about a mole per liter), the Gibbs plane is sufficiently close to the physical boundary between the metal and the adjacent solvent layer. Then the above *negative* values of ion Gibbs adsorptions at the p.z.c. and the linear slope in Eq. (11) points to the existence of an *ion-free layer* of the solvent near the metal surface, that is, a *compact layer,* the existence of which has been postulated long before these experiments. The factor in this relation, z_H, corresponds to *its thickness*, that is, to the distance of the closest approach of the centers of nonadsorbing ionic species to the metal surface, compared with the similar quantity for the solvent (Gibbs adsorption plane).

Surprisingly this important experimental information on the properties of the compact layer has been ignored in modern microscopic theories of the interfacial structure, while the thickness of the ion-free layer at the electrode surface could be calculated on the basis of the distribution functions for the solvent and ions.

Effects of the interfacial charging are of importance in various kinetic phenomena, in particular in the ion transport. The simplified treatment assumes that these effects can be adequately described by a single parameter, the capacitance C. However, the passage of a nonstationary current is accompanied by the time-variation of the electrolyte concentration outside the EDL region, and it must give an additional contribution to the change of the interfacial potential, Eq. (8a). Besides, there is generally an imbalance between the partial ion fluxes inside the diffusion layer and the time derivatives of the corresponding partial ionic charges, $\sigma_\pm$, Eq. (10). These factors result in the dependence of the nonstationary response of the system on all three interfacial parameters, C, $t_+^{m/s}$ and C_μ.

Interfacial thermodynamics has become one of the bases of the theory of the electrochemical phenomena determined by the charge transport, in particular in modeling the electrochemical impedance [25, 28]. The analysis of the results obtained within the framework of this approach has shown that for transport processes under usual electrochemical conditions, the characteristic timescales of the interfacial charging and of the diffusion process are well separated, and one can use a simplified treatment in which the interfacial charging effects are characterized by a single parameter, C. In particular, the relaxation after a potential step or the impedance in the proper frequency range may be represented as an in-series combination of the solution resistance and the interfacial capacitance. On the other hand, the interfacial contribution into the impedance arising from the perturbation of the electrolyte concentration outside of the EDL becomes essential for *thin-layer* systems in which the characteristic frequencies are overlapping or insufficiently separated.

2.1.4
Classical Models of Electrified Interfaces

2.1.4.1 Helmholtz Model

The thermodynamic treatment does not allow one to extract more detailed information on the interfacial structure. The already discussed experimental results for the ion-free solvent layer may be considered as an illustration of the simplest description of the interface in the Helmholtz model. The latter postulates two charged planes containing the extra electrons and the counterions, which are separated by a solvent ("compact") layer having thickness z_H, that is, the system represents a nanocondenser. Then the ratio of the electronic charge density at the electrode, $\sigma = Q/S$, and the interfacial potential difference, $\Delta\varphi^{m/s}$, gives the integral EDL (interfacial) capacitance (per unit surface area):

$$K_H = \frac{\sigma}{\varphi_H} \quad (13)$$

where

$$\varphi_H = \Delta\varphi^{m/s} - \Delta\varphi^{m/s}{}_{\sigma=0};$$
$$\Delta\varphi^{m/s} = E + \Delta\varphi^{ref} \quad (14)$$

Here, $\Delta\varphi^{ref}$ is the potential drop at the reference electrode. Another constant term, $\Delta\varphi^{m/s}{}_{\sigma=0}$, corresponds to the potential drop at the p.z.c. In the contemporary interpretation, its nonzero value is attributed to the potential variation inside the surface layer of the metal (affected by the adjacent solvent molecules) as well as to the polarization of the solvent layer, see Sect. 2.2. Equation (13) may also be rewritten as:

$$C_H = \frac{d\sigma}{d\varphi_H}; \quad C_H \equiv \left[\frac{d\left(\frac{\sigma}{K_H(\sigma)}\right)}{d\sigma}\right]^{-1} \quad (15)$$

containing the differential capacitance of the Helmholtz layer.

Equations (13)–(15) reproduce qualitatively the experimental findings for sufficiently concentrated solutions of surface-inactive electrolytes. In accordance with expectations, the capacitance values and their dependence on the interfacial potential are characteristic for each particular metal/solvent contact. One also observes a noticeable (sometimes, even strong) variation of the capacitance with the electrode charge. Equation (13) predicts a distorted (due to the dependence of K_H on σ) parabolic shape for $\gamma(E)$, with a maximum at p.z.c., see Eq. (3), also in conformity with experimental observations.

In electrochemical kinetics, this model corresponds to the Butler–Volmer equation widely used for the electrode reaction rate. The latter postulates an exponential (Tafel) dependence of both partial faradaic currents, anodic and cathodic, on the overall interfacial potential difference. This assumption can be rationalized if the electron transfer (ET) takes place between the electrode and the reactant separated by the above-mentioned compact layer, that is, across the whole area of the potential variation within the framework of the Helmholtz model. An additional hypothesis is the absence of a strong variation of the "electronic transmission coefficient", for example, in the case of adiabatic reactions.

The fundamental drawback of the classical Helmholtz picture is its inability to provide any explanation for the dependence of the electrode charge (or of the capacitance) on the electrolyte concentration, which is observed experimentally in sufficiently dilute solutions. Besides, there is a large group of electrokinetic phenomena that demonstrate a much greater extension of the EDL than it is assumed in the Helmholtz condenser model (see Sect. 5.4).

2.1.4.2 Gouy–Chapman–Stern–Grahame Model

Gouy [29] and Chapman [30] independently proposed an alternative treatment. It is based on an analytical solution of the nonlinear Poisson–Boltzmann equation for the electric potential created by a system of mobile point charges near a charged wall imitating an electrode surface. However, in the original form of this theory, most of its predictions were at variance with experimental data.

The situation changed radically after the suggestion of Stern [31] to combine the two approaches by Helmholtz and

Gouy–Chapman (GC), that is, to consider the whole interphase as consisting of two layers, a compact (or "inner" or "Helmholtz") and a "diffuse" one. The former corresponds to the aforementioned hypothesis of an ion-free layer of the solvent at the metal surface. The diffuse layer is located between the compact one and the bulk solution and the whole counterion charge is distributed inside this region in accordance with the GC theory.

The version of this theory, which is actually used in interpretation of all data for surface-inactive electrolytes, was proposed by Grahame [32]. The total potential drop across the interface is written as a sum of the compact- and diffuse-layer contributions:

$$\Delta\varphi^{m/s} \equiv E + \Delta\varphi^{ref} = \varphi_H(\sigma) + \varphi_{GC}(\sigma, c) \quad (16)$$

where both terms depend on the electrode charge density, σ, while only the second term depends on concentration c. The latter corresponds to the GC theory, for example, in the solution of a symmetrical single-charge electrolyte:

$$\varphi_{GC}(\sigma, c) = \left(\frac{2e}{\varepsilon L_B}\right)$$
$$\times \ln\left[\frac{\sigma}{\sigma^*} + \left(1 + \left(\frac{\sigma}{\sigma^*}\right)^2\right)^{1/2}\right] \quad (17)$$

Here,

$$\sigma^* = \frac{e}{2\pi L_B L_D} \quad L_D = \left(\frac{1}{8\pi L_B c}\right)^{1/2}$$

$$L_B = \frac{e^2}{\varepsilon k_B T} \quad (18)$$

ε is the bulk solvent dielectric constant, k_B the Boltzmann constant, L_D is the Debye screening length of the electrolyte solution and L_B is the Bjerrum length.

Division of each term in Eq. (16) by σ relates the integral capacitance of the whole interface, $K \equiv \sigma/(E - E_{\sigma=0})$, to those of the compact and diffuse layers, K_H in Eq. (13) and $K_{GC} \equiv \sigma/\varphi_{GC}(\sigma, c)$:

$$K^{-1} = K_H^{-1} + K_{GC}^{-1} \quad (19)$$

Differentiation of Eq. (16) gives a similar relation for the corresponding differential capacitances:

$$C^{-1} = C_H^{-1} + C_{GC}^{-1} \quad (20)$$

see Eqs. (2) and (15).

$$C_{GC}(\sigma, c) \equiv \frac{d\sigma}{d\varphi_{GC}} = \frac{\varepsilon}{2e} L_B$$
$$\times [\sigma^2 + (\sigma^*)^2]^{1/2} \quad (21a)$$

For aqueous solutions at room temperature

$$C_{GC}(\sigma, c) = 19.46[\sigma^2 + 137.8\,c]^{1/2} \quad (21b)$$

where C_{GC} is in units of $\mu F\,cm^{-2}$, σ in $\mu C\,cm^{-2}$, and c in M. In particular, at the p.z.c.

$$C_{GC}(0, c) = 228.5\,c^{1/2} \quad (21c)$$

One can rewrite Eqs. (17) and (21a) to represent the charge and the capacitance as a function of the potential drop within the diffuse layer, φ_{GC}:

$$\sigma = \sigma * \sinh\left(\frac{e\varphi_{GC}}{2k_B T}\right)$$

$$C_{GC} = \frac{\varepsilon}{4\pi L_D} \cosh\left(\frac{e\varphi_{GC}}{2k_B T}\right) \quad (22)$$

However, these widely quoted formulas are useless for real treatment of capacitance data since the potential drop, φ_{GC}, is unknown. Therefore in order to apply Eqs. (20) and (21a), one must first find the charge–potential relations for all the available solution compositions.

The key feature of Grahame's approach, which determines its success in interpretation of experimental data, is its semiphenomenological character. It postulates that the compact (Helmholtz)-layer capacitance, C_H, is independent of the concentration of the "surface-inactive electrolyte" (which does not penetrate noticeably into the compact layer) but avoids to accept a model hypothesis on its dependence on the electrode charge, $C_H(\sigma)$, that is, the latter is to be determined from experimental data. On the contrary, the properties of the diffuse layer are completely defined by Eqs. (17) and (21a), without any unknown parameter.

The ratio of the two contributions in Eq. (20) depends on the values of the electrode charge and the electrolyte concentration, as well as on the particular type of the media in contact. The diffuse-layer capacitance, Eqs. (21a) and (21b), is growing with the increase of the absolute value of the charge and with the concentration. Besides, it becomes almost independent of the concentration far from p.z.c. Near this point, $\sigma = 0$, the diffuse-layer capacitance shows a minimum, which is becoming progressively deeper at diluting the solution: $C_{GC}(0, c) = 228\,\mu\mathrm{F\,cm^{-2}}$ for $c = 1\,\mathrm{M}$ but only $7.2\,\mu\mathrm{F\,cm^{-2}}$ for 0.001 M aqueous solution of a 1,−1 electrolyte. The compact-layer capacitance, C_H, is within $20{-}60\,\mu\mathrm{F\,cm^{-2}}$ for most "mercury-like" metals (Hg, Pb, In, Cd, Bi, Sb, etc) in aqueous and nonaqueous solutions, with much greater values for some other systems, such as Ag and Au single-crystal faces, up to $100{-}150\,\mu\mathrm{F\,cm^{-2}}$ near p.z.c.

As a result, for sufficiently concentrated solutions, about 1 M or higher, the diffuse-layer contribution in Eq. (20) represents in most cases only a minor correction to the compact-layer term. Then, one can even use the simple Helmholtz model for qualitative interpretation of the data. In particular, far from the p.z.c., this approximation is acceptable for all concentrations. On the contrary, the diffuse layer determines the shape of the capacitance curves in dilute solutions near the p.z.c. where the theory predicts a profound minimum whose depth increases with diminishing the electrolyte concentration. The minimum is usually well developed to about 0.01 M concentrations while in systems with a high hump of $C_H(\sigma)$ near $\sigma = 0$ (such as for Ag or Au single-crystal faces) a deep diffuse-layer minimum should already appear at 0.1 M. All these qualitative predictions of the Gouy–Chapman–Stern–Grahame (GCSG) theory are in line with extensive experimental information for numerous metal–solution interfaces for surface-inactive electrolytes (i.e. for systems without specific adsorption of solute species, see following text) [22, 33, 34].

2.1.5
Diffuse and Compact Layer Properties

2.1.5.1 Grahame Model in Comparison with Capacitance Data for Liquid Electrodes

Let us now discuss how to verify this theory quantitatively as well as to extract the information on the interfacial structure with the use of the data for the Hg/H$_2$O-NaF system [35, 36] as an illustration. The measured capacitance curves for several electrolyte concentrations are presented in Fig. 1. One can notice the expected diffuse-layer minimum for dilute solutions but the whole shape of the curves is sufficiently complicated, in particular within the range of negative potentials.

The streaming electrode measurements show a constancy of the p.z.c. value, $E_{\sigma=0}$, for different concentrations, which is considered as the necessary condition for the absence of a specific ionic adsorption. In this case one can obtain the charge-potential relations for all concentrations by the integration of the corresponding

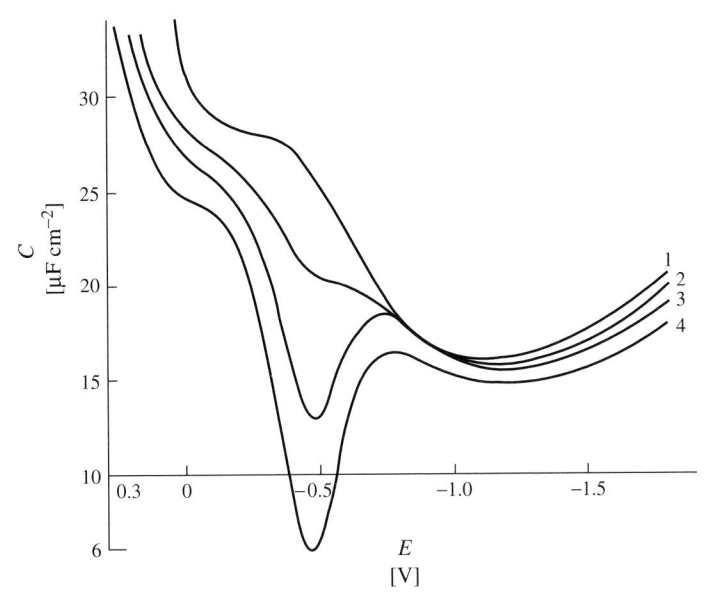

Fig. 1 Experimental capacitance curves for the mercury electrode in aqueous solutions of NaF. Concentrations: 0.916 M (1), 0.1 M (2), 0.01 M (3), 0.001 M (4) [35, 36].

capacitance curve, that is, one plots the capacitance versus the charge for all concentrations as in Fig. 2. These curves retain the diffuse-layer minimum at low concentrations but their behavior at high electrode charges has been strongly simplified, in particular all curves are practically identical at high negative charges.

Any justified comparison of measured data with theoretical predictions must take into account the experimental accuracy. However, the resulting error bars appear to depend strongly on the choice of the coordinate system, used for this treatment. In order to demonstrate the latter, let us assume for simplicity that the measured capacitance data in Fig. 1 have got the precision of $\pm 0.1 \, \mu F \, cm^{-2}$ for two highest concentrations, $\pm 0.2 \, \mu F \, cm^{-2}$ for the 0.01 M solution, and $\pm 0.5 \, \mu F \, cm^{-2}$ for 0.001 M, within the whole potential interval (in real measurements at low frequencies the dispersion also depends on the electrode charge and it may be even greater for millimolar solutions). The corresponding ranges of the dispersion are shown in Fig. 2 for two dilute solutions. Even though the initial dispersion in Fig. 1 was independent of the potential, the errors in Fig. 2 turned out to be markedly varying as a function of the electrode charge. They are close to those of the corresponding $C(E)$ curves at the zero charge and near the cathodic minimum. Then the errors are slightly enhanced at the dropping part of the cathodic branch, but they are several times smaller at lower electrode charges, where the capacitance varies very rapidly. The latter effect is due to a correlation of errors in the capacitance and the charge within this interval. It implies that even a considerable deviation between the theoretical prediction and experimental data in the initial $C(E)$ curves within this potential range may look rather small in the coordinates of Fig. 2.

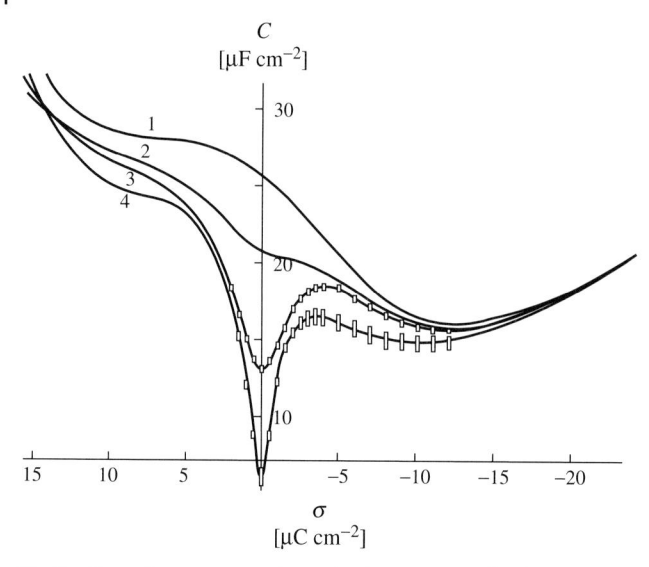

Fig. 2 Capacitance versus electrode charge curves for the mercury electrode in aqueous NaF solutions recalculated from the data in Fig. 1. The error bars were calculated from those in the capacitance versus potential curves (Fig. 1) that were assumed to have the precision of $\pm 0.1\,\mu\text{F cm}^{-2}$ for the two highest concentrations, $\pm 0.2\,\mu\text{F cm}^{-2}$ for the 0.01 M solution and $\pm 0.5\,\mu\text{F cm}^{-2}$ for 0.001 M, within the whole potential interval, as it is discussed in the text.

If one relies on the validity of the GC model for the diffuse-layer properties, one can use Eq. (20) to calculate the values of the compact-layer capacitance, C_H, from experimental data for C and theoretical formula (21a) for C_GC. This operation performed for all available values of the charge and the electrolyte concentration results in Fig. 3, in which the effect of the experimental dispersion is shown again for two dilute solutions.

According to Grahame's hypothesis, these curves for different concentrations should be coincident. One can notice first of all its brilliant confirmation in Fig. 3 for the two highest concentrations; a deviation is only seen at high positive charges, above $10\,\mu\text{C cm}^{-2}$. The proximity of the compact-layer capacitance curves is even more impressive for the 0.1 M and 0.01 M solutions, whose difference is inside the experimental error dispersion over the whole range of the electrode charges. This comparison enables one to conclude that the deviation of the data for the highest concentration at high positive charges is probably due to the violation of some assumption of the model, for example, dissolution of mercury or specific adsorption of cations. The concentration dependence of the diffuse-layer capacitance is probably too weak to explain this divergence, see in following text, even if the GC theory may be inapplicable quantitatively.

The curve for 0.01 M also demonstrates a marked increase of the experimental dispersion near the zero charge. This effect is even more pronounced for the 0.001 M solution in which the dispersion does not exceed $\pm 0.8\,\mu\text{F cm}^{-2}$, except for the range of very small charges where the

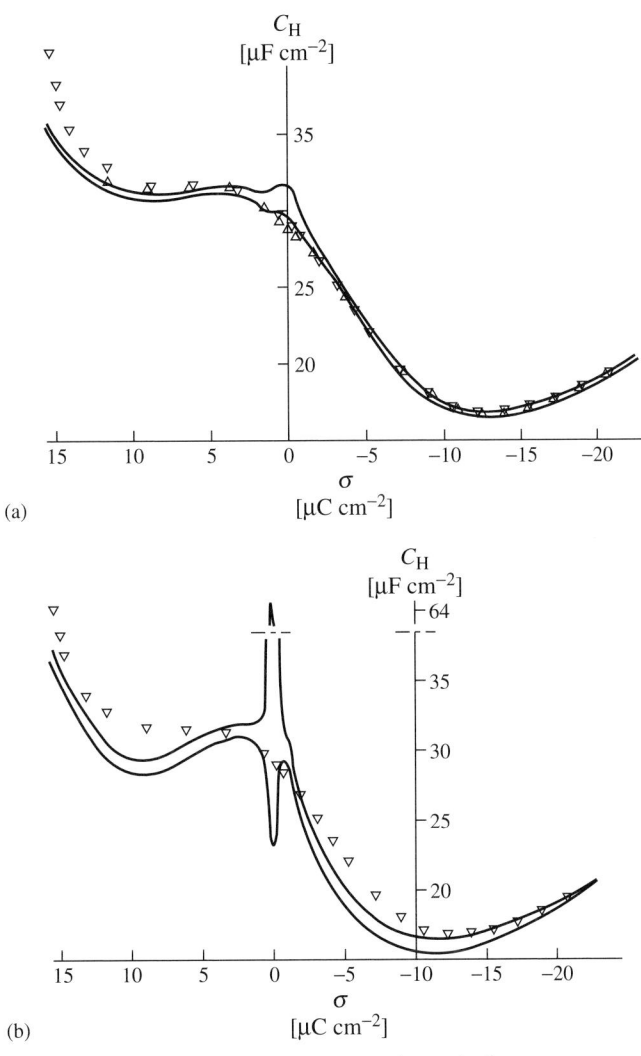

Fig. 3 Compact-layer capacitance versus electrode charge curves, $C_H(\sigma)$, for mercury electrode in aqueous NaF solutions calculated from the data in Fig. 2 with the use of Eqs. (20) and (21a), for each electrolyte concentration, 0.916 M (∇), 0.1 M (Δ), 0.01 M (solid line in Fig. 3a) and 0.001 M (solid line in Fig. 3b). Solid lines for lower concentrations were obtained with the use of the error bars in Fig. 2 for the corresponding curves. The error bars are small at the two higher concentrations and are not shown.

uncertainty in the C_H value is enormous. It means that the capacitance data for dilute solutions cannot be used immediately to determine the C_H value at zero electrode charge, since it is too sensitive even to small experimental errors. On the other hand, assuming that the compact-layer capacitance is a smooth function of the charge, one can interpolate its values for positive and negative charges, to obtain $C_H(0)$ for a dilute solution. Such a procedure in Fig. 3 results in the same value, about $29\,\mu\text{F}\,\text{cm}^{-2}$, for this quantity as for the other concentrations.

The curve for the 0.001 M solution deviates downward noticeably for moderate positive and negative charges. The most probable reason is the increase of experimental errors for such dilute solution related to the current passage across the solution.

An alternative and the most commonly used way to verify the Grahame model is based on Parsons–Zobel (PZ) plots [37], in which the inverse values of the experimentally measured capacitance, C^{-1}, are shown as a function of the inverse diffuse-layer capacitance given by Eq. (21a). The theory predicts a linear dependence with a unit slope for each value of the electrode charge. Such plots for the same experimental data are shown in Fig. 4, again with indication of the error dispersion of each point. All lines in the figure have the theoretical slope. Their extrapolation to the ordinate axis gives the value of C_H^{-1} for the corresponding electrode charge. The PZ method to check the validity of the Grahame model is more objective (since it suggests verifying the linearity of the plots and their slope), especially if the experimental dispersion is not taken into consideration. However, in most publications (especially for solid electrodes) this method has been used in a reduced form, where only the data for the p.z.c. are plotted, thus giving only $C_H(0)$. It means that the Grahame theory is

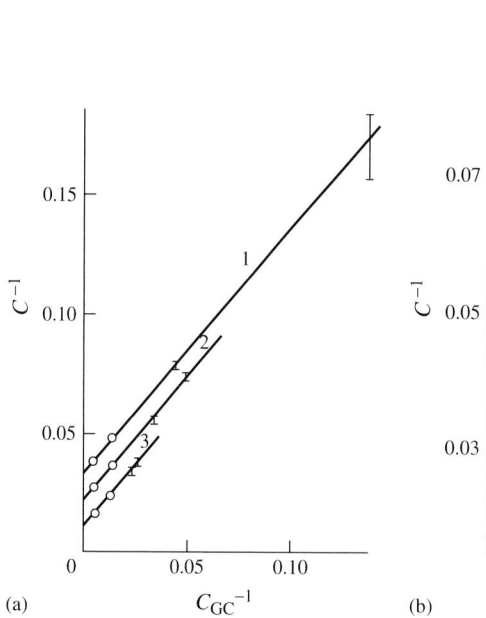

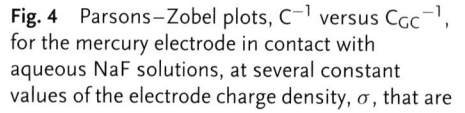

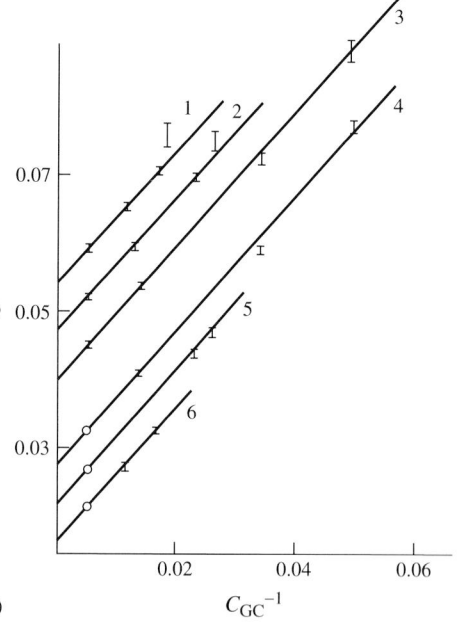

Fig. 4 Parsons–Zobel plots, C^{-1} versus C_{GC}^{-1}, for the mercury electrode in contact with aqueous NaF solutions, at several constant values of the electrode charge density, σ, that are indicated at each graph (in $\mu\text{C}\,\text{cm}^{-2}$): 0 (plot 1), +1 (plot 2), +2 (plot 3) in Fig. 4(a), −3 (plot 1), −2 (plot 2), −1 (plot 3), +1 (plot 4), +2 (plot 5), +3 (plot 6) in Fig. 4(b).

verified incompletely. Besides, it does not immediately give the properties related to individual parts of the interface, like the $C_H(\sigma)$ curve for the compact layer, or the characteristics of the diffuse layer.

The latter information may be extracted with the use of another coordinate system [38], δC^{-1} versus σ, where δC^{-1} is the difference of inverse experimental capacitance values for two electrolyte concentrations corresponding to the same electrode charge:

$$\delta C^{-1} \equiv [C(\sigma, c_1)]^{-1} - [C(\sigma, c_2)]^{-1} \quad (23a)$$

These curves can be plotted immediately on the basis of the experimental data for $C(\sigma, c)$ dependencies, see in the preceding text. For example, treating the data for Hg/H$_2$O-NaF interface in Fig. 2 leads to Fig. 5, in which the three lower concentrations, 0.1 M, 0.01 M and 0.001 M, were used for c_1, while c_2 was taken each time equal to the maximum concentration. The points thus found are shown together with the corresponding experimental dispersion.

On the one hand, it is sufficient to accept one of the Grahame's hypotheses, the one on the independence of the compact-layer capacitance of the electrolyte concentration, in order to relate the δC^{-1} value to the characteristics of the diffuse layer, that is, the change of its inverse capacitance $[C_{\text{diff}}(\sigma, c)]^{-1}$, with concentration, at a fixed value of the electrode charge:

$$\delta C^{-1} \equiv [C_{\text{diff}}(\sigma, c_1)]^{-1} - [C_{\text{diff}}(\sigma, c_2)]^{-1} \quad (23b)$$

On the other hand, if we also accept the second postulate of the Grahame theory, then $C_{\text{diff}}(\sigma, c)$ can be calculated with the use of the GC model, Eq. (21a). The corresponding *theoretical* curves are also shown in Fig. 5, in comparison with experimental ones.

Analysis of this figure leads to a surprising conclusion. It was accepted long ago that the capacitance data could only provide valuable information on the diffuse-layer properties in a close vicinity of the p.z.c., since this part of the interphase gives a minor contribution to the total interfacial capacitance outside this potential range, even in very dilute solutions. However, this reasoning does not take into account the very high precision of the capacitance measurements, which ensures a small dispersion of experimental points in Fig. 5 even for high electrode charges. This is especially true for the positive branch, where the dispersion is much less than the value itself. The points in Fig. 5 are limited to the range of $\pm 10\,\mu\text{F cm}^{-2}$ since the positive branch above this interval may be perturbed by the process of mercury dissolution. One should keep in mind that these *experimental* capacitance curves could be used to check *any* theory of the diffuse-layer structure within a *wide range of electrode charges*, since the GC model has *not* been used to obtain experimental points in these coordinates.

Figure 5 also demonstrates the results of this check for the GC model. One can see that it is able to reproduce adequately the data at the p.z.c. as well as within a wide range of the electrode charge, for both 0.1 M and 0.01 M solutions. One can see a noticeable discrepancy for the 0.001 M concentration, even for moderate electrode charges. However, the latter has probably a different origin since one should hardly expect that the mean-field type Poisson–Boltzmann theory fails at lower concentrations, while being applicable for higher ones!

These results represent a challenge for the modern statistical-mechanical theories of the interfacial structure, since the values of δC^{-1} in Eq. (23b) could be

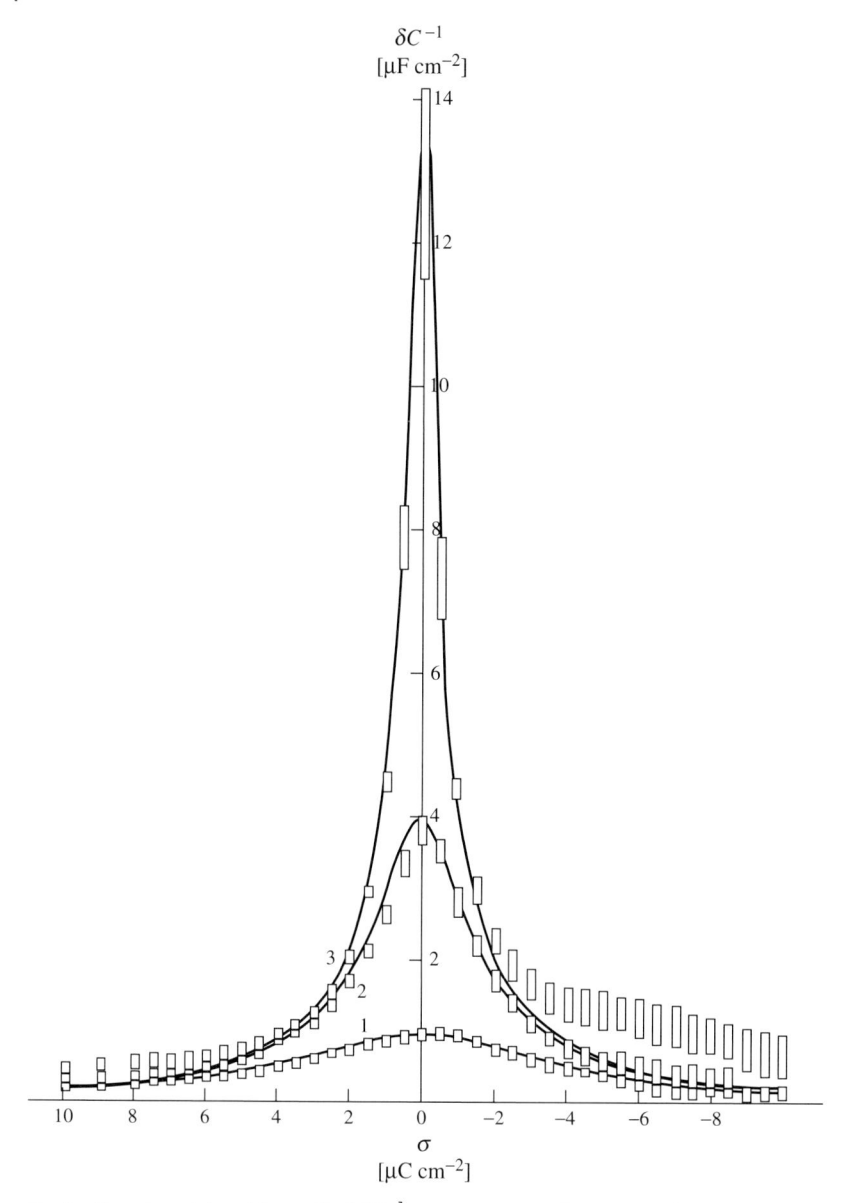

Fig. 5 Experimental and theoretical δC^{-1} versus σ curves for mercury/aqueous NaF solution [38], see Fig. 2, and the definitions in Eqs. (23a) and (23b). Error bars for the experimental curves correspond to those in Fig. 2.

immediately calculated from the theoretically derived ion concentration profiles. Although drastic deviations from the GC model have been predicted in these theories, no attempts have been made to subject them to the high precision experimental verification in accordance with Eq. (23b).

One should avoid, however, the mistake of interpreting the success of the GC passage through this test as its *complete* confirmation, in particular the proof of the validity of Eq. (21a) for the diffuse-layer capacitance at high electrode charges. In reality the theoretical lines in Fig. 5 based on this formula demonstrate a weak dependence of the diffuse-layer capacitance on the concentration (except for the p.z.c.), especially in dilute solutions (it is the effect, which is correctly reproduced by the model). However, the dominant part of the diffuse-layer capacitance has been *cancelled* (together with the compact-layer contribution) by the subtraction in Eq. (23b). Therefore the data in Fig. 5 cannot give any information on the *charge* dependence of the diffuse-layer capacitance far away from the p.z.c.

One has to keep in mind this impossibility to separate the compact-layer term and the concentration independent part of the diffuse-layer contribution in Eq. (20) at significant electrode charges. It results in an uncertainty in the values of the compact-layer capacitance, which is of primary importance in the modeling of the metal–solvent interfacial structure.

Recently the GC model has been modified to take approximately into account the screening effects for ions inside the diffuse layer [39, 40]. For this goal, the electrolyte concentration in the equations of the GC model (see previous text) were replaced by the electrolyte activity in the bulk solution. This treatment can be justified from the fundamental principles for a low charge density inside the diffuse layer (in the vicinity of the p.z.c.). On the other hand, the ion screening in this layer at high electrode charges is quite different compared with the bulk solution. Nevertheless this modification may extend the range of its applicability to higher ionic strengths.

2.1.5.2 Compact Layer Properties

The exposed variant of the theory has been applied to the study of numerous metal–solution interfaces, in particular for liquid metals, Hg, Ga, and their alloys. First of all for a particular metal/solvent boundary one has to ensure "no surface activity" of the solute components, that is, their absence inside the compact layer. Experimentally the necessary condition of this surface inactivity of an ionic species in the vicinity of the p.z.c. (which can be measured directly for liquid electrodes) is assumed to be the constancy of this potential, $E_{\sigma=0}$, for different electrolyte concentrations. Then the data are compared with the predictions of the Grahame model, as an additional check of the surface inactivity.

One should keep in mind that a detailed analysis has only been carried out for the Hg–H$_2$O–NaF interface (see previous text), in which both postulates (concentration independence of the compact-layer properties and the GC model for the diffuse layer) seem to be confirmed. For some other systems, the Grahame treatment was applied, to determine the compact-layer capacitance curve, $C_H(\sigma)$, from the data for the highest electrolyte concentration and to calculate on its basis the "theoretical" $C(E)$ curves for more dilute solutions, for their comparison with experimental ones, without taking into account the experimental dispersion. Moreover, in most cases the analysis was even reduced to constructing the PZ plot for the p.z.c. only. Generally the Grahame theory seems to provide at least a qualitative description of the results, even though certain "anomalies" have been observed for some systems (e.g. for liquid gallium [41]), see [38] for review.

Thus-found compact-layer capacitance curves turned out to be strongly dependent on the specific type of the metal (or alloy) and the solvent, be it water or a nonaqueous medium. On the other hand, the curves for different electrodes in contact with the same solvent are practically merging at high negative potentials (but not at positive ones). This experimental fact is still waiting for its interpretation.

The simplest interpretation of the compact-layer capacitance is represented by the Helmholtz model of the slab filled with a dielectric continuum and located between a perfect conductor (metal surface) and "the outer Helmholtz plane" considered as the distance of the closest approach of surface-inactive ions. Experimental determination of its thickness, z_H, may be based on Eq. (12). Moreover, its dielectric permittivity, ε_H, is often considered as a constant across the whole compact layer. Then its value can be estimated from the values of the compact-layer capacitance, for example, it gives about 6 or 10 (depending on the choice of z_H) for mercury–water interface, that is, a value that is much lower than the one in the bulk water, 80. This diminution was interpreted as a consequence of the "dielectric saturation" of the solvent in contact with the metal surface, its modified molecular structure or the effects of spatial inhomogeneity. The effective "dielectric permittivity" of the compact layer shows a complicated dependence on the electrode charge, which cannot be explained by the simple hypothesis of the saturation effects on one hand or by the unperturbed bulk-solvent nonlocal polarizability on the other hand.

Later studies revealed a more complicated structure of the compact layer whose properties are determined by the distribution of the metal electrons and discrete solvent molecules (see Sects. 2.1.7–2.1.9), so that the "continuous dielectric" consideration alone may become totally misleading. In particular, this simplified treatment identifies two physically different interfacial characteristics, the ion-free layer of the solvent and the region of a lowered dielectric permittivity (compared with the bulk-solvent one). The former manifests itself in Gibbs adsorption measurements of surface-inactive ions, Eq. (12); while the latter is related to the spatial correlation properties of the solvent polarization, see Sect. 2.1.7.

This distinction can be checked experimentally if the measurements are performed for different surface-inactive electrolytes but for the same electrode and solvent. According to the general theory [42–44] discussed in Sect. 2.1.7, the distance of the closest approach, which should be noticeably different for these electrolytes, has got almost no effect in the compact-layer capacitance if the ions do not penetrate into the region of the reduced dielectric response near the surface. This theoretical prediction turns out to be in conformity with experimental data [35, 37, 45, 46] for three mercury–aqueous solution interfaces for which the PZ plot at the p.z.c. gives practically identical values for the compact-layer capacitance, $C_H(0) \cong 29\,\mu F\,cm^{-2}$ (Fig. 6).

One should keep in mind that the theory predicts some effect of the ion properties on the compact-layer capacitance, which should manifest itself experimentally for systems with high values of this capacitance. Preliminary data for single-crystal faces testify in favor of this conclusion, but the theoretical relation has not been checked as yet. Interpretation of the observed values of the compact-layer capacitance and its dependence on the electrode

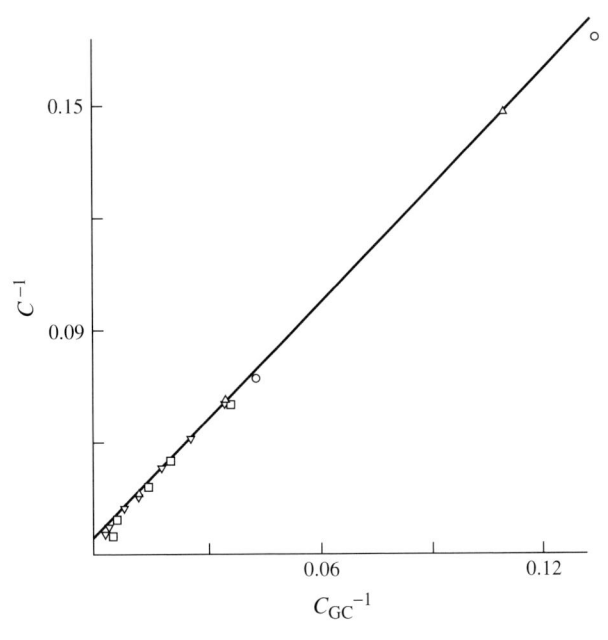

Fig. 6 Parsons–Zobel plot at the p.z.c. for mercury electrode in contact with various aqueous electrolyte solutions [38]. Capacitances are given in $\mu F\ cm^{-2}$. Experimental capacitance values: NaF (○) [35], NaH_2PO_4 (□) [37] and Na_2SO_4 (△, ▽) [45, 46].

charge requires a microscopic model of the interphase, as discussed in Sects. 2.1.8– 2.1.10.

2.1.5.3 Manifestations of the Diffuse Layer in Kinetics of Electrode Reactions

The existence of the EDL at an interface is of crucial importance for all interfacial electrochemical phenomena, in particular for the kinetics of electrode reactions. In most cases the principal factors influencing their rate are the overall potential difference across the interface (e.g. according to the Butler–Volmer equation for a simple ET reaction) and the specific adsorption of reactants, reaction products or intermediates. However, there are important cases in which the diffuse layer plays a dominant role [47, 48].

In a dilute solution the potential difference across the diffuse layer, φ_{GC}, attains significant values, being dependent on both the electrode charge and the electrolyte concentration, Eq. (17). Since the ET distance between the reactant and the electrode does not exceed a few angstroms under usual conditions, the reacting species must approach the compact layer or even enter it. Therefore the ET rate is only affected by a *fraction* of the overall potential difference, $\Delta\varphi^{m/s}$ in Eq. (16), equal to $\Delta\varphi^{m/s} - \psi_1$ where ψ_1 corresponds to the average potential at the point of the reactant localization, z^*. Besides, its concentration varies across the diffuse layer in accordance with the Boltzmann law.

Frumkin has combined this reasoning with his theory of the slow discharge [49] to derive the expression for the rate of an electrochemical reaction [50–52], for example, for the cathodic component of the faradaic current:

$$i_c = neSk_c^\circ c(z^*)$$
$$\times \exp\left(-\alpha(\Delta\varphi^{m/s} - \psi_1)\frac{ne}{k_BT}\right)$$
$$= neSk_c^\circ c^\circ \times \exp\left(-\alpha(\Delta\varphi^{m/s} - \psi_1)\right.$$
$$\left.\frac{ne}{k_BT} - \psi_1\frac{z_i e}{k_BT}\right) \quad (24)$$

Here, n is the number of transferred electrons ($n > 0$), S the surface area, k_c° the cathodic rate constant, α the cathodic charge-transfer coefficient, and z_i the charge valence of the reactant. This expression should be modified to take into account the adsorption energies of the reactant and the product if the ET takes place for a species inside the compact layer [11]. The value of the coordinate, z^*, which determines the ψ_1 potential is a priori unknown. In most studies it was postulated that the ET took place at the species located at the outer Helmholtz plane (the boundary between the compact and diffuse layers). Then, ψ_1 coincides with the potential drop within the diffuse layer, φ_{GC}, Eq. (17).

Equation (24) contains an additional factor,

$$K_{\psi 1} \equiv \exp\left(-(z_i - \alpha n)\psi_1\frac{e}{k_BT}\right) \quad (25)$$

compared to the conventional Tafel formula. In sufficiently concentrated solutions, for example, in the presence of the background electrolyte, the Frumkin correction may be neglected. On the contrary, it may radically change the shape of the polarization curve in dilute solutions near p.z.c., depending on the value of $z_i - \alpha n$. This effect originates from the linear dependence of the overall ($\Delta\varphi^{m/s}$) and diffuse-layer (ψ_1) potential drops on the electrode charge, σ, at small deviations from p.z.c., Eqs. (16), (13) and (17). Then the ratio of these potentials

$$\frac{\Delta\varphi^{m/s}}{\psi_1} \cong 1 + \frac{C_{GC}(0, c)}{C_H(0)} \quad (26)$$

determined by the ratio of the diffuse and compact-layer capacitances at the p.z.c. is about 1 (or even close to 1), see Eq. (21a).

The type of the effect on the polarization curve depends also on the sign of the coefficient, $z_i - \alpha n$. In most cases this sign is determined by the charge valence of the reactant, z_i. In the case of the electroreduction of a *cation* (e.g. for the hydrogen evolution or a metal deposition) $z_i > 0$, and the diffuse-layer factor $K_{\psi 1}$ changes in the same direction as the Tafel term, $\exp(-\alpha\Delta\varphi^{m/s}ne/k_BT)$. Therefore the former leads to a deviation from the straight line with the $2.3\, k_BT/\alpha ne$ slope in the Tafel coordinates, but the current-potential dependence remains *monotonous*. Besides, z_i and αn have opposite signs and for single-charged cations, the coefficients of the diffuse-layer effect represents a rather small correction to the Tafel dependence. Nevertheless it cannot be ignored, both for the quantitative treatment of the polarization curves and for the interpretation of the dependence of the current on the concentration of the indifferent electrolyte [53].

Quite a different situation takes place for the reduction of *anions*, or for the oxidation of *cations*. Then the diffuse-layer multiplier, $K_{\psi 1}$, varies in the *opposite* direction compared to the Tafel term. For sufficiently dilute solutions, the former factor dominates and the cathodic current *diminishes* at the negative shift of the electrode potential in the vicinity of the p.z.c. At higher electrode charges (positive or negative) the variation of the diffuse-layer potential becomes very slow (logarithmic function of the electrode charge, Eq. (17)) compared with the overall potential because the compact layer gives the principal contribution. Within these potential ranges the variation of the diffuse-layer factor only modifies the global tendency determined by the Tafel term, that is, the cathodic current increases for negative potential shift. As a whole, Eq. (24) predicts a *nonmonotonous* current-potential curve in dilute solutions, with a dropping branch in the vicinity of the p.z.c. The amplitude of this effect depends strongly on the charge of the reactant, z_i, and the electrolyte concentration, and the current of the anion reduction may almost vanish within a wide range of negative electrode potentials, $E - E_{\sigma=0} < 0$.

One must keep in mind that all this analysis was performed for systems without a specific adsorption of reacting species or indifferent electrolyte. The violation of this requirement leads to the necessity to introduce the energies of specific adsorption of reactants or their interaction with adsorbed species, into Eq. (24). Fortunately this problem is of less importance for the case of anion reduction, since the most interesting effects take place at the negatively charged electrode surface in which anion-adsorption effects are minimal.

The analysis in Eq. (24) does not take into account the transport limitations. Therefore these predictions of the diffuse-layer effects are only valid for the interval of the potentials in which the rate of the process is determined by the electron-transfer step. If this kinetic regime corresponds to high negative electrode charges, the $K_{\psi 1}$ factor in Eq. (25) varies slowly and the deviations from the Tafel behavior are rather weak. On the contrary, the anion electroreduction wave that starts at positive electrode charges may demonstrate a complicated curve: a usual behavior within this potential range, with approach to the limiting current at less positive potentials, then a deep minimum of the current within the range of small negative electrode charges.

Experimentally such nonmonotonous curves were observed for numerous reactions of the anion electroreduction: $S_2O_8^{2-}$, $S_4O_6^{2-}$, $Fe(CN)_6^{3-}$, $PtCl_4^{2-}$, and so on, where all above-mentioned qualitative predictions of the slow-discharge theory were confirmed [54]. For the quantitative treatment, it was suggested that corrected Tafel plots be used in coordinates: $\log i_c + (z_i e/2.3 k_B T)\psi_1$, $E - \psi_1$ [55], where one should observe a straight line with the slope $-\alpha n e/2.3 k_B T$. Besides, the data for different concentrations of the background electrolyte should fall on the same line [56]. These predictions have been confirmed (some curvature of corrected plots was found in some cases) by experimental studies of the persulfate, tetrathionate, ferricyanide, and some other anions [57].

An additional test of the theory is given by the comparison of the data for different electrode materials. Experimentally the reduction of the same anion at different metals leads to quite different patterns for the polarization curve, in particular, a drastic variation of the interval in which the diffuse-layer minimum is observed. According to the theory (24), if the reacting species do not enter the compact layer and the possible change of the electron-tunneling factor does not influence the reaction rate (e.g. for adiabatic electrochemical reactions [58]) the corrected Tafel plots must be independent of the electrode material. This prediction has also been confirmed experimentally for the persulfate reduction at Hg (amalgams), Bi, Sn, Pb, and Cd [59]. Similar results have been obtained for the persulfate reduction in some nonaqueous solvents [60, 61].

The exposed version of the model disregards an essential feature of this class of reactions, that is, the formation of ion associates between the reacting anion and background cations, which is especially pronounced for the multicharged anions. The ways to determine the charge of the discharging species in the bulk solution (which is generally different from the one inside the EDL) are analyzed in [62]. Some deviations from the Frumkin theory may also originate from the difference of the closest approaches of anions and cations to the electrode surface [47].

The Frumkin correction in Eq. (24) reduces the EDL structure effect to its single characteristic, ψ_1 potential, while real reactants represent a distribution of electric charges. A recent development has taken into account this distribution in the calculation of the work term, so that it depends on the whole profile of the potential across the EDL. Besides, the presence of the reactant perturbs this potential profile in the vicinity of the species (a kind of discreteness-of-charge effects [48] discussed in Sect. 2.1.11.3), due to both the field created by its charges and the cavity effects arising from the replacement of solvent molecules in the volume occupied by the solute species [63, 64].

The quantitative theory also has to take into account a variation of the reaction volume with the electrode charge [65]. Another weak point of the model (24) is the hypothesis of the Tafel dependence on the potential difference across the compact layer. Quantum-mechanical estimates [64] testify in favor of a considerable variation of the charge-transfer coefficient, α, over the wide potential range explored experimentally, or even its approach to zero, which corresponds to the activationless regime of the charge-transfer process [66]. Recent analysis which has included these factors [64, 65] has allowed one to provide a quantitative interpretation of experimental data for anion electroreduction at high overvoltages and has confirmed expectations of its proximity to the activationless region.

2.1.5.4 Electrokinetic Phenomena

Another area in which the existence of the diffuse part of the EDL plays a dominant role is represented by various *electrokinetic phenomena* [67, 68].

Some of them are observed at membranes separating two identical electrolyte solutions. For example, in *electroosmosis*, one can impose a difference between the electric potentials in these solutions, by means of immersed electrodes, which leads to the generation of an electric field across the membrane because of the current passage. As a response, one observes a solution flow across the membrane. For its qualitative interpretation, one considers pores of the membrane as a set of parallel cylindrical capillaries. An EDL is formed at the internal surface of each capillary, due to the ionization of surface groups or/and ion adsorption/association, accompanied by the accumulation of the counterions inside the diffuse layer. As soon as the electric field **E** is imposed along the capillary surface, the force $z_i e\,\mathbf{E}$ acts on each ion, including those within the diffuse layer. Owing to the mobility of the latter and the viscosity of the solvent, this force induces the movement of the solution along the capillary. The velocity of this flow (with respect to the electric field amplitude) in the stationary conditions is described by the Smoluchowski equation:

$$\frac{v}{E} = (4\pi)^{-1}\varepsilon\eta^{-1}\zeta \qquad (27)$$

where ε is the relative dielectric permittivity of the solvent, η the dynamic viscosity of the solvent, and ζ ("zeta potential") the potential in the "slip plane" inside the diffuse layer separating the hydrodynamically immobile (attached to the surface) and moving volumes of the solution. This potential is often identified to the potential of the outer Helmholtz plan, φ_{GC}, Eq. (16).

An inverse phenomenon (*streaming potential*), generation of an electric field inside the membrane, takes place if the solution passes through this porous medium due to an imposed hydrostatic pressure. This time it is the flow of the fluid inside the pores that induces the displacement of the mobile part of the EDL at the surface of capillaries, with respect to the charges attached to the surface. These dipoles create an electric field, which, under stationary conditions, prevents the farther displacement of the mobile charges. The resulting potential difference across the membrane, $\Delta\varphi$, is proportional to the excessive hydrostatic pressure, ΔP:

$$\frac{\Delta\varphi}{\Delta P} = (4\pi)^{-1}\varepsilon\eta^{-1}\Lambda^{-1}\zeta \qquad (28)$$

where Λ is the electrolyte conductivity, and the remaining notation is identical to that in Eq. (27).

The existence of the extended diffuse layer in dilute solutions is also of crucial importance in colloidal systems formed by particles whose size is comparable to the wavelength of light, that is, about 1 μm. For example, *electrophoresis*, the movement of colloidal particles with respect to the solution under the influence of an external electric field, originates from the displacement of the particle and the mobile part of its EDL in the opposite directions under the action of this field, since these components of the system possess charges of opposite signs. A related phenomenon occurs if the solid particles move with respect to the solution, for example, under the influence of the gravity force (sedimentation). The viscosity force results in a displacement of the mobile diffuse-layer charges with respect to the fixed ones at the surface, that is, to the creation of an electric dipole. The superposition of the fields of such dipoles gives an electric field oriented along the movement axis (*sedimentation potential*).

It should be noted that the *direction* of the induced effect depends on the *sign* of the diffuse-layer potential, so that one can determine from such experiments the sign of the "fixed" surface charge or even estimate its value.

The variation of the electrolyte concentration or the solution composition leads to drastic changes in the properties of the colloidal system, in particular to changes of its *stability*, which is primarily determined by the electrostatic repulsion between the particles. The latter is due to the formation of similar EDLs owing to the identical chemical nature of the surfaces. The repulsion is only efficient at long distances between the particles (at shorter distances the interaction between the particles becomes *attractive* due to van der Waals forces). This is why long-range diffuse layers are necessary for the colloid's stability. The increase of the electrolyte concentration strongly diminishes the Debye screening length, that is, the distance of the particle–particle interaction, with the coalescence and the subsequent sedimentation. Another scenario of the kind takes place upon change of the solution pH. In most cases the fixed charge of the particle's surface is formed at least partially by the proton-exchange equilibrium. The change of pH results in the modification of this surface charge, and the systems loses its stability at the moment when the surface charge passes zero and the repulsion between the particles disappears.

2.1.6
Solid Electrodes: Effects of Polycrystallinity and Surface Roughness

2.1.6.1 "Uniform" Model of Solid Electrodes. Surface Roughness

A brief outline of this complicated and still developing area is given in this section. A more detailed description may be found in reviews [22, 33, 34] as well as Chapter 2.4 of this volume [20].

Historically the interfacial properties of the solid metal–solution interface were initially studied for polycrystalline (PC) samples. At that time it was expected that the characteristics of all metals should be rather similar, saying nothing about the different single-crystal faces of the same metal.

The first studies of capacitance curves for numerous PC metals were qualitatively in accordance with these expectations; for carefully prepared smooth PC surfaces in the absence of the specific adsorption of ionic or neutral species, a minimum was observed in sufficiently dilute solutions, the depth of which increases for lower electrolyte concentrations. Therefore the potential of this minimum was interpreted as the p.z.c., $E_{\sigma=0}$, of the metal–solution interface.

The capacitance values at this minimum for solid metals turned out to be noticeably higher compared with those for the mercury electrode at the same electrolyte concentrations. Since the interfacial properties of different metals were assumed during that period to be similar, it was natural to attribute these higher experimental capacitance values for solid electrodes to a greater interfacial area.

On this basis, one even tried to estimate the *roughness factor* (RF) of the solid surface, R, that is, the ratio of the "real" (S) and the "apparent" (S_{app}) surface areas:

$$R \equiv \frac{S}{S_{app}}, \quad R \geq 1 \quad (29)$$

The real surface area S being unknown for solid electrodes, the charge density and the capacitance in this case were defined with the use of the apparent (or "geometrical") surface area, compare Eqs. (3) and (2):

$$\sigma_{app} \equiv \frac{Q}{S_{app}} = R\sigma$$

$$C_{app} \equiv S_{app}^{-1}\frac{dQ}{dE} = RC \quad (30)$$

In accordance with the Stern–Grahame model of the EDL structure the values of C are determined by both the diffuse and compact-layer properties, the latter being dependent on the metal properties. However, in very dilute solutions of a surface-inactive electrolyte the dominant contribution to C near the p.z.c. (at the capacitance minimum) is given by the diffuse layer, $C \cong C_{GC}(0, c)$. Therefore the ratio of capacitances in these conditions should be close to the RF for the surface of the solid metal M:

$$R \equiv \frac{C_{app}^{M}}{C^{Hg}} \quad (31)$$

It became clear later that this treatment had certain defects. The data for most dilute solutions (typically, within mM range) are especially sensitive to various experimental errors, while the capacitance values to be measured are very low. Another weak point of the reasoning is the assumption that the solid surface is *uniform*, which is hardly justified, as we see in the following text. One should also keep in mind that this consideration is inapplicable to the surface roughness, whose scale is smaller or comparable with the thickness of the diffuse layer, that is, with the Debye screening length in the solution, which is about 10 nm in mM solutions.

In a modified variant of this approach, the surface roughness was determined from the same ratio of capacitances (31) measured for the identical high negative charges [69]. Under these conditions the diffuse-layer contribution may be practically neglected while the difference between the compact-layer capacitances of *different metals* becomes rather small, compared to the vicinity of the p.z.c. or at positive electrode charges. The principal shortcoming of this estimate is a significant variation of the cathodic branch of the capacitance curve with the electrolyte concentration observed in most publications.

All this reasoning is based on the hypothesis that the separation (20) of the capacitance to unit area of the *real* surface, C, into the compact- and diffuse-layer contribution remains valid for solid metals (see [70] and earlier references therein). It gives for experimentally measurable values of the capacitance:

$$(C_{app})^{-1} = R^{-1}C_H^{-1}(\sigma) + R^{-1}C_{GC}^{-1}(\sigma, c) \quad (32)$$

One must keep in mind that the GC expression (21a) for the diffuse-layer capacitance, C_{GC}, contains the charge density per unit *real* surface area, so that it can only be calculated (except for its value for $\sigma = 0$) if the RF is known.

The attempts to apply the Grahame treatment in the *quantitative* manner encountered certain problems. First, the impedance data interpreted in terms of the in-series combination of the electrolyte resistance and the EDL capacitance, C, demonstrated a significant frequency dispersion of these elements (which implies that the equivalent circuit is more complicated). Besides, contrary to the theoretical predictions, there was a strong variation of the cathodic branch of the curve as a function of the electrolyte concentration, which was hardly possible to attribute to the specific adsorption of cations. The anodic branch was even more irregular.

As a result, within the initial period of the solid metal studies, the only quantitative treatment was made for the data at the diffuse-layer minimum. In accordance to Eq. (32), the corresponding PZ plots (dependence of C_{app}^{-1} versus C_{GC}^{-1}) for various solid metals had *lower* slopes than 1, which were interpreted as R^{-1}. The plots (if the concentration range was sufficiently broad) showed also a significant curvature in most cases. The latter was attributed to various experimental errors.

After having the RF determined one could calculate the charge density per unit "real" surface area, σ, by the integration of the $C(E) \equiv R^{-1}C_{app}(E)$ curve. Determining then C_{GC} in Eq. (32) one may find the compact-layer capacitance for all electrode charges. Finally, the "theoretical" capacitance curves could be found for lower electrolyte concentrations. However, such an analysis has almost never been performed, that is, only the values of R and the compact-layer capacitance for the p.z.c., $C_H(0)$, were extracted (see Ref. [34] for a review).

An alternative method to determine the RF and the compact-layer capacitance curve from experimental data for solid surfaces was proposed in [69]. It is based on the same model (32) of a uniformly charged rough surface. On the basis of this equation the compact-layer capacitance curve, $C_H(\sigma)$, was calculated for *several* values of the RF R, see Fig. 7, for a fixed electrolyte concentration. Similar to Fig. 3 for the mercury electrode (in which the

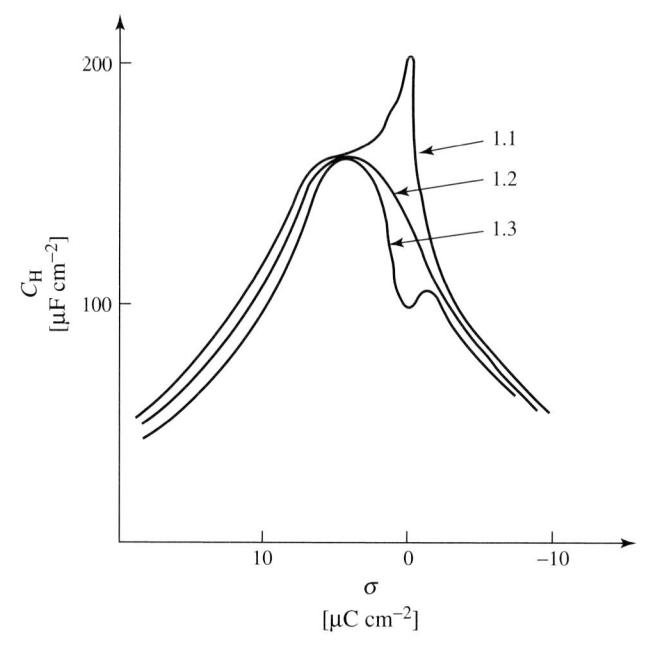

Fig. 7 The $C_H(\sigma)$ plots for a polycrystalline silver electrode calculated with the use of Eq. (32) for several values of the RF, R, indicated at each plot [69].

effect originated from the experimental errors in the capacitance values), the shape of the curve within the interval near the p.z.c. was especially sensitive to the choice of R. In particular, a value of R ($R = 1.2$ in Fig. 7) was found for which the whole curve was passing this interval *smoothly*. It was assumed that this procedure allowed one to determine both R and the $C_H(\sigma)$ curve. The value of R obtained in such a way was in accordance with the one found from the comparison of EDL capacitance values for PC samples of this metal (silver) and its single-crystal faces at high negative electrode charges.

One should note that one has to verify, whether these characteristics are the same for different electrolyte concentrations, like it was done in Fig. 3 for mercury. This requirement was mostly not verified or (if verified) not satisfied, especially for PC electrodes.

Equation (32) of this model implies that the surface roughness is realized by the spatial configuration whose scale is well above the thickness of the EDL, in particular the Debye screening length of the solution, L_D (18), to allow the interfacial structure to follow the curved surface without its especial distortion. Sometimes in earlier studies, (see [34] for review) the treatment of the capacitance data for solid metals was based on a different relation:

$$(C_{app})^{-1} = R^{-1} C_H^{-1}(\sigma_{app}) + C_{GC}^{-1}(\sigma_{app}, c) \quad (33)$$

Physically, it may simulate the situation of a solid electrode with a *microroughness* whose scale is well below the diffuse-layer thickness. Then the compact-layer capacitance per unit apparent surface area is increased by the RF, while the diffuse layer is formed outside this microrough interface and its structure is identical to the one for liquid metals.

Model (33) predicts a *unit* slope for the PZ plots. It is in contradiction with experimental data, at least for PC electrodes. However, it should be noted that recent theoretical studies of the EDL structure at rough (but uniform) surfaces have shown that these two models may be considered as limiting cases of a more general description, Sect. 2.1.6.4.

One must keep in mind that the increase of the EDL thickness upon a decrease of the electrolyte concentration is only valid for a close vicinity of the p.z.c. The GC formulas for the ion charge distribution show that outside this narrow interval of the electrode charges, the diffuse-layer thickness becomes independent of the ionic strength of the solution, see Eq. (21a) or (21b) as an illustration. This effect originates from the *nonlinear* response of the diffuse layer and makes the area of the applicability of Eq. (33) quite narrow.

2.1.6.2 Crystallographic Inhomogeneity Effects for Solid Electrode Surfaces

The analysis of models (32) and (33) demonstrates the necessity to be very careful in comparing the values of the RF of the same surface found by different experimental techniques, since they may be crucially dependent on the scale probed in the corresponding method.

Significant new information was provided by capacitance measurements at *single-crystal faces* of solid metals. It was found that there existed two groups of solid metals having qualitatively different properties. One of them includes numerous "mercury-like" metals or semimetals, like Pb, Bi, Sb, Sn, and Cd, for which the difference of the p.z.c. potentials between the single-crystal faces of the same metal is sufficiently small, within 100 mV or less. On the contrary, a great dispersion of p.z.c. values for different single-crystal faces (300 mV and greater) was revealed for Ag, Au, Cu, and so forth. Besides, for all face-centered cubic metals of this group, two basic crystal faces, (111) and (100), possess the most positive among those p.z.c. values, while even a relatively small change of the crystallographic orientation results in a rapid shift of the p.z.c. to the most negative values that are characteristic for most higher-order crystal planes.

This finding led to the development of a new concept emphasizing the *effects of crystallographic inhomogeneity* of the PC electrodes by Frumkin [71]. It postulates (in its simplest version) that the PC surface represents a combination of various single-crystal faces – each of them having their own EDL, the structure of which is identical to the one at the corresponding real single-crystal electrode (model of "independent electrodes" [72] or "independent diffuse layers (IDL)" [20]). It was noted [71] that at a fixed value of the PC electrode potential, E, the charge densities of individual faces were *different* and that this factor might change the shape of the $C(E)$ curve for the PC electrode near its diffuse-layer minimum.

Later the capacitance properties of a PC surface were analyzed on the basis of Eq. (34) [69, 73, 74], where the $C^{PC}(E)$ curve was represented as the sum of the capacitance curves for individual faces, $C^{(i)}(E)$, multiplied by their fractions of the total surface area, $\theta^{(i)}$:

$$C^{PC}(E) = \sum_i \theta^{(i)} C^{(i)}(E) \qquad (34)$$

Though this concept looks almost self-evident, one should emphasize that it results in quite different predictions on the EDL properties, compared to the aforementioned model of a rough but uniform surface (32) or (33):

- the diffuse-layer minimum of the $C^{(i)}(E)$ curve in dilute solutions is widened and not so deep, in comparison with the minimums for individual faces;
- for metals with a considerable variation of p.z.c. values among the faces the minimum of the $C^{(i)}(E)$ curve, E_{min}^{PC}, is close to the most electronegative among p.z.c. of the faces, for example, $E_{min}^{PC} \cong E_{\sigma=0}^{(110)}$ for Ag, Au, Cu;
- the formally constructed PZ plot, that is, the dependence of $(C_{min}^{PC})^{-1}$ on C_{GC}^{-1}, cannot be used to determine the electrode surface roughness or the compact-layer capacitance;
- this plot is expected to be curved (downward for lower concentrations) as a result of crystallographic inhomogeneity effects, if the concentration interval is sufficiently wide;
- the overall charge of the PC surface, σ^{PC}, is *nonzero* at E_{min}^{PC}, and its value depends on the electrolyte concentration;
- at the p.z.c. of the PC surface, $\sigma^{PC} = 0$, the charges of regions (faces) at the surface are *nonzero*, $\sigma^{(i)} \neq 0$, that is, this point corresponds simply to the compensation of the contributions of opposite signs into the overall PC surface charge;
- if the compact-layer capacitance curve of the PC surface, $C_H^{PC}(\sigma^{PC})$, is calculated with the use of the traditional scheme assuming the uniformity of the PC surface, Eq. (32), then this curve displays rapid nonmonotonous variation in the vicinity of the zero PC charge, $\sigma^{PC} = 0$, because of incorrect subtraction of the diffuse-layer contribution. Moreover, such calculated "compact-layer curves" are strongly dependent on the electrolyte concentration and have no relation to the ones obtained by the averaging of the compact-layer curves for individual faces with the same fractions $\theta^{(i)}$, similar to Eq. (34).

These strong "anomalies" were predicted for metals whose PC surfaces were composed by *low-index* single-crystal faces (e.g. (111),(100) and (110) for face centered cubic (fcc) lattices) having a strong difference between their p.z.c. (several hundred mV). For "mercury-like" PC metals, the predicted effects were less pronounced, even though the formal PZ plots were to be markedly curved.

It was also recognized [74] that there existed an alternative model of the PC surface, "EDL with a common/united diffuse layer". It means that if the size of the individual faces at the surface is sufficiently small, the diffuse layer is smooth along this heterogeneous surface, i.e., it is analogous to the one for a uniform surface but with the charge density in expression (21a) replaced by the *average* charge density of the heterogeneous surface:

$$\sigma^{PC} = \sum_i \theta^{(i)} \sigma^{(i)} \quad (35)$$

Then the interfacial capacitance of such PC surface should be described by the equation:

$$(C^{PC})^{-1} = (C_H^{PC})^{-1} + C_{GC}^{-1}(\sigma^{PC}, c) \quad (36a)$$

where "the compact-layer capacitance of the PC surface", C_H^{PC}, is given by the average of the compact-layer capacitances of the individual faces:

$$C_H{}^{PC} = \sum_i \theta^{(i)} C_H^{(i)}(\sigma^{(i)}) \qquad (36b)$$

The latter depends on the charge density at the corresponding face, $\sigma^{(i)}$, rather than on the average density σ^{PC}. Evidently this model predicted quite different effects:

- the coincidence of the p.z.c. of the PC surface with the potential of its $C^{PC}(E)$ curve, $E_{min}{}^{PC}$;
- the independence of its value on the electrolyte concentration;
- a straight line with a unit slope for the PZ plots at any fixed value, and so on.

In an attempt to choose an adequate treatment, the predictions for both models were compared with experimental data for PC samples of various metals. Qualitatively the conclusions of the model of "independent electrodes" turned out to be in agreement with experimental data for such metals as Ag, Au, Cu and so on, see reviews in Ref. [20, 34], contrary to the failure of the "common diffuse layer (CDL)" model, Eq. (36a). However, in quantitative terms the effects of crystallographic inhomogeneity were smaller than predicted by the model based on Eq. (34). For example, the capacitance at the minimum, $C_{min}{}^{PC}$, for PC Ag diminishes more rapidly than in its theoretical estimates. The experimental slope of the PZ plots calculated with no account for the surface heterogeneity effects is close to 1 for Bi, within a very wide range of the concentration, in contrast to the model analysis. The value of R found from the same plot is close to 1 for some PC Ag electrodes in rather concentrated solutions.

One of the possible reasons of this disparity may be an inappropriate choice of the fractions of the PC surface occupied by the individual faces, $\theta^{(i)}$, in Eqs. (34) and (35) since they have been postulated on an intuitive basis rather than determined from measurements. One should also keep in mind the necessity to distinguish between the capacitance values related to the *"real"* or *"apparent"* surface areas of the PC electrode, Eq. (30). In other words, one must take into account the RF of the PC surface, R^{PC}. Then the experimentally measured capacitance value, $C_{app}{}^{PC}(E)$, is given by Eq. (37) if the "model of independent electrodes" is used:

$$C_{app}{}^{PC}(E) = R^{PC} \sum_i \theta^{(i)} C^{(i)}(E) \qquad (37)$$

Here, similar to the previous relations, the sum of the fractions of the individual faces at the PC surface, $\theta^{(i)}$, is equal to 1. The capacitance curves $C^{(i)}(E)$ correspond to *perfect faces* without any roughness.

To avoid a misunderstanding, one must keep in mind that all conclusions derived on the basis of Eq. (34) remain valid for this case, too. In particular, it is impossible to obtain the RF of the PC surface, R^{PC}, from the slope of the PZ plots. On the other hand, one may determine both the RF and the fractions of the individual faces from the comparison of experimental capacitance curves for the PC surface and the individual faces (if the latter are known) [75]. These characteristics of the PC surface are found by minimizing the difference between the left- and right-hand side terms in Eq. (37) with respect to the fitting parameters, R^{PC} and $\theta^{(i)}$. This operation applied to the PC Ag electrode enabled one to reproduce properly its experimental capacitance curve on the basis of those for the three silver low-index faces, (111), (100) and (110). Moreover, it was found that the values of R^{PC} and $\theta^{(i)}$ obtained from the fitting for *each electrolyte concentration* turned out to be practically the same for all concentrations. The value of R^{PC} obtained in this way was close to the ratio of the C_{app}^{PC} and $C^{(i)}$ at high negative potentials, in accordance with earlier expectations, see previous text.

2.1.6.3 EDL Structure for NonUniform Electrode Surfaces

All this modeling was based on the hypothesis that the EDL structure is *one-dimensional*, that is, all profiles (concentrations, ion charge density, and electric potential) depend on a single coordinate normal to the surface, either within the whole diffuse layer in model (36a) or within the EDL at each single-crystal face in model (34) or (37). However, a rigorous theoretical analysis [76, 77] shows that these distributions are essentially perturbed near all contacts of the surface domains possessing different p.z.c. or/and compact-layer capacitance values. It results in a *nonuniform* distribution of the surface charge density. The width of this perturbed zone is determined by the screening properties of the electrolyte inside the diffuse layer, in particular it is close to the Debye screening length in the bulk solution, L_D, in the simplest approximation of the *linear* response. Then the criterion of the aforementioned models is in the comparison between the characteristic size of the individual uniform areas at the PC surface (e.g. single-crystal faces), y_*, and the Debye screening length, L_D. For example, for model (34) and (37) it has the form $y_* \gg L_D$.

This approximation of the linear response may be justified for heterogeneous surfaces with a small difference of the partial p.z.c. values, in particular for some mercury-like solid metals. However, the analysis is much more complicated [72, 78] in the case of a surface with a great difference of these p.z.c. values, for example, observed for single-crystal faces of many metals. Then the response of the EDL is *always* nonlinear, since the electron charge density is sufficiently high at least on some of the surface domains, since the electrode potential E is far from their p.z.c. values, $E_{\sigma=0}^{(i)}$. For these surface regions, the dominant contribution to the local capacitance is given by the *compact layer*, even if the electrolyte concentration is very low. For some other regions, the potential E may be close to their p.z.c., the EDL structure may be described within the linear-response model and their overall capacitance may be essentially determined by the diffuse-layer contribution. However, the latter leads to a strong reduction of the contributions of such regions into the overall capacitance of the heterogeneous surface, since the contributions of the regions are additive.

As a result the criteria of models (37) or (36a) depend mostly on the difference of the partial p.z.c. values and on the partial compact-layer capacitances. In particular, for the most interesting case of a strongly heterogeneous surface (i.e. having a significant difference between the partial p.z.c. values) the principal condition relates the characteristic size of the uniform regions, y_*, and the lengths $L_H^{(i)}$ derived from the values of the partial compact-layer capacitances, $C_H^{(i)}$, e.g., for the "model of independent electrodes" (Eqs. (34) and (37)):

$$y_* \gg \max\{\varepsilon L_H^{(1)}, \varepsilon L_H^{(2)}, \ldots\},$$
$$L_H^{(i)} \equiv (4\pi C_H^{(i)})^{-1} \quad (38a)$$

One should emphasize that this condition is sufficient even for very dilute solutions, in which the Debye screening length in the bulk, L_D, may strongly exceed the geometrical size, y_*, contrary to the intuitive expectations based on the linear-response approximation.

The criterion is slightly more complicated for systems with a *moderate* (but not small) difference of the partial p.z.c. values for which, the diffuse-layer minima for the different uniform regions as a function of the electrode potential overlap partially. Then, condition (38a) must be supplemented by another one; the geometrical size should also significantly exceed the shortest among the partial screening lengths *inside the diffuse layers* at each region:

$$y_* \gg \max\{L_D^{(1)}, L_D^{(2)}, \ldots\},$$
$$L_D^{(i)} \equiv (4\pi\varepsilon C_{GC}^{(i)})^{-1} \quad (38b)$$

determined by the diffuse-layer capacitances for the corresponding regions, $C_{GC}^{(i)}$. The latter depends on the local charge density for this region according to Eq. (21a).

The characteristic lengths $\varepsilon L_H^{(i)}$ vary widely as a function of the metal and solvent properties. For aqueous solutions, they are within 5 nm, or even smaller for metals with high values of the compact-layer capacitance for single-crystal faces at small electrode charges, such as Ag or Au. Therefore this analysis testifies in favor of the "model of independent electrodes" for such systems, in conformity with the experimental evidences mentioned above.

However, one should keep in mind that all this analysis is based on the assumption that the principal contribution to the PC capacitance is given by perfect single-crystal faces at its surface, which must have a sufficiently large size, Eq. (38a). One cannot a priori exclude a considerable contribution due to intermediate regions between well-defined uniform surface fragments, or even a completely amorphous surface structure. The corresponding model was first proposed for the *imperfect* single-crystal (111) face of Au in Ref [79], which was considered as a combination of perfect (111) regions and intermediate zones. Then in accordance with the approach of "independent electrodes", the overall capacitance was represented as the sum of the perfect $C^{111}(E)$ curve and the one of the PC Au electrode. Possible influence of various higher-order faces and surface defects on the capacitance properties of PC surfaces was discussed in [80]. Later, it was suggested to represent the overall PC capacitance as the sum of the contributions from all low-index faces (identical to Eq. (37)) *and* from the intermediate regions, the latter being modeled in accordance with Eq. (36). This treatment was extensively used to analyze the properties of Bi electrodes, see [20].

For metals with a great difference between the p.z.c. values of the faces, it is necessary to keep in mind the specific dependence of the p.z.c. values of the single-crystal faces as a function of their orientation; for example, for various fcc metals the low-index faces, (111) and (100), possess "singular" p.z.c. properties, while this characteristic rapidly approaches a much more positive value typical for (110) and higher-index faces, at a relatively small change of the crystallographic orientation. Therefore, one cannot exclude an alternative interpretation of the above-mentioned experimental observations for such PC electrodes. Namely, one can envisage that their surfaces do not contain *perfect* (111) or (100) faces but only perturbed ones. Their p.z.c. values are not much different from those for the (110) or (210) faces.

The possibility of getting a substantiated answer might lie in the use of modern methods of analysis to establish the PC surface structure more reliably [21, 22]. On the other hand, one must always keep in mind that each particular experimental technique often uses *different definitions* of the term "surface structure", in particular its "roughness".

2.1.6.4 Surface Roughness Effect on the Diffuse-Layer Capacitance

Characterizing roughness by a geometrical RF, as the ratio of the true surface area and the apparent cross-section area, Eq. (29), one distinguishes typically the cases of weak, moderate, and strong roughness, depending on whether $R - 1 << 1$, $R - 1 \approx 1$, $R - 1 >> 1$. Weak roughness is typical for single-crystalline electrodes, moderate roughness for polycrystalline electrodes, while the case of strong roughness is met for specially fabricated catalysts. However, as we have mentioned, the RF alone can describe the effect of surface corrugation only if the characteristic scales of roughness are much greater than the diffuse-layer thickness, while in the opposite limiting case its effect will not be seen. In the general case, one must develop a more general theory.

Such a theory was developed recently [81–83], however, with no account for possible "energetic inhomogeneity" of the surface, as discussed in the preceding text. Since, facets of a rough surface always expose different elementary faces of the crystal structure, such a theory has a limited applicability; one might expect it to be valid only for metals with a minor difference in p.z.c. Since this theory is discussed also in Chapters 2.2 and 2.4 of this volume, we only briefly outline its basic ideas. We will do it for the case of relatively small electrode charges, when the linearized Poisson–Boltzmann theory could be used [81]. The case of arbitrary large electrode polarizations, beyond the linearized Poisson–Boltzmann approximation, was considered in Ref. [83]; the solution obtained helps in explaining a number of experimentally observed potential-dependent effects. The latter is, however, more cumbersome for presentation, and we will explain the idea of the theory for the case of small electrode polarizations [82].

As it was shown in Ref. [81] within the linearized Poisson–Boltzmann approximation the diffuse double-layer capacitance takes the form,

$$C = \tilde{R}(\kappa) C_{GC} \quad (39)$$

where $\tilde{R}(\kappa)$ is not a roughness factor but a "*roughness function*" of the inverse Debye length, κ. This function varies between the obvious limits $\tilde{R}(0) = 1$ and the RF $R = \tilde{R}(\infty) > 1$. For the case of weak roughness, a relationship was derived between the roughness function and the height-height correlation function of the surface corrugation,

$$\tilde{R}(\kappa) = 1 + \frac{\kappa h^2}{(2\pi)^2} \int d\mathbf{K} g(\mathbf{K})$$
$$\times [\sqrt{(\kappa^2 + K^2)} - \kappa] \quad (40)$$

Here the "height" h stands for the mean square departure from flatness and $g(\mathbf{K})$ is the height–height correlation function

$$g(\mathbf{K}) = \frac{\langle |\xi_\mathbf{K}|^2 \rangle}{S_{app} h^2} \quad (41)$$

where S_{app} is the apparent surface area, $\xi(\mathbf{R})$ the corrugation amplitude ($<\xi(\mathbf{R})> = 0$) as a function of the lateral coordinate $\mathbf{R}$, $\xi_\mathbf{K} = \int d\mathbf{R} \xi(\mathbf{R}) \exp\{-i\mathbf{K}\mathbf{R}\}$ is its Fourier transform (FT).

For Gaussian height–height correlations, $g(\mathbf{K}) = \pi l^2 \exp(-l^2 K^2/4)$, one obtains

$$\tilde{R}(\kappa) = 1 + \frac{2\sqrt{\pi}h^2\kappa}{l} \exp\left(\frac{\kappa^2 l^2}{4}\right)$$
$$\times \left[1 - \Phi\left(\frac{\kappa l}{2}\right)\right] \quad (42)$$

and

$$R = 1 + \frac{2h^2}{l^2} \quad (43)$$

where $\Phi(x)$ is the probability function [84]. For this case, $\tilde{R}(\kappa)$ is plotted in Fig. 8.

As it was shown in Refs. [81, 82], the particular form of the corrugation has only a small effect on the roughness function, which is constrained between 1 and R.

In order to verify how critical the assumption of weak corrugation for this approach is to the calculation of the roughness function, the case of corrugation described by a model of rectangular grating was studied [82]. For this case, there is an exact analytical expression for the roughness function, obtained in the weak roughness approximation, and an exact numerical solution, obtained for any ratio between the height and the period of grating. The comparison of the two has shown that the analytically tractable approximation of weak roughness gives practically a universal description of the possible roughness effect on the diffuse-layer capacitance for nonfractal electrode surfaces.

The situation is more complicated for a self-affine fractal surface. The latter is characterized by the self-affine exponent, H, which is related to the Haussdorf fractal dimension $D = 3 - H$ [85], and a long-length cutoff, l, which characterizes the scales after which the surface is apprehended as a Euclidian one. As it was argued in Ref. [81], the roughness function for the Debye length shorter than the long-length cutoff scales as $\tilde{R}(\kappa) - 1 \propto (h/l)^2 (l\kappa)^{2(1-H)}$, at least for moderate fractal dimensions, that correspond to $H < 0.5$. It could have been interesting to verify this law.

However, it is more important to verify first the basic predictions of the theory

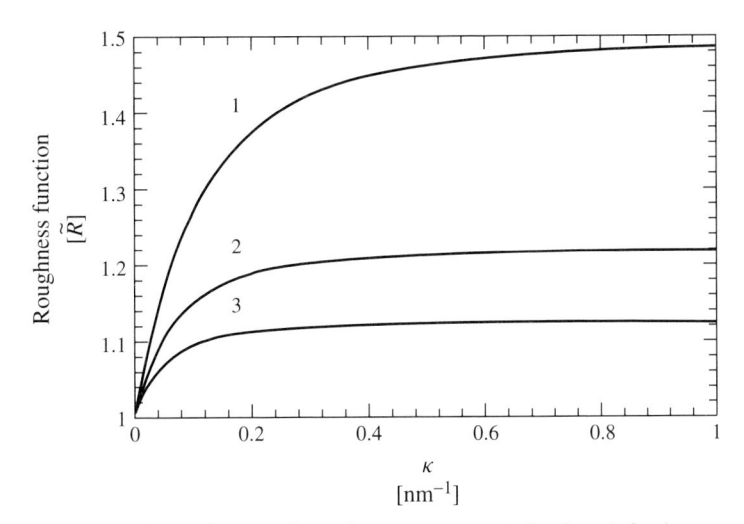

Fig. 8 Roughness function dependence on inverse Debye length for the random Gaussian surface. $\varepsilon = 80$, $h = 5$ nm, $l = (1) 10$, $(2) 15$, $(3) 20$ nm.

for nonfractal electrodes. Efforts in this direction have been made by the Tartu group [86] that have extracted $\tilde{R}(\kappa) = C/C_{GC}$ from their data and plotted it versus κ. It was found that for metals with close values of the p.z.c. for different crystal faces the trends of the theory are basically correct, but even for such metals the deviations from the nonequipotential character of the outer Helmholtz surface (assumed to be equipotential in Ref. [81]) are likely to be important for a quantitative comparison. For metals with strong variance of the p.z.c., the theory in its present form essentially fails. It needs an extension that would combine this theory with the models of energetic inhomogeneity described in the previous section. Note that Lust and coworkers compared the nonlinear variant of the theory also [83], that is, the so-called *potential-dependent roughness function*, with the values extracted from their data [87], and came essentially to the same conclusions. Further systematic data would be highly welcome, as well as the mentioned extension of the theory.

2.1.7
Semi-phenomenological Nonlocal Theory of the Double Layer

The dielectric response of the interface can be described in a unified manner in terms of the nonlocal electrostatic theory [88, 89]. Indeed, it was shown to be possible to express the electric properties of the interface through the dielectric function of the metal/solvent system, not applying a particular form of this function, for any structure of the interface. Such an approach allows revealing general properties of the double layer and expressing the parameters involved via the nonlocal dielectric function. We briefly describe the main points of this theory [43, 44, 90].

2.1.7.1 Basic Equations

At a given surface charge density, σ, of the electrode, the potential distribution along the z-normal to the electrode surface, $\phi_\sigma(z)$, obeys the Poisson equation $d^2\phi_\sigma(z)/dz^2 = -4\pi(\rho^\sigma_{ms} + \rho^\sigma_{ion})$, where ρ^σ_{ms} is the bound charge density of the metal and solvent in contact and ρ^σ_{ion} is the charge density of electrolyte ions. Subtracting from this equation the same equation at $\sigma = 0$, for the so-called *rationalized potential*, $\phi(z) = \phi_\sigma(z) - \phi_0(z)$, we get

$$\frac{d^2\phi(z)}{dz^2} = -4\pi(\rho_{ms} + \rho^\sigma_{ion} - \rho^0_{ion}) \quad (44)$$

where $\rho_{ms} = \rho^\sigma_{ms} - \rho^0_{ms}$. To close this equation let us introduce the nonlocal permeability of the metal/solvent system, $\chi(z, z')$, which relates the change in the bound charge density to the corresponding change of the electric field

$$\rho_{ms} = \frac{d}{dz}\int_{-\infty}^{\infty} dz' \chi(z, z') \frac{d\phi(z')}{dz'} \quad (45)$$

Then the equation for ϕ reads:

$$\frac{d}{dz}\int_{-\infty}^{\infty} dz' \varepsilon(z, z') \frac{d\phi(z')}{dz'} = -4\pi[\rho^\sigma_{ion} - \rho^0_{ion}] \quad (46)$$

where

$$\varepsilon(z, z') = \delta(z, z') + 4\pi \chi(z, z') \quad (47)$$

is the static nonlocal dielectric function of the metal/solvent system that keeps all the information about its structure [88–91].

Equation (46) is still not closed, as we still need an equation that relates the r.h.s. with ϕ. A particular closure is determined by a specific statistical-mechanical scheme for the ionic subsystem, the classical Poisson–Boltzmann scheme approximation (which results in the Gouy–Chapman–Stern (GCS) model) being the simplest one. Here, we will show the equation that follows in the case of the linearized Poisson–Boltzmann approximation, valid for $\sigma \to 0$ and moderately large concentrations c:

$$\frac{d}{dz}\int_{-\infty}^{\infty} dz' \varepsilon(z,z') \frac{d\phi(z')}{dz'} = \kappa_0^2 \theta(z-l) \quad (48)$$

Here, $\kappa_0 = \kappa\sqrt{\varepsilon}$ where ε is the dielectric constant of the bulk solvent and κ^{-1} is the Debye length; θ stands for the Heaviside unit step function, which reflects here the distance of closest approach, l, of ions to the electrode (l is calculated from the edge of the metal skeleton).

Because of the general properties of the kernel $\varepsilon(z,z')$, Eq. (48) possesses a unique solution. For the solution obtained, the capacity of the metal electrolyte interface is calculated via the formula,

$$\frac{1}{4\pi C} = \lim_{\sigma \to 0}\left[\frac{\phi(-\infty) - \phi(\infty)}{\sigma}\right]$$
$$= \frac{\phi(-\infty) - \phi(\infty)}{\kappa_0^2 \int_l^{\infty} dz \phi(z)} \quad (49)$$

Generally in order to find the solution, one must specify the particular form of $\varepsilon(z,z')$. Such solutions have been repeatedly reported in the literature (see e.g. [44, 92–96]). However, in the spirit of the semi-phenomenological approach, we are now interested in the solution for a general form of $\varepsilon(z,z')$. Such a solution can be obtained approximately [44] using the technique of matching asymptotic expansions [97], in the case of a dilute electrolyte when the Debye length is much greater than all the characteristic lengths of the nonlocal dielectric function. Referring the reader to Ref. [44], we show here only the result for the capacitance in the limit of low electrolyte concentrations:

$$\frac{1}{4\pi C} = \frac{1}{\varepsilon\kappa} + L_H + O(\kappa) \quad (50)$$

Here,

$$L_H = \int_{-\infty}^{\infty} dz \left\{ \left(\int_{-\infty}^{\infty} dz' \varepsilon^{-1}(z,z')\right) - \frac{\theta(z-l)}{\varepsilon} \right\} \quad (51)$$

where the inverse operator $\varepsilon^{-1}(z,z')$ is determined by

$$\int_{-\infty}^{\infty} dz' \varepsilon^{-1}(z,z') \varepsilon(z,z') = \delta(z,z') \quad (52)$$

To the accuracy of small $O(\kappa)$-terms Eq. (50) coincides with the Grahame ansatz at $\sigma = 0$. The Grahame ansatz is thus recovered in the low concentration limit, giving for the "compact-layer capacity" the expression

$$C_H = \frac{1}{4\pi L_H} \quad (53)$$

Hence, we see that C_H is determined primarily by the quantum-statistical properties of the metal-solvent systems rather than by the distance of solvent approach. The primitive local estimate gives an example when l is, indeed, of minor importance. Take a model of a film with low dielectric constant that separates an ideal metal and the bulk solvent:

$$\varepsilon(z,z') = \delta(z,z') \begin{cases} \infty & z < 0 \\ \varepsilon_* & 0 > z > z_* \\ \varepsilon & z > z_* \end{cases}$$

If $l > z_*$, this model gives $L_H = z_*(1/\varepsilon_* - 1/\varepsilon) + l/\varepsilon$, and for $\varepsilon \gg \varepsilon_*$ and $z_* \sim l$, $L_H \approx z_*/\varepsilon_*$, that is, it is determined by the parameters of dielectric inhomogeneity but not by the distance of closest approach of electrolyte ions.

An extension of the nonlocal dielectric theory to the range of moderate deviations from the p.z.c. was reported in Refs. [43, 90]. The results differ in replacing $1/\varepsilon\kappa$ by $1/4\pi C_d$, where C_d is the nonlinear differential capacitance of the diffuse layer (e.g. of GC type), and $\varepsilon(z, z')$ by the so-called *differential dielectric function* $\varepsilon(z, z', \sigma)$ (for details see Ref. [90]). Somehow the Grahame parameterization is again recovered, but the validity criteria now involve the value of σ. Critical is the requirement that the amount of the electrolyte charge, accumulated within the layer of thickness z_*, constitutes a small portion of the total charge σ. For aqueous electrolytes, the criterion limits the results to $\sigma < 10\,\mu\text{C/cm}^2$ and $c < 0.1$ M. This is the sufficient criterion. In fact the differences between the values of C_H as evaluated from small and high concentrations typically do not exceed 5%. The reason for that phenomenon is discussed in Ref. [88]. In general Grahame parameterization may eventually work at higher charges and concentrations.

2.1.8
"Dipolar" Models of the Compact Layer

Historically, molecular models of the compact layer were first to rationalize its nonlinear dielectric response. Starting with Ref. [98] and until the beginning of the 1980s, this type of models comprised the main direction of research in the theory of the electrochemical interface (for detailed review see Refs. [88, 99]). We will discuss several of the most important ones in the following text.

2.1.8.1 Two-State Model
This model [98] assumes the existence of two admissible orientations of solvent dipoles in the layer bounding the metal, "spin–up" (↑) and "spin–down" (↓), with a dipole pointing away or to the metal, respectively. Some residual energy is attributed to each state in the field of an uncharged electrode, $U_\uparrow$ and $U_\downarrow$. Boltzmann statistics relates then the occupation numbers of the states, $N_\uparrow$ and $N_\downarrow$: $k_B T \ln(N_\uparrow/N_\downarrow) = U_\downarrow - U_\uparrow + 2p\Phi/d$ where p is a permanent dipole molecule of the molecule, Φ is the potential drop across the layer of thickness d, and k_B is Boltzmann's constant. Here the dipole moments are assumed to have the same magnitude for both orientations, but the dipole moments are, in general, different from those in the gas phase or in the bulk of the solvent. The order parameter then reads $(N_\uparrow - N_\downarrow)/N = \tanh p\Delta\Phi/k_B T d$, where $\Delta\Phi = \Phi - \Phi_m$, $\Phi_m = d/2p(U_\uparrow - U_\downarrow)$, and $N = N_\uparrow + N_\downarrow$ is the total number of molecules in the layer per unit surface area. This is not yet a closed equation; the closure is reached by the assumption that Φ is a superposition of the potential drop in the absence of dipoles and the contribution due to dipoles:

$$\Phi = \frac{1}{\varepsilon_\infty}\left\{4\pi\sigma d - 4\pi pN \tanh\frac{p\Delta\Phi}{k_B T d}\right\} \quad (54)$$

where σ is the specific charge of the electrode and ε_∞ is the high frequency effective dielectric constant, related to the intramolecular polarizability of the molecules. The p.z.c., obtained from Eq. (54) at $\sigma = 0$, lies generally between 0 and Φ_m.

Equation (54) implicitly determines the value of potential for a given charge of the electrode. Its differentiation over σ gives the famous bell-shaped differential capacitance of the molecular layer with a maximum centered at Φ_m.

$$C_M = \frac{\varepsilon_\infty}{4\pi d} + \frac{Np^2}{k_B T d^2} \frac{1}{\cosh^2\left\{\frac{p\Delta\Phi}{k_B T d}\right\}} \quad (55)$$

The parametric dependence of C_M on σ is given by Eq. (55) combined with $\sigma(\Phi)$ dependence given by Eq. (54). According to Eq. (54) the maximum of capacitance is at $\sigma = \varepsilon_\infty \Phi_m / 4\pi d$. Electrochemists used to believe that, if there were a preferential orientation of water molecules on the metal surface, the molecule would look at the metal by its negative end. That means $U_\downarrow > U_\uparrow$. Hence, $\Phi_m < 0$. Since the maximum of the compact-layer capacitance is in the anodic but not in the cathodic range, historically this model was not accepted in electrochemistry. Nowadays we are less certain in that (see discussion in Sect. 2.1.10). Furthermore, since we now know that this is only one of the contributions to the inverse capacitance, this limitation may not be that critical. Indeed for small $|\Phi_m|$ the maximum of C_H may still be shifted to the anodic range by the basic slope of the σ-dependence of C_H (see Sect. 2.1.9.2). Somehow the two-state model [98, 100] has resulted in the understanding that humps on the capacitance curve may be associated with dipolar reorientations. The two-state model rationalized the value of the effective dielectric constant of the Helmholtz layer, $\varepsilon_*(\sigma) = \varepsilon_\infty + 4\pi Np^2/k_B T d [\cosh\{p\Delta\Phi/k_B T d\}]^2$.

Further modifications of the model [101, 102] were aimed at shifting the maximum towards the anodic range at the cost of an artificial assumption that the dipole moments in the up and down orientations are different (for a critical discussion of other modifications made in Refs. [101, 102] see Ref. [88]).

2.1.8.2 Multistate Models

A three-state model was suggested in Ref. [103] with a third state ($\rightarrow$) corresponding to orientation of a dipole parallel to the electrode surface. It was analyzed on a level similar to Ref. [101] and successfully applied for interpretation of the capacitance data for aprotic solvents.

A cluster model, essentially a four-state model, was suggested [104] and developed [105, 106] to interpret the capacitance data in associated solvents. According to this model the molecules in the bounding layer can exist in free states or in clusters. Up-and-down orientations are allowed for a free molecule or a cluster. Owing to constraints on the mutual orientations of molecules in a cluster, the effective dipole moment of a cluster is assumed to be smaller than that of a free molecule. The residual energy for a molecule in a cluster state is assumed to be lower than in a free state. In such a model there will be more clustered than free molecules at the p.z.c. However, charging the electrode shifts the equilibrium towards the occupation of free states due to the gain in polarization energy, that is, it causes the field-induced destruction of the clusters. Since the decay of clusters increases the net polarization of the layer, the rising branches of the $C_M(\sigma)$ curve emerge. Because of the interplay between the polarization of single molecules and clusters, and the cluster decay and formation, the $C_M(\sigma)$ curve may generally have a sophisticated form. Qualitative features of Grahame's curve for the compact-layer capacitance of the Hg–H$_2$O-NaF interface and its temperature dependence were reproduced in the elegant parameterization of Ref. [105] (discussed also in Refs. [88, 99]) with three fitting parameters and four parameters, which can be evaluated from independent data. A better fit was achieved in a more involved analysis of Ref. [106], but as noted in Ref. [99] the best fit would have been achieved for a dipole moment of water twice as large as its value in the gas phase, which may be a bit too much. Furthermore, the sign of the temperature derivative [105, 106] is opposite to the experimentally observed one.

The lesson from this four-state model is that the interplay between the field-induced destruction of the clusters and reorientations of the clusters and single molecules may be a source of a most peculiar $C_M(\sigma)$-behavior. The richest curve exhibits four maxima and three minima.

2.1.9
The Role of Metal Electrons

We now briefly describe the effects associated with the surface electronic profile of the metal and its influence on the response of the electrochemical interface to charging. For a detailed review, the reader is addressed to Ref. [107].

2.1.9.1 Surface Electronic Profile and its Response to Charging

How close do solvent molecules come to the metal? Where is the boundary of the metal? To answer these questions, we must go back to the structure of the metal surface. The existence of a tail of quasi-free electrons spilled out of the ionic skeleton of the metal was predicted in the early days of quantum mechanics [108]. In the 1970s, this picture was approved and detailed by the electron-density functional theory of the inhomogeneous electron gas (later leading to Nobel fame for Walter Kohn in 1998) applied to metal surfaces [109]. First calculations were performed in the so-called *jellium model* [110], in which the skeleton is treated as a homogeneous semi-infinite background of positive charge. Later on the calculations were modified to account for the discrete structure of the skeleton with the use of ionic pseudopotentials [111].

These predictions have been verified by experiments on low-energy helium atom scattering by metal surfaces [112]. The helium atom repels the Bloch electrons, because their wave function must become distorted to preserve orthogonality to the wave functions of the closed shells of helium. The repulsion potential appears in practise to be proportional to the local free electron density. Thus a helium atom scatters like a ping-pong ball from the electron cloud of the metal, and this allows to probe the distribution of electrons in the cloud, that is, the profile of the electronic tail and its lateral corrugation that usually follows the periodicity of the surface crystal plane [112].

How does the surface electronic profile deform with charging? Although there was no direct observation of the charge-induced profile modulation, a lot is known from the theory and indirect manifestation of this effect in field emission. For the jellium model, after the pioneering papers [110, 113, 114], the problem was studied in detail in Ref. [115] within the trial function version of the density functional formalism [116] and in Ref. [117, 118] within the more general Kohn–Sham scheme [119]. We summarize here the properties necessary for our further discussion.

If no discreteness of lattice charge is taken into account or if we speak about the laterally averaged electron density profile, $n(z)$, neglecting Friedel oscillations [110], it can be roughly approximated by a trial function

$$n_\sigma(z) = n_+ \begin{cases} 1 - 0.5\exp[\beta(\sigma)(z-\bar{z})], & z < \bar{z} \\ 0.5\exp[\beta(\sigma)(\bar{z}-z)], & z > \bar{z} \end{cases}$$

$$\bar{z} = -\frac{\sigma}{n_+} \quad (56)$$

where n_+ is the volume density of the positive charge of the ionic skeleton that occupies the half-space $z < 0$, σ is the net surface charge density ($\int_{-\infty}^{0}[n_\sigma(z) - n_+] + \int_{0}^{\infty} n_\sigma(z) = -\sigma$). According to Ref. [115], the expansion

$$\beta(\sigma) \approx \beta(0) + \tilde{a}\sigma + \tilde{b}\sigma^2 + \tilde{d}\sigma^3 \quad (57)$$

with the obtained analytical expressions for the coefficients $\beta(0), \tilde{a}, \tilde{b}, \tilde{d}$ as functions of n_+ reproduces well the direct numerical simulation of $\beta(\sigma)$. Figure 9 shows the thus calculated profile of $n_\sigma(z)$ for n_+ corresponding to Hg. The corresponding excess charge distribution $\Delta n_\sigma(z) = n_\sigma(z) - n_0(z)$ is shown in Fig. 10.

The important characteristic of the excess charge distribution is the position of its center of mass,

$$z_\sigma = -\frac{1}{\sigma}\int_{-\infty}^{\infty} dz\, z[n_\sigma(z) - n_0(z)] \quad (58)$$

which for the approximation (56) takes the form

$$z_\sigma = \frac{\bar{z}}{2} + \frac{n_+}{\sigma}\left[\frac{1}{\beta(0)^2} - \frac{1}{\beta(\sigma)^2}\right]$$

$$\approx z_0 + p\sigma + r\sigma^2 \quad (59)$$

again with definite expressions for z_0, p, r [115]. Thus-calculated z_σ plots appear to be close to direct Kohn–Sham calculations [117, 118].

$n(z)$ obeys an exact sum rule with no dependence on the particular approximation for the electron density functional. For example, the Theopilou sum rule derived for the metal vacuum interface reads $\int_{-\infty}^{0} dz\, z[n_\sigma(z) - n_0(z)] = \sigma^2/2n_+$. The substitution of a single parameter trial function (56) gives just an equation on β as a function of σ and n_+. This gives a qualitatively similar charge dependence, but quantitatively it is different from the

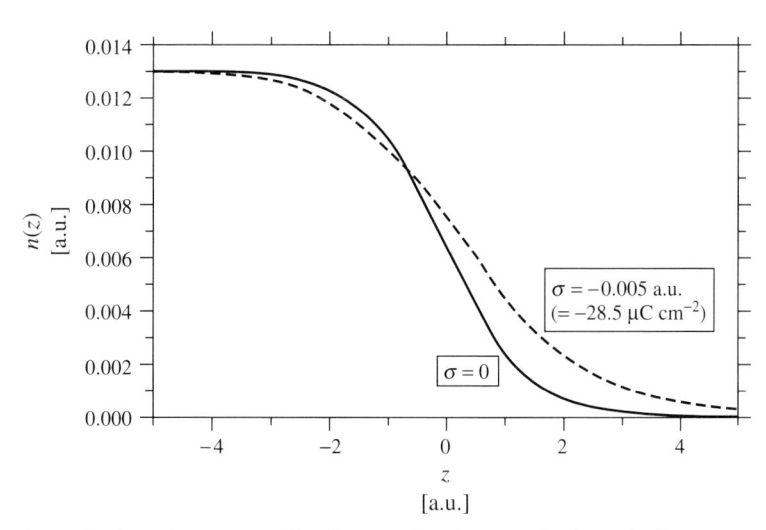

Fig. 9 Surface electronic profile of a neutral and negatively charged jellium calculated for the bulk electron density of Hg.

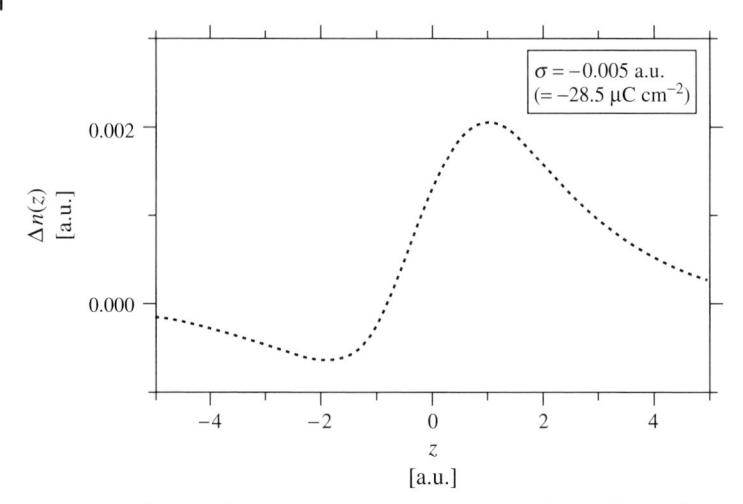

Fig. 10 Distribution of excess negative charge corresponding to the profile shown in Fig. 9.

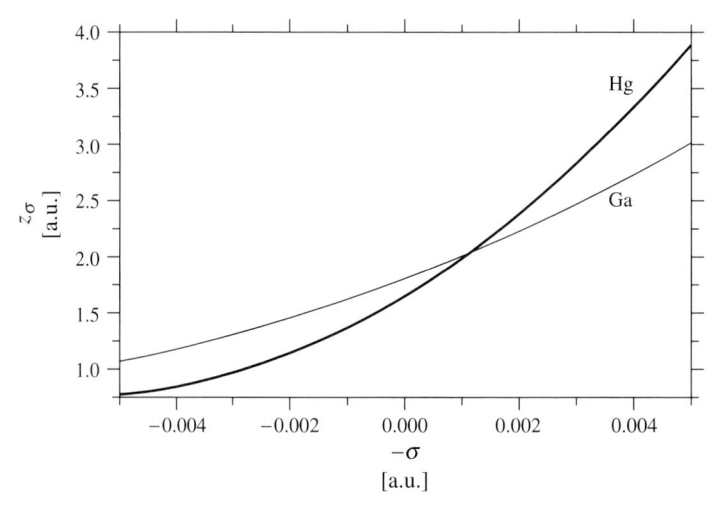

Fig. 11 The center of mass of excess charge distribution as a function of the net charge on jellium metals. At zero charge, it is located outside the jellium; with negative charging, it moves further out. 1 a.u. of surface charge = $5.710^{-3}\,\mu C\,cm^{-2}$.

density functional results, because neither is the trial function exact nor the density functional. The dramatic history of the employment of exact sum rules in electrochemistry is discussed in Ref. [107].

The z_σ^- and $\beta(\sigma)$-curves for bulk electron densities corresponding to Hg and Ga are shown in Figs. 11 and 12, respectively. For all the metals studied, $z_0 > 0$, $\tilde{a} > 0$, $p < 0$. This means that for negative charges the electronic profile broadens and the excess charge center-of-mass moves further outside the metal. These two features are also essential

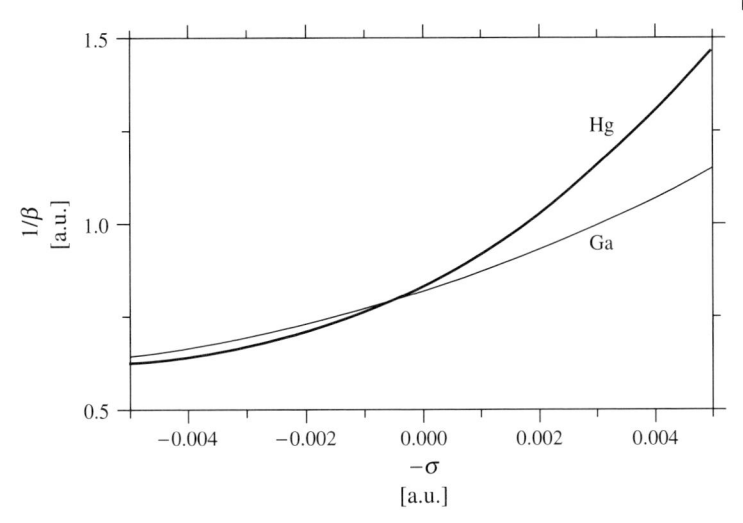

Fig. 12 The same graphs as in Fig. 11, but for the characteristic length of smearing of the surface electronic profiles. Smearing increases with negative charging.

for understanding the metal response to charging in situ, although the bounding medium may affect the values of $\beta(\sigma)$ and z_σ. However, the qualitative behavior of these functions remains the same, unless the solvent molecules chemisorb on the metal surface. In that case with negative charging of the metal the extra electrons may have to be localized within the skeleton half-space, thereby decreasing z_σ and $\beta(\sigma)^{-1}$.

2.1.9.2 Response to Charging of the Electrochemical Interface and the Compact-Layer Capacitance: Main Qualitative Effects

The reader interested in the history of the step-by-step comprehension of the role of metal electrons in the double-layer theory and various ideas in this field is referred to the reviews [107, 120–124]. In the following sections, we will quote only the references needed for understanding the simple but basic qualitative features discussed in this tutorial overview.

The profile of an ideally smooth interface is sketched in Fig. 13. The half-space $z < 0$ is occupied by the ionic skeleton of the metal. This can be described, roughly, in a jellium model, as a continuum of positive charge n_+ and the effective dielectric constant ε_b due to the polarizability of the bound electrons (this quantity is, with rare exceptions (Hg: $\varepsilon_b = 2$, Ag: $\varepsilon_b = 3.5$), typically close to 1 [125]). The gap $0 < z < a$ accounts for a nonzero distance of the closest approach of solvent molecules to the skeleton. The region of $a < z < a + d$ stands for the first layer of solvent molecules, while $z > a + d$ is the diffuse-layer region. $n(z)$ denotes the profile of the density of free electrons. This is, of course, an extremely crude picture, but it eventually helps to rationalize the results of the various theoretical models and simulations.

The integral K_H or differential C_H capacitances of the compact layer, related to each other as $1/C_H = (d/d\sigma)[\sigma/(K_H(\sigma))]$, are extractable from the overall capacitance

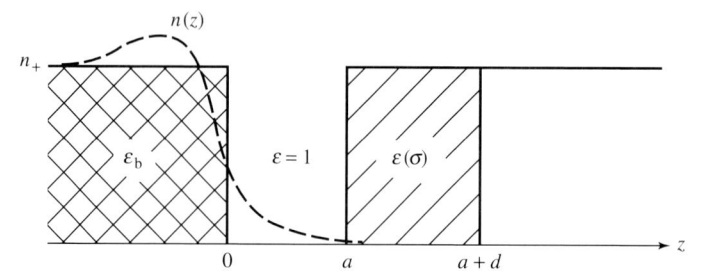

Fig. 13 A cartoon of a profile of a smooth electrochemical interface. The half-space $z < 0$ is occupied by the metal ionic skeleton that, within the jellium model, is described as a continuum of positive charge density ($n+$) and the dielectric constant due to bound electrons (ε_b), the value of which lies typically between 1 and 2. The gap accounts for a finite distance of closest approach of solvent molecules to the skeleton; the gap is determined by the balance of forces that attract the molecules to the metal and the Pauli repulsion of the closed shells of the molecules from the free electron cloud of the metal of density $n(z)$. The regions $a < z < a + d$ and $z > a + d$ correspond, respectively, to the first layer of solvent molecules (which can be roughly characterized by charge-dependent effective dielectric constant) and the diffuse-layer part.

of the interface via for example, the PZ plots [37] or other methods discussed in this chapter. To the accuracy of exponentially small terms, they can be written as:

$$\frac{1}{4\pi K_H} \simeq a - z_\sigma + \frac{d}{\varepsilon(\sigma)} \quad (60)$$

$$\frac{1}{4\pi C_H} \simeq a + \sigma \frac{da}{d\sigma} - z_* + \frac{d}{\varepsilon_*(\sigma)}$$

$$z_* = \frac{d(\sigma z_\sigma)}{d\sigma}, \quad \frac{1}{\varepsilon_*(\sigma)} = \frac{d}{d\sigma}\left[\frac{\sigma}{\varepsilon(\sigma)}\right] \quad (61)$$

Here the $d/\varepsilon(\sigma)$-contribution is due to the potential drop across the first layer of solvent molecules, where the effective dielectric constant $\varepsilon(\sigma)$ takes into account the nonlinear dielectric polarizability of the layer. Thus in order to understand the σ-dependence of the compact-layer capacitance, it is sufficient to understand the σ-dependence of $a - z_\sigma$ and $d/\varepsilon(\sigma)$.

Different microscopic models of the interface may be classified with regards to these dependencies (see Ref. [107]).

Here, we just refer to the behavior that explains the systematic trend for all s,p-metals [126] – the positive slope of the capacitance near the p.z.c., shown Fig. 14.

Such a behavior was first found in Refs. [127, 128] by density functional simulation. Here, for clarity, the degrees of freedom of the molecular reorientations in the first layer were frozen, that is, $\varepsilon(\sigma) = $ const. The peculiar $a(\sigma)$ behavior is determined by two main effects. The first effect is asymmetric with respect to p.z.c. That is, the position of the solvent molecules that are "sitting on the tail" of free electrons follows the deformation of the tail. With negative charging the profile swells out of the metal skeleton and a increases; with positive charging the profile moves back towards the metal and a decreases. The second effect is symmetric with respect to the p.z.c. Any capacitor with a flexible distance between the plates will contract upon charging regardless of the sign of the charge. The combination of these two effects results in the negative

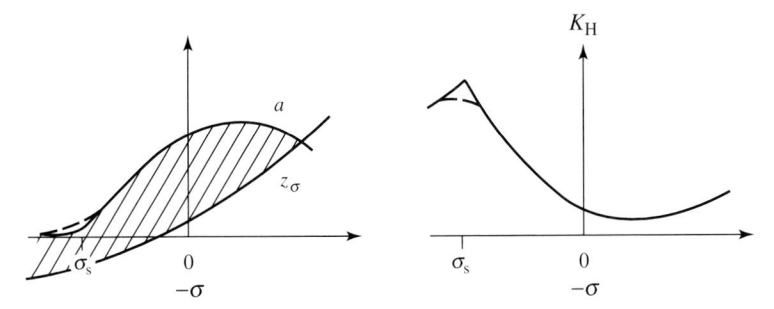

2.1 Electrochemical Interfaces: At the Border Line | 77

Fig. 14 A sketch of the charge dependence of the distance of closest approach of solvent molecules to the skeleton, a, the center of mass of the excess charge distribution, z_σ, and the corresponding plots for the integral capacitance of the compact layer, K_H. In the case of "soft landing" of the molecule onto the metal skeleton edge (dashed curves), the spike in the capacitance transforms into a hump.

slope of $a(\sigma)$ near the p.z.c. (which is typically steeper than the slope of z_σ) and the maximum in the cathodic range. At some positive charge, denoted here as σ_s, a will reach zero and will stop changing with further increase of σ, because the molecules will encounter the edge of the metal skeleton. The stop would not be abrupt (solid curve), but soft (dashed curve) if the skeleton were not a hard wall for solvent molecules. The simplest approximation for $a(\sigma)$ is

$$a(\sigma) = [a(0) - A\sigma - B\sigma^2],$$
$$A > 0, B > 0 \qquad (62)$$

The difference between $a(\sigma)$ and z_σ (Fig. 14) leads to $K_H(\sigma)$ plots that are in line with the data for simple metals (Figs. 15 and 16), (for review and references, see Ref. [88]) in contact with surface-inactive electrolytes; the slope at the p.z.c. is positive, the minimum lies in the cathodic range and a kind of a spike, which might be apprehended as a hump, takes place in the anodic range.

Of course there could be other reasons for the hump. In the molecular models of the compact layer (see Sect. 2.1.8), they were attributed to the nonlinear response of molecular reorientations that were so far frozen in order to make clear the mere effect of $a(\sigma)$ and z_σ. In terms of Eq. (60), that would mean that $\varepsilon(\sigma) \neq$ const. For the simplest two-state molecular models, it has a bell-shaped form, the maximum of which is expected in the cathodic range. The hump is, however, observed in the anodic range. Since the $(a(\sigma) - z_\sigma)$-determined "background" curve is properly asymmetric, a bell of $\varepsilon(\sigma)$ centered near the p.z.c. will be seen as a hump in the anodic range.

Is the smeared "stop-spike" for K_H (Fig. 14) what we see for a hump on the measured capacitance curve, or is it screened by molecular reorientations? We don't know that. Moreover, the specific adsorption of ions, which was assumed absent, can give rise to a hump too. Somehow this feature is interesting because of its quantum-mechanical origin and direct dependence on the nature of the metal. The more densely packed the metal surface, the more abrupt the stop, the sharper the spike. Open surfaces should favor "soft landing" and the hump would

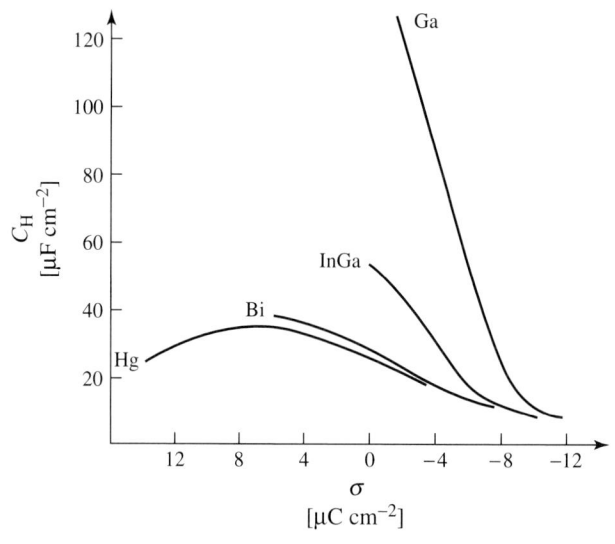

Fig. 15 Metal effect on the compact-layer differential capacity: the series Hg, Bi, InGa alloy, Ga. Data from Refs. [129–132].

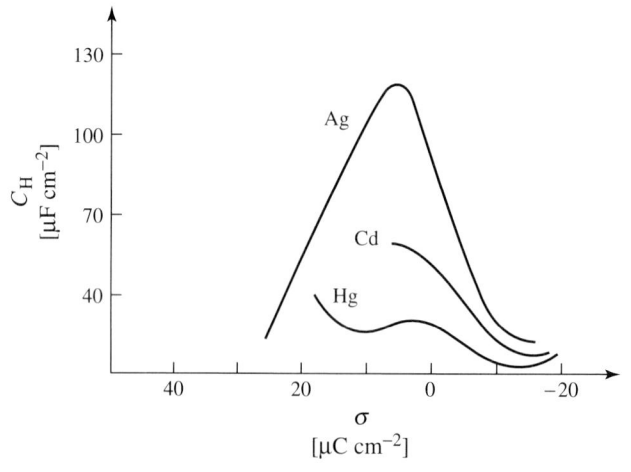

Fig. 16 Metal effect on the compact-layer differential capacity: the series Hg, Cd, polycrystalline Ag. Data from Refs. [35, 133, 134].

be smoother. For single-crystalline surfaces, the spikes should be sharper than for liquid metals, in which the interface undergoes fluctuations, and for uneven polycrystalline surfaces, in which inhomogeneous broadening may take place. The higher the bulk electron density of the metal, the denser the electronic tail, and the stronger the repulsion of the molecules from it. Higher positive charges will be needed to press the solvent to the skeleton and σ_s will move farther to the anodic range. This is what we see, if we compare Ga and Hg.

In contrast to what was suggested in an early "swing" and "slide" model [135], there is no simple relation between the slope of $K_H(\sigma)$ and n_+ (for details see Refs. [107, 126]).

Models of this "flexible capacitor" came across negative values of the calculated $C_H(\sigma)$. The reasons for that exotic effect (not necessarily an artifact!) have been discussed in detail in reviews [107, 121, 136].

2.1.9.3 Optical Response and the Interfacial Structure

In the early 1980s, phenomenological electrodynamics of smooth interfaces was developed with a capability to parameterize the main types of electromodulated optical signals. The signals were expressed through "irreducible integrals" of the nonlocal response function, which characterizes the interface in the optical frequency range. This had opened opportunities for model microscopic theories that could calculate or approximate the response function, specify the irreducible integrals, and rationalize their dependence on the charge of the metal and frequency (for review see [107, 137–140]).

At planar interfaces there are four main types of linear optical signals that are detected by different techniques. Three of them are related to reflection and one to attenuated total reflection. In reflection methods the basic measurable parameters are related to r_p and r_s – the complex amplitudes of the reflection coefficients of the light polarized parallel (p) and perpendicular (s) to the incidence plane, respectively (Fig. 17). These are reflection coefficient $R_{s(p)} = |r_{s(p)}|^2$, phase of reflected light $\delta_{s(p)} = (1/2i) \ln\{r_{s(p)}/r_{s(p)}^*\}$, ellipsometric parameters $\psi = \tan^{-1}|r_p/r_s|$ and $\Delta = \delta_p - \delta_s$. In attenuated total reflection, by measuring the frequency position of the dip in reflection coefficient as a function of the angle of incidence, one detects the dependence of the surface plasmon frequency on the wave-vector, $\bar{\omega}(K_\parallel)$.

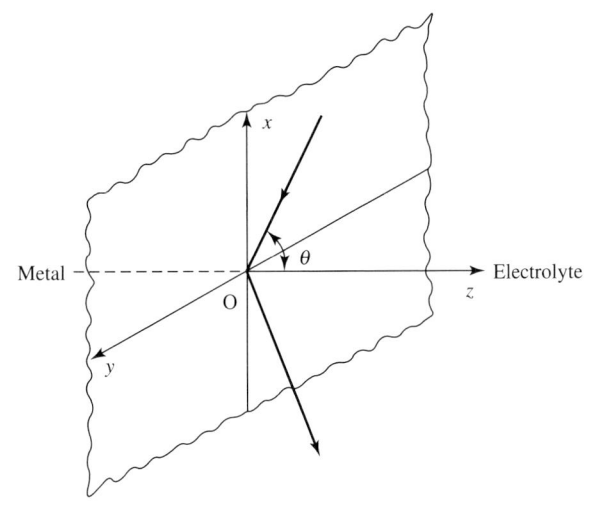

Fig. 17 The geometry of electromagnetic wave reflection experiments: (z, x) is the incidence plane, and θ is the angle of incidence.

The modulation technique is based on the application of a potential $E = E_0 + \delta E \sin(\Omega t)$, where $\delta E \to 0$ and Ω is smaller than all the characteristic frequencies at this interface. Measuring the Ω-harmonics of the signals, for example, $r_{s(p)} = r_{s(p)}|_{E=E_0} + (dr_{s(p)}/dE|_{E=E_0}) \, \delta E \sin(\Omega t)$, gives the precise value of the derivative over potential (if there is no singularity at $E = E_0$). The four types of modulation spectroscopy therefore give $d\ln(R_{s(p)})/dE$ (electroreflectance), $d\delta_{s(p)}/dE$ (electromodulation of reflected phase), $d\psi/dE$ and $d\Delta/dE$ (electromodulated ellipsometry), and $\bar{\omega}(K_\parallel, E)$ (potential-induced surface-plasmon shift). The potential modulation allows to separate a contribution of the microscopic interfacial layer from the signal. Though different equipment is needed to register these signals, a unified electrodynamic theory exists for their parameterization [137–140]. The interested reader is referred to Ref. [107] for a summary of results of this theory and to see how far can one proceed with microscopic models in the interpretation of the potential dependence of all the mentioned signals. Here, we will draw one simple example – electroreflectance of s-polarized light [141].

As it was shown in Ref. [141] in a Drude-like model of the metal and the model of the interface sketched in Fig. 13

$$\frac{d(\ln R_s)}{dE} = C 4\omega \frac{\sqrt{\varepsilon_\infty}}{c} \cos\theta \frac{1}{en_+}$$

$$\times \left\{ \left[(\varepsilon_\infty - 1)\frac{da}{d\sigma}\frac{1}{en_+} + \varepsilon_\infty - \mathrm{Im}\varepsilon_b \right] \right.$$

$$\left. \times \mathrm{Im}\varepsilon_M + (\mathrm{Re}\varepsilon_M - \varepsilon_\infty)\mathrm{Im}\varepsilon_b \right\}$$

$$\times \frac{1}{(\varepsilon_\infty - \mathrm{Re}\varepsilon_M)^2 + (\mathrm{Im}\varepsilon_M)^2} \quad (63)$$

where C is the capacitance of the metal–electrolyte interface, ε_∞ is the bulk optical dielectric constant of the solution, $\varepsilon_M(\omega)$ is the bulk optical dielectric constant of the metal, and $\varepsilon_b(\omega) - 1$ is the high frequency contribution to it, which comes from bound electrons of the skeleton. When $(da/d\sigma) = 0$, this formula reduces to the well-known Aspnes and McIntyre result [142], which tells us that the potential dependence of the s-polarized electroreflectance signal should follow the $C(E)$ plot. However, this is not observed experimentally. It is known that for a number of systems the signal may even change sign with the variation of E. Equation (63) can easily explain this effect by changing the sign of $(da/d\sigma) = 0$. Experimental data of Ref. [143, 144] were used to extract $da/d\sigma$ [141, 145] together with the data for optical constants of In and Pb from the literature.

Since the optical constants were not measured in situ in the same experiments the results may be considered as a rough evaluation only (for the same reason the independence of extracted values for $da/d\sigma$ on ω was not checked). Somehow the results shown in Fig. 18 seem to be in line with general expectations of the interfacial relaxation theory sketched in the upper corner of the figure; the meaningfulness of the obtained absolute values (for In, $1\,\mu C$ causes a variation of a by 0.26 a.u.) is discussed in Ref. [145].

The results for Pb [145], however, caused debates [107]. This electrode is not the best candidate for probing the plasma free electron properties. Single electron interband transitions for Pb lie in the low frequency part (~ 2 eV) of the spectrum. On the other hand, the data on potential dependent electroreflectance [144] were reported at one relatively low frequency (1.96 eV) where an interference between single electron and

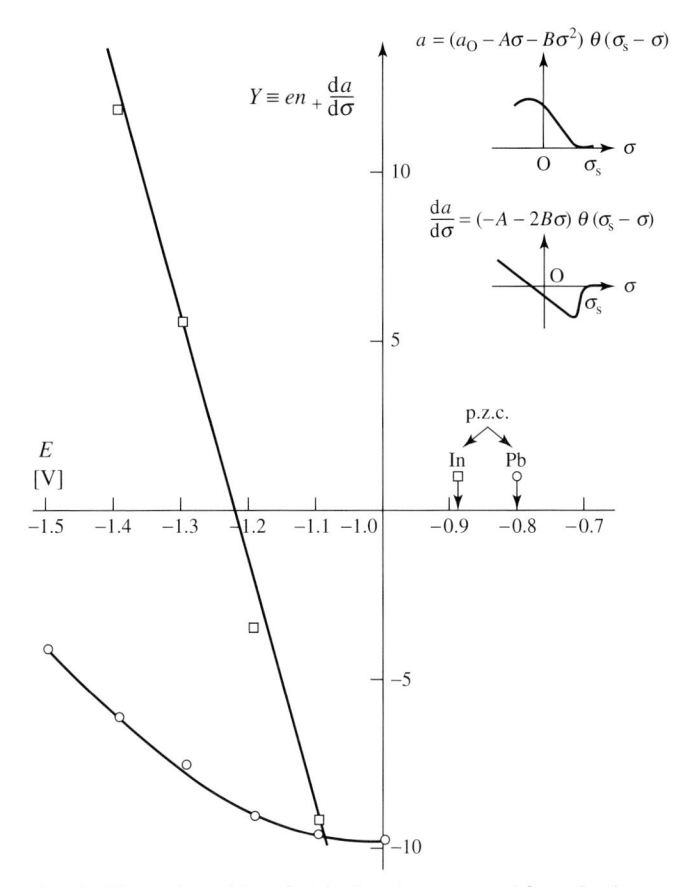

Fig. 18 The evaluated interfacial relaxation, extracted from the data on s–polarized electroreflectance, as a function of electrode potential. The insert plots show schematically, the shape of the $a(\sigma)$ and $da(\sigma)/d\sigma$ curves for parabolic approximation (Eq. (62)).

plasma contributions may be expected. For more on this interference, see in Ref. [107].

Simple jellium-model concepts also helped to understand the potential dependence of a nonlinear optical signal from the electrochemical interface – the second harmonic generation [146, 147], and of the frequency of surface plasmon resonance [148, 149]. In both cases, the effect of field-induced adsorption of anions at positive charges of the electrodes was rationalized. An application of the "inverted" Lang model [107] revealed the compression of the surface electronic profile under the adsorbed anions. The greater the field-induced adsorption, the stronger the compression, and this affects the signal in line with experimental observations.

In this section, we have focused on the qualitative features of the response of the metal–electrolyte interface to charging rather than on first principle calculations of the absolute values of the parameters that characterize the interface, about which one can learn more in Refs. [107, 124]. Such calculations, however, are often insufficiently paid back, due to the lack of precise knowledge of the solvent effect on the metal surface electronic properties. Refs. [150, 151] show what a "resistance" the steps towards "improving" the picture of this interaction encounter.

2.1.10
Effects of the Molecular Nature of Electrolyte Solutions

The "real" electrochemical double layer is not a simple arrangement of capacitors with concomitant potentials and fields, but it consists of "real" atomic and molecular ions and of a vast number of solvent molecules, all of which have a finite size and whose arrangement depends as much on the interatomic and intermolecular interactions as on external electric fields and potentials. In addition, molecular ions possess a very specific and asymmetric charge distribution. Owing to their (chemical) interactions with the metallic phase the charge of all ions can also depend to some extent on its distance from the metal surface (see previous text). Furthermore, polar solvent molecules possess a permanent dipole moment and higher electrostatic moments, that lead to lateral interactions between them, which in turn, render the simple picture of the solvent as a dielectric continuum or n-state dipole models (see Sect. 2.1.8.2) inappropriate; protic solvents such as water and alcohols have, in addition, the capability to form directional yet fluctuating hydrogen bonds that give rise to specific structural arrangements of the solvent molecules relative to each other and in the solvation sphere around ions. These arrangements, in turn, can either enhance or counteract the orientational influence due to the electronic structure and surface charge (or potential) of the metal.

The simplest model that goes beyond the Debye–Hückel and GC theories of point ions and takes into account the finite size of ions is the so-called *primitive model*. Here, ions are described as simple hard spheres and the solvent is approximated by a dielectric continuum. An even more special case is the so-called *restricted primitive model*, in which both ions have the same size.

Various attempts have been made in going beyond the primitive models towards *civilized models*. In the most simple of such extensions, called the *solvent-primitive model* [152], the solvent is still treated as a dielectric continuum. In addition, individual solvent molecules are treated as hard spheres. This takes into account the molecular nature of the solvent in but the crudest fashion. Replacing the description of the solvent from dielectric continuum by molecular point dipoles at the center of soft spheres representing solvent molecules (the Stockmayer model) leads to the *ion-dipole models*. Various extensions of these models incorporate higher-order electrostatic moments (up to octupole moments [153]) and molecular polarizability for the description of the solvent bound charge density.

In order to understand ionic distributions in the EDL a realistic description of hydration or, more generally, solvation phenomena is necessary, which in protic liquids implies an adequate description of the hydrogen-bond network. Theory and computer simulation of bulk liquids showed that the most efficient way to include these properties into the models is via *distributed charge models* in which the intramolecular charge distribution is represented by several point charges. The point charges are adjusted to reproduce experimental dipole and/or quadrupole moments of the molecule, the bulk structure of the liquid, or various other properties. For the ubiquitous solvent water, the simplest models treat water as a rigid three-site molecule in its gas phase geometry with partial charges on oxygen and hydrogen sites. Ions are described as charged soft spheres with a point charge at their center. Molecular and ionic size is maintained through Lennard–Jones interactions or similar potential functions in an attempt to describe short-range exchange correlation and long-range dispersion interactions. Hydrogen bonding and hydration are produced by these models through the balance between electrostatic and nonelectrostatic interactions. In addition to these simple rigid three-site models many more sophisticated flexible and polarizable water and ion models have been developed. The importance of the dipolar nature of the solvent and of the interactions between solvent and electrode were recognized in the double-layer model by Bockris, Devanathan, and Müller [154]. Water ''hydrates'' the electrode, which is regarded as a giant ion, and so contributes to the electric fields near the interface.

Molecular computer simulations are the method of choice when it comes to investigating molecular models like the one discussed earlier, both in the bulk and at interfaces. Contrary to analytical theories, simulation methods can treat a variety of different models on the same statistical-mechanical footing, without mathematically mandated approximations that cannot be rigorously justified and/or tested. In fact, computer simulations have been used as *computer experiments* to verify or falsify theoretical work.

In addition to explicitly accounting for the molecular nature of electrolyte solutions near the EDL, computer simulations can also be used towards realistic modeling of the metal part of the electrochemical interface. Simple models of the metal are hard walls with image interactions that account, in a very simplistic way, for the electronic properties of the metal. This model is nonspecific for the chemical identity of the metal. The jellium model incorporates the chemical identity of the metal in the form of the density of free electrons and ionic skeleton pseudopotentials; it models the *electronic spillover* effect into the liquid phase (see Sect. 2.1.9.1). Several interaction models employed in computer simulations are chemically specific, that is, they have been developed for a specific metal and/or surface geometry. These models may or may not be based on quantum-chemical calculations of the interaction between water molecules or ions and a small cluster (or an extended periodic slab) of ions. A widely used model has been the model of the Platinum(100)–water interface [155]. It consists of a flexible atomic representation of the metallic phase, of surface corrugation through site-dependent water-platinum interactions, and a description of the orientational anisotropy via specific *chemical* interactions between the oxygen atom of a water molecule and the platinum metal. This model has been extended to the (111) and other surfaces; a similar model has been used for mercury–water interactions [156, 157].

In spite of the fact that rather sophisticated models can be treated, several limitations of computer simulation methods such as Monte Carlo (MC) or Molecular Dynamics (MD) limit their use in interfacial electrochemistry:

- The spatial and concentration inhomogeneities of the EDL are large. This requires the simulation of large systems at high electrolyte concentrations. Dynamic processes near the interface

occur on a much longer time scale than the inherent fluctuations of the hydrogen-bond network and the hydration shell, or they are very rare, which mandates long simulation times. Even with present-day computer resources, the requirements of large system size and long simulation times are hard to reconcile.
- Simulations with molecular solvent description can only be performed in the concentration range above about 0.5 mol/liter, much higher than the concentrations of many electrochemical experiments.
- Simulations of the EDL structure around surface defects and thus the modeling of *real* interfaces is almost impossible, both because of the large system size involved and because of insufficient knowledge of the electronic structure around these defects, and thus the lack of suitable interaction potentials.
- MC and MD are methods of statistical mechanics. The quality of the results depends on the extent to which the phase space of the system can be sampled. At best the statistical error of a simulation decreases only slowly of the order of the square root of the computational effort; at worst, the phase space contains bottlenecks, making ergodicity impossible to achieve.
- Reliable and accurate potential energy surfaces on the basis of quantum chemistry for the interaction between molecules and ions with the metal surface are still very scarce. Especially, cluster calculations fail to reach the limit in which the cluster of metal atoms begins to exhibit metallic properties. Methods based on periodic slabs appear to be more promising for the future; to date, only few calculations have been performed [158].
- The use of classical simulations on the basis of the Born–Oppenheimer approximation is problematic when dealing with metals, even in the absence of electrochemical reactions involving the transfer of electrons, which can certainly not be approximated by classical mechanics.
- Ab initio simulation methods (like the Car–Parrinello method) are currently so much limited in system size (typically substantially less than 200 atoms) and simulation time (a few picoseconds), that they have only been applied to vacuum systems, for example, to the investigation of molecular and dissociative adsorption of water on a magnesium oxide surface [159, 160], and not to an entire metal–solution interface. However, these methods hold substantial promise for the future.

In spite of these shortcomings, simulation methods have nevertheless substantially contributed to our understanding of the double layer because of their ability to describe:

- Packing effects like the formation of distinct layers of solvent molecules and ions;
- The formation of a hydrogen-bond network in protic solvents such as water;
- Orientational structure and its change near charged and uncharged surfaces;
- Hydration of ions;
- Diffusion and reorientational dynamics;
- Solvent effects on the chemical dynamics of ion transfer, ET and several other electrochemical reactions.

One of the goals of applying molecular simulation to the electrochemical double layer is, of course a molecular interpretation of capacitance data (see Sects. 5 and 6). Because of the above-mentioned limitations of the method, progress in this field has been rather limited. Henderson and coworkers recently studied the properties of the electrochemical double layer both at high ionic concentrations (a model of the molten salt–metal interface) and at low ionic concentrations (modeling the double layer of an electrolyte solution) by MC calculations. They showed, with the primitive model, that at low temperatures the double-layer capacitance decreases with decreasing temperatures. This behavior was observed both for the low-concentration electrolyte solution [161] and for the molten salt case [162]. In their studies, they were only able to calculate the integral capacitance at the point of zero charge. GC theory without (GC) and with a Stern layer (GCS) as well as the mean spherical approximation (MSA), on the other hand, predicts a monotonous decrease of double-layer capacitance with temperature. The primitive model MC data showed that the capacitance versus temperature curve goes, at least at lower ion densities, through a maximum and that at high temperature, analytical *theories* and MC *experiment* agree well. The primitive model simulations thus indicate that ion size-dependent effects become important, thus limiting the range of applicability of the analytical theories of sizeless ions to the high-temperature regime.

At high electrolyte concentrations, the primitive model simulations clearly exhibit the importance of ion size in the form of oscillatory density and charge profiles in the vicinity of charged surfaces [162]. The MC adsorption isotherm $\Gamma = \int \sum_i [\rho_i(z) - \rho_i] dz$, where $\rho_i(z)$ is the density profile and ρ_i is the bulk value, z the distance from the charged surface and i runs over all ion species, shows drying at low density and exhibits a maximum in the range of the critical density of the bulk electrolyte [161]. With increasing surface charge density, ionic adsorption passes from negative to positive.

MC simulations of the solvent primitive model [163] show, not surprisingly, that the double-layer capacity increases with increasing electrolyte concentration and also with increasing solvent concentration. At the same time, with increasing solvent concentration, the ionic density profiles pass from nonoscillatory to oscillatory behavior, a common feature of dense systems. Furthermore, at low surface charge densities the contact value of the charge density increases and a passage between partial drying (at low solvent concentration) and partial wetting of the surface by the electrolyte is observed with increasing solvent density. The temperature dependence is such that the ion adsorption is negative at low temperature and positive at high temperature. The negative ion adsorption at low temperature becomes less pronounced with increasing particle density. The behavior can be explained by the fact that electrostatic ion–ion interactions dominate at low temperature; whereas the excluded volume effect of the solvent is more important at higher temperature.

Extending these studies to ion-dipole mixtures in which a background dielectric constant is replaced by fluctuating solvent point dipoles, leads to "*practical nonergodicity*" due to the formation of clusters and/or strings of particles [164].

Computer simulations of "realistic" models of water, ions, and metal surfaces also suffer from nonergodicity problems; however, they are able to paint a very detailed picture of the microscopic liquid

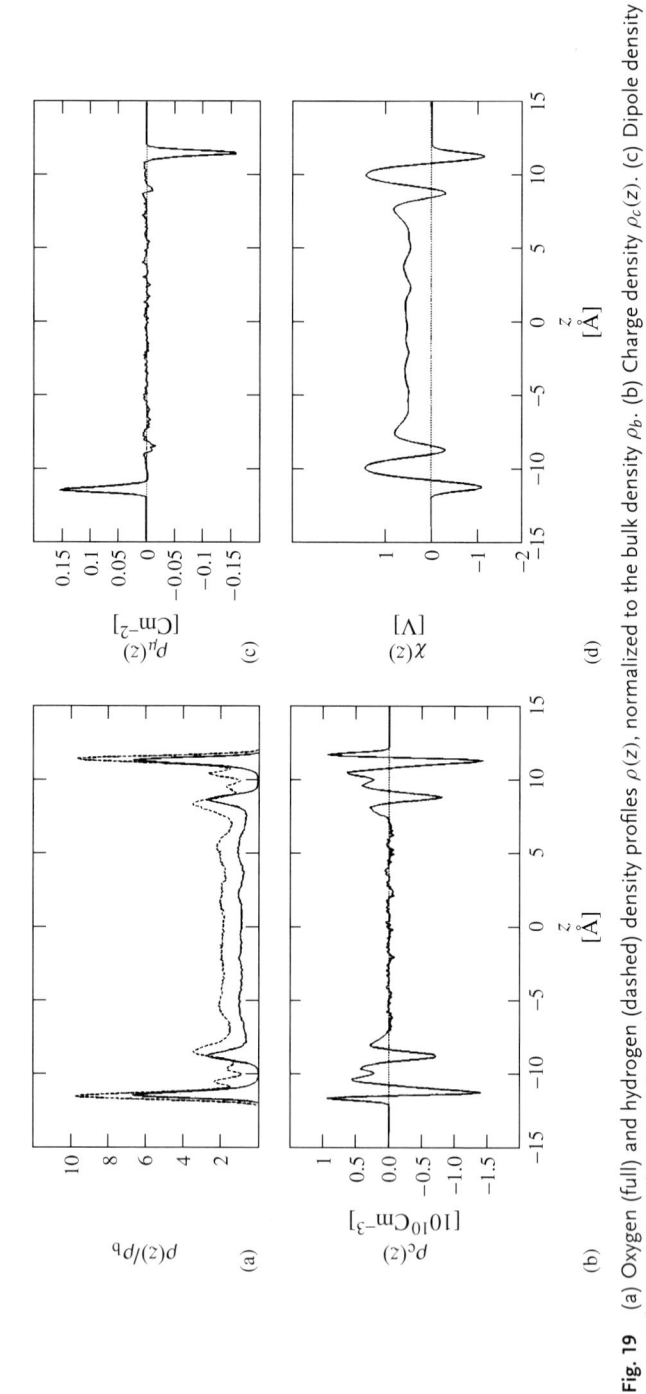

Fig. 19 (a) Oxygen (full) and hydrogen (dashed) density profiles $\rho(z)$, normalized to the bulk density ρ_b. (b) Charge density $\rho_c(z)$. (c) Dipole density $\rho_\mu(z)$. (d) Water contribution to the surface potential $\chi(z)$ calculated from the charge density $\rho_c(z)$ by means of Eq. (64). All data are taken from a 150 ps simulation of 252 water molecules between two mercury phases with (111) surface structure using Ewald summation in two dimensions for the long-range interactions.

structure near a metal electrode and near a liquid–liquid interface. Models of the interface range from simple image charge models [165] through jellium models [166] to complex potential energy functions derived from quantum-chemical calculations [167, 168] of clusters of metal atoms interacting with water molecules. These studies have been reviewed recently [169–171]. Only some key results are discussed in the following text, and the interested reader should consult the reviews and the references therein for further information.

Figure 19 shows as an example the data from a simulation of TIP4P water that is confined on both sides by a rhombohedral mercury crystal with (111) surface structure. Bosio and coworkers [172] deduced from their x-ray studies that a solid α-mercury lattice with a larger lattice constant in z direction may be used as a good structural model for liquid mercury. Thus the mercury phase was modeled as a rigid crystal, in order to simplify the simulations. The surface of such a crystal shows rather low corrugation. Figure 19a contains the oxygen and hydrogen density profiles. As a result of the significant adsorption energy of water on transition metal surfaces (typically of the order of 20 to $50\,\text{kJ}\,\text{mol}^{-1}$; see, e. g. Ref. [173]), pronounced density oscillations are observed next to the metal. Between three and four water layers have also been identified in most simulations near uncharged metal surfaces, depending on the model and on statistical accuracy. Beyond about 10–12 Å from the surface the density is typically constant and equal to the bulk value. In strong unscreened electric fields, several authors [174–177] report a phase transition towards a ferroelectric crystalline state in their simulations. However, the rather artificial character of these systems should be kept in mind.

The charge density profile and the dipole density profile can be calculated from the atomic density profiles and the orientational distributions. Figures 19b and 19c show the dipole and charge density, respectively, for water near the mercury surface. For both the first and the second water layers, a quadrupolar charge profile with alternating regions of positive, negative, and again positive charge density is observed. The form of the charge profiles originates from the fact that the centers of mass (or the oxygen atom) of the water molecules are mostly well localized, while the hydrogen density distribution is broadened as a result of librational motions and as a result of the hydrogen bonding between layers. Beyond the second layers the charge density approaches zero within the limits of statistical uncertainty. The dipole density profile indicates ordered dipoles in the adsorbate layer. The orientation is largely due to the anisotropy of the water–metal interaction potential in this particular model, which favors configurations in which the oxygen atom is closer to the surface. Most quantum-chemical calculations of water near metal surfaces to date predict a significant preference of "oxygen-down" configurations over "hydrogen-down" ones at zero electric field (e. g., Ref. [168, 178–183]). The dipole orientation in the second layer is only weakly anisotropic (see also Fig. 20). Solving the one-dimensional Poisson equation with the charge density profile $\rho_c(z)$ (Fig. 19d), the electrostatic potential drop near the interface can be calculated according to

$$\chi(z) = -4\pi \int_{-\infty}^{z} \rho_c(z') \cdot (z - z')\,\mathrm{d}z' \tag{64}$$

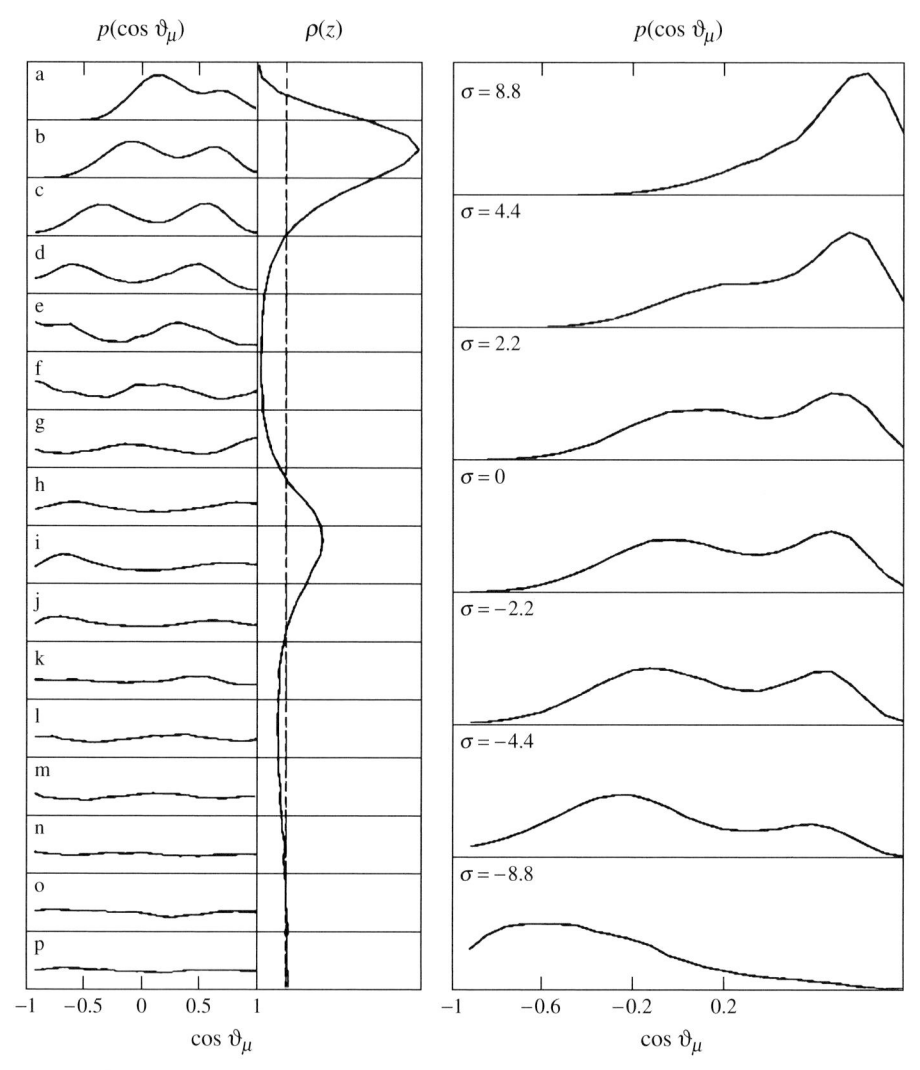

Fig. 20 Left side: Orientational distribution of the molecular dipole moment on uncharged Hg(111). $\cos(\theta_\mu)$ is the cosine of the angle between the water dipole vector and the surface normal that points into the water phase. Panels a to p on the left are sampled from the distance intervals that are indicated by the cuts through the density profile ρ scaled to its bulk value. Right side: Orientational distributions of adsorbed water molecules for various homogeneous surface charge densities, σ, which are given in units of µC cm^{-2} on the graphs. Data are from simulations of 700 TIP4P water molecules between Hg(111) surfaces.

It has been demonstrated in Ref. [184, 185] that the use of the Ewald summation is crucial to obtain the field-free bulk region of constant electrostatic potential in the center of the lamina.

Lateral density fluctuations are mostly confined to the adsorbed water layer. In the first layer the oxygen distribution shows the structure of the substrate lattice. In the second layer the distribution is more or less isotropic. As a consequence, oxygen motion is predominantly oscillatory rather than diffusive in the first layer. Consequently, the self-diffusion coefficient in the adsorbate layer is strongly reduced compared to the second or third layer [186–189]. It was concluded that the motion in the first layer is characteristic of a solid phase, while the motion in the second layer has more liquid-like character. The liquid–liquid water–mercury interface has also been studied [190]. The major difference of the water structure between the liquid–solid and the liquid–liquid interface is due to the capillary width of the liquid mercury surface, which smears out the water-density profiles considerably [191, 192].

The orientational structure of water near a metal surface has obvious consequences for the electrostatic potential across an interface, since any orientational anisotropy creates an electric field that interacts with the metal electrons. Hydrogen bonds are formed mainly within the adsorbate layer but also between the adsorbate and the second layer. The left side of Fig. 20 shows the orientational distribution of the molecular dipole moment, relative to the surface, normal in various distance ranges from the Hg(111) surface. Additionally, the oxygen density profile is plotted. The baselines between distribution functions cut through the density profile. The distribution function in each panel on the left side is for the subset of molecules that are located in the distance range between these lines on the right side. Over the first peak in the density profiles (panels a to d), there are almost no molecules whose dipole moment is perpendicular to the surface, as would be expected for an isolated molecule on the basis of the water–metal interaction potential. Within the adsorbate layer, there is a transition from the preference for orientations in which the dipoles point more or less into the solution (a and b) to one where a substantial fraction of the dipoles point more or less towards the surface (c and d). The orientational anisotropy ranges as far into the liquid phase as the density inhomogeneities do (roughly up to panel m), with increasingly less pronounced features. Slightly beyond the second maximum in the density profile the orientational distribution is isotropic, as it has to be the case for a bulk-like liquid. Thus, like in the case of nonpolar surfaces, the orientational distribution is governed by water–water interactions. All orientational distributions are rather wide and liquid-like, although the fact that the orientational preference changes within the adsorbate layer is sometimes attributed to ice-like structural elements near the interface (see Ref. [193] for a more detailed discussion).

An external electric field changes the orientational distribution and consequently the orientational polarization of the water molecules in the interfacial region. The effect of homogeneous [174, 175, 194–197] and inhomogeneous [198] electric fields on the orientational distribution near smooth model surfaces has been investigated using lattice models [197] and distributed point charge models for water. Also the field-induced changes on the more realistic Pt(100) surface were studied [176, 199, 200]. Because of the absence of free ions in these simulations of pure water the electric field is only screened by the water dipoles themselves; consequently a net electric field and a concomitant polarization persist through the lamina. At large field strengths, this has been found to lead to a field-induced phase transition to a crystalline water phase [174–176]. The right part of Fig. 20 shows the orientational distribution of the dipole moment vectors in the adsorbed water layer for various surface charge densities, σ. In the field-free case a wide bimodal-orientational distribution is observed. The overall preference of "oxygen down" bonding is manifest in the larger probability to find positive values of $\cos(\theta)$ (corresponding to angles smaller than $90°$ between dipole vector and the surface normal that points into the liquid phase) than to find negative ones. For positive surface charges, the average dipole moment (the first moment of the distribution) shifts with increasing surface charge density towards larger absolute values and the distribution becomes increasingly narrower. For low negative surface charge densities, the dipole-orientational distribution becomes more symmetric around the parallel orientation. At larger negative surface charge densities, the orientational distribution changes in such a way that the hydrogen atoms point preferentially to the surface. The effect of field-induced crystallization was observed at much higher surface charges [176]. The surface X-ray experiments by Toney and coworkers [201] give experimental evidence for voltage dependent ordering of water on a silver electrode. They observed a shift of the silver-oxygen distance with applied potential. However, they also made a debated conclusion a strong increase in local water density near the surface, while MD data suggest only an increase in correlation but not in overall packing density.

Several simulation studies of the potential of mean force of a single ion near a metal surface have been performed, with image charge models and quantum-chemical models of the metal [202–211]. With the image charge models, only the larger halide ions are contact adsorbed, whereas small ions such as Li^+ and F^- are not contact adsorbed, Cl^- being a borderline case. Calculations on the basis of quantum-chemical potentials [167, 191] show contact adsorption for almost all ions, although a detailed analysis of the local structures around these ions revealed characteristic differences between large and small ions.

The first simulation studies of full double layers with molecular models of ions and solvent were performed by Philpott and coworkers [212, 213] for the NaCl solution. The authors studied the screening of a negative surface charge by free ions in several highly concentrated NaCl solutions. A combination of (9-3) LJ potential and image charges was used to describe the metal surface. More recently, Spohr [214, 215] investigated the 2.2 molal NaCl and CsF solutions in the vicinity of a corrugated surface as a function of surface charge density. In a similar manner the interface between a 1 M KCl solution and a mercury electrode was studied by Dimitrov and Raev [216].

In the studies of Ref. [214, 215], which will be discussed in some more detail, water films consisting of 400 water molecules solvate 32 ions in the vicinity of the metal surface, with image charges and a full 2D Ewald treatment for the Coulomb interactions. Like in the studies by Philpott and coworkers, the solution is not necessarily electroneutral; the total charge in the solution is balanced by the image charges, which give rise to a surface charge density σ equal to the excess image charge divided by the area of the interface.

Figure 21 shows the ion density profiles near the metal surface for three surface charges. Beyond $z = 15$ Å all ion density profiles are identical within the limits of statistical errors. The oxygen density profile does not change much with surface

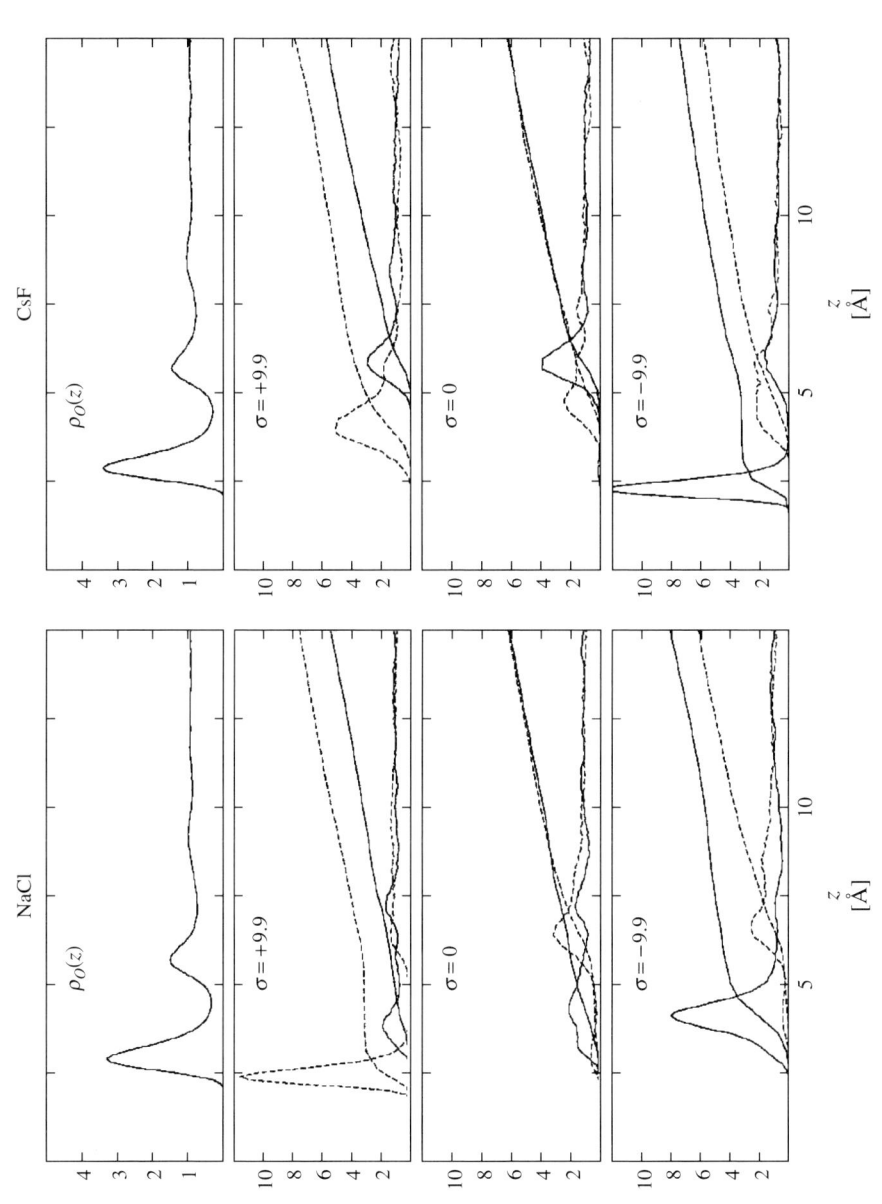

Fig. 21 Density profiles $\rho(z)$ and running integrals $N_{ion}(z)$ of the ion densities for cations (full lines) and anions (dashed lines) at three different surface charge densities in units of $\mu C\,cm^{-2}$ as indicated. Left: NaCl solutions; right: CsF solutions. The top graphs show, for reference, the corresponding oxygen density profile near the uncharged surface. Density profiles are normalized to the bulk densities corresponding to 2.2 molal solutions in each case.

charge and is repeated here only to provide a geometric reference for the ion positions. Together with the ion densities the running integrals of the densities, defined as

$$N_{\text{ion}}(z) = B_x B_y \int_0^z \rho(z')\,dz' \quad (65)$$

are plotted for cations (full lines) and anions (dashed lines). $B_x = B_y = 18\,\text{Å}$ are the box dimensions parallel to the interface and ρ is the particle number density.

Figure 21 (left) shows the ion density distributions of NaCl in the vicinity of the metal electrode. Near the uncharged electrode, there are no pronounced adsorption maxima. The density of Na^+ (full line) is slightly increased in the range around $z = 4.3\,\text{Å}$ between the first and second density maximum of water. The Cl^- density (dashed line) is significantly reduced up to about $z = 5\,\text{Å}$. There is no contact adsorption of Na^+, since no cations are found for $z < 2.8\,\text{Å}$, while water molecules can be found up to $z = 2\,\text{Å}$. Cl^- ions, on the other hand, can be found (with low probability) at distances below $2.3\,\text{Å}$.

At positive surface charge density, the Cl^- density exhibits a large maximum at very short distance from the electrode. The position of this maximum is closer to the electrode than that of the first water layer, thus giving a clear indication of contact adsorption of this anion. The Na^+ density near the electrode is slightly reduced as a result of the repulsion between the positively charged cations and the positively charged surface. The position of the first maximum of the Na^+ density profile is not shifted very much.

At negative surface charge density, the Na^+ density exhibits a large maximum at around $z = 4\,\text{Å}$. This position is very similar to the one at vanishing and positive surface charge densities. Obviously, with the models used in this study, Na+ does not contact-adsorb. The Cl^- density profile at $\sigma = -10\,\mu\text{C}\,\text{cm}^{-2}$ is similar to the one at $\sigma = 0$.

The monotonically increasing curves of the running integrals give the number of ions, whose distance from the electrode is smaller than z. The position at which the difference between the anion and cation running integral becomes 0 (for the uncharged surface) or ± 2 (for the charged surfaces) corresponds roughly to the thickness of the diffuse part of the double layer. This thickness is less than $10\,\text{Å}$ in all cases.

Figure 21 (right) shows the corresponding density profiles for the CsF solutions. At zero surface charge, there is very little contact adsorption of Cs^+ and no contact adsorption of F^-. The preferred position of the Cs^+ ions is in the second water layer, while F^- ions prefer the region between two water layers, similar to Na^+. At positive surface charge densities, the amount of F^- in the interlayer region increases, but no contact adsorption is observed. No Cs^+ cations are found in the first layer. At the negative surface charge density, Cs^+ forms a contact-adsorbed layer, similar to Cl^- at the positive surface charge. The thickness of the diffuse layer is also in the range of $10\,\text{Å}$, judging from the behavior of the running integral of the ion density.

No contact adsorption occurs in the simulations on uncharged electrodes. The small ions Na^+ and F^- do not adsorb directly on the electrode surface at moderate negative and positive surface charge densities, respectively. These ions form rather rigid hydration shells consisting of six or seven water molecules and favor the interlayer region between the first and second water layer, in which they can form stable hydration shells. However, the large ions Cs^+ and Cl^- exhibit contact adsorption when their interaction with the surface charge density becomes attractive.

From the ion density profiles, it is obvious that the surface charge is screened within less than 10 Å. Thus the thickness of the diffuse layer is of the same order of magnitude as the one derived from the Debye length ($r_D = 2.1$ Å at 2.2 mol/liter), in spite of the invalidity of the GC theory at these high concentrations. The results clearly show the correlation between the strength of the hydration forces and the preferred positions of the various ions relative to charged and uncharged electrode surfaces.

Solvent effects on chemical reactions are presently being studied intensively by molecular computer simulation methods. Outer sphere ET (e. g. Ref. [217–228]), ion transfer [229–232], proton transfer [233], and bondbreaking [234, 235] are among the reactions studied near electrodes and liquid–liquid interfaces. Much of this work has been reviewed recently [170, 236, 237] and will therefore not be discussed here. All studies involve the calculation of a free energy profile as a function of a spatial or a collective solvent coordinate.

2.1.11
Ionic Adsorption

2.1.11.1 Qualitative Aspects

In the general meaning of this term, *adsorption* of a solution component means any perturbation of its concentration within the interphase, $c_i(z)$, compared with its value in the bulk phase, c_i°. Its intensity is characterized by the integral value of this perturbation, which is called the *adsorption* of this component:

$$\Gamma_i = \int_{z_G}^{\infty} [c_i(z) - c_i^\circ] \, dz \quad (66)$$

Here, coordinate z is normal to the interface, and the integration is performed over the whole perturbed region. An important point in this formula is the lower integration limit, z_G, corresponding to the choice of the "phase boundary", whose attribution is not self-evident. This ambiguity of the definition (66) is removed by the thermodynamics of charged interfaces based on the Gibbs electrocapillary equation (67):

$$d\gamma = -\sum_j \Gamma_j \, d\mu_j \quad (67)$$

where γ is the interfacial energy, μ_j the electrochemical potential of the component j (the value of μ_j is identical in the bulk medium and within the interphase, in view of the thermodynamic equilibrium); summation includes all components of the phases in contact. The variations of different electrochemical potentials in Eq. (67) are not independent; in the bulk phase they are coupled by the Gibbs–Duhem identity. If the electrochemical potentials of all components j, except for component i and solvent s, are kept constant, Eq. (67) leads to the relation:

$$\Gamma_i^{(s)} \equiv \Gamma_i - \Gamma_s \frac{N_i}{N_s} = -\left(\frac{\partial \gamma}{\partial \mu_i}\right),$$

$$\text{keeping } \mu_j = \text{constant}, \quad j \neq i, s$$

$$(68)$$

The multiplier N_i/N_s in this equation represents the ratio of moles of components i and s in the bulk phase, or the ratio of their molar fractions. The resulting quantity, $\Gamma_i^{(s)}$, which is called the (Gibbs) adsorption of component i with respect to component s (or the relative interfacial excess of component i), is already independent of the choice of the "phase boundary", z_G. The latter is often chosen in the way to make the solvent adsorption zero, $\Gamma_s = 0$; then, both quantities become identical, $\Gamma_i^{(s)} \equiv \Gamma_i$. In dilute solutions the so-defined position of the separation plane is close to the physical boundary between the electrode and the adjacent solvent layer, being practically independent of the electrolyte concentration.

Let us consider first a *surface-inactive* solution, whose properties have already been outlined in Sects. 2.1.3 to 2.1.6. By definition, this term means that the compact layer is solely composed of solvent molecules. The distribution of components within the diffuse layer is determined mostly by the electrostatic forces. If the image forces (see Sect. 2.1.11.3) can be disregarded the concentrations of solute species, ionic or neutral, at the p.z.c. are identical to their bulk values, and the integral (66) over the diffuse layer vanishes. Inside the compact layer the concentration, $c_i(z)$, is zero, and the integration in (66) yields Eq. (12) for the Gibbs adsorption of surface-inactive components, $\Gamma_\pm = -z_H c$, which allows one to measure the thickness of the ion-free layer of the solvent, z_H.

For nonzero charges of the electrode, $\sigma \neq 0$, the distribution of ions across the diffuse layer is governed by the self-consistent electric potential, for example, given by the formulas of the GC theory given in Sect. 2.1.4.2. In particular, for a binary 1 : −1 electrolyte, this model yields for the adsorption of surface-inactive ions:

$$\Gamma_\pm = -z_H c^{-(1/2e)}$$
$$\times [\sigma_* \pm \sigma - (\sigma^2 + \sigma_*^2)^{1/2}] \quad (69)$$

F is the Faraday constant, c the electrolyte concentration, and σ_* is the characteristic charge defined by Eq. (18).

The estimates show that the charge-independent term in Eq. (69) is only noticeable for not too dilute solutions, about 1 M or higher. For lower electrolyte concentrations, it can be mostly disregarded. The other terms in Eq. (69) represent monotonous functions of the electrode charge, with the anion/cation contribution being dominant for sufficiently positive/negative values.

The behavior, that the electrolyte is surface–inactive with respect to an interface over a wide range of potentials including the p.z.c., represents a rare exception, while the majority of solute ion and neutral species demonstrate *specific adsorption*, that is, they significantly penetrate the compact layer. However, the existence of at least one surface-inactive electrolyte MA for a particular metal–solvent interface provides very important advantages for the further study of the *surface activity* of other species. First, the measurements in pure MA solutions allow one to determine the background characteristics of the metal–solvent interface, including the value of the p.z.c. and the compact-layer capacitance curve (see Sects. 2.1.2–2.1.4). Then, further studies (e.g. of the adsorption of anion A′) are performed in mixed electrolyte solutions, MA + MA′, possessing a constant ionic strength but a variable fraction of anions A′ [238, 239]. If the properties of these two anions inside the diffuse layer are sufficiently close, the variation of the interfacial energy, γ, with the fraction of anions A′ enables one to determine the *specific adsorption* of these anions, that is, its amount inside the *compact* layer, $\Gamma_{A'}^{(1)}$, rather than the overall interfacial excess of these species, $\Gamma_{A'}$, that appears in Eqs. (67) and (68).

Alternatively a study can be realized for binary electrolyte solutions, MA′, at varying concentrations [24] that provide, in favorable cases, the same information. However, the specific adsorption of A′ must vanish at sufficiently high (but still experimentally available) negative electrode potentials, so that the capacitance curves for the MA and MA′ solutions at identical concentrations coincide. This condition is necessary for the back-integration operation, thereby allowing the calculation of the charge-potential dependence for MA′ solutions. Besides, one has to subtract the *diffuse-layer* adsorption of A′, $\Gamma_{A'}^{(2)}$, from its overall adsorption, which is based on the formula (69) in which the electrode charge, σ, is replaced by the sum, $\sigma + \sigma_1$, $\sigma_1 = z_{A'} e \Gamma_{A'}^{(1)}$ being the charge of A′ inside the compact layer, and $z_{A'}$ the charge valence of anions A′. As one can conclude from the analysis in Sect. 2.1.5.1, the charge dependence predicted by this formula cannot be verified from existing experimental data for *high* electrode charges, so that the separation of the overall adsorption value into the diffuse and compact-layer contributions becomes ambiguous.

These measurements enable one to determine the *(specific) adsorption isotherm*, that is, the dependence of the interfacial excess of the species on their concentration in the bulk phase, $\Gamma_A^{(1)}(c_A)$. For a constant value of the bulk concentration, its adsorption also depends on the *electrical variable*, for example, on the electrode potential E. In mixed electrolytes, this process is also affected by the overall ionic strength, in particular via the diffuse-layer structure. In more complicated cases a combined specific adsorption is observed, that is, the compact layer contains (besides the solvent) species of different kinds, for example, both cations and anions.

Generally the specific adsorption at the surface of s,p-metals from protic solvents is more pronounced for anions than for cations. In particular, the alkali cations are almost surface-inactive, except for the interval of sufficiently high negative electrode charges, where the adsorption of Cs^+ is most noticeable (see also Sect. 2.1.1.10). However, even for this case the adsorption is *weak*, that is, the charge of their specific adsorption, σ_1, is less than the absolute value of the electrode charge, σ, that is, the charge of the diffuse layer has the opposite sign compared to the electrode charge, and the distribution of the average potential across the interphase is given by a monotonous function.

Anion adsorption possesses quite different features. For most anions, the specific adsorption is only absent for sufficiently high negative electrode charges. The diminution of this charge results in a rapid increase of the amount of anions inside the compact layer, even before the p.z.c. of the system is reached. For the zero charge of the electrode, the EDL is formed by the anions inside the compact layer and the countercharge in the diffuse layer. It means a shift of the corresponding electrode potential, that is, the p.z.c., in the negative direction, the value being correlated with the intensity of the anion specific adsorption. For their adsorption at the positively charged metal surfaces, one observes the *recharging effect*; the absolute value of the anion charge, σ_1, exceeds the electrode charge so that the diffuse layer should be charged *positively*. It leads to a nonmonotonous profile of the electric potential across the interphase, with important consequences for the phenomena determined by the diffuse-layer effects, Sect. 2.1.5.

Data on ionic adsorption are available for numerous systems, see for example, [34, 240–245]. One can see that the surface coverage by adsorbed ions (adsorption values with respect to their maximum level corresponding to the completely covered surface) is mostly much lower than 1 for mercury and "mercury-like" metals, even at high electrode charges. This limitation arises from a strong repulsive interaction of the electrostatic origin between adsorbed ions. On the other hand, the adsorption may approach a complete coverage in some other systems, for example, for halides adsorption at the silver surface [246], which implies that the ions have practically lost their charge in the transition to their adsorbed state, so that this process resembles the adsorption of neutral organic molecules, in which the complete coverage of the surface is often reached. Similar pronounced charge transfer from adsorbed ions to the metal takes place for many transition metals.

Despite of this charge transfer many of these systems may be treated as "ideally polarizable electrodes," if the adsorbed species are not transformed into a different component present inside the *bulk phase*. The latter condition is violated, for example, in the hydrogen adsorption at metals of the platinum group in which the adsorbed hydrogen atoms can be in equilibrium with protons in solution *and* hydrogen molecules in gas phase or hydrogen dissolved inside the metal. The latter system corresponds to *perfectly polarizable electrodes*, see Ref. [13] for further discussion.

Another possible complication may arise from a slow approach to the thermodynamic equilibrium leading to hysteresis phenomena, which are also typical for many transition metals.

2.1.11.2 Ion Adsorption Models

For metals of the mercury group, one can distinguish different regions in the ion adsorption isotherms. For sufficiently low bulk concentrations (which may be too low to be studied experimentally), the adsorption obeys the Henri isotherm, that is, the amount of ions inside the compact layer is proportional to the bulk concentration. At higher concentrations the adsorption increases much more slowly, approximately like the logarithm of the concentration. The behavior in both regions is influenced by the electrode potential and the ionic strength of the solution.

In the further discussion of the modeling of the adsorption isotherms, $\Gamma(c)$, it is assumed that the contribution of the diffuse-layer adsorption, $\Gamma^{(2)}$, is already subtracted or simply disregarded, that is, the symbol Γ will be used for the amount of *specifically adsorbed* species (called $\Gamma^{(1)}$ in Sect. 2.1.11.1).

Generally the expression for this dependence should be derived from the equality of the electrochemical potential of the ion in the adsorbed state and in the bulk solution:

$$\mu_{\text{ads}}(\Gamma) = \mu_{\text{sol}}(c) \quad (70a)$$

In most cases the solution is sufficiently dilute to use the ideal-solution approximation for the bulk value,

$$\mu_{\text{sol}}(c) \cong k_B T \ln(mc) + \text{constant} \quad (70b)$$

m being molar fraction of the surface-active ion in the mixed electrolyte, and c the overall concentration ($m = 1$ for the binary electrolyte). On the other hand the approximation for the $\mu_{\text{ads}}(\Gamma)$ dependence must take into account the interactions between the adsorbed ions, that is, the change of the energy to transfer an ion from the bulk solution to the adsorbed state as a function of the adsorption level.

The first theory of ion adsorption was proposed by Stern [247]. His results can be obtained from Eq. (70a) if the electrochemical potential of the ion in the adsorbed state is represented as the sum of the entropy term (with account of the saturation term) and the work to transfer the ion from the bulk solution into the adsorbed state in the average field of the EDL:

$$\mu_{ads} = k_B T \ln\left[\frac{\Gamma}{(\Gamma_{max} - \Gamma)}\right] + z_i e \varphi_{GC} + \text{constant} \quad (71)$$

$z_i e$ is the charge of the species. Stern assumed the location of the adsorption ion at the outer Helmholtz plane separating the compact and diffuse layer, so that the potential in the work term in Eq. (71) corresponds to the potential difference across the diffuse layer. The combination of Eqs. (70a) and (71) gives Stern's expression for the adsorption isotherm from a binary electrolyte solution for the case of a single ionic species being accumulated in the adsorbed state:

$$\frac{\Gamma}{(\Gamma_{max} - \Gamma)} = c \exp\left[\frac{-(\bar{W}_{ads} + z_i e \varphi_{GC})}{k_B T}\right] \quad (72)$$

where "the potential of specific adsorption" $\bar{W}_{ads}$ originated from the constant terms in Eq. (70b) and (71) which reflect the nonelectrostatic contributions to the energy of transfer into the adsorbed state, for example, the partial desolvation of the ion.

This model allows one to explain such qualitative features of the phenomenon as the drastic (exponential) dependence of Γ on the electrode charge, as well as the existence of the aforementioned two regions in the $\Gamma(c)$ variation. However, a more detailed experimental study has revealed serious deviations from the predictions of Eq. (72). For many ions, one can observe the effect of Esin and Markov [248], a stronger shift of the p.z.c. with the variation of the bulk concentration than it is predicted by the Stern theory. For zero electrode charge, the average potential in the adsorption plane (φ_{GC} in Eq. (72)) is equal to the electrode potential (at the p.z.c.), $E_{\sigma=0}$. Then the derivation of the Eq. (72) leads to an inequality:

$$\left(\frac{z_i e}{k_B T}\right) \frac{dE_{\sigma=0}}{d \ln c} < 1 \quad (73)$$

However, the experimental data show that the left-hand-site quantity exceeds significantly its upper limit, 1.

This "paradox" (which remains unresolved even if the adsorption plane is shifted into the compact layer) represented a strong stimulus for the development of the ion adsorption theory. It was shown [249] that it originated from the calculation of the work term in Eq. (71) as an *average* potential in the adsorption plane. The qualitative explanation of experimental observations can be achieved, if one takes into account that the transfer of the ion into the adsorption site is accompanied by a redistribution of the other adsorbed ions that creates a "hole" for it with the diminished repulsive potential. This inapplicability of the approximation of the uniform distribution of the adsorbed ion charge along the plane is called *discreteness-of-charge effects*, which must be taken into account in the theory of this phenomenon.

Another crucial factor is an adequate inclusion of these *diffuse-layer* and discreteness-of-charge effects into the model, which play a marked role if the overall electrolyte concentration is well below 1 M. It is why the adsorption parameters found by the interpretation of experimental data with the use of various isotherms derived mostly for uncharged species (Flory–Huggins, Frumkin, Temkin, etc) may strongly deviate from their proper values.

The aforementioned diffuse-layer and discreteness-of-charge effects have been taken into consideration in the model proposed by Grahame and Parsons [26, 250–252]. First, it was assumed (unlike in the Stern model) that the specifically adsorbed ions were located at the distance from the metal surface (in "the inner Helmholtz plane") ensuring their maximum bond strength, owing to the combination of forces of electrostatic and quantum-mechanical origins. It shows the need for the partial or even complete desolvation of the adsorbed species and its deep penetration into the compact layer. The position of this adsorption plane depends on all components of the system, metal, solvent, and adsorbed ion.

The second modification compared to the Stern theory was the introduction of "the discreteness-of-charge factor", λ. The electrostatic part of the work of ion transfer from the bulk solution into the adsorption plane across the EDL field in Stern's approximation may be represented as the sum of the contributions of the compact and diffuse layers:

$$W_{EDL} = z_i e(\varphi_1 - \varphi_{GC}) + z_i e \varphi_{GC}$$
$$= z_i e(\sigma + \sigma_1) K_O^{-1} + z_i e \varphi_{GC}$$
(74)

In the second expression in this formula, the potential difference across the external part of the compact layer is replaced by the sum of the electrode and specifically adsorbed charges, with the coefficient inversely proportional to the integral capacitance of "the external part of the compact layer", K_O, that is, between the inner (adsorption plane) and outer (diffuse-layer boundary) Helmholtz planes.

To take into account the adjustment of the distribution of the previously adsorbed ions inside the adsorption plane in order to minimize the total energy of the system after the insertion of an additional ion, the term in Eq. (74) dependent on the charge of the specific adsorption, σ_1, was diminished by "the discreteness-of-charge factor", λ, the value of which should lie between 0 and 1. In the original formulation by Grahame, this factor was expressed through the ratio of K_O and the capacitance of the whole compact layer, K_H in Eq. (19). Later, its value was treated as an unknown parameter to be found from the fitting of theoretical predictions to experimental data.

The use of the thus modified Eq. (74) and the equality of the electrochemical potentials (70a) lead to the adsorption isotherm:

$$\ln(mc) = \ln \Gamma + 2B\Gamma + z_i e \frac{\varphi_{GC}}{k_B T} + \ln \tilde{\beta}$$

$$\ln \tilde{\beta} = \ln \tilde{\beta}_o + z_i e \frac{\sigma}{k_B T} K_O \quad (75)$$

$$B = z_i^2 e^2 \frac{\lambda}{2 k_B T} K_O$$

Here the coefficient, B, is called *attraction constant*, in analogy with the corresponding factor in the Frumkin adsorption isotherm. However, it has got the opposite sign, that is, it reflects the *repulsion* between the adsorbed species, the amplitude of which is diminished owing to the relaxation of the adsorbed ion ensemble (factor λ). The entropy term for the adsorption state is taken without the "saturation term", compare Eqs. (71) and (72), since the charging degree for s,p-metals is mostly very low with respect to the complete coverage.

The expression (75) represents a *virial* isotherm for the $\Gamma(c)$ dependence, with a correction for the diffuse-layer effect, if the electrical variable is kept constant. Unlike the case of the Frumkin isotherm in which it was the *electrode potential*, the

Grahame–Parsons isotherm uses the electrode charge, σ, whose value influences the parameter, $\tilde{\beta}$, related to the standard adsorption energy. Equation (75) predicts that the isotherms for different electrode charges must be *congruent*.

This isotherm was used successfully to interpret experimental data for adsorption of various simple inorganic ions, including weak specific adsorption of cations. At the same time, significant deviations have been observed for some other systems, especially for low surface coverages as well as for the surface recharge by adsorbed anions. Moreover, in some cases in which the predictions of Eq. (75) were formally in accordance with experimental data, the values of the "discreteness-of-charge factor", λ, were found outside the expected interval, negative or exceeding 1.

Another shortcoming of this model is the absence of its derivation from the theory based on the analysis of the statistical-mechanical properties of the ensemble of adsorbed ions. Progress in the latter direction is discussed in the next subsection.

All these approaches disregard the effect of the gradual change of the dielectric properties of the compact layer as a consequence of the replacement of the solvent molecules in it by the adsorbed species. This approximation limits the applicability of such theories to moderate adsorption and to adsorbed species of a small size. In the opposite case of large ions (especially, those of organic type), this effect may become predominant even for sufficiently low adsorption charges. In the simplest approximation, one may accept a linear dependence of the compact-layer capacitance on the charge of the specific adsorption, such as in the model proposed by Kolotyrkin, Alexeyev, and Popov [253]. Further discussion on this development is given in Chapter 3.1 of this volume.

2.1.11.3 Image Potential for Ions in the Diffuse or Compact Layer

Within the framework of the GCSG treatment (Sect. 2.1.4) the ion concentration across the diffuse layer is governed by the Boltzmann distribution in the *average* electric field created by the electrode charge (plus the charge of the specifically adsorbed ions in the general case), and the ion countercharges in the diffuse layer:

$$c_i(z) = c_i^\circ \exp\left(\frac{-z_i e\varphi(z)}{k_B T}\right) \quad (76)$$

This formula implies that the presence of a particular ion at point z does not disturb this potential distribution. In reality, it is only valid as a zero order approximation. A more accurate formula contains an additional factor:

$$c_i(z) = c_i^\circ \exp\left(\frac{-z_i e\varphi(z)}{k_B T} - \frac{W_{im}(z)}{k_B T}\right) \quad (77)$$

where the "image potential" $W_{im}(z)$ represents the difference between the energies of the whole system if the additional ion of type i is added either at point z or in the bulk solution, as a source of the *external field* for the rest of the system. Since this quantity is related to a *single* ion, expression (77) contains the Boltzmann constant, k_B.

Within the diffuse layer the dominant contribution is given by forces of electrostatic origin, analogous to the Debye–Hückel ion atmosphere around each charge species in the bulk solution. It was assumed frequently on the intuitive ground that the image potential in Eq. (77) related to the location of the ion in the vicinity of the metal may be modeled by its expression for a charged species in the dielectric medium near its boundary with a conductor:

$$W_{im}^{\circ}(z) = -\frac{(z_i e)^2}{4\varepsilon z} \quad (78)$$

The negative sign in this formula means an attraction of the ion to the surface, in particular an accumulation of the ions inside the diffuse layer near the uncharged metal surface (at the p.z.c.). However, a more substantiated analysis has disproved this conclusion completely, see reviews [89, 254, 255] and references therein.

Let us consider first an idealized case of the uncharged metal in contact with the solvent without any electrolyte. The expression for the image potential within a wide range of the ion-metal distances (except for the close vicinity to the compact-layer surface) can be derived without specifying the particular form of the dielectric properties of the media in contact, solvent and metal, including their defect structure within the interphasial region (compact layer). The only assumption on these properties of the media and the interface is the existence of a *thin* layer within the interphase having a lowered value of the effective dielectric constant. The analysis of Sect. 2.1.4–2.1.6 shows that the capacitance data for surface-inactive electrolytes prove the existence of the compact layer, thereby justifying this assumption.

Correspondingly, the profile of the image potential:

$$W_{im}(z) \cong \frac{(z_i e)^2}{4\varepsilon z}\left[1 - E\left(\frac{2z}{\varepsilon L_H}\right)\right],$$

$$E(y) \equiv 2y \exp(y) \int_y^\infty z^{-1} e^{-z}\, dz \quad (79)$$

is only dependent on the *integral* characteristic of this "defective region", the compact-layer capacitance, C_H, defined by Eq. (20), via a characteristic length, L_H (which already appeared earlier, for example, in Eq. (38a)):

$$L_H \equiv (4\pi C_H)^{-1} \quad (80)$$

The value of this length can be found for any metal–solvent interface as soon as the capacitance data for a surface-inactive electrolyte are available; for example, it is about 0.3 Å for the Hg–water interface. As in other expressions containing L_H, this length is multiplied by the *bulk-solvent* dielectric constant, ε, which gives about 2.4 nm for this interface.

Equation (79) shows that Eq. (78) is only valid at distances from the surface much greater than this distance, $z \gg \varepsilon L_H$, where the amplitude of the image potential is negligible compared with the thermal energy, $k_B T$, that is, the concentration profile in Eq. (77) is not influenced by this factor. In the other interval of the distances, $z \ll \varepsilon L_H$ (but not too close to the compact layer) Eq. (79) may be approximated as:

$$W_{im}(z) = \frac{(z_i e)^2}{4\varepsilon z} \quad (81)$$

that is, the ion is *repelled* by the interface. This effect is explained by a stronger interaction of the ion with the charges induced inside the compact-layer region, which have got the *same* sign as the ion, see [254, 255] for a more detailed discussion. These two branches of $W_{im}(z)$ are separated by a minimum of purely electrostatic origin, whose depth depends on these characteristic parameters, ε and L_H, and the species charge, $z_i e$.

A similar analysis has been carried out for an ion near the contact between a charged metal and an electrolyte solution, in which the image potential profile is also a function of the electrode charge and the electrolyte concentration, $W_{im}(z; \sigma, c)$ [254–256]. Depending on these parameters, the profile, $W_{\text{im}}(z)$, may be a monotonous function (increasing or decreasing), or a curve with a minimum, see, for example, the numerical illustrations in Ref. [257].

For example, at the p.z.c. ($\sigma = 0$) the shape of the profile depends on the ratio of the Debye screening length in the bulk solution (32), L_D, and the product εL_H this ratio being identical to the ratio of the compact- and diffuse-layer capacitances, C_H and C_{GC} in Eq. (20).

For a sufficiently low electrolyte concentration (so that $C_H \gg C_{GC}$), the profile within the interval, $z \ll L_D$, is close to its form at vanishing concentration, Eq. (79), in particular there are the repulsion and attraction branches, Eqs. (81) and (78). For even larger distances from the metal surface (comparable with the Debye length), the interaction with the interface is screened:

$$W_{\text{im}}(z) = -\frac{(z_i e)^2}{4\varepsilon z} e^{-2z/L_D} \qquad (82)$$

More important changes take place in more concentrated solutions where $C_H \ll C_{GC}$ (which means $c \gg 0.1\,\text{M}$ for the Hg–water interface). The profile represents a monotonically decreasing curve (pure repulsion of the ion from the interface at *all* distances), in accordance with the formula:

$$W_{\text{im}}(z) = \frac{(z_i e)^2}{4\varepsilon z} e^{-2z/L_D} \qquad (83)$$

Thus the repulsive behavior at shorter distances is a general feature of the image potential profile at the p.z.c., whereas the large-distance variation depends on the ionic strength of the solution:

$$W_{\text{im}}(z) = \frac{(z_i e)^2}{4\varepsilon z} \frac{\varepsilon L_H - L_D}{\varepsilon L_H + L_D} e^{-2z/L_D},$$
$$z \gg \min(L_D, \varepsilon L_H) \qquad (84)$$

All these results correspond to the case that the ion is located within the diffuse layer far away from the outer Helmholtz plane. The conclusions change drastically for charged species inside the compact layer, in particular, for specifically adsorbed ions. This interphasial region represents a nonuniform dielectric medium whose effective dielectric constant, $\varepsilon(z)$, has got the values strongly reduced in comparison with those in the bulk solution. For such systems, the interaction of the species with the metal or the diffuse-layer ions represents a minor correction, while the dominant contribution to the image potential is given by the *short-range* interaction with the dielectric medium [258, 259]. For ions of a sufficiently small radius, r_o, one can use as a qualitative estimate a simple formula for the electrostatic contribution to the image potential:

$$W_{\text{el-stat}}(z) \cong \frac{(z_i e)^2}{2r_o} \left[\frac{1}{\varepsilon(z)} - \frac{1}{\varepsilon}\right] \qquad (85)$$

Because of lowered values of the effective dielectric constant inside the compact layer (independent of the physical reason leading to this reduction), the term given by Eq. (85) represents a strong repulsion of ions from this interphasial region. In reality, one should add the effect of the ion desolvation to enter this layer.

This consideration gives an explanation of the existence of surface-inactive ions for which the *attractive* contribution due to the short-range (electron exchange) metal surface–ion interaction is absent or insufficiently strong to compensate the aforementioned repulsive effect. It may also be responsible for the strong influence of the iodide adsorption on the hydrogen evolution, despite the absence of marked specific adsorption of H_3O^+ ions [260].

One can also derive analytical expressions for the variation of the image potential for a charged species inside the compact layer as a function of the bulk electrolyte concentration [254–256]. This contribution is relatively small compared with the short-range interactions, but its knowledge is of importance, since it represents the change of the single-ion (standard) adsorption energy with variation of the ionic strength of the solution in the adsorption isotherm, see Sect. 2.1.11.6.

2.1.11.4 Interaction Between Adsorbed Ions

The deviation of the adsorption isotherm at a uniform electrode surface from the linear behavior (Henri isotherm) is related to the interaction between the adsorbed species. A substantiated derivation of its form can only be made on the basis of an analysis of the statistical-mechanical properties of the whole ensemble of adsorbed ions, which, in turn, requires the knowledge of the interaction potential between the ions, U, as a function of their distance, R, along the surface. This quantity is defined as a difference between the energies of the system, when these two ions are fixed at distance R or are very far from each other.

There are numerous types of short-range interactions between adsorbed ions such as exchange-correlation effects, an overlap of their electronic cores, mutual polarization, interaction via the electrons of the metal, interaction between their solvation shells (which may partially be retained in the adsorbed state), changing of the local structure of the compact layer, and so on. All these forces decay rapidly with ion-ion distances exceeding the molecular scale, and the electrostatic contribution becomes dominant. A crucially important feature of this long-range contribution is a *surprisingly slow decay* as a function of R, and it leads to fundamental consequences for the correlation properties of this ensemble as well as for the form of the adsorption isotherm (see Sect. 2.1.11.6).

Intuitively, it seems to be "natural" to use an analog of Eq. (78) based on the same model of the "image charge" formed by the metal electrons as a response to the field of the adsorbed ion:

$$U^\circ(R) = (ze)^2 \left(\frac{1}{\varepsilon R} - \frac{1}{\varepsilon (R^2 + 4a^2)^{1/2}} \right)$$
$$\cong (ze)^2 \frac{2a^2}{\varepsilon R^3} \qquad (86)$$

Here, a is the distance between the center of the adsorbed charged species and the metal surface. Equation (86) predicts a rapid decrease of the interaction energy as a function of the distance between adsorbed ions, R, and it means that the ensemble of species with such interaction law should correspond to the one with *short-range* interactions. The uncertain point within this approach is to choose the value of the dielectric constant of the medium in Eq. (86), ε.

It is tempting to identify it with the value typical for the *compact* layer, which would mean that the interaction propagates along this interphasial region. However, it can be easily seen from simple calculations that the interaction energy must diminish exponentially (as a function of R), if the interaction between the charges is realized along the lines passing inside the compact layer, for example, in the model in which this layer is surrounded by *two conducting media* imitating the metal and the diffuse layer. As a result, this model cannot explain the very large values of the interaction constants observed experimentally for ion adsorption. The alternative attribution of ε to the bulk-solvent value, also leads to a very weak interaction.

It will be seen from the discussion that follows that the electrostatic interaction between adsorbed ions follows quite a different *functional* form. A detailed analysis of this problem may be found in reviews [89, 254, 255, 261].

Similar to Sect. 2.1.11.2, let us consider first the interaction of two charged species located at a small distance (atomic scale) from the metal–solvent interface. Once again, it turns out that the functional form of the $U(R)$ dependence is only influenced by the same characteristic length, εL_H, which had already appeared in Eq. (79) for the image potential, that is the product of the *bulk* dielectric constant of the solvent and the inverse value of the *compact-layer capacitance*, Eq. (80).

The binary interaction potential, $U(R)$, follows indeed the R^{-3} law within its long-distance limit:

$$U(R) \cong (z_i e)^2 \Delta^2 \frac{2\varepsilon L_H^2}{R^3} \quad (87)$$

but only for *very* large distances between the adsorbed ions, $R \gg \varepsilon L_H$. One should remember that this characteristic distance, εL_H (known for each system from independent capacitance experiments), is well beyond the atomic scale, for example, it is about 2.4 nm for any ion adsorbed at the Hg–water interface. The amplitude in Eq. (87) is several dozen times larger than the value predicted by Eq. (86).

The only parameter in the whole $U(R)$ expression that is unknown from independent experiments is the multiplier Δ^2, which depends on numerous factors, like, location of the adsorbed ion within the compact layer, the distribution of the dielectric properties across this layer (analogues of the ratio of the capacitances, K_H/K_O, in the Grahame–Parsons theory, Sect. 2.1.11.2), partial transfer of the electronic charge by the ion in its adsorbed state, and so on. In the simplest model of the compact layer as a uniform dielectric slab of thickness z_H and dielectric constant ε_H, this parameter is given by the formula:

$$\Delta = \frac{z_a}{z_i} \frac{a_i}{z_H} \quad (88a)$$

where a_i is the distance between the adsorbed ion and the metal surface (considered in this formula as a point charge), z_a and z_i are the charges of the ion in the adsorbed state and in the bulk solution. A somewhat more general model of the compact layer as a nonuniform dielectric slab whose local dielectric permittivity depends on the normal coordinate, $\varepsilon(z)$, results in the formula (88b) containing the ratio of the "effective thickness" for the overall compact layer, $L_H \equiv (4\pi C_H)^{-1}$, and that of its "inner part" between the bulk metal and the adsorption plane, L_I (again, the charge of the adsorbed ion is considered here as the *point* charge localized at the distance a from the surface) [262]:

$$\Delta = \frac{z_a}{z} \frac{L_I}{L_H}, \text{ where } L_H = \int_{-\infty}^{z_H} \frac{dz}{\varepsilon(z)},$$

$$L_I = \int_{-\infty}^{a_i} \frac{dz}{\varepsilon(z)} \quad (88b)$$

Similar expressions can easily be derived for any particular model of the metal–solvent interphasial structure on the basis of the general formula for this parameter [254, 261, 263], including a *spatial distribution* of the electronic charge of the ion in the adsorbed state. Generally, one may expect that its value should be between 0 and 1.

On the other hand, this parameter, Δ, can be found for any particular ion from the experimental data on its adsorption, see Sect. 2.1.11.6. It means that the form of the binary interaction potential and the correlation properties of this ion ensemble (Sect. 2.1.11.5) can be described without any unknown parameters.

Equation (87) only represents a limiting behavior at very large distances between adsorbed ions, R. There exists a wide range of intermediate distances in which the interaction potential follows quite a different functional law:

$$U(R) \cong (z_i e)^2 \Delta^2 \frac{2}{\varepsilon R},$$
$$\text{for } z_H \ll R \ll \varepsilon L_H \quad (89)$$

The crucial difference of this formula in comparison to Eqs. (86) and (87) is a *very slow* decrease of the interaction potential within this interval of R. The existence of this intermediate asymptotic behavior of $U(R)$ leads to crucial consequences for the correlation properties of the ion ensemble, Sect. 2.1.11.5.

One can demonstrate that this large increase of the intensity of the interaction is due to a specific distribution of the lines of the electric field created by an adsorbed ion. Because of a great difference between the effective dielectric constants in the compact layer and in the bulk solvent, the lines do *not* propagate along the compact layer. A fraction of these lines determined by the factor $1 - \Delta$, connect the ion with the charges at the metal surface. The remaining lines (their fraction being Δ) quit the compact layer and propagate outside of it at a distance R. The interaction given by Eqs. (87) and (89) is determined by the latter fraction of lines.

A similar analysis was carried out for the case of an electrochemical system in which the ion is adsorbed at the charged metal–electrolyte solution interface. The derived analytical expressions for the interaction potential become a bit more complicated [89], but all the above-formulated qualitative conclusions remain unchanged, including the universal functional form of $U(R)$ with a single unknown parameter, Δ, a high intensity of the interaction (compared to its estimates within oversimplified model (86)), existence of the "Coulombic" behavior of the potential (89) and so on.

The only qualitative modification is the appearance of the exponential decay at distances exceeding the screening length within the diffuse layer, which was assumed to be not too short (much longer than the compact-layer thickness). As a result the interval of intermediate distances in which the R^{-1} behavior is valid, Eq. (89), is determined by the inequalities, $z_H \ll R \ll R_*$, where R_* is the minimum of the distance εL_H, and the screening length inside the diffuse layer for the values of its charge and the ionic strength of the solution, in particular, the Debye screening length of the bulk solution for the zero charge of the diffuse layer.

The intensity of the interaction depends also on parameter Δ, whose experimentally found values for the adsorption at s,p-metals are not much smaller than 1 (see Sect. 2.1.11.6). For such systems, one can see that the "Bjerrum length", L_B, where the interaction energy is equal to the thermal energy belongs to the aforementioned interval of the intermediate distances, $z_H \ll L_B \ll R_*$. The importance of such a sequence of characteristic distances for the correlation properties of the system will be seen in the next section.

2.1.11.5 Statistical Mechanics of Adsorbed Ion Ensemble

During an extended period the structure of the two-dimensional (2D) ensemble formed by adsorbed ions located in the inner Helmholtz plane was *postulated* on intuitive grounds.

Most often, it was assumed that the ion distribution represents a *regular 2D lattice*, hexagonal or square, owing to the repulsive electrostatic forces between the ions. However, the theoretical studies (confirmed by experimental data) related to a similar 2D one-component system of charged species, electrons at the surface of liquid helium, have shown that such a regular lattice becomes unstable with respect to certain collective motions (dislocation formation, etc) at extremely low temperatures, that is, under the conditions in which the average interaction energy per ion is still much larger than the thermal energy (contrary to the standard stability criterion for 3D lattices). The application of this stability condition for 2D lattices to the ensemble of adsorbed ions shows that the hypothetical regular lattice *never* formed in these systems by electrostatic repulsion within the experimental temperature range, since the 2D density is limited by the repulsion between the cores of these ions.

Another hypothesis assumed that the binary interaction between ions is *weak*, so that the system could be considered as close to a 2D ideal gas. However, the analysis of the function, $U(R)$, has revealed that this energy exceeds the thermal one up to the distance about $R_B \sim 1$ nm, that is, it is *strong*.

The third popular approach was based on a cut-off-disk model. According to it the electrostatic repulsion between the ions leads to the formation of an area around each ion practically free from other adsorbed ions. The size of this area was identified with the average distance between the ions, so that it was dependent on the 2D density, that is, on the ion adsorption, Γ (as in the previous Sects. 2.1.11.2–2.1.11.4, this symbol is used for the amount of *specifically* adsorbed ions, unlike in Sect. 2.1.11.1). The results on the form of the $U(R)$ dependence in Sect. 2.1.11.4 confirm the existence of such a "cut-off-disk" but its size does *not* depend on Γ and is markedly different from the postulated values.

These assumptions on the distribution of ions in the adsorption plane were combined with some expression for the binary interaction potential, $U(R)$ (in most cases, given by Eq. (86)), to get the contribution from the interactions into the chemical potential of an adsorbed ion. However, the final results of such treatment can hardly be considered as reliable.

A substantiated analysis must be based on the dependence of the binary interaction energy on the distance between ions, $U(R)$, discussed in Sect. 2.1.11.4. Certain specific features of this system compared with usual 3D ones originate immediately from its *two-dimensional* character [264, 265], e.g., the absence of a long-range crystalline order at *any* finite temperature, slow decay of the autocorrelation function of the species velocity in the long time limit, and so forth.

Additional effects arise from the modification of the form of the binary interaction energy, $U(R)$, for the ions at the interface, compared to the one in the bulk solution. One of the most important effects is the great intensity of this interaction, for example, Eqs. (87) and (89), which may in certain cases be *stronger* than the same characteristic in the bulk phase (dielectric or solution). Another crucial feature is a *slow* decrease of this interaction, $U \sim R^{-1}$, in the intermediate range of the distances, $z_H \ll R \ll R_*$, with a much steeper diminution outside at greater distances, Eqs. (89) and (87), correspondingly. This combination makes it impossible to apply standard approaches of statistical mechanics derived separately for systems with a *strong short-range* or *weak long-range* binary interaction potential. Besides, one must take into account the screening properties of the *diffuse layer*, which depend markedly on the electrode charge and ion force of the solution.

The statistical-mechanical analysis of correlation properties of such an ion ensemble at the uniform metal surface was performed in Ref. [266], see also reviews [254, 255, 261, 267]. There are two characteristic values of the 2D density, $\Gamma = \Gamma_* \sim (\pi L_B R_*)^{-1}$ and $\Gamma = \Gamma_B \sim (\pi L_B^2)^{-1}$ corresponding to the intervals: 0.05–$0.5\,\mathrm{nm}^{-2}$ and 0.5–$2\,\mathrm{nm}^{-2}$, respectively.

Owing to the rapid decrease of the interaction potential at large distances, the system behaves like a 2D dilute gas with a weak interaction between the species if the surface coverage is extremely low, $\Gamma \ll \Gamma_*$. The virial expansion may be used to determine all energetic and correlation characteristics. However, the coefficients of this virial expansion turned out to be very large (compared with their estimates with the use of the Bjerrum length). As a result a noticeable repulsion between adsorbed ions in the intermediate distance range takes place already at very low ion densities, $\Gamma \sim \Gamma_*$.

Within this transitional interval of densities, $\Gamma \sim \Gamma_*$, the correlation properties of the system are subject to a gradual change. For lower coverages, the principal contribution to the average interaction energy is due to *pair* "collisions" of ions. On the contrary, at higher densities, $\Gamma \gg \Gamma_*$, new collective plasma-like excitations arise, leading to the screening of the interactions at the distance R_Γ. This new 2D screening length depends on the density of the system, $R_\Gamma \sim (\Gamma L_B)^{-1}$, that is, the increase of the coverage results in a more effective screening. Thus the correlation properties of this ensemble within this interval are close to those of a 2D one-component plasma, despite the absence of the R^{-1} behavior of $U(R)$ at very long distances. It would be of great interest in the future to perform experimental studies of these predictions with the use of small-angle scattering techniques, in particular of the anomalous behavior at small Fourier vectors owing to the plasma correlations.

At even higher system density, $\Gamma \sim \Gamma_B$, the interaction energy between a pair of adsorbed ions located at the average distance becomes comparable to the thermal energy, $k_B T$. Then the properties of the system correspond to those of a 2D irregular condensed state. Ultimately, when approaching the monolayer coverage determined by the short-range repulsion of ion cores or/and their solvating shells, a typical 2D "solid" is formed, which represents a crystal lattice possessing a quasi long-range ordering.

2.1.11.6 Ion Adsorption Isotherm

In a surprising contrast to this complicated picture for the correlation properties of the ensemble of adsorbed ions (Sect. 2.1.11.5), the same analysis has resulted in a very simple formula for the average interaction energy, that is, for the contribution to the chemical potential of adsorbed ions, μ_{int}, which is given as a linear function of the coverage over the whole range of surface coverages:

$$\mu_{\mathrm{ads}} = k_B T \ln \Gamma + \mu_{\mathrm{int}} + \mathrm{constant},$$

$$\mu_{\mathrm{int}} \cong 2v\Gamma k_B T \quad (90)$$

The coefficient in this linear relation serves as an analogue of the attraction constant in the Frumkin isotherm (with the minus sign) or the constant B in the Grahame–Parsons isotherm (75). However, unlike these models, in which this term as well as its dependence on the electrical variable (see following text) were introduced phenomenologically, Eq. (90) was derived on the basis of the binary interaction potential, $U(R)$, so that the constant, v, is related immediately to this characteristic:

$$v = \pi \int_{L_B}^{\infty} \left(1 - \exp\left(\frac{-U(R)}{kT}\right)\right) R \, dR \quad (91)$$

The introduction of the expression for the binary interaction potential discussed in Sect. 2.1.11.4 into this equation leads to the formula (92) for the "repulsion constant", v:

$$v = \Delta^2 \bar{v}, \quad \bar{v} = \frac{z_i^2 e^2}{2 k_B T (C_H + C_{GC})} \quad (92)$$

In this formula the coefficient, v, is represented as a product of the factor Δ^2 introduced in Sect. 2.1.11.4, and a function of the compact- and diffuse-layer capacitances (the GC expression for the latter, Eq. (21a), has been used in this derivation).

The combination of Eq. (70a) and (90) results in the adsorption isotherm (93) for a mixed electrolyte solution [267, 268]:

$$\ln(mc) = \frac{W_{ads}}{k_B T} + \ln \Gamma + 2v\Gamma \quad (93)$$

where W_{ads} is the single-ion adsorption energy corresponding to the energy of ion transfer to the surface in the absence of other adsorbed ions.

Similar to several other models, isotherm (93) has the *virial* form, with a linear dependence in the interaction term. The essential difference between them is related to another driving parameter of the system, which is called the *electrical variable* and which characterizes the state of the electrode while the bulk concentration of "active" species changes. The Frumkin isotherm is valid for a fixed *electrode potential*, the adsorption energy and the attraction "constant" as functions of this variable being modeled by the simplest approximation (linear or quadratic). In the theory of Grahame and Parsons, Eq. (75), it is the *electrode charge*.

According to the derivation of isotherm (93), the ions are transferred into the adsorbed state at a fixed value of the *electrode potential*, E, so that the parameters of the isotherm, W_{ads} and v, depend on this variable, as well as on the overall bulk concentration, c, in the case of the mixed electrolyte:

$$W_{ads} = W_{EDL} + W_{im}^c + W_{ads}^{\circ},$$

$$W_{EDL} = z_i e (E - \varphi_{GC})(1 - \Delta) + z_i e \varphi_{GC} \Delta \quad (94)$$

In other words, the single-ion adsorption energy is represented as the sum of the EDL term (compare Eq. (74)), the image potential as a function of the electrolyte concentration, $W_{im}^c = W_{im}(z; \sigma, c) - W_{im}(z; \sigma, 0)$ and the standard adsorption energy, W_{ads}°. The analytical expression for W_{im}^c is given in Ref. [256, 267, 268]. All these expressions including the capacitances in Eq. (92) and the diffuse-layer potential difference, φ_{GC}, in Eq. (94) depend on the electrode potential; these dependences correspond to *zero* specific adsorption so that they should be taken from experimental data for the *surface-inactive electrolyte*.

After this treatment the isotherm only contains *two* microscopic parameters of the adsorbed ion, whose values cannot be measured without experiments on its specific adsorption, W_{ads}° and Δ, which may generally be dependent on the electrode potential. The theoretical expression for Δ (in particular, its simplest form (88a)) shows that its value is expected to be between 0 and 1; it should diminish, if the ion is located closer to the electrode surface or/and at a greater charge transfer from the ion to the metal (i.e. if the adsorption is stronger).

Thus the derived isotherm has got a very simple final form, with a minimum number of fitting parameters, less than in the other approaches used to interpret data for ionic adsorption.

If the adsorption is measured as a function of the bulk concentration for different values of the electrode potential, $\Gamma\,(c, E)$, then the parameters in Eq. (93), W_{ads} and v, can be easily found by fitting for each potential, then their values may be used to determine the fundamental characteristics of the adsorbed ion, W_{ads}° and Δ, without any further uncertainties. However, the determination of experimental values of Γ on the basis of the *capacitance* (or electrode charge, or interfacial tension for liquid electrodes) measurements requires to perform the operation of the graphical differentiation of the curves for the interfacial energy, $\gamma\,(E, m, c)$ (for mixed electrolyte solutions), with respect to the solution composition, $\ln m$ or $\ln c$ (see Sect. 2.1.11.1).

This graphical differentiation can be avoided, owing to the particular form of the isotherm (93), which has allowed one to derive an analytical relation between γ and m [261, 267, 269]. Then the fitting first provides the values of W_{ads} and v, and from that W_{ads}° and Δ are obtained.

This method has been successfully applied to analyze the ion adsorption

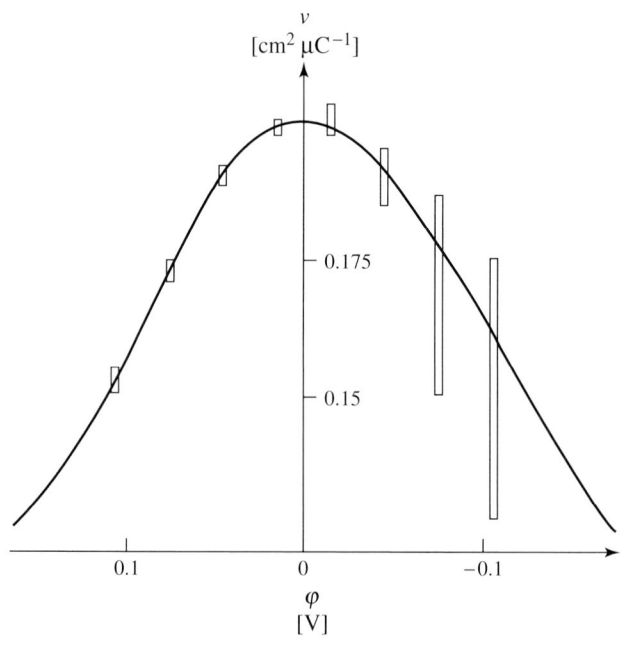

Fig. 22 Experimental values of the "repulsion constant", v, in isotherm (93) for iodide adsorption at Bi-water interface (points are shown with their dispersion) as a function of the electrode potential in the rational scale, $\varphi \equiv E - E_{\sigma=0}$. The theoretical curve (solid line) was calculated from Eq. (92) for $\Delta = 0.71$ with the use of experimental data for the compact- and diffuse-layer capacitances of the same interface in a surface-inactive electrolyte solution.

from mixed electrolyte solutions in the systems [261, 267, 270, 271], namely, Bi/H$_2$O−I$^−$, InGa/H$_2$O−I$^−$, InGa/H$_2$O −Br$^−$, Hg/ethylene glycole−Cs$^+$, Hg/ propylene glycole−Cs$^+$, Hg/H$_2$O−Cl$^−$, and so forth. These systems were chosen to verify the theoretical predictions for the form of the adsorption isotherm as well as for the dependence of the values of its parameters, $W_{ads}^°$ and Δ, on the properties of the metal, solvent, and adsorbed ion (both cations and anions).

The results of this treatment of the data for the first of these systems are presented in Figs. 22 and 23. One should note that the earlier application of the Grahame–Parsons isotherm to interpret these data was unsuccessful, despite a greater number of fitting parameters. The analysis with the use of isotherm (93) gave the values of the intermediate parameters, v and W_{ads}, which are shown in these figures as a function of the electrode potential in the rational scale, $\varphi \equiv E - E_{\sigma=0}$. The experimental dispersion in the thus obtained values of v is indicated for each point. It increases strongly at negative potentials since the coverages are much lower in this range and the precision of the determination of the *interaction* parameter becomes much lower.

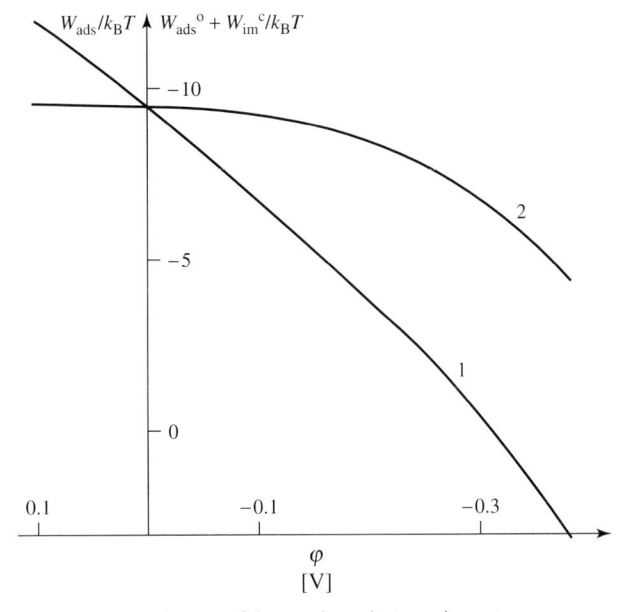

Fig. 23 Dependences of the total single-ion adsorption energy, W_{ads}/k_BT (curve 1), and of the sum of the "specific" and image contributions, $(W_{ads}^° + W_{im}^c)/k_BT$ (curve 2) for the system, Bi/H$_2$O−I$^−$, as a function of the electrode potential, $\varphi \equiv E - E_{\sigma=0}$. The values of W_{ads} were found from the fitting of isotherm (93) to experimental adsorption data for each electrode potential. Then the values of $W_{ads}^° + W_{im}^c$ were obtained with the use of Eq. (94) and the data for surface-inactive electrolyte solutions. The EDL term, W_{EDL}, was calculated for the same value of the adsorption parameter, $\Delta = 0.71$, see Fig. 22.

Then the value of Δ was determined from the experimental point for the most positive potential, with the use of the value of $\bar{v}$ found from Eq. (92) and experimental data for the Bi–water interface in the background electrolyte solution. This value was used to calculate, already without any unknown parameter, the whole theoretical curve for v (Fig. 22) as well as for $W_{ads}^{\circ} + W_{im}^{c}$ (Fig. 23) as functions of φ.

One can notice that a *constant* value of Δ is sufficient to reach an agreement with *all* experimental points for v, within their dispersion, which is very small for the positive potential range. Similarly the single-ion adsorption energy (after removal of its EDL component) is practically constant within the same range of potentials. Thus, in accordance with expectations, the microscopic characteristics of the adsorbed ion, I^-, do not change markedly in this region of electrode potentials, despite the change of the electric field created by the charged metal surface.

The data at high negative potentials show a significant reduction of the adsorption energy with the potential. It also corresponds to the qualitative expectations concerning the weaker bonding of an anion at negatively charged metal surface.

Similar results have been obtained for the other systems listed in the preceding text. In particular, the values of the "repulsion constant", v, were found within the limits of 0.6 and 3.5 nm^2 per adsorbed ion. These results represent an additional confirmation of the prediction of strong intensity and long-range of the binary interaction potential discussed in Sect. 2.1.11.4. One should keep in mind that the attempt to interpret the same experimental data with the use of the "single-image" expression (86) would result in the values of the ion-metal distance, a_i, well beyond the usual length of chemical bonds. It confirms the principal conclusion of Sect. 2.1.11.4 concerning the binary interaction potential, namely, that the dominant role is played by the interaction along the field lines passing across the *diffuse layer*, rather than those along the compact layer.

The values of v for each particular system demonstrate a marked variation with the potential, which can be described by Eq. (92) with a *constant* value of Δ, that is, this dependence on the electrode potential is given by the theoretical expression, $\bar{v}(E, c)$, which does not contain *any unknown parameter*.

As another success of the approach, one may notice a large variation between the values of v for different systems, up to a factor of 5, while the estimates of the principal microscopic parameter of the adsorbed ion, Δ, turned out to vary within a much more narrow range, from 0.55 to 1.0. Moreover, all these values fall inside the expected interval $\Delta < 1$. If one uses formula (88a) or (88b) to get a simple qualitative estimate of this parameter, one may conclude on the basis of these results that all these ions are located not too close to the metal surface, and that they retain most (if not all) of their electric charges in the adsorbed state, $z_a \sim z_i$.

Within the framework of the hypothesis that no charge transfer takes place in the course of the ion transition into the adsorbed state, the value of Δ for each system would allow one to determine the ratio of the capacitance between the metal surface and the adsorption plane to the overall compact-layer capacitance, Eq. (88b), similar to the Grahame–Parsons treatment in Sect. 2.1.11.2. (the Helmholtz–Stern model of the compact layer as a uniform dielectric slab does not correspond to the modern view of the interfacial structure, see previous sections, so that it is hardly possible to use Eq. (88a) to estimate the real distance between the adsorbed ion and the metal surface.) However, these adsorption data cannot give any argument in favor of or against such a partial charge transfer in these systems, in other words, the reduced value of this adsorption parameter, $\Delta < 1$, represents generally a combined effect of partial charge transfer from the adsorbed ion to the metal and of the location of the adsorbed ion inside the compact layer, see Eq. (88b) as an illustration.

The only exception is the case of the $Hg-H_2O-Cl^-$ interface [267, 271] in which the Δ value was close to 1. It allows one to conclude on the absence of a noticeable charge loss by this ion as well as on close values of the characteristic lengths, L_H and L_I in Eq. (88b), that is, of the corresponding capacitances. However, the latter does not mean that the physical location of the adsorbed ion is close to the outer Helmholtz plane. There are various indirect indications (in particular, presented in previous sections) that the effective dielectric permittivity, $\varepsilon(z)$, is quite nonuniform across the "compact layer", its lower values being within the "tail" of the metal electrons while the external part of the compact layer possesses a much higher permittivity (but still much lower than the one in the bulk solvent). Then the adsorbed ion can be localized at the distance of 0.1 to 0.2 nm from the ion skeleton of the metal, that is, deeply inside the compact layer having a thickness of about 0.3 nm, Eq. (12), but Δ can be close to 1.

The comparison of the Δ values for different systems have shown expected tendencies:

- this parameter for adsorbed iodide is greater at the Bi electrode than at InGa (the adsorption energy is greater for InGa);
- for the same electrode, InGa, the stronger adsorbed anion, iodide, has got a smaller value of this parameter, than bromide;
- in the case of a weakly adsorbed cation, cesium, there is no significant difference between ethylene and propylene carbonate solutions.

Despite the success in the interpretation of these experimental data for small-size inorganic ions, one should keep in mind the need for the further development of the theory, in particular to include the effect of the gradual change of the dielectric properties of the compact layer with the increase of the surface coverage by adsorbed ions.

2.1.12
Electric Field-Induced Modifications of the Surface Structure

Charging the electrode may not only affect the polarization of molecular dipoles on the surface, deform the surface electronic profile, and induce the adsorption of ions. It can also induce surface *reconstruction* and roughening transitions, that is, modify the atomic structure of the metal surface. In this section, we will briefly describe the principles of the statistical theory of this phenomenon [272–274].

In ultrahigh vacuum (u.h.v.), clean low-index single-crystal surfaces of gold and platinum are known to be reconstructed at low temperatures and unreconstructed at temperatures above the critical temperature [275]. Transition from the reconstructed to the high-temperature unreconstructed state is called *deconstruction*. This term does not necessarily imply that the surface structure at high temperature resumes the full 1×1 unreconstructed symmetry. Indeed the deconstruction is

Fig. 24 Top (upper) and side (lower) views on differently reconstructed (110) surfaces. The numbers indicate corresponding spin assignments of Ising model.

1 1 1 1 1 1 1 1 1

(1 × 1)

1 −1 1 −1 1 −1 1 −1 1

(1 × 2)

(1 × 3)

emersion of the electrode into u.h.v., in situ x-ray diffraction (XRD), electroreflectance, second harmonic generation, and scanning tunneling microscopy (for review see [276]). For instance, the 5×20 reconstructed Au(100) and the 1×23 reconstructed Au(111) surfaces are stable at sufficiently negative electrode potentials, but at positive potentials the reconstruction is lifted. It was noticed that anion adsorption definitely lifts the reconstruction, but it was hard to distinguish it from the mere effect of the positive charging, because ion adsorption is also a field-induced effect (see Sect. 2.1.11). Somehow at the metal electrolyte interface there is no need to go to high temperatures to deconstruct the surface. The reconstruction takes place by negative charging at room temperature.

Surfaces of Au(110) and Pt(110) in u.h.v. reconstruct to 1×2 or a mixture of 1×2 and 1×3 missing row structures (Fig. 24). In the electrochemical environment, the reconstructed state is stable for a range of electrode potentials smaller than a positive critical value. For potentials more positive than the critical value, the reconstruction is lifted. Potential variation reversibly changes the structure between the reconstructed and unreconstructed states. We will dwell on the missing row reconstruction in the following text, the easiest example for tutorial purposes, since it was rationalized by a relatively simple theory.

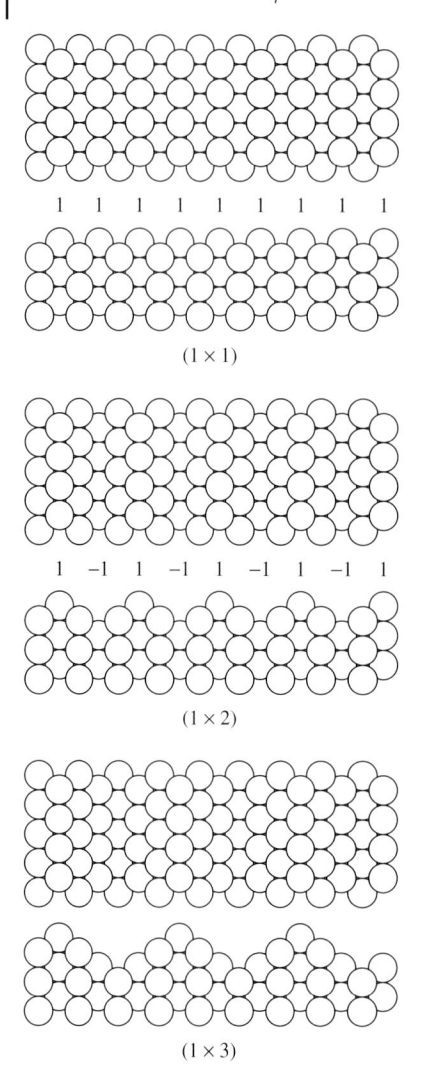

accompanied by the appearance of various kinds of imperfections, and the transition itself is not abrupt but passes through the phases of disorder, incommensurate superstructures and roughness.

Surface deconstruction has been discovered on single-crystalline electrodes in contact with electrolytes with the help of different experimental methods. These were the combination of cyclic voltammetry with ex situ analysis after

2.1.12.1 Electrochemical Missing Row Reconstruction: Facts and Speculations

Gao and coworkers [277] obtained scanning tunnelling microscopy (STM) images of Au (110) in 0.1 M $HClO_4$ aqueous solutions. At potentials more negative than −0.3 V versus. s.c.e., they report the appearance of 1 × 2 reconstruction (with small impurity of 1 × 3 and 1 × 4 blocks). Unreconstructed patterns emerge at potentials larger than zero; Hamelin [278] reported a threshold value of 0.1 V. Since the p.z.c. for this system is about −0.02 V [278], at zero charges the surface is not reconstructed in contrast to the same surface in u.h.v. at room temperature. The authors of Ref. [277] speculate, that this might be due to some adsorption of cations used to stabilize the reconstructed surface in u.h.v., which plays a role similar to negative charging. Between −0.3 and 0 V the interpretation of the images is unclear.

The same surfaces were studied by Magnussen and coworkers [279] in 0.1 M acid solutions: $HClO_4$, H_2SO_4 – pure and with traces of Cl^-. The Cl^--free solutions showed no dependence on the nature of electrolyte, but even small additions of Cl^- lifted reconstruction. The reconstructed state was observed at potentials from 0.05 down to −0.3 V, although less ordered than in u.h.v. A window of disordered and roughened states was found between 0.05 and 0.25 V, while at more positive potentials the 1 × 1 unreconstructed phase prevails. For slow variation of potential, the authors were able to follow the initial steps of reconstruction. They reported the transformation beginning near 0.2 V via a slight roughening of the step edges; further potential decrease induced roughening and zigzag step structure of the reconstructed surface with a simultaneous growth of anisotropic holes and islands one atom wide but up to 10 nm long in the 110 direction! When averaged over a big terrace, the concentration of islands was found to be nearly the same as the concentration of holes, in accordance with the idea that the reconstruction emerges by removing atoms from holes, which grow into missing rows, and redeposition of these atoms into the added rows.

Thus qualitatively the STM results of Behm's and Weaver's groups agree with each other. Simultaneously Ocko and coworkers [280] examined the same surfaces in 0.1 M perchloric acid and alkali halide salts solutions by a "more objective" in situ synchrotron diffraction method. For a meticulous discussion of this work, see the review [273]; here we just stress that the results, although less unambiguous in a variety of details, were qualitative similar to the results of STM studies.

2.1.12.2 Driving Force for Reconstruction

The structure and the shape of the metal crystal are determined by the net balance of forces. In the bulk of the noble metal, the stabilizing force is the compressive stress of s-p electrons balanced by repulsive interaction of the filled d-shells. The s-p electrons play the role of a "glue", which is as follows. They always "like" to spread out along the surface. However, at the (110) surface the external filed of the skeleton ions is very inhomogeneous. The electrons will have to form rather inhomogeneous distributions, which are energetically costly. This forces surface atoms to rearrange to form tightly-packed facets that mimic the lower energy (111) structure. On (110) surfaces, this is achieved by missing row (Fig. 24) reconstruction [281, 282]. The force that pushes the surface to reconstruct is stronger when more electrons are present in the surface layer. This happens when the surface is negatively charged.

The field-induced reconstruction was studied in the ground state Kohn–Sham electron density functional simulations for fixed atomic structures of the surface layers [283, 284]. Their energies are compared in order to find that configuration which has, at a given surface charge, the lower energy. It appears that in a range of negative charges the missing row configuration has lower energy at $T = 0$, and one may expect that the surface will be reconstructed. Physically, negative charging increases the density of electrons at the surface, which produces a tension that is relaxed by reconstruction, because the missing row structures allow for tighter packing of nearest neighbor atoms. Although no such calculations have been performed for the metal–electrolyte interface, in u.h.v. the comparison of 1×1 and 1×2 structures is available for (110) surfaces of Ag and Au. Subject to these calculations, one may approximate the difference in the ground state surface energies (per 1×1 surface unit cell)

$$\Delta E(\sigma) = E_{1 \times 2}(\sigma) - E_{1 \times 1}(\sigma) \qquad (95)$$

as

$$\Delta E(\sigma) \approx A(\sigma - \sigma_0) \qquad (96)$$

at least for σ close to the σ_0, the point where $E_{1 \times 2}(\sigma)$ and $E_{1 \times 1}(\sigma)$ curves intersect; for Au(110), $\sigma_0 \simeq 0$. A is a parameter, which we do not know for metal–electrolyte interface, but taking into account the u.h.v. density functional simulation, we may expect it to be no greater than 0.01 (eV cm^2) µC^{-1}.

2.1.12.3 Ising Model. Order–Disorder Transitions

Kornyshev and Vilfan [272–274] suggested the simplest way of describing the transitions between 1×1 and 1×2 states, based on an anisotropic 2D Ising model, and derived a closed expression for the temperature-surface charge phase diagram.

In this model an occupied topmost site is described by "spin" $s = 1$ and an empty site by $s = -1$ (Fig. 24). For an ideally 1×2 reconstructed surface, the $s = 1$ and $s = -1$ rows, oriented in y-direction (denoted as [1$\bar{1}$0]), alternate in x-direction, the one perpendicular to the rows ([001]). For an ideal 1×1 surface, all spins have $s = 1$ (or $s = -1$, which is equivalent). Let J_y and J_x and be the "spin interactions" of neighboring sites along the rows and perpendicular to the rows. The Ising–Hamiltonian then takes the form

$$H = - \sum_{i,j} (J_x s_{i,j} s_{i+1,j} + J_y s_{i,j} s_{i,j+1}) \qquad (97)$$

where i, j are the indices of the lattice sites determining their coordinates x_i and y_j. $J_y > 0$ for an anisotropic surface, and we will consider it to be constant. As for J_x, it could be positive or negative. If $J_x > 0$, the ground state is "ferromagnetically" ordered; if $J_x < 0$, it is "antiferromagnetically" ordered, using the lexicon of magnetic spin analogies. The transition between these two structures at $T = 0$ takes place when J_x changes the sign.

If we know $\Delta E(\sigma)$, it is easy to calculate J_x. Indeed the ground state energies of the two configurations calculated per unit cell of the reconstructed surface are $E_{1 \times 2} = E_0 + J_x - J_y$ and $E_{1 \times 1} = E_0 - J_x - J_y$, where E_0 is the part of the energy that does not depend on the spin state. Hence,

$$J_x = \frac{\Delta E}{2} \qquad (98)$$

J_y can also be obtained from the ground state calculations, since $J_y = E_0 - 1/2(E_{1 \times 2} + E_{1 \times 1})$. However, for further consideration, we will simply treat it as a scaling parameter, whose charge dependence is of minor importance.

At finite temperatures, we may use the well-known equation on the critical temperature of the anisotropic 2D Ising model [285]:

$$\sinh\left(\frac{|J_x(\sigma)|}{k_B T}\right)\sinh\left(\frac{J_y}{k_B T}\right) = 1 \quad (99)$$

Since, J_x is σ-dependent, this determines a critical line on the charge-temperature plane. Solving Eq. (99) for J_x and using Eq. (98), we obtain

$$\frac{|\Delta E(\sigma)|}{J_y} = F\left(\frac{k_B T}{J_y}\right) \quad (100)$$

where

$$F(x) \equiv x \ln\left[\cosh\left(\frac{1}{x}\right)\right] \quad (101)$$

Using, finally, Eq. (96), we obtain the equation for the critical line that determines the phase diagram:

$$\sigma = \sigma_0 + \frac{\text{sgn}(\sigma - \sigma_0)}{a} F\left(\frac{2k_B T}{J_y}\right) \quad (102)$$

where

$$a = \frac{A}{J_y} \quad (103)$$

The phase diagram corresponding to this function is shown in Fig. 25. This diagram tells us that moving at a given temperature from negative to positive charges, the system undergoes first a transition from reconstructed to a disordered state and then from a disordered state to the ordered unreconstructed state. The width of the disordered region follows from Eq. (103):

$$\Delta\sigma = 8\frac{k_B T}{A}\exp\left(-\frac{J_y}{k_B T}\right) \quad (104)$$

Taking as an estimate $A \sim 0.01\,\text{eV cm}^2\,\mu\text{C}^{-1}$ and $J_y \sim 4k_B T$, we get $\Delta\sigma \sim 0.5\,\mu\text{C cm}^{-2}$, but depending on the parameters, this value may be considerably larger. The origin of the disordered phase is clear. In order to have transitions, J_x must change sign at σ_0, but then $|J_x|$ must be small near σ_0, and there it cannot compete with the disordering effect of the temperature.

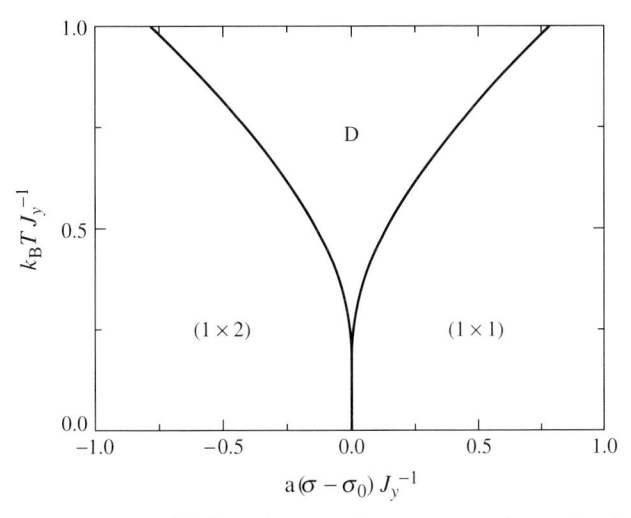

Fig. 25 Ising model phase diagram. The reconstructed (1 × 2) and unreconstructed (1 × 1) regions are separated by Ising disordered phase (D).

2.1.12.4 Roughening Transitions

While the Ising model of field-induced surface reconstruction is very convenient for tutorial purposes, it cannot describe roughening transitions that take place in the course of reconstruction-deconstruction transition. In order to take them into account, a more involved theory [272–274] was developed, based on the Vilfan and Villain approach to roughening transitions [286]. Kornyshev and Vilfan obtained the phase diagram by studying the excitation lines – Ising walls and roughening steps that break the long-range order. Each Ising domain wall is composed of two oppositely oriented steps, so that at both sides of the Ising domain wall the surface heights are the same. As long as opposite steps are bound into pairs, the surface behaves as flat. When the steps are no longer bound into pairs, thermal excitations causes the surface to become rough. Below the roughening transition, the surface can be flat but with Ising disorder whereas above the transition it is rough, because there is a random sequence of up-and-down steps. Strong anisotropy simplifies the calculation, because the steps and walls are oriented predominantly along the top atomic rows. One can imagine a step as a trajectory of a fermion moving in one-dimensional space, our x-axis, while the y-axis plays the role of time. The ground state energy of the corresponding quantum Hamiltonian is then equal to the free energy of the steps. In the flat phase the free energy is positive and infinitely long steps cannot be thermally excited. In the rough phase the free energy of a single step would be negative and the steps become thermodynamically stable. The steps repel each other and at the roughening transition the density of steps is infinitesimally small. Therefore it is enough to consider only one step, whose free energy is vanishing at the roughening transition temperature.

Kornyshev and Vilfan considered the statistical thermodynamics of such steps

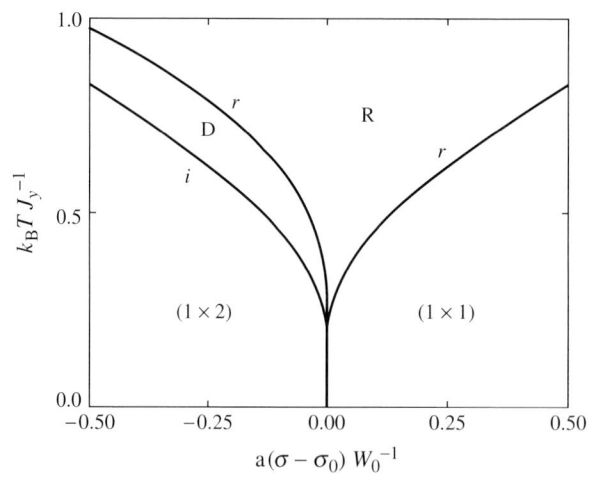

Fig. 26 Phase diagram of the Kornyshev and Vilfan theory exhibits a line of an Ising order-disorder transition and two lines of roughening transitions, with disordered, D, and rough, R, phases in between.

based on a free Fermion analogy, having again utilized the relationship of the charge dependence of the basic parameters of the theory with ΔE. Not going any further into the details of the theory (for tutorial presentation see the review article [273]), we draw in Fig. 26 the resulting phase diagram. While the Ising transition has a second order, the roughening transitions presumably are of Kosterlitz–Thouless universality class (see the discussion in Ref [273]).

In Ref. [273] the role of field-induced adsorption of anions together with a simple model of its incorporation into the theory is thoroughly discussed. Together with the analysis of the possible effects due to the presence of water [273], all this provides a framework for understanding the interesting phenomenon of electrochemically induced missing row reconstruction.

2.1.13
Soft Electrochemical Interfaces

Electrochemistry of soft interfaces is a rapidly progressing interdisciplinary field. The most important systems here are as follows:

- Polymer electrolyte–aqueous electrolyte interfaces in the porous interior of proton conducting or ion exchange membranes for separation, desalination, and fuel-cell applications [287].
- Interfaces between aqueous phase and the volumes confined by amphiphilic molecules [288]. *In vitro*, these refer to lipid vesicles and micelles, lipid lamellae, cubic and hexagonal phases, Langmuir–Blodgett (LB) films, which are important in colloid science and in extraction technology. *In vivo*, these are the surfaces of biological membranes.
- Oil–water interface [289, 290], important for oil extraction and catalysis.

The size of this chapter will not allow us to consider all these systems. We will therefore discuss the last one, well studied in electrochemistry, that is the interface between two immiscible electrolytes (ITIES). This is a biomimetic system and a model system for reaction kinetics, but also an actual medium for phase transfer catalysis. In this system, one of the solvents is usually water, and the other one is a low polar organic liquid (oil), such as, for example, nitrobenzene or dichloroethane. This interface separates "hydrophilic" and "hydrophobic ions", such as simple alkali-halides and organic ions, respectively. When two such salts are dissolved in this system, one composed of hydrophilic and the other one of hydrophobic ions, they form back-to-back EDLs and the interface can be polarized [289, 290]; standard electrochemical methods can be used for the characterization of this interface. The resulting electric field across the interface affects a set of phenomena that occur at the interface. A controllable variation of the potential drop across this interface can shed extra light on these phenomena.

First models of liquid–liquid interfaces in electrochemistry treated them as flat and sharp [289, 290]. The nonlinear capacitance due to two back-to-back ionic double layers at such an interface is given simply by [289]

$$\frac{1}{C_{GC}} = \frac{1}{C_1} + \frac{1}{C_2} \qquad (105)$$

$$C_i = \frac{\varepsilon_i \kappa_i}{4\pi} \left[1 + \frac{4\pi^2 \sigma^2}{(\varepsilon_i \kappa_i)^2}\right]^{(1/2)}, \ i = 1, 2$$
$$(106)$$

where σ is the surface density of the charge accumulated in each double layer per unit area of a flat interface.

However, these interfaces are known to be never ideally flat because of the thermal excitation of capillary waves. Depending on the interfacial tension between the two liquids, the interface between two solvents may occur to be somewhat diffuse. Stronger effect of diffuseness may come from the penetration of "hydrophobic" ions into water and "hydrophilic" ions into the organic phase, if the free energies of transfer of these ions between the two solvents are not much larger than k_BT. The overlap of the two space charge regions influences the interfacial double-layer capacitance. The two effects, the diffuseness of the interface and capillary waves are generally not unrelated. The short wave length excitations trigger local protrusions that contribute to the interface diffuseness, but the ion interpenetration reduces surface tension and affects long-range capillary waves. A rough estimate for the "width of the interface", d, is often used [291–293],

$$d^2 = d_0{}^2 + d_{cw}{}^2 \quad (107)$$

where d_0 and d_{cw} are the intrinsic and capillary wave contributions to the interface diffuseness. There is no unified theory that involves both phenomena; it is rather a subject for molecular dynamic simulation [294]. However, separately each effect has been rationalized [295, 296]. In the following text, we summarize the main results of these papers, in which one can read more about the history of the question and the previous achievements.

2.1.13.1 Capillary Waves at ITIES. Field Effect on Capillary Waves and Capillary Waves Effect on Capacitance

Measurements of optical [297, 298] and neutron reflectivity [299] and MD simulations (e.g. Ref. [294, 300]) show that the liquid–liquid interface is typically corrugated. The corrugation originates from thermal fluctuations of the interface, known as *capillary waves*. These fluctuations are described through the dependence of the local height of the interface, z, on the lateral coordinate $\mathbf{R} = (x, y)$, where the local height $z = \xi(\mathbf{R})$ is determined relative to the reference plane, $z = 0$, where the average value of the profile is zero ($<\xi(R)> = 0$).

The spectrum of capillary waves can be calculated starting with the functional of the free energy of two contacting liquids

$$F = F_0 + F_{cw}[\xi(\mathbf{R})] \quad (108)$$

where F_0 is the free energy of the system with a flat uncharged interface, while the second term stands for the excess energy due to surface corrugation:

$$F_{cw}[\xi(\mathbf{R})] = \frac{1}{2} \int d\mathbf{R}\gamma[(\nabla\xi(\mathbf{R}))^2 + k_{gr}^2\xi^2(\mathbf{R})] \quad (109)$$

Here, γ is the interfacial surface tension, $k_{gr}^2 = \Delta\rho g/\gamma$ is the small wave-vector gravitational cutoff, where g is the gravitational acceleration, and $\Delta\rho > 0$ is the density difference between the two liquids. After transformation to Fourier space,

$$\xi(\mathbf{k}) = \int d\mathbf{R}\xi(\mathbf{R})\exp[-i\mathbf{k}\mathbf{R}] \quad (110)$$

the free energy becomes a quadratic functional of the fluctuating variable $\xi(\mathbf{k})$,

$$F_{cw}[\xi(\mathbf{k})] = \frac{\gamma}{2} \int \frac{d\mathbf{k}}{(2\pi)^2} \xi(\mathbf{k})\xi(-\mathbf{k}) \times (k^2 + k_{gr}^2) \quad (111)$$

Note that this approximation is valid, strictly speaking, only if the amplitudes of capillary waves are small (i.e. when for each

Fourier component of surface fluctuation the inequality $|\xi(\mathbf{k})| < 2\pi/|\mathbf{k}|$ holds).

Further treatment is then trivial. For the Fourier transform of the height–height correlation function

$$g(\mathbf{k}) = \int d^2\mathbf{R} \langle \xi(\mathbf{R})\xi(0)\rangle \exp(-i\mathbf{k}\mathbf{R}) \quad (112)$$

with the help of the equipartition theorem one obtains [301]

$$g(\mathbf{k}) = \frac{k_B T}{\gamma(k^2 + k_{gr}^2)} \quad (113)$$

The mean square height of roughness, that is, the capillary wave contribution to the width of the interface, is expressed through the correlation function as

$$d_{cw}^2 = \frac{1}{2\pi} \int_0^{k_{max}} dk\, k\, g(\mathbf{k}) \quad (114)$$

For $g(\mathbf{k})$ given by Eq. (113),

$$d_{cw}^2 = \frac{k_B T}{2\pi \gamma} \ln\left(\frac{k_{max}}{k_{gr}}\right) \quad (115)$$

The upper wave-vector cutoff, k_{max}, is introduced to eliminate the divergence of the integral, which takes place for $g(\mathbf{k})$ given by Eq. (113), because in reality $g(\mathbf{k})$ decreases faster at large k. Various estimates of k_{max}^{-1} (proportional to the root mean square height [293], to the characteristic smearing of the interface [302], or to the largest bulk correlation length of the contacting liquids [303]) were proposed in the literature, but in any case, π/k_{max} cannot be less than the molecular diameter [304]. Gravity suppresses the long-range fluctuations and removes a divergence of the correlation function at small wave vectors.

Both d_{cw} and $g(\mathbf{k})$ can be measured in optical, neutron, and x-ray scattering experiments [297–299, 305], although it is more subtle to measure the height–height correlation spectrum than the integral characteristics of fluctuations, d_{cw}. Both experiments and MD [281, 284–287, 292], show that the macroscopic capillary wave theory works surprisingly well down to the nanometer range.

Time-resolved quasi-elastic laser scattering (QELS) experiments [306, 307] allow the measurement of the dispersion law (the relation between the frequency of capillary waves, ω, and their wavelength, $2\pi/k$). To derive the dispersion law theoretically, one must consider the dynamics of the two degrees of freedom – the position of the surface, $z = \xi(x, y, t)$, and the fluid velocity. In the limit of small viscosity, such a theory [291] gives

$$\omega^2 - 2i\omega\Delta(k)k^2 - \omega_c^2(k) = 0 \quad (116)$$

where

$$\omega_c^2(k) = \frac{k}{\rho_1 + \rho_2}\gamma(k^2 + k_{gr}^2),$$

$$\Delta(k) = 2k^2 \frac{\eta_1 + \eta_2}{\rho_1 + \rho_2} \quad (117)$$

ρ_1, ρ_2 and η_1, η_2 are the densities and viscosities of the two liquids in contact. In contrast to neutron and x-ray scattering experiments that allow the study of capillary fluctuations down to the nanometer scale, the QELS provides information on long wavelength fluctuations only, with wave lengths and frequencies of the order of 6.6×10^{-3} cm and 10 kHz, respectively.

Capillary waves at the interface of two immiscible electrolytes have been directly observed by time-resolved QELS [306, 307], investigated in MD computer simulations [294, 300] and described within the framework of a *phenomenological theory* [295, 308, 309]. The principles of this theory are clear, but the formalism, particularly for polarizations considerably far from the p.z.c. [295], is rather involved.

We refer the reader to Ref. [295], in which these are described in detail. Here, we only summarize the basic approach and the main results.

The interface is described as before, using the approximation of smooth corrugations. The two electrolyte solutions are characterized by the solvent dielectric constants, ε_1 and ε_2, and Debye lengths, κ_1^{-1} and κ_2^{-1}. In addition to the *interfacial tension* term, the free-energy functional now has two new terms. These are the *electrostatic energy*, F_e, and the term responsible for the *entropy of a dilute electrolyte*, F_s, so that

$$F = F_0 + F_{cw}[\xi(\mathbf{R})] + F_e[\xi(\mathbf{R})] + F_s[\xi(\mathbf{R})] \quad (118)$$

where

$$F_e[\xi(\mathbf{R})] = -\frac{\varepsilon_1}{8\pi} \int d\mathbf{R}$$
$$\times \int_{-\infty}^{\xi(\mathbf{R})} dz [\nabla \varphi_1(z,\mathbf{R})]^2 + e \int d\mathbf{R}$$
$$\times \int_{-\infty}^{\xi(\mathbf{R})} dz \varphi_1(z,\mathbf{R})(n_1^+(z,\mathbf{R})$$
$$- n_1^-(z,\mathbf{R})) - \frac{\varepsilon_2}{8\pi} \int d\mathbf{R}$$
$$\times \int_{\xi(\mathbf{R})}^{\infty} dz [\nabla \varphi_2(z,\mathbf{R})]^2 + e \int d\mathbf{R}$$
$$\times \int_{-\infty}^{\xi(\mathbf{R})} dz \varphi_2(z,\mathbf{R})(n_2^+(z,\mathbf{R})$$
$$- n_2^-(z,\mathbf{R})) + UQ \quad (119)$$

and

$$F_s[\xi(\mathbf{R})] = k_B T \int d\mathbf{R} \int_{-\infty}^{\xi(\mathbf{R})}$$
$$\times dz \left[n_1^+(z,\mathbf{R}) \log\left(\frac{n_1^+(z,\mathbf{R})}{n_1^0}\right) \right.$$
$$+ n_1^-(z,\mathbf{R}) \log\left(\frac{n_1^-(z,\mathbf{R})}{n_1^0}\right)$$
$$- (n_1^+ + n_1^- - 2n_1^0) \right] + k_B T$$
$$\times \int d\mathbf{R} \int_{\xi(\mathbf{R})}^{\infty} dz \left[n_2^+(z,\mathbf{R}) \right.$$
$$\times \log\left(\frac{n_2^+(z,\mathbf{R})}{n_2^0}\right) + n_2^-(z,\mathbf{R})$$
$$\times \log\left(\frac{n_2^-(z,\mathbf{R})}{n_2^0}\right)$$
$$- (n_2^+ + n_2^- - 2n_2^0) \right] \quad (120)$$

The notations are as follows: $\phi_i(z,\mathbf{R})$ and $n_i^\pm(z,\mathbf{R})$, the electrostatic potential and concentrations of positive and negative ions in the phase (i), respectively; U is the overall potential drop across the interface; Q and $-Q$ are the overall charges in the second and first phases, respectively. The UQ-term reflects the extra work needed to maintain the overall potential difference U if the latter is fixed [310].

This form of the functional assumes that the difference in the free energy of each ion transfer between water and organic phase is so large that each ion species is located either completely in water or completely in oil. Furthermore, formation of a spontaneous dipolar potential drop across the contact of the two solvents and its possible coupling with the ion distribution is not part of the model.

The functional depends on three types of fields, namely, surface fluctuations $\xi(\mathbf{R})$, electrostatic potential $\phi_i(z,\mathbf{R})$, and the ionic concentrations $n_i^\pm(z,\mathbf{R})$. Minimizing it with respect to ϕ_i and $n_i^\pm$ at a given $\xi(\mathbf{R})$, one obtains Poisson–Boltzmann equations that describe the distribution of the electrostatic potential and ionic concentrations. Substituting the results into the density functional, one can minimize that with respect to $\xi(\mathbf{R})$. This can be easily done in consistence with the already used harmonic approximation for $F_{cw}[\xi(\mathbf{k})]$ (Eq. 109). Then all characteristics of the interface are obtained in a closed form. For voltage drops smaller than $k_B T$, the results look relatively compact [308, 309], but for larger voltages they are very cumbersome [295].

The result for the electric capacitance reads

$$C = C_{GC}(V) R(\kappa_1, \kappa_2, V) \qquad (121)$$

where $V = eU/k_B T$, $C_{GC}(V)$ is the GC capacitance (Eqs. (105) and (106)) and $R(\kappa_1, \kappa_2, V)$ is the so-called roughness function, the factor in which all the deviations from GC are accumulated; when $\gamma \to \infty$, $R \to 1$. The general expression for $\tilde{R}(\kappa_1, \kappa_2, V)$ [295] is too cumbersome to repeat here, but its typical graph is shown in Fig. 27. This graph shows that the roughness function increases rapidly with V because the surface area increases due to a dramatic enhancement of the amplitude of capillary waves. The steepest wings in Fig. 27 should be ignored, because in this region the harmonic approximation breaks down, as they, essentially, reflect the electric field-induced destruction of the surface.

Indeed the potential-dependent FT of the height–height correlation function now reads

$$g(k, V) = \frac{k_B T}{\gamma[k^2 + k_{gr}^2] - \Delta f(k, V)} \qquad (122)$$

where $\Delta f(k, V)$ is a very complicated but elementary function of its arguments [295], which grows with $|V|$. For moderate $|V|$, it is proportional to V^2; for instance, for small wave vectors the expression is particularly simple,

$$\Delta f(k, V) \approx \frac{1}{2} k^2 V^2 C_{GC}^{(0)} (k \ll \kappa_1, \kappa_2) \qquad (123)$$

where $C_{GC}^{(0)} \equiv C_{GC}|_{\sigma=0}$.

The harmonic approximation breaks down when the denominator in Eq. (122) approaches zero, but the tendency towards dramatic enhancement of fluctuations is correct. This translates into the dependence of the mean square height of corrugation, d_{cw}^2, for which there is a full expression [295], but here we quote

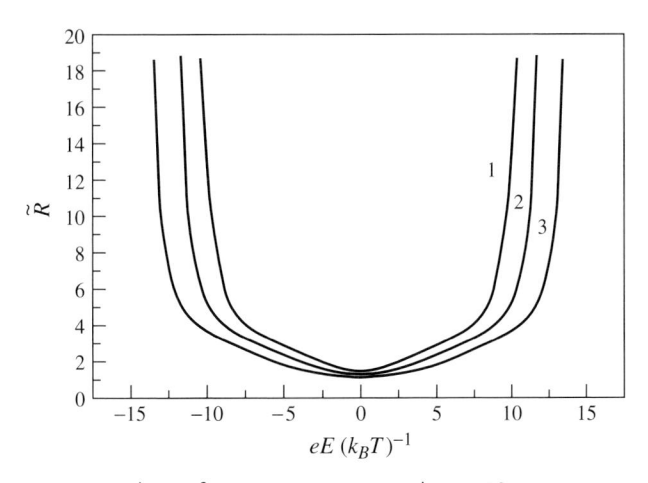

Fig. 27 Roughness function versus potential. $\varepsilon_2 = 10$; $\gamma = 30\,\mathrm{dyn\,cm^{-1}}$; κ_2^{-1}, nm: (1) 1; (2) 2; (3) 5.

only its simplified version for the case of moderate V:

$$d_{cw}^2 \approx (d_{cw}^{(0)})^2 + \frac{V^2 k_B T}{4\pi \gamma^2} C_{GC}^{(0)}$$
$$\times \ln \frac{\kappa_1 \kappa_2}{(\kappa_1 + \kappa_2) k_{gr}} \quad (124)$$

where $d_{cw}^{(0)}$ is the root mean square height at the p.z.c. Equation (124) shows unambiguously that charging the interface enhances surface corrugation. The general expression deviates from this law for $V > 4$ showing a stronger enhancement.

The dispersion law is also modified by charging the interface. It now reads [295, 308],

$$\omega^2 = -2i\omega\Delta(k) - \omega_c^2(k)$$
$$+ \frac{k}{\rho_1 + \rho_2} \Delta f(k, V) \quad (125)$$

that is, the charging reduces the frequency of capillary waves.

However, such a simple theory has its own limitations. One of them is that relaxation of the Debye plasma cannot follow the high frequency capillary waves, and their contribution to surface corrugation will not be "seen" by electrolyte ions. In Ref. [295], one can find a discussion of the yet unsolved dynamic aspects of the problem and suggestion for a correction, taking into account the incomplete adiabaticity.

2.1.13.2 Ion Penetration into an "Unfriendly Medium"

The treatment of interfacial capacitance in terms of the capacitance of two "back-to-back" GC double layers works fairly well for some electrolytes, while it fails for the majority of others [289, 311–316]. In contrast to the predictions of the simple GC theory, it was found that the capacitance of ITIES depends on the nature of the ions [311, 313–318]. Often the capacitance curves show a strong asymmetry as a function of the potential. The discrepancy between experimental results and the GC theory has stimulated theoretical work that went beyond the classical scheme. The first step in this direction has been done in Refs. [317–319], in which the "mixed boundary layer" was introduced and the effect of overlapping of the two space-charge regions on the double-layer capacitance has been considered. The authors used a quasi-chemical approximation (QCA) [317, 318] and MC simulations [319] in order to reveal the density profiles of the ions (the Poisson–Boltzmann equation has been solved numerically, assuming, for computational simplicity, that the dielectric constants of the two solvents are equal to each other and do not vary with the distance from the interface). Numerical studies of the capacitance [317–319] reveal the effect of the mixed layer. They, however, do not disclose general laws for the capacitance of this interface, leaving open the following questions:

- Which characteristics determine the observed capacitance dependence on the nature of ions?
- Which parameters determine the sign of the deviation of the capacitance from the Gouy–Chapman result and the asymmetry of the capacitance curves as a function of potential?
- What information on the ionic profiles and interfacial structure in general can we gain from the capacitance data?

Furthermore, smearing of the interface between two solvents should cause a smooth variation of dielectric properties across the interface (if it were sharp, similar effects would emerge due to nonlocal dielectric polarizability [92–94,

2.1 Electrochemical Interfaces: At the Border Line

91, 89]). It is important to understand how this will affect the capacitance curves.

A step towards answering these questions was made in an analytical theory, reported in Ref. [296]. The theory is based on a modified nonlinear Poisson–Boltzmann equation that takes into account the overlap of the two back-to-back double layers in the interfacial region and a smooth variation of dielectric properties across the interface. Similarly to [44], the analytical solution of this equation is obtained within the perturbation theory, which utilizes the smallness of the ratio of the "mixed layer" thickness to the Gouy lengths in the adjacent solutions. The solution is thus valid for *not too high* electrolyte concentrations and voltage drops across the interface. Note that the most interesting results have been observed in the range of low ionic concentrations [311–318].

The basic formula for capacitance, derived in Ref. [296], shows that the capacitance dependence on the nature of ions is controlled by *three integral parameters*. These parameters do not depend on the potential or ionic concentrations. They are expressed through the z-dependent profiles of the quantities, varying across the interface: (i) the short-range contributions to the free energy of ion transfer for all double layer ions and (ii) the effective dielectric constant of the solvent. These parameters are:

$$L_1 = \int_{-\infty}^{+\infty} dz [\exp(-g_1^+(z)) - \exp(-g_1^-(z))] \quad (126)$$

$$L_2 = \int_{-\infty}^{+\infty} dz [\exp(-g_2^+(z)) - \exp(-g_2^-(z))] \quad (127)$$

where $g_{1,2}^\pm(z)$ are the short-range energy profiles for the corresponding ions, and the combined parameter,

$$L_3 = \int_{-\infty}^{+\infty} dz \left[\frac{1}{2} \left(\frac{1}{\varepsilon_1} \exp(-g_1^+(z)) \right. \right.$$
$$+ \frac{1}{\varepsilon_1} \exp(-g_1^-(z)) + \frac{1}{\varepsilon_2} \exp(-g_2^+(z))$$
$$\left. \left. + \frac{1}{\varepsilon_2} \exp(-g_2^-(z)) \right) - \frac{1}{\varepsilon(z)} \right] \quad (128)$$

which is also affected by of the effective dielectric constant profile $\varepsilon(z)$, where $\varepsilon(z \to -\infty = \varepsilon_1)$ and $\varepsilon(z \to \infty) = \varepsilon_2$.

The effect of ion nature on these integral parameters can be clarified on the basis of a microscopic model of the interface or by MD simulations, giving $g_{1,2}^\pm(z)$ and $\varepsilon(z)$. On the other hand, these parameters can be extracted from the capacitance measurements, and their dependence on the ion nature can be traced. This suggests a new framework for the treatment of the capacitance data and establishes a relationship between experimental results and the microscopic structure of ITIEs. Here, we will not repeat the bulky expression for the capacitance, referring instead the reader to Ref. [296]. We show just one example here, a graph (Fig. 28) that demonstrates the effect on capacitance of the penetration of a hydrophobic ion into the aqueous phase. Generally the integral parameters L_1, L_2, and L_3 can take either positive or negative values depending on the shape of the free energy profiles and dielectric function. As a result the capacitance of ITIES can be either higher or lower than the GC capacitance of the two back-to-back double layers with a sharp interface (see Figs. 3–6 in Ref. [296]). Both types of behavior have been observed experimentally [313–315] and in numerical simulations [317, 319].

The effect of concentration on the capacitance curves has been studied using

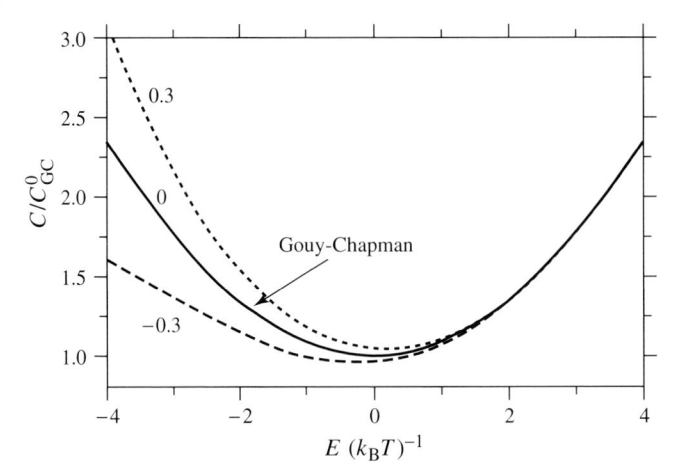

Fig. 28 The effect of penetration of a hydrophobic positive ion into the aqueous solution on the potential dependence of the normalized capacitance, C/C_{GC}^0. Curves correspond to the GC result obtained for $L_1 = L_2 = L_3 = 0$, and to the indicated values of the parameter $L_2^+ = \int_{-\infty}^{+\infty} dz[\exp(-g_2^+(z)) - \theta(z)]$ in nm. The figure is plotted for the case of no penetration of other ions through the interface, and no smearing of the dielectric profile. $\varepsilon_1 = 80$, $\varepsilon_2 = 10$, $\kappa_1 = \kappa_2 = 0.3$ nm^{-1}.

an equally concentrated solution in each phase. For this case, the results of Ref. [296] suggest that the deviations from the GC capacitance and differences between capacitance measured for different ions increase proportionally to the square root of the concentration, in line with experimental data [313].

There is a number of other interesting findings in Ref. [296], including the meaning of the apparent compact-layer contribution in the PZ plot. The latter is most transparent for small potential drops across the interface, where the expressions are most simple, that is:

$$\frac{1}{C} = \frac{1}{C_{GC}^0} + \frac{1}{C_H} \quad (129)$$

with the "compact-layer capacitance"

$$C_H = \frac{-1}{(4\pi L_3)} \quad (130)$$

Thus, C_H can be positive or negative, depending on the value of L_3. It should be stressed, that *there is no reason to expect in this system a true "compact layer", because the interface is permeable for both salts*. This result thus rationalizes a long-standing puzzle – the appearance of a compact-layer contribution in some systems and its absence in others. On the whole it clarifies the nature of the intercept of the PZ plot.

2.1.14
From Basic Concepts to Surface Electrochemistry

In this chapter, we have tried to give a consistent and comprehensive picture of the principles that describe the electrochemical interface, moving from long-standing but still valuable classical concepts of electrochemistry towards recent modern concepts based on condensed matter physics and statistical mechanics. To some extent the chapter followed the historic development of the ideas on the interfacial structure and properties.

Experimental interfacial electrochemistry is nowadays dominated by the so-called *"electrochemical surface science"* or *"electrochemistry with atomic resolution"*. Various methods and results of surface electrochemistry are thoroughly discussed in this and in subsequent volumes of the Encyclopedia. In terms of theoretical interpretation, they often ask for insight on the level of first principles calculations of electronic structure.

We did not review such calculations in any depth, the reason being two-fold; first, before one performs such calculations, one should become acquainted with the basic phenomenological concepts and principles of a global or "collective" description of the interface. Secondly, while first principle calculations in surface science have achieved a high level of sophistication, their combination with crude models of the neighboring solvent is controversial. At any rate, choosing the proper combination is currently a "fine art".

Using the methods of semi-empirical quantum chemistry can be as instructive as it can be, quite frequently, illusive, when it is disconnected from the real dynamics and the global properties of the interface. The development of theoretical interfacial electrochemistry in the near future will proceed in the direction of a combination of ab initio quantum chemical methods and MD. Progress and obstacles in this direction were recently reviewed [320]. MD of the solution part of the interface, based on empirical atom–atom interaction potentials, is often of great help for the verification of global models, rather than direct comparison with experiments. One obvious benefit of molecular simulation methods is the "natural" way in which the solvent is included, thus avoiding the use of "ad hoc" assumptions concerning the behavior of the dielectric constant, to name only one important aspect.

In general interfacial electrochemistry is an area, in which the conceptual physical models have to interact with chemical, quantum-mechanical "first principles" approaches. The former can guide the research, formulate questions and help rationalizing the results of the latter. Such a synthesis may, in a world of avalanche-like progress in computational resources, not lie in the distant future. It could be "tomorrow".

References

1. A. J. Bard, L. R. Faulkner, *Electrochemical Methods*, Wiley, New York, 1980 (1st ed.), 2001 (2nd ed.).
2. J. Goodisman, *Electrochemistry; Theoretical Foundations*, Wiley, New York, 1987.
3. J. Koryta, *Ions, Electrodes and Membranes*, Wiley, New York, 1991.
4. C. M. A. Brett, A. M. O. Brett, *Electrochemistry: Principles, Methods and Applications*, Oxford University Press, Oxford, 1993.
5. J. O'M. Bockris, S. U. M. Khan, *Surface Electrochemistry. A Molecular Level Approach*, Plenum Press, New York, 1993.
6. E. Gileadi, *Electrode Kinetics for Chemists, Chemical Engineers and Material Scientists*, VCH, Weinheim, 1993.
7. K. B. Oldham, J. C. Myland, *Fundamentals of Electrochemical Science*, Academic Press, San Diego, 1994.
8. R. G. Compton, G. H. W. Sanders, *Electrode Potentials*, Oxford University Press, Oxford, 1996.
9. Z. Galus, *Fundamentals of Electrochemical Analysis*, Ellis Horwood, Chichester, 1997.
10. C. H. Hamann, A. Hamnett, W. Vielstich, *Electrochemistry*, Wiley-VCH, Weinheim, 1998.
11. B. B. Damaskin, O. A. Petrii, G. A. Tsirlina, *Electrochemistry*, Khimiya, Moscow, 2001.

12. O. A. Petrii, G. A. Tsirlina, Electrode potentials, *Thermodynamics and Electrified Interfaces*, Encyclopedia of Electrochemistry (Eds.: E. Gileadi, M. Urbakh), Wiley-VCH, Germany, Vol. 1, Chap. 1. forthcoming.
13. B. B. Damaskin, O. A. Petrii, Specific adsorption, *Thermodynamics and Electrified Interfaces*, Encyclopedia of Electrochemistry (Eds.: E. Gileadi, M. Urbakh), Wiley-VCH, Germany, Vol. 1, Chap. 3. forthcoming.
14. O. A. Petrii, *Electrochim. Acta* **1996**, *41*, 2307.
15. M. A. Vorotyntsev, Yu. A. Ermakov, V. S. Markin et al., *Russ. J. Electrochem.* **1993**, *29*, 513.
16. M. A. Vorotyntsev, A. A. Rubashkin, J. P. Badiali, *Electrochim. Acta* **1996**, *41*, 2313.
17. A. N. Frumkin, *Electrocapillary Phenomena and Electrode Potentials*, Commerce typography, Odessa, 1919.
18. A. N. Frumkin, *Potentials of Zero Charge*, Nauka, Moscow, 1979.
19. S. Trasatti in *Comprehensive Treatise of Electrochemistry* (Eds.: J. O' M. Bockris, B. E. Conway, E. Yeager), Plenum Press, New York, 1980, p. 45, Vol. 1.
20. E. Lust, Double layers at single crystal and polycrystalline electrodes, *Thermodynamics and Electrified Interfaces*, Encyclopedia of Electrochemistry (Eds.: E. Gileadi, M. Urbakh), Wiley-VCH, Germany, Vol. 1, Chap. 2.4. forthcoming.
21. S. Trasatti, O. A. Petrii, *Pure Appl. Chem.* **1991**, *63*, 711.
22. S. Trasatti, E. Lust in *Modern Aspects of Electrochemistry* (Eds.: R. E. White, B. E. Conway, J. O' M. Bockris), Kluwer Academic/Plenum Publishers, New York, London, 1999, p. 1, Vol. 33.
23. D. C. Grahame, *J. Chem. Phys.* **1948**, *16*, 1117.
24. D. C. Grahame, B. A. Soderberg, *J. Chem. Phys.* **1954**, *22*, 449.
25. M. A. Vorotyntsev, J. P. Badiali, G. Inzelt, *J. Electroanal. Chem.* **1999**, *472*, 7.
26. D. C. Grahame, R. Parsons, *J. Am. Chem. Soc.* **1961**, *83*, 1291.
27. A. N. Frumkin, R. V. Ivanova, B. B. Damaskin, *Dokl. Akad. Nauk SSSR* **1964**, *157*, 1202.
28. B. M. Grafov, V. V. Elkin, *Sov. Electrochem.* **1990**, *26*, 1015.
29. G. Gouy, *J. Phys. Radium* **1910**, *9*, 457.
30. D. Chapman, *Philos. Mag.* **1913**, *25*, 475.
31. O. Stern, *Z. Elektrochem.* **1924**, *30*, 508.
32. D. C. Grahame, *Chem. Rev.* **1947**, *41*, 441.
33. A. Hamelin in *Modern Aspects of Electrochemistry* (Eds.: B. E. Conway, R. E. White, J. O' M. Bockris), Plenum Press, New York, 1985, p. 1, Vol. 16.
34. M. A. Vorotyntsev in *Modern Aspects of Electrochemistry* (Eds.: J. O' M. Bockris, B. E. Conway, R. E. White), Plenum Press, New York, 1986, p. 131, Vol. 17.
35. D. C. Grahame, *J. Am. Chem. Soc.* **1954**, *76*, 4819.
36. D. C. Grahame, Techn. Report no 14 to the office of Naval Research of Febr. 18, 1954. Deposited at the Library of Congress, Washington, D.C.
37. R. Parsons, F. G. R. Zobel, *J. Electroanal. Chem.* **1965**, *9*, 333.
38. M. A. Vorotyntsev, in *Advances in Science & Technology*, VINITI, Moscow, 1984, p. 3, Vol. 21.
39. R. Gonzalez, F. Sanz, *Electroanalysis* **1997**, *9*, 169.
40. B. B. Damaskin, *Sov. Electrochem.* **2001**, *37*, 745.
41. A. N. Frumkin, N. B. Grigoryev, *Sov. Electrochem.* **1968**, *4*, 533.
42. A. A. Kornyshev, M. A. Vorotyntsev, *Surf. Sci.* **1980**, *101*, 23.
43. A. A. Kornyshev, M. A. Vorotyntsev, *Can. J. Chem.* **1981**, *59*, 2031.
44. A. A. Kornyshev, W. Schmickler, M. A. Vorotyntsev, *Phys. Rev. B* **1982**, *25*, 5244.
45. B. B. Damaskin, N. V. Nikolaeva-Fedorovich, *Zh. Fiz. Khim.* **1962**, *36*, 1483.
46. R. Payne, *J. Electroanal. Chem.* **1975**, *60*, 183.
47. W. R. Fawcett, M. Opallo, *Angew. Chem., Int. Ed.* **1994**, *33*, 2131.
48. W. R. Fawcett in *Electrocatalysis* (Eds.: J. Lipkowski, P. N. Ross), Wiley-VCH, New York, 1998, p. 323.
49. A. N. Frumkin, *Z. Phys. Chem.* **1933**, *164*, 121.
50. A. N. Frumkin, G. M. Florianovich, *Dokl. Akad. Nauk SSSR* **1951**, *80*, 907.
51. G. M. Florianovich, A. N. Frumkin, *Zh. Fiz. Khim.* **1955**, *29*, 1827.
52. A. N. Frumkin, in *Advances in Electrochemistry and Electrochemical Engineering* (Ed.: P. Delahay) Intersci. Publ., New York, Vol. 1, 1961, p. 65; Vol. 3, 1963, p. 287.
53. L. I. Krishtalik, *Electrochim. Acta* **1965**, *13*, 1045.

54. P. Delahay, *Double Layer and Electrode Kinetics*, Intersci. Publ., New York, London, Sydney, 1965.
55. K. Asada, P. Delahay, A. Sundaram, *J. Am. Chem. Soc.* **1961**, *83*, 3396.
56. O. A. Petrii, A. N. Frumkin, *Dokl. Akad. Nauk SSSR* **1962**, *146*, 1121.
57. A. N. Frumkin, O. A. Petrii, N. V. Nikolaeva-Fedorovich, *Electrochim. Acta* **1953**, *8*, 177.
58. R. R. Dogonadze, A. M. Kuznetsov, M. A. Vorotyntsev, *Croat. Chim. Acta* **1972**, *44*, 257.
59. A. N. Frumkin, N. V. Nikolaeva-Fedorovich, N. V. Berezina et al., *J. Electroanal. Chem.* **1975**, *58*, 189.
60. D. I. Dzhaparidze, V. V. Shavgulidze, *Sov. Electrochem.* **1973**, *9*, 1390.
61. N. V. Fedorovich, M. D. Levi, A. V. Shlepakov, *Sov. Electrochem.* **1977**, *13*, 89.
62. R. R. Nazmutdinov, I. V. Pobelov, G. A. Tsirlina et al., *J. Electroanal. Chem.* **2000**, *491*, 126.
63. R. R. Nazmutdinov, G. A. Tsirlina, Yu. I. Kharkats et al., *J. Phys. Chem. B* **1998**, *102*, 677.
64. R. R. Nazmutdinov, G. A. Tsirlina, O. A. Petrii et al., *Electrochim. Acta* **2000**, *45*, 3521.
65. D. J. Gavaghan, S. W. Feldberg, *J. Electroanal. Chem.* **2000**, *491*, 103.
66. L. I. Krishtalik, *Charge Transfer Reactions in Electrochemical and Chemical Processes*, Plenum Press, New York, London, 1986.
67. R. J. Hunter, *Zeta Potential in Colloid Science. Principles and Applications*, Academic Press, London, New York, 1981.
68. J. Lyklema, *Fundamentals of Interface and Colloid Science*, Academic Press, London, Fundamentals, Vol. 1, 1993; Solid-liquid Interfaces, Vol. 2, 1995.
69. G. Valette, A. Hamelin, *J. Electroanal. Chem.* **1973**, *45*, 301.
70. D. Leikis, K. Rybalka, E. Sevastyanov et al., *J. Electroanal. Chem.* **1973**, *46*, 161.
71. A. N. Frumkin, *J. Res. Inst. Catalysis Hokkaido Univ.* **1967**, *15*, 61.
72. M. A. Vorotyntsev, *J. Electroanal. Chem.* **1981**, *123*, 379.
73. N. B. Grigoryev, *Dokl. Akad. Nauk SSSR* **1976**, *229*, 647.
74. I. A. Bagotskaya, B. B. Damaskin, M. D. Levi, *J. Electroanal. Chem.* **1980**, *115*, 189.
75. V. E. Kazarinov, M. A. Vorotyntsev, M. Ya. Kats in *Double Layer and Adsorption at Solid Electrodes*, Tartu University, 1981, p. 146, Vol. VI.
76. M. A. Vorotyntsev, *Sov. Electrochem.* **1981**, *17*, 472.
77. M. A. Vorotyntsev, *Sov. Electrochem.* **1981**, *17*, 835.
78. M. A. Vorotyntsev, *Sov. Electrochem.* **1981**, *17*, 162.
79. J. Lecoeur, *C. R. Acad. Sci. (Paris)* **1976**, *283C*, 651.
80. A. Hamelin, *Sov. Electrochem.* **1982**, *18*, 1413.
81. L. I. Daikhin, A. A. Kornyshev, M. I. Urbakh, *Phys. Rev. E* **1996**, *53*, 6192.
82. L. I. Daikhin, A. A. Kornyshev, M. I. Urbakh, *Electrochim. Acta* **1997**, *19*, 2853.
83. L. I. Daikhin, A. A. Kornyshev, M. I. Urbakh, *J. Chem. Phys.* **1998**, *108*, 1715.
84. M. Abramovitz, I. Stegun, (Eds.), *Handbook of Mathematical Functions*, Dover, New York, 1965.
85. B. B. Mandelbrot, *The Fractal Geometry of Nature*, Freeman, New York, 1982.
86. E. Lust, A. Jänes, V. Sammelselg et al., *Electrochim. Acta* **1988**, *43*, 373.
87. E. Lust, A. Jänes, V. Sammelselg et al., *Electrochim. Acta* **2000**, *46*, 373.
88. A. A. Kornyshev in *The Chemical Physics of Solvation* (Eds.: R. R. Dogonadze, E. Kalman, A. A. Kornyshev, J. Ulstrup), Elsevier, Amsterdam, 1988, p. 355, Part C Solvation Phenomena in Physical, Chemical and Biological Systems.
89. M. A. Vorotyntsev, A. A. Kornyshev, *Electrostatics of Media with the Spatial Dispersion*, Nauka, Moscow, 1993.
90. M. A. Vorotyntsev, V. Yu. Izotov, A. A. Kornyshev, *Sov. Electrochem.* **1983**, *19*, 364.
91. A. A. Kornyshev in *The Chemical Physics of Solvation* (Eds.: R. R. Dogonadze, E. Kalman, A. A. Kornyshev, J. Ulstrup), Elsevier, Amsterdam, 1985, p. 77, Part A Theory of Solvation.
92. M. A. Vorotyntsev, A. A. Kornyshev, *Sov. Electrochem.* **1979**, *15*, 560.
93. A. A. Kornyshev, M. A. Vorotyntsev, J. Ulstrup, *Thin Solid Films* **1981**, *75*, 105.
94. A. A. Kornyshev, J. Ulstrup, *Chem. Scr.* **1985**, *25*, 58.
95. A. A. Kornyshev, *J. Electroanal. Chem.* **1986**, *204*, 79.
96. A. A. Kornyshev, J. Ulstrup, *J. Electroanal. Chem.* **1985**, *183*, 387.

97. J. D. Cole, *Perturbation Methods in Applied Mathematics*, Plaisdell, Waltham, 1968.
98. R. G. Watts-Tobin, *Philos. Mag.* **1961**, *6*, 133.
99. W. R. Fawcett, *Isr. J. Chem.* **1979**, *18*, 3.
100. N. F. Mott, R. G. Watts-Tobin, *Electrochim. Acta* **1961**, *4*, 79.
101. S. Levine, G. M. Bell, A. L. Smith, *J. Phys. Chem.* **1969**, *73*, 3534.
102. I. L. Cooper, J. A. Harrison, *J. Electroanal. Chem.* **1975**, *66*, 85.
103. W. R. Fawcett, *J. Phys. Chem.* **1978**, *82*, 1385.
104. B. B. Damaskin, A. N. Frumkin, *Electrochim. Acta* **1974**, *19*, 173.
105. R. Parsons, *J. Electroanal. Chem.* **1975**, *59*, 229.
106. B. B. Damaskin, *J. Electroanal. Chem.* **1977**, *75*, 359.
107. A. A. Kornyshev, *Electrochim. Acta* **1989**, *34*, 1829.
108. Ya. I. Frenkel, *Z. Phys.* **1928**, *51*, 232.
109. M. D. Lang, *Solid State Phys.* **1973**, *28*, 225.
110. N. D. Lang, W. Kohn, *Phys. Rev. B* **1971**, *3*, 1215.
111. I. E. Inglesfield in *The Chemical Physics of Solid Surfaces and Heterogeneous Catalysis*, Elsevier, Amsterdam, 1981, p. 183, Vol. 1.
112. J. Harris, A. Liebsch, *Vacuum* **1983**, *33*, 655.
113. A. K. Theophilou, A. Modinos, *Phys. Rev. B* **1972**, *6*, 801.
114. M. B. Partenskii, Ya. G. Smorodinskii, *Fiz. Tverd. Tela* **1974** *16*, 644.
115. P. G. Dzhavakhidze, A. A. Kornyshev, G. I. Tsitsuashvili, *Solid State Commun.* **1984**, *52*, 401.
116. J. R. Smith, *Phys. Rev.* **1969**, *181*, 522.
117. P. Gies, R. Gerhardts, *Phys. Rev. B* **1985**, *31*, 6843.
118. P. Gies, R. Gerhardts, *Phys. Rev. B* **1986**, *33*, 982.
119. W. Kohn, L. J. Sham, *Phys. Rev.* **1965**, *140*, 1133.
120. W. Schmickler, D. Henderson, *Prog. Surf. Sci.* **1986**, *22*, 323.
121. V. I. Feldman, M. B. Partenskii, M. M. Vorobjev, *Prog. Surf. Sci.* **1986**, *23*, 3.
122. J.-P. Badiali, *Ber. Bunsen-Ges. Phys. Chem.* **1987**, *91*, 270.
123. S. Amokrane, J.-P. Badiali in *Modern Aspects of Electrochemistry* (Eds.: B. E. Conway, J.O'M. Bockris, R. E. White), Plenum Press, New York, 1992, p. 22, Vol. 1.
124. J. W. Halley, S. Walbran, D. Price in *Advances in Chemical Physics* (Eds.: I. Prigogine, S. A. Rice), Wiley, New York, 2001, p. 337.
125. M. A. Vorotyntsev, A. A. Kornyshev, *Sov. Electrochem.* **1984**, *20*, 1.
126. A. A. Kornyshev, M. A. Vorotyntsev, *J. Electroanal. Chem.* **1984**, *167*, 1.
127. J. W. Halley, B. Johnson, D. Price et al., *Phys. Rev. B* **1985**, *31*, 7695.
128. V. I. Feldman, A. A. Kornyshev, M. B. Partenskii, *Solid State Commun.* **1985**, *53*, 157.
129. A. V. Schlepakov, I. A. Bagotskaya, *Elektokhimija* **1982**, *18*, 26.
130. L. M. Doubova, I. A. Bagotskaya, *Elektokhimija* **1977**, *13*, 64.
131. E. K. Patjarv, U. V. Palm, *Elektokhimija* **1976**, *12*, 806.
132. R. Payne, *J. Am. Chem. Soc.* **1967**, *89*, 459.
133. G. Vallette, A. Hamelin, *J. Electroanal. Chem.* **1973**, *45*, 301.
134. V. A. Panin, K. V. Rybalka, D. I. Leikis, *Elekrokhimija* **1972**, *8*, 390.
135. D. Price, J. W. Halley, *J. Electroanal. Chem.* **1983**, *150*, 347.
136. M. B. Partenskii, V. Dorman, P. C. Jordan, *Int. Rev. Phys. Chem.* **1996**, *15*, 153.
137. A. M. Brodsky, M. I. Urbakh, *Usp. Fiz. Nauk* **1982**, *138*, 413.
138. A. M. Brodsky, L. I. Daikhin, M. I. Urbakh, *Usp. Khim.* **1985**, *54*, 3.
139. A. M. Brodsky, M. I. Urbakh, *Prog. Surf. Sci.* **1984**, *15*, 121.
140. A. M. Brodsky, M. I. Urbakh, *Electrodynamics of Metal Surfaces*, Nauka, Moscow, 1989.
141. A. A. Kornyshev, M. I. Urbakh, *J. Electroanal. Chem.* **1987**, *235*, 11.
142. J. D.-E. McIntyre in *Advances in Electrochemistry and Electrochemical Engineering* (Eds.: P. Delahay, C. N. Tobias), Wiley, New York, 1973, p. 61, Vol. 3.
143. A. B. Ershler, A. M. Foontikov, V. N. Alexeev et al., *Elektrochimiya* **1980**, *16*, 1171.
144. A. G. Foontikov, Dissertation, A. N. Frumkin Institute of Electrochemistry of the Academy of Sciences of the USSR, Moscow, 1981.
145. A. A. Kornyshev, M. I. Urbakh, *J. Electroanal. Chem.* **1988**, *257*, 305, **1987**, *235*, 11.
146. P. G. Dzhavakhidze, A. A. Kornyshev, A. Liebsch et al., *Phys. Rev. B* **1992**, *45*, 9339.
147. P. G. Dzhavakhidze, A. A. Kornyshev, A. Liebsch et al., *Electrochim. Acta* **1991**, *36*, 1835.

148. P. G. Dzhavakhidze, A. A. Kornyshev, A. Tadjeddine et al., *Phys. Rev. B* **1989**, *39*, 13 106.
149. P. G. Dzhavakhidze, A. A. Kornyshev, A. Tadjeddine et al., *Electrochim. Acta* **1989**, *34*, 1677.
150. A. A. Kornyshev, A. M. Kuznetsov, G. Makov et al., *J. Chem. Soc., Faraday Trans.* **1996**, *92*, 3997.
151. A. A. Kornyshev, A. M. Kuznetsov, G. Makov et al., *J. Chem. Soc., Faraday Trans.* **1996**, *92*, 4005.
152. D. Henderson, M. Lozada-Cassou, *J. Colloid Interface Sci.* **1986**, *114*, 180.
153. L. Blum, F. Vericat, D. Bratko, *J. Chem. Phys.* **1995**, *102*, 1461.
154. J. O' M. Bockris, M. A. Devanathan, K. Müller, *Proc. R. Soc.* **1963**, *A274*, 55.
155. E. Spohr, *J. Phys. Chem.* **1989**, *93*, 6171.
156. J. Böcker, R. R. Nazmutdinov, E. Spohr et al., *Surf. Sci.* **1995**, *335*, 372.
157. J. Böcker, E. Spohr, K. Heinzinger, *Z. Naturforsch.* **1995**, *50a*, 611.
158. T. Kramar, D. Vogtenhuber, R. Podloucky et al., *Electrochim. Acta* **1995**, *40*, 43.
159. W. Langel, M. Parrinello, *Phys. Rev. Lett.* **1994**, *73*, 504.
160. W. Langel, M. Parrinello, *J. Chem. Phys.* **1995**, *103*, 3240.
161. D. Boda, D. Henderson, K.-Y. Chan et al., *Chem. Phys. Lett.* **1999**, *308*, 473.
162. D. Boda, D. Henderson, K.-Y. Chan, *J. Chem. Phys.* **1999**, *110*, 5346.
163. D. Boda, D. Henderson, *J. Chem. Phys.* **2000**, *112*, 8934.
164. D. Boda, K.-Y. Chan, D. Henderson, *J. Chem. Phys.* **1998**, *109*, 7362.
165. R. M. Townsend, J. Gryko, S. A. Rice, *J. Chem. Phys.* **1985**, *82*, 4391.
166. W. Schmickler, E. Leiva, *Mol. Phys.* **1995**, *86*, 737.
167. J. Seitz-Beywl, M. Poxleitner, M. M. Probst et al., *Int. J. Quantum Chem.* **1992**, *42*, 1141.
168. S. Holloway, K. H. Bennemann, *Surf. Sci.* **1980**, *101*, 327.
169. P. A. Bopp, A. Kohlmeyer, E. Spohr, *Electrochim. Acta* **1998**, *43*, 2911.
170. I. Benjamin, *Modern Aspects Electrochem.* **1997**, *31*, 115.
171. E. Spohr in *Advances in Electrochemical Science and Technology* (Eds.: D. M. Kolb, R. Alkire), Wiley-VCH, Weinheim, 1999, pp. 1–75, Vol. 6.
172. L. Bosio, R. Cortes, C. Segaud, *J. Chem. Phys.* **1979**, *71*, 1979.
173. P. A. Thiel, T. E. Madey, *Surf. Sci. Rep.* **1987**, *7*, 211.
174. A. M. Brodsky, M. Watanabe, W. P. Reinhardt, *Electrochim. Acta* **1991**, *36*, 1695.
175. M. Watanabe, A. M. Brodsky, W. P. Reinhardt, *J. Chem. Phys.* **1991**, *95*, 4593.
176. X. Xia, M. L. Berkowitz, *Phys. Rev. Lett.* **1995**, *74*, 3193.
177. K. J. Schweighofer, X. Xia, M. L. Berkowitz, *Langmuir* **1996**, *12*, 3747.
178. R. R. Nazmutdinov, M. M. Probst, K. Heinzinger, *J. Electroanal. Chem.* **1994**, *369*, 227.
179. M. W. Ribarsky, W. D. Luedtke, U. Landman, *Phys. Rev. B* **1985**, *32*, 1430.
180. A. Ignaczak, J. A. N. F. Gomes, *J. Electroanal. Chem.* **1997**, *420*, 209.
181. H. Sellers, P. V. Sudhakar, *J. Chem. Phys.* **1992**, *97*, 6644.
182. H. Yang, J. L. Whitten, *Surf. Sci.* **1989**, *223*, 131.
183. M. Rosi, C. W. Bauschlicher Jr., *J. Chem. Phys.* **1989**, *90*, 7264.
184. S. E. Feller, R. W. Pastor, A. Rojnuckarin et al., *J. Phys. Chem.* **1996**, *100*, 17 011.
185. E. Spohr, *J. Chem. Phys.* **1997**, *107*, 6342.
186. E. Spohr, *Chem. Phys.* **1990**, *87*, 141.
187. K. Foster, K. Raghavan, M. L. Berkowitz, *Chem. Phys. Lett.* **1989**, *162*, 32.
188. K. Raghavan, K. Foster, M. L. Berkowitz, *Chem. Phys. Lett.* **1991**, *177*, 426.
189. K. Raghavan, K. Foster, K. Motakabbir et al., *J. Chem. Phys.* **1991**, *94*, 2110.
190. J. Böcker, Z. N. Gurskii, K. Heinzinger, *J. Phys. Chem.* **1996**, *100*, 14 969.
191. J. Böcker, R. R. Nazmutdinov, E. Spohr et al., *Surf. Sci.* **1995**, *335*, 372.
192. J. Böcker, E. Spohr, K. Heinzinger, *Z. Naturforsch.* **1995**, *50a*, 611.
193. M. L. Berkowitz, I.-C. Yeh, E. Spohr in *Interfacial Electrochemistry* (Ed.: A. Wieckowski), Marcel Dekker, New York, 1999.
194. J. Hautman, J. W. Halley, Y.-J. Rhee, *J. Chem. Phys.* **1989**, *91*, 467.
195. S. B. Zhu, G. W. Robinson, *J. Chem. Phys.* **1991**, *94*, 1403.
196. G. Aloisi, M. L. Foresti, R. Guidelli et al., *J. Chem. Phys.* **1989**, *91*, 5592.
197. G. Aloisi, R. Guidelli, *J. Chem. Phys.* **1991**, *95*, 3679.
198. A. A. Gardner, J. P. Valleau, *J. Chem. Phys.* **1987**, *86*, 4171.

199. G. Nagy, K. Heinzinger, *J. Electroanal. Chem.* **1990**, *296*, 549.
200. G. Nagy, K. Heinzinger, E. Spohr, *Faraday Discuss.* **1992**, *94*, 307.
201. M. F. Toney, J. N. Howard, J. Richter et al., *Nature* **1994**, *368*, 444.
202. D. A. Rose, I. Benjamin, *J. Chem. Phys.* **1991**, *95*, 6956.
203. D. A. Rose, I. Benjamin, *J. Chem. Phys.* **1993**, *98*, 2283.
204. E. Spohr, *Chem. Phys. Lett.* **1993**, *207*, 214.
205. L. Perera, M. L. Berkowitz, *J. Phys. Chem.* **1993**, *97*, 13 803.
206. E. Spohr, *Acta Chem. Scand.* **1995**, *49*, 189.
207. J. N. Glosli, M. R. Philpott, *J. Chem. Phys.* **1992**, *96*, 6962.
208. J. N. Glosli, M. R. Philpott, *J. Chem. Phys.* **1993**, *98*, 9995.
209. J. Seitz-Beyl, M. Poxleitner, K. Heinzinger, *Z. Naturforsch.* **1991**, *46a*, 876.
210. B. Eck, E. Spohr, *Electrochim. Acta* **1997**, *42*, 2779.
211. G. Toth, K. Heinzinger, *Chem. Phys. Lett.* **1995**, *245*, 48.
212. M. R. Philpott, J. N. Glosli, *J. Electrochem. Soc.* **1995**, *142*, L25.
213. M. R. Philpott, J. N. Glosli, S. B. Zhu, *Surf. Sci.* **1995**, *335*, 422.
214. E. Spohr, *J. Electroanal. Chem.* **1998**, *450*, 327.
215. E. Spohr, *Electrochim. Acta* **1999**, *44*, 1697.
216. D. I. Dimitrov, N. D. Raev, *J. Electroanal. Chem.* **2000**, *486*, 1.
217. D. A. Rose, I. Benjamin, *J. Chem. Phys.* **1994**, *100*, 3545.
218. X. Xia, M. L. Berkowitz, *Chem. Phys. Lett.* **1994**, *227*, 561.
219. J. B. Straus, G. A. Voth, *J. Phys. Chem.* **1993**, *97*, 7388.
220. J. B. Straus, A. Calhoun, G. A. Voth, *J. Chem. Phys.* **1995**, *102*, 529.
221. B. B. Smith, J. W. Halley, *J. Chem. Phys.* **1994**, *101*, 10 915.
222. D. A. Rose, I. Benjamin, *Chem. Phys. Lett.* **1995**, *234*, 209.
223. B. B. Smith, J. W. Halley, A. J. Nozik, *Chem. Phys.* **1996**, *205*, 245.
224. A. Calhoun, G. A. Voth, *J. Phys. Chem.* **1996**, *100*, 10 746.
225. Y. G. Boroda, A. Calhoun, G. A. Voth, *J. Chem. Phys.* **1997**, *107*, 8940.
226. I. Benjamin, *J. Phys. Chem.* **1991**, *95*, 6675.
227. B. B. Smith, J. T. Hynes, *J. Chem. Phys.* **1993**, *99*, 6517.
228. B. B. Smith, A. J. Nozik, *J. Phys. Chem.* **1997**, *101B*, 2459.
229. I. Benjamin, *J. Chem. Phys.* **1992**, *96*, 577.
230. I. Benjamin, *Science* **1993**, *95*, 6675.
231. O. Pecina, W. Schmickler, E. Spohr, *J. Electroanal. Chem.* **1995**, *394*, 29.
232. O. Pecina, W. Schmickler, E. Spohr, *J. Electroanal. Chem.* **1995**, *405*, 239.
233. O. Pecina, W. Schmickler, *J. Electroanal. Chem.* **1997**, *431*, 47.
234. A. Calhoun, M. T. M. Koper, G. A. Voth, *Chem. Phys. Lett.* **1999**, *305*, 94.
235. A. Calhoun, M. T. M. Koper, G. A. Voth, *J. Phys. Chem.* **1999**, *103B*, 3442.
236. I. Benjamin, *Chem. Rev.* **1996**, *96*, 1449.
237. I. Benjamin, *Annu. Rev. Phys. Chem.* **1997**, *48*, 407.
238. H. D. Hurwitz, *J. Electroanal. Chem.* **1965**, *10*, 35.
239. E. Dutkiewics, R. Parsons, *J. Electroanal. Chem.* **1966**, *11*, 100.
240. R. Payne, *Advances in Electrochemistry and Electrochemical Engineering*, Intersci. Publ., New York, 1970, p. 1, Vol. 7.
241. R. Payne, *J. Electroanal. Chem.* **1973**, *41*, 277.
242. B. B. Damaskin, R. V. Ivanova, *Usp. Khim.* **1979**, *48*, 1747.
243. R. Parsons, *J. Electroanal. Chem.* **1983**, *150*, 51.
244. A. Hamelin, T. Vitanov, A. Popov et al., *J. Electroanal. Chem.* **1983**, *145*, 225.
245. M. D. Levi, A. V. Shlepakov, B. B. Damaskin et al., *J. Electroanal. Chem.* **1982**, *138*, 1.
246. G. Valette, A. Hamelin, R. Parsons, *Z. Phys. Chem. N. F.* **1978**, *113*, 71.
247. O. Stern, *Z. Electrochem.* **1924**, *30*, 508.
248. O. Essin, B. Markov, *Acta Physicochim. URSS* **1939**, *10*, 353.
249. B. V. Ershler, *Zh. Fiz. Khim.* **1946**, *20*, 679.
250. D. C. Grahame, *J. Am. Chem. Soc.* **1958**, *80*, 4201.
251. J. M. Parry, R. Parsons, *Trans. Faraday Soc.* **1963**, *59*, 241.
252. C. V. D'Alkaine, E. R. Gonsalez, R. Parsons, *J. Electroanal. Chem.* **1971**, *32*, 57.
253. Ya. M. Kolotyrkin, Yu. V. Alexeyev, Yu. A. Popov, *J. Electroanal. Chem.* **1975**, *62*, 135.
254. M. A. Vorotyntsev, *Advances in Science & Technology*, VINITI, Moscow, 1988, p. 3, Vol. 26.
255. M. A. Vorotyntsev in *The Chemical Physics of Solvation* (Eds.: R. R. Dogonadze, E. Kalman, A. A. Kornyshev, J. Ulstrup), Elsevier, Amsterdam, 1988, p. 401, Part C

256. M. A. Vorotyntsev, K. Holub, *Sov. Electrochem.* **1984**, *20*, 256.
257. M. A. Vorotyntsev, E. M. Itskovich, *Sov. Electrochem.* **1979**, *15*, 114.
258. M. A. Vorotyntsev, *Sov. Electrochem.* **1978**, *14*, 781.
259. M. A. Vorotyntsev, *Sov. Electrochem.* **1978**, *14*, 783.
260. M. A. Vorotyntsev, V. S. Krylov, L. I. Krishtalik, *Sov. Electrochem.* **1979**, *15*, 636.
261. M. A. Vorotyntsev, *J. Res. Inst. Catalysis Hokkaido Univ.* **1982**, *30*, 167.
262. M. A. Vorotyntsev, *Sov. Electrochem.* **1980**, *16*, 1112.
263. M. A. Vorotyntsev, *Sov. Phys. Surf.: Phys., Chem., Mech.* **1989**, 39.
264. L. D. Landau, E. M. Lifshitz, L. P. Pitaevskii, *Statistical Physics*, Pergamon Press, New York, 1980.
265. A. Z. Patashinskii, V. L. Pokrovskii, *Fluctuation Theory of Phase Transitions*, Pergamon Press, Oxford, 1979.
266. M. A. Vorotyntsev, S. N. Ivanov, *Sov. Phys. JETP* **1985**, *61*, 1028.
267. M. A. Vorotyntsev in *Condensed Matter Physics Aspects of Electrochemistry* (Eds.: M. P. Tosi, A. A. Kornyshev), World Scientific, Singapore, 1991, p. 229.
268. M. A. Vorotyntsev, *Sov. Electrochem.* **1985**, *21*, 257.
269. M. A. Vorotyntsev, *Sov. Electrochem.* **1985**, *21*, 323.
270. M. A. Vorotyntsev, *Sov. Electrochem.* **1985**, *21*, 470.
271. A. V. Shcheglov, M. A. Vorotyntsev, B. B. Damaskin et al., *Sov. Electrochem.* **1985**, *21*, 1061.
272. A. A. Kornyshev, I. Vilfan, *Phys. Rev. B* **1993**, *47*, 10775.
273. A. A. Kornyshev, I. Vilfan, *Electrochim. Acta* **1995**, *40*, 109.
274. A. A. Kornyshev, I. Vilfan in *Microscopic Models of Electrode-Electrolyte Interfaces*, Electrochem. Soc. Proceedings Pennington, NJ, 1993 (Eds.: J. W. Halley, L. Blum), p. 205, Vol. 93–5.
275. P. Fery, W. Moritz, D. Wolf, *Phys. Rev. B* **1988**, *38*, 7275.
276. D. M. Kolb in *Structure of Electrified Interfaces* (Eds.: J. Lipkowski, Ph. Ross), 1993, VCH, New York, p. 65.
277. X. Gao, A. Hamelin, M. J. Weaver, *Phys. Rev. B* **1991**, *44*, 10983.
278. A. Hamelin, *J. Electroanal. Chem.* **1992**, *329*, 3350.
279. O. M. Magnussen, J. Wiechers, R. J. Behm, *Surf. Sci.* **1993**, *289*, 139.
280. B. M. Ocko, G. Hegelson, B. Schardt et al., *Phys. Rev. Lett.* **1992**, *69*, 3350.
281. V. Heine, L. D. Marks, *Surf. Sci.* **1986**, *165*, 65.
282. K.-M. Ho, K.-P. Bohnen, *Phys. Rev. Lett.* **1987**, *59*, 1833.
283. C. L. Fu, K. M. Ho, *Phys. Rev. Lett.* **1989**, *63*, 1617.
284. K.-P. Bohnen, K.-M. Ho, *Electrochim. Acta* **1995**, *40*, 129.
285. B. M. McCoy, T. T. Wu, *The Two-Dimensional Ising Model*, Harvard University Press, Cambridge, 1973.
286. I. Vilfan, J. Villain, *Surf. Sci.* **1991**, *257*, 368.
287. G. Pourcelli, C. Gavach in *Proton Conductors* (Ed.: Ph. Colomban), Cambridge University Press, Cambridge, 1992, p. 294.
288. M. Daoud, C. E. Williams, (Eds.), *Soft Matter Physics*, Springer, Berlin, 1999.
289. H. H. J. Girault, D. H. Schiffrin in *Electroanalytical Chemistry* (Ed.: A. J. Bard), Marcel Dekker, New York, 1989, p. 1, Vol. 15.
290. A. G. Volkov, D. W. Deamer, D. I. Tanelian et al., *Liquid Interfaces in Chemistry and Biology*, Wiley, New York, 1989.
291. C. A. Croxton, *Statistical Mechanics of the Liquid Surface*, Wiley, New York, 1980.
292. M. P. Gelfand, M. E. Fisher, *Physica A* **1990**, *166*, 1.
293. F. Buff, R. Lovett, F. Stillinger, *Phys. Rev. Lett.* **1965**, *15*, 621.
294. I. Benjamin in *Modern Methods for Multidimensional Dynamics Computations in Chemistry* (Ed.: D. L. Thompson), World Scientific, Singapore, 1998, p. 101.
295. L. I. Daikin, A. A. Kornyshev, M. Urbakh, *J. Electroanal. Chem.* **2000**, *483*, 68.
296. L. I. Daikin, A. A. Kornyshev, M. Urbakh, *J. Electroanal. Chem.* **2001**, *500*, 461.
297. D. Beaglehole, *Phys. Rev. Lett.* **1987**, *58*, 1434.
298. L. T. Lee, D. Langevin, J. Meunier et al., *Prog. Colloid Polym. Sci.* **1990**, *81*, 209.
299. L. T. Lee, D. Langevin, B. Farnoux, *Phys. Rev. Lett.* **1991**, *67*, 2678.
300. K. J. Schweighofer, I. Benjamin, *J. Electroanal. Chem.* **1995**, *391*, 1.

301. S. A. Safran, *Statistical Thermodynamics of Surfaces. Interfaces and Membranes*, Addison-Wesley, Reading, 1994.
302. D. Beaglehole, *Phys. Rev. Lett.* **1987**, *58*, 1434.
303. J. D. Weeks, *J. Chem. Phys.* **1977**, *67*, 3106.
304. R. Evans, *Adv. Phys.* **1979**, *28*, 143.
305. B. M. Ocko, X. Z. Wu, E. B. Sirota et al., *Phys. Rev. Lett.* **1994**, *2*, 242.
306. Z.-H. Zhang, I. Tsuyumoto, S. Takashi et al., *J. Phys. Chem. A* **1997**, *101*, 4163.
307. Z.-H. Zhang, I. Tsuyumoto, T. Kitamori et al., *J. Phys. Chem. B* **1998**, *102*, 10 284.
308. L. I. Daikin, A. A. Kornyshev, M. Urbakh, *Chem. Phys. Lett.* **1999**, *309*, 137.
309. L. I. Daikin, A. A. Kornyshev, M. Urbakh, *Electrochim. Acta* **1999**, *45*, 685.
310. L. D. Landau, E. M. Lifshitz, *Electrodynamics of Continuous Media*, Pergamon Press, New York, 1984.
311. H. H. Girault in *Modern Aspects of Electrochemistry* (Eds.: J.O'M. Bockris et al.), Plenum Press, New York, 1993, p. 1, Vol. 25.
312. Z. Samec, *Chem. Rev.* **1988**, *88*, 617.
313. C. M. Pereira, A. Martins, M. Rochas et al., *J. Chem. Soc., Faraday Trans.* **1994**, *90*, 143.
314. H. H. Girault, D. J. Schiffrin, *J. Electroanal. Chem.* **1983**, *150*, 43.
315. D. Homolka, P. Hajkova, V. Marecek et al., *J. Electroanal. Chem.* **1983**, *159*, 233.
316. C. Yufei, V. J. Cunnane, L. Murtomaki et al., *J. Chem. Soc. Faraday Trans.* **1991**, *87*, 107.
317. C. M. Pereira, W. Schmickler, F. Silva et al., *Chem. Phys. Lett.* **1997**, *268*, 13.
318. C. M. Pereira, W. Schmickler, F. Silva et al., *J. Electroanal. Chem.* **1997**, *436*, 9.
319. T. Huber, O. Pecina, W. Schmickler, *J. Electroanal. Chem.* **1999**, *467*, 203.
320. J. W. Halley, S. Walbran, D. L. Price, *Adv. Chem. Phys.* **2000**, *116*, 337.

2.2
Electrical Double Layers: Theory and Simulations

Wolfgang Schmickler
Abteilung Elektrochemie, University of Ulm, Ulm, Germany

2.2.1
Introduction

Electrical double layers exist in many systems. They form whenever two conducting phases meet at an interface that is impermeable to the charge carriers. Then the application of a potential difference between the two phases leads to the buildup of charges of equal magnitude and opposite sign at the interface. The resulting charge distribution is the same as in a capacitor, and therefore much research has been focused on the resulting capacity. Such double layers may also form in systems in which charge carriers can cross the interface, for example, at the contact between two metals with different work functions. In this case, an external voltage gives rise to a current across the interface, but often when the current is small and the concentration of charge carriers is high, the double layer remains undisturbed. Finally, double layers even exist at an interface in which one of the two adjoining phases does not conduct current, but carries an intrinsic immobile excess charge on its surface – a situation that occurs in membranes and in biological systems. This excess charge is then compensated by a charge layer in the conducting system.

A general review of double layer theory covering all these cases would certainly be desirable, but is beyond the scope of this chapter. Here we focus on electrochemical systems, and even within this realm we have to restrict ourselves to two important cases: metal–solution and liquid–liquid interfaces. Of course, semiconductor–solution interfaces are just as important in electrochemistry, but they are of little interest to electrochemical theory: their double-layer properties are dominated by the space-charge region of the semiconductor, which are covered in textbooks on semiconductor physics. Therefore, in the remainder of this chapter we shall treat the above-mentioned two systems in turn, and cover both theory and simulations, as suggested by the title. We shall assume throughout that there is no specific adsorption of ions at the interface. While it is doubtful if there is any system for which this assumption is strictly valid, specific adsorption is different for each system, so there is no point in including it in a general theory. So the models presented in this chapter form the basis of double-layer theory, to which the effects of specific adsorption must be added for each individual system.

2.2.2
Gouy–Chapman Theory and its Extensions

Statistical double-layer theory started at the beginning of the last century with the works by Gouy [1] and Chapman [2], two of the most influential papers ever written in electrochemistry. Their theory, which we will briefly review below, explains the double-layer capacity for metal–solution interfaces at low electrolyte concentrations quite well. Unfortunately, further progress has been slow, and during the last decade, there has been more work in simulations of the double layer than in proper theory.

The Gouy–Chapman (GC) theory is treated in all good textbooks, so we only summarize the main points. It considers a planar metal electrode in contact with an ionic solution, and starts from the following assumptions:

- the metal is a perfect conductor, and its excess charge is distributed evenly on the surface;
- the solvent is a dielectric continuum, characterized by a dielectric constant ε;
- the ions are point particles, whose distribution is determined by the Poisson–Boltzmann equation.

For the case of a z–z electrolyte, explicit expressions for the differential capacity and for the profile of the electrostatic potential can be derived. We give the expression for differential capacity C per unit area:

$$C_{GC} = \frac{d\sigma_M}{d\phi} = \frac{\varepsilon\varepsilon_0}{L_D} \cosh \frac{ze_0(\phi - \phi_{pzc})}{2kT} \quad (1)$$

where ϕ is the electrode potential and ϕ_{pzc} is the *potential of zero charge* (pzc), at which the charge density σ_M on the electrode vanishes. n_0 is the concentration of the ions in the bulk of the solution, k_B is the Boltzmann constant, and T is the temperature. The Debye length L_D, which is familiar from the Debye–Hückel theory for electrolyte solutions, is given by

$$L_D = \left(\frac{\varepsilon\varepsilon_0 kT}{2(ze_0)^2 n_0}\right)^{1/2} \quad (2)$$

The GC capacity has a pronounced minimum at the *pzc*, and rises rapidly on both sides (see Fig. 1). It compares well with experimental data for low electrolyte concentrations and for low to medium charge densities, where the structure of the system, the size and chemical nature of the solution, and the electronic properties of the metal, do not matter.

Considering the simplicity of the GC theory, it works surprisingly well – so well that it still serves as the reference point for all further advances in double layer theory, and new results are often stated as corrections to the GC theory.

A cursory review of the experimental data shows that the GC theory overestimates the capacity for high electrolyte

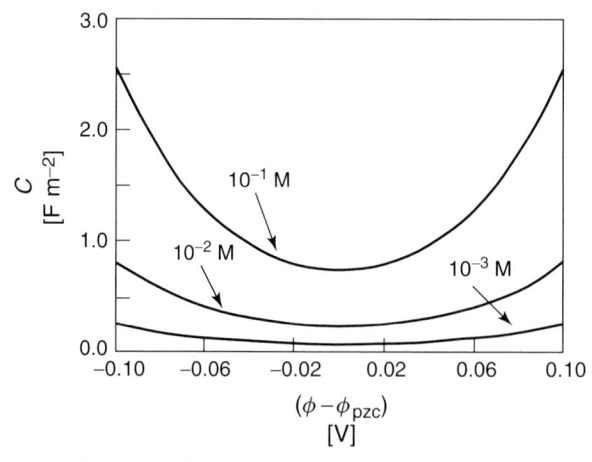

Fig. 1 The Gouy–Chapman capacity for several ionic concentrations of a 1:1 electrolyte.

concentrations and for high charge densities. In a real solution, the ions and the solvent molecules have a finite size, which prevents their centers from coming arbitrarily close to the metal surface, and which also limits the amount of charge that can be packed into the first few layers. To account for these effects, Stern [3] suggested that the first layer of the solution is free of ions, and that this *inner* or *compact layer* contributes a term to the capacity, which is in series with the GC term:

$$\frac{1}{C} = \frac{1}{C_H} + \frac{1}{C_{GC}} \quad (3)$$

where the *Helmholtz capacity* C_H is attributed to the inner layer; C_H is sometimes called the *inner layer* or *compact layer* capacity, but since Eq. (3) also holds in models that do not assume the existence of an inner layer, we shall not use that terminology. In the absence of specific adsorption, the Helmholtz capacity is expected to be independent of the electrolyte concentration, and also of the nature of the ions. Experimentally it can be obtained by measuring the capacity, at a given charge density, for several electrolyte concentrations, and plotting the inverse of the experimental capacity versus the calculated GC capacity; this is known as a Parsons and Zobel plot [4]. This should result in a straight line, and the Helmholtz capacity can be obtained from the intercept. For some metals, in particular for mercury, this procedure works well, while for others, such as gold [5], the plot is curved, which may indicate that there is always at least weak specific adsorption on these metals; of course, it may also indicate that Eq. (3) does not hold for these systems.

The Helmholtz capacity always depends strongly on the charge density σ_M; different metals may have quite different capacities, but typically there is a maximum, also known as *the hump*, near the *pzc* (see Fig. 2). Since the strength of the electric field increases on both sides of the *pzc*, the occurrence of the hump can be explained by dielectric saturation. Watts-Tobin [6] elaborated this idea into

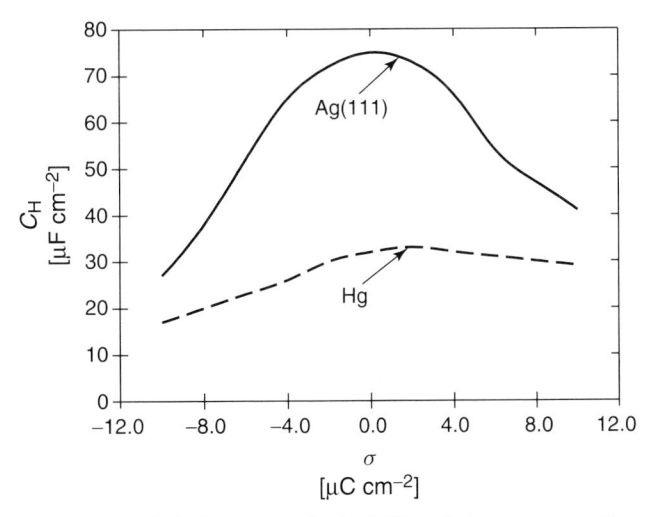

Fig. 2 The Helmholtz capacity for Ag(111) and Hg in contact with an aqueous solution; after Ref. [77] with permission from the author.

a quantitative model: he treated the inner layer as a two-dimensional lattice of dipoles, which could be oriented either parallel or antiparallel to the surface normal. In effect, this is equivalent to the Ising model, which was evaluated in the mean field approximation (MFA). As expected, this model predicts a capacity hump at the *pzc*, but otherwise it does not agree well with the Helmholtz capacity of mercury, on which the early theories were focused, or of any other metal. In mercury, the Helmholtz capacity rises again at charge densities above the hump. This increase was explained [7] by the presence of two water species in the inner layer: single water molecules and water clusters composed of two molecules, which have a smaller dipole moment. A high electric field can break up the clusters, and the capacity increases again since the average dipole moment becomes larger. Parsons [8] worked out the statistical mechanics of this four-state model (two orientations for each water species), and by an appropriate choice of the system parameters he obtained reasonable agreement with the data for mercury. Later this model was elaborated by other researchers, who added further details and increased the number of adjustable parameters, so that the mercury data could be fitted perfectly; such work has been critically reviewed by Guidelli [9]. However, if this kind of model were to represent the data for other metals, all parameters would have to be fitted again, so the physical significance of these models was dubious from the beginning.

Modern theories have abandoned the idea that the water dipoles can take up only two orientations, though it occasionally still crops up in the literature. A significant conceptual progress was achieved in the work of Guidelli [10, 11]. He modeled the inner layer as a two-dimensional lattice, in which each point is occupied by one water molecule, which can form hydrogen bonds with its nearest neighbors and take up a fairly large number (25) of orientations. The statistical mechanics of this ensemble was treated in the quasi-chemical approximation, which is much better than the mean-field approximation that was employed in the Ising-like models discussed above.

However, all these early monolayer models suffer from two defects: the connection of the inner layer to the rest of the solution is not clear and the electronic structure of the metal is neglected – it just acts as a perfect conductor. Nowadays, it is generally recognized that there are three contributions to the double-layer capacity on metals:

1. the capacity of the diffuse layer, which is described by the GC theory;
2. a contribution from the boundary layer of the solution, which extends over a distance of several solvent diameters toward the bulk;
3. the electronic response of the metal surface to the high field at the interface.

A treatment of these effects requires more advanced methods than the GC theory does; therefore, we shall briefly review the methods that have been used in the double-layer theory before discussing the latter two contributions in greater detail.

2.2.3
Theoretical Methods and Principles of Computer Simulations

Double-layer theory is complicated by the fact that it involves the theory of liquids and of solids, and of the interactions between them. While the theory of the bulk of these phases is quite well developed, the treatment of boundary effects is notoriously difficult, and the interaction between the two phases is almost intractable. At present, we understand the basic features of the double layer, but quantitative calculations, even for simple systems, are beyond our capabilities.

Below we briefly review the methods that have been used in double-layer studies; they comprise both theoretical methods and simulations. The latter have become quite popular in recent years, since they allow the treatment of more complicated models than can be handled by pure theory.

2.2.3.1 Integral Equations

The structure of a bulk fluid can be characterized by the *pair distribution functions* $g_{ij}(1, 2)$, which give the probability of finding a particle of species i at the position $\mathbf{r}_1$ with orientational coordinates Ω_1 and a particle of species j at the position $\mathbf{r}_2$ with orientational coordinates Ω_2, while all other particles are distributed statistically. The arguments $(1, 2)$ are shorthand for $\mathbf{r}_1, \Omega_1, \mathbf{r}_2, \Omega_2$. These distribution functions are normalized such that they tend to unity as the separation between the two particles goes to infinity. The quantity $h_{ij}(1, 2) = g_{ij}(1, 2) - 1$ is called the *total correlation function*; it vanishes for infinite separation and characterizes the interaction between the two particles.

In principle, these correlation functions should be calculated from the interaction potentials $v_{ij}(1, 2)$ between the particles. However, even for an ensemble of hard spheres, this is an intractable problem, let alone for systems with realistic interactions. Therefore various schemes have been devised to calculate the correlation functions from approximate relations between these functions. For this purpose, the Ornstein–Zernike relation is often used as a starting point. It is based on the following idea: the correlations between two particles can be decomposed into two sources, that is, the direct interaction between the particles and the interaction mediated by other particles in the vicinity. Therefore a *direct correlation function* $c_{ij}(1, 2)$ is introduced through

$$h_{ij}(1, 2) = c_{ij}(1, 2) + \sum_k \frac{\rho_k}{\omega}$$
$$\times \int c_{ik}(1, 3) h_{ik}(3, 2) \, d3 \quad (4)$$

where ρ_k is the density of the species k and ω is a normalization factor, which is unity for most systems of interest. So far, Eq. (4) is but a definition for the direct correlation $c_{ij}(1, 2)$. However, the range of the direct correlation is of the same order of magnitude as that of the interaction potentials $v_{ij}(1, 2)$ and shorter than that of the total correlation function, which contains the mediated interactions. This fact has been used as a basis for approximations based on diagrammatic expansion schemes. They start from the exact expansion:

$$c(1, 2) = h(1, 2) - \beta v(1, 2)$$
$$- \log[h(1, 2) + 1] + d(1, 2) \quad (5)$$

where the subscripts are omitted to simplify notations. Here, $d(1, 2)$ is the sum of all *bridge diagrams* in the diagrammatic expansion and $\beta \equiv 1/kT$. Unfortunately, the combination of Eqns. (4) and (5) is intractable because it involves an infinite series of many-dimensional integrals of products of $h(1, 2)$. Neglecting the bridge functions results in the *hypernetted chain approximation* (HNC). This is an example of a *closure relation*, because together with Eq. (4) it forms a closed system of equations for the unknown functions $c_{ij}(1, 2)$ and $h_{ij}(1, 2)$, which can be solved at least in principle. In the special case in which the interaction is small, $\beta v_{ij}(1, 2) < 1$, the total correlation functions h_{ij} are also small, and the HNCs simplify to the *mean spherical approximation* (MSA):

$$c_{ij}(1, 2) = -\beta v_{ij}(1, 2) \qquad (6)$$

which results in a linear theory. This is a useful approximation for weakly interacting particles in the vicinity of the *pzc*, where the Coulomb forces are weak.

A better approximation, which has become quite popular recently, is the *reference hypernetted chain* (RHNC) approximation. It makes use of a *reference system* consisting of a fluid of neutral hard spheres, which is the only potential for which the *bridge function* $d(1, 2)$ in Eq. (2) is known with sufficient accuracy. If we denote by the subscript 0 a property of the reference system and set $h = h_0 + \Delta h$, $c = c_0 + \Delta c$, $v = v_0 + \Delta v$, and $d = d_0 + \Delta d \approx d_0$, Eq. (5) can be written as

$$\Delta c(1, 2) = \Delta h(1, 2)$$
$$- \log[g_0 + \Delta h(1, 2)]$$
$$+ \log g_0 - \beta \Delta v(1, 2) \qquad (7)$$

where $g_0 \equiv h_0 + 1$ is only a function of the separation $r \equiv |\mathbf{r}_1 - \mathbf{r}_2|$ between the two particles. The pair-interaction potential v_0 of the reference system is just the spherically symmetrical part of $v(1, 2)$. Equation (7) is exact except for the replacement of the true bridge function $d(1, 2)$ of the system under study by the bridge function $d_0(1, 2)$ of the reference system. The RHNC has been applied to several simple models of water [12, 13]; the results are all numerical.

The MSA, HNC, RHNC, and similar schemes were first developed to treat bulk fluids, and then applied to interfaces by using the following trick: one spherical particle from the ensemble is singled out and its radius is taken to be infinitely large or, if that limit is not tractable numerically, much larger than the radii of all other molecules. This large particle represents the surface.

2.2.3.2 Computer Simulations

Even comparatively simple models turn out to be difficult to treat by the methods of statistical mechanics. In such cases, computer simulations may be used to explore the consequences and predictions of the model, even though they do not give the same detailed understanding as analytical methods do. This is not the place to describe these methods in any detail; the book by Allen and Tildesley [14] provides an excellent introduction to the field. There are three different methods.

Monte Carlo simulations The basic method works in the following way: starting from a given configuration, a new configuration is tentatively chosen at random from a certain region of phase space, for example, by moving one particle a certain distance. If the energy of the new configuration is lower than that of the previous one, it is accepted, and a new Monte Carlo (MC) step is initiated. Otherwise a random number r is chosen from the interval (0, 1). If $r < \exp -\Delta E/kT$, where ΔE is the change in energy, the new configuration is accepted, otherwise it is discarded. It can be shown that averaging over a large number of configurations that have been produced in this way leads to correct thermodynamic quantities [15]. There are various variants of this method; they mainly provide information about the structure and thermodynamics of an ensemble, but a few have been designed to investigate the dynamics as well.

Molecular dynamics The orbits of the particles in an ensemble are obtained from Newton's second law: at each step the forces on all particles, and hence their accelerations, are calculated, and from these their movements during a small time step, in which the forces can be taken as constant, are obtained. Then the forces are calculated again, and the system moves another step forward. During an initialization period, the system must be equilibrized to the desired temperature, and then it can be sampled to obtain statistical properties. Obviously, this method gives both the dynamics and the thermodynamics of the ensemble.

Car–Parinello method In ordinary simulations the electronic structure of the particles is taken as fixed, and the molecules interact through given potentials. In the Car–Parinello method [16], at each step of the simulation the electronic structure of the system is recalculated, usually through density-functional methods [17, 18], and from these the forces on the particles are obtained. In principle, this is the most exact but also the most time-consuming method, so that only small ensembles can be considered. Sometimes this method is simplified by restricting the electronic structure calculations to a part of the ensemble.

All of these methods have been applied to electrochemical systems, and we shall bring them up again as we discuss specific models.

Real systems contain a practically infinite number of particles, while the ensembles that are used in simulations contain typically a few hundred, at best a few thousand, particles. In order to mimic an infinite system, cyclic boundary conditions can be imposed. In electrochemical investigations they are usually imposed in the two directions parallel to the electrode surface. In the perpendicular direction the dimensions should be so large that a region with bulklike properties is obtained, so that bulk and surface properties can be contrasted.

2.2.4
Electronic Structure Calculations

In spite of the explosion of computing power during the last decade, it is still impossible to calculate the electronic structure of semi-infinite solids. Three different approaches have been taken to deal with this problem.

Simple model systems A good method in such situations is a simplification of the model so as to make it mathematically tractable. The art of the theorist consists in keeping the most relevant features of the system and discarding the others. Since in the electric double layer the electronic properties of the metal surface play the major role, the jellium model [18], which disregards the details of the atomic structure and focuses on the distribution of the electrons, has been widely applied. While this model does not give quantitative results for particular systems, it has greatly improved our understanding of the electronic effects; we will discuss it later.

Slab calculations At a perfect surface the atomic structure of a solid is periodic in the direction parallel to the surface. This symmetry makes it possible to calculate the electronic structure of slabs that are infinite in two directions and consist of a few layers of atoms. Typically, about 10 layers are sufficient to obtain a region with bulklike properties in the center and two surface regions. By their very nature such calculations are limited to structures that are periodic parallel to the surface; the unit cell must not be too large – with present-day computers, it must not contain more than a few atoms – or the calculations become computationally too expensive.

Cluster calculations The conceptually simplest method consists in replacing the infinite system by a large cluster of particles. Such calculations can be performed with several packages, from both commercial and public domain, and have therefore become quite popular. However, even in relatively big clusters a large fraction of the atoms lie at the surface. Therefore the results of such calculations have to be interpreted with care – particularly, if the system is charged or has an uneven distribution of charges, since charges tend to accumulate on surfaces. In addition, the results obtained often depend on the size of the cluster, so that one has to focus on relative rather than on absolute values.

Nowadays such quantum calculations are usually based on *density functional formalism* [17], in which the electronic density, and not the wave functions, is used as the basic variable. This formalism results in much faster programs, so that larger systems can be treated than that with the conventional approaches.

2.2.5
Models for the Solution

The surface properties of a phase generally differ from the bulk properties. Therefore the GC theory, which is based on macroscopic bulk concepts such as the dielectric constant, requires corrections. Indeed, this was the idea behind the monolayer models discussed earlier – their main fault was that they were too simple: boundary layers extend beyond the first layer of molecules that is in contact with the metal. Furthermore, these models completely neglected the effect of the metal, which was treated as a perfect conductor without any electronic structure. In this section we will discuss a few models for the solution before we turn to the metal surface and then to the metal–solution interface.

2.2.5.1 Field-theoretical Approach to the Double Layer

The GC theory is based on macroscopic electrostatics, while much of the present-day work is aimed at a description on a molecular level. There is, however, an elegant alternative based on the methods of statistical field theory, which was developed by Badiali and associates [19, 20]. This leads to a coarse-grained description in terms of the particle-distribution functions. By setting up a suitable model Hamiltonian, a wide variety of phenomena can be described.

Badiali and coworkers started by applying this method to a combination of the primitive model, in which the ions are considered as point charges and the solution is treated as a dielectric continuum, and a simple model for the metal in which the charge distribution is taken as uniform in the direction parallel to the surface. In this model the Hamiltonian depends only on the charge distributions $e_0 q(r)$ in the double layer and $e_0 q_0(z)$ in the metal, and can be written in the form

$$\beta H[q(r)] = \int dr\, dr' \frac{\beta e_0^2 [q_0(z) + q(r)][q_0(z') + q(r')]}{8\pi \varepsilon |r - r'|}$$
$$+ \sum_{n=2}^{\infty} \frac{a_n}{\rho_b n!} \int [q(r)]^n \, dr \quad (8)$$

where $\beta = 1/k_B T$, and ρ_b is the bulk concentration of the ions. The first term is the electrostatic energy in the systems, and all other terms, in particular structural and entropic terms, have been written as an expansion in powers of $q(r)$. The coefficients a_n are dimensionless quantities.

The simplest way to treat this Hamiltonian is the MFA. The statistical average $\langle q(z) \rangle$ is then the value of $q(r)$ for which the functional derivative $\delta H[q(r)]/\delta q(r)$ vanishes. For the Hamiltonian of Eq. (8) this gives

$$\beta V(z) + \sum_{n=1}^{\infty} \frac{a_n}{\rho_b} \langle q(z)^{(n-1)} \rangle = 0 \quad (9)$$

where $V(z)$ is the average electrostatic potential. If one assumes that the average charge distribution is related to the average potential via a Boltzmann distribution and that the metal is a classical conductor (i.e. $q_0(z) \propto \delta(z)$), then all the coefficients a_n can be determined, and one obtains the nonlinear GC theory.

So far, this is nothing but an ingenious method for obtaining a well-known result. However, this approach can be extended in various ways by including other effects. Stafiej and coworkers discuss explicitly ion–ion interactions, chemical interactions with the wall, and nonlocality. The main value of this method lies in the fact that the calculations are transparent and help in understanding the incorporated effects.

2.2.5.2 Ensembles of Hard Spheres

In the GC theory the size of the molecules was disregarded, so it is a natural improvement to treat them as hard spheres. Actually, there are two different models that use hard spheres: in the *primitive model* the ions are considered as charged hard spheres, while the solvent is a dielectric continuum; in the *civilized* or *nonprimitive* model both the ions and the solvent molecules are considered to be spheres, but the latter have a dipole moment at their center. The primitive model is mainly of academic interest: ions and solvent molecules are roughly of the same size and should be treated on the same footing. However, one prediction from the primitive model deserves attention: it suggests that at low temperatures the interfacial capacity should increase with temperature, while the GC theory and most model calculations predict a decrease. In this context, it is useful to define an *effective* or *reduced* temperature through $T^* = kT\sigma_i (4\pi\varepsilon\varepsilon_0)/e_0^2$, where σ_i is the diameter of the ions, which is assumed to be equal for the two species. Low temperature means $T^* < 1$, so that aqueous solutions are high-temperature systems because of their high dielectric constant. Somewhat surprisingly, molten salts are low-temperature systems by this definition. This result is not only based on simulations within the primitive model [21] but also on more general arguments based upon pressure balance at the interface [22]; it may explain why the capacity of gold electrodes in contact with a frozen, highly concentrated aqueous solution has been found to increase with temperature [23].

We now focus our attention on the civilized model. In the electrochemical context the basic system consists of hard spheres in contact with a charged hard wall; there are three types of hard spheres: solvent molecules of diameter σ_s, which have a dipole moment μ at their center, and ions with a diameter σ_i and ions with a charge number ± 1 – note that both kinds of ions have the same diameter. The integral equations for this model have been solved analytically, but in the MSA [24, 25] only, so that the results are valid for small excess-charge densities on the electrode, that is, in the vicinity of the *pzc*.

The solution is implicit and consists of a pair of nonlinear algebraic equations, which can be solved numerically. In the limit of low concentrations, where the Debye inverse length $\kappa = 1/L_D$ tends to zero, explicit results can be derived. The distribution of the ions $g_i(x)$ and of the dipoles $g_s(x)$ as a function of the separation x from the electrode show oscillations, which are caused by the packing of the particles (see Fig. 3); they die out far from the surface.

The interfacial capacity C is of particular interest since it is measurable. For low ionic concentrations, its inverse can be developed into a power series in κ; the first two terms are

$$\frac{1}{C} = \frac{1}{\varepsilon\varepsilon_0}\left[\frac{1}{\kappa} + \frac{1}{2}\left(\sigma_i + \frac{\varepsilon-1}{\lambda}\sigma_s\right)\right] \quad (10)$$

The next term is of the order of κ. Within the MSA, the dielectric constant ε is related to the dipole moment μ and the density ρ_s of the solvent through

$$\frac{\varepsilon-1}{\varepsilon} = \frac{4\pi\mu^2}{3kT}\frac{9}{\lambda^2(\lambda+1)^2}\rho_s \quad (11)$$

and the parameter λ is obtained from the dielectric constant via

$$\lambda^2(1+\lambda)^4 = \varepsilon \quad (12)$$

The first term in Eq. (10) is just the GC capacity at the *pzc*; the second term is independent of the ionic concentration, and can be identified with the Helmholtz capacity. However, in this model the Helmholtz capacity is not caused by a single monolayer of solvent with special properties, like in the Stern model, but results from an extended boundary layer. It depends on the dielectric properties of the solvent and on the diameters of the particles. Since $\lambda \ll \varepsilon$, the influence of the ions on the capacity is predicted to be small. This is in line with the experimental

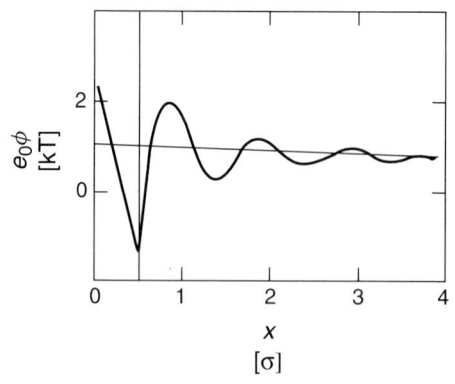

Fig. 3 Oscillations in the electrostatic potential in an ensemble of hard spheres caused by the packing of the particles against the electrode; the straight line is the prediction of the GC theory. (Data taken from Ref. [24].)

observation that the Helmholtz capacity is practically independent of the nature of the ions, provided they are not specifically adsorbed. So this prediction is one of the successes of the theory. The second term in the Helmholtz capacity represents the effect of the solvent. The parameter λ accounts for the fact that the dielectric properties at the interface differ from those of the bulk; roughly speaking, the effective dielectric constant in this region is much reduced.

For aqueous solutions, with $\sigma_s \approx 3\,\text{Å}$ and $\varepsilon \approx 80$, Eq. (10) predicts a Helmholtz capacity of the order of $16\,\mu\text{F cm}^{-2}$ at the *pzc*; this is much smaller than any experimental values. The contribution of the metal is missing in this model. We will see later that the electronic polarizability of the metal surface increases the Helmholtz capacity.

These results for the ensemble of hard spheres are now more than 20 years old; all the same, it has not been possible to extend them into the nonlinear region away from the *pzc* in a rigorous way, though heuristic extensions have been suggested [26]. Even computer simulations have met with little success: for realistic values of the dipole moments and the diameters, the ensemble tends to get stuck in metastable states away from equilibrium, so that it is impossible to obtain thermodynamic averages. Therefore, Henderson and coworkers [27] have performed MC simulations for an ensemble in which both the dipole moment of the solvent and the charges on the ions have been scaled down to a fraction of their values for aqueous solutions. Such systems do reach equilibrium; the particle profiles show a strong layering of the particles, similar to that observed in the MSA, but it is not possible to extrapolate the results to realistic charges and dipole moments.

2.2.6
The Role of the Metal

2.2.6.1 Qualitative Considerations

In concentrated solutions the electric field at the electrode surface is so high that it distorts the electronic density of the metal, and thereby changes the surface potential χ. It is natural to define a contribution of the metal to the interfacial capacity through

$$\frac{1}{C_m} = \frac{\partial \chi}{\partial \sigma_M} \qquad (13)$$

According to Le Chatelier's principle, this metal capacity is negative: a positive charge density increases the potential drop between the metal and the solution, and the surface electrons react so as to make it smaller. Thus a positive charge density on the metal surface gives rise to an electric field directed outward from the surface, which pulls the electrons back into the metal and thereby decreases the surface dipole moment. Conversely, a negative charge pulls the electrons toward the solution and increases the surface potential (see Fig. 4). Since the metal capacity is independent of the electrolyte concentration, it contributes to the Helmholtz capacity, which can thus be decomposed into parts pertaining to the metal and to the solvent:

$$\frac{1}{C_H} = \frac{1}{C_m} + \frac{1}{C_{sol}} \qquad (14)$$

However, since the metal and the solvent interact at the interface, these two contributions are not independent. Since the metal capacity is negative, it makes the total capacity, which must be positive in any reasonable model, larger. In essence, the metal surface possesses a high electronic polarizability, which increases the double-layer capacity.

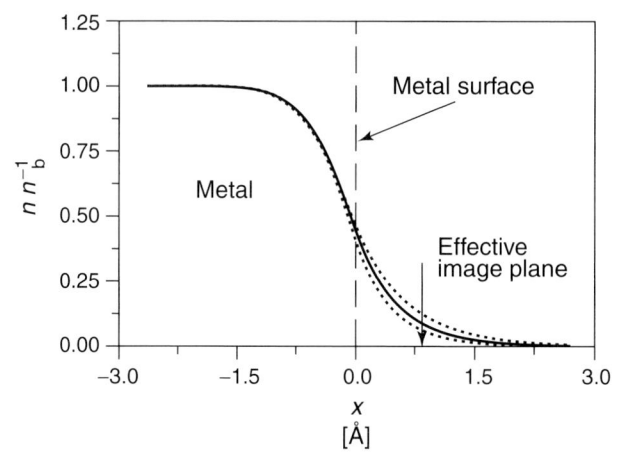

Fig. 4 Normalized electronic density profile at the surface of jellium with a bulk density of 15.3 atomic units. The upper dashed curve is for a surface charge density $\sigma_M = -0.1\,\text{C m}^{-2}$, the lower dashed curve for $\sigma_M = 0.1\,\text{C m}^{-2}$. The full curve is for an uncharged surface. The arrow gives the position of the effective image plane for an uncharged surface.

Trasatti [28] has noticed that in aqueous solutions the Helmholtz capacity C_H of the simple sp metals, taken at the *pzc*, correlates with their electronic densities (see Fig. 5). This indicates that the surface polarizability of simple metals increases with their electronic densities, a trend that can be explained by the jellium model presented below.

2.2.6.2 The Jellium Model

As early as 1928, Rice [29] attempted to estimate the contribution of the metal to the interfacial capacity within the framework of the Thomas–Fermi theory. However, he obtained a positive contribution of the metal to the capacity, which was practically the same for all metals. Only some 50 years later, Kornyshev and coworkers [30] showed that Thomas–Fermi-like models are unsuitable for double-layer theory because they do not give a realistic description of the work functions and the surface potentials of metals.

A major breakthrough in our understanding of the electric double layer came with the introduction of the jellium model, which had already been used extensively in the theory of metal surfaces [31, 32], into electrochemistry [34, 36]. In this model the lattice of positively charged metal ions is represented by a constant positive background charge, which drops abruptly to zero at the surface. The electrons are modeled as an inhomogeneous electron gas, which interacts with the positive background. Explicit calculations for this model are usually based on density-functional theory, which has already been mentioned above. The quantum-mechanical self-interactions of the electron gas – the exchange and the correlation energies – are treated in the *local density approximation*, which is based on the following idea: these interactions are known for an electron gas of constant electronic density. As a first approximation, one can assume that in an

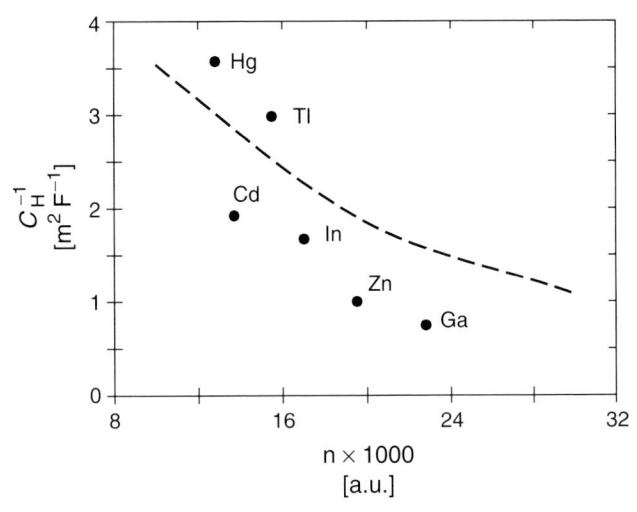

Fig. 5 Inverse Helmholtz capacity of aqueous solution in contact with sp metals; experimental values from Trasatti [28]. (The dashed line is based on a model calculation by Schmickler and Henderson [26].)

inhomogeneous gas the exchange and correlation energies at each point take on the same values that they would have in a homogeneous gas with the same density. While this approximation is simple, it has the advantage that it fulfills a number of sum rules that the exact expression must also obey. Most importantly, it gives good results for the electronic structures of surfaces, and only for complicated interactions must one resort to better mathematical treatments such as the generalized gradient approximation.

The electronic polarizability of the jellium surface, which is the relevant quantity for the interfacial capacity, can be expressed in terms of the *effective position of the image plane*, which has the meaning that its name suggests: a test charge placed in front of the jellium surface experiences an image force as if the image plane were a distance x_{im} in front of the surface (see Fig. 4). In a metal capacitor the effective positions of the plates lie at a distance x_{im} in front of the geometrical surface. Thus a positive value of x_{im} indicates that the effective plate separation is smaller than the physical separation between the two surfaces. Typical values for x_{im} are of the order of 0.5 Å; for a normal plate capacitor such a small shift of the apparent plate position is quite negligible. However, for high ionic concentrations the double layer can be considered as a capacitor with a plate separation of a few Ångstrøms, so that a shift of this order of magnitude has a noticeable effect.

The relation between the position of the image plane and the metal capacity is given by

$$x_{im} = -\varepsilon_0 \frac{\partial \chi}{\partial \sigma_M} = -\varepsilon_0 \frac{1}{C_m} \quad (15)$$

As one might expect, the polarizability increases with the electronic density (see Fig. 6), which qualitatively explains the corresponding increase of the Helmholtz capacity, though for jellium in vacuum

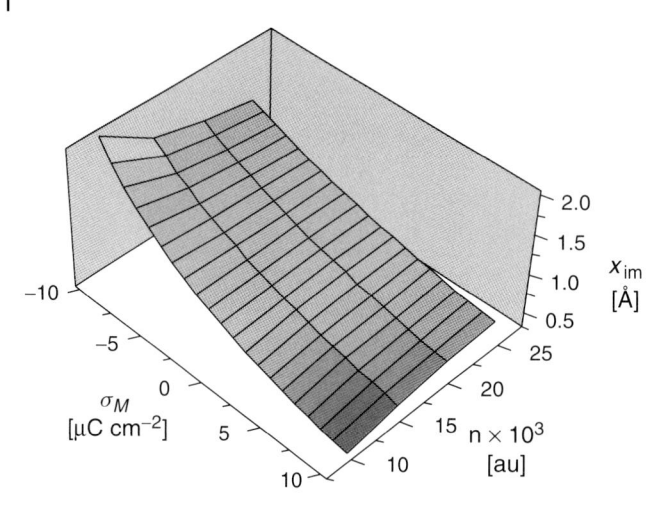

Fig. 6 Variation of the effective position of the image plane with the electronic density and with the surface charge density, calculated within the jellium model.

this dependence is relatively weak. It also depends on the surface charge density, and is much higher when there is a considerable excess of electrons on the surface.

As already mentioned, the distribution of electrons at the jellium surface entails a surface dipole moment. Therefore, at the *pzc* a water molecule situated within the electronic tail experiences a positive field that tends to orient the molecule with its oxygen end toward the metal surface. Conversely, the water molecules tend to enhance the electronic spillover. As a result, the work function for jellium in contact with water is lower than in the vacuum. Model calculations for this effect have been performed by several authors [34, 37, 39–41]. While the results depend on the details of the model, it is generally agreed that the larger the overall change in the dipole potential, the higher the electronic density.

Jellium is a simple, structureless model for a metal, and can therefore only reproduce trends and orders of magnitudes for polycrystalline materials. In an improved version the positive background charge is replaced by a lattice of pseudopotentials so that single crystals can be represented. Usually, the pseudopotentials are averaged parallel to the surface in order to keep the calculations one-dimensional. Such models predict sizable differences for the surface polarizabilities, measured in terms of x_{im}, of different single-crystal surfaces, and may thereby explain why the interfacial capacity varies with the crystal orientation [40, 42]. However, the calculated values depend more strongly on the orientation than do the experimental values; this indicates that the contributions of the metal and of the solvent are not independent.

2.2.6.3 Calculations for Metal Clusters

Modern quantum-chemical methods account for the chemical properties of the atoms much better than jellium-type models can, but they are usually limited

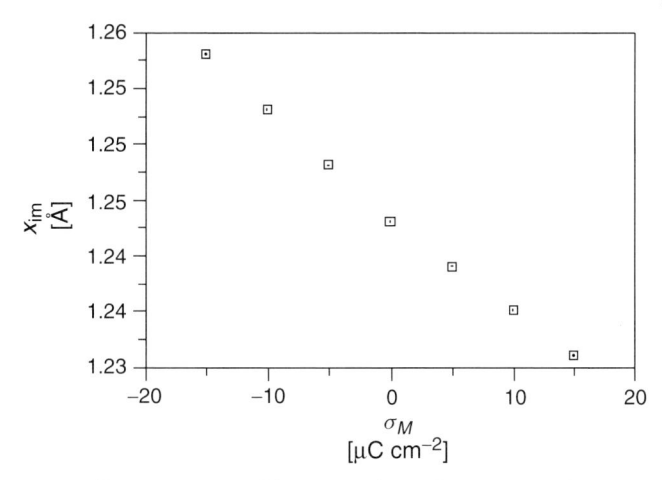

Fig. 7 Effective position of the image charge for a mercury cluster as a function of the surface charge density. (Data taken from Ref. [44].)

to treating a finite number of atoms. It is therefore natural to model an electrode surface by a large cluster of metal atoms. However, even in a large cluster most of the atoms lie near the surface, so that the charge distribution differs considerably from that in a semi-infinite metal. Therefore an early attempt [43] to estimate the metal contribution to the capacity by quantum-chemical calculations for a lithium cluster met with mixed success: the values obtained for the surface polarizability were quite large – in the case of Li(111) even excessively large. A new attempt has recently been undertaken by Nazmutdinov and coworkers [44]. These authors suggested that the surface polarizability should not be calculated from the change in this surface dipole potential, but from a renormalization of the density matrix. They chose mercury as a model system and obtained values for the effective position of the image plane that increase toward negative charge densities (see Fig. 7) and are quite similar to those obtained from jellium-type calculations.

2.2.7 The Metal–Solution Interface

The models presented in the previous two sections dealt with just one side of the interface. Here, we discuss models for the whole metal–solution interface. They can be constructed by combining models for the solution and for the metal, but a few attempts have also been made to treat the interface as a whole. We present several approaches in increasing order of complexity.

2.2.7.1 Jellium in Contact with Model Solutions

The simplest molecular model for an electrolyte solution is an ensemble of hard spheres treated in the MSA. This can be combined with jellium to obtain a model for the whole interphase [46–48]. The hard sphere model has been solved at the PZC only, so the combined model is restricted to this point. It is natural to consider the jellium surface as a hard wall for the electrolyte and add the contributions of the hard-sphere electrolyte and jellium to

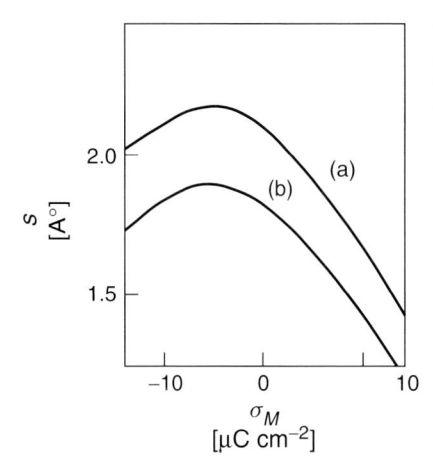

Fig. 8 Metal–solvent separation "s" as a function of the surface charge density for two different choices of the pseudopotentials [52, 53].

the inverse capacity. However, in this way one obtains negative interfacial capacities for metals with higher electronic densities, which are clearly unphysical. Obviously, the interaction of the metal with the solution affects the response to an external field. A simple remedy was proposed by Schmickler and Henderson [47, 48]: they introduced a barrier that repels the metal electrons from the solution and reduces the surface polarizability. The height of the barrier was chosen such that the model gives reasonable values for the capacities of the sp metals – in this way they obtained the broken line in Fig. 5. Obviously, this ad hoc procedure is not quite satisfactory, even though the experimental trend is well reproduced.

The main problem is the interaction between the metal and the solution. Several groups have tried to improve on the simple hard-wall model by constructing semiempirical interaction potentials. A particular issue in these approaches has been whether the average distance s between the metal surface and the first layer of solvent molecules changes with the charge density. If it does change, it makes an additional contribution:

$$\frac{1}{C_{\text{shift}}} = \frac{\sigma_M}{\varepsilon_0} \frac{\partial s}{\partial \sigma_M} \quad (16)$$

to the inverse capacity. Halley's group [49–51] proposed that this is a major contribution, which determines the form of the capacity charge characteristics. This idea was taken up by Amokrane and Badiali [52, 53], who constructed a metal–water potential composed of an attractive, long-range dispersion force and a short-range repulsion. They performed explicit calculations for Ag(111), modeled as jellium with pseudopotentials, and water, and obtained a surprisingly strong variation of the water–metal separation s with the surface charge density σ_M (see Fig. 8). The separation is smaller for high charge densities, irrespective of their sign, because the electrostatic pressure pushes the water toward the surface. This effect is less marked for negative charge densities, where the electrons spill out further and exert an outward pressure on the solvent. The absolute values of the $s - \sigma_M$ characteristics depend on the pseudopotentials for the jellium–water interaction, which determine the repulsive part of the potential, but the shape of these

Fig. 9 Solvent capacity for aqueous solutions according to the phenomenological theory of Amokrane and Badiali [52, 53]. The full lines are from the experimental data for a solution of KPF$_6$, the dashed curves for NaF.

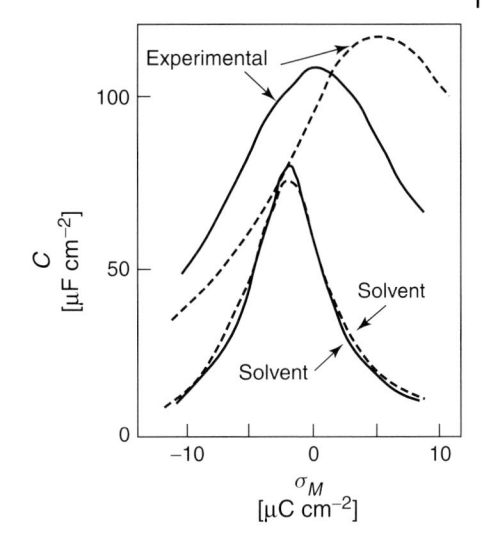

curves is the same for reasonable choices of these pseudopotentials.

To arrive at a complete double-layer model, Amokrane and Badiali proceeded in a semiempirical manner. Assuming that the contributions of the metal and the solvent to the inverse capacity are additive (see Eq. 14), they obtained the solvent capacity from experimental data for the Helmholtz capacity of Ag(111). In this way they obtained a solvent capacity for water, which shows a pronounced maximum near the pzc (see Fig. 9); this curve can be fitted to a model in which the solvent is represented by a layer of dipoles.

This procedure can be carried further by taking experimental data for the Helmholtz capacities of other metals and extracting the metal contribution by assuming that the solvent part is not affected by the metal. In this way, Amokrane and Badiali arrive at a consistent interpretation of all experimental data. However, their theory hinges on the calculations for Ag(111). In fact, the variation of the calculated metal capacity is so strong that it practically determines the solvent capacity in the sense that different experimental capacity data give almost the same solvent capacities.

Bérard and coworkers [54] have applied the RHNC to a simple water model and combined this with the jellium model. The solution and the jellium were assumed to interact electrostatically. This leads to a coupled set of equations for the density profiles of the electrons and the particles in the solution, which was solved self-consistently through an iterative procedure. A critical issue in this coupling is the distance of closest approach of the solvent molecules to the jellium edge, modulated by the repulsion of the inhomogeneous electron gas from the orbitals of the solvent molecules. Bérard and coworkers [54] allow the solvent spheres to come into contact with the jellium edge. Moreover, owing to the presence of the solvent, the electron density spills further out of the metal, dropping to 0.1% of the bulk value at 3 Å. At an uncharged metal, the resulting strong coupling causes the solvent dipoles in the layer contacting the metal to have a small net orientation normal to the metal

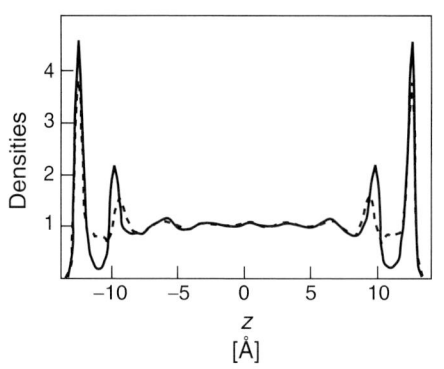

Fig. 10 Oxygen (solid line) and hydrogen (dashed line) densities as a function of the distance from the center of the simulation box. The densities have been normalized to unity in the bulk. (Data taken from Ref. [56].)

surface, with the positive ends out. This is in agreement with the results of several computer simulations that we will discuss below.

2.2.7.2 Molecular Dynamics Simulations

Realistic models for solvents such as water are too complicated to be treated by anything but computer simulations, and molecular dynamics has usually been the method of choice. Most studies have been limited to pure water in contact with a metal surface, but recently a few simulations have been performed for concentrated aqueous solutions. We will not attempt to survey this very active field, but give a few representative results.

There are a variety of semiempirical models for water; most of them model the molecule by a few point charges with a Lennard–Jones or similar potential superimposed. For example, the popular rigid SPC/E model [55] consists of three point charges of $-0.8476e_0$ on the oxygen and $0.4238e_0$ on the hydrogen sites (e_0 is the proton charge) and a Lennard–Jones potential centered at the oxygen atom. The resulting dipole moment is somewhat larger than that of the isolated water molecule in the gas phase in order to account for the polarizability. Other water models have a built-in polarizability, but such complications increase the computer time.

The interaction of water with the metal has been modeled in various ways: hard-wall, Lennard–Jones, and Morse potentials have been used, as well as potentials derived from ab initio cluster calculations. Whatever interaction is chosen, the presence of the metal always induces a layering of the water molecules, similar to that seen in the hard-sphere ensemble. As an example, we show the density profiles for the oxygen and hydrogen atoms obtained by Foster and coworkers [56] in Fig. 10. The corresponding simulations were performed in the absence of an external field, that is, for the *pzc*, and in the *z*-direction the water was contained between two identical walls, while periodic boundary conditions were applied in the xy directions. As a result of the layering effect, the density profiles exhibit a series of oscillations, whose amplitude diminishes toward the bulk. In the first two layers the profiles for oxygen and hydrogen differ somewhat: on an average, the oxygen atoms are a little closer to the electrode surface than the hydrogen atoms. Therefore the dipole moment of the water molecules has a small net orientation along the direction of the surface normal, which produces a surface dipole potential χ. The magnitude of this effect

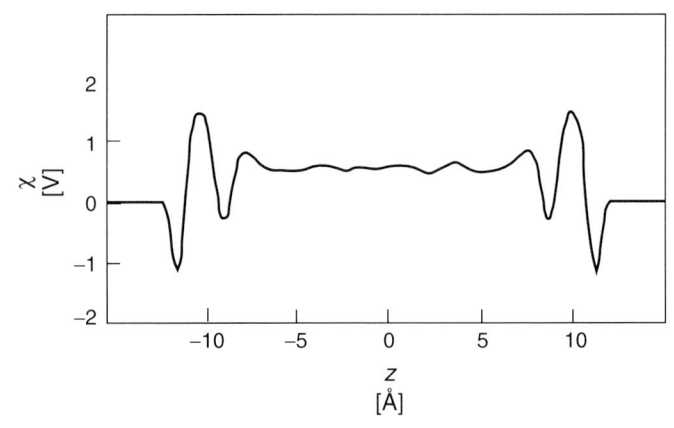

2.2 Electrical Double Layers: Theory and Simulations

Fig. 11 Surface potential χ as a function of the distance from the center of the simulation box. (Data taken from Ref. [44].)

depends on the interaction between the metal and water; Fig. 11 shows the dipole potential for a simulation of water in contact with a mercury electrode. The total potential drop between the electrode surface and the bulk is of the order of several tenths of an electron volt, which agrees qualitatively with estimates of Trasatti [57] based on experimental data.

The detailed structure of water at the surface is often discussed in terms of the bilayer model proposed by Doering and Madey [58] to explain water adsorption on metal surfaces in the vacuum at low temperatures. As the name suggests, this model proposes two layers of water that are hydrogen-bonded to each other. Obviously, in the liquid state and at ambient temperatures such a bilayer must be strongly disturbed, though a vestige seems to appear in some simulations [59]. Similar, but more icelike, structures have been suggested by Krämer and coworkers [60].

When the electrode surface is charged or, equivalently, an electric field is introduced into the simulation box, the water molecules tend to align with the field; the extent to which this happens depends on the water model employed and on the interaction of the first layer with the metal surface. Xia and Berkowitz [61] suggest that at very high fields water undergoes a phase transition to an icelike ordered structure.

The simulations reported above have been performed for pure water. Modeling the double layer requires the introduction of ions, which are usually represented as charged spheres with a Lennard–Jones potential. Given the limited size of the ensembles that can be handled, a statistically meaningful number of ions results in a high concentration; typically of the order of 2 M. At present, a few such simulations have been reported [62, 63], and more can be expected in the future as computing power grows. The results obtained so far are still tentative, but generally seem to be in line with physicochemical intuition and the notions developed by Grahame [64, 65] and Parsons [66], though a few details are surprising. Thus, the solvation of the ions plays a central role in the double-layer structure. In particular, the fact that the anions tend to be less strongly solvated than the cations induces an asymmetry with respect to the *pzc*. As

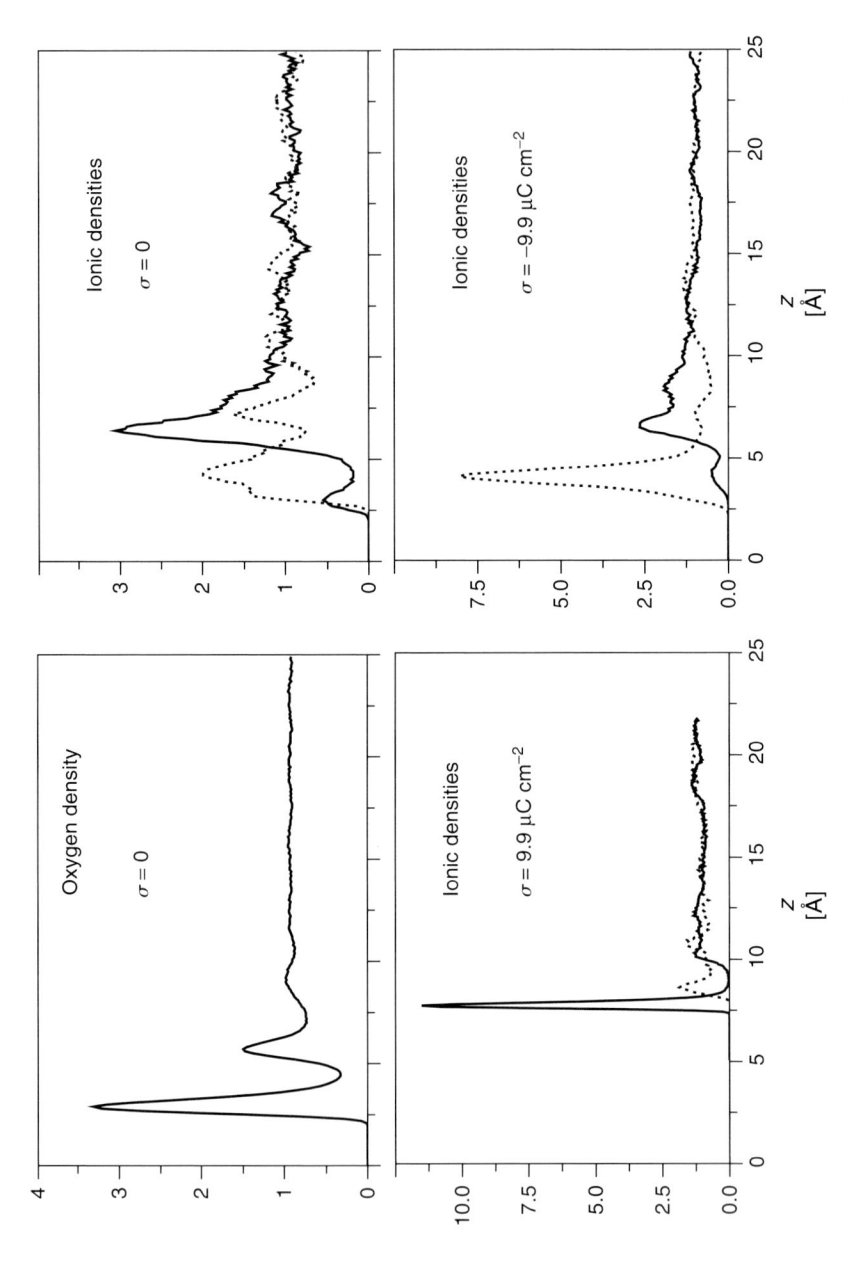

Fig. 12 Distribution of the ions at an idealized metal surface with various surface charge densities. Dotted lines: cations (Na$^+$), full lines: anions (Cl$^-$). For comparison, the figure (a) shows the distribution of the oxygen atoms at the pzc.

an example, we show in Fig. 12 the density profiles of Na$^+$ and Cl$^-$ obtained by Spohr [63] at the *pzc* and at surface charge densities of $\sigma = \pm 9.9\,\mu\text{F}\,\text{cm}^{-2}$. At the positively charged surface the first peak for the weakly solvated anions is closer to the metal (in this case mercury) surface than the peak for the cation is at the negative surface. However, at the *pzc* the sodium ions are closer to the surface because their solvation shell can be accommodated between the first and second water layers.

The total potential drop at the interface is determined both by the distribution of the ions and by the orientation of the water; these two contributions tend to cancel each other, so there is a complicated interplay. While it is still too early to obtain the interfacial capacity from such simulations, first estimates give a reasonable order of magnitude [63].

2.2.7.3 Car–Parinello-type Simulations

In the simulations reported above the interactions between the various components were mostly based on semiempirical potentials. In contrast, Price, Halley, and their collaborators attempt to model the whole interface in the spirit of the Car–Parinello method [16]. One of the first systems investigated was the interface between a copper electrode and water [67]. For this purpose these authors set up a simulation cell with approximate dimensions of $42\,\text{Å} \times 15\,\text{Å} \times 15\,\text{Å}$. Each cell contained a slab of copper atoms that were five-layers thick, the two surfaces having (100) structure. The remaining space was filled with water molecules. Cyclic boundary conditions were applied in all directions. Obviously, an ab initio, all electrons calculation is quite impossible for such a system, and may not even be desirable. Instead, Halley and collaborators used a mixture of pseudopotentials and interaction potentials derived from quantum-chemical calculations.

The system is investigated by molecular dynamics, performing a self-consistent calculation for the electronic subsystem at each step. The presence of the solvent is found to have a pronounced effect on the electronic distribution on the metal surface. In particular, the simulations clearly show the formation of image charges on the metal surface.

This and similar studies from this group provide a very detailed description of the interface, to which we cannot give full justice in this limited space. Of course, such comprehensive simulations are presently of a preliminary nature, but they do indicate the important physical processes that occur at the interface. We may expect more work of this type to be performed in the future.

2.2.8
Rough Electrodes

Most of the theoretical work has been restricted to perfect single-crystal electrodes. However, in practical applications such as fuel cells, in which the electrode acts as a catalyst, it is desirable to maximize the surface area by using rough electrodes. Very rough electrodes can be modeled as fractals [68], but a double-layer theory for such electrodes has remained elusive. The opposite limit of small roughness has been treated by Daikhin and coworkers [69] at the GC level.

These authors start by defining a *roughness function* $\tilde{R}(\kappa)$ through the relation

$$C = \tilde{R}(\kappa)C_{\text{GC}} \quad (17)$$

where C is the actual capacity, C_{GC} the GC value, and κ is the Debye inverse length. Effectively, the Debye length $1/\kappa$ defines the yardstick with which the surface area is measured. Therefore,

$$\lim_{\kappa \to 0} \tilde{R}(\kappa) = 1 \quad \text{and} \quad \lim_{\kappa \to \infty} \tilde{R}(\kappa) = R \tag{18}$$

where R is the geometrical roughness of the electrode, which is defined as the ratio between the true and the apparent surface areas. If the z-direction is taken as perpendicular to the average surface, then, for a moderately rough electrode, the true surface is given by a single-valued function:

$$z = \zeta(x, y) \tag{19}$$

When the roughness is small it can be treated as a perturbation, and the roughness function can be expressed through the Fourier transform (FT) of the surface. This theory is simple only near the *pzc*; away from the *pzc* the uneven distribution of the surface charge must also be considered: the charge accumulates on the surface protrusions. Then the roughness function becomes quite involved, and depends also on the electrode potential [70].

2.2.9
Liquid–Liquid Interfaces

In a certain sense, interfaces between two liquids are simpler than those between a metal and a liquid, since they do not involve solid-state properties. However, for a long time the structure of liquid–liquid interfaces has remained a matter of controversy. Essentially, there are two different views: one holds that the interface is sharp and contains a compact layer of solvent molecules into which the ions cannot penetrate. The other view posits the

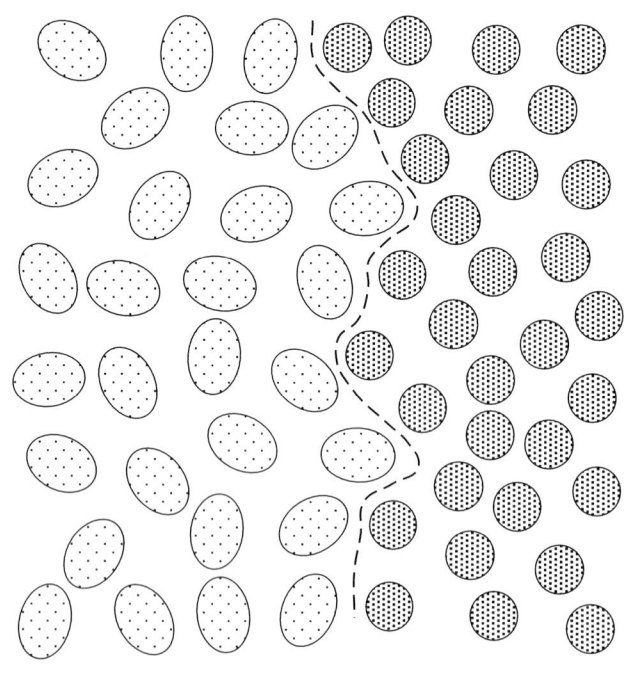

Fig. 13 Schematic picture of the interface between two immiscible liquids.

existence of an extended boundary layer in which the two solutions mix.

During the last few years, the latter view has received more supporting evidence. Already the early experimental work of Girault and Schiffrin [71], who determined the surface excess of water at the interface with 1,2-dichloroethane, had indicated the existence of a mixed boundary layer. Recent X-ray scattering experiments [72] indicate an average interfacial width of the order of 3 to 6 Å. These experiments are in line both with model calculations based on the density functional formalism [73] and with computer simulations [74, 75]. Accordingly, the interface is best visualized as rough on a molecular scale as indicated in Fig. 13.

So far, there have been no molecular dynamics simulations with a larger number of ions for these interfaces, and the double-layer theory rests on analytical calculations and on simulations of the lattice gas model (LGM).

2.2.9.1 Verwey–Niessen Theory

The basic system consists of two immiscible solvents, each of which contains a salt that is badly solvable in the other solvent. When a potential difference is applied between the two solutions, opposing space charge layers form on both sides of the interface and give rise to a capacity. The simplest model consists of two diffuse double layers back to back, which are described by the GC theory – in the context of liquid–liquid interfaces this model is known as the Verwey–Niessen theory [76].

The interfacial capacity per unit area at the interface between two solutions labeled (1) and (2) is defined as

$$C = \frac{d\sigma}{d(\phi_2^\infty - \phi_1^\infty)} \quad (20)$$

where ϕ_i^∞ is the value of the potential in the indicated phase far from the interface. The capacity can be decomposed into a series combination of two GC capacities C_i, one for each interface:

$$C_i = \frac{\varepsilon_i \varepsilon_0}{L_D^i} \cosh \frac{z_i e_0 (\phi^s - \phi_i^\infty)}{2kT} \quad (21)$$

where ϕ^s is the potential at the idealized, flat interface (cf. Eq. 1). The latter must be calculated from the Poisson–Boltzmann equation [77].

The Verwey–Niessen model gives a reasonable first approximation, but at a closer glance significant deviations are observed even at low ionic concentrations, where this theory is expected to hold. Surprisingly, at low ($\leq 10^{-2}$ M) concentrations the experimental capacity is often *higher* than that predicted by Verwey–Niessen theory – this is just the opposite to the behavior of metal–solution interfaces, where the capacity is always *lower* than the GC value.

These deviations were first explained by the presence of a compact, ion-free layer at the interface; this is known as the modified Verwey–Niessen model. Obviously, the presence of an ion-free layer can only reduce the capacity, so the theory had to be modified further. For a few systems a consistent interpretation of the experimental capacity was achieved [78–80] by combining this model with the so-called *modified Poisson–Boltzmann* (MPB) theory [81], which attempts to correct the GC theory by accounting for the finite size of the ions and for image effects, while the solvent is still treated as a dielectric continuum. The combined model has an adjustable parameter, so it is difficult to judge whether the agreement with experimental data is significant. The existence of a compact solvent layer is just a postulate, and is difficult to reconcile with the extended boundary layer discussed in the previous section.

2.2.9.2 Capillary Waves

Since liquid–liquid interfaces are rough, the ideas for rough electrodes presented above can be applied, though with some modifications. Again, if this roughness is not too large, the interface can be represented by a function $\zeta(x, y)$ at a particular moment of time. The two-dimensional Fourier components $A(\mathbf{q})$ of this surface function are known as *capillary waves*; $\mathbf{q}$ is a two-dimensional wave vector. Thus, a rough surface can be considered as a superposition of capillary waves, which are present at all wavelengths. Of course, these cannot be observed as traveling waves, but only as a fluctuating surface roughness.

The energy stored in a capillary wave is proportional to the surface area it creates. This entails a prediction for the distribution of the amplitudes:

$$\langle A(\mathbf{q}) A^*(\mathbf{q}) \rangle = \frac{kT}{\gamma |\mathbf{q}|^2} \quad (22)$$

where the angular brackets denote thermal expectation values. This relation has indeed been observed in computer simulations of liquid–liquid interfaces [74, 75].

The concept of capillary waves can be used to explain how the surface roughness increases the interfacial capacity beyond the Verwey–Niessen value. For this purpose, Pecina and Badiali [82] have solved the linear Poisson–Boltzmann equation across the interface between two solutions with different dielectric constants and Debye lengths separated by a corrugated surface. A major difficulty is the boundary condition at the rough interface, which these authors handle by a perturbation expansion treating the roughness as a small parameter. Following Daikhin and coworkers [69] (see previous section), Pecina and Badiali express the deviation from the Verwey–Niessen theory through a *roughness function* $\tilde{R}(L_D^i, \varepsilon_i)$, which here depends on the dielectric constants of the two solutions and on their Debye lengths. Explicit expressions for the roughness function can be given for simple surface corrugations. While Pecina and Badiali consider small deviations from the *pzc*, the nonlinear case has been treated by Urbakh and coworkers [83].

The concept of capillary wave explains the increase of the interfacial capacity beyond the Verwey–Niessen limit, and thus helps in understanding the structure of the interface. A quantitative interpretation of experimental data within this model is, however, difficult since there are other causes for the deviations from Verwey–Niessen theory besides the surface roughness.

2.2.9.3 Lattice Gas Model

The Ising model is one of the most versatile models in physics and chemistry, and in the equivalent form of the LGM it has also been applied to liquid–liquid interfaces. In this case it is based upon a three-dimensional, typically simple cubic lattice. Each lattice site is occupied by one of a variety of particles. In the simplest case the system contains two kinds of solvent molecules, and the interactions are restricted to nearest neighbors. If we label the two types of solvent molecules S_1 and S_2, the interaction is specified by a symmetric 2×2 matrix w_{ij}, where each element specifies the interaction between two neighboring molecules of type S_i and S_j. Whether the system separates into two phases or forms a homogeneous mixture depends on the relative strength of the cross-interaction w_{12} with respect to the self-interaction terms w_{11} and w_{22}, which can be expressed through the combination

$$w = w_{12} - (w_{11} + w_{22})/2 \qquad (23)$$

If w/kT is large, the system separates into two phases: a phase 1, which contains mainly solvent molecules S_1 and has only a low concentration of S_2, and a phase 2, which is mainly composed of S_2. If w is small, entropy wins and the systems forms one phase. If the system separates, it can be extended to a model for the interface between two solutions by introducing ions. In the basic case the system contains a salt composed of cations K_1^+ and anions A_1^-, which is preferentially solvated by the solvent S_1, but badly solvable in solution 2, and a salt $K_2^+ A_2^-$ that is preferentially dissolved in solvent 2. This can be achieved by choosing suitable interaction parameters between the ions and the two solvents.

In addition to the nearest-neighbor interaction, each ions experiences the electrostatic potential generated by the other ions. In the literature this has generally been equated with the macroscopic potential ϕ calculated from the Poisson–Boltzmann equation. This corresponds to a mean-field approximation, in which correlations between the ions are neglected. The better this approximation, the lower the concentrations of the ions.

The lattice gas can be treated both by analytical approximations such as the mean-field or the quasi-chemical approximation, or by MC simulations. Both methods complement each other: the

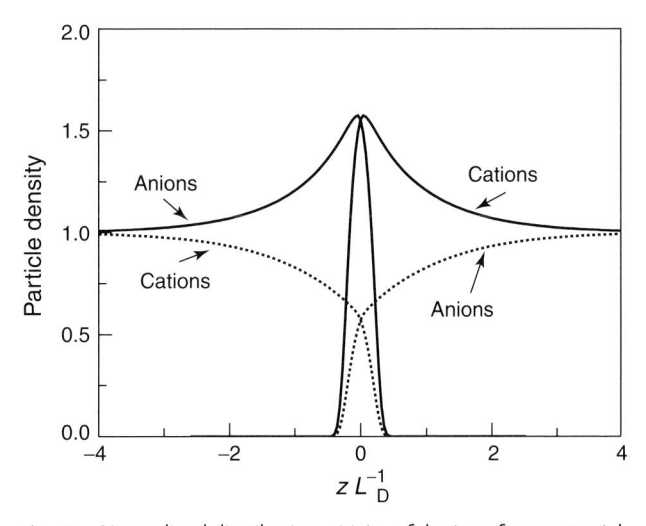

Fig. 14 Normalized distribution $c(x)/c_0$ of the ions for a potential drop of $e_0 \Delta\phi/kT = 1$. The full lines give the distribution of the majority ions; in the region $x < 0$ these are the anions, in $x > 0$ these are the cations. The dashed lines give the distributions of the counterions. The calculations have been performed in the quasi-chemical approximation for equal Debye lengths and dielectric constants in both solutions [84].

simple analytical approximations work best at low ionic concentrations, while the simulations, because of the finite ensemble size, are more suitable for higher concentrations. The quasi-chemical approximation has been employed by Pereira and coworkers [84] and MC simulations have been employed by Huber and coworkers [85]. Both works give similar results. The most important effect is that the roughness of the interface entails an overlap of the double-layer regions pertaining to the two solvents (see Fig. 14). This overlap decreases the average distance between the two opposing charges, and hence it increases the capacity. At higher ionic concentrations the finite size of the particle becomes noticeable, an effect that makes the capacity smaller. Hence, depending on the concentrations and on the various interaction parameters, the capacity can be larger or smaller than that predicted by the Verwey–Niessen theory; a few examples are shown in Fig. 15. Further details of the LGM can be found in a recent article by Schmickler [86].

2.2.10
Conclusion

All electrochemical processes take place in the double-layer region. Therefore, electrochemists need to know the distribution of the various particles and of the electrostatic potential and the charge near the interface in detail. After more than a century of double-layer theory, and more than 50 years after the seminal paper by Grahame [64], it is pertinent to ask how far we have succeeded. Of course, the following remarks are the personal view of the author.

The hard-sphere model, introduced in the 1980s, was a major step forward: it demonstrated the existence of an extended boundary layer at the solution side of the interface and gave a good estimate for the contribution of the solution to the interfacial capacity. However, it was solved within the rough MSA, which holds only for small excess charges. Within the integral-equations approach, further progress has been slow. So our

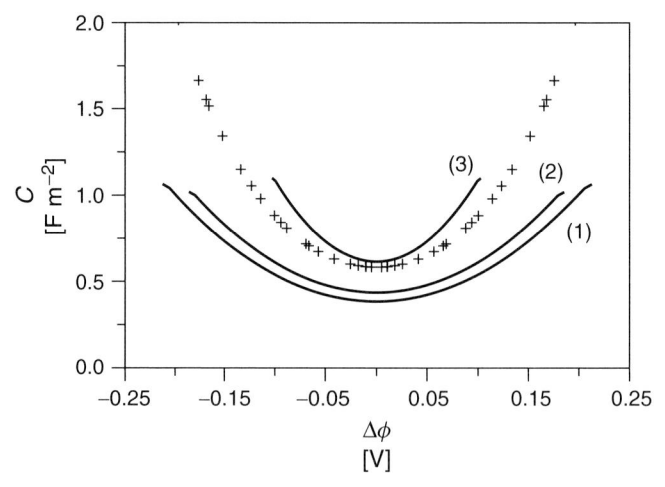

Fig. 15 Interfacial capacity in the lattice gas model for various ion–solvent interactions. The data were obtained from an MC simulation [85]. The crosses denote the Gouy–Chapman capacity.

main knowledge of the distribution of particles in the solution rests on computer simulations. With increasing computing power, it has become possible to include ions in the ensemble. To a large extent, these simulations support the notions presented in Grahame's review: small cations are less likely to approach the metal surface than anions, the first layers of water have a more rigid structure, and the water dipoles are oriented by the electric field. However, the simulations do provide additional insight into the water structure and the solvation of ions near the interface. This is true both for metal–solution and for liquid–liquid interfaces.

Another advance in the 1980s was the recognition that the metal surface contributes to the interfacial capacity. This explains why the capacity varies so strongly with the nature of the metal. Calculations based on the jellium model gave an estimate for the order of magnitude of the metal contribution, and reproduced the main trends. But again, further progress has been slow, and so far the methods of modern computational chemistry have not contributed much to our understanding of the double layer.

So the advances during the last 50 years have been mainly in our qualitative understanding of the double layer. Given the complexity of the interface, this is no small achievement. However, quantitative results are clearly desirable. Most probably they will come from further advances in simulation techniques and computing power, while the contribution of proper theory will diminish. This trend toward more numerical works may be regrettable, but is unavoidable. In any case, there are many open questions in double-layer modeling, and it will stay an active field of research for years to come.

Acknowledgments

I would like to thank Dr. E. Spohr, Jülich, for useful discussions on simulations, and for sending me the data of Fig. 10.

References

1. G. Gouy, *Compt. Rend.* **1910**, *149*, 654.
2. D. L. Chapman, *Philos. Mag.* **1913**, *25*(6), 475.
3. O. Stern, *Z. Elektrochem.* **1924**, *56*, 508.
4. R. Parsons, F. G. R. Zobel, *J. Electroanal. Chem.* **1965**, *9*, 333.
5. D. Eberhardt, E. Santos, W. Schmickler, *J. Electroanal. Chem.* **1996**, *419*, 23.
6. R. J. Watts-Tobin, *Philos. Mag.* **1961**, *6*, 133.
7. B. B. Damaskin, A. N. Frumkin, *Electrochim. Acta* **1974**, *19*, 173.
8. R. Parsons, *J. Electroanal. Chem.* **1975**, *21*, 229.
9. R. Guidelli in *Trends in Interfacial Electrochemistry* (Ed.: A. F. Silva), Riedel, Dordrecht, 1986, pp. 387–452.
10. R. Guidelli, *J. Electroanal. Chem.* **1981**, *123*, 59.
11. R. Guidelli, *J. Electroanal. Chem.* **1986**, *197*, 77.
12. D. R. Bérard, G. N. Patey, *J. Chem. Phys.* **1992**, *97*, 4372.
13. G. M. Torrie, P. G. Kusalik, G. N. Patey, *J. Chem. Phys.* **1988**, *88*, 7826.
14. M. P. Alleen, D. J. Tildesley, *Computer Simulations of Liquids*, Clarendon Press, Oxford, 1987.
15. K. Binder, *Monte Carlo Methods in Statistical Physics*, Topics in Current Physics, Springer, Berlin, 1986, Vol. 7.
16. R. Car, M. Parinello, *Phys. Rev. Lett.* **1985**, *55*, 2471.
17. N. H. March, *Electron Density Theory of Atoms and Molecules*, Academic Press, London, 1992.
18. S. Lundquist, N. H. March, (Ed.), *Theory of the Inhomogeneous Electron Gas*, Plenum Press, New York, 1983.
19. J. Stafiej, J. P. Badiali, Z. Borkowska, *J. Electroanal. Chem.* **1995**, *395*, 1.
20. J. Stafiej, M. Dymitrowska, J. P. Badiali, *Electrochim. Acta* **1996**, *41*, 2107.
21. D. Boda, D. Henderson, *J. Chem. Phys.* **2000**, *112*, 8934.

22. D. Henderson, *J. Chem. Phys.* **2000**, *112*, 6716.
23. A. Hamelin, S. Röttgermann, W. Schmickler, *J. Electroanal. Chem.* **1987**, *230*, 281.
24. S. L. Carnie, D. Y. C. Chan, *J. Chem. Phys.* **1980**, *73*, 2949.
25. L. Blum, D. Henderson, *J. Chem. Phys.* **1981**, *74*, 1902.
26. W. Schmickler, D. J. Henderson, *J. Chem. Phys.* **1984**, *80*, 3381.
27. D. Boda, D. Henderson, *J. Chem. Phys.* **1998**, *109*, 7362.
28. S. Trasatti, *J. Electroanal. Chem.* **1981**, *123*, 2031.
29. O. K. Rice, *Phys. Rev.* **1928**, *31*, 105.
30. A. A. Kornyshev, W. Schmickler, M. A. Vorotyntsev, *Phys. Rev. B* **1982**, *25*, 5244.
31. J. R. Smith, *Phys. Rev.* **1969**, *181*, 522.
32. N. D. Lang, W. Kohn, *Phys. Rev. B* **1970**, *1*, 4555.
33. N. D. Lang, W. Kohn, *Phys. Rev. B* **1971**, *3*, 1215.
34. J. P. Badiali, M. L. Rosinberg, J. Goodisman, *J. Electroanal. Chem.* **1983**, *143*, 73.
35. J. P. Badiali, M. L. Rosinberg, J. Goodisman, *J. Electroanal. Chem.* **1983**, *150*, 25.
36. W. Schmickler, *J. Electroanal. Chem.* **1983**, *150*, 19.
37. W. Schmickler, D. J. Henderson, *J. Chem. Phys.* **1984**, *80*, 3381.
38. W. Schmickler, D. J. Henderson, *Prog. Surf. Sci.* **1986**, *22*, 323.
39. W. Schmickler, *Chem. Phys. Lett.* **1983**, *99*, 135.
40. E. Leiva, W. Schmickler, *Proc. Indian Acad. Sci.* **1986**, *97*, 267.
41. M. I. Rojas, E. P. M. Leiva, *J. Electroanal. Chem.* **1991**, *303*, 55.
42. E. Leiva, W. Schmickler, *Surf. Sci.* **1993**, *291*, 226.
43. W. H. Mulders, J. H. Sluyters, J. H. van Lenthe, *J. Electroanal. Chem.* **1989**, *261*, 273.
44. R. R. Nazmutdinov, M. Probst, K. Heinzinger, *Chem. Phys. Lett.* **1994**, *222*, 101.
45. J. Böcker, R. E. Nazmutdinov, E. Spohr et al., *Surf. Sci.* **1995**, *333*, 372.
46. J. P. Badiali, M. L. Rosinberg, F. Vericat et al., *J. Electroanal. Chem.* **1985**, *158*, 82.
47. W. Schmickler, D. J. Henderson, *J. Phys. Chem.* **1984**, *80*, 3381.
48. W. Schmickler, D. J. Henderson, *J. Phys. Chem.* **1985**, *82*, 2925.
49. D. Price, J. W. Halley, *J. Electroanal. Chem.* **1983**, *150*, 347.
50. J. W. Halley, B. Johnson, D. Price et al., *Phys. Rev.* **1985**, *B31*, 7695.
51. J. W. Halley, B. Johnson, D. Price et al., *Phys. Rev.* **1988**, *38*, 9357.
52. S. Amokrane, J. P. Badiali, *Electrochim. Acta* **1989**, *34*, 39.
53. S. Amokrane, J. P. Badiali, *J. Electroanal. Chem.* **1989**, *266*, 21.
54. D. R. Bérard, M. Kinoshita, X. Ye et al., *J. Chem. Phys.* **1994**, *101*, 6271.
55. H. J. C. Berendsen, J. R. Grigeria, T. P. Straatsma, *J. Phys. Chem.* **1987**, *91*, 6269.
56. K. Foster, R. Raghavan, M. Berkowitz, *Chem. Phys. Lett.* **1989**, *162*, 32.
57. S. Trasatti in *Trends in Interfacial Electrochemistry* (Ed.: A. F. Silva), Reidel, Dordrecht, 1986.
58. D. L. Doering, T. E. Madey, *Surf. Sci.* **1982**, *123*, 305.
59. J. P. Valleau, A. A. Gardner, *J. Chem. Phys.* **1987**, *86*, 4162.
60. A. Krämer, M. Vossen, F. Forstmann, *J. Chem. Phys.* **1997**, *106*, 2792.
61. X. Xia, M. L. Berkowitz, *Phys. Rev. Lett.* **1995**, *74*, 3193.
62. M. R. Philpott, J. N. Glosli, S. B. Zu, *Surf. Sci.* **1995**, *335*, 422.
63. E. Spohr, *Electrochim. Acta* **1999**, *44*, 1697.
64. D. C. Grahame, *Chem. Rev.* **1947**, *41*, 441.
65. D. C. Grahame, R. Parsons, *J. Am. Chem. Soc.* **1961**, *83*, 1291.
66. R. Parsons in *Comprehensive Treatise of Electrochemistry* (Eds.: J. O'M. Bockris, B. E. Conway, E. Yeager), Plenum Press, New York, 1980, Vol. 1.
67. D. L. Price, J. W. Halley, *J. Chem. Phys.* **1995**, *102*, 6603.
68. B. B. Mandelbrot, *The Fractal Geometry of Nature*, Freeman, San Francisco, 1982.
69. L. I. Daikhin, A. A. Kornyshev, M. Urbakh, *Phys. Rev. E* **1996**, *53*, 6192.
70. A. Kornyshev, private communication.
71. H. H. Girault, D. H. Schiffrin, *J. Electroanal. Chem.* **1983**, *150*, 43.
72. A. M. Tikhonov, D. M. Mitrinovic, M. Li et al., *J. Phys. Chem. B* **2000**, *200*, 6336.
73. D. Henderson, W. Schmickler, *J. Chem. Soc., Faraday Trans.* **1996**, *92*, 3839.
74. I. Benjamin, *Chem. Rev.* **1996**, *96*, 1449.
75. I. Benjamin, *Annu. Rev. Phys. Chem.* **1997**, *48*, 407.
76. E. J. W. Verwey, K. F. Niessen, *Philos. Mag.* **1939**, *28*, 435–446.

77. W. Schmickler, *Interfacial Electrochemistry*, Oxford University Press, New York, 1996.
78. T. Wandlowski, K. Holub, V. Marecek et al., *Electrochim. Acta* **1996**, *40*, 1897–2887.
79. Q. Cui, G. Zhu, E. Wang, *J. Electroanal. Chem.* **1994**, *372*, 15.
80. Q. Cui, G. Zhu, E. Wang, *J. Electroanal. Chem.* **1995**, *383*, 7.
81. C. W. Outhwaite, L. B. Bhuyiyan, S. Levine, *J. Chem. Soc., Faraday Trans. 2* **1980**, *76*, 1388.
82. O. Pecina, J. P. Badiali, *Phys. Rev E* **1998**, *58*, 6041.
83. M. Urbakh, A. A. Kornyshev, L. I. Daikin, *J. Electroanal. Chem.* **2000**, *483*, 68.
84. C. M. Pereira, W. Schmickler, A. F. Silva et al., *Chem. Phys. Lett.* **1997**, *268*, 13.
85. T. Huber, O. Pecina, W. Schmickler, *J. Electroanal. Chem.* **1999**, *467*, 203.
86. W. Schmickler The lattice gas and other simple models for liquid-liquid interfaces in *Liquid Interfaces in Chemical, Biological and Pharmaceutical Applications* (Ed.: A. G. Volkov), Marcel Dekker, New York, 2000.

2.3
Electrochemical Double Layers: Liquid–Liquid Interfaces

Alexander G. Volkov
Oakwood College, Huntsville, Alabama

Vladislav S. Markin
University of Texas, Dallas, Texas

2.3.1
Polarizable and Nonpolarizable Liquid Interfaces

The electrical double layer (edl) at the oil–water interface is a heterogeneous interfacial region that separates two bulk phases of polarized media and maintains a spatial separation of charges. EDLs at such interfaces determine the kinetics of charge transfer across phase boundaries, stability and electrokinetic properties of lyophobic colloids, mechanisms of phase transfer or interfacial catalysis, charge separation in natural and artificial photosynthesis, and heterogeneous enzymatic catalysis [1–5].

The interface between two immiscible electrolyte solutions (ITIES) can be either polarizable or nonpolarizable, depending on permeability to charged particles. If the interface is relatively impermeable it is called *polarizable*, otherwise it is called *nonpolarizable* or reversible. Planck [6] introduced a rigorous thermodynamic concept called a *completely polarizable interface*, defined as "an interface whose state is completely determined by the charge that has passed through it beginning from a given instant". Planck's definition assumes that direct current cannot flow through the interface. There can be only the transitive current in the boundary layer. However, the definition says nothing about how charged particles from different phases are distributed in this transitional layer. The distribution of particles in the boundary layer can have any configuration and the tales of this distribution for different particles can even overlap in the boundary layer (Fig. 1a, c). At a completely polarizable interface, charge transfer between bulk phases becomes impossible.

A different concept of an *ideally polarizable interface* was formulated by Koenig [7] and represents a particular case of a completely polarizable interface (Fig. 1b, d). Two homogeneous phases α and β are separated by a transitional layer whose properties gradually change from the properties of phase α to those of phase β. A Koenig's surface, which is impermeable to all charged particles (both ions and electrons), is drawn in the transitional layer. This surface represents an infinitely thin and infinitely high-energy barrier for all the charged particles involved. It is this assumption that distinguishes an ideally polarizable interface from a completely polarizable surface, which makes it irrelevant that charged particles are contained in each phase. Koenig's definition cannot be described thermodynamically because it suggests a definite structure of the interfacial region and employs nonthermodynamic assumptions about the existence of an infinitely high-energy barrier for all charged particles. The term "ideally polarizable interface" was introduced by Grahame and Whitney [8], whereas Koenig used Planck's term. Frumkin demonstrated that a completely polarized electrode, unlike an ideally polarizable one, can include systems with local adsorption charge transfer. Although local charge transfer exists at the interface and Koenig's surface cannot be drawn, there is no actual charge transfer between the phases, so the electrode is completely polarizable according to Planck, but nonideally polarizable according to Koenig.

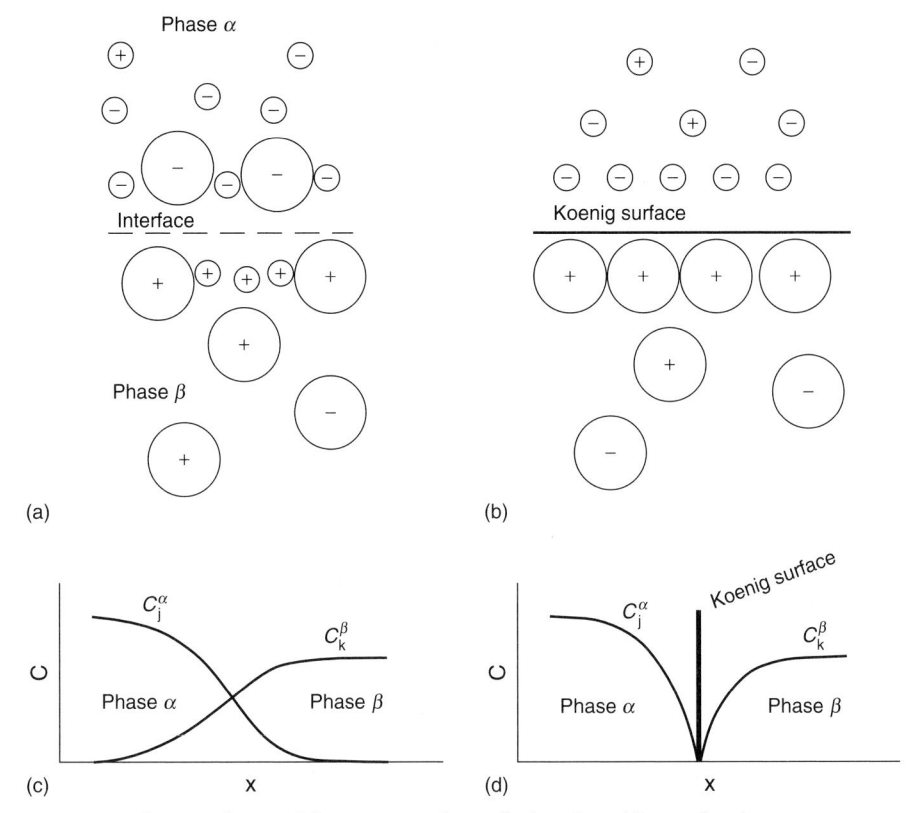

Fig. 1 An arbitrary scheme of the structure of a perfectly polarizable interface between two immiscible liquids (a, b, c, d) and an ideally polarizable interface (b, c).

If phases α and β contain at least one common ion that can freely pass across the interface, the interface may be called *reversible* or *nonpolarizable* (Fig. 2). Although the latter term is commonly accepted, it does not correctly describe the state of the system. A reversible (nonpolarizable) interface can be partially polarizable or completely nonpolarizable. A completely nonpolarizable interface containing at least one common ion can pass a high current in either direction without causing a deviation of the interfacial potential difference from the equilibrium value. Although, in practice we encounter neither ideally polarizable nor completely nonpolarizable interfaces, under certain conditions the properties of some interfaces are very close to being ideal.

2.3.2
Verwey–Niessen Model

Verwey and Niessen first described the EDL at ITIES as two noninteracting diffuse layers, one at each side of the interface [9]. Both solvents were assumed to be structureless media with macroscopic dielectric permittivities, and potential distribution in the EDL was defined by Gouy–Chapman theory.

Gavach and coworkers [10] extended the Verwey–Niessen model (MVN) by introducing an ion-free transition layer

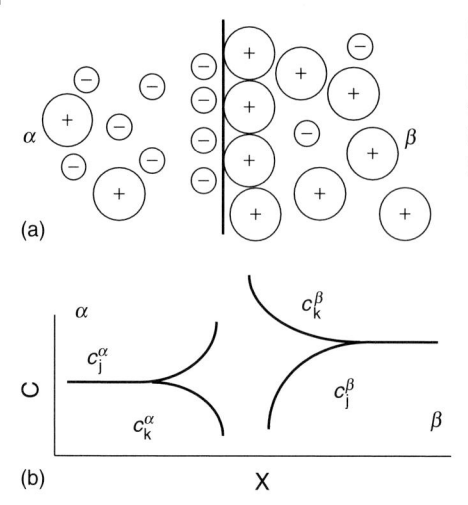

Fig. 2 Scheme of the structure of a nonpolarizable interface between two immiscible electrolyte solutions (a) and the distribution of the concentrations of ions (b) capable of passing from one phase to another.

at the ITIES. This is directly analogous to the compact layer (or the inner Helmholtz layer) in classical electrochemistry. Stern theory was extended to ITIES, and the final model is referred to as the modified Verwey–Niessen model.

In the MVN model, the EDL consists of two diffuse ion layers back-to-back, which produce a compact inner layer between the two phases (Fig. 3). Dielectric permittivity of the medium at any point in the diffuse layer is assumed to be constant and equal to the bulk phase value ε. The compact layer or inner Helmholtz layer is located between $-\delta^\alpha$ and $+\delta^\beta$. In a more detailed analysis, the dielectric permittivities in both parts of the compact layer are ε_h^α and ε_h^β.

In the MVN model, the boundary of the compact layer (called the outer Helmholtz plane) is at the distance of closest ion approach to the interface. For ions with different radii, a few outer Helmholtz planes can be introduced as necessary. In the absence of specific adsorption, the compact layer located between two nearest outer Helmholtz planes disposed on different sides of an ideal interface is usually called the ion-free layer.

The maximum electrical potential in the compact layer $\Delta_\beta^\alpha \phi_i$ includes a dipolar potential $\Delta_\beta^\alpha g$, which is shown schematically as a narrow region at the sharp interface. A dipolar layer can be located not only in the compact layer but can also occupy part of the diffuse layer. The amplitude and sign of $\Delta_\beta^\alpha g$ can differ from the total interfacial potential. Figure 4 illustrates four possibilities for potential distribution at ITIES. Generally, the dipolar potential depends on the total interfacial potential $\Delta_\beta^\alpha \phi$.

The interfacial potential difference consists of the sum of the potential drops:

$$\Delta_\beta^\alpha \phi \equiv \phi^\alpha - \phi^\beta = \phi_d^\beta - \phi_d^\alpha + \Delta_\beta^\alpha \phi_h \quad (1)$$

Charge density at each side of the EDL is usually called the interfacial free charge density, and depends on the model chosen for the interphase. In an ideal polarized electrode, the free charge equals the total thermodynamic charge in the Lippmann equation of electrocapillarity if all components of an interface have the same charges as they have in the bulk phases.

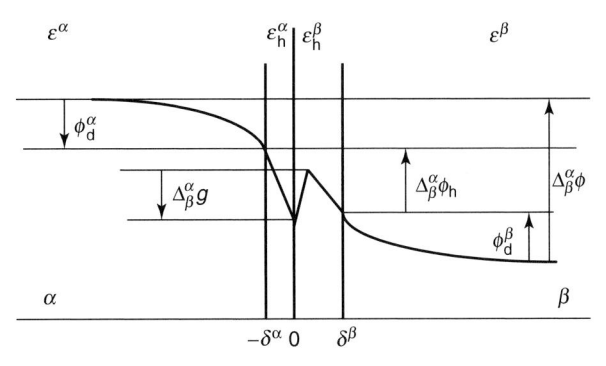

2.3 Electrochemical Double Layers: Liquid–Liquid Interfaces

Fig. 3 Distribution of the electric potential in the electric double layer at the interface between two immiscible electrolyte solutions.

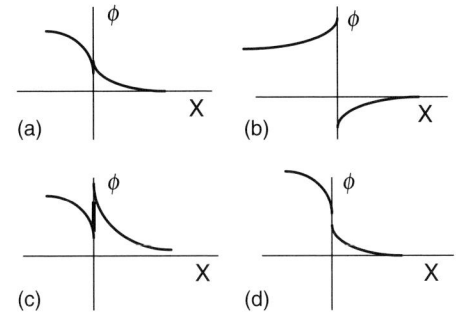

Fig. 4 Different profiles of Galvani potential differences at the interface between two immiscible electrolyte solutions.

We will denote surface charge density in the diffuse layers as q_d^α, q_d^β and in the compact layer as q_h. From electroneutrality of the interphase, we have:

$$q_d^\alpha + q_h + q_d^\beta = 0 \quad (2)$$

At the potential of the free zero charge (fzcp), the compact layer has not adsorbed ions and the charges of the diffuse layers are equal to zero:

$$\phi_d^\alpha = \phi_d^\beta = 0 \text{ and } \Delta_\beta^\alpha \phi_h = \Delta_\beta^\alpha \phi^{fzcp} \quad (3)$$

The dipolar potential $\Delta_\beta^\alpha \phi_h$ also needs the index *fzcp* because of its dependence on total interfacial potential. Usually, the dependence is weak and Eq. (3) is used to determine the dipolar potential.

From the Gouy–Chapman theory for a 1 : 1 electrolyte, it follows that:

$$q_d = -\frac{2\varepsilon_0 RT}{F} \varepsilon \kappa \sinh \frac{F\phi_d}{2RT} \quad (4)$$

where κ is the Debye length,

$$\kappa = \sqrt{\frac{2F^2 c^0}{\varepsilon \varepsilon_0 RT}} \quad (5)$$

and c^0 is the bulk electrolyte concentration in the corresponding phase.

If a Helmholtz layer does not have charges, potential drops in the two diffuse layers have a simple relation:

$$\varepsilon^\alpha \kappa^\alpha \sinh \frac{F\phi_d^\alpha}{2RT} + \varepsilon^\beta \kappa^\beta \sinh \frac{F\phi_d^\beta}{2RT} = 0 \quad (6)$$

The relation between potentials of two diffuse layers depends on a parameter:

$$\eta^{\alpha/\beta} = \frac{\varepsilon^\beta \kappa^\beta}{\varepsilon^\alpha \kappa^\alpha} \equiv \sqrt{\frac{\varepsilon^\beta c^{0\beta}}{\varepsilon^\alpha c^{0\alpha}}} \quad (7)$$

which is the reverse ratio of the diffuse-layer capacitance. There is a direct proportional dependence between ϕ_d^α and ϕ_d^β with coefficient $\eta^{\alpha/\beta}$ when the potential drops are small:

$$\frac{\phi_d^\alpha}{\phi_d^\beta} = -\eta^{\alpha/\beta} \quad (8)$$

From Eqs. (7) and (8), it follows that the potential drop ϕ_d in the diffuse layer will be less if the dielectric permittivity ε or electrolyte concentrations c^0 are increased or, if the diffuse double-layer capacitance C_d is increased. At the contact between aqueous and organic phases, virtually the entire drop of the potential occurs in the organic phase. However, with increasing interfacial potential, the situation changes dramatically. When potentials are large enough, Eq. (6) can be presented in a different approximation:

$$\phi_d^\alpha = -\phi_d^\beta - \frac{RT}{F} \sin \phi_d^\beta \cdot \ln \eta^{\alpha/\beta} \quad (9)$$

Here, the coefficient of proportionality is equal to one.

From Eqs. (1) and (6), it is possible to find the dependence of the potential drop in both diffuse layers $\Delta_\beta^\alpha \phi - \Delta_\beta^\alpha \phi_h$ on potential drops ϕ_d^α and ϕ_d^β in each diffuse layer:

$$\phi_d^\beta = \frac{RT}{F}$$
$$\times \ln \frac{(\varepsilon^\beta \kappa^\beta / \varepsilon^\alpha \kappa^\alpha) + \exp[F(\Delta_\beta^\alpha \phi - \Delta_\beta^\alpha \phi_h)/2RT]}{(\varepsilon^\beta \kappa^\beta / \varepsilon^\alpha \kappa^\alpha) + \exp[-F(\Delta_\beta^\alpha \phi - \Delta_\beta^\alpha \phi_h)/2RT]} \quad (10)$$

$$\phi_d^\alpha = \frac{RT}{F}$$
$$\times \ln \frac{(\varepsilon^\alpha \kappa^\alpha / \varepsilon^\beta \kappa^\beta) + \exp[F(\Delta_\beta^\alpha \phi - \Delta_\beta^\alpha \phi_h)/2RT]}{(\varepsilon^\alpha \kappa^\alpha / \varepsilon^\beta \kappa^\beta) + \exp[-F(\Delta_\beta^\alpha \phi - \Delta_\beta^\alpha \phi_h)/2RT]}$$
$$= -(\Delta_\beta^\alpha \phi - \Delta_\beta^\alpha \phi_h) + \frac{RT}{F}$$
$$\times \ln \frac{(\varepsilon^\beta \kappa^\beta / \varepsilon^\alpha \kappa^\alpha) + \exp[F(\Delta_\beta^\alpha \phi - \Delta_\beta^\alpha \phi_h)/2RT]}{(\varepsilon^\beta \kappa^\beta / \varepsilon^\alpha \kappa^\alpha) + \exp[-F(\Delta_\beta^\alpha \phi - \Delta_\beta^\alpha \phi_h)/2RT]} \quad (11)$$

The spatial distribution of potential near the interface is accompanied by changes in electrolyte concentrations that produce surface excesses of ions. In each phase, the surface excess Γ_i of an ion i can be divided into two components in the inner layer Γ_i^h and in the diffuse layer Γ_i^d:

$$\Gamma_i = \Gamma_i^h + \Gamma_i^d \quad (12)$$

The charge of each contacting phases is equal to:

$$q = \sum_i z_i F \Gamma_i \quad (13)$$

For a binary electrolyte, the ion excess in the diffuse layer can be written as:

$$\Gamma_i^d = \frac{2c^0}{\kappa} \left[\exp\left(-\frac{z_i F \phi_d}{2RT}\right) - 1 \right]$$
$$= \frac{2c^0}{\kappa^\alpha} \left[\sqrt{1 - \frac{q^2}{8\varepsilon_0 \varepsilon RT c^0}} + \frac{z_i q}{8\varepsilon_0 \varepsilon RT c^0} - 1 \right] \quad (14)$$

where c^0 is the electrolyte concentration in the bulk phase.

The interfacial capacitance C consists of the capacitance of the compact layer C_h and capacitance of two diffuse layers C_d^α and C_d^β. Differentiating Eq. (13) with respect to charge q^α of phase α, and assuming that the drop of the potential in the compact layer $\Delta_\beta^\alpha \phi_h$ does not depend on an electrolyte concentration, we have:

$$\frac{1}{C(q,c)} = \frac{1}{C_h(q)} + \frac{1}{C_d^\alpha(q,c)} + \frac{1}{C_d^\beta(q,c)} \quad (15)$$

Diffuse layer capacitance depends on potential and electrolyte concentration:

$$C_d = F\sqrt{\frac{2\varepsilon_0 \varepsilon c}{RT}} \cosh \frac{F\phi_d}{2RT}$$

$$= \frac{F}{2RT}\sqrt{8\varepsilon_0 \varepsilon RTc + q^2} \quad (16)$$

In classical electrochemistry, the capacitance usually depends on surface charge q (Fig. 6).

2.3.3
The Electrocapillary Equation

A comprehensive thermodynamic theory of the electrocapillary phenomena at polarizable and nonpolarizable liquid interfaces was developed by Markin and Volkov [11, 12] using Hansen's method [13]. Electrochemical processes occurring at the interface between two immiscible liquids are traditionally described on the basis of Gibbs thermodynamics of surfaces. However, Hansen's method, by extending and generalizing Gibbs method gives us a better understanding of the nature of interfacial phenomena and provides us with an improved method for describing them. Consider two phases in contact, α and β with r_h neutral components, whose chemical potentials are designated by μ_h, r_c types of cations and r_a types of anions with chemical potentials μ_i. Using Hansen's reference system and choosing the solvents α and β as reference substances, one can write down Hansen's rendition of Gibbs adsorption equation:

$$d\gamma = -s^s_{(\alpha,\beta)}\,dT + \tau^s_{(\alpha,\beta)}\,dp$$
$$- \sum_{h \neq \alpha,\beta} \Gamma_{h(\alpha,\beta)}\,d\mu_h - \sum_i \Gamma_{i(\alpha,\beta)}\,d\bar{\mu}_i \quad (17)$$

Here $s^s_{(\alpha,\beta)}$, $\tau^s_{(\alpha,\beta)}$ and $\Gamma_{h(\alpha,\beta)}$ are the surface excesses of the entropy, volume, and substance h in the reference system, when the absolute excesses α and β are zero. The terms corresponding to different components of the system are divided into two summations, one for neutral and the other for charged particles, the reference substances α and β being excluded from the sum of neutral particles. Individual components of the system, both charged and neutral, can be present either in both phases or in only one. The electrocapillary equation in its final form, which includes the electromotive force of the measuring cell, can be written as:

$$d\gamma = -\left(s^s_{(\alpha,\beta)} - \left(\frac{1}{z_{j'}F}\right)Q^\alpha s^*_{R_\alpha}\right.$$
$$\left. - \left(\frac{1}{z_{m'}F}\right)Q^\beta s^*_{R_\beta}\right)dT$$
$$+ \left(\tau^s_{(\alpha,\beta)} - \left(\frac{1}{z_{j'}F}\right)Q^\alpha v^*_{R_\alpha}\right.$$
$$\left. - \left(\frac{1}{z_{m'}F}\right)Q^\beta v^*_{R_\beta}\right)dp$$
$$- \sum_{h \neq \alpha,\beta} \Gamma_{h(\alpha,\beta)}\,d\mu_h - \sum_{k \neq k'} \left(\frac{1}{\bar{v}_{j'k}}\right)$$
$$\times \Gamma_{k(\alpha,\beta)}\,d\mu_{j'k}$$

$$-\sum_{j\neq j'}\left(\frac{1}{v_{jk'}^+}\right)\Gamma_{j(\alpha,\beta)}\,d\mu_{jk'}$$

$$-\sum_{m\neq m'}\left(\frac{1}{v_{l'm}^-}\right)\Gamma_{m(\alpha,\beta)}\,d\mu_{l'm}$$

$$-\sum_{l\neq l'}\left(\frac{1}{v_{lm'}^+}\right)\Gamma_{l(\alpha,\beta)}\,d\mu_{lm'} - \left(\frac{1}{z_{k'}v_{j'k'}^-}\right)$$

$$\times\left(z_{k'}\Gamma_{k'(\alpha,\beta)} + \sum_{j\neq j'}z_j\Gamma_{j(\alpha,\beta)}\right)$$

$$\times\,d\mu_{j'k'} - \left(\frac{1}{z_{lk'}v_{l'm'}^+}\right)$$

$$\times\left(z_{l'}\Gamma_{l'(\alpha,\beta)} + \sum_{m\neq m'}z_m\Gamma_{m(\alpha,\beta)}\right)$$

$$\times\,d\mu_{l'm'} - Q^\alpha\,dE_{\beta(m')}^{\alpha(j')} \quad (18)$$

Note the symmetry of the last term with respect to the transposition of the α and β indexes:

$$Q^\alpha\,dE_{\beta(m')}^{\alpha(j')} = Q^\beta\,dE_{\alpha(j')}^{\beta(m')} \quad (19)$$

Equation (18) contains $r_h + r_c + r_a - 1$ independent differentials of intensive variables, whose number is equal to the number of the degrees of freedom of the system.

The impedance or electrocapillary properties of the ITIES can be measured with the cell mentioned in Sch. 1.

Silver–silver chloride electrodes are usually used as the reference electrodes (RE), water is used as phase α, and nitrobenzene or 1,2-dichloroethane as phase β. The interface between phases 1 and 2 is the polarized oil–water interface serving as a working electrode (interface). The common cation A^+ is usually the terabutylammonium ion, and B^- is tetraphenylborate or dicarbollylcobaltate.

The potential difference measured in the cell (I) consists of the sum of two interfacial potential drops:

$$-E = \Delta_\beta^\alpha\phi + \Delta_\alpha^{Ag}\phi_{RE_1} - \Delta_\alpha^{Ag}\phi_{RE_2} + \Delta\phi_D \quad (20)$$

$\Delta\phi_D$ is the Nernst–Donnan potential difference between the nonaqueous fraction and the aqueous solution of RE_2 with the common ion A^+:

$$\Delta\phi_D = -\Delta_\beta^\alpha\phi_{A^+}^0 + \frac{RT}{F}\ln\frac{a_2}{a_3} \quad (21)$$

and $\Delta_\beta^\alpha\phi_{A^+}^0$ is the standard potential difference for A^+ known from literature data or calculated theoretically.

By suitable selection of the Cl^- concentrations and ionic strengths of the electrolytes into which $Ag/AgCl^-$ electrodes are immersed, we obtain:

$$\Delta_\alpha^{Ag}\phi_{RE_1} - \Delta_\alpha^{Ag}\phi_{RE_2} = \frac{RT}{F}\ln\frac{a_1}{a_3} \quad (22)$$

Substituting (7.18) and (7.19) into (7.17), we obtain:

$$d\Delta_\beta^\alpha\phi = -dE - d\Delta_\beta^\alpha\phi_D + \frac{RT}{F}d\ln\frac{a_3}{a_1} \quad (23)$$

With Eq. (23), the electrocapillarity equation at the flat oil–water interface can be

RE$_1$	H$_2$O (α) M$^+$Cl$^-$	1	A$^+$B$^-$ (β) oil	2	A$^+$C$^-$ (α) H$_2$O	3	RE$_2$
	a_1		a_2		a_3		
	$\Delta_\alpha^{Ag}\phi_{RE_1}$		$\Delta_\beta^\alpha\phi$		$\Delta\phi_D$	$-\Delta_\alpha^{Ag}\phi_{RE_2}$	

Scheme 1

written as follows:

$$d\gamma = -\sum_i \Gamma_i\, d\mu_i - Q_\alpha\, dE$$

$$+ Q_\alpha \frac{RT}{F} d\ln \frac{a_2 a_1}{a_3^2} \quad (24)$$

where Q is the charge that must be supplied to each side (α or β, correspondingly) of the polarizable interface when expanding it by a unit area in order to maintain the potential difference between the phases [14]. Figures 5 to 7 illustrate dependencies of interfacial tension and capacitance on the potential difference between two immiscible electrolyte solutions.

At the ideally polarizable interface, the ions of each group are separated by the Koenig barrier, which rules out a possible overlap between the groups (Fig. 1b). In this case, the free charges q_α and q_β of the phases are equal to the thermodynamic charges: $q_\alpha = Q_\alpha = -q_\beta = -Q_\beta$.

2.3.4 Zero-Charge Potentials

Potentials of zero charge of the interface can be found reliably by the same independent methods that are used at the metal–water interface. These include finding the differential capacitance minimum of the electric double layer, from electrocapillary curves, with a flowing-electrolyte electrode, with the vibrating boundary method, with radiotracers, or by measuring the second harmonic

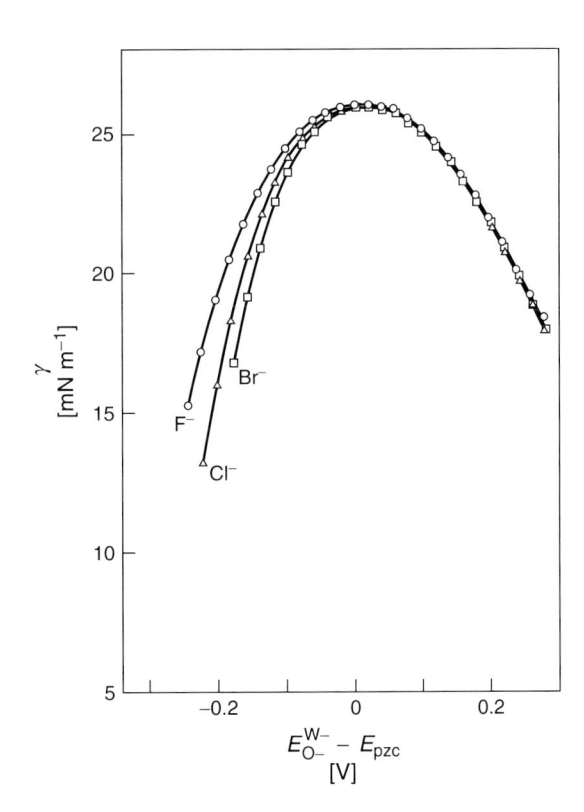

Fig. 5 Comparison of electrocapillary curves for the interface between nitrobenzene solution of 0.1 M hexadecyltrimethylammonium tetraphenylborate and aqueous solution of 0.05 M LiF (○), LiCl (∇), and LiBr (□). (From Ref. [15], reproduced by permission of The Chemical Society of Japan.)

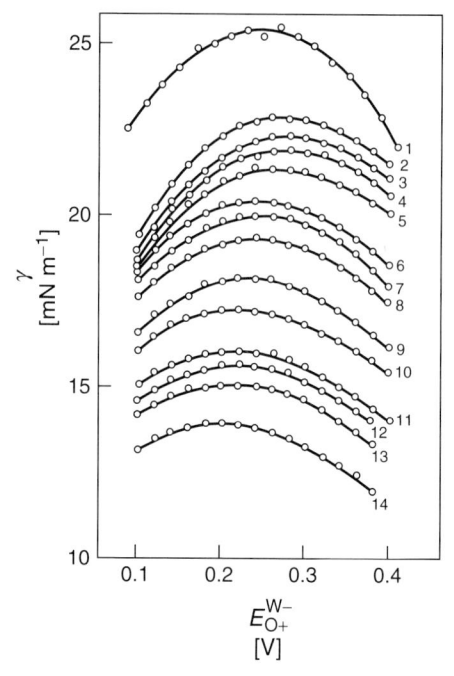

Fig. 6 Electrocapillary curves at 25 °C for the interface between nitrobenzene solution of 0.1 M tetrapentylammonium tetraphenylborate and aqueous solution of 0.05 M LiCl in the presence of x mmol dm^{-3} C12E4: $x = 0$ (curve 1), 1 (curve 2), 2 (curve 3), 5 (curve 4), 7 (curve 5), 10 (curve 6), 15 (curve 7), 20 (curve 8), 30 (curve 9), 40 (curve 10), 50 (curve 11), 70 (curve 12), 80 (curve 13), and 100 (curve 14). (From Ref. [16], reproduced by permission of The Chemical Society of Japan.)

generation. Potentials of the zero free charge at nitrobenzene/water and 1,2-dichloroethane/water interfaces, obtained from the differential capacitance minimum of the electric double layer in solutions of a surface-inactive electrolyte do not necessarily correspond to the potentials of thermodynamic zero charge (Fig. 7). They can depend on electrolyte concentration (Fig. 8) when the capacitance of the compact layer is affected by surface charge as a result of nonlinear double-layer properties.

The potential difference between two immiscible electrolyte solutions can be written as the sum:

$$\Delta_\beta^\alpha \phi = \Delta_\beta^\alpha \phi_h + \phi_d^\alpha + \phi_d^\beta \quad (25)$$

where $\Delta_\beta^\alpha \phi$ and ϕ_d are the potential drops across the compact and diffuse layers, respectively. In the linear approximation and in the absence of specific adsorption, the electric double layer is equivalent to three capacitors in series:

$$\frac{d\Delta_\beta^\alpha \phi}{dq} = \frac{d\Delta_\beta^\alpha \phi_h}{dq} + \frac{d\phi_d^\alpha}{dq} + \frac{d\phi_d^\beta}{dq} \quad (26)$$

One can use the following approximation when modeling the electric double layer:

$$\frac{d^2 \Delta_\beta^\alpha \phi}{dq^2} = \frac{d^2 \Delta_\beta^\alpha \phi_h}{dq^2} + \frac{d^2 \phi_d^\alpha}{dq^2} + \frac{d^2 \phi_d^\beta}{dq^2} \quad (27)$$

where

$$\Delta_\beta^\alpha \phi'' = -\frac{C_\phi''}{C^3}; \quad \Delta_\beta^\alpha \phi_h'' = -\frac{C_{h\phi_h}'}{C_h^3};$$

$$\phi_\alpha'' = -\frac{C_{d\phi_\alpha}^{\alpha'}}{(C_d^\alpha)^3}; \quad \phi_\beta'' = -\frac{C_{d\phi_\beta}^{\beta'}}{(C_d^\beta)^3} \quad (28)$$

We find the position of the minimum in the C vs. $\Delta_\beta^\alpha \phi$ curve as:

$$\Delta_\beta^\alpha \phi'' = \Delta_\beta^\alpha \phi_h'' + \phi_d^{\alpha''} + \phi_d^{\beta''} = 0 \quad (29)$$

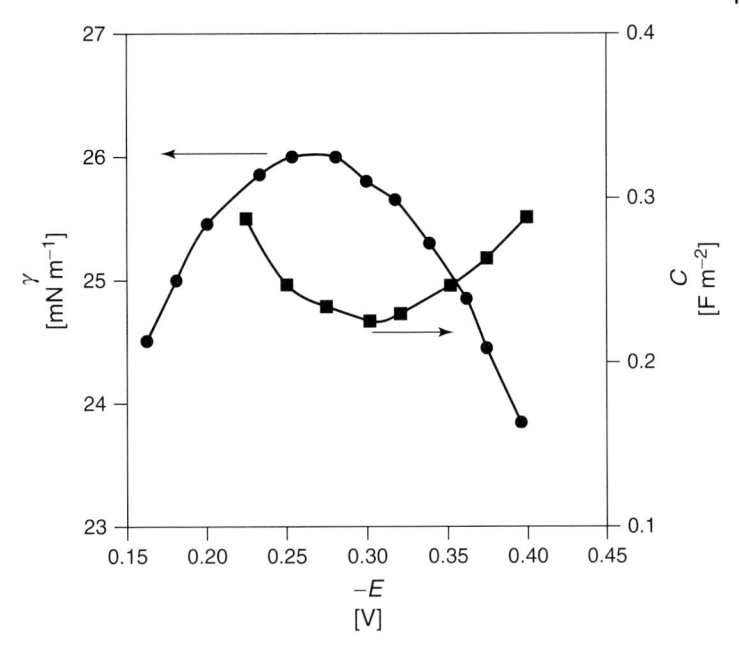

Fig. 7 Interfacial tension and differential capacity as functions of applied potential difference. The medium: 0.1 M LiCl in water, 0.1 M tetrabutylammonium tetraphenylborate in the nitrobenzene at 25 °C. (Experimental points are taken from Refs. [17, 18].)

It follows from the Gouy–Chapman diffuse-layer theory that if $q_\alpha = q_\beta = 0$ then, $(\phi_d^\alpha)'' = (\phi_d^\beta)'' = 0$. Therefore, $\Delta_\beta^\alpha \phi'' = 0$ applies only when $\Delta_\beta^\alpha \phi_i'' = 0$, that is, when the compact layer capacitance is independent of surface charge. It has been shown that C_c generally depends on potential and charge, and $\Delta_\beta^\alpha \phi_i'' \neq 0$. As the value of $\Delta_\beta^\alpha \phi_i''$ in the region of the potential of zero free charge increases, the value of q corresponding to the minimum in the C vs. $\Delta_\beta^\alpha \phi$ curve will also increase [14].

For the water-nitrobenzene system, one observes a strong dependence of capacitance of the compact double layer on surface charge, even in the absence of specific adsorption. This leads to a dependence of the potential minimum in the capacitance curve on electrolyte concentration (Fig. 8). Therefore, a correct determination of the zero free-charge potential for the nitrobenzene-water system in the presence of a binary surface-inactive electrolyte is possible only at base-electrolyte concentrations less than 0.01 M. For the water-nitrobenzene system, this quantity $\Delta_{nb}^w \phi_{pzfc} = -0.29$ V and for the water-1,2-dichloroethane system $\Delta_{de}^w \phi_{pzfc} = -0.27$ V.

For many systems, the maximum of the electrocapillary curve is located in the region of ideal polarizability of the electrode, and the maximum potential corresponds to the potential of zero total (thermodynamic) and zero free charge. At the mercury–water interface, the region of ideal polarization is a few volts when soluble mercury salts are absent (2.2 V), but at the interface between two immiscible

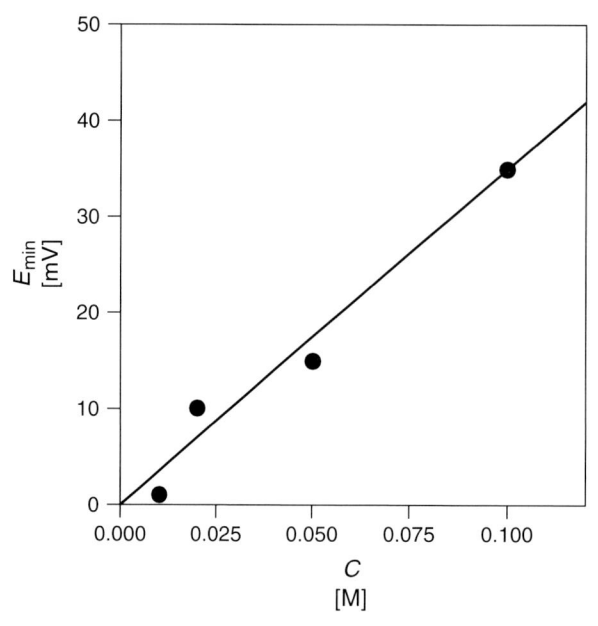

Fig. 8 Displacement of the differential-capacity minimum on the potential axis as a function of electrolyte concentration in the water-nitrobenzene system. The medium: 0.1 M LiCl in water and 0.1 M tetrabutylammonium tetraphenylborate in nitrobenzene. (Constructed from data of Samec and coworkers [19].)

liquids, the region is about a tenth of that value. Under conditions in which the current flow does not significantly alter the compositions of the liquid phases and in which the equilibrium at the interface is not disturbed by the current, the electrocapillary curves yield the potential of zero total charge. However, under realistic conditions in which the region of ideal polarization is narrow, the maximum of the electrocapillary curve corresponds to the potential of zero total charge, rather than zero free charge. This closely resembles systems consisting of amalgams and liquid electrolyte (LE) solutions that are described by Frumkin [20].

The structure of the electric double layer at ITIES has been investigated using such common electrochemical methods of capacitance measurements as impedance, galvanostatic, and potentiodynamic techniques. More recently, it has become possible to measure electric double-layer capacitance at the ITIES using a four-electrode potentiostat or two-electrode potentiostat. Commonly studied systems are nitrobenzene/water and 1,2-dichloroethane/water, in which the organic phase has a relatively high dielectric permittivity and a high dissociation constant.

In typical investigations, cyclic current-potential curves are usually determined before impedance measurements in order to find the potential range in which the contribution of faradaic impedance is low. The electric double-layer capacitance is measured in the potential window between extremes in the cyclic voltamogram. Over this range of potentials, the water–dichloroethane interface has properties close to those of an ideally polarizable electrode.

A "diffuse" picture of the compact layer has been considered [19]. According to this hypothesis, adsorbed ions can penetrate into the compact layer, which consists of solvent dipoles, analogous to nonlocalized electron gas penetration from a metal electrode to an aqueous electrolyte solution. Here, a penetrating ion can act as a hydrophobic anion in the organic phase and potential distribution $\phi(x)$ near to the point of zero charge can be calculated using the Poisson–Boltzmann equation:

If $x < x_2^w$, $\varepsilon = \varepsilon_w$:

$$\phi'' = (\kappa_w)^2(\phi - \Delta_{org}^w\phi) \quad (30)$$

If $x_2^w < x < x_2^{org}$, $\varepsilon = \varepsilon_h$:

$$\phi'' = (\kappa_h)^2 \phi \quad (31)$$

If $x > x_2^{org}$, $\varepsilon = \varepsilon_{org}$:

$$\phi'' = (\kappa_{org})^2(\phi - \Delta_{org}^w\phi) \quad (32)$$

In each of these three areas, the solution can be written as

$$\phi(x) = A_+ e^{\kappa x} + A_- e^{-\kappa x} \quad (33)$$

where six coefficients A_+ and A_- can be determined from the six boundary conditions:

$$\begin{aligned}
&\phi(-\infty) = \Delta_{org}^w\phi \\
&\phi(\infty) = 0 \\
&\phi(x_2^w - 0) = \phi(x_2^w + 0) \\
&\phi(x_2^{org} - 0) = \phi(x_2^{org} + 0) \\
&\varepsilon_w \phi'(x_2^w - 0) = \varepsilon_h \phi'(x_2^w + 0) \\
&\varepsilon_h \phi'(x_2^{org} - 0) = \varepsilon_{org} \phi'(x_2^{org} + 0)
\end{aligned} \quad (34)$$

As a result, the equation for the capacitance of the electric double layer is:

$$C = \frac{q^w}{\Delta_{org}^w\phi} = -\frac{(\kappa_w)^2 \varepsilon_w \varepsilon_{org}}{\Delta_{org}^w\phi}$$

$$\times \int_{-\infty}^{x_2^w} (\phi - \Delta_{org}^w\phi)\,dx = \kappa_w \varepsilon_w \varepsilon_{org}$$

$$\times \frac{\left[\begin{array}{c}\left(1 + \dfrac{\varepsilon_h \kappa_h}{\varepsilon_{org}\kappa_{org}}\right)\exp(\kappa_h \delta) \\ + \left(1 - \dfrac{\varepsilon_h \kappa_h}{\varepsilon_{org}\kappa_{org}}\right)\exp(-\kappa_h \delta)\end{array}\right]}{\left(1 + \dfrac{\varepsilon_w \kappa_w}{\varepsilon_h \kappa_h}\right)\left(1 + \dfrac{\varepsilon_h \kappa_h}{\varepsilon_{org}\kappa_{org}}\right)}$$

$$\times \exp(\kappa_h \delta) + \left(1 - \dfrac{\varepsilon_h \kappa_h}{\varepsilon_{org}\kappa_{org}}\right)$$

$$\times \left(1 - \dfrac{\varepsilon_w \kappa_w}{\varepsilon_h \kappa_h}\right)\exp(-\kappa_h \delta) \quad (35)$$

where $\delta = x_2^{org} - x_2^w$. It should be noted that the size of a hydrophobic penetrating anion is large in relation to the compact part of the EDL and it is not clear if the Poisson–Boltzmann equation can be used in this case.

2.3.5
Specific Adsorption at Liquid–Liquid Interfaces

Ions can be adsorbed specifically if the main contribution to their interaction with the interface (ions, dipoles) is caused by non-coulombic short-range forces. *Specific adsorption* cannot be explained using only the theory of diffuse double layer. Specifically adsorbed ions penetrate to the compact layer and form a compact or loose monolayer. The surface passing through the centers of specifically adsorbed ions is usually called the *inner Helmholtz plane*. If several kinds of specifically adsorbed ions are present, each ion can have its own inner Helmholtz plane.

It was found while studying the mechanism of electron transfer (ET) across the interface between two immiscible liquids that specific anion adsorption occurs at the octane–water interface in the presence of metalloporphyrins. This adsorption increases in the order of Cl^-, Br^-, I^-, and is caused by coordinative bonding of the anions as ligands of the porphyrin metal atoms.

Specific ion adsorption at the polarizable nitrobenzene–water interface containing monolayers of phosphatidylcholine, phosphatidylserine, octaethylene glycol monodecyl ethers, tetraethylene glycol monodecyl ether, and hexadecyltrimethylammonium (HTMA$^+$) was studied in detail. HTMA$^+$ exhibited no specific adsorption in the potential range in which the aqueous phase was positive [21], whereas a strong adsorption occurred in the potential range in which the electric potential in the aqueous phase was negative with respect to the nitrobenzene (Fig. 9).

Another example of specific ion adsorption was discussed in terms of the formation of interfacial ion pairs between ions in the aqueous and the organic phase. The contribution of specific ionic adsorption to the interfacial capacitance can be calculated using the Bjerrum theory of ion-pair formation. The results show that a phase boundary between two immiscible electrolyte solutions can be described as a mixed solvent region with varying penetration of ion pairs into it, depending on their ionic size. The capacitance increases with increasing ionic size in the order $Li^+ < Na^+ < K^+ < Rb^+ < Cs^+$. Yufei et al. [22] found that significant specific ion adsorption occurs at the interface between two immiscible electrolytes and the potential dependence of the capacitance is strongly influenced by the adsorption isotherms owing to the interfacial ionic association.

When there is specific adsorption of ions dissolved in phase α the condition of electroneutrality can be written as

$$q^\beta = -q^\alpha = -(q_i^\alpha + q_d^\alpha) \quad (36)$$

where q_i is the charge of the inner Helmholtz plane. The separation of q_α from q_i^α and q_d^α cannot be done without introducing a model of the interface.

Although capacitance of the interface can be calculated as before using the equation

$$\frac{d\Delta_\beta^\alpha \phi}{dq} = \frac{d\Delta_\beta^\alpha \phi_i}{dq} + \frac{d\phi_d^\alpha}{dq} + \frac{d\phi_d^\beta}{dq} \quad (37)$$

the second term in the right side of the equation is not the diffuse-layer capacitance since the charge of a diffuse layer in phase a is equal to

$$q_d^\alpha = -q_\beta - \sum_j q_i^{\alpha,j} \quad (38)$$

The diffuse-layer capacitance is equal to

$$C_d = -\frac{dq_\alpha}{d\phi_d^\alpha} \quad (39)$$

and one can write

$$C^{-1} = (C^i)^{-1} + (C_d^\alpha)^{-1}\left(1 + \frac{d\sum_j q_i^{\alpha,j}}{dq^\beta}\right)$$
$$+ (C_d^\beta)^{-1} \quad (40)$$

The drop of a potential in the compact layer depends on the surface charge q_α and the charge of the inner Helmholtz plane $q_i^{\alpha,j}$.

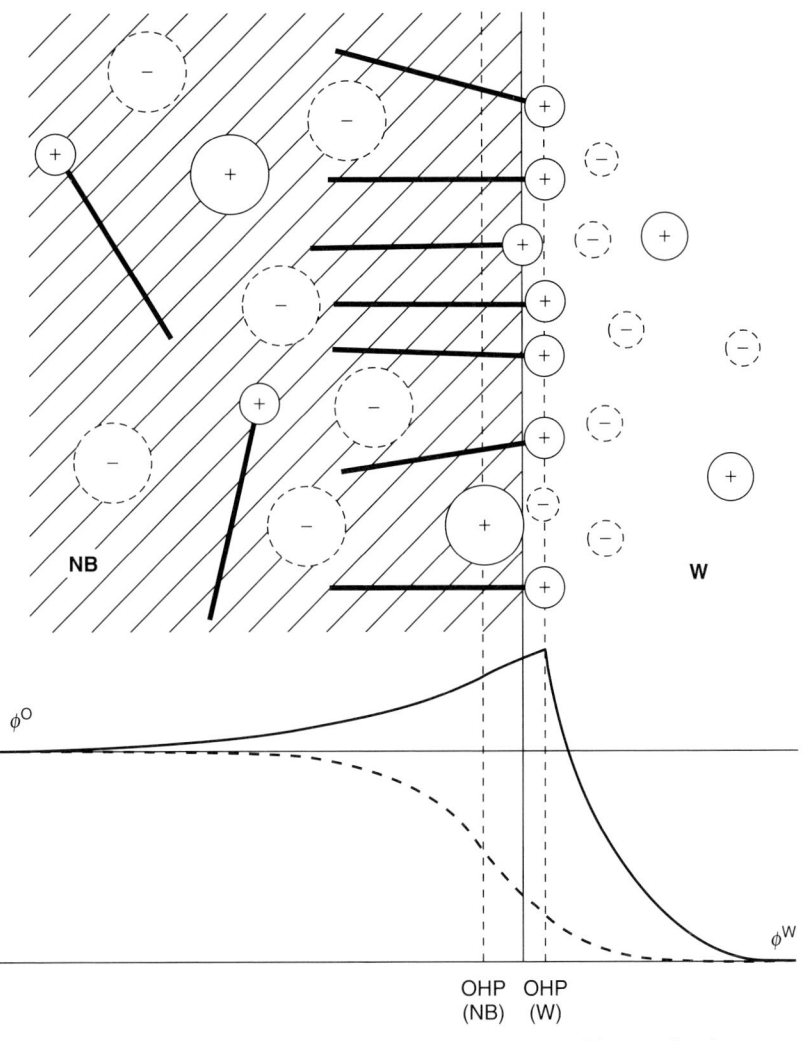

Fig. 9 Schematic representation of the double-layer structure of the interface between nitrobenzene and aqueous solutions in the presence of the specific adsorption of hexadecyltrimethylammonium ions [21]. (Reproduced by permission of the Chemical Society of Japan.)

It shows that

$$d\phi_i = \left[\frac{\partial \phi_i}{\partial q^\beta}\right]_{q_i^{\alpha,j}} dq^\beta + \sum \left[\frac{\partial \phi_i}{\partial q_i^{\alpha,j}}\right]_{q^\beta, q_i^{\alpha,k}} dq_i^{\alpha,j} \quad (41)$$

or

$$\frac{d\phi_i}{dq^\beta} \equiv \frac{1}{C_i}$$

$$= \left[\frac{\partial \phi_i}{\partial q^\beta}\right]_{q_i^{\alpha,j}} + \sum \left[\frac{\partial \phi_i}{\partial q_i^{\alpha,j}}\right]_{q^\beta, q_i^{\alpha,k}} \frac{dq_i^{\alpha,j}}{dq^\beta} \quad (42)$$

Here the index $q_i^{\alpha,j}$ means that all $q_i^{\alpha,j}$ are constant except one.

In the absence of specific adsorption the capacitances of the interface, compact and diffuse layers are always positive. The capacitance of the compact layer in the presence of specific adsorption can be either positive or negative.

2.3.6
Adsorption Isotherm

Traditional models for calculation of adsorption isotherms are based on the assumption that surface-active compounds at the interface can substitute for adsorbed molecules of one solvent but cannot penetrate the second phase. Although these models are useful for metal–water interfaces, recent interest has focused on the surface chemistry of amphiphilic compounds that can penetrate both phases and replace adsorbed molecules of both solvents, for example, water and oil. Amphiphilic molecules consist of two moieties with opposing properties: a hydrophilic polar head and a hydrophobic hydrocarbon tail. We present here, a theoretical analysis of the generalized Frumkin adsorption isotherm for amphiphilic compounds.

The interface between two immiscible liquids may be considered to be a surface solution of surfactant in a special kind of solvent. In order to calculate the entropy of such a solution, we will adopt a simplified lattice model and use lattice statistics, a widely used method for describing surface solutions. The transition from three-dimensional (3-d) to two-dimensional (2-d) geometry may cause errors in statistical formulas, if some peculiarities of 2-d solutions are overlooked.

The main difficulty when dealing with a monolayer of a surfactant, one can consider this monolayer as a 2-d system. The solvent molecules do not form a monolayer, but rather a multilayer. Therefore, the transition from 3-d- to 2-d-geometry should be specified. Consider molecules of both solvents that are substituted by a surfactant (Fig. 10). Suppose that these molecules can be assembled into columns consisting of m_o molecules of oil and m_w molecules of water. Suppose that one column of oil molecules matches the n_w molecules of water. This match of 1 oil column and n_w water columns will be considered in what follows as a quasi-molecule of solvent Q. These quasi-molecules constitute a "monolayer" of solvent. They consist of m_o oil molecules and $n_w m_w$ water molecules.

Designate the molecules of surfactant in the bulk as A, and in the monolayer as B. At the interface, aggregation of surfactant molecules can take place, B ⇔ rA, such as dimerization of porphyrin or pheophytin molecules at the octane–water interface. Let the surfactant B replace p quasi-molecules at the interface. Therefore, one can write

$$pQ + rA = B + p(\text{oil}) + pn_w (\text{water}) \tag{43}$$

The chemical potentials for (7.76) are:

$$p\mu_Q^s + r\mu_A^b = \mu_B^s + p\mu_o^b + pn_w\mu_w^b \tag{44}$$

Taking the 2-d solution as ideal, we have:

$$\mu_Q^s = \mu_Q^{0,s} + RT \ln X_Q^s \tag{45}$$

$$\mu_B^s = \mu_B^{0,s} + RT \ln X_B^s \tag{46}$$

In the bulk phase, we have:

$$\mu_A^b = \mu_A^{0,b} + RT \ln X_A^b \tag{47}$$

$$\mu_o^b = \mu_o^{0,b} + RT \ln X_o^b \tag{48}$$

$$\mu_w^b = \mu_w^{0,b} + RT \ln X_w^b \tag{49}$$

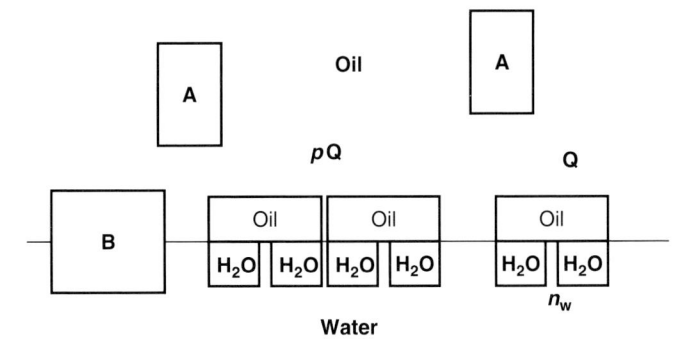

2.3 Electrochemical Double Layers: Liquid–Liquid Interfaces

Fig. 10 Structure of oil–water interface with adsorbed monolayer of amphiphilic surfactant B.

In all these equations, X designates the mole ratio of corresponding substances. Substituting these equations into (44), one obtains:

$$p\mu_Q^{0,s} + r\mu_A^{0,b} - \mu_B^{0,s} - p\mu_o^{0,b} - pn_w\mu_w^{0,b}$$
$$+ RT \ln \frac{(X_A^b)^r}{(X_o^b)^p (X_w^b)^{pn_w}} = RT \ln \frac{X_B^s}{(X_Q^s)^p} \quad (50)$$

Using the standard Gibbs free energy of adsorption

$$\Delta_b^s G^0 = \mu_B^{0,s} - r\mu_A^{0,b} + p\mu_o^{0,b}$$
$$+ pn_w\mu_w^{0,b} - p\mu_Q^{0,s} \quad (51)$$

one obtains the adsorption isotherm:

$$\frac{X_B^s}{(X_Q^s)^p} = \frac{(X_A^b)^r}{(X_o^b)^p (X_w^b)^{pn_w}} \exp\left(-\frac{\Delta_b^s G^0}{RT}\right) \quad (52)$$

We considered the 2-d solution of surfactant B in the solvent of quasi-particles Q, in which the mole ratios were defined as

$$X_B^s = \frac{N_B^s}{N_B^s + N_Q^s}; \quad X_Q^s = \frac{N_Q^s}{N_B^s + N_Q^s} \quad (53)$$

Some authors prefer another set of definitions when real particles in the interface are considered. The equation for this state with real particles A, O, W becomes

$$X_A^s = \frac{N_A^s}{N_A^s + N_o^s + N_w^s} \quad (54)$$

$$X_O^s = \frac{N_o^s}{N_A^s + N_o^s + N_w^s} \quad (55)$$

$$X_w^s = \frac{N_w^s}{N_A^s + N_o^s + N_w^s} \quad (56)$$

and, we can obtain:

$$X_B^s = \frac{X_A^s}{X_A^s + X_o^s}; \quad X_Q^s = \frac{X_o^s}{X_A^s + X_o^s} \quad (57)$$

The adsorption isotherm can then be presented in the form

$$\frac{X_A^s}{(X_Q^s)^p}(X_A^s + X_o^s)^{p-1}$$
$$= \frac{(X_A^b)^r}{(X_o^b)^p (X_w^b)^{pn_w}} \exp\left(-\frac{\Delta_b^s G^0}{RT}\right) \quad (58)$$

In the past, the adsorption isotherm was presented in terms of the fraction θ of the surface actually covered by the adsorbed surfactant.

If we introduce η as the ratio of areas occupied in the interface by the molecules of surfactant and oil, the mole fractions in surface solution can be presented as follows:

$$X_B^s = \frac{\Theta}{\Theta + \eta(1-\Theta)}; \quad X_o^s = \frac{\eta(1-\Theta)}{\Theta + \eta(1-\Theta)} \quad (59)$$

The adsorption isotherm takes the form

$$\frac{\Theta}{\eta^p(1-\Theta)^p}[\Theta + \eta(1-\Theta)]^{p-1}$$
$$= \frac{(X_A^b)^r}{(X_o^b)^p(X_w^b)^{pn_w}} \exp\left(-\frac{\Delta_b^s G^0}{RT}\right) \quad (60)$$

In this isotherm, the mole fractions X_A^b, X_o^b, X_w^b of the components in the bulk solution are presented. In the general case, they must be substituted with activities:

$$\frac{\Theta}{\eta^p(1-\Theta)^p}[\Theta + \eta(1-\Theta)]^{p-1}$$
$$= \frac{(a_A^b)^r}{(a_o^b)^p(a_w^b)^{pn_w}} \exp\left(-\frac{\Delta_b^s G^0}{RT}\right) \quad (61)$$

If the molecules B can interact as pairs in the adsorbed layer and the energy of each new particle is proportional to its concentration, then their chemical potential, μ_B^s, instead of Eq. (46), should be presented as:

$$\mu_B^s = \mu_B^{s,0} + RT \ln X - 2aRTX \quad (62)$$

where a is so-called attraction constant. Then, after some algebra we obtain, instead of Eq. (60), the isotherm:

$$\frac{\Theta[\eta-(\eta-1)\Theta]^{p-1}}{\eta^p(1-\Theta)^p} \exp(-2a\Theta)$$
$$= \frac{(X_A^b)^r}{(X_o^b)^p(X_w^b)^{pn_w}} \exp\left(-\frac{\Delta_b^s G^0}{RT}\right) \quad (63)$$

Recall that η was introduced as the ratio of areas occupied in the interface by the molecule of surfactant to the same of oil, and p was introduced as the number of columns of oil, which could be supplanted with one molecule of surfactant. Therefore, p is a relative size of the surfactant molecule in the interfacial layer. It is reasonable to suppose that:

$$\eta = p \quad (64)$$

If the concentration of surfactant in the solution is not high and the mutual solubility of oil and water is low, then we can use the approximation $X_o^b = X_w^b = 1$, so that the general Eq. (63) simplifies to:

$$\frac{\Theta[p-(p-1)\Theta]^{p-1}}{p^p(1-\Theta)^p} \exp(-2a\Theta)$$
$$= (X_A^b)^r \exp\left(-\frac{\Delta_b^s G^0}{RT}\right) \quad (65)$$

This is the final expression for the isotherm that we will call the amphiphilic isotherm [23]. It is straightforward to derive classical adsorption isotherms from the amphiphilic isotherm (65).

1. The Henry isotherm, when $a=0$, $r=1$, $p=1$, $\Theta \ll 1$:

$$\Theta = X_a^b \exp\left(-\frac{\Delta_b^s G^0}{RT}\right) \quad (66)$$

2. The Freundlich isotherm, when $a=0$, $p=1$, $\Theta \ll 1$:

$$\Theta = (X_a^b)^r \exp\left(-\frac{\Delta_b^s G^0}{RT}\right) \quad (67)$$

3. The Langmuir isotherm, when $a=0$, $r=1$, $p=1$:

$$\frac{\Theta}{1-\Theta} = X_a^b \exp\left(-\frac{\Delta_b^s G^0}{RT}\right) \quad (68)$$

4. The Frumkin isotherm, when $r=1$, $p=1$:

$$\frac{\Theta}{1-\Theta}\exp(-2a\Theta) = X_a^b \exp\left(-\frac{\Delta_b^s G^0}{RT}\right) \tag{69}$$

Therefore, the amphiphilic isotherm (65) could be considered as a generalization of the Frumkin isotherm, taking into account the replacement of some solvent molecules with larger molecules of surfactant. Of course, the amphiphilic isotherm includes all the features of the Frumkin isotherm and displays some additional ones. To elucidate them, it will be convenient to change the variable X_A^b to the relative concentration $y = X_A^b/X_A^b(\Theta = 0.5)$, where $X_A^b(0.5)$ is the concentration corresponding to the surface coverage $\theta = 0.5$:

$$y = \frac{\Theta[p - (p-1)\Theta]^{p-1}}{(p+1)^{p-1}(1-\Theta)^p}\exp(a - 2a\Theta) \tag{70}$$

This equation gives the coverage fraction θ as a function of relative concentration y, while a and p are the parameters of this isotherm, the first being the attraction constant and the second, the size of surfactant. These parameters play an important role because their effect on the shape of amphiphilic isotherm is very strong.

Amphiphilic isotherm (65) analysis can be used for the determination of the interfacial structure. An amphiphilic molecule, which consists of two moieties with opposing properties such as a hydrophilic polar head and a hydrophobic hydrocarbon tail, should be used as an analytical tool located at the interface. Pheophytin a is a well-known surfactant molecule that contain a hydrophobic chain (phytol) and a hydrophilic head group. The value of p, less than 1.0, indicates that adsorbed molecules of n-octane are parallel to the interface between octane and water [23] (Fig. 11). Substitution of one adsorbed octane molecule requires about 4 to 5 adsorbed pheophytin or chlorophyll molecules. These are supported by molecular dynamic studies in the systems decane/water, nonane/water, and hexane/water. The structure of both water and octane at the interface is different from the bulk. Adsorbed at the interface, octane molecules have a lateral orientation at the interface, as it shown in Fig. 11.

2.3.7
Image Forces

Any charge near phase boundary interacts with it because of the different polarization of two phases. The force of this interaction can be found by solution of the Laplace equation, but it is more convenient to use the method of images. This method employs a fictitious charge ("image"), which together with the given charge creates the right distribution of electric potential in the phase [24]. All ions interact with all images giving rise to the so-called "image forces." The image forces are largely responsible for both positive and negative adsorption. Important contribution to this effect is given by the change of the size of solvation shells of ions in the boundary layer. This has an impact on the planes of closest approach of ions and dipolar molecules to the phase boundary.

The electrostatic Gibbs free energy for an ion in the vicinity of a boundary between two liquids with dielectric constants ε_1 and ε_2 (Fig. 12) is determined by the Born ion solvation energy and by the interaction with its image charge. In research dealing with the energy of image forces and with the interactions of charges at the oil–water interface, approximate models of the interface are often employed, which are based on the traditional description of the interface between two local dielectrics. In the oil–water system, the force of attraction (or repulsion) of charge in the

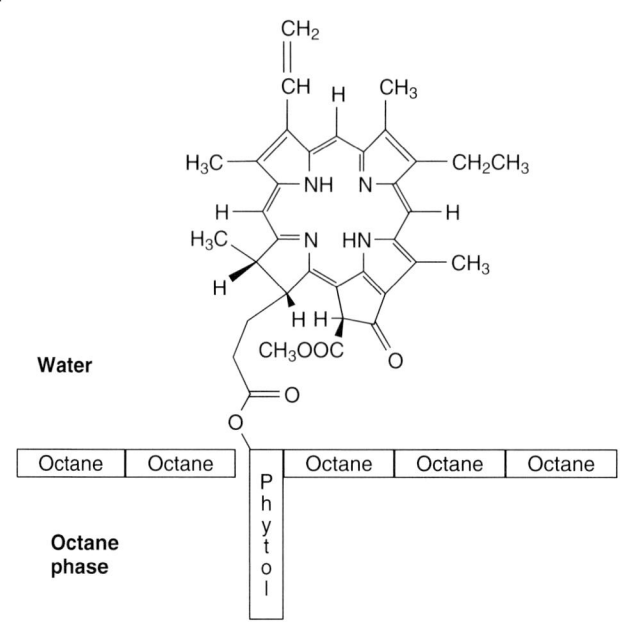

Fig. 11 The amphiphilic compound pheophytin a, which can substitute one-sixth of the lateral oriented adsorbed molecule of octane, is a tool for the evaluation of the interfacial structure.

organic (β) phase with its image in the aqueous phase (α) sitting on the same axis perpendicular to the dividing plane is given by:

$$F(h) = -\frac{\varepsilon_\alpha - \varepsilon_\beta}{\varepsilon_\alpha + \varepsilon_\beta} \frac{(ze_0)^2}{16\pi \varepsilon_0 \varepsilon_\beta h^2} \quad (71)$$

where h is the distance from the interface. If $\varepsilon_\alpha > \varepsilon_\beta$, the charge in the nonpolar phase is attracted to its image, but if $\varepsilon_\alpha < \varepsilon_\beta$ there is repulsion between the charge in an organic phase and its image. From Eq. (71) it follows that charges in the organic phase are attracted to the oil–water interface. Image forces attract the diffuse layer on the organic side, making it thinner, and repel and thicken the aqueous diffuse layer.

The Kharkats–Ulstrup model [25] incorporates the finite radius of an ion a, which is assumed to have a fixed spherically uniform charge distribution and can continuously pass between the two phases. For a point charge at long distances from the interface, electrostatic Gibbs free energy can be written as

$$\Delta G = \frac{(ze_0)^2}{32\pi \varepsilon_0 \varepsilon_\alpha a} \left(4 + \frac{\varepsilon_\alpha - \varepsilon_\beta}{\varepsilon_\alpha + \varepsilon_\beta} \frac{2a}{h} \right) \quad (72)$$

where a is an ionic radius, h is the distance from the interface and ze is the charge of an ion.

The potential ϕ for the spherically symmetrical charge distribution is

$$\phi = \begin{cases} \dfrac{ze_0}{4\pi \varepsilon_0 \varepsilon_\alpha r_1} \\ \quad + \dfrac{ze_0}{4\pi \varepsilon_0 \varepsilon_\alpha r_1'} \dfrac{\varepsilon_\alpha - \varepsilon_\beta}{\varepsilon_\alpha + \varepsilon_\beta} & \text{(in region } \alpha) \\ \dfrac{ze_0}{4\pi \varepsilon_0 \varepsilon_\beta r_1} \dfrac{2\varepsilon_\beta}{\varepsilon_\alpha + \varepsilon_\beta} & \text{(in region } \beta) \end{cases} \quad (73)$$

Fig. 12 The relative arrangement of a spherical ion and a planar interface between two immiscible liquids with dielectric permittivities ε_α and ε_β: (a) the ion-center distance from the interfacial plane P exceeds the ion radius; (b) the ion partially penetrates the interface, $h < a$.

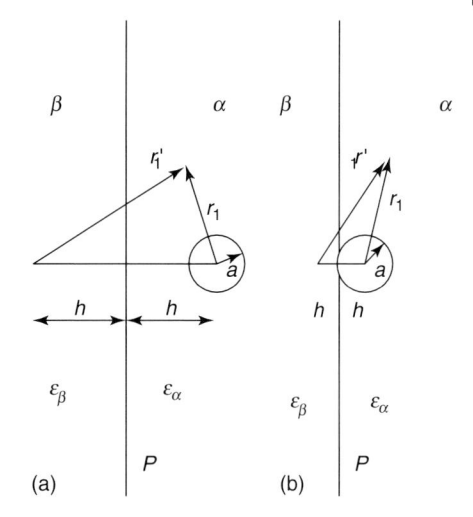

Using standard procedures of calculations, Kharkats and Ulstrup obtained equations for the electrostatic part of the free Gibbs energy of a finite-size ion.

When $h > a$

$$\Delta G = \frac{(ze_0)^2}{32\pi \varepsilon_0 \varepsilon_\alpha a} \left\{ 4 + \left(\frac{\varepsilon_\alpha - \varepsilon_\beta}{\varepsilon_\alpha + \varepsilon_\beta} \right) \frac{2a}{h} \right.$$

$$+ \left(\frac{\varepsilon_\alpha - \varepsilon_\beta}{\varepsilon_\alpha + \varepsilon_\beta} \right)^2$$

$$\left. \times \left[\frac{2}{1 - (2h/a)^2} + \frac{a}{2h} \ln \frac{2h+a}{2h-a} \right] \right\} \quad (74)$$

The first term in (74) is the Born solvation energy, and the second is the interaction with the image. For a charge located in region β, the electrostatic free energy is obtained from (74) by exchanging $\varepsilon_\alpha \Leftrightarrow \varepsilon_\beta$.

Kharkats and Ulstrup [25] obtained the expression for the electrostatic free energy in region $0 \leq h \leq a$:

$$\Delta G = \frac{(ze_0)^2}{32\pi \varepsilon_0 \varepsilon_\alpha a} \left\{ \left(2 + \frac{2h}{a} \right) \right.$$

$$+ \left(\frac{\varepsilon_\alpha - \varepsilon_\beta}{\varepsilon_\alpha + \varepsilon_\beta} \right) \left(4 - \frac{2h}{a} \right) + \left(\frac{\varepsilon_\alpha - \varepsilon_\beta}{\varepsilon_\alpha + \varepsilon_\beta} \right)^2$$

$$\times \left[\frac{(1+h/a)(1-2h/a)}{1+2h/a} \right.$$

$$\left. + \frac{a}{2h} \ln \left(1 + \frac{2h}{a} \right) \right] \right\}$$

$$+ \frac{(ze_0)^2}{16\pi \varepsilon_0 \varepsilon_\beta a} \left(\frac{2\varepsilon_\beta}{\varepsilon_\alpha + \varepsilon_\beta} \right)^2 \left(1 - \frac{h}{a} \right) \quad (75)$$

If $h = 0$ and center of an ion is at the interface, electrostatic Gibbs energy will be equal to

$$\Delta G(h=0) = \frac{(ze_0)^2}{4\pi \varepsilon_0 (\varepsilon_\alpha + \varepsilon_\beta) a} \quad (76)$$

The Figure 13 illustrates the effect of image forces on the electrostatic Gibbs energy of an ion at the oil–water interface.

Image forces play a significant role in electric double-layer effects. The excess surface charge density is

$$q = ec_0 \int_{h_a}^{h_c} \left\{ \exp \left[-\frac{\Delta g(h)}{k_B T} \right] - 1 \right\} dh \quad (77)$$

where

$$\Delta g(h) = \Delta G(h) - \frac{(ze_0)^2}{8\pi \varepsilon_0 \varepsilon_\beta a} \quad (78)$$

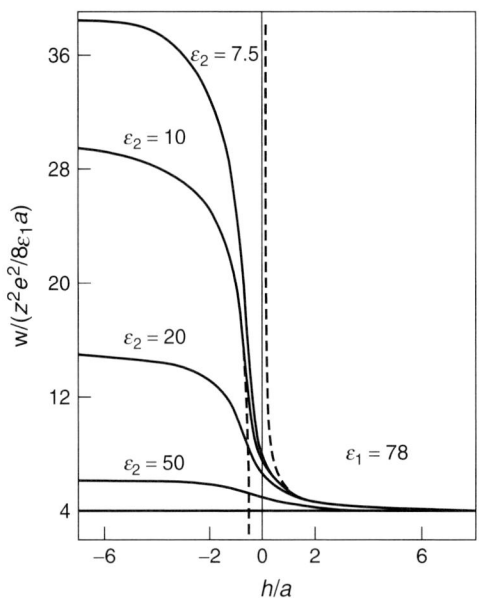

Fig. 13 Electrostatic Gibbs energy profiles for ion transfer across the ITIES boundary. Solid lines: finite-size ion profiles in units of $(ze)^2/\varepsilon_1 a$, $\varepsilon_1 = 78$ and different values of ε_2. Dashed lines: profile for the point change model in the same units, $\varepsilon_1 = 78$, $\varepsilon_2 = 10$ [25]. (Reproduced by permission of Elsevier Sequoia S. A.)

$\Delta G(h)$ is the electrostatic contribution to the Gibbs energy of solvation, and h_a and h_c are the distances of closest approach of the anion and the cation. If image forces are taken into account, the diffuse layer capacitance can be calculated by the equation:

$$C_d = C_d^{GC} \exp\left[\frac{\Delta g(h_0)}{2k_B T}\right] \quad (79)$$

where C_d^{GC} is the Gouy–Chapman diffuse layer capacitance without image terms. As noted by Kharkats and Ulstrup [25], Eq. (79) always gives diffuse layer capacitance corrections toward higher values.

The first step to statistically correct the Gouy–Chapman theory for the diffuse double layer used a restricted primitive electrolyte model. This model considers ions to be charged hard balls of identical radii in a structureless dielectric continuum with constant dielectric permittivity. There are three main approaches to creating a statistical theory with this model. The first approach uses the Gouy–Chapman theory and the average electrical field in the diffuse layer is calculated from the Poisson equation. The structure and physical properties of the double layer were then calculated from the restricted primitive electrolyte model. As a result, the modified Poisson–Boltzmann theory (MPB) was developed.

The second statistical approach incorporates correlation functions and integral equations from the theory of liquids. In this case, the well-known uniform Ornstein–Zernike equations were modified to calculate the ion distribution function near a charged interface. Modified in this way, the Ornstein–Zernike equations became nonuniform and were solved using hyperneted chain approximations (HCA) incorporating a mean spherical approximation (MSA) for correlation functions (HCA/MSA). A third statistical approach used to describe the EDL was achieved by using the integral Born–Green–Ivon equations.

These correlation function approaches resulted in the theory of ionic plasma in semi-infinite space. The next advance in describing the double layer was achieved by upgrading the restricted primitive electrolytes model to a "civilized" or "nonprimitive" model. This was accomplished by addition of hard spheres with embedded point dipoles to the restricted "primitive" model. This theory takes into account the presence of long-range Coulomb interactions, image effects, and external electrical fields. The new "nonprimitive" model became the basis for the theory of ion-dipole plasma. The serious mathematical difficulties encountered in describing the "nonprimitive" models necessitated the development of less statistically rigorous intermediate models. In these intermediate models, the first monolayer of solvent molecules at the charged interface was considered as a complex of discrete particles and an electrolyte outside this layer was described by the Gouy–Chapman theory. This intermediate approach was able to provide good theoretical agreement with experimental results.

The concept of constant dielectric permittivity in the double layer is also of concern. Different factors such as high ionic concentration or strong electric fields can affect the dielectric permittivity. In addition, description of the dielectric properties of solutions by a *single parameter* – the dielectric permittivity – also came under criticism and led to the development of nonlocal electrostatics.

Nonlocal electrostatics takes into consideration discrete properties of solvent molecules. It assumes that fluctuations of solvent polarization are correlated in space. This means that the average polarization at each point is correlated with the electric displacement at all other points and therefore, uses the solvent dielectric function that couples polarization vectors throughout the solvent.

2.3.8
Modified Poisson–Boltzmann (MPB) Model

The popular Poisson–Boltzmann equation considers the mean electrostatic potential in a continuous dielectric with point charges and is therefore, an approximation of the actual potential. An improved model and mathematical solution resulted in the MPB equation [26]. This equation is based on a restricted primitive electrolyte model that considers ions as charged hard spheres with diameter d in a continuous uniform structureless dielectric medium of constant dielectric permittivity ε. The sphere representing an ion has the same permittivity ε. The model initially was developed for an electrolyte at a hard wall with dielectric permittivity ε_w and surface charge density σ. The charge is distributed over the surface evenly and continuously.

This theory takes into account the finite size of ions, the fluctuation potential, and image forces in the electrolyte solution next to a rigid electrode, but it still an approximation. The MPB theory begins with the Poisson equation for the mean electrostatic potential ψ in solution:

$$\frac{d^2\psi}{dx^2} = -\frac{1}{\varepsilon_0\varepsilon}\sum_i z_i e_0 n_i^0 g_{0i}(x) \quad (80)$$

where ε_0 is the electrical constant, z_i is the charge number of ion i, e_0 is the elementary charge, n_i^0 is the bulk number density, and $g_{0i}(x)$ is the wall-ion distribution function representing the ratio of the local ion density to the ion density of the bulk electrolyte solution. The distance x is measured from the electrode, so that the distance of closest approach of an ion i is d/2. According to Kirkwood, the distribution function is given by the following equation:

$$g_{0i}(x) = \varsigma_i(x) \exp\left[-\frac{1}{kT} z_i e_0 \psi(x) + \eta_i\right] \quad (81)$$

The factor $\varsigma_i(x)$ accounts for the excluded volume of the ion, η_i is the fluctuation potential that takes into consideration the ionic atmosphere around an ion created by other ions. The fluctuation potential can then be presented as

$$\eta_i = -\frac{1}{kT}\int_0^{z_i e_0} \lim_{r\to 0}(\phi_i^* - \phi_{ib}^*)d(z_i e_0) \quad (82)$$

where

$$\phi_i^* = \phi_i - \frac{z_i e_0}{4\pi\varepsilon\varepsilon_0 r} \quad (83)$$

ϕ_i is the fluctuation potential created by the ion i and its atmosphere, r is the distance from the center of the ion. The difference ϕ_i^* is the potential created by the ionic atmosphere only and ϕ_{ib}^* is the value of this potential in the bulk electrolyte solution. There are several methods for calculating ϕ_i, and if the image effect is taken into account, it can be presented as:

$$\phi_i = B_0\left[\frac{1}{r}\exp(-\kappa r) + \frac{f}{r^*}\exp(-\kappa r^*)\right],$$

$$\text{if } r > d, x > 0 \quad (84)$$

where f is the image factor determined by the difference between two dielectric constants of the electrolyte solution and the electrode:

$$f = \frac{\varepsilon - \varepsilon_w}{\varepsilon + \varepsilon_w} \quad (85)$$

The local Debye–Huckel parameter at a given point is

$$\kappa^2 = \frac{e_0^2}{\varepsilon_0\varepsilon kT}\sum_i z_i^2 n_i^0 g_{0i}(x) \quad (86)$$

In Eq. (84), B_0 is a constant, and r^* is the distance from the mirror image in the interface. Hence Eq. (84) takes into account both the ionic atmosphere and image forces.

If $r < d$ and $x > 0$, the equation for the fluctuation potential should be written as

$$\nabla^2 \phi_i = -\frac{d^2\psi}{dx^2}, \quad r < d, x > 0 \quad (87)$$

The term $\varsigma(x)$ can be presented as a power series expansion in the bulk density

$$\varsigma_i(x) = 1 + B(x) + \frac{4\pi d^3}{3}\sum_j n_j^0 \quad (88)$$

where

$$B(x) = \pi\sum_j n_j^0 \begin{cases} \int_{d/2}^{x+d}[(x'-x)^2 - d^2] \\ \times g_{0j}(x')\,dx', \\ \quad \frac{d}{2} \leq x \leq \frac{3d}{2} \\ \int_{x-d}^{x+d}[(x'-x)^2 - d^2] \\ \times g_{0j}(x')\,dx', \quad x \geq \frac{d}{2} \end{cases} \quad (89)$$

Now, we can combine these equations and obtain the modified Poisson–Boltzmann equation for the diffuse layer:

$$\frac{d^2\psi}{dx^2} = -\frac{1}{\varepsilon_0\varepsilon}\sum_i z_i e_0 n_i^0 \varsigma_i(x)\exp\left\{-\frac{z_i e_0}{kT}\right.$$

$$\left.\times\left[L_1(\psi) + \frac{z_i e_0}{8\pi\varepsilon_0\varepsilon d}(F - F_0)\right]\right\},$$

$$x > \frac{d}{2} \quad (90)$$

Since the distance of the maximal approach of ions to the interface cannot be less then $d/2$, the usual linear dependence of potential in the compact layer can be introduced:

$$\psi(x) = \psi(0) + x \frac{d\psi(0)}{dx} \quad (91)$$

Additional functions in Eq. (173) are

$$\zeta_i(x) = 1 + \frac{4\pi d^3}{3} i \sum_i n_i^0$$

$$+ \pi \sum_i n_i^0 \int_{\max\left(\frac{d}{2}, x-d\right)}^{x+d}$$

$$\times [(x'-x)^2 - a^2] g_{0i}(x') \, dx' \quad (92)$$

$$L_1(\psi) = \frac{F}{2} [\psi(x+d) + \psi(x-d)]$$

$$- \frac{F-1}{2d} \int_{x-d}^{x+d} \psi(x') \, dx' \quad (93)$$

$$F = \frac{\left\{\begin{array}{l} 2(1+\kappa x)\exp(-\kappa d) \\ +2(1-\kappa d)\exp(-\kappa x) \\ +2\kappa^2 \, dx \int_x^a r^{-1} \exp(-\kappa r) \, dr \\ + \frac{f}{\kappa x}[\exp(-\kappa d) \\ +2\kappa(d-x)\exp(-\kappa x) \\ -\exp(-\kappa(2x+d))] \end{array}\right\}}{\left\{\begin{array}{l} 2[\exp(-\kappa d)(1+\kappa(x+d)) \\ +\exp(-\kappa x)] \\ -\frac{f}{\kappa x}[2\kappa x \exp(-\kappa x)+(1+\kappa d)] \\ \times (\exp(-2\kappa x)-1)\exp(-\kappa d)] \end{array}\right\}},$$

$$\frac{d}{2} \leq x \leq d \quad (94)$$

$$F = \frac{(1+f\delta_2)}{(1+\kappa d)(1-f\delta_1)}, \text{ if } x \leq d \quad (95)$$

$$F_0 = \lim_{x\to\infty} F = \frac{1}{1+\kappa_0 d} \quad (96)$$

$$\delta_2 = \frac{1}{2\kappa x} \exp[\kappa(d-2x)] \sinh \kappa d \quad (97)$$

$$\delta_1 = \frac{(\kappa d \cosh \kappa d - \sinh \kappa d)\delta_2}{(1+\kappa d)\sinh \kappa d} \quad (98)$$

$$\kappa_0^2 = \lim_{x\to\infty} \kappa^2 = \frac{e_0^2}{\varepsilon_0 \varepsilon kT} \sum_i z_i^2 n_i^0 \quad (99)$$

The boundary conditions for these equations are

1. $\psi(x), \psi'(x) \to 0$, if $x \to \infty$
2. $\psi'(0+) = -\delta/\varepsilon_0\varepsilon$ (100)
3. $\psi(x), \psi'(x)$ are continuous at $x > 0$

This set of equations completes the mathematical formulation of the problem, but the equations are very cumbersome and can only be solved numerically. The major difficulty is related to the finite ion size. If we consider the limiting case of point charges, the equation for potential in the double layer would be:

$$\frac{d^2\psi}{dx^2} = -\frac{1}{\varepsilon_0\varepsilon} \sum_i n_i^0 z_i e_0$$

$$\times \exp\left\{-\frac{z_i e_0}{kT}\left[\psi(x) + \frac{z_i e_0}{8\pi\varepsilon_0\varepsilon}\right.\right.$$

$$\left.\left.\times \left\{\kappa_0 - \kappa + \frac{f}{2x}\exp(-2\kappa x)\right\}\right]\right\}$$
(101)

In comparison with the Gouy–Chapman theory, the MPB equation for point charges contains two improvements in that the Debye–Huckel parameter κ depends on distance, and the ion image is screened.

In solving the MPB equations, different estimates of $\zeta_i(x)$ were considered. Depending on the type of $\zeta_i(x)$, the equations were called MPB1, MPB2, MPB3 and so on.

The MPB theory overcomes certain limitations of the Gouy–Chapman model by taking into account ionic volume, the ionic atmosphere, and image forces. Results of both theories coincide at low electrolyte concentrations and small surface charges, but beyond this limiting case, there is a very significant qualitative and quantitative discrepancy. For instance, the MPB theory predicts a thinner diffuse ion layer and higher capacitance than the Gouy–Chapman model. Near a charged interface, MPB predicts considerable adsorption of co-ions. The major difference is that the decaying potential distribution of the Gouy–Chapman model transforms into the decaying-oscillating solution of the MPB theory, with oscillations beginning at high electrolyte concentrations when $\kappa_0 d > 1.24$. This oscillation was also predicted by Stilinger and Kirkwood [27] who estimated a different value $\kappa_0 d > 1.03$. Other modern theories predict oscillatory behavior of the electrostatic potential, but under different conditions. In the HCA theory, the condition is $\kappa_0 d > 1.23$, which is very close to MPB theory. The Bogolyubov–Born–Green–Ivon theory puts the condition at $\kappa_0 d > 1.72$. Therefore, oscillatory behavior is a consistent finding of all theories. It is worthwhile to mention that damped oscillation in the MPB theory is due to a fluctuating potential and its existence does not depend on the excluded volume.

One of the major drawbacks of MPB theory is that it does not take into account the discreteness of a solvent. In the diffuse layer, the discreteness of solvent molecules strongly influences ion–ion interactions at small distances. Furthermore, an understanding of the structure and properties of the compact layer is impossible without an adequate model of solvent molecules and their interactions with the electrode and between themselves.

Application of MPB model to ITIES was made first by Torrie and Valleau [28]. Cui and coworkers [29] applied the modified Poisson–Boltzmann theory (MPB4) to the ITIES and found that MPB4 describes the experimental results at the nitrobenzene/water and dichloroethane/water phase boundaries. It reproduces the Monte Carlo (MC) calculation more accurately than the MVN theory for 1:1 electrolytes over a wide range of conditions including variations in the image forces, ion size, the inner-layer potential distribution, electrolyte concentration, surface charge density and solvent effects. They used the equation for electrostatic potential from the MPB4 model for the inner layer at ITIES when $x < a/2$:

$$\psi(x) = \psi(0) + x\psi'(0) - x\lambda c |q^{aq}|^{1/2} \frac{q^{aq}}{\varepsilon_h}$$

(102)

where $\psi(x)$ is the electrostatic potential at coordinate x, $\psi(0)$ is the electrostatic potential at the surface, $\psi'(0)$ is the derivative of $\psi(0)$, c is the electrolyte concentration, qaq is the surface charge density of the aqueous phase, λ is the parameter determined by fitting the experimental values of the Galvani potential differences and ε_h is the dielectric permittivity of the compact layer. The third term in Eq. (102) represents the dependence on electrolyte concentration, surface charge density, and the solvent effect on the inner-layer potential distribution. These effects can be ascribed to ionic penetration into the opposite phase and ion–ion correlations across the interface. Cui et al. [29] obtained good agreement between theoretical calculations and experimental data and came to the conclusion that the structure of the 1,2-dichloroethane–water interface is similar to that of the water/nitrobenzene interface, except that the effects of the image force and ion size are more pronounced and the inner-layer potential drop plays a more important role.

References

1. A. G. Volkov, D. W. Deamer, D. I. Tanelian et al., *Liquid Interfaces in Chemistry and Biology*, John Wiley & Sons, New York, 1998.
2. A. G. Volkov, D. W. Deamer, D. I. Tanelian et al., *Prog. Surf. Sci.* **1996**, *53*, 1–136.
3. O. S. Ksenzhek, A. G. Volkov, *Plant Energetics*, Academic Press, San Diego, 1998.
4. A. G. Volkov, D. W. Deamer, (Eds.), *Liquid–Liquid Interfaces: Theory and Methods*, CRC Press, Boca Raton, London, Tokyo, 1996, p. 421.
5. A. G. Volkov, (Ed.), *Liquid Interfaces in Chemical, Biological, and Pharmaceutical Applications*, Marcel Dekker, New York, 2001.
6. M. Planck, *Ann. Phys.* **1891**, *44*, 385–428.
7. F. O. Koenig, *J. Phys. Chem.* **1934**, *38*, 111–128.
8. D. C. Grahame, R. W. Whitney, *J. Am. Chem. Soc.* **1942**, *64*, 1548–1552.
9. E. J. W. Verwey, K. F. Nielsen, *Philos. Mag.* **1939**, *28*, 435–446.
10. C. Gavach, P. Seta, B. d'Epenoux, *J. Electroanal. Chem.* **1977**, *83*, 225–235.
11. V. S. Markin, A. G. Volkov, *Prog. Surf. Sci.* **1989**, *30*, 233–356.
12. V. S. Markin, A. G. Volkov, *Electrochim. Acta* **1989**, *34*, 93–107.
13. R. S. Hansen, *J. Phys. Chem.* **1962**, *66*, 410–415.
14. A. G. Volkov, *Langmuir* **1996**, *12*, 3315–3319.
15. T. Kakiuchi, M. Kobayashi, M. Senda, *Bull. Chem. Soc. Jpn.* **1988**, *61*, 1545–1550.
16. T. Kakiuchi, T. Usui, M. Senda, *Bull. Chem. Soc. Jpn.* **1990**, *63*, 2044–2050.
17. T. Kakiuchi, M. Senda, *Bull. Chem. Soc. Jpn.* **1983**, *56*, 1322–1326.
18. T. Osakai, T. Kakutani, M. Senda, *Bull. Chem. Soc. Jpn.* **1984**, *57*, 370–376.
19. Z. Samec, V. Mareček, D. Homolka, *J. Electroanal. Chem.* **1985**, *187*, 31–51.
20. A. N. Frumkin, *Zero Charge Potentials*, Nauka Publ., Moscow, 1979.
21. T. Kakiuchi, M. Kobayashi, M. Senda, *Bull. Chem. Soc. Jpn.* **1987**, *60*, 3109–3115.
22. C. Yufei, V. J. Cunnane, D. J. Schiffrin et al., *J. Chem. Soc., Faraday Trans.* **1991**, *87*, 107–114.
23. V. S. Markin, A. G. Volkov in *Liquid–Liquid Interfaces. Theory and Methods* (Eds.: A. G. Volkov, D. W. Deamer), CRC Press, Boca Raton, London, Tokyo, 1996, pp. 63–75.
24. L. D. Landau, E. M. Lifshitz, *Electrodynamics of Continuous Media*, 2nd ed., Pergamon Press, New York, 1984.
25. Yu. Kharkats, J. Ulstrup, *J. Electroanal. Chem.* **1991**, *308*, 17–26.
26. C. W. Outhwaite, L. B. Bhuiyan, S. Levine, *J. Chem. Soc., Faraday Trans. 2* **1980**, *76*, 1388–1408.
27. F. H. Stilinger, J. G. Kirkwood, *J. Chem. Phys.* **1960**, *33*, 1282–1290.
28. G. M. Torrie, J. P. Valleau, *J. Electroanal. Chem.* **1986**, *206*, 69–79.
29. Q. Cui, G. Y. Zhu, E. K. Wang, *J. Electroanal. Chem.* **1995**, *383*, 7–12.

2.4
Electrical Double Layers. Double Layers at Single-crystal and Polycrystalline Electrodes

Enn Lust
University of Tartu, Tartu, Estonia

2.4.1
General Aspects

Remarkable progress has been made in the last few years in electrochemistry on single-crystal surfaces [1–10]. This parallels with the progress on the surface science and it has been partly simulated by developments in that field especially regarding the preparation and characterization of surfaces. The most widely used techniques for the determination of the structure of single-crystal surfaces are the low-energy electron diffraction (LEED), scanning tunneling microscopy (STM), atomic force microscopy (AFM), X-ray scattering (SXRS), and other methods [4–6, 8–10]. Auger electron spectroscopy (AES) is used to characterize the chemical composition or cleanliness of surfaces. To the first approximation, the low Miller indexes reflect the symmetry of the bulk structure and they have lower surface energy than vicinal surfaces (the high index planes). However, LEED, AFM, STM and SXRS studies have revealed two important features of atomic surface structure: relaxation and reconstruction [4–6, 8–11] (see also Chaps. 14–21 in Vol. 3).

2.4.1.1 The Potential of Zero Charge and Electric Double-layer Structure

The electric double layer (edl) is fundamental for electrochemistry because the rate and mechanism of the various electrochemical reactions (hydrogen evolution, corrosion and corrosion inhibition by surfactants, metal deposition and dissolution, and so forth) depend on the metal–electrolyte phase boundary structure (see also Vol. 2). A large number of techniques have been used for the experimental study of the edl structure and for the determination of the potential of zero charge (pzc) $E_{\sigma=0}$, which is a fundamental parameter in electrochemistry [1]. However, the greatest success at solid polycrystalline (PC) and single-crystal electrodes has been achieved mainly by impedance, optical, scrape, and surface tension methods, and therefore in this review more attention has been paid to the data obtained using these methods [1–5, 12, 13].

The charged particles may or may not cross the interface between two phases and in this respect, interphases may be divided into the two limiting types: nonpolarizable and polarizable, respectively [12]. The analysis by Frumkin et al. [14] in 1970 led to the conclusion that it is possible to treat thermodynamically not only the polarizable but also the nonpolarizable electrode cases, and to define the charge density $Q = \sigma + A_{H^+}$, where σ is the surface charge density and A_{H^+} is the fraction of charge that has crossed the interface moving from one phase to the other. Thus, it has been proposed to call the potential at which $Q = 0$ the potential of zero total charge (PZTC) and the potential at which $Q = \sigma = 0$ the potential of zero free charge (PZFC) (see also Chap. 1 in Vol. 1). A PZTC has been observed and measured for Pt-group metals, and this is due to the chemisorption of hydrogen atoms. In all other cases the PZFC is usually observed and for simplicity will be termed as a pzc, denoted by $E_{\sigma=0}$.

2.4.1.2 Differential Capacitance of Electrode–Electrolyte Solution Interface

For (ideally) polarizable metals (i.e. electrodes having a large energy barrier for charge transfer) with a sufficiently broad double-layer region (such as Hg, Ag, Au, Bi, Sn, Pb, Cd, Tl, and others [1]), $E_{\sigma=0}$ can be obtained from measurements of the edl capacitance in dilute solutions, where it is detected by a pronounced minimum in the capacitance–potential (C, E) curve. The related differential capacitance C, defined as

$$C = \left(\frac{\partial \sigma}{\partial E}\right)_{T,p} = -\left(\frac{\partial^2 \gamma}{\partial E^2}\right)_{T,P,\mu_i} \quad (1)$$

can be measured by a great variety of techniques [1, 5, 9] (see also Chaps. 1–9 in Vol. 3). In Eq. (1), γ is the interfacial (for liquid electrodes) or superficial (specific surface) work (for solid electrodes).

It is usually assumed that the edl at the ideally polarizable metal–electrolyte interfaces is splitted into two parts: the inner layer and the diffuse layer, and can be represented by two capacitances in series [1, 5, 7, 9]:

$$C^{-1} = C_i^{-1} + C_d^{-1} \quad (2)$$

Here, C_i is the inner-layer capacitance, independent of the surface-inactive electrolyte concentration and C_d is the diffuse (Gouy) layer capacitance, expressed according to the Gouy–Chapman theory for a z,z-type electrolyte by

$$C_d = \frac{d\sigma}{d\phi_d} = \frac{|z|F}{2RT}\sqrt{4A^2 c + \sigma^2} \quad (3)$$

where $\phi_d = 2RT(|z|F)^{-1}\text{arcsh}(0.5\sigma A^{-1} c^{-1/2})$ is the potential drop in the diffuse layer and $A = \sqrt{2\varepsilon_0 \varepsilon_d RT}$; ε_0 is the permittivity of vacuum and ε_d is the dielectric constant of the diffuse layer, usually taken equal to the macroscopic dielectric permittivity of the solvent. Thus, at $E_{\sigma=0}$, the value of C_d decreases linearly with $\sqrt{c}$. According to the Gouy–Chapman–Stern–Grahame (GCSG) model [1], in a surface-inactive electrolyte solution, the potential of the minimum in C, E curve for $c_{el} \leq 10^{-2}$ M would correspond to the condition $\sigma = 0$, that is, to $E_{\sigma=0}$ value. Exact values of $E_{\sigma=0}$ in the absence of adsorption can be obtained by linear extrapolation of E_{min} as a function of the electrolyte concentration at $c_{el} \leq 1 \times 10^{-2}$ M [5, 15, 16].

Models put forward in the 1960s to 1970s by several authors for describing the metal–solution interface did not take into account the metal contribution and consider the solid as a perfect conductor, so that the potential drop in the metal layer is charge-independent and the potential drop at the metal–solution interface is just given by the solvent molecules and free ionic charges. Although early theories considered the penetration of the electric field into the metal, this idea was subsequently dropped for a long time [1, 9, 17–19]. Penetration of the electric field into the solid has been found to be important for semiconductors, and its possible relevance to semimetals such as Bi and Sb was suggested [1, 5, 9, 20, 21]. In the past 20 years different models have been proposed, describing the traditional so-called *inner-layer capacity* as $1/C_i = 1/C_M + 1/C_{dip} + 1/C_H$, where C_M, C_{dip}, and C_H are the capacitances of the metal phase, the dipole (solvent) layer, and the Helmholtz layer, respectively. However, it is impossible experimentally to separate the overall capacity of inner layer into its components [1–5] (see also Chap. 2.3 in Vol. 1).

2.4.1.3 Electrical Double-layer Structure at Solid Electrodes

During several decades, the quantitative analysis of the edl structure, adsorption of ions and molecules on the ideally polarizable solid electrodes was carried out on the basis of relationships, developed for the liquid Hg surface without any consideration of the structure of the solid surface. But the theoretical analysis of the electrochemical characteristics for the PC solid electrodes, even for simple metals, is considerably complicated in comparison with that for Hg. One of the reasons is both the differences between properties of individual faces of the metal and the influence of various defects of the surface structure [1–10]. In the beginning of the 1970s, by a careful experimental study of the adsorption of pyridine on the solid bismuth drop electrode of known PC surface structure and of tetraalkylammonium ions on PC Zn, the phenomenon of splitting the adsorption–desorption maximum was discovered [1, 20, 21] (Fig. 1). Already the first investigations showed that splitting the maxima was due to the PC nature of the surface, and also to the nature of the adsorbing particle and the solvent, to the surface charge density σ, and to the solution concentration. On account of that, the edl and adsorption parameters (e.g. $E_{\sigma=0}$, interaction parameter in the adsorption (Frumkin's) isotherm, maximal Gibbs adsorption Γ_{max}, adsorption equilibrium constants, and so forth), obtained for the PC electrodes, often may be apparent [1–5, 7, 15, 16, 20–22].

It may be noted that for a PC surface, there are complications in defining the PZC because one does not have zero-charge density of all crystal faces simultaneously, exposed on PC electrode. The work function is different for various homogeneous regions (planes), but if the grains are in contact, the electrochemical potential of the electron has a single value. Thus, the Volta or outer potential is different for various faces, so different surface charge densities are present. Similarly, the surface structure of solid metal makes the electronic structure more complicated than that for a liquid metal or for a single-crystal plane. Certainly, the use of the density properties depending only on distance perpendicular to the interface

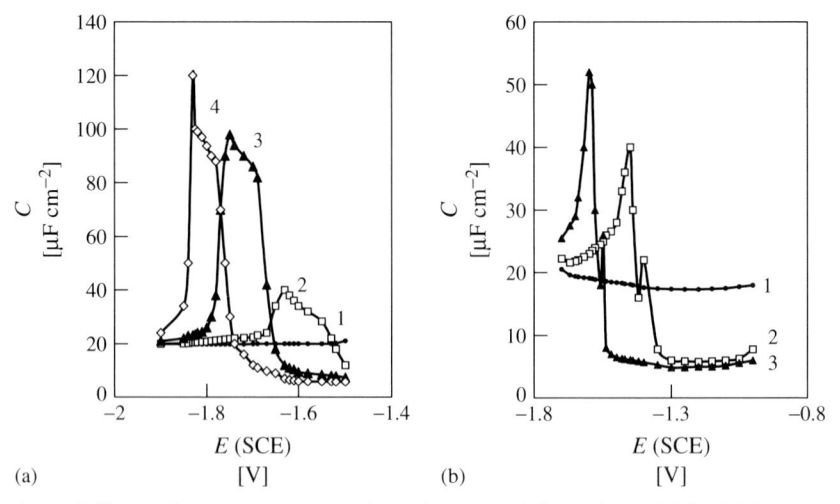

Fig. 1 Differential capacitance versus electrode potential dependence: (a) for PC Zn in 0.1 M KI containing TBAI (M): (1) 0, (2) 3×10^{-4}, (3) 3×10^{-3}, (4) 3×10^{-2}; (b) for PC Bi in 0.1 M KF containing pyridine (M): (1) 0, (2) 0.1, (3) 0.2 (updated from Refs. [1, 20, 21]).

2.4.1.4 Surface Roughness of Solid Electrodes

Surface roughness is an important property in electrochemistry of solid electrodes as the most of edl and adsorption characteristics, as well as kinetic parameters, are extensive quantities and are referred to the apparent unit (flat cross-section) area of electrode surface [1–5]. The examination of the working area of solid electrodes is a difficult matter owing to the irregularities at a submicroscopic level. For the determination of the real surface area of the solid electrodes, different in situ and ex situ methods have been proposed and used, which are discussed in Refs. [5, 23]. The in situ methods more commonly used in electrochemistry to obtain the surface roughness

$$R = \frac{S_{\text{real}}}{S_{\text{geom}}} \quad (4)$$

(where S_{real} and S_{geom} are the real (working) surface and geometrical area, respectively) are (1) differential capacitance measurements in the region of ideal polarizability, Parsons–Zobel (PZ) plot method [24], Valette–Hamelin approach [25, 26], and other similar methods [1, 5, 7]; (2) mass transport under diffusion control with assumption of homogeneous current distribution; (3) adsorption of ratioactive organic compounds hydrogen, oxygen, or metal monolayers; (4) microscopy (optical, electron, STM, and AFM); and (5) quartz microbalance, as well as a number of ex situ methods [23]. It is to be noted that depending on the irregularity-to-probe size ratio, either the entire surface or only a fraction of it is accessible to a particular measurement. Only when the size of the molecule or ion, used as a probe particle, is smaller than the smallest surface irregularity, the entire surface can be evaluated. But each method is applicable to a limited number of electrochemical systems so that a universal method of surface area measurements is not available for the time being. Thus, it is useful to stress again that the value of R depends on the method used [1–5, 15, 16, 22, 23].

2.4.1.5 The Applicability of the Gouy–Chapman–Stern–Grahame Model to Solid Electrodes

2.4.1.5.1 Shape of C, E Curves
Dependence of the C, E curves for a solid metal on a method of electrode surface preparation was reported long ago [1–5, 7, 20–22, 25, 26]. In addition to the influence of impurities and faradaic processes, variation in the surface roughness was pointed out as a possible reason for the effect. For the determination of R, it was first proposed to compare the values of C of the solid metal (M) with that of Hg, that is, $R = C^M/C^{Hg}$. The data at $E_{\sigma=0}$ for the most dilute solution were typically used for such a comparison to eliminate the influence of possible differences in the inner-layer capacities. However, C_i of different solid metals and Hg has shown such a large variation that this approach can hardly be considered as appropriate. It is to be noted that the error in C, which for solid electrodes is much higher than for liquid electrodes, increases with the decrease of c_{el}; further, as shown later (Sect. 2.4.1.5.2), the effects of the surface crystallographic inhomogeneity on C values also prove especially appreciable in the dilute electrolytes [1–5, 7, 15, 16, 22].

Frumkin was the first to give a qualitative consideration of the electrochemical properties of PC electrodes. He noted that the charge σ_j at individual faces j may be different at $E = $ const and this may change the form of the capacitance curve near the diffuse-layer capacitance minimum [1]. Important results were obtained in a pioneering paper by Valette and Hamelin [25, 26]. They compared experimental capacitance curves for a PC Ag electrode and Ag faces. It was found that the capacitance of a PC Ag electrode can be obtained by the superposition of the corresponding C_j, E curves for individual faces exposed at the PC surface:

$$C_{PC}(E) = \sum_j \theta_j C_j(E) \quad (5)$$

where θ_j is the fraction of the face on the PC electrode surface. A weighed sum of C, E curves for the faces was found to be similar to the C, E curve for a PC electrode, and thus all main Ag faces are exposed on the surface. The diffuse-layer capacitance minimum potential E_{min}^{PC} for a PC Ag electrode was only slightly less negative (30 mV) than the PZC of the Ag(110) face (i.e. for the face with the more negative value of $E_{\sigma=0}$). The diffuse-layer capacitance minimum for PC Ag was wider and less deeper than for the Ag faces.

Thereafter, the influence of the crystallographic inhomogeneity of PC and monocrystalline electrodes (with various surface defects) has been discussed for various metals in many papers [1–5, 7, 15, 16, 22] (Fig. 2). Bagotskaya et al. [27] showed that integration of the partial C, E curves from $E_{\sigma=0}$ of each face to the minimum potential of the C_{PC}, E curve, E_{min}^{PC}, on the PC electrode, taking into account the fraction of plane, gives $\sigma_{Ag}^{PC} = -0.04$ C m^{-2} at E_{min}^{PC}. At E_{min}^{PC}, the Ag(110) plane has a positive σ and other planes have a negative σ, thus, at the PC Ag surface there are no surface regions with $\sigma = 0$. The same conclusions hold for PC Au ($\sigma_{Au}^{PC} = -0.03$ C m^2), PC Bi ($\sigma_{Bi}^{PC} = -0.01$ C m^{-2}), and for other PC electrodes [5].

Mathematical simulation of C_{PC}, E curves shows that the shape of the diffuse-layer capacitance minimum depends on the difference of $E_{\sigma=0}$ for individual faces, exposed on the PC electrode surface, and their fractions, as well as on the shape of partial C_j, E curves (Fig. 3). The results of experimental capacitance studies at two-plane model PC Bi electrodes were in agreement with these conclusions [5, 15, 16]. Thus it has been shown that the potential of the diffuse-layer capacitance minimum for a PC electrode with noticeable difference in $E_{\sigma=0}$ values for various planes does not correspond to the PZC of the whole surface and $\sum_j \theta_j \sigma_j \neq 0$ at E_{min}.

2.4.1.5.2 Parsons and Zobel Plot Method

Substantial contributions to the interpretation of the experimental data for solid electrodes have been made by Leikis et al. [28] and by Valette and Hamelin [25, 26]. Both approaches are based on the same model: the value of $E_{\sigma=0}$ and of the inner-layer capacitance per unit "true" surface area, C_i, are assumed to be constant over the whole PC surface and the value of E_{min}^{PC} was identified by $E_{\sigma=0}$. Thus, the GCSG model is considered as applicable to the capacitance characteristics related to the unit of "true" surface area, which differ by a factor R^{-1} from those per unit "apparent" surface area

$$C_{real}(E) = R^{-1} C_{geom}(E) \text{ and}$$
$$\sigma_{real}(E) = R^{-1} \sigma_{geom}(E) \quad (6)$$

$$[C_i(\sigma_{real})]^{-1} = [C_{real}(\sigma_{real})]^{-1} - [C_d(\sigma_{real})]^{-1} \quad (7)$$

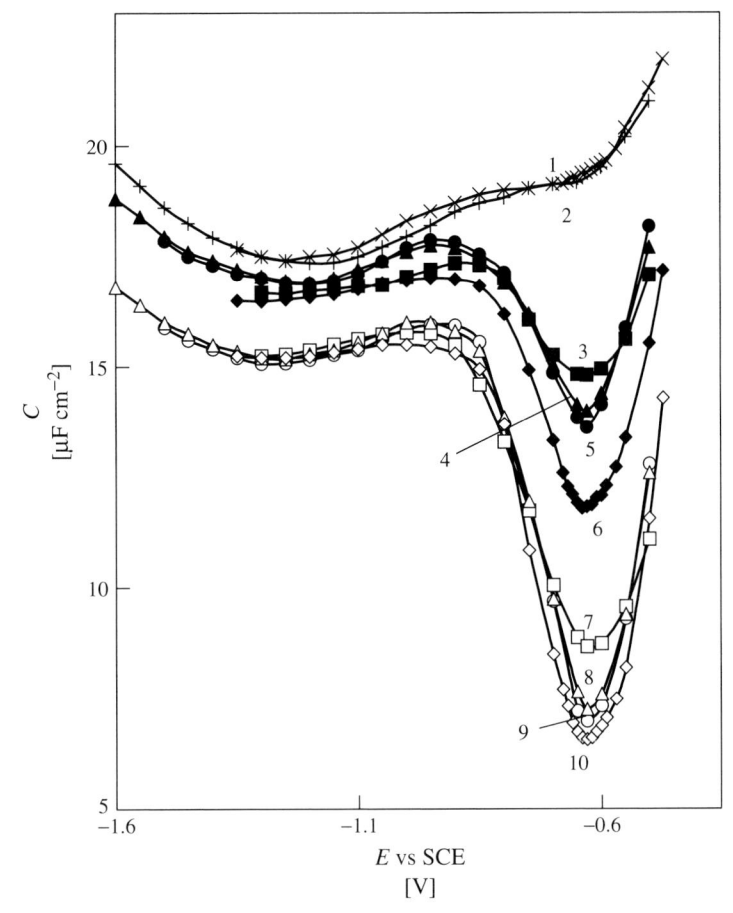

Fig. 2 C, E curves ($v = 210$ Hz) for $(Bi_{DE})^R$ (1, 4, 8), $Bi(111)^{EP}$ (2, 6, 10), $(Bi_{PC})^{MP\ CE}$ (3, 7), and $Bi(111)^{CE}$ (5, 9) in aqueous solution of NaF, M: (1, 2) – 0.1; (3–6) – 0.01; (7–10) – 0.002.

where $C_d(\sigma)$ is the diffuse-layer capacitance obtained according to the Gouy theory [1–5, 7]. Extrapolation of C_{geom}^{-1} versus C_d^{-1} to $C_{geom}^{-1} = 0$ provides the inner-layer capacitance as $R^{-1}C_{i,geom}^{-1}$. The inverse slope of the PZ plot at E_{min}^{PC} is identified by R. In the absence of ion-specific adsorption and for ideally smooth surfaces, these plots should be linear with the unit slope. However, data for Hg and single-crystal face electrodes have shown that the test is somewhat more complicated [1–5, 7, 29]. The PZ plots for Hg–surface-inactive electrolyte solution interfaces at $\sigma = 0$ and $\sigma < 0$, albeit usually linear, exhibit reciprocal slopes that are somewhat greater than unity. The main reason for this is the experimental error in measuring C, as well as the hyperbolic form of C in the GCSG model [15, 16, 22, 29].

The error in C_i is the total differential of Eq. (2):

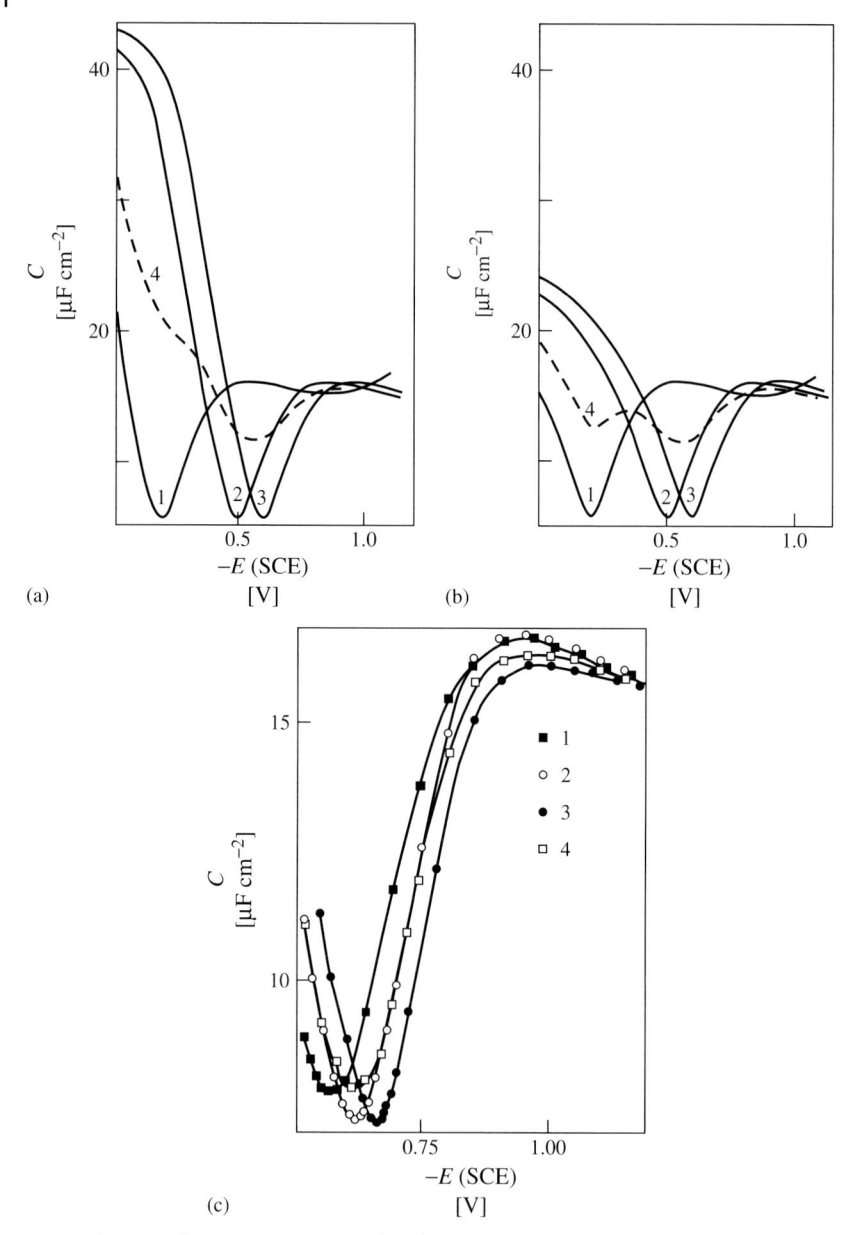

Fig. 3 Theoretical C, E curves (1, 2, and 3) for single-crystal planes and for a model PC surface (4) calculated by the superposition of the C, E curves at $E =$ const with $\theta_1 = \theta_2 = \theta_3 = 1/3$, (1) faces with high hydrophilicity; (2, 3) faces with weak hydrophilicity. (1, 2) $\Delta E_{\sigma=0}(\max) = 0.4$ V and (3) $\Delta E_{\sigma=0}(\max) = 0.09$ V.

$$dy = dC_i = \frac{1}{x^2} \cdot \frac{1}{C^2} dC \quad (8)$$

where dC, for a given value of σ, includes the experimental error in the determination of C and the error in the integration of the C, E curves. When $x (x = C^{-1} - C_d^{-1})$ is small, dC_i is large and tends to infinity; when x is large, dC_i is small and tends to dC. For a given positive x, the smaller the C_d, the more an error in C affects C_i. The same error for Bi, Cd, and Ag at fixed c_{el} causes the error in C_i to increase in the same order of metals since the value of C_i increases. The same experimental error entails a larger uncertainty in C_i for the lowest c_{el} and σ. At $|\sigma| \gg 0$, the uncertainty in C does not bear on C_i since x is large [5, 15, 16, 22, 29].

In the case of liquid Hg, the uncertainty in the measurements may be induced by possible errors connected with (1) experimental measurement of C, (2) preparation of solutions of an exact c_{el}, (3) incomplete dissociation of electrolytes, (4) slight specific adsorption of anions, and (5) deviations of diffuse layer from GC theory. In the case of solid electrodes, in addition to the above-mentioned reasons, sources of inaccuracy are the possible erratic preparation of the electrodes with the same geometrical surface area and the same crystallographic orientation [1–5, 15, 16, 22, 29]. Studies with wedge-shaped, two-faced Bi electrodes show that with increasing $\Delta E_{\sigma=0}$ of different faces exposed at a model PC electrode surface, the deviation of the PZ plot from linearity increases and the value of R also increases [15, 16] (Fig. 4). This was explained by the fact that R is a complex quantity, being $R = f_R f_{CR}$, where f_{CR} is a factor of crystallographic inhomogeneity of the PC electrode surface [15, 16, 22, 25–30] and

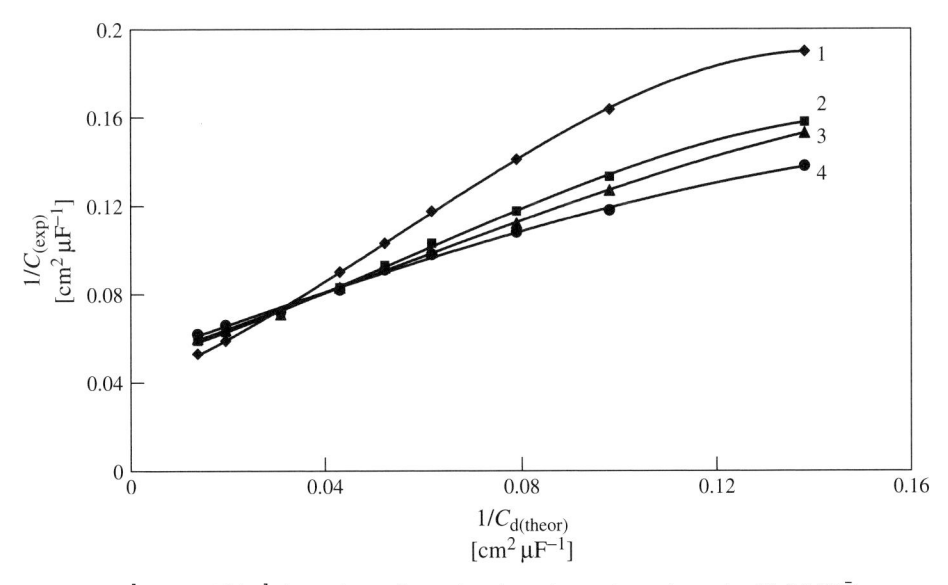

Fig. 4 $(C_i)^{-1}$ versus $(C_d)^{-1}$ dependence for wedge-shaped two-plane electrode: (1) $0.5(01\bar{1}) + 0.5(\bar{1}0\bar{1})$; (2) $0.8(01\bar{1}) + 0.2(111)$; (3) $0.3(01\bar{1}) + 0.7(111)$; and (4) $0.5(01\bar{1}) + 0.5(111)$ with $\Delta E_{\sigma=0} = 15$ mV (1) and $\Delta E_{\sigma=0} = 70$ mV (2–4) (C_i, inner-layer capacitance; C_d, diffuse-layer capacitance) (updated from Refs. [15, 16]).

f_R was assumed to be the actual surface roughness factor independent of c_{el} and σ. The more dilute the solution and the larger the difference between $E_{\sigma=0}$ of individual planes (i.e. the homogeneous regions) exposed at a PC surface the higher the values of f_{CR} have been established. Thus, f_{CR} decreases as $|\sigma|$ increases. A comparison of the data for F^-, BF_4^-, and ClO_4^- solutions shows that R increases in the order $F^- < BF_4^- < ClO_4^-$ with increasing weak specific adsorption [5, 15, 16, 22].

Thus, the experimental value of R depends on the surface charge density and sometimes on the electrolyte concentration c_{el}. The real R cannot depend on σ and on c_{el}. Experimental PZ plots for single-crystal faces in the region of $E_{\sigma=0}$ ($-3 < \sigma < 3$ µC cm^{-2}) often show slopes increasing with $|\sigma|$, that is, the apparent R decreases as $|\sigma|$ rises [1–5, 7, 15, 16, 22]. These findings indicate that the experimental R (the inverse slope value of the PZ plot) is not a real measure of the actual R. However, the practically unit slope of the C^{-1}, C_d^{-1}-plots for Hg, Bi, Cd, Sb, Ag, and for other systems with correlation coefficients better than 0.996 provides convincing evidence both for the validity of the GC theory and for the lack of experimentally detectable deviations of the roughness factor from unity. Slopes of C^{-1}, C_d^{-1} plots much lower than unity very near $E_{\sigma=0}$ ($-0.5 \leq \sigma < 1$ µC cm^{-2}) can be interpreted as deviations from the simple GC theory caused by the geometrical roughness and energetical inhomogeneity of the solid electrode surface [2–5, 15, 16, 22, 25–29].

2.4.1.5.3 Surface Roughness and Shape of Inner-layer Capacitance Curves
In 1973, Valette and Hamelin [25, 26] proposed another method to determine the roughness factor R of solid PC surfaces and to test the GCSG theory on the basis of Eqs. (6) and (7). For each c_{el}, a set of C_i, σ curves was calculated for various R values and the optimum value of R was selected on the basis of the assumption that near E_{min} the C_i, σ curve must be smooth. The experimental values of R were found to increase as c_{el} decreased from 1.40 to 1.80. This was explained by the fact that R is a complex quantity, being $R = f_R f_{CR}$. Later, the influence of the crystallographic inhomogeneity of PC and monocrystalline electrodes (with various surface defects) was discussed in many papers [1–5, 7, 15, 16, 22, 25–29] (Fig. 5).

2.4.1.5.4 Polycrystalline Electrode Models
The edl models for PC electrodes may be classified in two groups [5, 7, 10, 27, 30]. The first group considers the PC electrode surface as one consisting only of relatively large homogeneous (monocrystalline) regions with linear parameter $y^* \gg 10$ nm, which corresponds to macropolycrystallinity. Within these surface regions, both the compact and the diffuse layers at different homogeneous areas can be viewed as independent ones, and accordingly

$$C_{PC}^{app} = R \sum_j \frac{X_j C_{ij} C_{dj}}{C_{ij} + C_{dj}} \quad (9)$$

where C_{ij} and C_{dj} are the inner- and diffuse-layer capacities for the plane j, respectively. Therefore, this is the so-called model of independent diffuse layers (IDL) (Fig. 6).

In terms of the second group of models, the PC surface consists only of very small crystallites with linear parameter y^*, whose sizes are comparable with the edl parameters in moderate electrolyte concentrations. For such electrodes, the

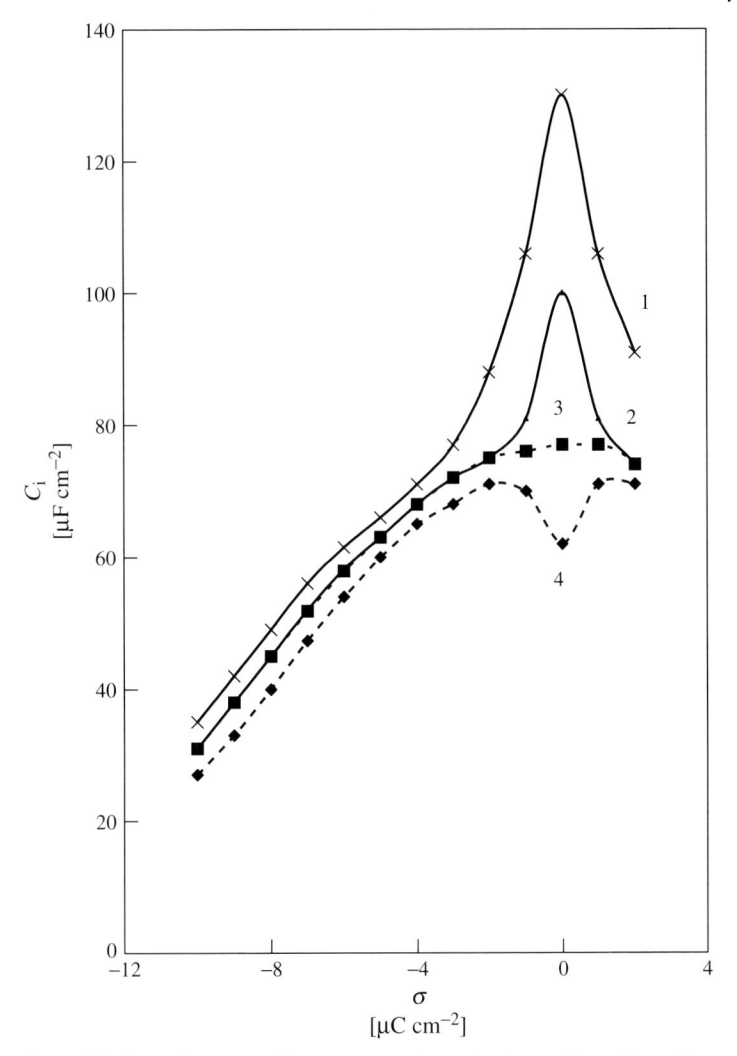

Fig. 5 The inner-layer capacitance versus electrode charge dependence for the PC Cd electrode at different values of roughness factor: (1) 1; (2) 1.08; (4) 1.15; (3) extrapolation by an approximate equation determined between -1.0 and $-10\ \mu C\ cm^{-2}$.

compact layers at different monocrystalline areas are considered to be independent, but the diffuse layer is common for the entire surface of a PC electrode and depends on the total medium charge density $\bar{\sigma}_{PC} = R \sum_j X_j \sigma_j$, averaged over the PC electrode surface, and capacitance C_{PC}^{app} can be obtained by the equation

$$C_{PC}^{app} = C_d(\bar{\sigma}) R \sum_j X_j C_{ij}$$

$$\times \left(C_d(\bar{\sigma}) + R \sum_j X_j C_{ij} \right)^{-1} \quad (10)$$

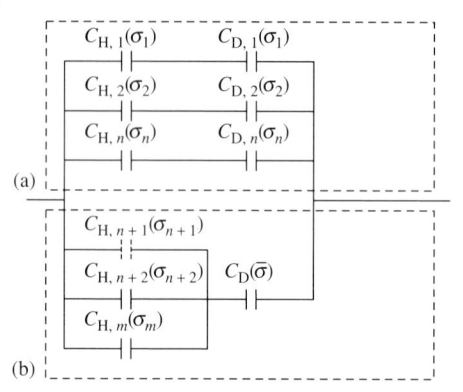

Fig. 6 Theoretical model of the electrical double layer at an ideally polarizable electrode with a PC surface structure. (a) Model of independent diffuse layers (Eq. 9); (b) model of common diffuse layer (Eq. 10).

This model is known as the model of the common diffuse layer (CDL) [27]. As shown in Ref. [30], both models can describe only some limiting cases and the exposition for the total capacity of the PC electrode (equivalent circuit) investigated depends on the relationship between the three lengths: (1) the characteristic size of the individual planes at a PC electrode surface y^*; (2) the effective screening length in the bulk of the diffuse layer near the face j $l_{dj} \equiv l_d(\sigma_j) = l_d/\sqrt{1 + (\sigma_j/\sigma^*)^2}$, where l_d is the Debye screening length and $\sigma^* = \varepsilon RT/2\pi F l_d$; (3) εl_{ij}, where ε is the bulk dielectric constant of the solvent and the length $l_{ij} = (1/4)\pi C_{ij}$ is determined by the capacity of the compact layer of the face j. According to the theoretical analysis, the CDL model is valid for PC electrodes with very small grains ($y^* < 5 \div 10$ nm) first of all with a moderate difference in $E_{\sigma=0}$ for the faces ($\Delta E_{\sigma=0} = 0.1 \div 0.15$ V) and for very dilute electrolyte solutions ($c \leq 0.01$ M) near the point of total zero charge. Thus, if the value of y^* is much less than two of the three lengths: $y^* \ll \varepsilon l_{i1} \sim \varepsilon l_{i2}, l_{d1}, l_{d2}$, the CDL model is valid. For other cases, the model of IDL would be valid [30].

The real PC electrode usually consists of the comparatively large homogeneous surface regions ($y^* > 10$ nm), and between the large homogeneous areas there are aggregates, which consist of many very small crystallites, whose sizes are probably smaller than the effective Debye screening lengths. Therefore, the structure of edl on a real PC electrode can be expressed by the relations

$$C_{PC}^{app} = R \sum_{j=1}^{n} \frac{X_j C_{ij} C_{dj}}{C_{ij} + C_{dj}} + X_{m+1}$$

$$\times \left[\frac{C_d(\bar{\sigma})R \sum_{j=n+1}^{m} X_j C_{ij}}{C_{dj}(\bar{\sigma}) + R \sum_{j=n+1}^{m} X_j C_{ij}} \right] \quad (11)$$

$$X_{m+1} + \sum_{j=1}^{n} X_j = 1, \quad \sum_{j=n+1}^{m} X_j = 1,$$

$$m > n$$

The results of computer simulation of many experimental C_{PC}^{exp}, E curves for various bismuth solid drop electrodes show that 10 to 30% of the whole surface of bismuth solid drop electrode is covered with small crystallites ($y^* < 10$ nm), and the standard deviation $\bar{\Delta}(\Delta C)$ of C_{PC}^{exp}, E curves from C_{PC}^{app}, E curves is smaller if

Eq. (11) is used instead of Eqs. (9) or (10). The data for the wedge-shaped two-plane electrode show that the share of small crystallites at the surface of this electrode is not more than 5 to 10% [15, 16, 22].

2.4.1.5.5 Debye Length–dependent Surface Roughness Model

Daikhin et al. [31] have shown that the competition between the Debye length and the characteristic sizes of roughness would modify the GC result for the diffuse-layer capacitance. The limiting value of the capacitance at short Debye length should follow the Gouy result, but with surface area value replaced by $S_{real} = RS_{geom}$. Thus, the diffuse layer simply follows every bump or dip of the electrode, the surface of which looks flat at the Debye scale. In the limit of long l_D (dilute solutions), the roughness cannot be manifested in the capacitance that would obey the native Gouy expression and the analogy with macroscopic capacitance can be used. The crossover between these two limits can be simulated as

$$C = \tilde{R}(\kappa) C_{GC} \quad (12)$$

where the roughness function, $\tilde{R}(\kappa)$, varies between $\tilde{R}(0) = 1$ and $\tilde{R}(\infty) = R > 1$; κ is the Gouy length (the inverse Debye length). It was demonstrated that the PZ plot method is not the most convenient tool for the characterization of surface roughness and the more informative would be the plot of $\tilde{R}(\kappa)$ versus κ, where

$$\tilde{R}(\kappa) \approx (C_{exp}^{-1} - C_i^{-1})^{-1} C_{GC}^{-1} \quad (13)$$

and C_i is evaluated from the measurements at high electrolyte concentration according to Valette–Hamelin approach [25, 26]. In the case of weak nonfractal surface roughness the two limiting approximations are valid: (1) At high electrolyte concentrations, when l_D is shorter than the smallest characteristic length of surface inhomogeneity, $\tilde{R}(\kappa)$ can be expressed as

$$\tilde{R}(\kappa) \approx R\left(\frac{1 - \langle H^2 \rangle}{2R\kappa^2}\right) \quad (14)$$

where $\langle H^2 \rangle$ is the mean square curvature of surface [31–33]. Equation (14) shows that the roughness function $\tilde{R}(\kappa)$ approaches the geometrical roughness factor R at small l_D. With the increase of l_D it decreases with respect to R, and the correction is proportional to the square of l_D. (2) In the range of large Debye lengths (low concentrations), $\tilde{R}(\kappa)$ can be expressed as

$$\tilde{R}(\kappa) \approx 1 + \frac{\kappa h^2}{L} - \kappa^2 h^2 \quad (15)$$

where h is the height of the characteristic size of roughness in the z-direction (h denotes the root-mean-square departure of the surface from flatness); L is the length that is in the order of the maximal correlation length l_{max}. At very low electrolyte concentrations ($l_D > l_{max}$), the roughness of surface is not detectable in the capacitance and the first correction to the flat surface result is linear in κ. Intermediate to these two limiting cases, the specific form of the roughness function depends on the morphology of the interface. The various geometrical surface layer models – sinusoidal corrugation, random Gaussian roughness, and periodical system of linear defects and rectangular grating – have been analyzed [31–33].

Experimental results of electrochemically polished, cut, electrochemically etched, and chemically etched Bi, Sb, and Cd electrodes (Fig. 7) demonstrate that $\tilde{R}(\kappa)$ rises with the increase of surface roughness, in agreement with AFM data and conclusions of Refs. [32, 33].

The nonlinear version of roughness theory applicable for arbitrary large charges

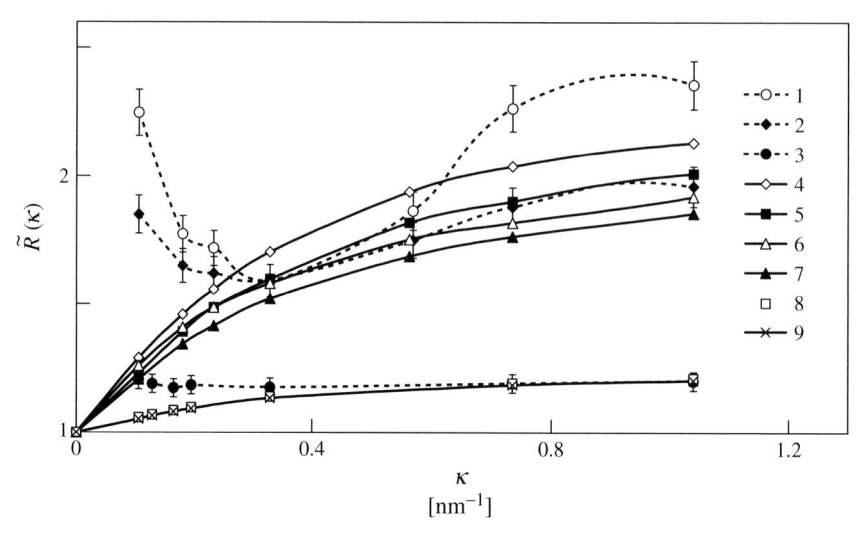

Fig. 7 Roughness function versus inverse Debye length plots of Bi(111)^{CE} (1); Bi(111)^{ECE} (2); Bi(DE)^R (3); and for various theoretical simulation models: sinusoidal corrugation (SC) – $h = 30$, $l_{SC} = 50$ (4); random Gaussian roughness (RGR) – $l_{RGR} = 50$, $h = 30$ (5); SC – $h = 5.0$, $l_{SC} = 2.5$ (6); RGR – $h = 2$, $l_{RGR} = 1$ (7); SC – $h = 0.85$, $l_{SC} = 2.5$ (8); and RGR – $h = 0.9$, $l_{RGR} = 2.0$ (9) (all parameters in nm) (l, characteristic lateral correlation length; h, root-mean-square height of surface roughness).

has been worked out in Ref. [34]. Far from $E_{\sigma=0}$ the effects of nonlinear screening have been found, which lead to the charge dependence of PZ plot, observed experimentally in many works [1–5, 7, 25–33]. Thus, the roughness factor of PC surface depends on the effective diffuse-layer thickness $\kappa_{\text{eff}}(E_r) = \kappa \cosh(e\beta E_r/2)$, where $e\beta E_r$ is the dimensionless electrode potential, $E_r = E_x - E_{\sigma=0}$ is the rational electrode potential, $\beta = (k_B T)^{-1}$ (k_B is the Boltzmann constant and e is the elementary charge); E_x is an electrode potential. According to theoretical calculations, $\tilde{R}(\kappa, E)$ is minimal close to $E_{\sigma=0}$ and rises to the limiting value at moderate and high σ. According to Ref. [34], for a fixed value of κ, the effect of surface structure on the capacitance increases with E_r. However, in contrast to the case of low E_R, $\tilde{R}(\kappa, E)$ could be a non-monotonous function of κ with a maximal value $\tilde{R}_{\max} > R$, and the crossover occurs at $E_R \approx 65$ mV ($T \approx 298$ K).

This model has been used for interpretation of the experimental capacitance data of rough Cd electrodes (Fig. 8). In the region of moderate E_r, the roughness function rises with the decrease of the effective Debye length l_{eff}. At higher E_r, the surface roughness function levels off to the constant value, that is, to the so-called geometrical surface roughness value. In the region of $E_{\sigma=0}$ there are deviations from the theory caused by the energetic inhomogeneity of electrode surface. It is to be noted that the new effective Debye length–dependent surface roughness theory needs future corrections taking into account the role of crystallographic inhomogeneity of the geometrically rough electrode surface, that is, the dependence of the local surface charge density on the crystallographic

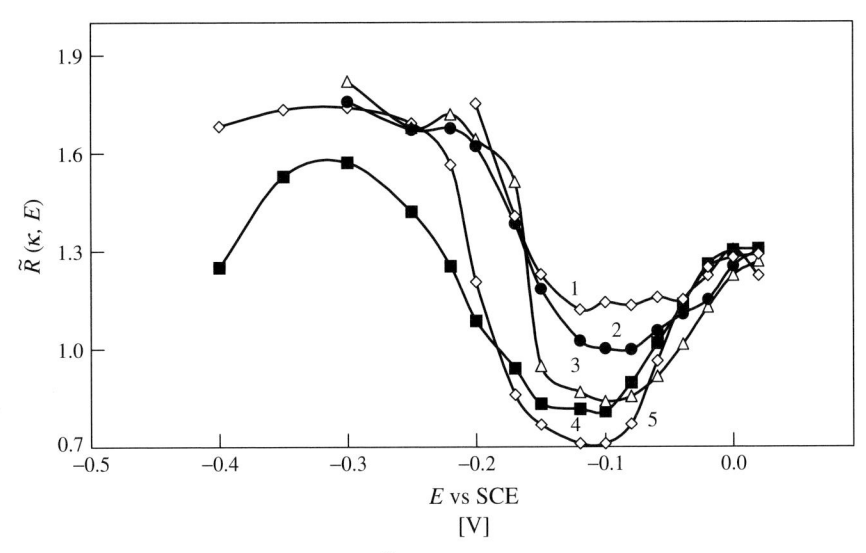

Fig. 8 Experimental roughness function $\tilde{R}(\kappa, E)$ versus E curves for electrochemically etched Cd(0001) for the fixed electrolyte (NaF + H_2O) concentrations: (1) 0.05 M; (2) 0.02 M; (3) 0.007 M; (4) 0.002 M; (5) 0.001 M.

structure of energetically homogeneous regions exposed on the surface of the macropolycrystalline electrode.

2.4.1.5.6 Electric Double Layer and Fractal Structure of Surface

Electrochemical impedance spectroscopy (EIS) in a sufficiently broad frequency range is a method well suited for the determination of equilibrium and kinetic parameters (faradaic or non-faradaic) at a given applied potential. The main difficulty in the analysis of impedance spectra of solid electrodes is the "frequency dispersion" of the impedance values, referred to the constant phase or fractal behavior and modeled in the equivalent circuit by the so-called *constant phase element* (CPE) [5, 15, 16, 22, 35, 36]. The frequency dependence is usually attributed to the geometrical nonuniformity and the roughness of PC surfaces having fractal nature with so-called self-similarity or self-affinity of the structure resulting in an unusual fractal dimension of the interface according to the definition of Mandelbrot [35–37]. Such a structural nonuniformity may result in a nonuniform distribution of σ at the electrode surface due to the different $E_{\sigma=0}$ of the different grains existing at the electrode surface. The fractal carpet model [35, 36] is representative of this approach. The impedance of a fractal electrode is

$$Z = R_S + Z_0(j\omega)^{-\alpha} \quad (16)$$

where Z_0 is a preexponential factor (analogous to the inverse of the capacitance of the edl $(1/C)$), $\omega = 2\pi \nu$ is the angular frequency, $j = \sqrt{-1}$ and α is a dimensionless parameter with a value usually between 0.5 and 1. The CPE angle φ is related to α by

$$\varphi = \frac{\pi}{2}(1 - \alpha) \quad (17)$$

The value $\alpha = 1$ corresponds to ideal capacitive behavior. The so-called fractal dimension D introduced by Mandelbrot [37] is a quantity that attains a value between

2 and 3 for a fractal structure and reduces to 2 when the surface is flat. According to Ref. [35, 36], D is related to α by

$$\alpha = \frac{1}{(D-1)} \quad (18)$$

The CPE model has been used to study PC Au, Cd, Ag, Bi, Sb and it has been found that for electrochemically polished electrodes the surface roughness is very small compared with mechanically polished surfaces [5, 15, 16, 22, 35, 36] (see also Chap. 8 in Vol. 3).

2.4.2
Experimental Aspects. Short Review of Data

2.4.2.1 Silver

Ag crystallizes in the face-centered cubic (fcc) system. Various procedures for the preparation of the Ag electrode surface have been used [5, 7]. Vitanov et al. [2, 3, 38] have used electrodes electrochemically grown (EG) in a Teflon capillary (so-called *quasi-perfect* Ag(100) and Ag(111) planes). Valette and Hamelin [25, 26] have used electrochemically polished (EP) Ag single crystals (so-called *real planes*), and Trasatti et al. [39] have used chemically polished (CP) Ag crystal faces. $E_{\sigma=0}$ is found to depend on the crystallographic orientation of Ag faces (Table 1), increasing with the atomic density of the faces (see also Sect. 2.4.3.1). The specific adsorption of anions increases in the order $PF_6^- \leq BF_4^- \leq F^- < ClO_4^-$ for Ag(100) and Ag(111) and in the order $BF_4^- \leq PF_6^- < ClO_4^- < F^-$ for Ag(110). The activity of ClO_4^- on Ag electrodes increases in the sequence Ag(110) < Ag(100) < Ag(111), but for F^- in the reverse order. This was explained by the decrease in hydrophilicity in the order Ag(111) < Ag(100) < Ag(110) because the adsorption of the large structure-breaking ClO_4^- anion from H_2O at Ag is mainly caused by the squeezing-out effect [4, 5, 38].

The inner-layer capacitance $C_i^{\sigma=0}$ increases in the sequence Ag(111) < Ag(100) < Ag(110) as the reticular density of EP planes decreases, in good agreement with the conclusions of Leiva and Schmickler [18], and with the data of quantum-chemical calculations of water adsorption at metal clusters [40]. For EG electrodes, C_i varies in the reverse order and the values of $C_i^{\sigma=0}$ for *quasi-perfect* Ag planes are remarkably higher than those for real Ag planes [2–5].

The contradiction in the data might be explained by various concentrations of surface defects that are claimed to be minimal on EG Ag(111) and Ag(100), and somewhat higher on EP and CP electrodes. EG Ag(100), Ag(111), and Cd(1000) were found to be free of screw dislocations and atomically smooth. However, the surface structure of real Ag single-crystal faces was observed by STM to be very smooth and atomically perfect too. For thermally grown and thereafter CP Ag surfaces, the correct LEED pattern has been obtained [8]. The large difference of $E_{\sigma=0}$ values existing for the various Ag faces should cause a marked dependence of $E_{\min}$ on the surface structure [1–5, 7, 15, 16, 22, 25, 26]. However, the $E_{\sigma=0}$ values obtained by different groups demonstrate a surprisingly good agreement (Table 1; see also Chaps. 14–20 in Vol. 3).

The temperature dependence of the edl parameters has been determined for real and quasi-perfect Ag planes [4, 5]. The electronic structure of Ag single-crystal face $H_2O + NaF$ interfaces has been studied by Chao et al. by ellipsometry and differential capacitance, which is discussed in Ref. [5].

Tab. 1 Electrical double-layer parameters for Ag, Au, and Cu in aqueous solutions

Electrode	Electrolyte	$E_{\sigma=0}$ [V] versus SHE	R	Atomic density [cm^{-2}]	Method	References
PC Ag^{EP}	NaF	−0.70 to −0.72	1.2	—	Impedance	5, 25
"	"	−0.66 ± 0.02	—	—	Open circuit scrape	5, 13
Ag(111)^{EG}	NaF	−0.45 ± 0.01	1.02	1.380 × 10^{15}	Impedance	38
Ag(111)^{EP}	NaF	−0.454 ± 0.010	1.07 ± 0.10	—	"	2–5, 26
Ag(111)^{CP}	NaF	−0.450 ± 0.016	—	—	"	3, 39
Ag(100)$^{EP, EG, CP}$	NaF	−0.621 ± 0.005	1.15	1.195 × 10^{15}	"	2–5, 26, 38, 39
Ag(110)^{CP}	NaF, NaBF$_4$, KPF$_6$	−0.734 ± 0.005	1.10–1.20	0.844 × 10^{15}	"	39
Ag(311)^{CP}	NaF, KPF$_6$	−0.664 ± 0.005	—	0.724 × 10^{15}	"	"
Ag(331)^{CP}	NaF, KPF$_6$	−0.670 ± 0.010	—	—	"	"
Ag(210)^{EP}	NaF	−0.750 ± 0.010	—	0.536 × 10^{15}	"	2–5
Au(111)	0.05–0.2 M NaF	0.56	—	—	"	2
Au(100)	"	0.33	—	—	"	10
Au(110)	"	0.19	—	—	"	2
Au(210)	"	0.11	—	—	"	"
Au(311)	"	0.25	—	—	"	"
Au(111) (22 × √3)	0.1 M HClO$_4$	0.55	—	—	Immersion	5
Au(111) (1 × 1)	"	0.47	—	—	Impedance	6
Au(100) "hex"	"	0.54	—	—	"	"
Au(100) (1 × 1)	"	0.32	—	—	"	"
PC Cu	0.1 M NaF	−0.50	—	—	Open circuit scrape	1, 13
"	0.050 M. Na$_2$SO$_4$ + TBN$^+$	−0.64 ± 0.05	—	—	Impedance	45
"	0.050 M. Na$_2$SO$_4$ + TBN$^+$	−0.33 ± 0.02^a	—	1.772 × 10^{15}	"	"
Cu(111)	0.001–0.04 M KClO$_4$	−0.20 ± 0.01	1.0	—	"	44
"	0.050 M. Na$_2$SO$_4$ + TBN$^+$	−0.46 ± 0.05^a	—	1.535 × 10^{15}	"	45
Cu(100)	0.001–0.04 M KClO$_4$	−0.54 ± 0.01	1.0	—	"	44
"	0.05 M Na$_2$SO$_4$ + TBN$^+$	−0.63 ± 0.05^a	—	1.084 × 10^{15}	"	45
Cu(110)	0.005 M NaClO$_4$	−0.69 ± 0.01	—	—	"	46

Note: Abbreviations: TBN$^+$: tetrabutylammonium ion; SHE: standard hydrogen electrode; PC: polycrystalline; EP: electrochemically polished; CP: chemically polished; EG: electrochemically grown.
^{a} By back integration method.

The edl at PC Ag–aqueous solution interface has been discussed by Leikis et al., Valette and Hamelin, Vorotyntsev, and Trasatti and Lust [2–5, 7, 25–28]. A diffuse-layer minimum was found at $E_{min} = -0.96 \pm 0.02$ V (SCE), but E_{min} and C_i do not correspond to $E_{\sigma=0}$ and $C_i^{\sigma=0}$ (see also Sect. 2.4.1.5). E_{min} values for PC Ag–nonaqueous electrolyte interface, depending on the solvent nature, have been discussed in Ref. [5].

2.4.2.2 Gold

Au crystallizes in the same system (fcc) as Ag. Experimental data obtained by cyclic voltammetry (CV), electrochemical impedance, STM, AFM, SXRS, and other techniques (see also Chaps. 14–20 in Vol. 3) show that the edl and interfacial properties depend strongly on the crystallographic structure of the electrode surface, the surface charge density, as well as the chemical composition of the solvent and the electrolyte [2–11] (Table 1). Most of the results before 1980 were obtained with gold faces isolated by a noncontaminating resin and cycled before use in dilute Na_2SO_4 or H_2SO_4 solutions until a stable, reproducible CV was obtained. From 1980 onward, most of the results were obtained after a flame annealing treatment of Au planes, applied in a way similar to that originally used for ordered Pt electrodes [41]. The surface cleanliness achieved by this method allowed investigations to be made in a meaningful way already during the first or initial scans of the CV.

Over the past 15 years it has been demonstrated by a variety of in situ and ex situ techniques that flame-annealed Au faces are reconstructed in the same way as the surfaces of samples prepared in ultrahigh vacuum (UHV) [2–11] and that the reconstructed surfaces are stable even in contact with an aqueous solution if certain precautions are taken with respect to the applied potential and the electrolyte composition. The surface reconstruction of Au(110) is more rapid than that of Au(111) and Au(100). A comprehensive review of reconstruction phenomena at single-crystal faces of various metals has been given by Kolb [6] and Gao et al. [42].

The edl at a PC Au–solution interface has been studied mainly by Clavilier and Nguyen Van Houng, Hamelin, Borkowska et al., and others [1–7, 43]. A diffuse-layer minimum was found at $E_{min} = -0.04 \pm 0.02$ V (SCE) for PC Au|H_2O (NaF, $KClO_4$, $HClO_4$), but E_{min} obtained for PC Au does not correspond to $\bar{\sigma}_{PC} = \sum_i \theta_i \sigma_i = 0$ and C_i at E_{min} does not correspond to $C_i^{\sigma=0}$ [2–5, 7]. The value of E_{min} depends noticeably on the solvent nature and E_{min} (versus bis-biphenylchromium (BBCr(I/0)) scale) rises in the order of solvents MeOH < H_2O < DMSO < dimethylformamide < propylene carbonate < acetonitrile [5, 43].

2.4.2.3 Copper

Cu crystallizes in the fcc system. The experimental data for single-crystal Cu|H_2O and for PC Cu are controversial [2–5, 7, 44, 45]. The first studies with Cu(111), Cu(100), Cu(110), and PC Cu in surface-inactive electrolyte solutions show a capacitance minimum at E less negative than the positive limit of ideal polarizability of Cu electrodes. More reliable values of $E_{\sigma=0}$ for Cu single-crystal faces have been obtained by Lecoeur and Bellier [44] with EP Cu(111) and Cu(100) (Table 1). Foresti et al. [46] have found $E_{\sigma=0} = -0.93 \pm 0.01$ V (SCE) for the Cu(110)–aqueous solution interface, and the validity of the GCSG model has been verified. As for Zn single-crystal electrodes, reliable values of $E_{\sigma=0}$ have been obtained indirectly from the dependence of the adsorption–desorption peak potential E^{max} of TBN$^+$ on the crystallographic orientation [45]. Like other sd metals, $\sigma_{PC} \neq 0$ on the PC Cu–H$_2$O interface at E_{min} [2–5, 7, 25–27].

In situ AFM studies in aqueous H$_2$SO$_4$ and HClO$_4$ solutions show that Cu(111) exhibits the correct hexagonal structure. Cu(100) shows a correct structure and spacing at $E = -0.63$ V (SCE) in 0.1 M HClO$_4$, but the Cu(110) surface in dilute HClO$_4$ (0.003 M) at $E = -0.13$ V (SCE) (160 mV more negative than the rest potential) exhibits a correct unreconstructed structure (see also Chaps. 15–17 in Vol. 3). The strength of the Cu–O bond will be lower on the Cu(111) face than on the Cu(100) and Cu(110). These results suggest that a specific Pourbaix pH–E phase diagram is needed to describe the behavior of each low index face of Cu [47].

2.4.2.4 Lead

Pb crystallizes in the fcc system and electrochemically polished Pb(111), Pb(100), Pb(110), and Pb(112) faces have been studied by impedance. The potential of the minimum in C, E curves was independent of c_{NaF}, as well as of $c_{Na_2SO_4}$. The dependence of $E_{\sigma=0}$ on the atomic density of the faces is weak, and the value of $E_{\sigma=0}$, becoming less negative as the atomic density decreases (Table 2), is in contradiction to that for other fcc metals [2, 3, 5, 7]. The inner-layer capacitance at $\sigma = 0$ increases in the sequence Pb(100) $\leq$ Pb(110) $\leq$ Pb(112) < PC Pb < Pb(111), which has been explained by the increasing hydrophilicity in the same order [48].

EDL data for PC Pb have been obtained in aqueous, glacial acetic acid, MeOH, EtOH, and dimethylformamide surface-inactive electrolyte solutions. At E_{min} the value of $\sigma_{PC} \approx 0$ and thus $E_{min} \approx E_{\sigma=0}$ [1–5, 7, 48].

2.4.2.5 Tin

Sn crystallizes in the tetragonal system. $E_{\sigma=0}$ increases only slightly with the atomic density of faces Sn(110) < Sn(100) < Sn(111) [49, 50] (Table 2), in

Tab. 2 Electrical double-layer parameters for Pb [5, 48] and Sn [5, 49, 50] in various solvents

Electrode	Solution	$E_{\sigma=0}$ [V] versus SHE	R
PC Pb	0.01 M NaF/H$_2$O	-0.60 ± 0.02^I	1.10
,,	KF/MeOH	-0.43 ± 0.02^I	1.16
,,	CH$_3$COONa/CH$_3$COOH	-0.58 ± 0.02^I	1.15
,,	NaClO$_4$/FM	-0.61 ± 0.02^{IM}	
Pb(111)	0.01 M NaF/H$_2$O	-0.62 ± 0.01^I	1.05
Pb(100)	0.01 M NaF/H$_2$O	-0.59 ± 0.01^I	1.10
Pb(110)	0.01 M NaF/H$_2$O	-0.58 ± 0.01^I	1.10
Pb(112)	0.01 M NaF/H$_2$O	-0.58 ± 0.01^I	1.15
PC Sn	NaClO$_4$ + (HClO$_4$)	-0.43 ± 0.02^I	–
Sn(001)	Na$_2$SO$_4$/H$_2$O	-0.37 ± 0.01^I	1.1
Sn($\bar{1}$10)	Na$_2$SO$_4$/H$_2$O	-0.37 ± 0.01^I	1.1
Sn(110)	Na$_2$SO$_4$/H$_2$O	-0.38 ± 0.01^I	1.1

Note: Abbreviations: I: impedance method; IM: immersion method; PC: polycrystalline; FM: formamide; MeOH: Methanol.

accordance with the general trend observed for sd and sp metals [2–5, 7]. The applicability of the GCSG theory for monocrystalline and PC Sn has been verified by Khmelevaya and Damaskin [49, 50]. At $E_{\min}$, $\sigma_{PC} \approx 0$ for PC Sn has been obtained.

2.4.2.6 Zinc

Zinc crystallizes in the hexagonal close-packed (hcp) system. As the zinc dissolution takes place at potentials very close to $E_{\sigma=0}$, the differential capacitance curves in the region of $E_{\sigma=0}$ in pure surface-inactive electrolyte solutions can be determined directly for the Zn(11$\bar{2}$0) face only, for which $E_{\sigma=0}$ is more negative than the steady state potential [51, 52] (Table 3). The values of $E_{\sigma=0}$ obtained by the salting-out method for Zn(11$\bar{2}$1) and by the dependence of $E_{\min}$ on c_{KCl} at $c_{TU} = $ const are in good agreement with the values of $E_{\sigma=0}$ obtained from the C, E curves for pure KCl solutions. $E_{\sigma=0}$ values for other faces have been obtained by the dependence of $E_{\min}$ in the C, E curves on c_{KCl} in solutions in the presence of thiourea at $c_{TU} = 3 \times 10^{-1}$ M [51, 52]. The maximum difference between the $E_{\sigma=0}$ values for the various faces of Zn is about 90 mV, and this is in good agreement with earlier estimates [1–5, 7] (see also Sect. 2.4.3.1).

The edl at PC Zn–H$_2$O interfaces has been studied in many works, but the situation is somewhat ambiguous and complex. The values of $E_{\min}$ depend on the surface preparation method. The value of $E_{\sigma=0}$ for PC Zn lies between $E_{\sigma=0}$ values for single-crystal planes, and at $E_{\min}$, $|\sigma_{PC}| \leq 0.01$ C m^{-2} [1–5, 7, 51, 52].

2.4.2.7 Cadmium

Cadmium crystallizes in the hcp system. Korotkov et al. [53] have established that in surface-inactive electrolyte solutions (NaF, LiClO$_4$) the $E_{\sigma=0}$ of Cd (11$\bar{2}$0) is slightly shifted to more negative potentials compared with the $E_{\sigma=0}$ of polished PC Cd. On a scraped surface of Cd(11$\bar{2}$0), $E_{\sigma=0}$ is more negative than on a polished one, and this difference can be related to the greater inhomogeneity of the scraped surface. Vitanov, Popov et al. [54, 55] have studied Cd(0001), EG in a Teflon capillary. A linear PZ plot with $R = 1.09$ has been found at $\sigma = 0$. A slight dependence of C_i, σ-curves on c_{el} has been found. The temperature dependence of the edl parameters has been studied for Cd(0001). The positive value of $\partial E_{\sigma=0}/\partial T$ (0.33 ± 0.03 mV K^{-1}) is taken to indicate that the H$_2$O dipoles are oriented with their negative end toward the metal surface. C_i has been found to be higher for Cd(0001) than for PC Cd [54, 55].

Electrochemically polished and chemically treated Cd(0001), Cd(10$\bar{1}$0), Cd(11$\bar{2}$0), Cd(10$\bar{1}$1), and Cd(11$\bar{2}$1) electrodes have been studied by impedance [15, 16, 22], and the main parameters obtained are given in Table 3. The value of $E_{\min}$ for PC Cd [1–5, 7] lies between the $E_{\sigma=0}$ values for single-crystal planes, and at $E_{\min}$, $\sigma_{PC} \leq 0.01$ µC cm^{-2}. $E_{\sigma=0}$ for a cut Cd(0001)^c electrode is in good agreement with that for electrochemically polished Cd(0001). For chemically treated Cd electrodes, a remarkable dependence of $E_{\min}$ on c_{NaF} has been observed [5, 22], characteristic of PC surface. The inner-layer capacitance of EP Cd faces increases as the atomic density decreases, in good agreement with Leiva–Schmickler model and quantum-chemical calculations [18, 40]. It has been suggested that hydrophilicity increases in the order Cd(0001) < Cd(10$\bar{1}$0) < Cd(11$\bar{2}$0). The same order has been proposed on the basis of data of organic compound adsorption.

Tab. 3 Electrical double-layer parameters of Zn and Cd in various solvents

Electrode	Solution	$E_{\sigma=0}$ [V] versus SHE	R	α^a	D^b	Atomic density [cm^{-2}]	Reference
PC Zn	NaClO$_4$ (pH = 7)	−0.92	—	—	—	—	51, 52
,,	0.1 M KCl + TU	−0.91	—	—	—	—	,,
Zn(0001)	,,	−0.78 ± 0.02	—	—	—	1.630 × 10^{15}	,,
Zn(10$\bar{1}$0)	,,	−0.84 ± 0.02	—	—	—	1.315 × 10^{15}	,,
Zn(11$\bar{2}$0)	,,	−0.87 ± 0.02	—	—	—	—	,,
Zn(11$\bar{2}$0)	3 × 10^{-3} M KCl	−0.85 ± 0.01	—	—	—	—	,,
PC Cd	KF/H$_2$O	−0.73 ± 0.02	—	0.94	2.06	—	1, 5, 7
,,	LiClO$_4$/DMF	−0.600 ± 0.01	—	—	—	—	5, 7
,,	LiClO$_4$/i-PrOH	−0.75 ± 0.01	—	—	—	—	,,
Cd(0001)^{EG}	NaF/H$_2$O	−0.75 ± 0.01	1.09	—	—	1.308 × 10^{15}	54, 55
Cd(0001)^{EP}	,,	−0.73 ± 0.01	1.06	0.98	2.01	—	22
Cd(10$\bar{1}$0)^{EP}	,,	−0.77 ± 0.01	1.12	0.97	2.03	1.038 × 10^{15}	,,
Cd(11$\bar{2}$0)^{EP}	,,	−0.76 ± 0.01	1.1	—	—	0.692 × 10^{15}	53
Cd(11$\bar{2}$0)^{EP}	,,	−0.78 ± 0.01	1.22	0.98	2.02	—	22
Cd(10$\bar{1}$1)^{EP}	,,	−0.760 ± 0.015	1.24	0.94	2.07	0.544 × 10^{15}	,,
Cd(11$\bar{2}$1)^{EP}	,,	−0.750 ± 0.015	1.25	0.95	2.06	0.442 × 10^{15}	,,

^{a}Fractional exponent, dimensionless parameter.
^{b}Fractal dimension, dimensionless parameter.
Note: Abbreviations: PC: polycrystalline; EG: electrochemically grown; EP: electrochemically polished; DMF: N,N-dimethylformamide; i-PrOH: iso-propanol; TU: thiourea.

The inverse capacitance of metal phase, shown in Fig. 9, obtained according to the Amokrane–Badiali model [15, 16, 19, 22], depends on the crystallographic structure as well as on the chemical nature of the metal studied.

2.4.2.8 Bismuth

Bi crystallizes in a rhombohedral system. Electrochemically polished Bi single-crystal faces were first used for edl studies by Frumkin, Palm and coworkers [1, 20, 21]. The E_{min} (Table 4) was observed to be shifted 30 mV toward more negative potentials compared with Bi solid drop electrode and EP PC Bi. Later, electrochemically polished Bi(01$\bar{1}$), Bi(2$\bar{1}\bar{1}$), Bi(001), and Bi(101) faces have been studied [5, 15, 16]. The difference between $E_{\sigma=0}$ for the above faces and the Bi(111) face is comparatively big and this can be explained by the different surface states of the Bi faces. E_{min} for PC Bi lies between $E_{\sigma=0}$ values for various planes with $\sigma_{PC} \leq 0.01\ \mu C\ cm^{-2}$.

The admittance of Bi single-crystal faces and PC Bi in acetonitrile and various alcohols including LiClO$_4$ solutions has been studied, and $E_{\sigma=0}$ has been found to depend on the nature of the solvent. Maximal $\Delta E_{\sigma=0}$ for various faces decreases in the order of solvents: acetonitrile > H$_2$O > methanol > 2-propanol > ethanol [5, 15, 16].

C_i values increase as the atomic density of the face decreases, except for Bi(111), in good agreement with the theoretical calculations by Leiva and Schmickler, predicting the lowest interfacial capacitance for the most densely packed planes [18]. The capacitance of a metal

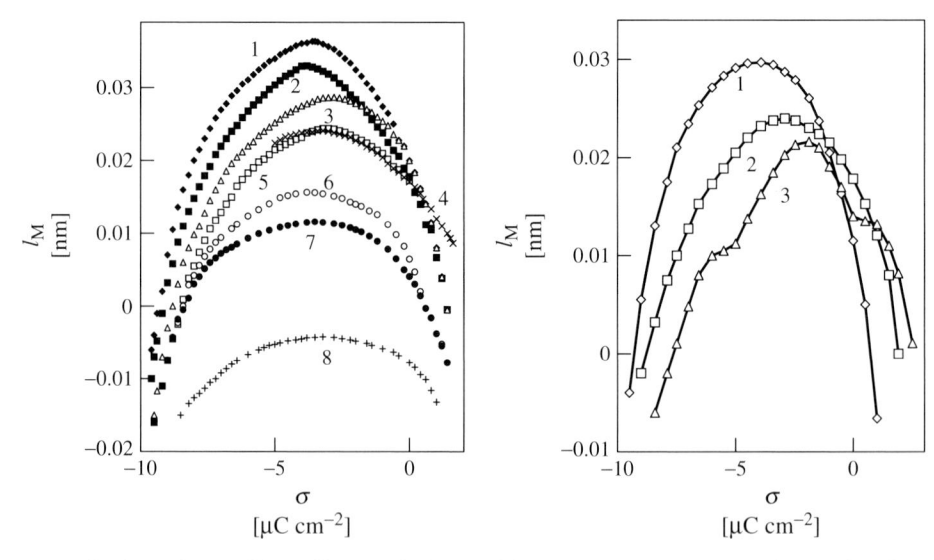

Fig. 9 The dependence of the "effective" thickness of the potential drop region in the metal surface layer, calculated by the Amokrane–Badiali model [19] (at constant capacitance of solvent monolayer, C_S), on the charge density for aqueous solution on different electrodes (a): 1, Sb(111); 2, Sb(001); 3, Bi(111); 4, Hg; 5, Bi(001); 6, Cd (0001); 7, Cd(11$\bar{2}$0); and 8, Ga; and for Bi(001) in various solvents (b): 1, methanol; 2, H$_2$O; 3, acetonitrile (updated from Ref. [15, 16]).

Tab. 4 Electrical double-layer parameters of Bi and Sb [5, 15, 16, 22] electrodes in various solvents

Electrode	Solvent	$E_{\sigma=0}$ [V] versus aq. SHE	R	D	Atomic density [cm^{-2}]
Bi(111)	KF/H$_2$O	-0.434 ± 0.005	1.04	2.03	5.6×10^{15}
Bi($\bar{1}0\bar{1}$)	,,	-0.34 ± 0.01	1.07	2.05	4.5×10^{15}
Bi(001)	,,	-0.35 ± 0.01	1.06	2.05	5.3×10^{15}
Bi(01$\bar{1}$)	,,	-0.35 ± 0.01	1.05	2.04	6.4×10^{15}
Bi(2$\bar{1}\bar{1}$)	,,	-0.33 ± 0.015	1.14	2.06	7.4×10^{15}
Bi(111)	LiClO$_4$/AN	-0.25 ± 0.01	1.06	2.04	–
Bi($\bar{1}0\bar{1}$)	,,	-0.18 ± 0.01	1.01	2.04	–
Bi(001)	,,	-0.17 ± 0.01	0.98	2.03	–
Bi(01$\bar{1}$)	,,	-0.18 ± 0.01	1.08	2.06	–
Bi(2$\bar{1}\bar{1}$)	,,	-0.17 ± 0.015	1.03	2.05	–
Bi(111)	LiClO$_4$/MeOH	-0.28 ± 0.01	1.05	–	–
Bi(001)	,,	-0.21 ± 0.01	1.07	–	–
Bi(01$\bar{1}$)	,,	-0.23 ± 0.01	1.10	–	–
Bi(111)	LiClO$_4$/EtOH	-0.21 ± 0.01	1.03	2.04	–
Bi(001)	,,	-0.18 ± 0.01	1.08	2.07	–
Bi(01$\bar{1}$)	,,	-0.20 ± 0.01	1.07	–	–
Bi(2$\bar{1}\bar{1}$)	,,	-0.22 ± 0.01	1.12	–	–
Bi(111)	LiClO$_4$/i-PrOH	-0.21 ± 0.01	1.02	–	–
Bi(001)	,,	-0.170 ± 0.015	1.07	–	–
Bi(01$\bar{1}$)	,,	-0.220 ± 0.015	1.03	–	–
SbDE^R	KF; NaF/H$_2$O	-0.17 ± 0.01	1.10–1.14	–	–
PC Sb	NaF/H$_2$O	-0.16 ± 0.02	1.03	2.05	–
SbDE^R	HClO$_4$/MeOH	-0.02 ± 0.025	1.08	–	–
Sb(111)	KF/H$_2$O	-0.22 ± 0.01	1.06	2.04	–
Sb(001)	,,	-0.13 ± 0.01	1.04	2.04	–
Sb(01$\bar{1}$)	,,	-0.15 ± 0.01	1.18	2.08	–
Sb(2$\bar{1}\bar{1}$)	,,	-0.10 ± 0.015	1.25	2.10	–
Sb(111)	LiClO$_4$/EtOH	-0.02 ± 0.01	1.01	–	–
Sb(001)	,,	0.05 ± 0.01	1.04	–	–

Note: Abbreviations: AN, acetonitrile; MeOH, methanol; EtOH, ethanol; i–PrOH, isopropanol; SbDE^R, an antimony drop electrode with remelted surface; PC, polycrystalline.

phase, C_M, has been calculated according to the Amokrane–Badiali model [19]. The thickness of a thin metal surface layer $l_M = 1/4\pi C_M$ as a function of σ decreases in the order Bi(111) > PC Bi > Bi(101) > Bi(2$\bar{1}\bar{1}$) > Bi(01$\bar{1}$) > Bi(001), in agreement with other data. The value of l_M^{max} rises in the order AN < H$_2$O < MeOH < EtOH (Fig. 9b), which has been explained by slight deviation of experimental systems from the Amokrane–Badiali model [19].

The influence of the surface pretreatment of Bi single-crystal faces has been studied, and a noticeable dependence of $E_{\sigma=0}$ on the surface structure has been established [5, 15, 16, 22].

2.4.2.9 Antimony

Sb crystallizes in the same rhombohedral system as Bi. Sb(111) exhibits an appreciably more negative value of $E_{\sigma=0}$ compared with the other planes (Table 4). The difference $\Delta E_{\sigma=0}$ for Sb(001), Sb(01$\bar{1}$), and Sb(2$\bar{1}\bar{1}$) is not more than 0.05 V and $E_{\sigma=0}$ decreases as the atomic density of the surface increases. $E_{\sigma=0}$ for the Sb solid drop electrode has an intermediate value, and $\sigma_{PC} \leq 0.01$ C m^{-2} at E_{min}^{PC} [5, 15, 16, 22].

The C_i values for Sb faces and PC Sb are noticeably lower than those for Bi. The closest-packed faces show the lowest values of C_i (except Sb(111)), in good agreement with the Leiva–Schmickler model [18]. The anomalous position of Sb(111) and Bi(111) is probably caused by a more pronounced influence of the capacitance of the metal phase compared with other Sb and Bi faces (Fig. 9a).

The edl structure of Sb(111)–EtOH and Sb(001)–EtOH interfaces has been investigated by impedance. As for H$_2$O, $\Delta E_{\sigma=0}$ in EtOH for Sb(001) and Sb(111) is higher than for Bi(111) and Bi(001). $C_i^{\sigma=0}$ for Sb single-crystal faces in EtOH are lower than for Bi, which has been explained by the lower lyophilicity of Sb, as well as by the higher thickness of the thin metal layer [5, 15, 16, 22].

2.4.2.10 Iron

Fe crystallizes in the body-centered cubic (bcc) system. Fe(100) and Fe(111) single-crystal faces were grown at 750 ÷ 780 °C from FeBr$_2$ in pure H$_2$ atmosphere and reduced for 1 h at $E = -0.95$ V (SCE) in the working solution (pH = 2.5). A diffuse-layer capacitance minimum was observed with E_{min} practically independent of c_{el} (Table 5). The PZ plot was linear with R somewhat higher than unity. The inner-layer capacitance decreases from Fe(111) to Fe(100) as the atomic density of the face increases [56].

The PC Fe electrodes with electrochemically polished (cathodically pretreated for 1 h) surfaces show a diffuse-layer minimum at $E_{min} = -0.94$ V (SCE), independent of $c_{H_2SO_4}$ [57]. A dependence of the PZC on pH has been observed [5].

The edl at the renewed Fe–LiClO$_4$ interface has been studied in various

Tab. 5 Electrical double-layer parameters of Fe [5, 56–58] and Ni [5, 59] in various solvents

Electrode	Solution[a]	$E_{\sigma=0}$ [V] versus aq. SHE	R	Atomic density [cm^{-2}]
PC Fe	KF (0.01 M)/	-0.70 ± 0.02	–	–
"	NaClO$_4$/H$_2$O	-0.64	–	–
Fe(111)	NaClO$_4$/H$_2$O	-0.69 ± 0.01	1.43	0.707×10^{15}
Fe(100)	NaClO$_4$/H$_2$O	-0.72 ± 0.01	1.33	1.222×10^{15}
PC Fe	LiClO$_4$/TMU	-0.56	5.0	–
"	LiClO$_4$/DMF	-0.41	–	–
"	LiClO$_4$/DMAA	-0.31	–	–
"	LiClO$_4$/N-MP	-0.61	–	–
"	LiClO$_4$/HMPA	-0.56	4.8	–
PC Ni	H$_2$SO$_4$/H$_2$O (pH = 3.0)	-0.24	–	–
Ni(100)	HClO$_4$/H$_2$O	-0.39 ± 0.02	1.0	1.614×10^{15}
Ni(110)	HClO$_4$/H$_2$O	-0.53 ± 0.02	1.0	1.141×10^{15}

[a]TMU: tetramethylurea; DMF: dimethylformamide; DMAA: N,N,-dimethylacetamide; MPF: N-methyl-N-2-pyridylformamide; HMPA: hexamethylphosphoramide.

nonaqueous aprotic solvents by Safonov et al. [58]. The E_{min} in C, E curves depends on time; for Fe|1,1,3,3-tetramethylurea (TMU) + LiClO$_4$ interface, $E_{min} = -0.8$ V (SCE in H$_2$O) just after surface renewal and -0.35 V after 15 min. The PZ plot at E_{min} was linear, with R equal to 5.0 in TMU and 4.8 in hexamethylphosphoramide. The reconstruction of the surface is probable during the experiment with a decrease in surface roughness with time [58].

A less pronounced dependence of C_i on E for an Fe electrode than that for Hg has been observed, and the values of C_i for Fe are remarkably lower than that for Hg. Thus, the interfacial potential drop is substantially higher for Fe than for Hg and, accordingly, strong chemisorption of solvent molecules, weakly depending on E, is probable at an Fe–electrolyte interface [58].

2.4.2.11 Nickel

Ni crystallizes in the fcc system. A diffuse-layer minimum has been observed for the Ni(100) and Ni(110) single-crystal faces in H$_2$O/Na$_2$SO$_4$ system, with E_{min} independent of ν and c_{el}, and $R \approx 1$ (Table 5). In the case of Ni(111) and PC Ni the capacitance also decreases with dilution, but no deep minimum was observed in the C, E curves. Very low C_i values have been obtained. The difference between sp and sd metals has been explained by a different strength of the interaction between the metal and solvent molecules [59].

$E_{\sigma=0}$ of PC Ni is located in the negative potential range and depends strongly on the solution's pH [1, 5]. Widely scattered E_{min} values have been reported by different authors at similar solution pH (Table 5). However, the values obtained by scraped method are in reasonable agreement with impedance data [5, 13].

2.4.2.12 Pt-group Metals

Pt and Pt-group metals (poly- and single crystals) have long been among the most intensively studied systems in electrochemistry; nevertheless, reliable $E_{\sigma=0}$ values have been determined only recently [1, 5, 8, 14, 60–67].

The analysis by Frumkin et al. [1, 14] in 1970 led to the conclusion that in the case of nonpolarizable electrodes, two values of zero-charge potential can be defined: $E_{\sigma=0}$ and $E_{Q=0}$ (see Sect. 1). The values of $E_{\sigma=0}$ obtained by the scrape method [13] are in reasonable agreement with the Frumkin et al. data (Table 6). Results for other Pt-group metals are due to Frumkin and coworkers [1, 14] (Table 6). An Rh electrode with the surface renewed in closed circuit has been used by Lazarova [60]. In 0.005 M Na$_2$SO$_4$, $E_{\sigma=0} = -0.09 \pm 0.02$ V (SCE), while in 0.5 M Na$_2$SO$_4$, pH 2.5, $E_{\sigma=0} = -0.22$ V has been reported with $\partial E_{\sigma=0}/\partial \text{pH} = 55$ mV.

It is to be stressed that in the case of PC Pt-group metals, the crystallographic structure of the surface probably exerts a very pronounced influence, so that the experimental PZC values do not correspond to the condition $\bar{\sigma}_{PC} = 0$.

The surface electrochemistry of Pt single-crystal electrodes has been exhaustively studied using cyclic voltammetry [5, 8–12, 61–67]. Phenomena of a step reconstruction and step coalescence have been observed [61]. For Pt(111)–H$_2$O interface, prepared by the flame annealing method, a double-layer charging has been observed only in a very narrow potential region ($0.1 < E < 0.35$ V (SCE) in 0.05 M H$_2$SO$_4$), which depends on the chemical

Tab. 6 Potentials of zero charge of Pt-group metal single-crystal faces in aqueous solutions

Electrode	Electrolyte	$E_{\sigma=0}$ [V] versus SHE	$E_{Q=0}^{local}$ [V] versus SHE	Adsorption state (Terrace)	Method	Atomic density (cm^{-2})	References
Pt(111)	0.1 M HClO$_4$	–	0.34	(111)	Cyclic voltammetry	1.503×10^{15}	68
	0.1 M HClO$_4$	–	0.29	–	"	–	63
	0.1 M HClO$_4$	–	0.27	–	"	–	66
Pt(110)	0.1 M HClO$_4$	–	0.16	(110)-(1×2)$^{(a)}$	"	0.920×10^{15}	68
	0.15 M HClO$_4$	–	0.18	(110)	"	–	66
Pt(100)	0.1 M HClO$_4$	–	0.25	(100)-1D	"	1.302×10^{15}	68
Rh(111)	0.1 M HClO$_4$	–	0.16–0.17	(111)	"	1.599×10^{15}	"
Ir(111)	0.1 M HClO$_4$	–	0.13	(111)	"	1.574×10^{15}	"
PC Pt	0.3 M HF + 0.12 M KF (pH 2.4)	0.185	0.235	–	Titration	–	1, 14
"	NaClO$_4$ (pH = 1.0)	0.24	–	–	Scrape	–	1, 13
"	NaClO$_4$ (pH = 7.0)	0.02	–	–	"	–	"
"	NaClO$_4$ (pH = 11.0)	−0.30	–	–	"	–	"
PC Pd	0.05 M Na$_2$SO$_4$ + 0.001 M H$_2$SO$_4$ (pH 3)	0.10	0.26	–	Titration	–	1, 14
"	NaClO$_4$ (pH 7.0)	0.00	–	–	Scrape	–	5, 13
PC Rh	0.3 M HF + 0.12 M KF (pH 2.4)	−0.005	0.085	–	Titration	–	1, 14
"	NaClO$_4$ (pH 1.0)	0.17	–	–	Scrape	–	5, 13
"	NaClO$_4$ (pH 7.0)	−0.02	–	–	"	–	"
PC Ir	0.3 M HF + 0.12 M KF (pH 2.4)	−0.01	–	–	Titration	–	1, 14
"	NaClO$_4$ (pH 1.0)	0.21	–	–	Scrape	–	5, 13
"	NaClO$_4$ (pH 7.0)	0.52	–	–	"	–	"
PC Pt	LiClO$_4$ AN	0.14	–	–	Impedance	–	69, 70
PC Pt	LiClO$_4$/DMSO	0.09	–	–	"	–	"
PC Pd	LiClO$_4$/AN	0.32	–	–	"	–	"
PC Pd	LiClO$_4$/DMSO	0.24	–	–	"	–	"

Note: Abbreviations: AN: acetonitrile; DMSO: dimethylsulfoxide.

nature of the anion [5, 41, 61] (see also Chaps. 14–17 and 19 in Vol. 3).

For Pt(111) in the aqueous solutions of hydrofluoric acid, Wagner and Moylan [62] have estimated $E_{\sigma=0} = 0.22$ V (versus reversible hydrogen electrode) by comparing voltammetric curves and high-resolution electron energy loss spectroscopic data of water + H$^+$ coadsorption from the gas phase. Iwasita and Xia [63] prepared Pt single crystals according to the Clavilier et al. method [41]. According to the data of Fourier transform infrared (FTIR) reflection adsorption spectra, the water orientation changes from hydrogen down to oxygen down at 0.29 V (SHE), and this has been taken to indicate that $E_{Q=0}$ for Pt(111) is close to this value.

Hamm et al. [64] treated a Pt(111) surface in a UHV chamber by sputtering and annealing until the surface was clean and well ordered. A very narrow potential region, $0.13 < E < 0.28$ V (SCE), is left to bare double-layer charging with a capacitance $C = 110 \pm 13$ µF cm^{-2}. Thus, only in this region, a Pt(111) surface is free from surface adlayers. Thereafter the immersion method (IM) was used, and the Q, E-plot has been found to go through zero at $E \geq 0.80$ V (SHE). Assuming a small dependence of C on E, a somewhat more positive $E_{\sigma=0} \sim 1.1$ V (SHE) has been estimated. However, it is to be noted that these values of $E_{\sigma=0}$ correspond to the potential region, where the Pt surface is oxidized [5].

Clavilier et al. [65] have studied CO adsorption on electrochemically faceted Pt(111) and Pt(110) electrodes, and from the charge transients, these authors have provided a "definite" determination of $E_{Q=0}$. However, electrochemically faceted Pt(111) electrodes have a PC surface structure, and thus the value of $E_{Q=0}$ for such electrodes lies between $E_{Q=0}^{(111)}$ for terraces and $E_{Q=0}^{(110)}$ for steps [5]. Gomez et al. have studied the stepped Pt(111) surfaces [66]. The dependence of the PZTC for Pt(111) surfaces in acidic aqueous solution having increasing densities of the ordered monoatomic steps in the (111) to (110) and (111) to (100) zones is evaluated from CO "charge-displacement" measurements, with the objective of elucidating the influence of the electrochemical double layer on the large step-induced changes in surface potential known for the clean uncharged surface in UHV [67]. The PZTC values in 0.1 M HClO$_4$ and 0.5 M H$_2$SO$_4$ solutions decrease noticeably upon increasing the (110) step density, whereas smaller effects are found for the (100) steps. The location of the $E_{Q=0}$ values within the so-called *hydrogen* region, however, complicates the interpretation of the $E_{Q=0}$ – step density dependences owing to the presence of faradaic charge associated with potential-dependent hydrogen adsorption [66].

A detailed study of N$_2$O reduction on a variety of single-crystal Pt-group metal electrodes shows that N$_2$O reduction current maxima occurred exactly at Clavilier's $E_{Q=0}$. Therefore the N$_2$O reduction data have been used for obtaining the $E_{Q=0}$ values (Table 5). The importance of a local value of $E_{Q=0}$ has been emphasized, especially with respect to reconstructing metal surfaces such as Pt(100) and Pt(110), which can be prepared in a variety of crystallographic states [5, 65–68].

Pt and Pd electrodes with a renewed surface obtained by cutting with a ruby knife under the working nonaqueous solution have been studied by Petrii et al. [69, 70] and by other investigators (Table 6). A more detailed discussion of data obtained is presented in Ref. [5].

Electrical double layer at solid metal alloy–electrolyte interface have been studied, but one of the features of alloys is that their bulk and surface compositions are, as a rule, substantially different because one of the components is more surface-active than the other. It is to be noted that the surface composition of solid alloys can change with time as well as through surface process (selective oxidation or dissolution, surface migration and diffusion of components). Therefore, for the more detailed discussion the future experimental studies are inevitable; however, a short review of experimental data is given in Ref. [5].

2.4.3
General Correlations

2.4.3.1 Zero-charge Potential and Reticular Density of Planes

Figure 10 shows the graphical dependence of the PZC (or PZTC for Pt planes) on the crystallographic orientation of the surface for fcc metals. The plots exhibit a typical pattern with minima and maxima, which fall at the same angle for all fcc metals, and $E_{\sigma=0}$ values are linearly correlated with the coordination number of surface metal atoms [2–5].

$E_{\sigma=0}$ values are correlated with the density of atoms on the given surface and are given in Fig. 11. The PZC is more positive for dense surfaces and more negative for open surfaces. Pb also crystallizes in the fcc system and therefore the same dependence of $E_{\sigma=0}$ on the crystallographic orientation should be expected, but $E_{\sigma=0}$ varies exactly in the other direction [48]. A possible explanation can be sought in the surface mobility of Pb atoms at room temperature, which may lead to extensive surface reconstruction phenomena [5]. The data for Ni(001), for Ni(110) [59], and for Pt planes ($E_{Q=0}$) [68] are in good correlation with those for Au, Ag, and Cu planes.

Zn and Cd both crystallize in the hcp system. For both metals, the basal (0001) face is the most dense, and as for fcc metals, $E_{\sigma=0}$ is more positive than for the other faces [15, 16, 22, 48] (Fig. 11). For Sn planes the variation of $E_{\sigma=0}$ with the crystal face is practically negligible [49, 50].

Being semimetals, Bi and Sb show anomalies in the correlation of $E_{\sigma=0}$ with the surface atomic density [15, 16, 22], explained in terms of face-specific space charge effects. However, the definite dependence of $E_{\sigma=0}$ on the reticular density of plane has been established in a good agreement with general tendency.

The data discussed above indicate that the difference in $E_{\sigma=0}$ between the most dense and the most open surface depends on the "softness" of the metal surface, which to a first approximation can be measured by the melting point of the given metal [2–5]. $\Delta E_{\sigma=0}$ is highest for Au ($T_f = 1063\,°C$) and lowest for Sn ($T_f = 231.9\,°C$).

2.4.3.2 Potential of Zero Charge, Work Function, Interfacial Parameter, and Inner-layer Capacitance

According to the basic concepts of electrochemistry (see also Chap. 1 in Vol. 1), the electrode potential E measured with respect to the constant reference electrode can be expressed as

$$E(\text{vs. ref.}) = \Phi + \delta\chi_M + \delta\chi_S + \Delta\chi_{\text{ion}} + \text{const} \quad (19)$$

where $\delta\chi_M = \chi_M^{(S)} - \chi_M$ is the perturbation of the electron distribution at the surface of the metal (χ_M is the surface potential of the bare metal surface in vacuum), $\delta\chi_S = \chi_S^{(M)} - \chi_S$ is due to the effect of any solvent dipolar layer preferentially oriented at the phase boundary (χ_S is the surface potential at the free surface of the

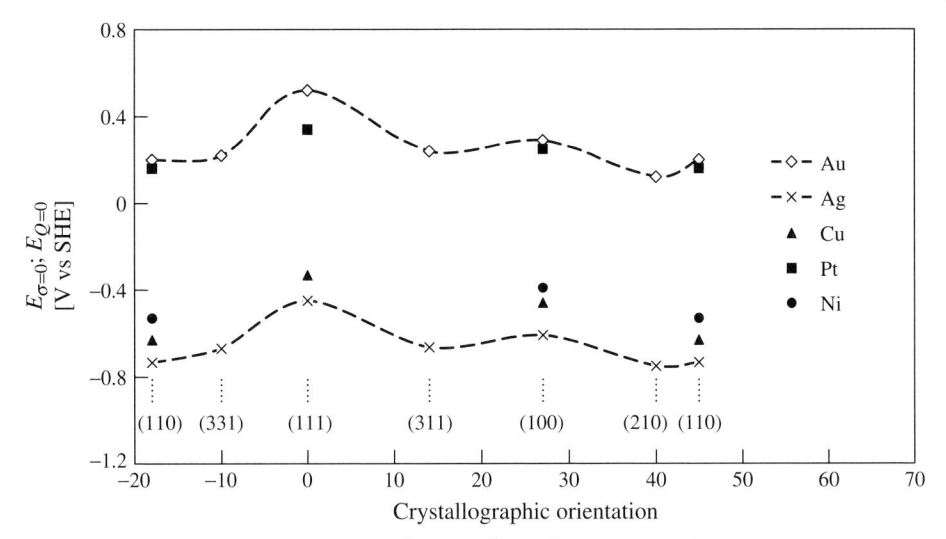

Fig. 10 Dependence of the pzc, $E_{\sigma=0}$, on the crystallographic orientation for the metals (Cu, Ag, Au, and Ni), which crystallize in the fcc system. For Pt electrodes, $E_{Q=0}$ values have been used. (Updated from Ref. [5].)

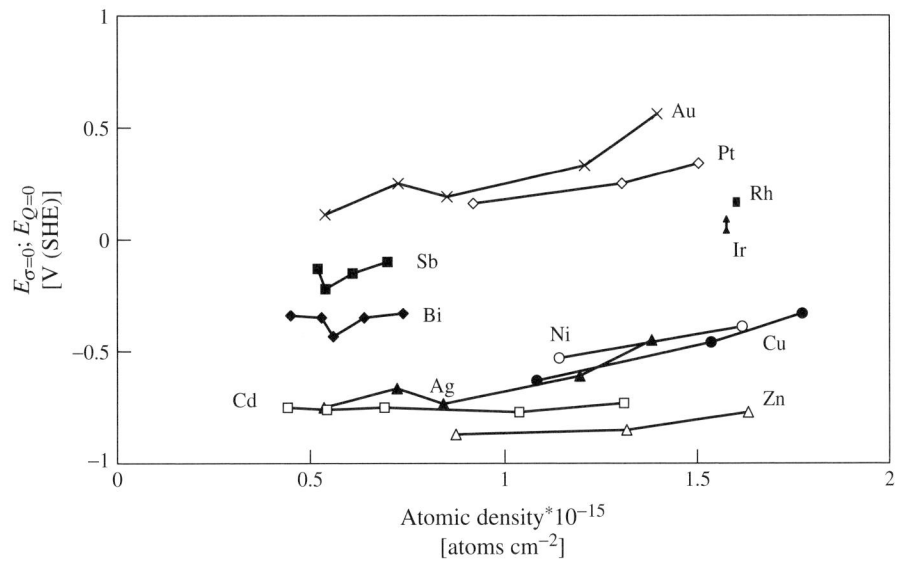

Fig. 11 The dependence of zero-charge potential (PZTC values for Pt metals) on the atomic density for various single-crystal electrodes (metals noted in figure) (updated from Ref. [5]).

solvent), and $\Delta\chi_{ion}$ accounts for any possible charge separation across the interface. At the PZC in surface-inactive electrolyte solutions, the ionic specific adsorption is absent ($\Delta\chi_{ion} = 0$) and Eq. (19) modifies to

$$E_{\sigma=0}(\text{vs. ref.}) = \Phi + \delta\chi_M + \delta\chi_S + \text{const}$$
$$= \Phi + \chi + \text{const} \quad (20)$$

Since the two perturbation terms ($\delta\chi_M$ and $\delta\chi_S$) are specific to the given interface and are experimentally inseparable, they have grouped into a single quantity X and, according to Trasatti, is called the *interfacial parameter* [1, 4, 5, 71, 72]. Thus, the difference in $E_{\sigma=0}$ for different metals can be written as $\Delta E_{\sigma=0} = \Delta\Phi + \Delta X$ if the same reference electrode is used.

The main problem in $E_{\sigma=0}$ versus Φ correlation is that the two experimental quantities are as a rule obtained in different laboratories with different techniques. More than with $E_{\sigma=0}$, the problem is with the selection of values for Φ. The values of $E_{\sigma=0}$ have been obtained more recently and are usually known with accuracy to ± 0.01 V or better, the measurements of Φ are rather occasional and have an accuracy from ± 0.05 to ± 0.25 eV [4, 5]. Thus, experimental data do not ensure an appropriate accuracy for ΔX and the uncertainty may outweigh the value itself. The best way to proceed is to plot $E_{\sigma=0}$ versus Φ for many electrodes and to derive information about ΔX term from graphical correlations using statistical analysis.

Figure 12 shows a dependence of $E_{\sigma=0}$ on Φ. For PC sp metals, the average work function values, defined as $\langle\Phi^M\rangle = \sum_j \theta_j \Phi_j^M$, have been used [5, 73]. Hg is taken as a reference surface and a straight

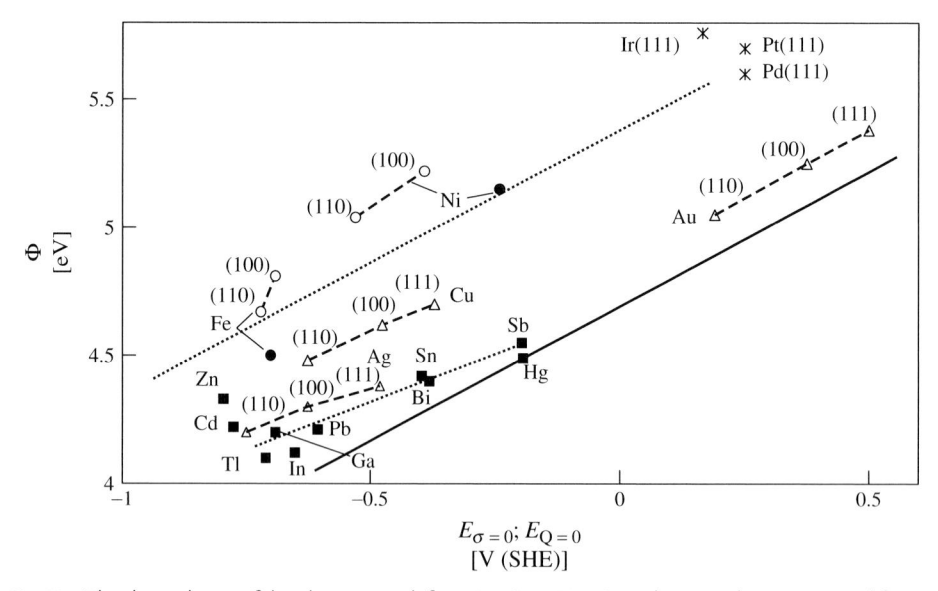

Fig. 12 The dependence of the electron work function (into UHV) on the zero-charge potential for various PC (filled marks) and monocrystalline (open marks) electrodes (metals and faces noted in the figure). (———) Straight line of the unit slope through the point for Hg. For Pt-metals potentials of zero total charge are shown.

line of unit slope is drawn through its point. The data show that all metals lie on the left of the straight line, thus all of them have an X term more negative than that of Hg and there are no metals on whose surface the contact with water produces a lower $\Delta\Phi$ than on Hg. The horizontal distance (along the $E_{\sigma=0}$ axis) of each metal point from the line of unit slope measures ΔX with respect to Hg. Thus, ΔX measures the relative value of the contact potential difference at the metal–water interface [4, 5].

All sp metals, except Zn, can be apparently gathered in a single group and there is a clear trend within the group for $E_{\sigma=0}$ to become more negative as Φ decreases. Thus, ΔX increases as Φ decreases. Zn has much higher value of ΔX than the other sp metals with comparable work function. Since $X \equiv \delta\chi^M + \delta\chi^S$, the major contribution is probably in $\delta\chi^S$.

In view of the large heterogeneity effect on the value of $E_{\sigma=0}$ for sd metals, the data for single-crystal planes are shown in Fig. 12. It is remarkable that each metal forms a separate group in which the faces are aligned along apparently parallel straight lines with the same slope as for sp metals. ΔX varies in the sequence Au < Ag < Cu and for the same metal with the crystal face (111) < (100) < (110). These relations are in good agreement with the work function drop values, caused by the adsorption of H_2O molecules, obtained from the UHV data [4, 5].

It is to be noted that in many cases, the preparation of a single-crystal face is not followed by a check of its perfection by means of appropriate spectroscopic techniques, and therefore we actually have "nominal" single-crystal faces. The only values of ΔX of high reliability are, in principle, those for singular planes, specifically Au(111), for which "congruent" pairs of data exist for Φ and $E_{\sigma=0}$ [4, 5].

$E_{\sigma=0}$ values for PC d-metals are not very reliable, except Fe and Ni, and therefore the points for PC Fe and Ni only as well as for single-crystal planes are included in Fig. 12 as broadly representative of d-metals. ΔX values for these electrodes are higher than those for sp and sd metals.

Pt metals have never found a definite position in $E_{\sigma=0}$, Φ correlation. Recently, with the improvement achieved in the preparation and control of surfaces, the PZTC has been estimated indirectly ($0.22 < E_{Q=0} < 0.35$ V (SHE)) close to the PZC of PC Pt. In view of the heterogeneity of Pt surfaces, this closeness may be puzzling [5]. However, if the PZTC estimated around 0.30 V (SHE) is taken for Pt(111) as $E_{Q=0}$, the point of Pt would be located further from the line for d-metals, with a high value of ΔX that is not justified by the known behavior of the Pt surface [5].

The value of $E_{\sigma=0}$ for Pt(111), obtained directly by the IM [64], is much more positive than others reported [62, 63, 68]. It is in the right direction with respect to a PC surface, even though it is an extrapolated value and probably corresponds to the oxidized Pt(111) surface [5]. If this value is taken as PZTC and 6.1 eV assumed to be Φ for Pt(111), then a linear dependence seems to be valid for Hg, Au(111), and Pt(111), and ΔX increases in the order Hg < Au(111) < Pt(111). The value of $E_{\sigma=0}$ given by Frumkin et al. [1, 14] would imply $\Phi = 5.4$ eV for the PC Pt, which is within the usual range of experimental data. However, there remains the puzzling aspect that while $E_{Q=0}$ refers to a surface with hydrogen adsorbed on it, Φ values do not include contributions from adsorbed H [5].

Data for single-crystal planes of sp metals have not been plotted in Fig. 12 as there are no correct Φ values in the literature. For that reason the so-called electrochemical work function values Φ^* have been obtained according to Frumkin et al. and Trasatti's conception discussed in Refs. [1, 71, 72]. According to the data in Fig. 13, there is a good linear correlation between Φ^* and $E_{\sigma=0}$ values for single-crystal planes of sp metals, with the same slope as for PC sp metals. The dependence of ΔX^* on the crystallographic orientation of plane is weak, but ΔX^* seems to increase with the decrease of the reticular density of planes, thus in the same direction as for fcc sd metals.

ΔX and $\Delta \Phi$ measure change in the structure of the surface layers of a metal and of a solvent as the two phases are brought in contact (see also Chap. 14 in Vol. 3). The reorientation of a solvent molecule is possible if there are orienting forces acting on it, which in the absence of an electric field ($\sigma = 0$) can only be the short-range chemical interactions. So, ΔX can also be interpreted in terms of metal–water interaction, and the concept of "hydrophilicity" has been introduced [1, 4, 5, 71, 72]. However, a quantitative comparison (of ΔX with $\Delta \Phi$) with reliable recent data is possible only for Ag(110), Cu(110), and Pt(111), and the gas phase data confirm the sequence observed electrochemically. Thus, $\Delta \Phi$ varies in the order Au < Ag < Cu ($\Delta \Phi$ for PC Au is compared with ΔX for the (110) face); also, $\Delta \Phi$ decreases Cu(110) > Pt(111) >

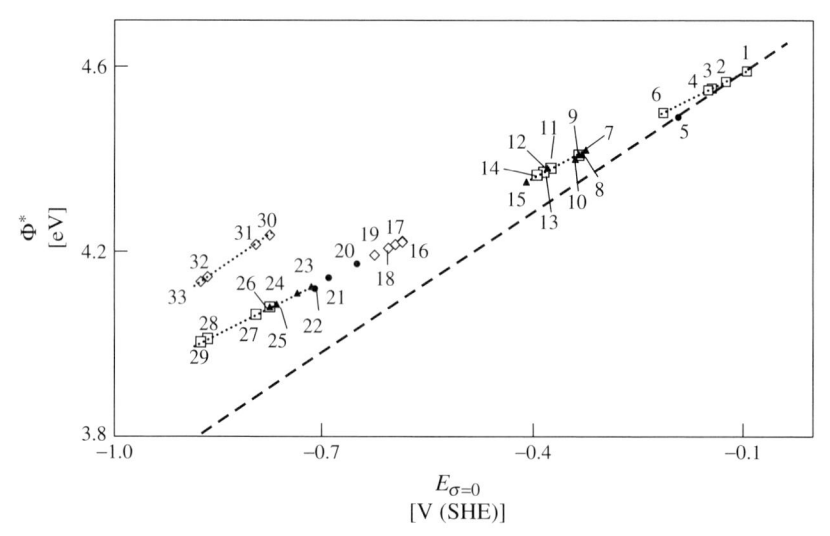

Fig. 13 Dependence of the electrochemical work function Φ^* on the pzc $E_{\sigma=0}$ for various single-crystal and PC electrodes: 1, Sb($2\bar{1}\bar{1}$); 2, Sb(001); 3, Sb($01\bar{1}$); 4, PC-Sb; 5, Hg; 6, Sb(111); 7, Bi($2\bar{1}\bar{1}$); 8, Bi($\bar{1}0\bar{1}$); 9, Bi($01\bar{1}$) and Sn(001); 10, Bi(001); 11, Sn(110); 12, PC-Bi; 13, Sn($1\bar{1}0$); 14, PC-Sn; 15, Bi(111); 16, Pb(110) and Pb(112); 17, Pb(100); 18, PC-Pb; 19, Pb(111); 20, In(Ga); 21, Ga; 22, Tl(Ga); 23, Cd(0001); 24, PC-Cd; 25, Cd($10\bar{1}0$); 26, Cd($11\bar{2}0$) and Zn(0001); 27, PC-Zn; 28, Zn($10\bar{1}0$); 29, Zn($11\bar{2}0$); 30, Zn(0001); 31, PC-Zn, 32, Zn($10\bar{1}0$); 33, Zn($11\bar{2}0$). The values of Φ^*1–29 have been calculated according to the equation $E_{\sigma=0}$ (versus SHE, V) $= 1.27 \Phi^*(eV) - 5.91$; and 30–33 according to equation $E_{\sigma=0}$ (versus SHE, V) $= \Phi^*(eV) - 5.01$ (updated from Ref. [71]).

Ag(111) [4, 5]. According to the data of thermal desorption spectroscopy (TDS), the Ag–H$_2$O interactions are weaker than H$_2$O–H$_2$O interactions, and the activity of Cu(110) is higher than that of (111) and (100) faces. The M–H$_2$O interactions are stronger on Cu than on Ag. A surface of Pt(111) is more reactive than that of sd metals, but the high-temperature TDS peak for Pt(111) is very close to that of Cu(110). More complex spectra are exhibited by Ni(110), which is known to be easily oxidized and thus tends to react strongly with water molecules [5, 74].

A high value of ΔX entails an appreciable modification of the surface structure as a consequence of strong interactions between the phases and vice versa, and thus the inner-layer capacitances have to correlate with ΔX (or $\Delta \Phi$) values. A good correlation between the experimental inner-layer thickness $l_i = 4\pi C_i$ and ΔX has been demonstrated for various electropolished PC and single-crystal sp and sd metal electrodes in H$_2$O medium (Fig. 14). The decrease of l_i with the increase of ΔX appears to be a general occurrence [5, 71], and for electropolished Ag electrodes, l_i decreases in the sequence (111) > (100) > (110) as the hydrophilicity of surface decreases. The point of Au(100) is also located near the general correlation [5, 37].

A linear correlation between l_i and electrochemical ΔX^* values for sp single-crystal planes (Fig. 14), in correlation to the general behavior for PC sp metals and electropolished Ag and Au planes, can be observed. For Bi(111) and Sb(111) there are small deviations toward higher values of l_i, caused by the different semimetallic properties of these planes [5, 71].

On the other hand, the same values of C_i cannot fit Fig. 14 if the values of ΔX estimated by Valette are used [26]. The same is the case for the values of C_i reported by Popov et al. [38] for

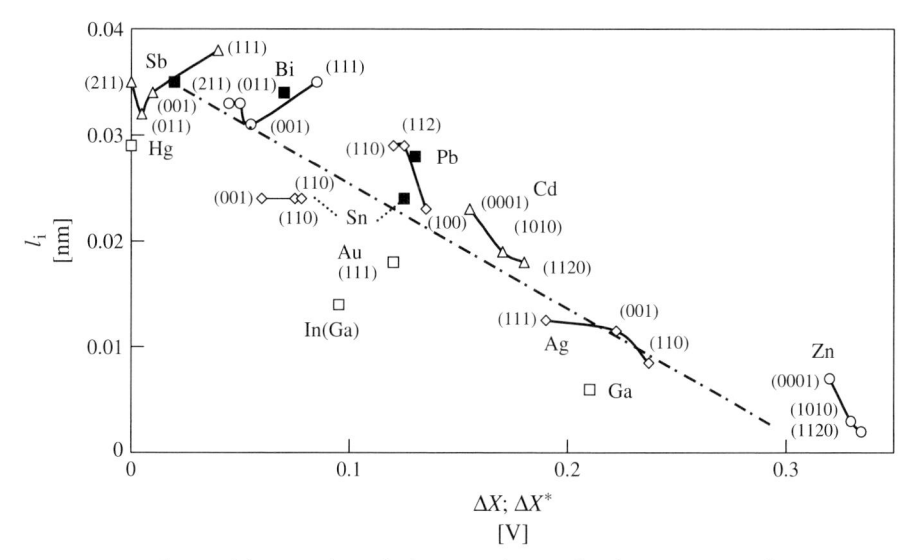

Fig. 14 Dependence of the inner-layer thickness on the interfacial parameter ΔX for various polycrystalline sp metals (filled marks) and single-crystal sp and sd metal electrodes, noted in figure (updated from Refs. [5, 71]). For single-crystal sp metals, ΔX^* values have been used.

single-crystal faces grown in a Teflon capillary. For electropolished Cd, higher capacitances for the faces with more negative PZC have been obtained in good agreement with the data for EP Ag [4, 5, 22]. For the Cd electrodes grown in a Teflon capillary, a reverse order of capacitance values has been reported [54, 55]. This indicates that differently prepared Ag and Cd single crystals behave not only quantitatively but also qualitatively differently [4, 5].

2.4.3.3 Data of Quantum-chemical Calculations

Quantum-chemical calculations, molecular dynamics (MD) simulations, and other model approaches have been used to describe the state of water on the surface of metals. The decrease in Φ at the surface of Ag(110) has been successfully reproduced by a jellium/point dipole model by assuming a disordered water structure at the interface [75]. In the case of hydrophilic surfaces, the first layer (or two) of water is strongly oriented by the forces emanating from the surface, but the bulk structure is soon recovered in a few layers after some very disorganized layers (see also Chap. 2.3 in Vol. 1).

MD simulations have been used for H_2O at Pt(100) and Pt(111), as well as at Ag(111). The structure of water is predicted to conform to a hexagonal pattern and the metal–H_2O interaction is probably stronger for the (111) than for the (100) surface. However, on the basis of the extended Hückel theory, different conclusions in favor of the (100) face have been made [5, 76, 77] (see also Chap. 2.3 in Vol. 1).

Water reorientation is usually predicted by theoretical models, but for Ag(111), MD simulations do not confirm the noticeable increase in water population near a charged surface claimed by Toney et al. on the basis of surface X-ray scattering experiments, which proposed that H_2O molecules are oriented with H toward a negatively charged surface [5, 77, 78]. This picture is not confirmed by far-IR spectroscopy results [79] according to which, although H_2O molecules change the orientation with charge, they always point O atom toward the surface. The importance of the presence of other molecules for the interaction of a water molecule with a metal surface is seen clearly in calculations. The adsorption energy calculated for H_2O on Hg is -32.2 kJ mol^{-1}, which is less than that for hydrogen bonding, and therefore Hg behaves as a hydrophobic surface [80].

According to quantum-chemical calculations, the metal–H_2O interaction has been proposed to be in the sequence Hg < Ag(100) < In < Cu(100) [81]. Compared to the experimental data, it appears that the positions of In and Ag(100) are exchanged. The complete neglect of differential overlap method predicts for any given metal a weaker interaction on the more dense surface [40]. Thus, the predicted sequence is (111) < (100) < (110) for fcc metals and (0001) < (1100) for hcp metals. However, for the most compact surfaces, the calculated sequence is Hg < Ag(111) < Cu(111) ≈ Zn(0001) < Au(111) < Cd(0001). It is difficult to accept that Zn can be less hydrophilic than Au or Cd, and also that Au can be more reactive than Cu. More recent calculations gave the other order of metals: Ag < Au < Cu. This confirms the position of Cu, but Au still appears to be more reactive than Ag [5].

Thus, the data obtained are controversial on the whole, theoretical calculations provide only a general insight into the

2.4.3.4 Adsorption of Organic Compounds and Hydrophilicity of Electrode

Adsorption at electrodes is considered to be a solvent replacement reaction: B(sol) + nH$_2$O(ad) ↔ B(ad) + nH$_2$O(sol), where B is an adsorbing substance replacing n water molecules on the electrode surface [1, 4, 5, 71, 72]. Adsorption will affect the PZC since water dipoles are replaced by adsorbate dipoles. On the other hand, the Gibbs energy of adsorption for a given adsorbate B can be divided into several contributions: $\Delta G_A^0 = G_{M-B} - G_{B-S} - G_{M-S}$, where S stands for solvent and G is the bond strength. If the same adsorbate is studied on different metals in the same solvent, then $G_{B-S} \approx$ const. Furthermore, if only physical adsorption occurs, $G_{M-B} \approx$ const. Under similar circumstances, ΔG_A^0 is only a function of G_{M-S}, hence it is expected to be correlated with ΔX, as well as C_i values.

Data in Fig. 15 show a nice linear correlation for various normal alcohols, with $\Delta(\Delta G_A^0)$, decreasing as ΔX increases ($\Delta(\Delta G_A^0)$ is the difference between the Gibbs energy of adsorption of an organic compound at a metal–electrolyte interface, ΔG_A^0, and the increase of Gibbs energy of organic compound adsorption, caused by the addition of electrolyte into the solution, ΔG_{sol}^0 ($\Delta(\Delta G_A^0) = \Delta G_A^0 - \Delta G_{sol}^0$). Thus the more hydrophilic metals (or faces) adsorb less. However, there are two aspects: (1) The slope of the correlation depends marginally on the nature of the adsorbate, that is, it is a property of the adsorbed layer structure. (2) The correlation is valid for both PC and single-crystal electrodes, which suggests that common

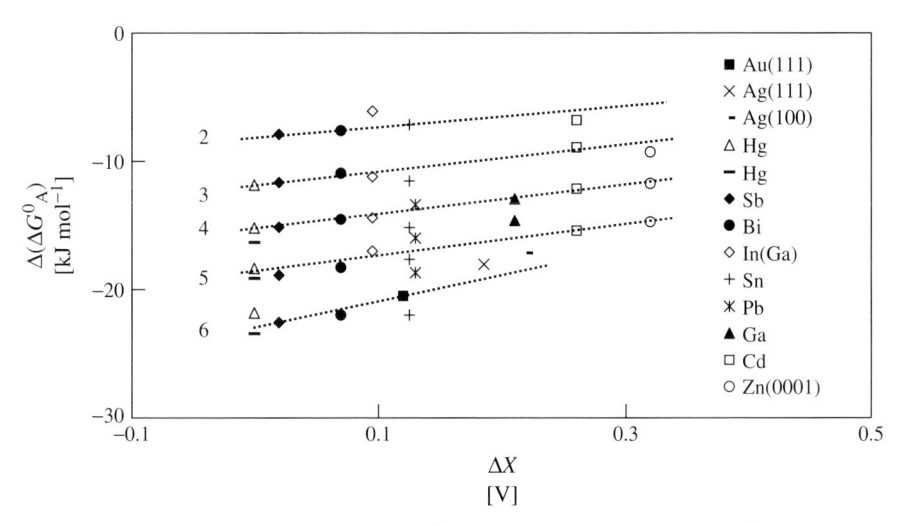

Fig. 15 Dependences of the difference $\Delta(\Delta G_A^0)$ for various alcohols on the interfacial parameter ΔX for various metals. The number of each plot is equal to the number of carbon atoms of given aliphatic compound (updated from Refs. [5, 71]). $\Delta(\Delta G_A^0) = \Delta G_A^0 - \Delta G_{sol}^0$ where ΔG_A^0 is the Gibbs energy of adsorption of an organic compound at a metal–electrolyte interface and ΔG_{sol}^0 is the increase of Gibbs energy of organic compound adsorption, caused by the addition of electrolyte into the solution.

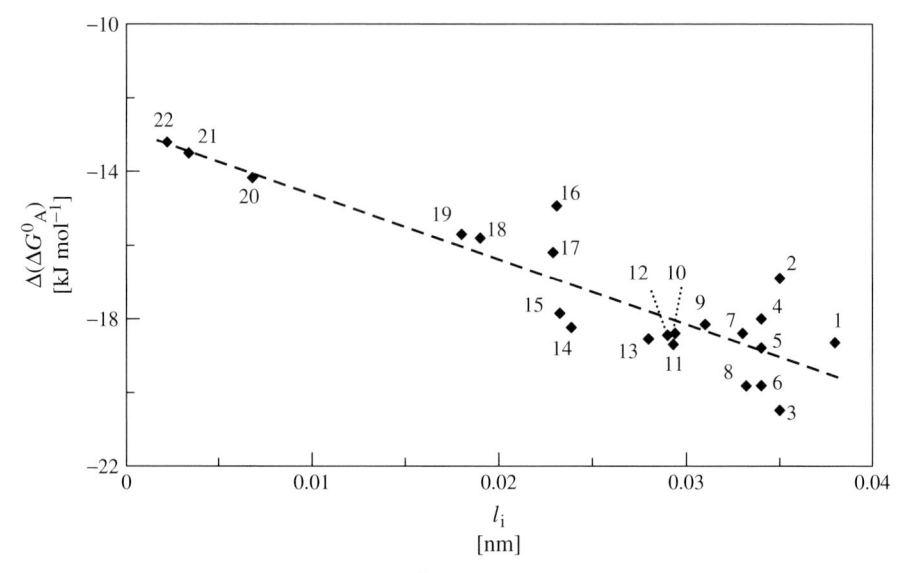

Fig. 16 Dependence of the difference $\Delta(\Delta G_A^0)$ for cyclohexanole adsorption on the thickness of inner layer at $\sigma = 0$ in the solution of surface-inactive electrolyte for various electrodes: 1, Sb(111); 2, Bi(111); 3, Sb(2$\bar{1}\bar{1}$); 4, PC-Bi; 5, Bi($\bar{1}$0$\bar{1}$); 6, Sb(001); 7, Bi(2$\bar{1}\bar{1}$); 8, Bi(01$\bar{1}$); 9, Bi(001); 10, Pb(001); 11, Pb(110); 12, Hg; 13, PC-Pb; 14, Sn(001); 15, Pb(111); 16, PC-Cd; 17, Cd(0001); 18, Cd(10$\bar{1}$0); 19, Cd(11$\bar{2}$0); 20, Zn(0001); 21, Zn(10$\bar{1}$0); 22, Zn(11$\bar{2}$0) (updated from Ref. [71]).

factors are behind this phenomenon. ΔX and water–metal interaction strength are parallel, and these data confirm the hydrophilicity sequence Hg < Au < Ag and the crystal-face sequence for fcc metals (111) < (100) < (110). Therefore the hydrophilicity scales of Popov et al., Silva et al., and Valette cannot be sustained on the basis of sound experimental and theoretical arguments [5].

This order of metals is in good agreement with the data in Fig. 16, where the dependence of $\Delta(\Delta G_A^0)$ on the inner-layer thickness is presented. To the first approximation, this plot can be considered linear and the adsorption activity of organic compound at the electrode–solution interface decreases as the thickness of double layer decreases, thus as the hydrophilicity of electrode increases [71]. Data for Bi(111) and Sb(111) planes show a noticeable deviation from the general plot, which is mainly caused by the more pronounced influence of the capacitance of metal phase of these electrodes in comparison with other Bi and Sb single-crystal planes.

References

1. A. N. Frumkin, *Zero Charge Potentials*, Nauka, Moscow, 1979.
2. A. Hamelin in *Modern Aspects of Electrochemistry* (Eds.: B. E. Conway, R. E. White, J. O'M. Bockris), Plenum Press, New York, 1985, pp. 1–102, Vol. 16.
3. A. Hamelin, T. Vitanov, E. Sevastyanov et al., *J. Electroanal. Chem.* **1983**, *145*, 225–264.
4. S. Trasatti, L. M. Doubova, *J. Chem. Soc., Faraday Trans.* **1995**, *91*, 3311–3325.
5. S. Trasatti, E. Lust in *Modern Aspects of Electrochemistry* (Eds.: R. E. White, B. E. Conway, J. O'M. Bockris), Kluwer Academic/Plenum Publishers, New York, London, 1999, pp. 1–216, Vol. 33.

6. D. M. Kolb in *Structure of Electrified Interfaces* (Eds.: J. Lipkowski, P. N. Ross), VCH, New York, 1993, pp. 65–102.
7. M. A. Vorotyntsev in *Modern Aspects of Electrochemistry* (Eds.: J. O'M. Bockris, B. E. Conway, R. E. White), Plenum Press, New York, 1986, pp. 131–222, Vol. 17.
8. R. Adzic in *Modern Aspects of Electrochemistry* (Eds.: R. E. White, J. O'M. Bockris, B. E. Conway), Plenum Press, New York, 1990, pp. 163–236, Vol. 21.
9. J. O'M. Bockris, S. U. M. Khan, *Surface Electrochemistry. A Molecular Level Approach*, Plenum Press, New York, 1993, pp. 59–210.
10. A. Hamelin in *Nanoscale Probes of the Solid/Liquid Interface* (Eds.: A. A. Gewirth, H. Siegenthaler), Kluwer, Dordrecht, 1995, pp. 285–306.
11. G. A. Somorjai, *Chemistry in Two Dimensions: Surfaces*, Cornell University Press, Ithaca, 1991.
12. R. Parsons in *Comprehensive Treatise of Electrochemistry* (Eds.: J. O'M. Bockris, B. E. Conway, E. Yeager), Plenum Press, New York, 1980, pp. 1–44, Vol. 1.
13. R. S. Perkins, T. N. Andersen in *Modern Aspects of Electrochemistry* (Eds.: J. O'.M. Bockris, B. E. Conway), Plenum Press, New York, 1969, pp. 203–282, Vol. 5.
14. A. Frumkin, O. Petrii, B. Damaskin, *J. Electroanal. Chem.* **1970**, *27*, 81–100.
15. E. Lust, K. Lust, A. A.-J. Jänes, *Russ. J. Electrochem.* **1995**, *31*, 807–821.
16. E. Lust, A. Jänes, K. Lust et al., *Electrochim. Acta* **1997**, *42*, 771–783.
17. M. A. Vorotyntsev, A. A. Kornyshev, *Elektrokhimiya* **1984**, *20*, 3–47.
18. E. Leiva, W. Schmickler, *J. Electroanal. Chem.* **1987**, *143*, 73–88.
19. S. Amokrane, J. P. Badiali in *Modern Aspects of Electrochemistry* (Eds.: R. E. White, J. O'M. Bockris, B. E. Conway), Plenum Press, New York, 1991, pp. 1–95, Vol. 22.
20. B. B. Damaskin, U. V. Palm in *Itogi Nauki i Tekhniki. Elektrokhimiya* (Eds.: Yu. M. Polukarov), VINITI, Moscow, 1977, pp. 99–143, Vol. 12.
21. A. N. Frumkin, V. V. Batrakov, A. I. Sidnin, *J. Electroanal. Chem.* **1972**, *39*, 225–228.
22. E. Lust, A. Jänes, K. Lust et al., *Electrochim. Acta* **1997**, *42*, 2861–2878.
23. S. Trasatti, O. A. Petrii, *Pure Appl. Chem.* **1991**, *63*, 711–734.
24. R. Parsons, F. G. R. Zobel, *J. Electroanal. Chem.* **1965**, *9*, 333–348.
25. G. Valette, A. Hamelin, *J. Electroanal. Chem.* **1973**, *45*, 301–319.
26. G. Valette, *J. Electroanal. Chem.* **1989**, *269*, 191–203.
27. I. A. Bagotskaya, M. D. Levi, B. B. Damaskin, *J. Electroanal. Chem.* **1980**, *115*, 189–209.
28. D. Leikis, K. Rybalka, E. Sevastyanov et al., *J. Electroanal. Chem.* **1973**, *46*, 161–167.
29. A. Hamelin, L. Stoicoviciu, *J. Electroanal. Chem.* **1989**, *271*, 15–26.
30. M. A. Vorotyntsev, *J. Electroanal. Chem.* **1981**, *123*, 379–387.
31. L. I. Daikhin, A. A. Kornyshev, M. Urbakh, *Phys. Rev. E* **1996**, *53*, 6192–6199.
32. E. Lust, A. Jänes, V. Sammelselg et al., *Electrochim. Acta* **1998**, *43*, 373–383.
33. E. Lust, A. Jänes, V. Sammelselg et al., *Electrochim. Acta* **2000**, *46*, 185–191.
34. L. I. Daikhin, A. A. Kornyshev, M. Urbakh, *J. Chem. Phys.* **1998**, *108*, 1715–1723.
35. L. Nyikos, T. Pajkossy, *Electrochim. Acta* **1985**, *30*, 1533–1539.
36. T. Pajkossy, *J. Electroanal. Chem.* **1994**, *364*, 114–125.
37. B. B. Mandelbrot, *The Fractal Geometry of Nature*, Freeman, San Francisco, 1982.
38. T. Vitanov, A. Popov, E. S. Sevastyanov, *J. Electroanal. Chem.* **1982**, *142*, 289–297.
39. M. Bacchetta, S. Trasatti, L. Doubova et al., *J. Electroanal. Chem.* **1986**, *200*, 389–396.
40. An. M. Kuznetsov, R. R. Nazmutdinov, M. S. Shapnik, *Electrochim. Acta* **1989**, *34*, 1821–1828.
41. J. Clavilier, R. Faure, G. Gainet et al., *J. Electroanal. Chem.* **1980**, *107*, 205–209.
42. X. Gao, G. J. Edens, A. Hamelin et al., *Surf. Sci.* **1993**, *296*, 333–351.
43. Z. Borkowska, G. Jarzabek, *J. Electroanal. Chem.* **1993**, *353*, 1–17.
44. J. Lecoeur, J. P. Bellier, *Electrochim. Acta* **1985**, *30*, 1027–1033.
45. H. Hennig, V. V. Batrakov, *Elektrokhimiya* **1979**, *15*, 1833–1837.
46. M. L. Foresti, G. Pezzatini, M. Innocenti, *J. Electroanal. Chem.* **1997**, *434*, 191–200.
47. J. R. LaGraff, A. A. Gewirth in *Nanoscale Probes of the Solid/Liquid Interface* (Eds.: A. A. Gewirth, H. Siegenthaler), Kluwer, Dordrecht, 1995, pp. 83–101.
48. L. P. Khmelevaya, A. V. Chizhov, B. B. Damaskin, *Elektrokhimiya* **1978**, *14*, 1304–1305.

49. L. P. Khmelevaya, B. B. Damaskin, *Elektrokhimiya* **1981**, *17*, 1721–1725.
50. L. P. Khmelevaya, Thesis, Moscow State University, **1982**, p. 4.
51. V. V. Batrakov, B. B. Damaskin, *J. Electroanal. Chem.* **1975**, *65*, 361–372.
52. Yu. P. Ipatov, V. V. Batrakov, *Elektrokhimiya* **1976**, *12*, 1174–1178.
53. A. P. Korotkov, E. B. Bezlepkina, B. B. Damaskin et al., *Elektrokhimiya* **1985**, *22*, 1298–1304.
54. R. Naneva, V. Bostanov, A. Popov et al., *J. Electroanal. Chem.* **1989**, *274*, 179–183.
55. I. A. Popov, N. Dimitrev, R. Naneva et al., *J. Electroanal. Chem.* **1994**, *376*, 97–100.
56. V. V. Batrakov, N. I. Naumova, *Elektrokhimiya* **1979**, *13*, 551–555.
57. L. E. Rybalka, D. I. Leikis, A. G. Zelinskii, *Elektrokhimiya* **1976**, *12*, 1340–1341.
58. V. A. Safonov, L. Yu. Komissarov, O. A. Petrii et al., *Elektrokhimiya* **1987**, *23*, 1375–1381.
59. J. Arold, J. Tamm, *Elektrokhimiya* **1989**, *25*, 1417–1418.
60. E. M. Lazarova, *Elektrokhimiya* **1982**, *18*, 1654–1655.
61. A. Rodes, J. Clavilier, *J. Electroanal. Chem.* **1993**, *348*, 247–264.
62. F. T. Wagner, T. E. Moylan, *Surf. Sci.* **1988**, *206*, 187–202.
63. T. Iwasita, X. Xia, *J. Electroanal. Chem.* **1996**, *411*, 95–102.
64. U. W. Hamm, D. Kramer, R. S. Zhai et al., *J. Electroanal. Chem.* **1996**, *414*, 85–89.
65. J. Clavilier, R. Albalat, R. Gómez et al., *J. Electroanal. Chem.* **1992**, *330*, 489–497.
66. R. Gómez, V. Climent, J. M. Feliu et al., *J. Phys. Chem. B* **2000**, *104*, 597–605.
67. M. J. Weaver, *Langmuir* **1998**, *14*, 3932–3936.
68. G. A. Attard, A. Ahmadi, *J. Electroanal. Chem.* **1995**, *389*, 175–190.
69. O. A. Petrii, I. G. Khomchenko, *J. Electroanal. Chem.* **1980**, *106*, 277–286.
70. E. Yu. Alakseyeva, V. A. Safonov, O. A. Petrii, *Elektrokhimiya* **1984**, *20*, 945–950.
71. E. Lust, A. Jänes, K. Lust et al., *J. Electroanal. Chem.* **1997**, *431*, 183–201.
72. S. Trasatti, *J. Electroanal. Chem.* **1971**, *33*, 351–378.
73. D. R. Lide, (Ed.), *CRS Handbook of Chemistry and Physics*, 76th ed., CRS Press, New York, 1995,1996, 12-122–12-123.
74. B. W. Callen, K. Griffiths, U. Memmert et al., *Surf. Sci.* **1990**, *230*, 159–174.
75. M. I. Rojas, E. P. M. Leiva, *Surf. Sci.* **1990**, *227*, L121–L124.
76. K. Heinzinger, *Pure Appl. Chem.* **1991**, *63*, 1733–1742.
77. K. J. Scweighofer, X. Xia, M. L. Berkowitz, *Langmuir* **1996**, *12*, 3747–3752.
78. J. G. Gordon, O. R. Melroy, M. F. Toney, *Electrochim. Acta* **1995**, *40*, 3–8.
79. A. E. Russell, A. S. Lin, W. E. O'Grady, *J. Chem. Soc., Faraday Trans.* **1993**, *89*, 195–198.
80. R. R. Nazmutdinov, M. Probst, K. Heinzinger, *J. Electroanal. Chem.* **1994**, *369*, 227–231.
81. R. R. Nazmutdinov, M. S. Shapnik, O. I. Malyucheva, *Elektrokhimiya* **1991**, *27*, 1275–1278.

2.5
Analyzing Electric Double Layers with the Atomic Force Microscope

Hans-Jürgen Butt
Universität Siegen, Siegen, Germany

2.5.1
Introduction

Electric double layers are studied by a variety of methods. These methods give information about different properties of the double layer. One method to study the properties of electric double layers is surface force measurements. They directly inform about the distance dependency of the potential and ion concentration. In addition, surface force measurements allow determining the surface charge or potential. A brief introduction to surface force measurements is given in the second chapter.

With the introduction of the atomic force microscope (AFM), the measurement of surface forces was greatly facilitated. The AFM was invented in 1986 by Binnig, Quate, and Gerber [1]. It has become one of the major tools to analyze surfaces. Usually the AFM is used to image the topography of surface (Fig. 1). This can be done in vacuum, gaseous, or liquid environment. For imaging, the sample is scanned underneath a tip, also called a "probe," which is mounted to a cantilever spring. The sample is scanned by a piezoelectric translator. While scanning, the force between the tip and the sample is measured by monitoring the deflection of the cantilever. A topographic image of the sample is obtained by plotting the deflection of the cantilever versus its position on the sample. Alternatively, it is possible to plot the height position of the piezo translator. This height is controlled by a feedback loop that maintains a constant force between tip and sample.

Image contrast arises because the force between tip and sample is a function of both tip–sample separation and the material properties of the tip and the sample. To date, in most applications the AFM was used to image surface topography in the contact mode. Then, image contrast is obtained from the very short-range repulsion that occurs when the electron orbitals of tip and sample overlap (Born repulsion). However, further interactions between tip and sample can be used to investigate material properties on a nanometer scale.

Soon after the invention of the AFM, it was realized that by taking force-versus-distance measurements, valuable information about the surfaces could be obtained [2, 3]. These measurements are usually known as "force measurements." The technique of force measurements with the AFM is described in detail in the third chapter. Force measurements with the AFM were first driven by the need to reduce the total force between tip and sample in order to be able to image fragile, biological structures [4, 5]. Therefore it was obligatory to understand the different components of the force. In addition, microscopists tried to understand the contrast mechanism of the AFM to interpret images correctly. Nowadays most force measurements are done by surface scientists, electrochemists, or colloidal chemists who are interested in surface forces per se. Excellent short [6] or comprehensive [7] reviews about surface force measurements with the AFM have appeared. Also an older review about surface force measurements in aqueous electrolyte exists [8]. This overview focuses on electrostatic double-layer forces.

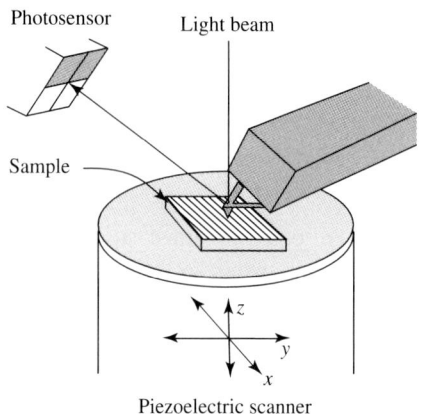

Fig. 1 Schematic drawing of an atomic force microscope.

First measurements of the electrostatic double-layer force with the AFM were done in 1991 [9, 10]. The electrostatic double layer depends on the surface charge density (or the surface potential) and the ionic strength. A brief introduction to the theory of the electrostatic force is given in Chap. 4. The electrostatic double-layer force is in many cases responsible for the stabilization of dispersions. An AFM experiment can be regarded as directly probing the interaction between a sample surface and a colloidal particle (or the AFM tip). Since the AFM tip is relatively small, this interaction can be probed locally. The lateral spacial resolution can be of the order of few nanometers.

2.5.2
Surface Forces in Aqueous Electrolyte

2.5.2.1 DLVO Forces

At the end of the 19th century, it was well known that many colloids in aqueous medium coagulate after the addition of salt. It was even known that di- or trivalent ions are much more efficient in destabilizing dispersions than monovalent ions [11]. The explanation for this behavior was eventually given in a quantitative way with the DLVO theory, named after Derjaguin, Landau, Verwey, and Overbeek [12, 13]. The DLVO theory uses a similar approach as in the Gouy–Chapman (GC) [14, 15] and Debye–Hückel [16] models of the electric double layer. To explain the coagulation of sols after the addition of salts, the DLVO theory takes explicitly into account the interaction between colloidal particles. The interaction is assumed to consist of two contributions: the van der Waals attraction and an electrostatic double-layer repulsion. At low salt concentration, the repulsion is strong enough to keep the colloidal particles apart. With increasing salt concentration, the electrostatic repulsion is more and more screened, and at a certain concentration, the van der Waals attraction overcomes the repulsive electrostatic barrier and coagulation sets in.

Van der Waals forces arise from correlations between time-dependent dipoles in atoms or molecules (for a review see Refs. [17–19]). The time dependence occurs either because a permanent dipole is changing its orientation in space or because the dipole itself is changing with time. The dominant term in the van der Waals force is usually the second frequency-dependent term, also known as the dispersion force. Since all materials are polarizable, a van der Waals force occurs between all atoms and all surfaces. The force between two atoms is relatively weak and short-ranged, decreasing proportional to D^{-7} for small separations and D^{-8} for larger separations where the force is "retarded" owing to weaker correlations between the atoms. Calculation of the forces in many atom systems is complex because the forces between individual atoms are not additive, and the modern treatment of van der Waals forces, known as Lifshitz theory, is a combination of quantum electrodynamic theory and spectroscopic data [20]. Fortunately, assuming additivity is a reasonable approximation in many cases, some simple equations can be derived.

The electrostatic double-layer force is the other contribution to the DLVO force. This force arises because of surface charges at interfaces. Water has a high dielectric constant. Thus, surface dissociation or adsorption of a charged species in water is very common. The surface charge is balanced by dissolved counterions that are attracted back to the surface by the electric field, but spread away from the surface to increase the entropy. Together the ions and charged surface are known as the electric double layer. When another surface approaches, the double layer is perturbed, and the resulting force is known as the double-layer force. When the approaching surface charges have the same sign, the concentration of ions between the surfaces always increases and results in a repulsive force. The origin of the surface charge is dissociation, so for finite dissociation constants, the surface charge is influenced by the interaction with another surface. This phenomenon is called *charge regulation*, and results in less repulsive forces, or even attractive forces, if one of the surface charges is reversed. For a more complete description of double-layer forces see Refs. [18, 19].

2.5.2.2 Early Surface Force Measurements

The development of the DLVO theory stimulated an interest in measuring forces between the surfaces to verify the theory (review Ref. [21], see also Ref. [22]). One of the earliest direct force measurements were made between two polished glass bodies [23, 24]. One glass body was fixed, the other was mounted to a spring. The distance between the two glass surfaces and the deflection of the spring were measured. Multiplication of the deflection with the spring constant gave the force. Using these simple early devices and after overcoming severe problems, it was possible to verify the theoretically predicted van der Waals force for glass interacting with glass across gaseous medium [25, 26]. Derjaguin, Rabinovich, and Churaev measured the force between two thin metal wires. In this way they could determine the van der Waals force for metals [27].

In these early measurements two problems became obvious: the minimal distance accessible ($\approx$20 nm) was rather limited and experiments could not be done in liquid environment. In more recent experiments these problems could be overcome and the minimal distance could be decreased to $\approx$2 nm with improved devices [28, 29]. Historically, however, improvements were made by using other types of devices.

One inherent problem of all techniques to measure surface forces directly is surface roughness. The surface roughness over the interacting areas limits the distance resolution and the accuracy of how well zero distance (contact) is defined. Practically, zero distance is the distance of closest approach because if a protrusion is sticking out of one surface, the other surface cannot approach any further. Two methods can be used to reduce this problem of surface roughness: either atomically smooth surfaces are chosen or the interacting areas are reduced. The first approach was chosen with the development of the surface forces apparatus (SFA) and the measurement and analysis of surface interaction forces (MASIF), while the advantages of the AFM are small interacting areas.

2.5.2.3 The Surface Forces Apparatus

The development of the SFA (Fig. 2) was a big step forward because it allowed to measure directly the force law in liquids and vapors at ångström resolution level [30–32]. The SFA contains two crossed atomically smooth mica cylinders of approximately 1 cm radius between which the interaction forces are measured. One mica cylinder is mounted to a piezoelectric translator. With this translator, the distance is adjusted. The other mica surface is mounted to a spring of known and adjustable spring constant. The separation between the two surfaces is measured by means of an optical technique using multiple beam interference fringes. Knowing the position of one cylinder and the separation to the surface of the second cylinder, the deflection of the spring and the force can be calculated.

With the SFA, the main predictions of the DLVO theory were verified. In particular the electrostatic double-layer force was analyzed for different salts and under

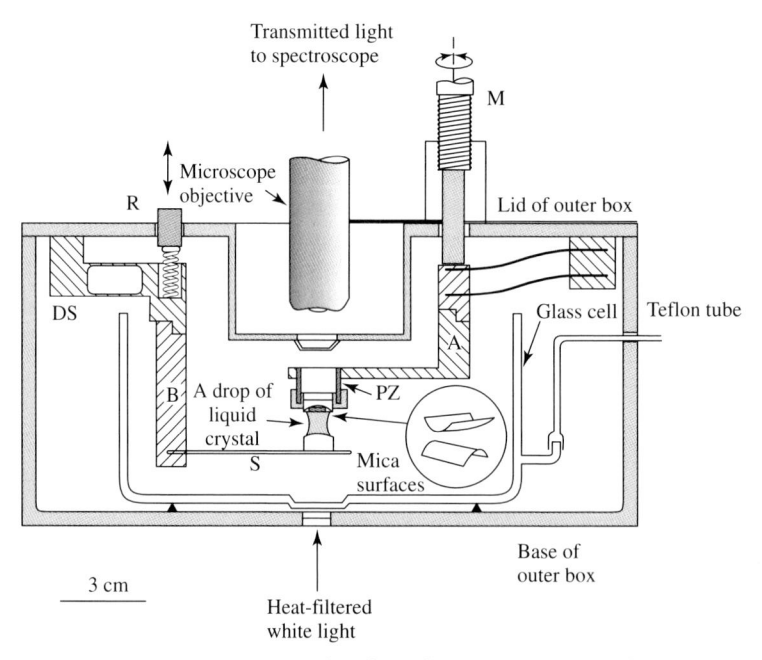

Fig. 2 Apparatus to measure surface forces between two mica surfaces (SFA). (The figure is reproduced with kind permission from J. Klein [33].) This particular device was built to study the behavior of liquid crystals confined in the narrow gap between the mica surfaces.

different conditions. Also, limits of the DLVO theory became obvious. The DLVO theory treats the intervening medium as continuous, so it is not surprising that the model breaks down when the liquid medium between two surfaces is only few molecular diameters in width. At high salt concentrations and small separations, an additional monotonic repulsive force of about 1 to 3 nm range was detected in aqueous electrolyte [34]. It has been attributed to the energy required to remove the water of hydration from the surface, or the surface-adsorbed species (secondary hydration), presumably because of strong charge–dipole, dipole–dipole or H-bonding interactions. Because of the correlation with the low (or negative) energy of wetting of these solids with water, it was termed *hydration force* [35–38]).

2.5.2.4 The Osmotic Stress Method

In parallel, another important (although less direct) technique for measuring forces between macromolecules or lipid bilayers was developed, namely, the osmotic stress method [39–41]. A dispersion of vesicles or macromolecules is equilibrated with a reservoir solution containing water and other small solutes, which can freely exchange with the dispersion phase. The reservoir also contains a polymer that cannot diffuse into the dispersion. The polymer concentration determines the osmotic stress acting on the dispersion. The spacing between the macromolecules or vesicles is measured by X-ray diffraction (XRD). In this way, one obtains pressure-versus-distance curves. The osmotic stress method is used to measure interactions between lipid bilayers, DNA, polysaccharides, proteins, and other macromolecules [36]. It was particularly successful in studying the hydration force between lipid bilayers and biological macromolecules.

2.5.2.5 The Bimorph Surface Force Apparatus

Despite the new insights achieved with the SFA, there are also severe problems and limits. It is difficult and slow to operate, only mica and few other materials [42, 43] provide suitable substrates, elastic deformation has to be taken into account. With the invention of a bimorph surface force apparatus, known as MASIF, some of these limitations could be overcome [44, 45]. In this case the force between two glass or polymer spheres of typically few millimeter diameter is measured. One sphere is therefore attached to a piezoelectric bimorph that serves as a force sensor.

With the MASIF, the existence of another force component, namely, the hydrophobic force [46], which is not described by the DLVO theory and which was already observed with the SFA [47], was confirmed. The hydrophobic force is a monotonic attractive force of longer range (100 nm) and greater magnitude than van der Waals forces. These occur in aqueous medium between surfaces that generally exhibit high water contact angles (e.g. hydrocarbon and fluorocarbon surfaces). At present there is no clear understanding of the mechanism of these forces. It is easy to conceive of a negative hydration force, for example, the disruption of water H-bonding near a noninteracting surface, with a similar range as the hydration force. However, it is difficult to conceive of one with the long measured range. At present the focus of understanding these long-ranged forces is on the role of dissolved gas.

Even with the MASIF, a severe limit remained: that of surface roughness. Glass and, in particular, polymer surfaces are not atomically smooth over areas of millimeter or even 10 µm diameter. This limit could be drastically reduced with the AFM. In the AFM the interacting areas are typically 10 to 50 nm in diameter. Hence, all samples whose surface is smooth, clean, and homogeneous over such small areas can be quantitatively analyzed. This increased the number of accessible materials enormously.

Small interacting areas provide another principal advantage. When two solid surfaces come into contact, the surfaces deform. Even under zero load, the adhesion between two surfaces leads to the formation of a contact area rather than a contact point. It also leads to an indentation of the two interacting surfaces. For a sphere and a flat planar surface (or two crossed cylinders), this central indentation δ at negligible load is proportional to $\delta \propto \sqrt[3]{R/E^2}$, where R is the radius of the sphere and E is Youngs modulus [48, 49]. Using typical values for the SFA ($R = 1$ cm, E of mica $\approx$60 GPa [50]), the MASIF ($R = 1$ mm, E of silica $\approx$60 GPa), and the AFM ($R = 50$ nm, E of silicon nitride $\approx$150 GPa), the estimated central indentations at zero loads are in the order of 14 nm, 6 nm, and 0.1 nm, respectively. This was calculated with the theory of Johnson, Kendall and Roberts [48, 49] using a surface energy of $\gamma = 0.05$ N m^{-1} in all the cases. Hence, in a usual AFM measurement, the elastic deformation can be neglected, while in SFA and MASIF experiments, deformations have to be taken into account.

2.5.3
Force Measurements with the AFM

2.5.3.1 The Practice of Force Measurements

In a force measurement the sample is moved up and down by applying a voltage to the piezoelectric translator, onto which the sample is mounted, while measuring the cantilever deflection. The direct result of such a force measurement is the cantilever deflection, Δz_c, versus position of the piezo, Δz_p, normal to the surface (Fig. 3). To obtain a force-versus-distance curve, Δz_c and Δz_p have to be converted into force and distance. The force, F, is obtained by multiplying the deflection of the cantilever with its spring constant K: $F = K \cdot \Delta z_c$. The tip–sample separation D is calculated by adding the deflection to the position.

The deflection of the cantilever is usually measured using the optical lever technique. Therefore a beam from a laser diode is focused onto the end of the cantilever and the position of the reflected beam is monitored by a position sensitive detector (PSD). Often the back of the cantilever is covered with a thin gold layer to enhance its reflectivity. When a force is applied to the probe, the cantilever bends and the reflected light beam moves through an angle equal to twice the change of the end-slope $\tan \alpha$. For a cantilever with a rectangular cross section of width w, length L, and thickness h, the change of the end-slope (Fig. 4) is given by

$$\tan \alpha = \frac{6FL^2}{Ewh^3} \quad (1)$$

E is the elastic modulus of the cantilever material. F is the force applied to the end of the cantilever in normal direction. The signal detected with the optical lever technique is proportional to the end-slope of the cantilever. The deflection of the cantilever is given by

$$\Delta z_c = \frac{4FL^3}{Ewh^3} = \frac{2}{3} L \tan \alpha \quad (2)$$

Hence, the deflection is proportional to the signal. One should, however, keep in

Fig. 3 Schematic deflection-versus-piezo translator position curve (Δz_c versus Δz_c). At large separation, no force acts between tip and sample (a). The cantilever is not deflected. When approaching the surface, it was assumed that a repulsive force is acting (b) and the cantilever is bent upwards. At a certain point, the tip often jumps onto the sample. This happens when the gradient of attractive forces, for example, the van der Waals forces, exceeds the spring constant of the cantilever. After the jump-in, the tip is in contact with the sample surface (c). When retracting the tip, often an adhesion force is observed (d) and the tip has to be pulled off the surface. The deflection-versus-piezo translator position curve has to be transformed as described in the text to obtain a force-versus-distance curve (F versus D).

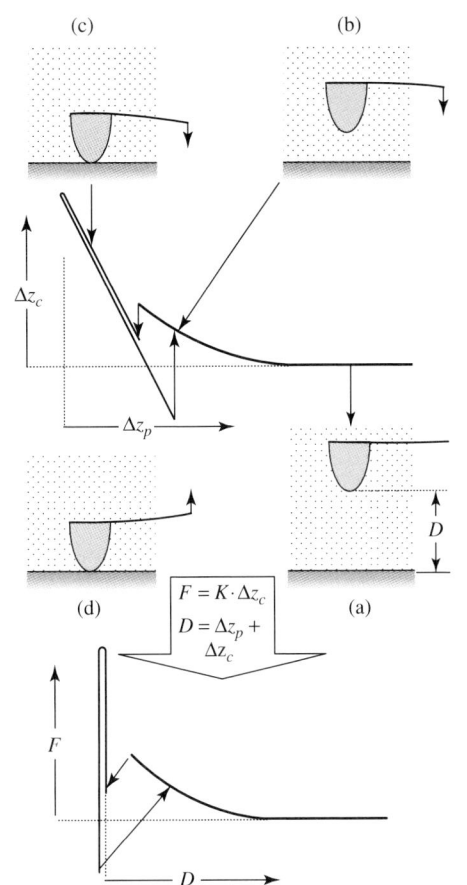

Fig. 4 Schematic diagram of a cantilever with rectangular cross section.

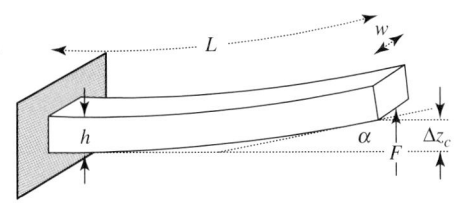

mind that these relations only hold under equilibrium condition. If the movement of the cantilever is significantly faster than that allowed by its resonance, frequency Eqs. (1) and (2) are not valid anymore and the signal is not necessarily proportional to the deflection.

The resolution of the optical lever technique is approximately $10^{-13}/\sqrt{t}$ [102], with t being the time for measuring a pixel of the force curve in seconds. The result is in meter. With typically $t = 0.1$ ms, the deflection resolution is 0.1 Å. In practice the deflection sensitivity is often limited by thermal cantilever vibrations, which are $\sqrt{4k_BT/3K}$ [103]. Here, k_B and T are the Boltzmann constant and the absolute temperature, respectively. With typical spring constants between 0.01 to 1 N m^{-1}, the amplitude of thermal noise is between 7 to 0.7 Å at room temperature.

2.5.3.2 The Cantilever

An important property is the spring constant of the cantilever K. The spring constant can in principal be calculated from the material properties and dimensions of the cantilever. For a cantilever with rectangular cross section, it is

$$K = \frac{F}{\Delta z_c} = \frac{Ewh^3}{4L^3} \quad (3)$$

The cantilever is in fact a key element of the AFM and its mechanical properties are largely responsible for its performance. A good cantilever should have a high sensitivity. High sensitivity in Δz_c is achieved with low spring constants or low ratio h/L. Hence, in order to have a large deflection at small force, cantilevers should be long and thin. In addition, the design of a suitable cantilever is influenced by other factors:

- With the optical lever technique, deflections down to 0.1 Å can be detected. It is useless to make cantilevers so soft that thermal fluctuations exceed this value by more than, say, a factor 10. Hence, the lowest useful spring constant is in the order of 0.02 to 0.1 N m^{-1} at room temperature.
- External vibrations, such as vibrations of the building, the table, or the noises that are usually in the low-frequency regime, are less transmitted to the cantilever, when the resonance frequency of the cantilever [104, 105]

$$\nu_r = 0.163 \frac{h}{L^2} \sqrt{\frac{E}{\rho}} \quad (4)$$

(ρ is the density of the cantilever material) is as high as possible. A high resonance frequency is also important to be able to scan fast because the resonance frequency limits the time resolution [104, 105]. In liquids the resonance frequency is reduced by a typical factor 3 to 6 because of the liquid that is dragged with the cantilever [105–107].

- Cantilevers have different top and bottom faces. The top portion is usually coated with a layer of gold to increase its reflectivity. Therefore, any temperature change leads to a bending of the cantilever as in a bimetal. In addition, adsorption of substances or electrochemical reactions in liquid environment slightly change the surface stress of the two faces. These changes in surface stress are in general not the same on the bottom and the top. Any difference in surface stress $\Delta\sigma$ will lead to a bending of the cantilever [108, 109]

$$\Delta z_c \approx \frac{4L^2 \Delta\sigma}{Eh^2} \quad (5)$$

Practically, these changes in surface stress lead to an unpredictable drift of the cantilever deflection that disturbs force measurements. To reduce drift, the ratio h/L should be high.

Hence, the optimal design of a cantilever is a compromise between different factors. Depending on the application, the appropriate dimensions and materials are chosen. Cantilevers for AFMs are usually "V" shaped to increase their lateral stiffness. They are typically 100 to 200 μm long (L), each arm is about 20 μm wide (w) and 0.5 μm thick (h). The spring constant of "V"-shaped cantilevers is often approximated by that of a rectangular bar of twice the width of each arm [104]. A more detailed analysis of the mechanical properties of cantilevers and calculations of spring constants are given in Refs. [105, 110–113].

In principle, the spring constant can be calculated. The thickness and modulus of elasticity of cantilevers are, however, not easy to determine, so it is desirable to measure spring constants. Several methods have been described [105, 114, 115] but none appears to be simple, reliable, and precise at the same time. In the most common methods, tungsten or gold beads are attached to the end of the cantilever spring. The spring constant can be calculated by measuring a change in the resonant frequency due to the additional mass at the end of the spring [116]. Torii and coworkers determine the spring constant of microfabricated cantilevers by pressing them against a calibrated macroscopic reference cantilever [117]. Sader and coworkers determined the spring constant of cantilevers with a rectangular cross section from the resonance frequency and the quality factor [118]. Another promising approach is to use thermal cantilever noise to determine the spring constant [119].

2.5.3.3 The Probe: Tip or Colloidal Particle

In the AFM, the two interacting surfaces are the planar sample surface and the surface of the AFM probe. For microscopy the most commonly used probes are the sharp microfabricated silicon nitride or silicon tips that provide a high lateral resolution. These probes also provide high resolution when measuring surface forces, but introduce the problem that the surface geometry is difficult to determine in the 10-nm regime [120–122]. This would, however, be necessary since typical radii of curvature are from 5 to 60 nm. In addition, the surface chemistry of silicon nitride tips, which are most frequently used for force measurements, is rather complex [51, 52, 123, 124].

Alternatively, spherical particles of typically 2 to 20 µm diameter are attached to the end of cantilevers. This can be achieved either by gluing them with a water-resistant polymer glue [9, 10] or by melting [97] or sintering them onto the cantilever (Fig. 5). With particles of defined radius, the force can be analyzed more quantitatively. In particular, the surface

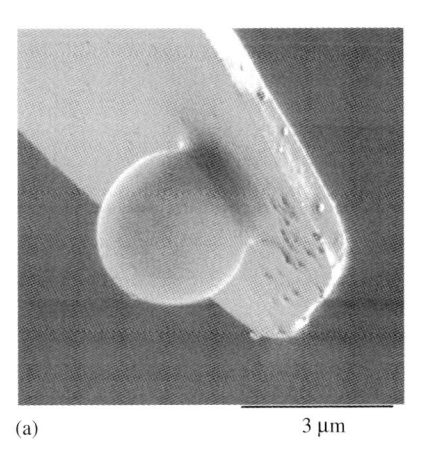

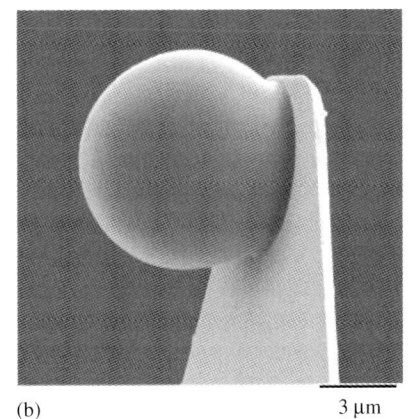

Fig. 5 Scanning electron micrographs of a silica microsphere glued with polymer glue to the end of a silicon nitride cantilever (a). The polystyrene particle shown to the right was sintered at 120 °C to the cantilever (b).

charge and den Hamaker constant can be extracted. The number of materials that can be brought into the shape of microspheres with sufficient smoothness is, however, limited. Most frequently, silica microspheres are used. Silica spheres can be made with different diameters. They have a low surface roughness of less than 1 Å (as determined by AFM) and their surface chemistry is well known. Different kinds of polymer particles are commercially available. They are, however, much rougher than silica. Their roughness can be decreased by annealing roughly at the glass transition temperature. Still, the surface roughness is significantly higher than silica. Zirconia microspheres were produced by annealing zirconia powder [83].

While the preceding paragraph concentrated on the importance of knowing the geometry of the tip, it is equally important to know the geometry of the sample. Again, the most convenient geometry is a smooth sphere or a planar surface. Problems due to sample roughness are greatly reduced in the AFM compared to other surface force techniques since the sample needs only to be smooth on a scale comparable to the radius of curvature at the end of the tip. Several materials such as mica, oxidized silicon, graphite (HOPG), MoS_2, and so on are available with sufficient smoothness over the required areas (Table 1). In addition, the surface of many materials can be smoothed by template stripping. Therefore the material, say a polymer, is melted on a smooth surface [76]. This can be a mica or a silicon wafer surface. After cooling and right before an experiment, the surface is peeled off and the freshly exposed polymer surface is used for the force measurement. The same technique can also be used for gold or other materials that can be sputtered or evaporated [125, 126]. First, the material is deposited onto mica. Then, a steel plate is glued on top. Finally, the steel plate with the deposited material is cleaved off the mica surface. Since mica is extremely inert, chances are high that cleavage really happens at the mica surface. The surface of the deposited material now exposed can then be used in a force experiment.

2.5.3.4 An Instability: The Jump-in

Implicit in the measurement of surface forces with a "spring force" is the assumption that the two are equal and opposite and thus the spring is not in motion. However, close above the surface, the spring often cannot provide a sufficient force to counterbalance the surface force. These regimes are characterized by rapid motion of the cantilever, and cause difficulty for determining surface forces. The onset of this unstable regime ("jump-in") is characterized by the point at which the gradient of the attractive force exceeds the spring constant plus the gradient of the repulsive force [127]. Several techniques have been employed to access the unstable regime: (1) the application of a force feedback (electric or magnetic) to balance the surface force [33, 128–131]; (2) measurement of the time-dependent force and displacement in the unstable regime, then calculation of the velocity and acceleration to yield the surface force according to Newton's equation of motion [132]; (3) use a high approaching speed so that the hydrodynamic drag F_H, which for a sphere approaching a flat with a velocity v increases with decreasing distance as $F_H = 6\pi \eta v R^2 / D$ (η is the viscosity of the liquid) [133], compensates for the attraction [134].

2.5 Analyzing Electric Double Layers with the Atomic Force Microscope

Tab. 1 Measurements of electrostatic double-layer forces with the AFM

Probe	Sample	Refs.
Silicon nitride tip	Silicon nitride	51, 52
Silicon nitride tip	Mica	4, 10, 51
Silicon nitride tip	SiO_2	53
Silicon nitride tip	Al_2O_3	54
Silicon nitride tip	Graphite	55
Silicon nitride tip	PEEK	56
Silicon nitride tip	LB film	57
Silicon nitride tip	Silane monolayer on SiO_2	58
Silicon nitride tip	Lipid bilayer	59, 60
Silicon nitride tip	Purple membrane	54, 61, 62
Silicon nitride tip	Oocysts	63
SiO_2 sphere	SiO_2	9, 52, 64–69
SiO_2 sphere	Mica	68, 70
SiO_2 sphere	Al_2O_3	71, 72
SiO_2 sphere	TiO_2	73, 74
SiO_2 sphere	Carbon	75
SiO_2 sphere	Polypropylene	76, 77
SiO_2 sphere	Ultrafiltration membrane	78
Glass sphere	SiO_2	10, 79
Amine-modified glass sphere	Amine-modified glass	80
LB film on glass sphere	LB film on glass	81
Al_2O_3 particle	Mica	10
TiO_2 particle	TiO_2	82
ZrO_2 sphere	ZrO_2	83–85
ZnS sphere	Mica	86
ZnS sphere	PbS	87
ZnS sphere	ZnS sphere	88
Gold sphere	Gold	64, 89–93
Thiol monolayer on gold-coated tip	Thiol monolayer on gold sample	94–96 (review)
Polystyrene sphere	Polystyrene	97, 98
Polystyrene sphere	Mica	98
Protein-coated probe	Protein-coated sample	99, 100
Bacteria-coated tip	Mica/glass	101

Note: AFM measurements of electrostatic double-layer forces in aqueous electrolyte. Force measurements between surfactant layers are not included. They are discussed in the text. PEEK: Poly(etheretherketone); LB film: Langmuir–Blodgett film.

Another instability occurs when tip or sample surfaces are not perfectly rigid but are deformed by the tip. For that case of elastic deformation, Pethica and Sutton showed that at sufficiently small separations, typically 1 to 2 Å, the tip and the sample would jump together, irrespective of apparatus construction [135]. A similar jump occurs for nonelastic deformation. In this case the jump depends on the rate of the plastic deformation [136].

2.5.3.5 AFM-related Techniques

Commercial AFMs are made for imaging. For force measurements, certain features, such as the lateral scanning capability and the feedback electronic that regulates the height, are not required. Hence, if one plans to use the device solely for force measurements, it is significantly less expensive to build homemade setups [137–139]. In addition, some features of commercial AFMs are not perfectly suited for measuring force curves. One such feature is the split photodiode, which has an extremely high sensitivity but a limited dynamic range. In addition, the sensitivity depends on the precise shape of the reflex and the AFM needs to be calibrated for each experiment [140]. These disadvantages were overcome in some homebuilt devices by replacing the spilt photodiode with a linear position-sensitive device [132, 141]. With position-sensitive devices, the accessible force range is highly increased. This is particularly important when measuring long-range forces during the approach and the adhesion force during the retraction in one experiment. With microfabricated tips, long-range forces in aqueous electrolyte are of the order of 0.1 to 1 nN. Adhesion forces can be 100 times higher. To have a sufficient sensitivity as well as the capability to measure the strong adhesion force, a position-sensitive device is necessary.

The position of the sample is adjusted by the piezoelectric translator. Piezoelectric crystals show creep and hysteresis, which affect the accuracy of the distance determination [142]. One possibility to overcome this problem is to use piezoelectric translators with integrated capacitive position sensors, which are commercially available [143]. In the same setup, another deficit of commercial AFMs was overcome. Standard fluid cells of commercial AFMs are small and manually difficult to access. In addition, they consist of different materials (glass, steel, silicon, etc.) that are difficult to clean. In self-made devices the fluid cell can be made of one or few materials (such as Teflon and quartz) that can be cleaned thoroughly, for example, with hot sulfuric or nitric acid.

2.5.4
Theory of Measuring Surface Forces with the AFM in Aqueous Electrolyte

2.5.4.1 Derjaguins Approximation

The force between two surfaces depends on both the material properties and the geometry of the surfaces. Derjaguin [144] approximated the influence of arbitrary geometry on the interaction potential $V(D)$ by reducing it to the simple geometry of two flat surfaces:

$$V(D) = \int_D^\infty V_A(z)\,dA \qquad (6)$$

$V(D)$ is the interaction potential that depends on the distance between the interacting surfaces D. $V_A(z)$ is the potential energy per unit area between two flat surfaces that are separated by a distance z. dA is the increase of the cross-sectional area of the two surfaces with increasing separation distance.

According to Eq. (6), the fundamental property of the material is the interaction energy, V_A, (or force, $F_A = -dV_A/dz$) per unit area. The most useful measurement is the one between two surfaces of known geometry, so that V_A can be determined and thus the interaction energy of all geometries can be calculated.

The problems of unknown surface geometry and chemistry can partly be solved by attaching smooth micrometer-sized spheres to the cantilever. For a sphere of radius R, and a flat, a simple equation relates the measured force, F, to the interaction energy per unit area, V_A:

$$V_A(D) = \frac{F(D)}{2\pi R} \qquad (7)$$

From this relationship, it can also be seen that a larger radius results in a higher force, and thus greater sensitivity in V_A. Naturally, this increase in V_A sensitivity comes at the expense of reduced lateral resolution. Another advantage of using spheres as probes is the possibility to make a variety of probes by attaching particles of different chemical composition to the cantilevers. Relation (7) is derived from Eq. (6) by assuming a parabolic shape of the probe. It can also be used for microfabricated tips since they often show a roughly parabolic shape. In this case, R is the radius of curvature of the tip at its end.

2.5.4.2 The Electrostatic Double-layer Force

When two charged surfaces approach each other at some point, the electric double layers start to overlap and a force, also called *disjoining pressure*, begins to act [13, 145]. This electrostatic double-layer force decays roughly exponentially. When the surface potentials of both surfaces are below approximately 25 to 50 mV, the force between a spherical tip and a flat sample can be approximated in an analytical form [146–149]:

$$F_{el} = \frac{4\pi R \sigma_1 \sigma_2 \lambda_D}{\varepsilon \varepsilon_0} e^{-D/\lambda_D} \quad (8)$$

with the Debye length

$$\lambda_D = \sqrt{\frac{\varepsilon \varepsilon_0 k_B T}{2 n_\infty e^2}}$$

Here, σ_1 and σ_2 are the surface charge densities of probe and sample, ε_0 is the permittivity of free space, ε is the dielectric constant, e is the unit charge. The Debye length λ_D is determined by the salt concentration; n_∞ is the bulk concentration of a monovalent salt in molecules per volume. For more detailed calculations that are also valid at higher potentials and that explicitly take into account the geometry, see Refs. [150–154].

The requirement of low potentials is not the only limit of Eq. (8). In addition, Eq. (8) is restricted to distances roughly larger than the Debye length. For $D \leq \lambda_D$, boundary conditions influence the force. The force is calculated from the potential distribution between the two surfaces [18, 19]. In continuum theory the potential distribution is determined from the Poisson–Boltzmann equation, which is a second-order differential equation. To solve this equation, certain boundary conditions have to be assumed. Two boundary conditions are often used: either it is assumed that upon approach the surface charges remain constant (constant charge) or that the surface potentials remain constant (constant potential). These boundary conditions have a strong influence on the electrostatic force at distances smaller than λ_D. Two surfaces with constant charge of equal sign always repel each other for $D \to 0$. Two surfaces with constant potential are attracted for $D \to 0$, even when the surface potentials have the same sign (except for the hypothetical case that the potentials are precisely equal in magnitude and sign) [155].

To demonstrate the effect of boundary conditions, Fig. 6 shows the force between a microsphere of 3-µm radius interacting with a flat surface versus distance. The force was calculated with the so-called nonlinear Poisson–Boltzmann equation, which is also valid for potentials above 50 mV. A salt concentration of 1 mM was chosen resulting in a Debye length $\lambda_D = 9.6$ nm. At close distances, the force calculated with constant charge boundary conditions is much higher than the force calculated assuming constant potential. For $D \gg \lambda_D$, the force calculated with

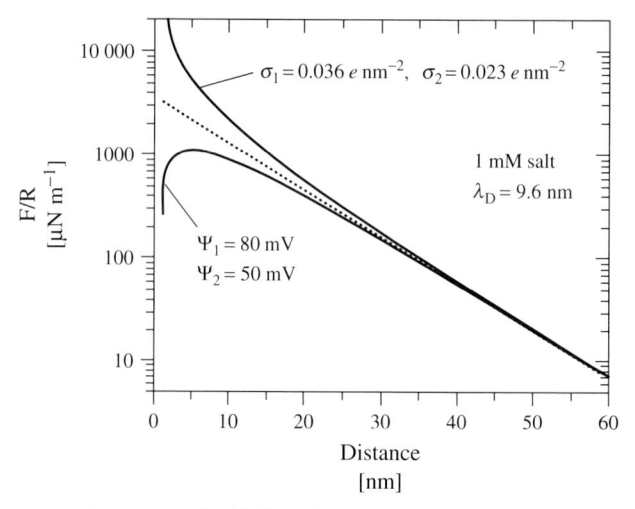

Fig. 6 Electrostatic double-layer force acting on a sphere of 3-µm radius in water containing 1 mM monovalent salt. The force was calculated with the nonlinear Poisson–Boltzmann equation and the Derjaguin approximation for constant potentials ($\Psi_1 = 80$ mV, $\Psi_2 = 50$ mV) and for constant surface charge ($\sigma_1 = 0.0058$ C m^{-2} $= 0.036\ e$ nm^{-2}, $\sigma_2 = 0.0036$ C m^{-2} $= 0.023\ e$ nm^{-2}). The surface charge was adjusted by $\sigma_{1/2} = \varepsilon\varepsilon_0 \Psi_{1/2}/\lambda_D$, so that at large distances both lead to the same potential. In addition, the force expected with approximation (8) is plotted as a dotted line.

constant charge and constant potentials are equal and surface charges and potentials Ψ can be converted according to $\sigma_{1/2} = \varepsilon\varepsilon_0 \Psi_{1/2}/\lambda_D$. Also the force calculated with approximation (8) is plotted. It lies between the two extreme cases.

Which boundary condition is more realistic depends on the materials used. In addition, the electrolyte and the speed of the approach might have an influence. Prica and coworkers measured force curves between zirconia, which showed a constant charge behavior [84]. Also the force between two surfaces coated with densely packed carboxylic groups followed constant charge conditions [94]. In other cases, constant potential conditions are more appropriate. Most cases, however, lay between the two extremes. Then often, a charge regulation model is applied [156].

In this model the surface charge is caused by the dissociation of ions from surface groups. The dissociation constant and as a consequence, the surface charge depends on the potential. This also explains the often observed dependency of the surface charge on pH and the salt concentration. The charge regulation model is successfully applied to many materials, especially those in which the surface charge is mainly determined by pH, such as oxides, silicon nitride [51, 52, 123, 124], mica [34, 157–159], and biological materials.

2.5.4.3 Other Surface Forces in Aqueous Electrolyte

With any surface force–measuring device, the total force between two surfaces is determined. The origin of all surface forces is the interaction between electric charges. Under normal circumstances it is practical, however, to separate the total force into several components and take the components as being additive (for a critique see Ref. [160]). These components are the van der forces, the electrostatic double-layer force, the hydration repulsion (between hydrophilic surfaces), and the hydrophobic attraction (between hydrophobic surfaces). To analyze the electrostatic double-layer force, one needs to separate it from all other components. This is usually done by assuming a certain distance dependency for the van der Waals force and the electrostatic force. In addition, it can be identified by the influence of salt.

The van der Waals potential per unit area is given by

$$V^A = -\frac{H}{12\pi D^2} \quad (9)$$

Using Eq. (7), the force between a sphere interacting with a plane can be calculated:

$$F_{vdW} = -\frac{HR}{6D^2} \quad (10)$$

For $D \gg 10$ nm, retardation effects have to be considered, and the force decays with D^{-3} instead of D^{-2}. H is the so-called Hamaker constant that depends on the dielectric properties of the interacting materials and the medium between. Detailed calculations showed that Derjaguins approximation leads to errors that are usually less than 10% [161]. For large distances the precise shape of the probe has to be considered [19, 162].

The van der Waals force is usually attractive. It is always attractive between two identical materials immersed in any third material. For hydrocarbons immersed in water, the Hamaker constant is about $0.2 - 1 \times 10^{-20}$ J, for many oxides in water, it is $0.8 - 5 \times 10^{-20}$ J, and for metals in water, it is $30 - 40 \times 10^{-20}$ J. Great progress has been made during the last few years in calculating Hamaker constants from spectroscopic data. A list of Hamaker constants for different combinations of the three media involved is given in Refs. [163, 164]. There are some cases in which the van der Waals force is repulsive [165]. A repulsive van der Waals force was, for example, detected between a silicon nitride tip and a silicon oxide surface in diiodomethane [166]. Also the van der Waals interaction between many solids and vapors in water is often repulsive and causes the adsorption of water to solid surfaces.

Hydration forces [35–38] are relatively short-ranged, so that at salt concentrations below 0.1 M, they can easily be distinguished from the longer range electrostatic and van der Waals forces. They are repulsive and, except for the case of molecularly smooth surfaces and low salt concentrations where force oscillations were observed [167], decay exponentially with distance:

$$V^A = Ae^{-D/\lambda_0} \quad (11)$$

Characteristic decay lengths λ_0 determined with the SFA, the osmotic stress method, or the AFM range from 0.2 to 1.4 nm. Typical amplitudes are $A = 10^{-3}$ to 10 J m^{-2}. In contrast to the electrostatic double-layer force, hydration forces tend to become stronger and longer ranged with increasing salt concentration, especially for divalent cations.

Hydrophobic forces occur between surfaces that are hydrophobic and show a contact angle around 90° or higher. They are attractive and often lead to a jump-in at distances between 5 and 100 nm. They can easily be avoided by using hydrophilic probes. For contact angles below 70 to 80°, they are usually much smaller then DLVO force and can be neglected [168].

Microfabricated tips of silicon nitride or silicon oxide are naturally hydrophilic. One should, however, keep in mind that they might contaminate and as a result become hydrophobic.

2.5.5
AFM Force Measurements of the Electric Double Layer

2.5.5.1 Between Solid Surfaces
With the AFM, DLVO forces were measured between several materials that are of interest especially in colloidal science, for example, glass, silica, and silicon nitride [4, 9, 10, 51, 53, 64–69], gold [64, 89–91], zinc, and lead sulfide [86–88], titanium oxide [73, 74, 82], zirconia [83–85], alumina [10, 71, 72], different polymers [56, 76, 77, 97, 98, 169–171], LB and other thin organic layers [57, 58, 94, 95, 172], surfactant adsorbed to solid surfaces [92, 173–175], biological membrane structures [54, 61, 62], adsorbed protein layers [99], or even whole cells [63, 101]. In all cases the force could be well described by DLVO theory; independently measured surface potentials (or charge densities) and Hamaker constants agreed with AFM results. Only at short separation below 1 to 5 nm, differences occurred [10, 65, 79, 176], particularly at high ionic strength. These are usually attributed to hydration forces.

As one example, the force-versus-distance between a silica particle and a titania flat is shown in Fig. 7. Like in many publications, the force is scaled by division through the radius of the silica microsphere. According to Eq. (7), the scaled force, F/R, is equal to $2\pi V_A$. Force curves were recorded at different pH values ranging from pH 8.8 for the top curve to pH 3.0 for the bottom curve. The surface charges of both materials are mainly determined by the pH. Silica has

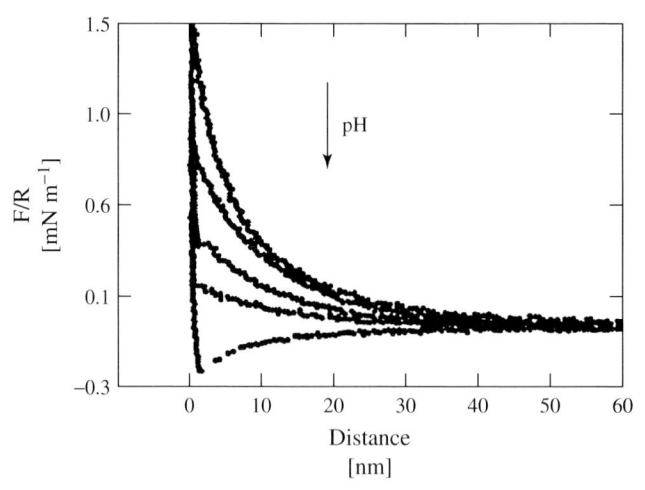

Fig. 7 Force between a silica (SiO_2) microsphere of 2.5-μm radius and a titania (TiO_2) crystal versus distance. The force is scaled by the radius of the sphere. The curves were recorded at pH values of 8.8, 7.2, 6.3, 5.3, and 3.0 from top to bottom with 1 mM KNO_3 background electrolyte. (The figure is reproduced with kind permission from C. J. Drummond [73].)

an isoelectric point around pH 3.0, while the isoelectric point of titania is pH 5.6. As a consequence, at high pH, where both materials are negatively charged, an electrostatic repulsion is observed. The repulsion decreases as the pH decreases, and at pH 3.0, that is, below the isoelectric point of titania, there is an electrostatic attraction as well as a van der Waals force resulting in an overall attraction between the two surfaces.

2.5.5.2 High Surface Potentials

With classical surface force–measuring techniques, it is impossible to analyze the electrostatic double layer at surface potentials significantly higher than 0.1 V. The reason is that high surface potentials can only be obtained by applying an external potential to a metallic or conducting sample. All materials that are accessible to the SFA or MASIF are not conducting. Therefore it was a particular challenge to measure electrostatic double-layer forces at high surface potentials with the AFM.

In first experiments, Ishino and coworkers [177] coated a microfabricated AFM tip with gold and used it as working electrode. They indeed observed a variation of the force with the applied potential when measuring the interaction between the gold-coated tip and a stearic acid monolayer. An electrostatic repulsion was observed at negative applied potentials and attractive forces were measured at positive potentials. Since the stearic acid monolayer was negatively charged, such a behavior is expected.

A more convenient setup consists of a conducting sample, which serves as a working electrode, and an insulating probe. Raiteri and coworkers and Döppenschmidt and coworkers [55, 178] measured the force between a gold, platinum, or graphite sample and a silicon nitride tip.

Campbell and Hillier determined the force between a silica sphere as probe and carbon [75]. Figure 8 shows the result of Hillier and coworkers [179] who measured the force between a gold sample and a silica sphere attached to the end of a cantilever. Silica is negatively charged at the pH $\approx$ 5.5 used in these experiments. As expected, the silica microsphere is repelled at negative potentials of the gold. When changing the potential to more positive values, the repulsion decreased and finally attractive forces were observed. For separations larger than 10 nm, the force curves could be described by DLVO theory. For smaller separations, measured force curves deviated significantly from calculated force-versus-distance curves. Hillier and coworkers attribute this discrepancy to several factors: an overestimation of the Hamaker constant, the roughness of the two interacting surfaces, and hydration effects.

Although the results of Hillier and coworkers qualitatively agree with results of Raiteri and coworkers [178] obtained with silicon nitride tips, there are also some discrepancies. First, the force observed between silica and gold changed almost linearly with the applied potential in a range from -0.7 to $+0.4$ V_{SCE}. Raiteri and coworkers observed a change of the force only in a narrow potential range of approximately 0.4 V. Above and below this voltage range, the force saturated. This saturation behavior is also predicted by DLVO theory. Second, Hillier and coworkers [179] always measured decay lengths that agreed with the calculated Debye length. Raiteri and coworkers observed the same for potentials that lead to a repulsive force. In the attractive regime, however, the decay length was significantly larger than the Debye length. These two questions are still open.

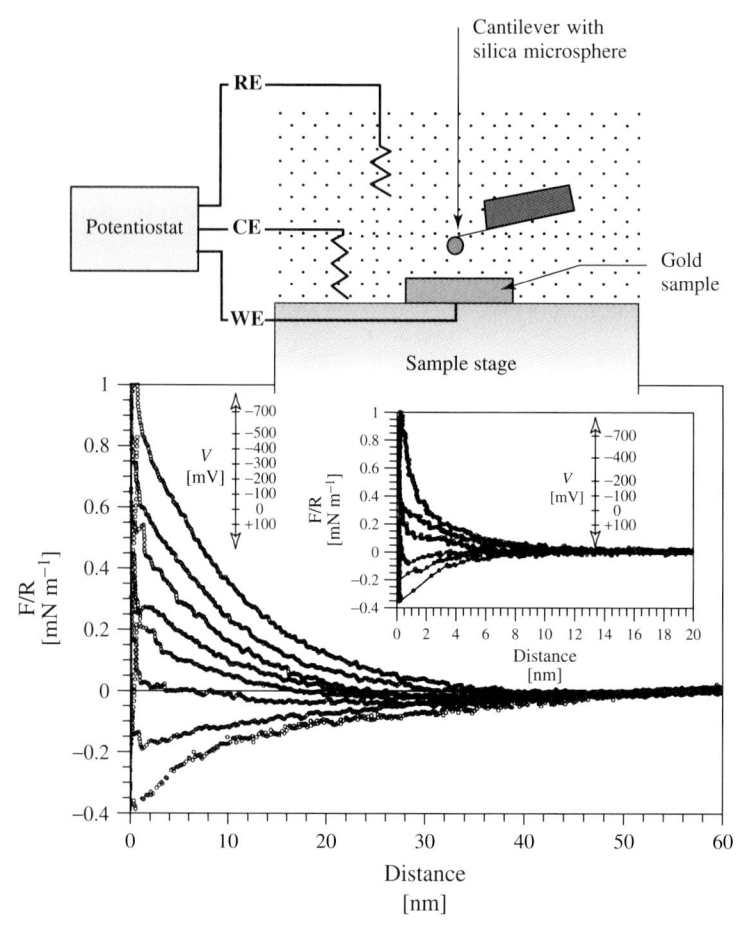

Fig. 8 Force between a silica sphere and gold electrode in an aqueous electrolyte solution of 1 mM KCl and pH ≈ 5.5 as a function of the applied potential at the gold electrode. The curves correspond to, from top to bottom, electrode potentials of −0.7, −0.5, −0.4, −0.3, −0.2, −0.1, 0, and +0.1 V (versus SCE). Inset: Force curves recorded in a 10 mM KCl solution. (The figure is reproduced with kind permission from A. C. Hillier [179].)

Also, preliminary experiments are done for the symmetric case of gold interacting with gold [93]. Both gold surfaces were shortcut and served as working electrodes. In this case a repulsion was observed at high positive potentials and high negative potentials. In between, around the point of zero charge, the repulsion was reduced. This behavior agrees with predictions of continuum theory.

2.5.5.3 Between Deformable Surfaces

While the electric double layer on a solid surface is relatively well understood and theories are able to account for colloidal stability and coagulation kinetics quite well, there has been much less success in understanding the double-layer structure at liquid–liquid or liquid–gas interfaces. This is despite the fact that the stability of emulsions or dispersion of particles and gas bubbles play a central role in many industrial processes such as flotation or the deinking of paper. With the AFM or with AFM-related setups, the electric double layers at such deformable interfaces can be analyzed. It is possible to measure the force between a solid microsphere and an oil drop (or another immiscible liquid) or a bubble in aqueous medium. These experiments are, however, more difficult to perform and to interpret. First, since the interface is deformable, it is difficult to determine zero distance. In principle, the shape of the interface can be calculated using the Laplace equation [180, 181]. This is, however, not trivial and in many cases practically impossible. Second, the tip or particle can even penetrate into the bubble or oil drop. In this case, a three-phase contact line is formed and the capillary force completely dominates the interaction. Electrostatic double-layer forces can only be detected before a three-phase contact is formed.

First experiments were done between differently treated silica particles and an air bubble [132]. As expected, they showed that a hydrophilic particle is repelled by an air bubble, while a hydrophobic particle jumps into the bubble. These early experiments were not very precise and the distance resolution was below 10 nm. The first precise experiments were done by Ducker and coworkers [182] who observed an electrostatic double-layer force that decayed with the expected Debye length. This observation that the electric double layer decays exponentially with the Debye length even at air–water interfaces was later confirmed with devices that were equipped with capacitatively calibrated piezo translators (Fig. 9). Ducker and coworkers detected, however, an unexpected attractive force between a presumably hydrophilic silica particle and a bubble. Fielden and coworkers confirmed that between hydrophilic particles and air bubbles only repulsive forces are acting [183] and their experiments can be considered as the first accurate measurements. As expected from adsorption experiments, a hydrophilic particle is repelled by an air bubble and a stable water film remains on the particle surface. The previously observed attraction was probably due to contamination. In contrast, a hydrophobic particle immediately forms a three-phase contact and snaps into the bubble (or vice versa: the bubble engulfs the particle). Before contact, the electrostatic double-layer force

Fig. 9 Decay lengths of the exponentially decaying component versus one over the square root of the salt concentration. The dashed line corresponds to the calculated Debye length. The figure summarizes results of measurements with hydrophilic and hydrophobic silica microspheres in solutions with different concentrations of SDS (sodium dodecylsulfate) or DTAB (from Ref. [143]).

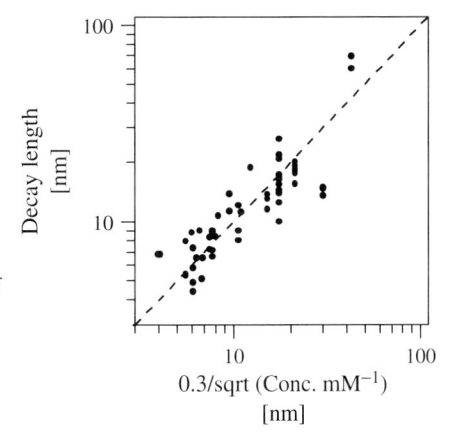

decaying with the Debye length could be observed in both cases.

Ducker and coworkers [182] and Fielden and coworkers [183] performed their experiments with commercial AFMs. For routine measurements this is not convenient because the liquid cells of commercial instruments are small and difficult to access manually. In addition, a thorough cleaning procedure is hampered by the fact that the cell consists of many different materials. Therefore, Preuss and coworkers built their own device to measure the force between particles and bubbles [143, 184–186]. With this device, the influence of different surfactants on the interaction of hydrophobic and hydrophilic particles with bubbles was studied.

As one example, the force between a hydrophilic silica particle and an air bubble at different concentrations of dodecyltrimethylammonium bromide (DTAB) is shown in Fig. 10. Without surfactant, the particle is repelled by the air bubble. At distances above 5 nm, the electrostatic repulsion dominates. The reason being the negative surface charges on the silica surface and at the water–air interface [187–190]. Even at close distance, a stable water remains on the particle surface and no three-phase contact is formed. Adding even small amounts of the cationic surfactant DTAB changes the interaction drastically. At concentrations between 0.1 mM and typically 5 mM DTAB (critical micellar concentration is $\approx$16 mM), no repulsion was observed. When the particle comes into contact with the air–water interface, it jumps into the bubble and a three-phase contact is formed. Such a behavior can be explained with the strong adsorption of long-chain alkyltrimethylammonium ions to silica [191]. At a concentration of 0.1 mM, DTA^+ forms a monolayer on the silica surface. This reverses its surface charge from negative to positive and it makes the surface hydrophobic.

When increasing the DTAB concentration to 5.4 mM, a small electrostatic repulsion was observed before the jump-in. The repulsion was at least partly due to electrostatic repulsion since it decayed exponentially with the Debye length. The reason is probably that at DTAB concentrations above 5 mM also, the air–water interface becomes positively charged because of adsorbed DTAB. This leads to an electrostatic repulsion between the positively charged air–water interface and the positively charged silica surface. Above approximately 6 mM DTAB, the jump-in disappeared and only repulsive force were observed.

The interaction of a particle with an oil drop is relatively similar to that with an air bubble [192–194]. Both interfaces, the oil–water and the water–gas, are negatively charged around neutral pH. Therefore an electrostatic repulsion is observed when interacting with a silica particle. As expected from DLVO theory, this repulsion decays roughly exponentially with a decay length equal to the Debye length [195].

2.5.5.4 Imaging Charge Distributions

Besides measuring the distance dependency of a force, it is sometimes interesting to know how the force changes from one place on the sample to another. To obtain this lateral information, Senden, Drummond, and Kékicheff [196] imaged a silicon nitride surface at low force. When imaging at very low forces (about 0.1 nN), the tip did not touch the surface because electrostatic repulsion kept the tip several nanometers above the surface. The tip "rode" on top of the double layer. In this way a constant electrostatic force image is obtained. When increasing the force, the

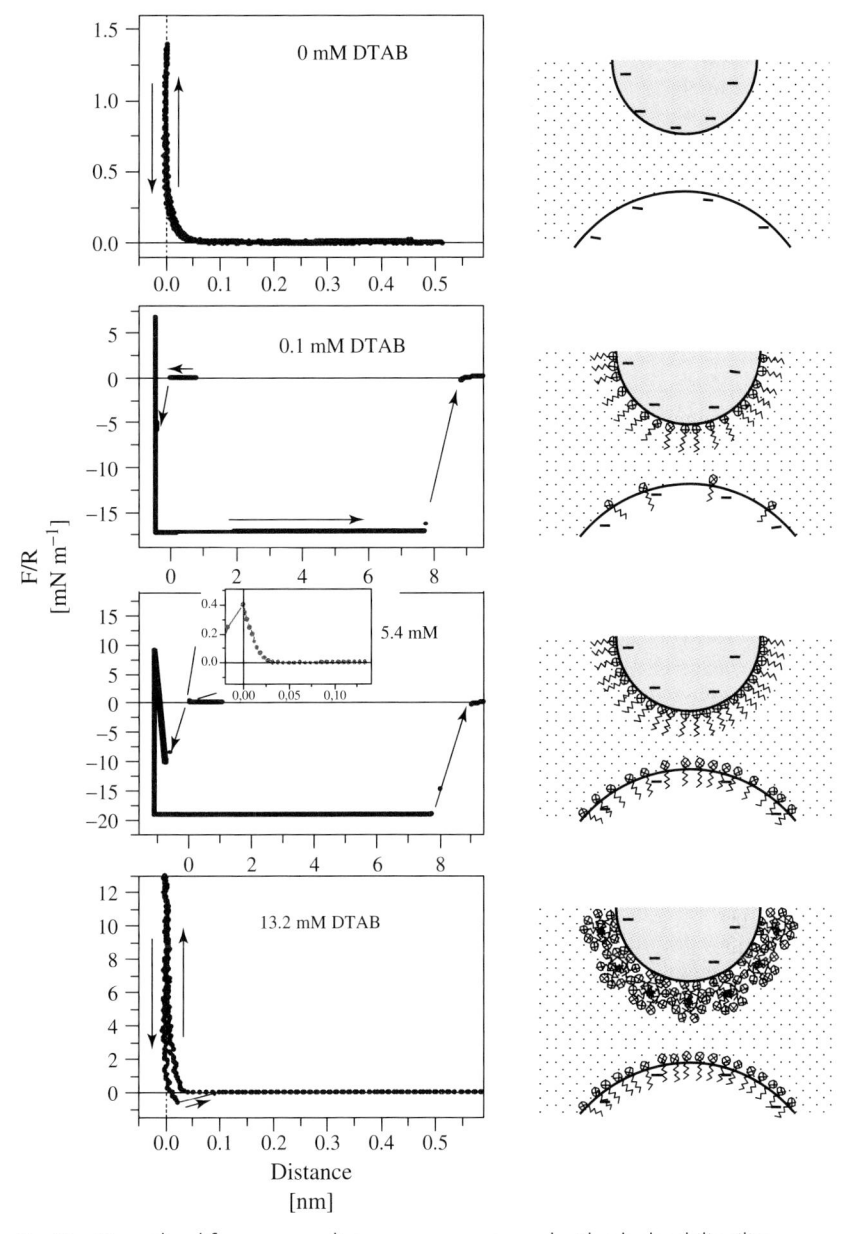

2.5 Analyzing Electric Double Layers with the Atomic Force Microscope | 245

Fig. 10 Normalized force-versus-distance curve measured with a hydrophilic silica particle in aqueous electrolyte with no added DTAB, 0.1 mM DTAB, 5.4 mM DTAB, and 13.2 mM DTAB (from Ref. [143]). The electrolyte contained 0.3 mM KCl, the pH was around 5.5. The insert at 5.4 mM shows the electrostatic repulsion before the jump-in. The particle radius R was 2.5 μm.

tip came into contact with the sample and a topographic image was obtained.

Using the electrostatic repulsion mode, Manne and coworkers studied the structure of surfactants adsorbed to solid surfaces. They imaged the cationic surfactant cetyltrimethylammonium bromide (CTAB) adsorbed to a hydrophobic graphite surface [197]. When imaging with a very low force, they observed parallel stripes spaced about 4.2 nm apart, that is, about twice the length of the CTA^+ ion. The adsorbed stripes were generally observed in three orientations. On the basis of observations like these, the authors proposed that CTAB forms hemicylindrical hemimicelles on graphite. Manne and coworkers and others extended this work to other solids such as silica [198], mica [175], or gold [199] and to other surfactants such as sodium dodecyl sulfate (SDS) or dodecyldimethylammoniopropanesulfonate (DDAPS) [200]. As an example, Fig. 11 shows mica surfaces covered with micells or hemimicells of two gemini surfactants. AFM images taken in the electrostatic repulsion mode allowed to analyze the supramolecular structure of surfactants on solid surfaces and greatly advanced our knowledge about surfactant adsorption. The advantage over contact mode is the significantly less destructive interaction between tip and sample. Burgess and coworkers [201] could even visualize changes in the structure of SDS adsorbed to gold(111) upon variation of the potential. At negative and neutral surface charge, SDS forms a hemimicellar structure. At positive surface charge, a condensed monolayer is formed.

In the described experiments the images are not solely images of the charge distribution but they are a mixture of topography and charge density. Heinz and Hoh [203] developed a protocol, called D minus D mapping, on how the two contributions can be separated. Therefore they take isoforce images at different salt

Fig. 11 To study the structure surfactants on surface, these AFM images (150 × 150 nm) were recorded in electrostatic repulsion mode. They show two different gemini surfactants, $(C_mH_{2m+1})[N^+(CH_3)_2](CH_2)_n[N^+(CH_3)_2](CH_2)_j$ with m-n-j of 18-3-1 (a) and 12-4-12 (b), in aqueous medium on mica. The concentrations were 3.0 and 2.2 mM, respectively. Both are above the critical micellar concentration. (The figure is reproduced with kind permission from S. Manne [202].)

concentrations to remove topography and isolate electrostatic contributions to the tip–sample interaction.

To get more quantitative information on the surface charge density, whole force curves must be taken at each individual point of the sample. In this way, differences between the charge densities of different regions on a sample can be obtained. This not only allows to distinguish regions on the basis of their electric properties but also opens the possibility to measure quantitative charge densities with a standard AFM tip. Usually, it is practically impossible to measure local charge densities with standard tips directly. The surface chemistry of silicon nitride and the size of the tip are not known accurately enough to allow quantitative measurements. To circumvent this problem, the sample is adsorbed on a substrate with known charge density. Then, force curves are taken on the substrate and on the deposited material. This substrate is used to calibrate the tip and then the charge density of the deposited material can easily be calculated. In this way the author determined the charge density of purple membranes deposited on alumina [54]. Rotsch and coworkers verified the positive surface charge of dimethyldioctadecylammonium bromide (DODAB) patches on negatively charged mica (Fig. 12). At neutral pH they observed a repulsive force between the negatively charged mica and the negatively charged silicon nitride tip. On a DODAB patch an electrostatic attraction was observed, indicating that the patch is positively charged.

Taking many force curves can nowadays be done automatically with commercial instruments. Usually it requires 0.1 to 1 s to take one force curve. To obtain say 64 × 64 force curves lasts 7 to 70 min. To increase the speed, Miyatani and coworkers developed a procedure in which the cantilever is moved in a sinusoidal form rather than a triangular pulse [205].

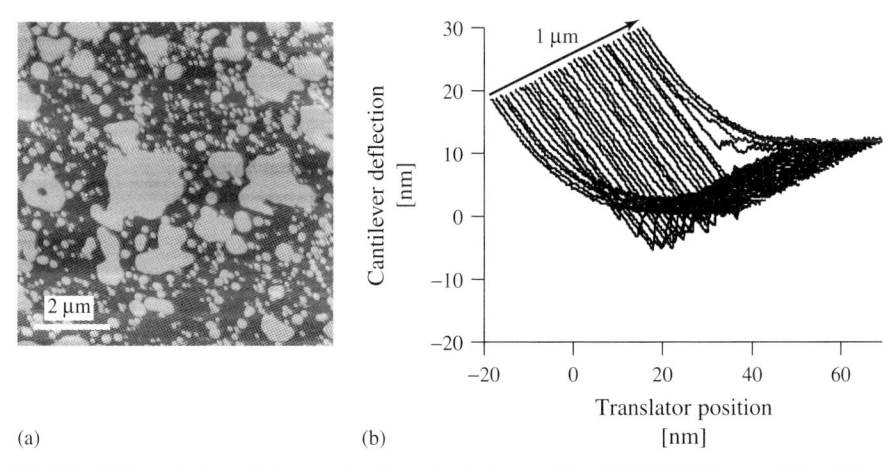

Fig. 12 AFM contact mode image showing a typical assembly of DODAB bilayer patches (a). The patches were self-assembled from a vesicle suspension and imaged in pure water. To the right, a series of approaching force curves are shown that were taken along a line across one of the patches in 5 mM LiNO$_3$ (b). The different electrostatic interaction – repulsive on the mica substrate at the beginning and end of the line and attractive on the DODAB patch – is visible. (The figure is reproduced with kind permission from M. Radmacher [204].)

This method is, however, limited by the increasing viscous drag when the cantilever approaches a sample [134].

2.5.6
Conclusion and Outlook

Since the first measurements of the electrostatic double-layer force with the AFM not even 10 years ago, the instrument has become a versatile tool to measure surface forces in aqueous electrolyte. Force measurements with the AFM confirmed that with continuum theory based on the Poisson–Boltzmann equation and applied by Debye, Hückel, Gouy, and Chapman, the electrostatic double layer can be adequately described for distances larger than 1 to 5 nm. It is valid for all materials investigated so far without exception. It also holds for deformable interfaces such as the air–water interface and the interface between two immiscible liquids. Even the behavior at high surface potentials seems to be described by continuum theory, although some questions still have to be clarified. For close distances, often the hydration force between hydrophilic surfaces influences the interaction. Between hydrophobic surfaces with contact angles above 80°, often the hydrophobic attraction dominates the total force.

In the future, more fundamental investigations are probably going to focus on the interaction at close distances where only few molecular layers are present between the interacting surfaces. Then, however, the separation into electrostatic, van der Waals, and hydration forces might not be useful anymore. Also the investigation of shear forces (friction) or the change of interaction under shear is a promising area of research and many important fundamental and applied questions can be addressed. Apart from these fundamental issues, the AFM has become a routine tool to measure interaction force for specific particles or surfaces relevant to practical applications.

References

1. G. Binnig, C. F. Quate, C. Gerber, *Phys. Rev. Lett.* **1986**, *56*, 930–933.
2. A. L. Weisenhorn, P. K. Hansma, T. R. Albrecht et al., *Appl. Phys. Lett.* **1989**, *54*, 2651–2653.
3. E. Meyer, H. Heinzelmann, P. Grütter et al., *Thin Solid Films* **1989**, *181*, 527–544.
4. A. L. Weisenhorn, P. Maivald, H.-J. Butt et al., *Phys. Rev. B* **1992**, *45*, 11 226–11 232.
5. J. H. Hoh, J. P. Revel, P. K. Hansma, *Nanotechnology* **1991**, *2*, 119–122.
6. S. Manne, H. E. Gaub, *Curr. Opin. Colloid Interface Sci.* **1997**, *2*, 145–152.
7. B. Cappella, G. Dietler, *Surf. Sci. Rep.* **1999**, *34*, 1–104.
8. H.-J. Butt, M. Jaschke, W. Ducker, *Bioelectrochem. Bioenerg.* **1995**, *38*, 191–201.
9. W. A. Ducker, T. J. Senden, R. M. Pashley, *Nature* **1991**, *353*, 239–241.
10. H.-J. Butt, *Biophys. J.* **1991**, *60*, 1438–1444.
11. H. Freundlich, *Z. Phys. Chem.* **1910**, *73*, 385–423.
12. B. Derjaguin, L. Landau, *Acta Physicochim. URSS* **1941**, *14*, 633–662.
13. E. J. W. Verwey, J. T. G. Overbeek, *Theory of the Stability of Lyophobic Colloids*, Elsevier Publishing, New York, 1948.
14. G. Gouy, *J. Phys. (Paris)* **1910**, *9*, 457–468.
15. D. L. Chapman, *Philos. Mag. (London)* **1913**, *25*, 475–481.
16. P. Debye, E. Hückel, *Phys. Z.* **1923**, *24*, 185–206.
17. J. Mahanty, B. W. Ninham, *Dispersion Forces*, Academic Press, New York, 1976.
18. J. N. Israelachvili, *Intermolecular and Surface Forces*, 2nd ed., Academic Press, London, 1992.
19. R. J. Hunter, *Foundations of Colloid Science I+II*, Clarendon Press, Oxford, 1995.
20. I. E. Dzyaloshinskii, E. M. Lifshitz, L. P. Pitaevskii, *Adv. Phys.* **1961**, *10*, 165–209.
21. P. M. Claesson, T. Ederth, V. Bergeron et al., *Adv. Colloid Interface Sci.* **1996**, *67*, 119–183.
22. N. V. Churaev, *Adv. Colloid Interface Sci.* **1999**, *83*, 19–32.

23. B. V. Derjaguin, A. S. Titijevskaia, I. I. Abricossova et al., *Discuss. Faraday Soc.* **1954**, *18*, 24–41.
24. J. T. G. Overbeek, M. J. Sparnaay, *Discuss. Faraday Soc.* **1954**, *18*, 12–24.
25. W. Black, J. G. V. de Jongh, J. T. G. Overbeek et al., *Trans. Faraday Soc.* **1960**, *56*, 1597–1608.
26. G. C. J. Rouweler, J. T. G. Oberbeek, *Trans. Faraday Soc.* **1971**, *67*, 2117–2121.
27. B. V. Derjaguin, Y. I. Rabinovich, N. V. Churaev, *Nature* **1978**, *272*, 313–318.
28. R. G. Horn, D. T. Smith, *J. Non-Cryst. Solids* **1990**, *120*, 72–81.
29. J. Keldenich, G. Peschel, *Colloid Polym. Sci.* **1991**, *269*, 1064–1070.
30. D. Tabor, R. H. S. Winterton, *Nature* **1968**, *219*, 1120–1121.
31. J. N. Israelachvili, D. Tabor, *Proc. R. Soc. London, A* **1972**, *331*, 19–38.
32. J. N. Israelachvili, G. E. Adams, *J. Chem. Soc., Faraday Trans. 1* **1978**, *74*, 975–1001.
33. S. A. Joyce, J. E. Houston, *Rev. Sci. Instrum.* **1991**, *62*, 710–715.
34. R. M. Pashley, *J. Colloid Interface Sci.* **1981**, *83*, 531–546.
35. G. Cevc, *J. Chem. Soc., Faraday Trans.* **1991**, *87*, 2733–2739.
36. S. Leikin, V. A. Parsegian, D. C. Rau et al., *Annu. Rev. Phys. Chem.* **1993**, *44*, 369–395.
37. L. Fischer, *J. Chem. Soc., Faraday Trans.* **1993**, *89*, 2567–2582.
38. J. Israelachvili, H. Wennerström, *Nature* **1996**, *379*, 219–225.
39. D. M. LeNeveu, R. P. Rand, V. A. Parsegian, *Nature* **1976**, *259*, 601–603.
40. V. A. Parsegian, R. P. Rand, N. L. Fuller et al., *Methods Enzymol.* **1986**, *127*, 400–416.
41. V. A. Parsegian, N. Fuller, R. P. Rand, *Proc. Natl. Acad. Sci. U.S.A.* **1979**, *76*, 2750–2754.
42. J. P. Chapel, *Langmuir* **1994**, *10*, 4237–4243.
43. A. Grabbe, R. G. Horn, *J. Colloid Interface Sci.* **1993**, *157*, 375–383.
44. J. L. Parker, *Langmuir* **1992**, *8*, 551–556.
45. J. L. Parker, *Prog. Surf. Sci.* **1994**, *47*, 205–271.
46. H. K. Christenson, The long-range attraction between macroscopic hydrophobic surfaces, in *Modern Approaches to Wettability* (Eds.: M. E. Schrader, G. Loeb), Plenum Press, New York, 1992, pp. 29–51.
47. J. Israelachvili, R. Pashley, *Nature* **1982**, *300*, 341–343.
48. K. L. Johnson, K. Kendall, A. D. Roberts, *Proc. R. Soc. London, A* **1971**, *324*, 301–313.
49. K. L. Johnson, *Contact Mechanics*, Cambridge University Press, Cambridge, 1985.
50. L. E. McNeil, M. Grimsditch, *J. Phys.: Condens. Matter* **1993**, *5*, 1681–1690.
51. T. J. Senden, C. J. Drummond, *Colloids Surf., A* **1994**, *94*, 29–52.
52. B. V. Zhmud, A. Meurk, L. Bergström, *J. Colloid Interface Sci.* **1998**, *207*, 332–343.
53. X. Y. Lin, F. Creuzet, H. Arribart, *J. Phys. Chem.* **1993**, *97*, 7272–7276.
54. H.-J. Butt, *Biophys. J.* **1992**, *63*, 578–582.
55. A. Döppenschmidt, H.-J. Butt, *Colloids Surf., A* **1999**, *149*, 145–150.
56. P. Weidenhammer, H. J. Jacobasch, *J. Colloid Interface Sci.* **1996**, *180*, 232–236.
57. T. Ishino, H. Hieda, K. Tanaka et al., *Jpn. J. Appl. Phys.* **1994**, *33*, 4718–4722.
58. T. Arai, D. Aoki, Y. Okabe et al., *Thin Solid Films* **1996**, *273*, 322–326.
59. Y. F. Dufrene, T. Boland, J. W. Schneider et al., *Faraday Discuss.* **1998**, *111*, 79–94.
60. H. Mueller, H.-J. Butt, E. Bamberg, *J. Phys. Chem.* **2000**, in press.
61. D. J. Müller, A. Engel, *Biophys. J.* **1997**, *73*, 1633–1644.
62. D. J. Müller, D. Fotiadis, S. Scheuring et al., *Biophys. J.* **1999**, *76*, 1101–1111.
63. R. F. Considine, D. R. Dixon, C. J. Drummond, *Langmuir* **2000**, *16*, 1323–1330.
64. W. A. Ducker, T. J. Senden, R. M. Pashley, *Langmuir* **1992**, *8*, 1831–1836.
65. L. Meagher, *J. Colloid Interface Sci.* **1992**, *152*, 293–295.
66. M. W. Rutland, T. J. Senden, *Langmuir* **1993**, *9*, 412–418.
67. G. Hüttl, D. Beyer, E. Müller, *Surf. Interface Anal.* **1997**, *25*, 543–547.
68. G. Toikka, R. A. Hayes, *J. Colloid Interface Sci.* **1997**, *191*, 102–109.
69. B. V. Zhmud, A. Meurk, L. Bergström, *Colloids Surf., A* **2000**, *164*, 3–7.
70. Y. Kanda, T. Nakamura, K. Higashitani, *Colloid Surf., A* **1998**, *139*, 55–62.
71. S. Veeramasuneni, M. R. Yalamanchili, J. D. Miller, *J. Colloid Interface Sci.* **1996**, *184*, 594–600.
72. I. Larson, C. J. Drummond, D. Y. C. Chan et al., *Langmuir* **1997**, *13*, 2109–2112.
73. I. Larson, C. J. Drummond, D. Y. C. Chan et al., *J. Phys. Chem.* **1995**, *99*, 2114–2118.
74. K. Hu, F. R. F. Fan, A. J. Bard et al., *J. Phys. Chem. B* **1997**, *101*, 8298–8303.

75. S. D. Campbell, A. C. Hillier, *Langmuir* **1999**, *15*, 891–899.
76. L. Meagher, R. M. Pashley, *Langmuir* **1995**, *11*, 4019–4024.
77. L. Meagher, R. M. Pashley, *J. Colloid Interface Sci.* **1997**, *185*, 291–296.
78. W. R. Bowen, N. Hilal, M. Jain et al., *Chem. Eng. Sci.* **1999**, *54*, 369–375.
79. R. H. Yoon, S. Vivek, *J. Colloid Interface Sci.* **1998**, *204*, 179–186.
80. M. Mizukami, M. Kurihara, *Prog. Colloid Polym. Sci.* **1997**, *106*, 266–269.
81. R. Ishiguro, D. Y. Sasaki, C. Pacheco et al., *Colloids Surf., A* **1999**, *146*, 329–335.
82. I. Larson, C. J. Drummond, D. Y. C. Chan et al., *J. Am. Chem. Soc.* **1993**, *115*, 11 885–11 890.
83. H. G. Pedersen, L. Bergström, *J. Am. Ceram. Soc.* **1999**, *82*, 1137–1145.
84. M. Prica, S. Biggs, F. Grieser et al., *Colloids Surf., A* **1996**, *119*, 205–213.
85. S. Biggs, *Langmuir* **1995**, *11*, 156–162.
86. D. T. Atkins, R. M. Pashley, *Langmuir* **1993**, *9*, 2232–2236.
87. G. Toikka, R. A. Hayes, J. Ralston, *J. Chem. Soc., Faraday. Trans.* **1997**, *93*, 3523–3528.
88. G. Toikka, R. A. Hayes, J. Ralston, *Langmuir* **1996**, *12*, 3783–3788.
89. S. Biggs, M. K. Chow, C. F. Zukoski et al., *J. Interface Colloid Sci.* **1993**, *160*, 511–513.
90. S. Biggs, P. Mulvaney, C. F. Zukoski et al., *J. Am. Chem. Soc.* **1994**, *116*, 9150–9157.
91. I. Larson, D. Y. C. Chan, C. J. Drummond et al., *Langmuir* **1997**, *13*, 2429–2431.
92. S. Biggs, P. Mulvaney, *J. Chem. Phys.* **1994**, *100*, 8501–8505.
93. R. Raiteri, M. Preuss, M. Grattarola et al., *Colloids Surf., A* **1998**, *136*, 191–197.
94. K. Hu, A. J. Bard, *Langmuir* **1997**, *13*, 5114–5119.
95. D. V. Vezenov, A. Noy, L. F. Rozsnyai et al., *J. Am. Chem. Soc.* **1997**, *119*, 2006–2015.
96. A. Noy, D. V. Vezenov, C. M. Lieber, *Annu. Rev. Mater. Sci.* **1997**, *27*, 381–421.
97. M. E. Karaman, L. Meagher, R. M. Pashley, *Langmuir* **1993**, *9*, 1220–1227.
98. Y. Q. Li, N. J. Tao, J. Pan et al., *Langmuir* **1993**, *9*, 637–641.
99. W. R. Bowen, N. Hilal, R. W. Lovit et al., *J. Colloid Interface Sci.* **1998**, *197*, 348–352.
100. H. Mueller, H.-J. Butt, E. Bamberg, *Biophys. J.* **1999**, *76*, 1072–1079.
101. Y. L. Ong, A. Razatos, G. Georgiou et al., *Langmuir* **1999**, *15*, 2719–2725.
102. C. A. J. Putman, B. G. De Grooth, N. F. Van Hulst et al., *J. Appl. Phys.* **1992**, *72*, 6–12.
103. H.-J. Butt, M. Jaschke, *Nanotechnology* **1995**, *6*, 1–7.
104. T. R. Albrecht, S. Akamine, T. E. Carver et al., *J. Vac. Sci. Technol.* **1990**, *A8*, 3386–3390.
105. H.-J. Butt, P. Siedle, K. Seifert et al., *J. Microsc.* **1993**, *169*, 75–84.
106. G. Y. Chen, R. J. Warmack, T. Thundat et al., *Rev. Sci. Instrum.* **1994**, *65*, 2532–2537.
107. F. J. Elmer, M. Dreier, *J. Appl. Phys.* **1997**, *81*, 7709–7714.
108. T. Thundat, R. J. Warmack, G. Y. Chen et al., *Appl. Phys. Lett.* **1994**, *64*, 2894–2896.
109. H.-J. Butt, *J. Colloid Interface Sci.* **1996**, *180*, 251–260.
110. J. E. Sader, L. White, *J. Appl. Phys.* **1993**, *74*, 1–9.
111. J. E. Sader, *Rev. Sci. Instrum.* **1995**, *66*, 4583–4587.
112. J. N. Neumeister, W. A. Ducker, *Rev. Sci. Instrum.* **1994**, *65*, 2527.
113. M. Sasaki, K. Hane, S. Okuma et al., *Rev. Sci. Instrum.* **1994**, *65*, 1930–1934.
114. Y. I. Rabinovich, R. H. Yoon, *Colloids Surf., A* **1994**, *93*, 263–273.
115. T. A. Senden, W. A. Ducker, *Langmuir* **1994**, *10*, 1003–1004.
116. J. P. Cleveland, S. Manne, D. Bocek et al., *Rev. Sci. Instrum.* **1993**, *64*, 403–405.
117. A. Torii, M. Sasaki, K. Hane et al., *Meas. Sci. Technol.* **1996**, *7*, 179–184.
118. J. E. Sader, J. W. M. Chou, P. Mulvaney, *Rev. Sci. Instrum.* **1999**, *70*, 3967–3969.
119. J. L. Hutter, J. Bechhoefer, *Rev. Sci. Instrum.* **1993**, *64*, 1868–1873.
120. P. Siedle, H.-J. Butt, E. Bamberg et al., *Inst. Phys. Conf. Ser.* **1992**, *130*, 361–364.
121. S. Xu, M. F. Arnsdorf, *J. Microsc.* **1994**, *173*, 199–210.
122. J. A. DeRose, J. P. Revel, *Microsc. Microanal.* **1997**, *3*, 203–213.
123. M. Grattarola, G. Massobrio, S. Martinoia, *IEEE Trans. Electron Devices* **1992**, *39*, 813–819.
124. D. L. Harame, L. J. Bousse, J. D. Shott et al., *IEEE Trans. Electron Devices* **1987**, *34*, 1700–1707.
125. H.-J. Butt, T. Müller, H. Gross, *J. Struct. Biol.* **1993**, *110*, 127–132.
126. M. Hegner, P. Wagner, G. Semenza, *Surf. Sci.* **1993**, *291*, 39–46.

127. N. A. Burnham, R. J. Colton, H. M. Pollock, *J. Vac. Sci. Technol., A* **1991**, *9*, 2548–2556.
128. G. L. Miller, J. E. Griffith, E. R. Wagner, *Rev. Sci. Instrum.* **1991**, 705.
129. P. J. Bryant, H. S. Kim, R. H. Decken et al., *J. Vac. Sci. Technol., A* **1990**, *8*, 2502.
130. D. A. Grigg, P. E. Russel, J. E. Griffith, *Ultramicroscopy* **1992**, *42*, 1504.
131. J. Mertz, O. Marti, J. Mlynek, *Appl. Phys. Lett.* **1993**, *62*, 2344–2346.
132. H.-J. Butt, *J. Colloid Interface Sci.* **1994**, *166*, 109–117.
133. D. Y. C. Chan, R. G. Horn, *J. Chem. Phys.* **1985**, *83*, 5311–5324.
134. O. I. Vinogradova, H.-J. Butt, G. Yakubov et al., *Rev. Sci. Instrum.* **2001**, *72*, 2330–2339.
135. J. B. Pethica, A. P. Sutton, *J. Vac. Sci. Technol., A* **1988**, *6*, 2490–2494.
136. H.-J. Butt, A. Döppenschmidt, G. Hüttl et al., *J. Chem. Phys.* **2000**, *113*, 1194–1203.
137. G. J. C. Braithwaite, A. Howe, P. F. Luckham, *Langmuir* **1996**, *12*, 4224–4237.
138. V. S. J. Craig, A. M. Hyde, R. M. Pashley, *Langmuir* **1996**, *12*, 3557–3562.
139. R. M. Pashley, M. E. Karaman, V. S. J. Craig et al., *Colloids Surf., A* **1998**, *144*, 1–8.
140. N. P. Costa, J. H. Hoh, *Rev. Sci. Instrum.* **1995**, *66*, 5096–5097.
141. M. Pierce, J. Stuart, A. Pungor et al., *Langmuir* **1994**, *10*, 3217–3221.
142. S. M. Hues, C. F. Draper, K. P. Lee et al., *Rev. Sci. Instrum.* **1994**, *65*, 1561–1565.
143. M. Preuss, H.-J. Butt, *Langmuir* **1998**, *14*, 3164–3174.
144. B. Derjaguin, *Kolloid Z.* **1934**, *69*, 155–164.
145. D. Y. C. Chan, D. J. Mitchell, *J. Colloid Interface Sci.* **1983**, *95*, 193–197.
146. V. A. Parsegian, D. Gingell, *Biophys. J.* **1972**, *12*, 1192–1204.
147. R. Hogg, T. W. Healy, D. W. Fuerstenau, *Trans. Faraday Soc.* **1966**, *62*, 1638–1651.
148. H.-J. Butt, *Biophys. J.* **1991**, *60*, 777–785.
149. H.-J. Butt, *Nanotechnology* **1992**, *3*, 60–68.
150. S. L. Carnie, D. Y. C. Chan, J. S. Gunning, *Langmuir* **1994**, *10*, 2993–3009.
151. H. Ohshima, *J. Colloid Interface Sci.* **1994**, *162*, 487–495.
152. J. Stankovich, S. L. Carnie, *Langmuir* **1996**, *1996*, 1453–1461.
153. A. V. Nguyen, G. M. Evans, G. J. Jameson, *J. Colloid Interface Sci.* **2000**, *230*, 208–212.
154. V. G. Levadny, M. L. Belaya, D. A. Pink et al., *Biophys. J.* **1996**, *70*, 1745–1752.
155. E. Barouch, E. Matijevic, *J. Chem. Soc., Faraday Trans. 1* **1985**, *81*, 1797–1817.
156. D. E. Yates, S. Levine, T. W. Healy, *J. Chem. Soc., Faraday Trans. 1* **1974**, *70*, 1807–1818.
157. P. M. Claesson, P. Herder, P. Stenius et al., *J. Colloid Interface Sci.* **1986**, *109*, 31–39.
158. P. J. Scales, F. Grieser, T. W. Healy, *Langmuir* **1990**, *6*, 582–589.
159. S. Nishimura, H. Tateyama, K. Tsunematsu et al., *J. Colloid Interface Sci.* **1992**, *152*, 359–367.
160. B. W. Ninham, V. Yaminsky, *Langmuir* **1997**, *13*, 2097–2108.
161. P. Johansson, P. Apell, *Phys. Rev. B* **1997**, *56*, 4159–4165.
162. C. Argento, R. H. French, *J. Appl. Phys.* **1996**, *80*, 6081–6090.
163. L. Bergström, *Adv. Colloid Interface Sci.* **1997**, *70*, 125–169.
164. H. D. Ackler, R. H. French, Y. M. Chiang, *J. Colloid Interface Sci.* **1996**, *179*, 460–469.
165. A. Milling, P. Mulvaney, I. Larson, *J. Colloid Interface Sci.* **1996**, *180*, 460–465.
166. A. Meurk, P. W. Luckham, L. Bergström, *Langmuir* **1997**, *13*, 3896–3899.
167. J. N. Israelachvili, R. M. Pashley, *Nature* **1983**, *306*, 249–250.
168. R. H. Yoon, D. H. Flinn, Y. I. Rabinovich, *J. Colloid Interface Sci.* **1997**, *185*, 363–370.
169. L. Meagher, V. S. J. Craig, *Langmuir* **1994**, *10*, 2736–2742.
170. J. Nalaskowski, S. Veeramasuneni, J. D. Miller, Interaction forces in the flotation of colemanite as measured by atomic force microscopy, in *Innovations in Mineral and Coal Processing* (Eds.: S. Atak, G. Önal, S. Celik), Brookfield, Rotterdam, 1998, pp. 159–165.
171. S. Nishimura, M. Kodama, H. Noma et al., *Colloids Surf., A* **1998**, *143*, 1–16.
172. V. Kane, P. Mulvaney, *Langmuir* **1998**, *14*, 3303–3311.
173. C. J. Drummond, T. J. Senden, *Colloids Surf., A* **1994**, *87*, 217–234.
174. M. E. Karaman, R. M. Pashley, N. K. Bolonkin, *Langmuir* **1995**, *11*, 2872–2880.
175. W. A. Ducker, E. J. Wanless, *Langmuir* **1999**, *15*, 160–168.
176. S. Veeramasuneni, M. R. Yalamanchili, J. D. Miller, *Colloids Surf., A* **1998**, *131*, 77–87.
177. T. Ishino, H. Hieda, K. Tanaka et al., *Jpn. J. Appl. Phys.* **1994**, *33*, L1552–L1554.
178. R. Raiteri, M. Grattarola, H.-J. Butt, *J. Phys. Chem.* **1996**, *100*, 16700–16705.

179. A. C. Hillier, S. Kim, A. J. Bard, *J. Phys. Chem.* **1996**, *100*, 18 808–18 817.
180. S. J. Miklavcic, R. G. Horn, D. J. Bachmann, *J. Phys. Chem.* **1995**, *99*, 16 357–16 364.
181. D. J. Bachmann, S. J. Miklavcic, *Langmuir* **1996**, *12*, 4197–4204.
182. W. A. Ducker, Z. Xu, J. N. Israelachvili, *Langmuir* **1994**, *10*, 3279–3289.
183. M. L. Fielden, R. A. Hayes, J. Ralston, *Langmuir* **1996**, *12*, 3721–3727.
184. M. Preuss, H.-J. Butt, *Int. J. Miner. Process* **1999**, *56*, 99–115.
185. M. Preuss, H.-J. Butt, *J. Colloid Interface Sci.* **1998**, *208*, 468–477.
186. S. Ecke, M. Preuss, H.-J. Butt, *J. Adhes. Sci. Technol.* **1999**, *13*, 1181–1191.
187. S. Usui, H. Sasaki, H. Matsukawa, *J. Colloid Interface Sci.* **1981**, *81*, 80–84.
188. R. H. Yoon, J. L. Yordan, *J. Colloid Interface Sci.* **1986**, *113*, 430–438.
189. P. Saulnier, J. Lachaise, G. Morel et al., *J. Colloid Interface Sci.* **1996**, *182*, 395–399.
190. H. J. Schulze, C. Cichos, *Z. Phys. Chem. (Leipzig)* **1972**, *251*, 252–268.
191. L. G. T. Eriksson, P. M. Claesson, J. C. Eriksson et al., *J. Colloid Interface Sci.* **1996**, *181*, 476–489.
192. P. Mulvaney, J. M. Perera, S. Biggs et al., *J. Colloid Interface Sci.* **1996**, *183*, 614–616.
193. B. A. Snyder, D. E. Aston, J. C. Berg, *Langmuir* **1997**, *13*, 590–593.
194. G. E. Yakubov, O. I. Vinogradova, H.-J. Butt, *J. Adhes.* **2000**, *14*, 1783–1799.
195. P. G. Hartley, F. Grieser, P. Mulvaney et al., *Langmuir* **1999**, *15*, 7282–7289.
196. T. J. Senden, C. J. Drummond, P. Kékicheff, *Langmuir* **1994**, 358–362.
197. S. Manne, J. P. Cleveland, H. E. Gaub et al., *Langmuir* **1994**, *10*, 4409–4413.
198. S. Manne, H. E. Gaub, *Science* **1995**, *270*, 1480–1482.
199. M. Jaschke, H.-J. Butt, H. E. Gaub et al., *Langmuir* **1997**, *13*, 1381–1384.
200. W. A. Ducker, L. M. Grant, *J. Phys. Chem.* **1996**, *100*, 11 507–11 511.
201. I. Burgess, C. A. Jeffrey, X. Cai et al., *Langmuir* **1999**, *15*, 2607–2616.
202. S. Manne, T. E. Schäffer, Q. Huo et al., *Langmuir* **1997**, *13*, 6382–6387.
203. W. F. Heinz, J. H. Hoh, *Biophys. J.* **1999**, *76*, 528–538.
204. C. Rotsch, M. Radmacher, *Langmuir* **1997**, *13*, 2825–2832.
205. T. Miyatani, M. Horii, A. Rosa et al., *Appl. Phys. Lett.* **1997**, *71*, 2632–2634.

2.6
Comparison Between Liquid State and Solid State Electrochemistry

Ilan Riess
Technion-Israel Institute of Technology, Haifa, Israel

Abstract

We discuss the similarities and differences between liquid-state electrochemistry (LSE) and solid-state electrochemistry (SSE). Although based on the same thermodynamic principles, the properties of these cells are quite distinct. Differences exist in the bulk conduction mechanism, partially in electrode reaction and in cell construction and morphology. This also leads to differences in applications.

2.6.1
Introduction

LSE, the classical electrochemistry, is concerned with electrochemical cells (ECs) based on liquid ionic-conductors (liquid electrolytes (LEs)). Solid-state electrochemistry is concerned with ECs in which the ionic conductor (electrolyte) is a solid. Both fields are based on common thermodynamic principles. Yet, the finer characteristics of ECs in the two fields are different because of differences in the materials properties, conduction mechanisms, morphology and cell geometry. Differences that come immediately to mind are: (1) The lack of electronic (electron/hole) conduction in most LEs, while electronic conduction exists to some extent in all solid electrolytes (SEs). (2) In LEs both cations and anions are mobile, while in SEs only one kind of ions is usually mobile while the other forms a rigid sublattice serving as a frame for the motion of the mobile ion. An example is AgI (at $T > 149\,^{\circ}C$), where only Ag^+ ions are mobile. (3) In liquid-state cells one has the option to stir and replenish the LE. This is not possible, of course, when using SEs. Contrary to SSE in which the electrode can be complex, in LSE it is usually the LE that has a complicated composition, containing many different kinds of ions coming from dissolution of different compounds. These differences and others will be discussed in more detail in the coming sections. In Sect. 2 we discuss the geometry of liquid and solid ECs; in Sect. 3 material properties, in particular, charge carriers, their nature, concentration and conduction mechanism; in Sect. 4 the theoretical current-voltage relations in the bulk and the relevant boundary conditions. In Sect. 5 experimental methods for characterizing materials properties and overpotentials are discussed, and in Sect. 6 electrode processes. Two short sections follow, Sect. 7 in which the use of the Galvani potential is considered, and Sect. 8 in which catalysis is briefly considered. A summary is given in Sect. 9.

LSE is a well-established field, for which there are many good, comprehensive text books and we mention only four: Bockris and Ready [1], Bard and Faulkner [2], Oldham and Myland [3] and Gileadi [4]. In SSE, being a younger field, the information is scattered in many review papers on different topics, books containing review chapters [5–7], books dedicated to particular topics: on solid-state batteries [8], solid-oxide fuel cells [9], and solid-state sensors [10]. A text book that summarizes the state of the art up to year 1981 is the book by Rickert [11]. A recent book (in German) was published by Maier. It presents a broad overview [12]. Chemical reactions in ionic solids are closely related to SSE as they are based on interdiffusion of ions and sometimes also on charge transfer. The driving forces are similar. The reactions can be driven by both chemical potential differences and electrochemical potential differences. A recent book on solid-state reaction is by Schmalzried [13]. Properties of many relevant ionic materials as well as an excellent summary of the theory of point defect in solids can be found in the book by Kröger [14]. An outstanding contribution to the field up to the year 1977 was made by Wagner, whose results are quoted in the reviews and books written later and mentioned in the preceding text.

2.6.2
Electrochemical Cell Geometry

2.6.2.1 Cells Based on LEs

An EC based on a LE is shown schematically in Fig. 1. It is composed of a liquid that conducts ions and two current-carrying electrodes, an anode and a cathode (E_1, E_2). An electrode can be composed of more than one piece of one kind of metal as when using a sacrificial anode. Additional electrodes (E_{ref}) that do not carry currents are added in most cases to serve as reference electrodes in voltage measurements and analysis of cell operation.

The LE serves not only as a medium for selective transport of ionic charge but also, in many cases, as the source and sink for the reactants and products involved in the electrochemical reaction. The electrodes are then chemically inert, serving only for the supply and removal of electrons. Furthermore, the LE tolerates the addition of other materials that contribute mobile ions which, however, do not participate in the electrical current under steady state conditions. The main use of this feature is in adding a supporting electrolyte, the importance of which will be discussed in the theoretical Sect. 4.

In other arrangements, the electrodes provide material or serve as a sink. An electrode that serves as a source of material is, for example, a metal anode that dissolves and liberates cations into the LE. A metal electrode serves as a sink when it is oxidized to form, for instance, an oxide layer or when electroplating occurs. This is also the case when insertion takes place as hydrogen into a palladium electrode. Material can be exchanged also with a gas phase. This is the case when bubbling hydrogen on a Pt electrode.

A key limitation of the existing LE ECs is that the operation temperature must be kept low, compatible with the vapor pressure of the liquid. (An exception is the use of a few high-temperature, molten salt LEs, for example, molten carbonates) [1, 15, 16]. While low-operating temperature has obvious advantages, it may also introduce limitations because of two very different reasons. (1) high temperatures allow to overcome electrode overpotentials,

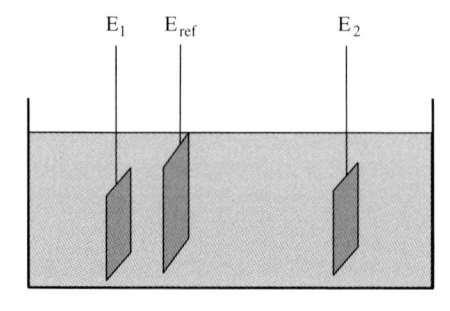

Fig. 1 Schematics of a liquid-state electrochemical cell. LE – liquid electrolyte; E_1, E_2 – current-carrying electrodes; E_{ref} – reference electrode.

and (2) the heat generated as a by-product of the electrochemical reaction in hot cells can be used for heating much more efficiently. This is the case in high-temperature solid-oxide fuel cells in which the hot exhaust gases can be used for cogeneration of electricity by gas turbine driven electrical generators. A second key limitation of LE-based ECs is that they cannot be used in thin-film devices and miniature cells in microelectronics, while solid-state cells are being developed for these applications.

2.6.2.2 Cells Based on SEs

Solid electrochemical cells are more restricted in the structure because of the solid nature of the ionic conductor. Schematic configurations of solid-state cells are presented in Fig. 2(a–c). Figure 2(a) shows an all-solid cell with the SE, SE(M$^+$), and the electrodes, E$_1$, E$_2$, being solids. These electrodes serve both as source and sink for electrons and for material. An example is the battery Li$_x$C|SE(Li$^+$)|Li$_y$TiS$_2$, where the anode is graphite intercalated with Li, and the cathode is TiS$_2$, a layered compound that allows intercalation of Li into it, serving both as a conductor and a source/sink of Li$^+$ ions and electrons. The SE conducts Li$^+$ ions, and can be, for example, a glass made of Li$_2$O−Li$_2$SO$_4$−B$_2$O$_3$ [8]. In this case, the electrodes materials serve for the supply and removal of both charge and matter. They must allow the diffusion (intercalation) of the material into them and they must also be electronic (electron/hole) conductors. The SE serves solely as a membrane that facilitates the selective conduction of certain ions (e.g. Li$^+$), but not that of electron/holes and of other ions

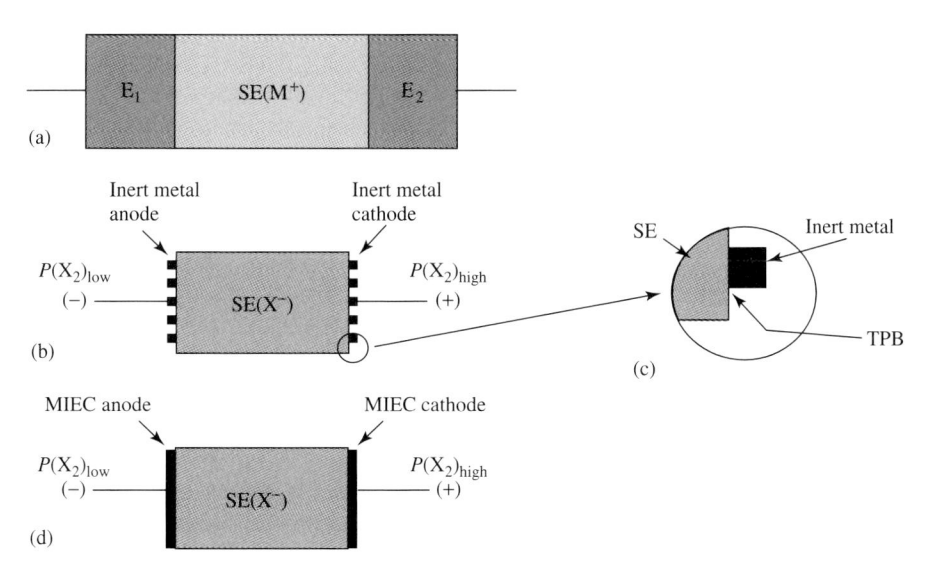

Fig. 2 Schematics of solid-state electrochemical cells. (a) Cell with two, solid, current-carrying electrodes E$_1$ and E$_2$, SE(M$^+$) − solid electrolyte conducting the cation M$^+$. (b) Cell with electrodes composed of a porous metal layer for electronic charge exchange and the gas phase, X$_2$, for material exchange. P_{X_2} −partial pressure of the gas X$_2$. SE(X$^-$) − solid electrolyte conducting the anion X$^-$. (c) The triple phase boundary (TPB) enlarged. (d) Cell with mixed − ionic − electronic conducting electrodes for supply of electronic charge permitting diffusion of ions through them, enabling material exchange with the gas phase.

that may come from the electrodes. SEs usually do not tolerate the addition of a second type of mobile ions to serve as supporting electrolytes. All this is quite different from the case of LSE. There are additional differences. Ions have to cross the SE/electrode interface. The mobility of ions in SEs is, for most cases, lower than in LE, and is realized usually only at elevated temperatures. Because of the low mobility, SEs contain high concentrations (>1 atomic%) of mobile ions. The highest conductivity obtained at elevated T reaches ~ 1 S/cm for some materials (e.g. for Ag^+ in AgI at $T > 149\,°C$ and for O^{2-} in Bi_2O_3 at $\sim 800\,°C$ [11]). However, for most SEs the conductivity is at least one order of magnitude lower.

Figure 2(b) exhibits a solid electrochemical cell with porous, inert electrodes in contact with a gas phase at each electrode that interacts with the electrode (by electron charge transfer) and with the SE. An example is the fuel cell arrangement, $H_2,H_2O,Pt|YSZ|Pt,O_2$. The SE yttria stabilized zirconia (YSZ) (the oxide: $(Y_2O_3)_x(ZrO_2)_{1-x}$, $x \sim 0.1$) is an oxygen-ion conductor. One electrode is exposed to a high concentration of oxygen (pure oxygen or air) and the other to a fuel such as H_2. (In modern fuel cells, Pt for the electrodes is replaced by other metals, mixed ionic electronic conductor (MIECs) or cermets [9].) In the cell mentioned here, each electrode is a mixture of phases, one (Pt) supplying the electronic charge and the second one (the gas phase) serving as the source or sink of material. The active electrode area is around the triple phase boundary (TPB) in which the three phases: gas, metal, and SE involved in the electrode reaction, meet (see Fig. 2c). The length of the TPB is the length of the pore edges in the porous metal. The width of this area varies from case to case and depends on the rates of the different and competing elementary steps in the electrode reaction. The geometry of this cell is again different from that of a LE-based cell.

A modification of the cell shown in Fig. 2(b) is presented in Fig. 2(d) in which nonporous MIECs are used as electrodes. In a fuel cell, oxygen has then to diffuse from the gas phase through the MIEC cathode to reach the SE and, correspondingly, the fuel and the exhaust gas have to diffuse through the anode. (In reality, the MIEC electrodes are porous, which increases the electrode active area). The MIEC can be either a semiconductor or a metal. An example of a continuous cathode is: YSZ|Ag|air, with the Ag layer as the MIEC electrode. This is possible because silver is permeable to oxygen at elevated temperature [17, 18].

2.6.2.3 Comparing the Two Groups of Cells

In LSE salt bridges are used to connect electrically (by selective ion motion) but to separate physically and to a good approximation also chemically, two compartments. Owing to ion leaks, these are useful only for low-current measurements, in particular, open circuit voltage measurements. In SSE, salt bridges are not needed. SEs usually conduct only one kind of ions. Let us consider the Daniel cell $Cu|LE(Cu^{2+},SO_4^{2-})|Bridge|LE(Zn^{2+},SO_4^{2-})|Zn$. The bridge is required to stop mixing of the two solutions. It allows a slow passage of ions, thus facilitating a continuous electrical circuit. If the bride would be absent, this would destroy the cell as mixing would occur followed by electroplating of Cu onto the Zn electrode. On the other hand, in SSE, if one would build a Cu, Zn cell, one would use a SE that conducts only one kind of ions, say Cl^-. This could be $SrCl_2$ [19]. The cell would then be: CuCl$_2$,Cu|SrCl$_2$|Zn,ZnCl$_2$. This cell would function without the need of a bridge. The SE (SrCl$_2$) serves as the separating membrane. We notice that the phases CuCl$_2$ and ZnCl$_2$ are added to the electrodes in order to fix the chemical potential of Cl$_2$ there.

If a concentration cell is considered, then in LSE the cell would be, for example: Zn|LE(Zn^{2+},Cl$^-$)|Bridge(K$^+$,Cl$^-$)|LE(Zn^{2+},Cl$^-$)|Zn, with the Cl$^-$ concentration different in the two LEs. In SSE one would use a cell such as: Cl$_2$($P^I_{Cl_2}$),Zn|SE(Cl$^-$)|Zn,Cl$_2$($P^{II}_{Cl_2}$) with different partial pressures, P_{Cl_2}, of Cl$_2$. No bridge is again needed since the SE serves as the separating membrane. (For this cell to work properly $P^I_{Cl_2}$ and $P^{II}_{Cl_2}$ should be kept below the equilibrium pressure of Zn/ZnCl$_2$, otherwise the chloride ZnCl$_2$ would be formed at the electrodes, and both P_{Cl_2}-s would obtain the (same) equilibrium value and the voltage would vanish). In SSE the metal electrodes in concentration cells are, in many cases, inert. A typical example is the cell: O$_2$($P^I_{O_2}$),Pt|SE(O^{2-})|Pt,O$_2$($P^{II}_{O_2}$).

When the source of material is a solid electrode, it is consumed under current. Therefore the lifetime of the electrode is limited. In SSE, care must be taken to press the electrode onto the SE continuously. At the other solid electrode, space must be left for the electrode to grow. In contrast, in LSE the LE stays always in good contact with the electrode. Furthermore, when the source of material is the LE, then material can be supplied indefinitely since the LE can be replenished. The geometry of LSE cells closest to all solid cells is when the LE is soaked in a porous solid.

A continuous supply of material is possible in SSE and in LSE, when it comes from the gas phase. In SSE this can also be achieved by supplying a liquid to an electrode.

One expression of the flexibility of the geometry in LSE, in contrast to SSE, is the possibility to use a dropping mercury electrode.

2.6.3
Charge Transport in Electrolytes

2.6.3.1 Charge Transport in LEs

In LSE charge is transported by ions in the LE and by electrons in the metallic electrodes. The LE is usually a solvent that contains an ionic solute. In many cases, the solvent is water and the solute a salt, acid or base. A different example is molten carbonates in which the salt is molten and the liquid serves also as the source for ions [1, 15, 16]. In some cases, the water in an ionic solution also takes part in the reaction. For example, the reaction: Cr(s) + 4H$_2$O(l) → CrO$_4^{2-}$(aq) + 8H$^+$(aq) + 6e$^-$(s) that can take place at a Cr electrode. In SSE on the other hand, the SE is always chosen to be chemically stable (in the steady state).

In LEs there are at least two kinds of mobile ions, the cations and anions. They have usually comparable mobilities and both may contribute to the electrical current. There is rarely an electron current in a LE, though a few exceptions exist (e.g. sodium metal dissolved in liquid ammonia contributes solvated electrons [20]). Electrons can also propagate in a liquid by hopping between a redox couple say: M$^+$ + M^{2+} → M^{2+} + M$^+$, which is equivalent to the propagation of an electron from left to right [21]. This is analogous to hopping conduction in an impurity band in solids. The rate of electron hopping in a liquid is low due to the relatively large average distance between the ions for common concentrations and in view of the large distance of closest approach due to the cloud of solvating molecules. We can therefore safely consider LEs as ionic conductors with negligible electronic conduction.

The motion of ions in a LE is impeded. The ion or the ion and its solvating sphere move in a viscous liquid. The force of retardation can be calculated according to Stokes equation and the corresponding diffusion coefficient or mobility is related to the viscosity.

Protons can propagate by two mechanisms. One is by the viscous flow of a complex $H^+(H_2O)_n$ where $n = 1, 2, 3, 4$, the second is by the Grotthuss mechanism, hopping of a proton from one hydronium H_3O^+ to a nearby water molecule. The latter mechanism has some similarity to hopping conduction of ions in solids as discussed in the following text.

2.6.3.2 Charge Transport in SEs

SEs are those solid ionic conductors for which ionic conductivity, σ_{ion}, is significant and the electronic conduction, σ_{el}, is small. There is no exact definition of what is a significant ionic conductivity, but usually $\sigma_{ion} > 10^{-4}$ S/cm is considered significant with the lower limit applying to cells with thin-layer SEs. Electronic conduction in solids is, usually, present to a higher degree than in LEs. SEs are those MIECs for which $\sigma_{el} \ll \sigma_{ion}$. The reasons for low electronic conduction in ionic solids are: (1) The energy gap is large and intrinsic electron and hole concentrations are very low. (2) The occupied electron states in the conduction band and hole states in the valence band are localized rather than delocalized. Therefore, the electron and hole mobilities are thermally activated and low. The lack of electron conduction in LE can be discussed in the same terms. In these terms, the quasi–energy gap of the disordered system is very large and the electron states highly localized with negligible mobility.

The ions and electrons in SEs and in MIECs polarize the ionic neighborhood. Therefore, the ions, electrons, and the polarized surrounding form the corresponding small polarons [22]. This would be analogous to polarization of an ionic liquid. However, the details are different and in the sphere of solvating molecules that exists in LSE, neutral molecules with an electric dipole are attracted to the ion and also change orientation. For SEs with a high concentration of mobile ions (>10 ion%), for example, Ag^+ ion in AgI at $T > 149\,°C$, the interaction between the mobile ions forces local relaxation within the sea of these ions, which affects the conductivity and its dependence on frequency for ac signals [23].

The high concentration of ions both in LEs and in SEs affect the width of the space charge that exists at the boundary of the electrolyte near the contact interface with the electrode. Any abrupt change in the concentration of charged species must result in a space charge [24, 25]. The width of that space charge is roughly the Debye length and for typical LEs, it is only 0.3 to 10 nm; while for typical SEs, it is 0.3 to 1 nm. The detailed characteristics of the space charge are not identical for LEs and SEs. One difference arises from the fact that in SEs usually only one type of ions is mobile. There is therefore a difference between enriching and depleting that ion, and the profile of the space charge when it is enriched or depleted is different [12]. The other differences are due to the ordered structure of the SE and the absence of a solvating cloud in SEs, as discussed in Sect. 6.

In SEs, ionic and electronic charge transport is associated with deviations from the normal state. This can be best understood by considering crystalline solids. Then conduction is associated with point defects [14]. The perfect crystal is defined as that with perfect lattice periodicity, at zero temperature. In semiconductors, electrons in the conduction band are thus defects formed by excitation into states that are empty at $T = 0$. Conduction in LEs can also be formulated as a deviation from a standard liquid state [26]. However, this is uncommon and is not discussed further here.

The electronic charge carriers can be generated, under equilibrium, in three ways:

1. By thermal excitation across the band gap.
2. By extrinsic doping with donors or acceptors.
3. By change of stoichiometry. The ionic defects formed in this way act as native donors or acceptors.

Of course, any combination of these three causes may be realized simultaneously in any given material. In addition to electrons/holes generated under thermal equilibrium, they may also be injected from electrodes and excited by optical pumping under nonequilibrium conditions. As mentioned before, the mobility of the electrons and holes is thermally activated as they form small polarons.

For ionic conduction to exist in a crystal, an ion has to move out of its normal position either into an interstitial site or find an existing defect in the form of a vacancy. The conductivity increases therefore linearly with the defects concentration, in the dilute limit (as long as interaction between the defect can be neglected). The defects can be generated, under equilibrium, in three ways, similar to those for electronic defects:

1. By thermal excitation. The result, in an elementary solid, is a vacancy and an interstitial. In a binary-ionic compounds, say MX (a compound composed of cations M and anions X), thermal excitation may result in one of five combinations of a vacancy, an interstitial and/or a misplaced ion [14]. The most widely discussed defects are Frenkel pairs, an interstitial of, for instance, a cation and the vacancy left behind, $M_i + V_M$, and Schottky pairs of vacancies of the cation M and anion X: $V_M + V_X$. We use here, the Kröger–Vink notation of point defects. Charge is measured relative to the perfect crystal. A dot denotes a positive relative charge, a prime denotes a negative relative charge and x denotes a zero relative charge [14]. Thus, the Frenkel pairs in AgBr, interstitial silver ions and silver ion vacancies, are, in the Kröger–Vink notation: $Ag_i^{\bullet} + V_{Ag}'$.
2. By doping. Charged ionic defects can be generated also by doping with charged impurities. For substitutional doping, they must be aliovalent. For example, doping substitutionally ZrO_2 with CaO forming the defect Ca_{Zr}'' generates also ionic defects in the form of charged oxygen vacancies, $V_O^{\bullet\bullet}$. The later are mobile at elevated temperatures. We have mentioned before that doping may result also in generating electrons or holes. Whether doping will result in ionic defects (called self compensation) or electronic ones depends on the materials involved, and can be predicted if the nature of the point defects and their concentrations in the host are known in detail [27].

3. By change of stoichiometry. For binary compounds, say oxides, the oxygen vacancies can be generated by reducing the oxygen partial pressure, P_{O_2}, of the gas in equilibrium with the oxide, at elevated temperatures. This yields for instance, CeO_{2-y}, $y > 0$. Cation vacancies or oxygen interstitials can be formed under high P_{O_2} as is the case in Cu_2O, which contains then V_{Cu}^x, $V_{Cu}^{\backslash}$ and $O_i^{\backslash\backslash}$ [28, 29].

The propagation of ionic defects makes use of the existing point defects in various combinations [30], provided they form a continuous, percolating path. Let us consider two simple but important paths: (1) Vacancy propagation by the hopping of a nearby ion (of the correct type) into the vacant site. This leaves a new vacancy behind. For example $M_M^x + V_M^{\backslash\backslash} \rightarrow V_M^{\backslash\backslash} + M_M^x$ describes the motion of a cation with valence 2+ from left to right or, equivalently, the motion of a vacancy $V_M^{\backslash\backslash}$ from right to left. (2) The other example is that of interstitial cation moving via interstitial sites. In all cases, the mobility is thermally activated with the thermal energy required to overcome the potential energy (or rather the enthalpy) barriers between allowed sites in the conduction path.

Another conduction mechanism in solids that has to be considered, besides electron and hole conduction in the corresponding energy bands and ion conduction, is electron hopping in an impurity band [31]. For example, let D be the impurity then, $D_i^{\bullet} + D_i^{\bullet\bullet} \rightarrow D_i^{\bullet\bullet} + D_i^{\bullet}$ ($D_i^{\bullet}$, $D_i^{\bullet\bullet}$ are the interstitial cations in the Kröger–Vink notation) describes the effective motion of an electron from left to right. This occurs in addition to any conduction due to electrons in the conduction band or holes in the valence band [27]. The hopping conductivity is proportional to the product of concentrations: $[D_i^{\bullet}][D_i^{\bullet\bullet}]$.

The conduction-electron conductivity is proportional to the concentration of electrons in the conduction band, which, under equilibrium, is proportional to the ratio of concentrations $[D_i^{\bullet}]/[D_i^{\bullet\bullet}]$, as can be deduced from the mass action law for the reaction $D_i^{\bullet\bullet} + e^{\backslash} \leftrightarrow D_i^{\bullet}$. The concentration of the electrons in the conduction band may be quite low, though both the ionic defects concentrations may be high. Then, the hopping conductivity may be higher than the conduction-electron conductivity. Not only impurities but also native defects with variable charged states allow hopping conduction in a defect band [32].

In SSE MIECs play an important role [33]. They are characterized by the relative concentrations of: (1) mobile ions (fixed or variable charge), (2) electron and holes, (3) immobile charged impurities (fixed or variable charge). The most widely considered MIECs (including SEs), both theoretically and experimentally, are those with a high concentration of one kind of mobile ionic defects (N_{ion}) and a low concentration of electrons (n) and holes (p): $n, p \ll N_{ion}$. A typical example is YSZ with ~ 5 anion% mobile oxygen vacancies, a low but nonnegligible hole concentration at $T \sim 1000\,°C$ for $P_{O_2} \sim 1$ atm and a low but significant electron concentration at $T \sim 1000\,°C$ for $P_{O_2} \sim 10^{25}$ atm. The electron and hole conductivities in YSZ are low at $T \sim 1000\,°C$ for intermediate P_{O_2}. In Ag_2S, on the other hand, the concentration ratio is again $n, p \ll N_{ion}$ but due to a high electron mobility, the electron conductivity is higher than the ionic one [11]. Another defect model is $p \ll n = z_{ion}N_{ion}$, where $z_{ion}q$ is the charge of the mobile ionic defect (q is the elementary charge). This is the case for example in CeO_{2-x} where $p \ll$

$n = 2[V_O^{\bullet\bullet}]$. In this case, the electron conductivity is higher than the ionic one due to a significantly higher electron mobility.

In amorphous condensed phases, the interpretation of the conduction is rather similar to that in crystalline materials if one views the amorphous phase as a crystalline solid that is highly disordered. In this view, the solid contains a high concentration of defects with at least one kind being mobile. This explains why glasses exhibit, in many cases, a higher ionic conductivity than the corresponding crystalline solid with the same composition.

Polymer ion and electron conductors are of much current interest in SSE because of their ease of production, low cost and relative high conductivity at room temperature. The conduction mechanism for ions is different from that in inorganic solids and is based either on segmental motion in the polymer and hopping of the ion (in a remote similarity to the Grotthuss mechanism for proton conduction in water) or on a LE embedded in the polymer. Electron conduction in polymers can be discussed in terms used for conduction in inorganic semiconductors. Single-phase MIECs polymer with significant conductivities do not exist yet. Quasi-MIECs are formed by blending two polymers, one conducting ions and the other conducting electrons/holes [34].

2.6.3.3 Summarizing the Differences Between Transport in LEs and SEs

The key differences between conduction in LEs and in SEs are

1. There is usually only one kind of mobile ions in SEs, as compared to at least two in LEs.
2. There is an electronic conductivity, σ_{el}, present in SEs that has to be taken into consideration, even if small ($\sigma_{el} \ll \sigma_{ion}$), whenever ambipolar motion of different charge carriers occurs, as is the case in chemical diffusion and under conditions in which the driving force for ionic motion is small. The existence of electronic conduction in addition to an ionic one yields a very important class of materials: MIECs.
3. The concentration of ionic defects in SEs must be high compared to that in LEs in order to obtain comparable ionic conductivity because the mobility of the ions in SEs is usually lower than in LEs. Even so, in most cases, the SEs will exhibit a high conductivity only at elevated temperatures.
4. The path for ion motion and the temperature dependence of ionic conductivity in SEs and LEs are very different.

2.6.4 Theoretical Current Voltage Relations for Ideal ECs (No Electrode Polarization)

2.6.4.1 General Current Density Equations

The current density equation for the ions in the bulk of the LE or SE are the same:

$$j_i = -\frac{\sigma_{ion}}{z_{ion}q}\nabla\tilde{\mu}_{ion} \qquad (1)$$

where $\tilde{\mu}_{ion}$ is the electrochemical potential of the ions, z_{ion} the charge of the ion and q is the elementary charge of a proton. For SEs the charge considered is the effective one in the Kröger–Vink notation (though absolute charges could also be discussed [35]). The convention used here is the one used in SSE referring the various quantities to a single particle rather than to a mole as done in LSE. Thus, the chemical potential is per particle, the charge is counted in elementary charges, q, rather than in Faraday units and one uses the Boltzmann constant rather than the gas constant.

For LEs one has to write at least two current-density equations, one for each mobile ionic specie. If a supporting electrolyte is added to the LE, two more current-density equations have to be added for the additional ions. The gradient in the latter two equations vanishes under steady state conditions because the electrodes are blocking for the ions of the supporting electrolyte. On the other hand, for SEs, one has to consider the current density of electrons and/or holes. For electrons:

$$j_e = \frac{\sigma_e}{q}\nabla\tilde{\mu}_e \qquad (2)$$

where $\tilde{\mu}_e$ is the electrochemical potential of the electrons. $\tilde{\mu}_e$ is equal to the redox potential of a redox couple, and it is equal to the Fermi level in solids.

Quasi-coupling between the ionic and electronic currents is sometimes introduced by adding a term $L_{ie}\nabla\tilde{\mu}_e$ to Eq. (1) and a term $L_{ei}\nabla\tilde{\mu}_i$ to Eq. (2). However, it was shown that these corrections are not really necessary (except perhaps in metals [33]) if the current density of all mobile species, as well as interactions between defects are all taken into consideration [36].

2.6.4.2 Half Cells

In LSE it is common to discuss an EC as if separated into halves, and to consider the potential at each electrode with respect to a third standard reference electrode. Furthermore, many of the ECs discussed in LSE contain supporting electrolytes. This is quite different from SSE, in which supporting electrolytes are absent. Thus, in LEs the voltage as well as the Nernst theoretical voltage can be expressed in terms of ion concentration in the LE. In SSE the analysis is done in terms of the chemical potentials of the materials and electrochemical potentials of the ions and electrons/holes involved in the cell reaction, and we shall use this approach here also. Reference electrodes are also used in SSE, but they are not universal standard ones, but rather chosen ad hoc for each particular system.

2.6.4.3 Supporting Electrolyte

Let us first examine the effect of a supporting electrolyte in a LEs and its absence in SEs. The supporting electrolyte does not participate in the electrochemical reaction, by definition. Thus, $j_{ion} = 0$ (where j is the current density) for the ions of the supporting electrolyte. Hence, by Eq. (1)

$$\nabla\tilde{\mu}_{ion} = 0, \quad j_{ion} = 0 \qquad (3)$$

An electrochemical potential, $\tilde{\mu}$, can be expressed in terms of a chemical potential, μ, and the inner (Galvani) electrical potential φ,

$$\tilde{\mu} = \mu + zq\varphi \qquad (4)$$

where zq is the charge of the relevant specie. In the limit of relative high concentrations of the supporting electrolyte, gradients in the chemical potential of its ions can be neglected even when a voltage is applied to the electrodes, and/or there exist a difference in the composition of the electrodes,

$$\nabla\mu_{ion} = 0 \qquad (5)$$

Combining Eqs. (3–5) yields:

$$\nabla\varphi = 0 \qquad (6)$$

that is, the Galvani potential is uniform, the inner electric field vanishes in the LE due to the supporting electrolyte. In this case, the other (minority) ions flow under a chemical potential gradient only.

In SSE, on the other hand, the mobile ions are those participating in the electrochemical reaction. The concentrations of the mobile ions can be high as is the case with $Ag_i^{\bullet}$ (interstitial silver ion, in the Kröger–Vink notation) in Ag_2S and with $V_O^{\bullet\bullet}$ in YSZ. Eq. (5) holds for these ions. However, their current does not vanish, in general, and Eqs. (3) and (6) do not hold. Instead, Eqs. (1) and (5) yield,

$$j_{ion} = -\sigma_{ion}\nabla\varphi \qquad (7)$$

that is, the ionic current through the SE is driven by an electric field: $-\nabla\varphi$.

In other types of MIECs, such as CeO_{2-x} ($p \ll n = z_{ion}N_{ion}$), Eq. (5) does not hold. Thus, Eq. (7) is not true anymore and one has to revert to Eq. (1) [37].

2.6.4.4 Boundary Conditions

The voltage measured on an EC is given by the difference in the electrochemical potential of the electrons at the two electrodes connected to the voltmeter,

$$-qV = \Delta\tilde{\mu}_e \qquad (8)$$

The Nernst voltage, V_{th}, is, in SSE, a theoretical quantity, having the dimension of voltage, that reflects a chemical potential difference. For a given chemical component, X, with mobile ions of charge z_{ion}, V_{th} is defined as,

$$-qV_{th} = \frac{\Delta\mu_X}{z_{ion}} \qquad (9)$$

where $\Delta\mu_X$ is the difference in the chemical potential of the component X at the two electrodes. Thus, for example, for YSZ one can consider $\Delta\mu(O) = \frac{1}{2}\Delta\mu(O_2)$ and $z_{ion} = -2$, while for Ag_2S one would consider $\Delta\mu(Ag)$ and $z_{ion} = 1$.

2.6.4.5 Relations Between Open-Circuit Voltage and the Nernst Voltage

$\Delta\mu_X$ is related to $\Delta\tilde{\mu}_{ion}$ and $\Delta\tilde{\mu}_e$. Let the mobile ion be, for example, $X_i^{\backslash\backslash}$. Then,

$$X_i^{\backslash\backslash} \longleftrightarrow X + 2e^{\backslash} \qquad (10)$$

If equilibrium prevails locally at each electrode, then,

$$\Delta\mu_X = \Delta\tilde{\mu}_{ion} + z_{ion}\Delta\tilde{\mu}_e \qquad (11)$$

with $z_{ion} = -2$ for $X_i^{\backslash\backslash}$. When the ionic current of $X_i^{\backslash\backslash}$ vanishes, then by Eq. (1), $\nabla\tilde{\mu}_{ion} = 0$ and the voltage measured on the cell is equal to the Nernst voltage (cf. Eqs. (8), (9), and (11)), despite the existence of an electronic conductivity,

$$V(j_{ion} = 0) = V_{th} \qquad (12)$$

As we shall see shortly, $V(j_i = 0)$ need not be equal to the open-circuit voltage of the cell.

When μ_X refers to a dilute concentration or an ideal gas, for example, oxygen at $P_{O_2} \leq 1$ atm, then Boltzmann statistics holds, and

$$\Delta\mu_X = \Delta\mu_X^0(T) + \frac{1}{2}k_BT\ln\left(\frac{[X_2]_1}{[X_2]_2}\right) \qquad (13)$$

where k_B is the Boltzmann constant, $[X_2]_1$ and $[X_2]_2$ the concentrations of X_2 near the two electrodes. The Nernst voltage depends then on the concentrations $[X_2]_1$ and $[X_2]_2$. From Eqs. (9) and (13),

$$-qV_{th} = \frac{RT}{2z_{ion}q}\ln\left(\frac{[X_2]_1}{[X_2]_2}\right) \qquad (14)$$

We notice that V_{th} does not depend on the concentration of ions in the SE (which for YSZ, for example, is fixed and uniform to a good approximation). μ_X can be fixed by a single phase, for example, O_2 gas with a given P_{O_2} or Ag metal or it can be fixed by a reaction in the electrode. For example, P_{O_2} may be fixed by the couple of reacting solids Fe/FeO when in equilibrium. $\mu(Na)$ can be fixed by $SnO_2/Na_2SnO_3/O_2$ under a P_{O_2} value for which SnO_2 is stable or it can be fixed by $CO_2/Na_2CO_3/O_2$ under a P_{O_2} value for which CO_2 is stable [38]. μ_X may also be fixed by an auxiliary EC [33].

The open-circuit voltage is the voltage when the total current through the EC vanishes. In SSE the total current, I_t, may contain a nonnegligible electronic contribution, I_{el} and

$$I_t = I_{ion} + I_{el} \qquad (15)$$

Under open circuit conditions, $I_t = 0$ and the partial currents $I_{ion} = -I_{el}$ need not vanish. Therefore, the voltage for $I_t = 0$ is different from the voltage for $I_{ion} = 0$. It can be shown that when a single ionic species is mobile then [37],

$$V_{oc} < V(j_{ion} = 0) = V_{th},$$

$$V_{oc} \longrightarrow V(j_{ion} = 0) \text{ only when}$$

$$I_{el} \longrightarrow 0. \qquad (16a)$$

Eq. (16a) holds whether the electron/hole transport is via the conduction/valence band or via defect band hopping.

When variable charge, mobile ions exist, for example, Cu^+ and Cu^{++}, then there are different Nernst voltages, one for each z_{ion} value (see Eq. 9). We neglect now any electronic current. Then, the open-circuit voltage is not equal to either of the Nernst voltages. For the example with $z_{ion} = 1$ and 2 [33, 39],

$$V_{th}(z_{ion} = 2) < V_{oc} < V_{th}(z_{ion} = 1) \qquad (16b)$$

In LSE the Nernst voltage is, in principle, defined similarly (with respect to a standard-reference electrode) and called the equilibrium potential. However, in practice one usually refers to the open-circuit voltage of a half cell with respect to a standard electrode. The open-circuit voltage is equal to the Nernst voltage when it is measured on purely ionic conductors, with fixed ionic charges and with negligible electronic conductivity.

The way the Nernst voltage is calculated in LSE is by considering the electrode reaction under open-circuit condition, rather as in SSE from $\Delta\mu_X$ (Eq. 9). Let us consider, for example, the cell, $Ag|AgCl|LE(H_2O,H^+,Cl^-)|Pt,H_2$, and calculate the Nernst voltage using first the LSE approach and then the SSE approach.

The voltage of the cell is, according to Eq. (8),

$$-qV = \tilde{\mu}_e(Pt) - \tilde{\mu}_e(Ag) \qquad (17)$$

$\tilde{\mu}_e(Ag)$ is fixed by the electrode reaction: $Ag + Cl^- \leftrightarrow AgCl + e^-$, which under equilibrium yields,

$$\tilde{\mu}_e(Ag) = \mu^0(Ag) + \tilde{\mu}(Cl^-)_{Ag,AgCl}$$
$$- \mu^0(AgCl) \qquad (18)$$

where $\tilde{\mu}(Cl^-)_{Ag,AgCl}$ is the value of the electrochemical potential of Cl^- near the interface Ag/AgCl and μ^0 denotes the standard chemical potential at the given temperature, T, which for solids is equal to the chemical potential. The measurement of the open-circuit voltage should be done under steady state. Then the Cl^- current must vanish in the two, purely ionic conductors AgCl and $LE(H_2O,H^+,Cl^-)$ placed in series. Thus, $\tilde{\mu}(Cl^-)$ is uniform throughout the cell and we may remove the subscript Ag/AgCl and we may add the subscript LE.

$\tilde{\mu}_e(Pt)$ is fixed by the electrode reaction $H_2 \leftrightarrow 2H+ + 2e^-$, which, under equilibrium, yields,

$$\tilde{\mu}_e(Pt) = \tfrac{1}{2}\mu(H_2) - \tilde{\mu}(H^+)_{Pt,LE} \qquad (19)$$

where $\tilde{\mu}(H^+)_{Pt,LE}$ is the electrochemical potential of the protons near the $H_2,Pt/LE$ interface. $\tilde{\mu}(H^+)$ turns out to be uniform in the LE, since both $\tilde{\mu}(Cl^-)$ and $\mu(HCl)$

are uniform in the LE, thus the subscript Pt,LE can be replaced by LE. Combining Eqs. (17), (18) and (19) yields,

$$-qV_{oc} = \tfrac{1}{2}\mu(H_2) - \mu^0(Ag) + \mu^0(AgCl)$$
$$- \tilde{\mu}(H^+)_{LE} - \tilde{\mu}(Cl^-)_{LE} \quad (20)$$

Under equilibrium one can make the following substitution:

$$\tilde{\mu}(H^+)_{LE} + \tilde{\mu}(Cl^-)_{LE} = \mu(HCl)_{LE}$$
$$= \mu(H^+)_{LE} + \mu(Cl^-)_{LE} \quad (21)$$

in Eq. (20), then,

$$-qV_{oc} = \tfrac{1}{2}\mu(H_2) - \mu^0(Ag) + \mu^0(AgCl)$$
$$- \mu(H^+)_{LE} - \mu(Cl^-)_{LE} \quad (22)$$

or when expressed in terms of activities,

$$-qV_{oc} = \tfrac{1}{2}\mu^0(H_2) - \mu^0(Ag) + \mu^0(AgCl)$$
$$- \mu^0(H^+)_{LE} - \mu^0(Cl^-)_{LE}$$
$$+ k_B T \ln \frac{a(H_2)^{1/2}}{a(H^+)_{LE} a(Cl^-)_{LE}} \quad (23)$$

For low ion concentrations $a(H^+)_{LE}$ and $a(Cl^-)_{LE}$ may be replaced by the corresponding (uniform) concentration $[H^+]$ and $[Cl^-]$ in the LE.

The argumentation in the SSE approach would be as follows: There is a selective membrane for Cl^- ions, that is, an electrolyte, made of two parts connected in series: AgCl and the LE. The Nernst voltage is determined by $\Delta\mu(Cl_2)$ according to Eq. (9),

$$2qV_{th} = \mu(Cl_2)_{Pt,LE} - \mu(Cl_2)_{Ag,AgCl} \quad (24)$$

$\mu(Cl_2)_{Ag,AgCl}$, the chemical potential of Cl_2 at the Ag/AgCl interface is governed by the reaction: $2Ag + Cl_2 \leftrightarrow 2AgCl$. Under equilibrium,

$$\tfrac{1}{2}\mu(Cl_2)_{Ag,AgCl} = \mu^0(AgCl) - \mu^0(Ag) \quad (25)$$

$\mu(Cl_2)_{Pt,LE}$, the chemical potential of Cl_2 at the LE/Pt,H_2 interface, is fixed by the reaction of H_2 and HCl in the LE: $H_2 + Cl_2 \leftrightarrow 2HCl \leftrightarrow 2H^+ + 2Cl^-$. Under equilibrium,

$$\tfrac{1}{2}\mu(Cl_2)_{Pt,LE} = \mu(H^+)_{Pt,LE} + \mu(Cl^-)_{Pt,LE}$$
$$- \tfrac{1}{2}\mu(H_2) \quad (26)$$

Combining Eqs. (25) and (26) and noticing that $\mu(H^+)$ and $\mu(Cl^-)$ are uniform in the LE, as mentioned before, yields,

$$-qV_{th} = \tfrac{1}{2}\mu(H_2) - \mu^0(Ag) + \mu^0(AgCl)$$
$$- \mu(H^+)_{LE} - \mu(Cl^-)_{LE} \quad (27)$$

V_{th} equals V_{oc} of Eq. (22) since the electronic component in the combined electrolyte AgCl/LE vanishes (see Eq. (16a)). This demonstrates that the two definitions of the Nernst voltage are indeed equal when the membrane is a pure ionic conductor. This would not be so if electronic conduction would be present (as can be the case in SSE), as then $V_{th} \neq V_{oc}$ (Eq. 16a). Different values for the Nernst voltage defined as V_{oc} rather than V_{th} would be obtained also when the LE contains variable charge-mobile ions, for example, Cu^+ and Cu^{++}, as then $V_{th} \neq V_{oc}$ as can be seen in Eq. (16b).

2.6.4.6 Chemical Diffusion Coefficient

Chemical diffusion of a single component occurs, in ionic solids, by the motion of ions. This must be accompanied by an equal current of electronic charge to avoid charge accumulation. Thus diffusion in ionic solids with a single type of mobile ions is an ambipolar motion of ions and electrons (or holes). This motion is governed by Eqs. (1) and (2) and the condition $I_t = 0$. The chemical diffusion coefficient is then [12, 40],

$$D_{\text{chem}} = \frac{k_B T}{z_{\text{ion}} q} \frac{\sigma_{\text{ion}} \sigma_{\text{el}}}{\sigma_{\text{ion}} + \sigma_{\text{el}}} \frac{1}{c_X} \frac{\partial \ln(a_X)}{\partial \ln(c_X)} \quad (28)$$

where c_X is the total, local, concentration of material X (or defects), which, in ionic solids, may be equal to the concentration of the mobile ions (or defects), a_X is the local activity of the component X. D_{chem} need not be uniform. It is dominated by the smaller of the conductivities with $D_{\text{chem}} \propto \sigma_{\text{el}}$ for $\sigma_{\text{el}} \ll \sigma_{\text{ion}}$ and $D_{\text{chem}} \propto \sigma_{\text{ion}}$ for $\sigma_{\text{ion}} \ll \sigma_{\text{el}}$.

2.6.4.7 I–V Relations

The ionic and electronic currents in the bulk of a MIEC are governed by the current-density equations of the kind of Eqs. (1) and (2), the boundary values of μ_X, of the mobile species, at the two electrodes and the voltage on the EC, Eq. (8). The $I-V$ relations are actually two relations: I_{ion} vs μ_{X1}, μ_{X2}, and V, and I_{el} vs μ_{X1}, μ_{X2}, and V. These relations are not universal but depend on the defect model of the MIEC.

For an EC based on an MIEC with high and uniform concentrations of mobile ions ($n, p \ll N_{\text{ion}}$) and thus a uniform ionic conductivity,

$$I_{\text{ion}} = \frac{V_{\text{th}} - V}{R_{\text{ion}}} \quad (29)$$

where R_{ion} is the resistance of the MIEC to ionic current. The inner electric field, $-\nabla \varphi$ is uniform, in the steady state, as can be deduced from Eq. (7) and the uniformity of j_{ion} and σ_{ion}.

The electronic current, I_{el}, is not linear in V. The electron and hole concentrations are not uniform and are dependent on the boundary values, μ_{X1}, μ_{X2}, and V. In particular, for MIECs with one type of minority electronic carriers, for instance, electrons, ($p \ll n \ll N_{\text{ion}}$) [33, 37],

$$I_e = -S\sigma_e^1 \frac{V_{\text{th}} - V}{L}$$
$$\times \frac{e^{-q(V_{\text{th}}-V)/k_B T} - e^{-qV_{\text{th}}/k_B T}}{1 - e^{-q(V_{\text{th}}-V)/k_B T}} \quad (30)$$

where σ_e^1 is the conductivity in the MIEC at the end, where $V \equiv 0$, L is the length and S the cross-sectional area of the MIEC. In SSE $1/k_B T$ is normally denoted as β. We shall not use this notation here to avoid confusion with the notation β of the symmetry factor in LSE.

A different defect model for a wide group of MIECs is one with comparable concentration of electrons (or holes) and mobile ions, ($p \ll n = z_{\text{ion}} N_{\text{ion}}$). The inner electric field $-\nabla \varphi$ in this case is not uniform and the $I-V$ relations are different from the previous ones [33, 37],

$$I_{\text{ion}} = 2S v_{\text{ion}} (n_1 - n_2) \frac{k_B T}{V_{\text{th}}} \frac{V_{\text{th}} - V}{L} \quad (31)$$

and

$$I_e = -2S v_e (n_1 - n_2) \frac{k_B T}{V_{\text{th}}} \frac{V}{L} \quad (32)$$

where v is the mobility, n_1 and n_2 the concentration of electrons (and of the mobile-ionic defects) at the two ends of the MIEC. When the electrodes are reversible, n_1 and n_2 are then fixed by $\mu_{X,1}$ and $\mu_{X,2}$ also under current, I_{ion} is linear in V and I_e is proportional to V. On the other hand, when the electrodes block the ionic current, (e.g. when full blocking occurs $I_{\text{ion}} = 0$, hence $V = V_{\text{th}}$) n_1 and/or n_2 depend then on V and the $I_e - V$ relations are not linear.

2.6.5 Methods for Characterizing ECs

2.6.5.1 Methods for Characterizing LEs and SEs/MIECs

The key parameters of interest in LEs and SEs/MIECs in ECs for the purpose of discussing the $I-V$ relations are the nature of the charge carriers, their concentrations and mobilities.

In LSE the concentrations of the mobile ions are usually known and fixed by the composition of the LE. This is certainly true for strong electrolytes where the concentration of the mobile ions is equal to the concentration of the corresponding material added to the solution, times the stoichiometry coefficient.

In SSE on the other hand, the nature and concentration of mobile defects can be guessed a priori with reasonable accuracy only when they are introduced by doping, provided the dopants are fully ionized. When the mobile defects are generated by thermal excitation, for example, Frenkel pairs, or by deviation form stoichiometry, they have then to be identified and their concentration measured. The experimental methods used can be direct ones, that is, spectroscopic ones, such as characteristic ultra-violet (UV), visible (VIS), and infrared (IR) absorption; electron spin resonance (ESR); and nuclear magnetic resonance (NMR). The methods can be indirect ones such as conductivity measurement, which is linear in the defect concentration (for low concentration), thermogravimetry (TGA) that follows the changes in the stoichiometry by weight changes, and by coulometric titration that follows the stoichiometry changes by electrical measurements and can be done more accurately than TGA. Other methods used, follow changes in stoichiometry such as pressure measurements, IR measurements and chemical-titration measurements of the gases evolving from the sample [41]. The indirect methods require a model to relate the measured quantities to the defect nature and concentrations. In particular, in order to decide whether the defects are charged or neutral, it is useful to combine coulometric-titration measurements with in situ conductivity measurements [28, 29]. Hall coefficient measurements to determine the charge-carrier concentration are difficult because of the low mobilities found in ionic solids both for ionic and electronic charge carriers. Only few Hall measurements were done on ionic solids to determine the electron/hole concentrations and we are aware of only work done by Funke that measured the ion Hall coefficient in an SE [42, 43]. The mobility can be determined from the electrical conductivity once the defect concentration has been determined, or directly from a drift experiment.

In SSE one has to determine both the ionic conductivity, σ_{ion}, and the electronic conductivity, σ_{el}. For that purpose, one can use blocking electrodes that are permeable to either ions or electrons, and measure the $I-V$ relation when a steady state is reached. This is the basis for the Hebb–Wagner polarization method for determining σ_{el} [33, 44–46]. In the Hebb–Wagner method, the ion-blocking electrode is an inert conductor, for example, graphite or platinum. The blocking occurs either because a continuous-inert electrode can block the access of material to the SE or because the gas phase contains a limited amount of the relevant material. The counter electrode in the Hebb–Wagner experiment is a reversible one. It fixes the composition in the MIEC being tested, near the electrode/MIEC contact. This allows to relate the measured local σ_{el} to a well-defined composition of the MIEC. An example of such a cell is: $(-)$Ag|AgBr|C$(+)$, which allows the determination of the distribution of σ_{el} throughout the AgBr. The voltage applied has to be kept below the decomposition voltage. When this voltage is exceeded the

MIEC decomposes, in analogy to electrolysis of an LE in LSE.

The partial ionic or electronic currents and conductivities can also be measured by setting the driving force for the other partial current to zero. If one examines the Hebb–Wagner method, it becomes apparent that this nullification also occurs there. Because of the extreme resistance of the ion-blocking electrode to ionic current, the driving force for ionic motion in the MIEC vanishes ($\nabla\tilde{\mu}_{ion} = 0$).

A simple way to make the driving force for electrons vanish ($\nabla\tilde{\mu}_e = 0$) is to short circuit the electrodes [47]. This sets the difference $\Delta\tilde{\mu}_e = 0$ and for most common MIECs, it also sets $\nabla\tilde{\mu}_e = 0$.

Another method for separating σ_{ion} from σ_{el} is by their different dependence on the chemical potential, μ_X, applied to the electrodes. This method works well when $\sigma_{ion} \gg \sigma_{el}$ or vice versa, and allows to identify which conductivity is being measured [22].

The chemical diffusion coefficient, D_{chem}, can be determined using time-dependent electrochemical methods. The best known is the galvanostatic-intermittent-titration-technique (GITT) [40]. In this measurement, ions are pumped through an SE into (or out of) an MIEC for a short time. The pumping is stopped and the relaxation of the cell open-circuit voltage is followed. The time dependence of the cell voltage during pumping and during relaxation, both yield D_{chem}. To overcome contact overpotential a modified arrangement was suggested in which the chemical potential of the diffusing specie is monitored as it arrives at the remote end [48].

Through the dependence of D_{chem} on σ_{ion} and σ_{el} (Eq. 28) D_{chem} can be used to determine the minority conductivity provided $c_X^{-1}\partial \ln(a_X)\partial \ln(c_X)$ is measured in addition (using coulometric titration in combination with EMF measurements [33]).

2.6.5.2 Methods for Characterizing Elements in the EC

Working ECs are commonly characterized using ac-impedance measurements. This allows to identify the different contribution to the impedance by the various elements of the cell [49–51]. This method, first developed for LSE, is also extensively used in SSE. As a result of the different geometry in LSE and SSE the processes and their sequence are different. Yet the analysis can use the same electrical elements, namely, resistors, capacitors, and Warburg elements (for simulating diffusion). In both cases non Faradaic currents may exist, which originate from charging double layers without a chemical reaction taking place.

Voltammetry is commonly used in LSE not so in SSE probably because of the difficulty in the interpretation of the results.

2.6.6
Electrode Processes and Overpotentials

2.6.6.1 General

The details of electrode processes in LSE and SSE are different, as mentioned before. To examine the electrode processes in SSE let us examine a few examples. The EC, (+)Ag|AgI|Ag$_2$S(−) represents the EC shown in Fig. 2(a). AgI is an SE that conducts Ag$^+$ ions (at $T > 149\,°C$ [11]). The electrodes Ag and Ag$_2$S are a source or sink for both the material (silver) and for the electrons. In the oxidation reaction of Ag at the Ag/AgI interface, silver atoms or ions have to leave the silver metal and enter the AgI lattice as ions. There is charge transfer of ions across the interface and there may also be electron transfer to form the ions there. At the AgI/Ag$_2$S interface, Ag$^+$ ions are transferred from the AgI lattice into the Ag$_2$S lattice.

For porous gas electrodes, shown schematically in Fig. 2(b), there can be many different steps involved. For example, the overall cathodic reaction on a Pt electrode on YSZ is: $O_2(gas) + 4e^-(Pt) \leftrightarrow 2O^{2-}(YSZ)$. One can suggest that the overall reaction consists of the following series of elementary steps: diffusion of O$_2$ in the gas phase toward the TPB, O$_2$ adsorption on the Pt, O$_2$ dissociation, electron transfer to O_{ad} to form O_{ad}^{2-}, diffusion of O_{ad}^{2-} along the Pt electrode surface toward the SE (YSZ), and then entering of the ion into the YSZ. There can be other series of steps that yield the same overall reaction [52].

A different electrode is one that is not porous but a continuous MIEC and the source or sink for material is not the MIEC but the gas phase, as shown schematically in Fig. 2(d), Then diffusion of material in the form of ions through the MIEC, as well as charge transfer of ions at the MIEC/SE interface has also to be considered.

It is evident from those examples that electrode processes in SSE involve elementary steps that may be limited by diffusion, others by slow-charge transfer of electrons in an electronation or de-electronation step, and yet others by a slow transfer of ions across an interface. The latter step is not common in LSE. On the other hand convection exits in LSE but not in SSE.

2.6.6.2 Inner (Galvani) Electrical Potential

2.6.6.2.1 Inner Electrical Potential in LSE

The analysis of elementary steps that involve charge transfer, whether of electrons or of ions, require a detailed discussion of the inner (Galvani) electrical potential, φ. The distribution $\varphi(\vec{r})$ and the concentration of the charges that fix $\varphi(\vec{r})$ therefore deserve an over view. In LSE the LE is terminated abruptly at the contact with the electrode (and at the upper side in contact with the atmosphere). Charge is transferred, also under equilibrium, from the metal electrode into the LE (or vice versa) leaving behind the opposite charge and thus forming a double layer. In the metal the excess or missing charge (in the "sea" of electrons) concentrates near the M/LE interface, in a very thin layer, less than a mono atomic–layer wide. In the LE the charge is spread out a little more over a length determined by the Debye length, λ_D (which is the inverse of the Debye reciprocal length, κ commonly used in LSE). λ_D is about 0.3 to 10 nm wide for most LEs due to the high concentration of the mobile ions.

In the LE the local dielectric constant at the interface may be significantly modified as a result of orientation of the water-molecule dipole in the electric field of the double layer. Furthermore, ions can flow to the interface and be adsorbed there, in the double-layer region.

The explanation for the charge transfer on contact of the metal and the LE, under equilibrium, depends on the approach taken. Of course, the different explanations are equivalent. From the SSE point of view, the charge in the double layer (for zero applied voltage) originates from a difference in two properties between the electrode material and the LE, the work function and the chemical potential of the mobile component. For inert metals it is only the difference in the electron work functions that matters. Electrons are transferred until the electrochemical potential of the electrons in the metal and the nearby LE is the same. Then the potential of zero charge, used in LSE, equals the difference in the work functions (divided by the elementary charge, q). When the electrode can be dissolved (e.g. a Zn electrode) then material and electrons are transferred, until the chemical potential of the material as well as the electrochemical potentials of the corresponding ions and of the electrons all are uniform across the interface.

At the free surface of the LE facing the gas phase a double layer can also be formed, in analogy to the free surface of a solid, discussed in the following text. Our main concern is the interface between the LE and the metal electrons.

2.6.6.2.2 Inner Electrical Potential in SSE

In SSE the situation is somewhat more complex because of the rigid structure of the solid. The Galvani potential distribution is shown in Fig. 3. Let us start with the free surface of a solid. A semiconductor may have surface states for the electrons because of the breaking of the translational symmetry at the surface. Electrons or holes trapped at these states cause a double layer between the trapped charges and the opposite charge left behind in the bulk of the semiconductor and distributed nearby [53]. Due to quantum effects, electrons can be found outside the free surface. This also contributes to a double layer.

In SEs and MIECs of interest the contributions to surface charges come mainly from mobile ions that exist in large concentrations rather than from electrons or holes, when they are the minority charge carriers. These ions can accumulate or be depleted near the surface forming a charged layer there (the "core" in Fig. 3a) [54, 55]. The position of the "core" may vary to be in between the first and second atomic layer, within the first atomic layer or on the first atomic layer. The opposite ionic charge left behind in the bulk is distributed nearby. For a core centered at x_c (in Fig. 3), the opposite

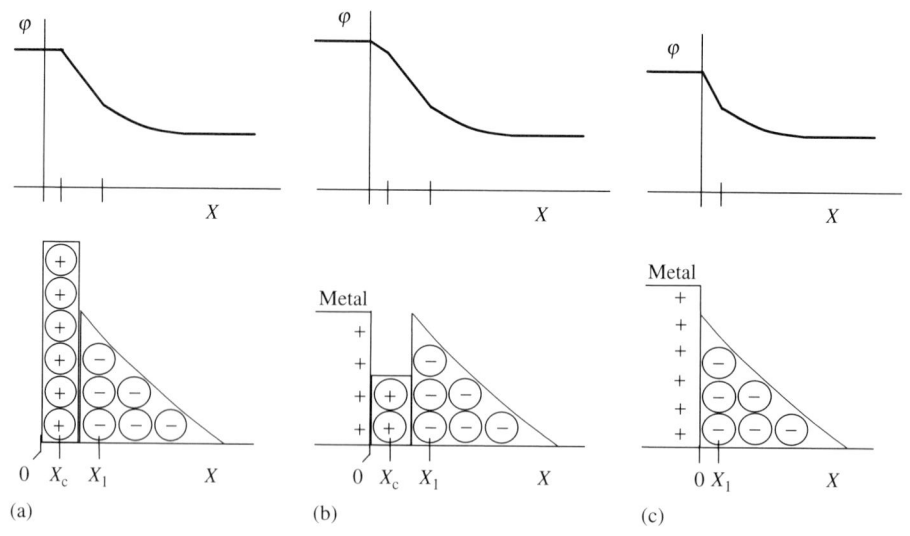

Fig. 3 Distribution of charge and Galvani potential, φ, at a solid interface. x_c - the position of the core. In (b) and (c) the metal is left of $x = 0$. (a) Free surface; (b) metal/MIEC interface that contain a core and (c) metal/MIEC interface without a core.

charge is distributed beyond x_1 to the right due to the finite size of the ions. In the double layer formed one layer is thin, of a mono-atomic width and the opposite charge is spread out over a few atomic distances.

When an inert metal electrode is applied on the SE or MIEC the difference in the work functions between the metal and the SE/MIEC, contributes an additional charge transfer that charges the metal surface and changes the charges in the SE/MIEC (i.e. in the core and the space charge). This modifies the electric potential distribution as shown schematically in Fig. 3(b). The charge in the metal, which is compensated in the SE/MIEC by a change in distribution of, mainly, majority ionic species, is electronic. The potential distribution is different and determined by redistribution of the charged species. In particular an electric field appears in the region between the surface of the metal and the core (between $x = 0$ and x_c in Fig. 3b). This region has a width of the order of 0.1 nm. The field for $x > x_c$ is also modified. When the electrode applied is not inert (e.g. an Ag electrode on AgI), material transfer as well as electron transfer occurs. This modifies the concentration of ionic defects in the core and in the space charge in the SE/MIEC. The distribution of φ looks, however, qualitatively as shown in Fig. 3(b).

A simplified situation exists when one can neglect the core and the corresponding space charge that exists in the free surface of a solid (shown in Fig. 3a). Then a space charge is formed only due to the contact between the SE/MIEC and the metal electrode. This situation is shown in Fig. 3(c) for an inert metallic electrode. This last case is similar to the situation in the metal-LE double layer in LSE mentioned before, when no ions are adsorbed in the double-layer region. When ions do exist in this region in LSE then Fig. 3(b) is the appropriate analogous presentation and the core is the analogue of the adsorbed-ions layer.

When applying a voltage on the EC a part η of this voltage appears across the interface and modifies the Galvani potential drop there. The modification is generated by a change of the double-layer charge.

2.6.6.3 I–V Relations for Charge-Transfer Processes in LSE

2.6.6.3.1 The Butler–Volmer Equation for an Elementary Step
Charge-transfer I–V relations are given in LSE by the Butler–Volmer type equation. For an elementary step in which a species with charge zq is transferred,

$$I = I_0(e^{(1-\beta)zq\delta(\Delta\varphi)/k_B T} - e^{-\beta zq\delta(\Delta\varphi)/k_B T}) \quad (33a)$$

where $\Delta\varphi$ is the difference in the Galvani potential across the interface, $\delta(\Delta\varphi)$ is the deviation of $\Delta\varphi$ from the equilibrium value $\Delta\varphi_{eq}$

$$\delta(\Delta\varphi) = \Delta\varphi - \Delta\varphi_{eq} \quad (33b)$$

z is unity for an electron or an ion of unit charge, $z > 1$ for an ion with a larger charge. In Eq. (33a) the concentration of the species involved are assumed not to be altered under current. I_0 depends on the concentration of these species [56].

In these I–V relations β is not necessarily an integer. This is due to the form of the potential barrier the species have to overcome. A potential barrier may exhibit a maximum as shown in Fig. 4(a), or be in the form of just one slope as shown in Fig. 4(b), which might represent a step potential or be one side of a minimum in the potential shown in Fig. 4(c).

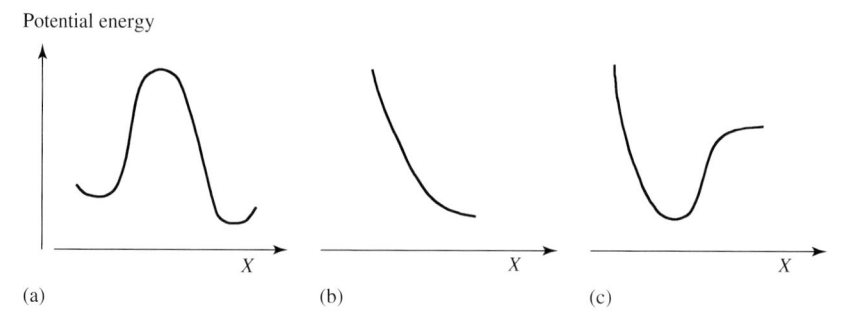

Fig. 4 Potential energy barriers of different forms.

Potential barriers of type (1) may have a contribution from the electrical potential energy as shown in Figs. 5(b) and (c). This then leads to expressions of the form of Eq. (33a) with $0 < \beta < 1$. The parameter β is associated with the symmetry of the potential maximum, therefore β is denoted as the symmetry factor.

Let us first see why a potential barrier of the form of Fig. 4(a) arises in LSE. For ion transfer between the solution and the metal electrode, say a solvated proton, the position, sufficiently removed from the metal electrode, where it is fully surrounded with water molecules, is of relative low energy. The position in which the proton is attached to the metallic electrode with most of the water molecule surrounding it removed, is also of relative low energy due to the interaction of the proton with the metal. The positions in between these two ones are of higher energy forming a maximum in the potential energy. The barrier may have a contribution of electrical potential energy, which is linear in position x. The subdivision into different energy contributions looks qualitatively then, as shown in Figs. 5(b) and (c).

The transfer of a proton from left to right over the barrier in Figs. 4(b), 5(b) and (c) requires thermal activation and the transfer from right to left also requires thermal activation but not necessarily the same energy. Under equilibrium the two transfers (which depend on the barrier heights, local concentrations of reactants and products and reaction constants) exhibit the same rate. The electrical potential contribution is, under equilibrium, $\Delta\varphi_{eq}$ (Fig. 5b). Under current a deviation $\delta(\Delta\varphi)$ from the equilibrium value $\Delta\varphi_{eq}$ is generated. Part $-\beta\delta(\Delta\varphi)$ modifies one side barrier height and a part $(1-\beta)\delta(\Delta\varphi)$ modifies the other side barrier height (Fig. 5c).

When the charge transferred is an electron, Eq. (33a) still holds. An example for such a reaction is: $H_{LE}^+ + e_M^- \leftrightarrow H_{ad,M}$, where: H_{LE}^+ is a proton in the LE, e_M^- an electron in the metal electrode, and $H_{ad,M}$ an hydrogen atom adsorbed on the metal electrode. It is not obvious that this equation holds for an electron transfer as an electron can easily tunnel through a potential barrier that is a few Å wide and only ~ 1 eV high, as is typically the case at the Metal/LE interface [57, 58]. However, the barrier that the electron has to cross is not the only factor to be considered. For tunneling to take place the initial and final states of the electron have to be degenerate, that is, have the same energy. Furthermore, the state the electron is transferred to must be empty

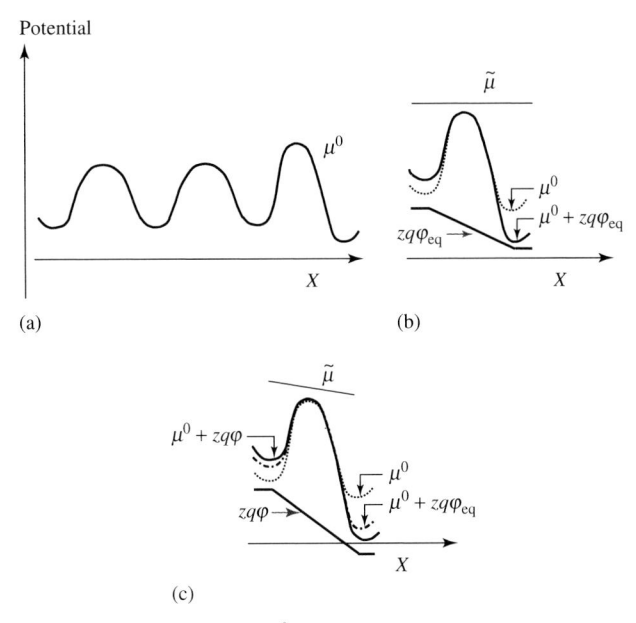

2.6 Comparison Between Liquid State and Solid State Electrochemistry

Fig. 5 Potential energies (μ^0, $\tilde{\mu}$, $zq\varphi$) in SSE. (a) Inside the SE (or MIEC); (b) at the SE/electrode interface under equilibrium and (c) at the SE/electrode interface under current.

to be able to receive the electron. It turns out that the position of the proton and hydrogen atom with the right energy for electron tunneling lies at a relative high energy and the motion toward that position forms a potential barrier similar to the one shown in Figs. 5(b) and (c). This results in Eq. (33a) also for electron transfer [56].

Charge transfer to a neutral species that approaches the metal electrode from beyond the outer Helmholtz plane, for example, O_2 dissolved in the LE, may also follow the Butler–Volmer Eq. (33a). This is expected to be different from the I–V relation for an electron transfer to O_2 already adsorbed onto the metal (see in the following text).

2.6.6.3.2 The Butler–Volmer Equation for Multistep Reactions
An over all process at an interface consists usually of a series of elementary steps some transferring charge and some not. The products of one step are the reactants for the following step. When the charge transfer occurs across exactly the same potential difference, $\Delta\varphi$ then the I–V relations can be related through I_0 to the concentrations of the species that exist in the first and the last step in the series. Taking one elementary step to be rate determining (rds) [56],

$$I = I_0(e^{\alpha_a q \delta(\Delta\varphi)/k_B T} - e^{-\alpha_c q \delta(\Delta\varphi)/k_B T}) \quad (34a)$$

where,

$$\alpha_c = \frac{\gamma_c}{\nu} + r\beta,$$

$$\alpha_a = \frac{n_t - \gamma_c}{\nu} + r\beta \quad (34b)$$

where, n_t is the total number of elementary charges transferred, γ_c is the number of elementary charges transferred before the rds, r is the number of elementary charges transferred in each rds, ν is the number of times the rds has to be repeated to complete one reaction and β is the symmetry factor of the rds. Thus, for example, $r = 1$ for an electron being transferred; $r = 2$ for an O^{2-} ion being transferred; and $r = 0$ if the rds is a chemical reaction without charge transfer.

2.6.6.4 I–V Relations for Charge-Transfer Processes in SSE

2.6.6.4.1 General
Equation (33a) for an elementary charge-transfer step has been applied in SSE [51, 59–61] as well as Eqs. (34a) and (34b) for a multistep reaction that contains charge transfer [62–65]. This application to SSE is justified for ion transfer across an interface. However, it may be problematic in two cases: (1) The transfer of an electron may not be associated with a symmetry factor $0 < \beta < 1$, but rather with $\beta = 0$ or 1. (2) The more complicated morphology may not allow the application of Eq. (34a) to a multistep reaction because the charge transfer elementary steps do not occur across exactly the same interface.

2.6.6.4.2 Ion Transfer
Let us first consider an ion being transferred from one solid into the other. The transfer is driven by a difference in the electrochemical potential of the ions across the interface. Except for a possible difference in the electric field, φ, a difference in the chemical potential of the ions may also exist. The chemical potential, μ, is composed of the standard one, μ^0, and the activity term:

$$\mu = \mu^0 + kT \ln(a) \quad (35)$$

For macroscopic considerations the standard chemical potential can be represented by the average over a volume of atomic size. However, for charge transfer at an interface the detailed distribution of the energies on an atomic scale must be considered, as these exhibit potential barriers that the ion has to overcome [66].

To be specific let us consider the case of dilute defect concentration, [i], so that the chemical potential is:

$$\mu = \mu^0 + kT \ln[i] \quad (36)$$

where $kT \ln[i]$ is contributed by the (configurational) entropy and $\mu^0 = u + Pv - Ts^0$ (where s^0 is the standard entropy per particle) is equal the enthalpy per particle up to the constant $-Ts^0$, u is the energy per particle, P the pressure and v the volume per particle. On the atomic scale, μ^0 varies mainly due to variations in $u(x)$. Variation of μ^0 vs. x are shown schematically in Fig. 5(a). The changes within a crystalline solid are periodic, on an atomic scale. The periodicity ends at the interface. A potential energy maximum occurs at the boundary as follows: the energy of the ion exhibits a local minimum at the normal site inside the solid and also at an interstitial site, the ion is at a local energy minimum outside as an adsorbed ion, and the energy exhibits a maximum as the ion moves between such two minima. The energy exhibits a maximum as the ion enters the solid, since it has to squeeze in between the ions of the solid in the outermost atomic layer before it reaches a normal or interstitial site inside. The concentration [i] is not uniform either. Under equilibrium conditions, $\tilde{\mu} = \mu + zq\varphi$ is uniform with variations in $kT \ln[i]$ compensating for a nonuniformity in μ^0 and $zq\varphi$ (see Fig. 5b). Figure 5(c) shows $\mu^0(x) + zq\varphi(x)$, $zq\varphi(x)$ and $\tilde{\mu}(x)$ under current with $\nabla \tilde{\mu} \neq 0$. The moving ion has to overcome a potential barrier of the type shown in Fig. 5(c), and therefore, the $I-V$ relations have to follow Eq. (33a).

2.6.6.4.3 Electron Transfer

For electrons the distributions of $\mu^0(x)$, $\mu^0(x) - zq\varphi(x)$, and $\tilde{\mu}(x)$ are qualitatively similar to those for ions as presented schematically in Fig. 5. However, the electron can tunnel through the potential barrier if the initial and final states are degenerate. In LSE a potential of the shape of Figs. 5(b) and (c) arose due to the motion of the ions and atoms involved in the electronation and de-electronation process. In SSE, however, the motion of the ions is much more restricted. One expects, for instance, that an ion/atom adsorbed on a metal electrode or an ion/atom at the interface between the two solids (e.g. SE/metal electrode) will only oscillate about a minimum. Thus the electron transfer is over a short distance of ~ 1 Å, while the adsorbed ion/atom/molecule hardly moves.

Tunneling to a degenerate state, in SSE, requires thermal excitation only in one tunneling direction. To see this let us consider Fig. 6, which exhibits the conduction band of a metal, the Fermi level, ε_F, the upper discrete energy level, ε_a, in an adsorbed atom/ion occupied with an electron that is to be transferred to the metal and the energy potential barrier, E_p, the electron has to tunnel through. In the case shown in Fig. 6(a), the energy of the electron has to be raised at least by the amount $\varepsilon_F - \varepsilon_a$ to reach degeneracy with an empty state in the conduction band of the metal to allow tunneling. On the other hand, in the case shown in Fig. 6(b), the electron can tunnel into a degenerate, empty state in the conduction band of the metal without thermal excitation. It is expected that the potential barrier, E_p, and the difference in energy of the electron in the metal and in the atom/ion, depend on a gradient in the electrical potential (not shown). The $I-V$ relations for this elementary step are,

$$I = I_0 f(\delta(\Delta\varphi))$$
$$\times \left[\frac{1-p_r}{1-p_{r,eq}} - \frac{p_r}{p_{r,eq}} e^{-q\delta(\Delta\varphi)/k_B T}\right] \text{ or }$$

$$I = I_0 f(\delta(\Delta\varphi))$$
$$\times \left[\frac{1-p_r}{1-p_{r,eq}} e^{q\delta(\Delta\varphi)/k_B T} - \frac{p_r}{p_{r,eq}}\right]$$
(37a)

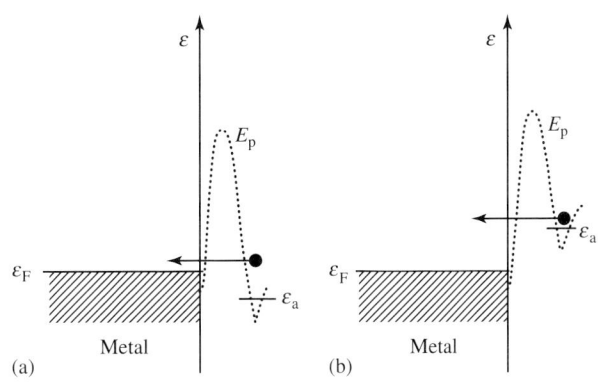

Fig. 6 Tunneling of an electron through a potential energy barrier into a metal. (a) Thermally assisted tunneling and (b) tunneling not requiring thermal excitation.

where $f(\delta(\Delta\varphi))$ yields the dependence of the tunneling current on the distortion in the potential barrier by $\delta(\Delta\varphi)$, p_r is the probability that the acceptor level is occupied by an electron and $p_{r,eq}$ is p_r under equilibrium. The relevant expression depends on the transfer direction being thermally activated relative to the polarity of the applied voltage. Eq. (37a) also holds when the electron is transferred from the metal to the atom/ion. The temperature dependence is given by the exponential term $\exp(\pm q\delta(\Delta\varphi)/k_B T)$. Neglecting the effect of $\delta(\Delta\varphi)$ on f and p_r, Eq. (37a) reduces to,

$$I = I_0(1 - e^{-q\delta(\Delta\varphi)/k_B T}) \text{ or}$$
$$I = I_0(e^{q\delta(\Delta\varphi)/k_B T} - 1) \quad (37b)$$

Eq. (37a) is different from Eq. (33a) as one term in the brackets does not depend on $\delta(\Delta\varphi)$.

2.6.6.4.4 Further Discussion of Ion and Electron Transfer

It is questionable if the rate of an electron transfer in SSE, from a metal to an atom/ion (or vice versa), can be slower than an ion transfer at the electrode. The electron transfer is by tunneling. For a given atom/ion as the acceptor (or donor), four factors govern the tunneling rate. First, availability of the electrons (or of an empty site for the reverse transfer). Electrons are ample in a metal. Then comes the distance for tunneling, the energy barrier height, and the energy $\Delta\varepsilon$ required to achieve degeneracy. Let us first compare the electron tunneling process in SSE with the one in LSE, in which the electron tunneling process is different and can be rate determining. In LSE the ion is solvated and the cloud of solvating molecules keeps the ion quite far from the metal electrode. In SSE, both the ion and the atom are adsorbed onto the metal electrode. This is true whether it is the free metal/gas interface or the metal/SE interface. Thus the tunneling distance for an electron is significantly smaller in SSE (except when in LSE the electron transfer is also to a species directly adsorbed onto the metal electrode. However, the concentration of uncharged adsorbed acceptors is expected to be relatively low). The energy barrier heights are expected to be similar in SSE and LSE; however, $\Delta\varepsilon$, the energy required to achieve degeneracy is lower than the energy barrier height. We therefore expect that electron tunneling at the electrode in SSE is faster than in LSE.

For comparing the transfer rate of ions and electrons in SSE across a potential barrier, let us consider an example of oxygen transfer from the gas phase into YSZ, first adsorbing onto a metal electrode with electron transfer and diffusion to the TPB occurring there:

$$\begin{aligned} O_2 &\longrightarrow 2O_{ad} \longrightarrow 2O^-_{ad} \\ &\longrightarrow 2O^{2-}_{ad} \longrightarrow 2O^{2-}_{ad,TPB} \\ &\longrightarrow 2O^{2-}(YSZ) \end{aligned} \quad (38)$$

We notice (see Fig. 6) that electron transfer in one direction does not require thermal excitation at all and in the opposite direction it needs only to be excited by $\Delta\varepsilon$, a fraction of the energy potential barrier E_p, to reach a degenerate state. $\Delta\varepsilon$ is expected to be also lower than the potential barrier that ions have to cross, since the later should be similar to the total barrier height in E_p of the electrons. The high concentration of electrons in the metal, the short tunneling distance of ~0.1 nm, the barrier height of ~1 eV, and the relative low value of $\Delta\varepsilon$ should result in a high value of the exchange current I_0 for electron transfer and I_0 should be higher than that for ion transfer. Thus if charge transfer is found experimentally to be the rate-determining step in SSE then it is, most likely, an ion transfer that is rate-determining not an electron transfer [60].

The electron as well as ion-transfer currents depend not only on the rate per adsorbed atom/ion but also on the atom/ion coverage. If this is low it lowers both exchange currents. There can be two reasons for low coverage: either another elementary step, which governs the supply of the atom/ion, is the rate determining step, or the equilibrium coverage is low. The latter has been suggested for gold cathodes on YSZ for oxygen transfer [67]. In this case a following rate- determining charge-transfer step is, most likely, an ion-transfer step.

The expression for multistep reactions assumes that all charge transfers occur at a single interface across the same $\Delta\varphi$. In view of the complexity of the interface in SSE the multistep reaction can take place spread over different sites with different Galvani potential differences. If so, Eq. (34a) cannot be applied. For instance, let us consider the overall reaction in which a neutral adsorbed oxygen atom on a metal electrode obtains two electrons and finally ends inside the SE (for instance, YSZ) at an interstitial site: $O_{ad} + 2e^- \to O_i^{2-}$(SE). Let us assume that the series of elementary steps involved are: (1) electronation of the adsorbed oxygen atom: $O_{ad} + e^-_M \to O_{ad}^-$, followed by (2) diffusion along the metal, (3) transfer of the ion into the SE to an interstitial site, $O_{ad}^- \to O_i^-$(SE), and (4) the adding of an electron there: O_i^-(SE) $+ e^-$(M) $\to O_i^{2-}$(SE). It is obvious that charge transfer occurs at four different locations (though all are associated with the same electrode reaction) with different $\Delta\varphi$ values and Eq. (34a) cannot be applied.

It was argued before that in SSE the transfer of electrons is usually rather rapid and that a charge-transfer rds is thus due to ion transfer. Then there is a negligible difference $\Delta\tilde{\mu}_e$ at the interface. An overpotential is still measured because of the way overpotentials, η, at the electrodes are measured [59, 60, 62, 68]. η is not measured by inserting an inert metallic electrode into the SE and measuring η with respect to the working electrode, that is, at the two sides of the interface. Such a measurement would yield $\eta = -\Delta\tilde{\mu}_e/\tilde{q}$. Rather, the measurement is done between a working electrode and a reference electrode put on top of the SE exposed to the same material (for instance, gas) as the working electrode, thus adopting a well-defined chemical potential μ_X. The overpotential then yields changes in the electrochemical potential of the ions between the working and reference electrodes.

The way the measurement of the overpotential is arranged $\delta(\Delta\varphi)$ at the rds is equal to the measured potential between the working electrode and the reversible reference electrode, $\delta(\Delta\varphi) = \eta$. Then Eq. (33a) can be written as,

$$I = I_0(e^{(1-\beta)zq\eta/k_BT} - e^{-\beta zq\eta/k_BT}) \quad (39a)$$

Eq. (34a) as,

$$I = I_0(e^{\alpha_a q\eta/k_BT} - e^{-\alpha_c q\eta/k_BT}) \quad (39b)$$

and Eq. (37b) as

$$I = I_0(1 - e^{-q\eta/k_BT}) \text{ or}$$
$$I = I_0(e^{q\eta/k_BT} - 1) \quad (39c)$$

2.6.6.4.5 Relating the Overpotential to a Difference in a Chemical or Electrochemical Potential
If one wants to allocate to $\delta(\Delta\varphi)$ a relation to a total driving force, this is not trivial. A total driving force is associated with a gradient in the electrochemical potential of electrons, ions, or of the

chemical potential of neutral species. For example, for electrons, it is the difference in the electrochemical potential of the electrons, $-\delta\tilde{\mu}_e/q = -\delta(\Delta\mu_e)/q + \delta(\Delta\varphi)$, which is not necessarily $\delta(\Delta\varphi)$. However, in some cases $\delta(\Delta\mu_e) = 0$ and $-\delta\tilde{\mu}_e/q = \delta(\Delta\varphi)$, as is the case in LSE when the electron is exchanged between a metal and a redox couple in the LE, the redox potential of which is fixed by a high concentration of the redox couple, as then $\delta(\Delta\mu_e) = 0$ also under current. In SSE, on the other hand, $\delta(\Delta\mu_e) = 0$, is not necessarily the case for electrons injected into an SE and electron exchange between a metal and adsorbed species.

$\Delta\varphi$ may represent $\delta\tilde{\mu}_{ion}$. For a fixed concentration of ionic defects at the interface $\delta\tilde{\mu}_{ion} = zq\delta(\Delta\varphi)$. Furthermore, if electron transfer is fast and $\delta\tilde{\mu}_e = 0$, one can express $zq\delta(\Delta\varphi)$ in terms of $\delta\mu_X = \delta\tilde{\mu}_{ion}$. This yields [51, 60, 68],

$$I = I_0(e^{(1-\beta)\delta\mu_X/k_BT} - e^{-\beta\delta\mu_X/k_BT}) \quad (40)$$

2.6.6.5 Conclusions for Charge-Transfer Processes

To summarize these results for SSE: If a Butler–Volmer $I-V$ relation is observed experimentally with a noninteger α_a and α_c, then the rds is most likely an ionic charge transfer. It is important to measure the $I-V$ relations for both direction of the current in order to determine accurately α_c and α_a. If α_a and α_c are integers then the rds may be one of the following: (1) an electron transfer and $\beta = 0$ or 1; or more likely (2) an ion transfer with $z = 2$ and $\beta = 0.5$; or (3) the rds is not a charge transfer one. We have noticed that in SSE a multistep reaction can be spread over close but different interfaces in which case Eqs. (34a) and (39b) cannot be applied. It is safe then to calculate each elementary step separately, determine the parameters associated with each step and use for the charge transfer Eq. (33a) or Eq. (37b) and Eq. (39a) or Eq. (39c). $\delta(\Delta\varphi)$ can, in certain cases, be related to $\delta\tilde{\mu}_e$, $\delta\tilde{\mu}_{ion}$ or $\delta\mu_X$ across the interface.

In LSE the specie adsorbed on the metal electrode, M, are in many cases neutral ones. Ions are solvated in the LE. For example, in the process Cl^- (LE) $\to Cl_{ad,M} + e_M^-$ the adsorbed specie is neutral Cl. In SSE charged specie are expected to be adsorbed in key reactions, for example, O_{ad}^- on Pt is obtained in the reaction: $O_{ad} + e_M^- \to O_{ad}^-$. Furthermore, adsorbed species with different charge may coexist.

We have mentioned that reference electrodes used for electrode characterization in SSE (and LSE) are chemically active ones. For determining the distribution of $\tilde{\mu}_e$ in the SE or MIEC, inert metal electrodes are used [46, 69, 70].

2.6.6.6 Diffusion

Diffusion in both LSE and SSE is governed by the gradient of the concentrations of the relevant species. Owing to the more complex morphology of electrodes in SSE, the paths for diffusion in SSE are more diverse as mentioned before.

2.6.7 Galvani Potential Distribution versus Electrochemical Potential Distribution

In LSE it is customary to follow the distribution of the Galvani potential, φ, along the EC. φ changes within the bulk of a single phase (except when a supporting electrolyte is used) as well as at interfaces. The drop at interfaces exists also under equilibrium. A detailed discussion of the distribution of φ requires the knowledge of the specific changes at interfaces for the materials involved.

In SSE it is customary to follow the distribution of the electrochemical potential of the electrons, $\tilde{\mu}_e$, and of the ions, $\tilde{\mu}_{ion}$. This has the advantage that under equilibrium, $\tilde{\mu}_e$ and $\tilde{\mu}_{ion}$ are uniform. Furthermore, under current for a reversible electrode with respect to electron transfer: $\Delta\tilde{\mu}_e = 0$ and for a reversible electrode with respect to ion transfer: $\Delta\tilde{\mu}_{ion} = 0$ at the interface. $\Delta\tilde{\mu}_e$ can be directly related to the measurable voltage, V, and $\Delta\tilde{\mu}_{ion} + z\Delta\tilde{\mu}_e$ to the chemical potential difference at the electrodes, $\Delta\mu_X$, that can also be measured.

2.6.8
Heterogeneous Catalysis in SSE

Catalysis plays an important role in SSE as it does in LSE. At low temperatures, relevant to LSE, the catalysts are relative few, usually, precious metals, such as Pt. At elevated temperatures, normally encountered in SSE, the reactions are accelerated and a wider range of materials can serve as catalysts many of them quite inexpensive.

ECs can be used to change the rate of chemical reactions. Two modes of operation are known. In the first one, a reactant is supplied via the SE to the working electrode (from a counter electrode that serves as a reservoir of the material). That reactant reacts with other reactants supplied directly to the working electrode via the gas phase. The state of the material that is transferred through the SE (i.e. its chemical potential) is controlled by the applied voltage [71]. This controls also the reaction rate at the working electrode.

In the second mode of operation, all the reactants are supplied via the gas phase in the working electrode compartment. The working electrode acts as a catalyst. The catalytic activity of the electrode is controlled by the voltage applied to the EC. The supply of ions through the SE is usually negligible and does not contribute significantly the reaction. Instead the ions cover the electrode material and thus change the catalytic activity of the electrode. The applied voltage governs that coverage. This is the so-called NEMCA effect developed by Vayenas and coworkers [72–74]. For example, a cell used for the oxidation of CH_4 will be: $(-)$air,Pt$|$YSZ$(O^{2-})|$Pt,CH_4,$O_2(+)$ where CH_4 reacts with O_2 supplied together with it. The rate of reaction is governed by the voltage on the EC, which controls the coverage of the working Pt electrode by oxygen coming through the YSZ SE.

2.6.9
Summary

Although based on the same thermodynamic principles, LSE and SSE are different in many aspects. This arises because of difference in material properties, morphology, and the geometry of the ECs. The differences are in bulk properties of the ionic conductors and in the elementary steps of electrode processes. Differences arise also in applications. Solid state cells have an advantage where dry cells are required, in microelectronics and at elevated temperatures. Liquid-state devices are richer in the variety of ions, which can be transported in a large range of LEs, and have significant advantage in large production electrochemical plants where high volumes of materials have to be processed.

References

1. J. O'M. Bockris, A. K. N. Ready, *Modern Electrochemistry*, Plenum/Rosetta, N.Y., 1970, Vol. 1, 2.
2. A. J. Bard, L. R. Faulkner, *Electrochemical Methods, Fundamentals and Application*, John Wiley & Sons, New York, 1980.

3. K. B. Oldham, J. C. Myland, *Fundamentals of Electrochemical Science*, Academic Press, New York, 1994.
4. E. Gileadi, *Electrode Kinetics for Chemists, Chemical Engineers and Materials Scientists*, VCH, New York, 1993.
5. P. G. Bruce, (Ed.), *Solid State Electrochemistry*, Cambridge University Press, New York, 1995.
6. P. J. Gelling, H. J. M. Bouwmeester, (Eds.), *The CRC Handbook of Electrochemistry*, CRC Press, Boca Raton, 1997.
7. T. Takahashi, (Ed.), *High Conductivity Solid Ionic Conductors. Recent trends and Applications*, World Scientific, Rive Edge, 1989.
8. C. Julien, G.-A. Nazri, Solid state batteries: materials, design and optimization, in *Electronic Materials: Science and Technology Series* (Ed.: H. L. Tuller), Kluwer Academic Publisher, Norwell, 1994.
9. N. Q. Minh, T. Takahashi, *Science and Technology of Ceramic Fuel Cells*, Elsevier, Amsterdam, New York, 1995.
10. K. S. Goto, *Solid State Electrochemistry and its Applications to Sensors and Electronic Devices*, Elsevier, Amsterdam, New York, 1988.
11. H. Rickert, *Solid State Electrochemistry*, Springer-Verlag, New York, 1982.
12. J. Maier, *Festkörper-Fehler und Funktion*, Teubner Studienbücher, Stuttgart, 2000.
13. H. Schmalzried, *Chemical Kinetics of Solids*, VCH, New York, 1995.
14. F. A. Kröger, *The Chemistry of Imperfect Crystals*, North Holland, New York, 1974, Vol. II.
15. G. Mamantov in *Materials for Advanced Batteries* (Eds.: D. W. Murphy, J. Broadhead, B. C. H. Steele), Plenum Press, New York, 1979, pp. 111–122.
16. A. Bélanger, F. Morin, M. Gauthier et al., in *Materials for Advanced Batteries* (Eds.: D. W. Murphy, J. Broadhead, B. C. H. Steele), Plenum Press, New York, 1979, pp. 211–222.
17. T. A. Ramanarayanan, R. A. Rapp, *Metall. Trans.* **1972**, *3*, 3239.
18. J. Van herle, A. J. McEvoy, *Phys. Chem. Solids* **1994**, *55*, 339.
19. S. Fujitu, K. Koumoto, H. Yanagida, *Solid State Ionics* **1986**, *18, 19*, 1146.
20. K. B. Oldham, J. C. Myland, *Fundamentals of Electrochemical Science*, Academic Press, New York, 1994, p. 9.
21. J. O'M. Bockris, A. K. N. Ready, *Modern Electrochemistry*, Plenum/Rosetta, N.Y., 1970, Vol. 2, p. 983.
22. H. L. Tuller, Mixed conduction, in *Nonstoichiometric Oxides* (Ed.: Toft-Sørensen), Academic Press, New York, 1981, pp. 271–335.
23. K. Funke, *Prog. Solid State Chem.* **1993**, *22*, 111.
24. W. Shockley, *Bell Syst.Tech. J.* **1949**, *28*, 435.
25. I. Riess *Solid State Ionics* **1994**, *69*, 43.
26. J. Maier, *Angew. Chem., Int. Ed. Engl.* **1993**, *32*, 313.
27. Y. Tsur, I. Riess, *Phys. Rev. B* **1999**, *60*, 8138.
28. O. Porat, I. Riess, *Solid State Ionics* **1994**, *74*, 229.
29. O. Porat, I. Riess, *Solid State Ionics* **1995**, *81*, 29.
30. J. R. Manning, *Diffusion Kinetics for Atoms in Crystals*, van Nostrand, New York, 1968, pp. 2–9.
31. N. F. Mott, E. A. Davis, *Electronic Processes in Non-Crystalline Materials*, Clarendon Press, Oxford, 1971, Chap. 6.
32. F. Lange, M. Martin, *Ber. Bunsen-Ges. Phys. Chem.* **1997**, *101*, 176.
33. I. Riess Electrochemistry of mixed ionic electronic conductors, in *CRC Handbook of Solid State Electrochemistry* (Eds.: P. J. Gellings, H. J. M. Boumeester), CRC Press, Boca Raton, 1997, pp. 223–268.
34. I. Riess, *Solid State Ionics*, **2000**, *136–137*, 1119.
35. J. Maier, *J. Am. Ceram. Soc.* **1993**, *76*, 1212.
36. H.-D. Weimhöfer, *Solid State Ionics* **1990**, *40/41*, 530.
37. I. Riess, *J. Phys. Chem. Solids* **1986**, *47*, 129.
38. J. Maier, M. Holzinger, W. Sitte, *Solid state Ionics* **1994**, *74*, 5.
39. J. Maier, *J. Am. Ceram. Soc.* **1993**, *76*, 1218.
40. W. Weppner, R. A. Huggins, *Annu. Rev. Mater. Sci.* **1978**, *8*, 269.
41. I. Riess, D. S. Tannhauser, Control and measurement of stoichiometry in oxides, in *Transport in NonStoichiometric Compounds*, Materials Science Monographs 15 (Ed.: J. Nowotny), Elsevier Scientific Publishing Company, New York, 1982, pp. 503–517.
42. K. Funke, R. Hackenberg, *Ber. Bunsen-Ges. Phys. Chem.* **1972**, *76*, 883.
43. J. Sohége, K. Funke, *Ber. Bunsen-Ges. Phys. Chem.* **1984**, *88*, 657.
44. M. H. Hebb, *J. Chem. Phys.* **1952**, *20*, 185.
45. C. Wagner, *Z. Elektrochem.* **1956**, *60*, 4.
46. I. Riess, *Solid State Ionics* **1996**, *91*, 221.

47. I. Riess, *Solid State Ionics* **1991**, *44*, 207.
48. K. D. Becker, H. Schmalzried, V. von Wurmb, *Solid State Ionics* **1983**, *11*, 213.
49. J. R. Macdonald, D. R. Franceschetti, *J. Chem. Phys.* **1978**, *68*, 1614.
50. J. R. Macdonald, (Ed.), *Impedance Spectroscopy*, John Wiley & Sons, New York, 1987.
51. S. B. Adler, J. A. Lane, B. C. H. Steele, *J. Electrochem. Soc.* **1996**, *143*, 3554.
52. I. Riess, J. Schoonman in *CRC Handbook of Solid State Electrochemistry* (Eds.: P. J. Gellings, H. J. M. Boumeester), CRC Press, Boca Raton, 1997, pp. 269–294.
53. N. W. Ashcroft, N. D. Mermin, *Solid State Physics*, Holt-Saunders Intl., Philadelphia, 1976, pp. 367–371.
54. R. N. Ediriweera, P. R. Mason, *Solid State Ionics* **1985**, *17*, 213.
55. J. Maier in *Recent Trends in Superionic Solids and Solid Electrolytes* (Eds.: S. Chandra, A. Laskar), Academic Press, New York, 1989, p. 137.
56. J.O'M. Bockris, A. K. N. Ready, *Modern Electrochemistry*, Plenum/Rosetta, N.Y., 1970, Vol. 1, 2, Chaps. 8 & 9.
57. R. H. Fowler, L. Nordheim, *Proc. R. Soc. London, Ser. A* **1928**, *119*, 173.
58. S. M. Sze, *The Physics of Semiconductor Devices*, Wiley, N.Y., 1969.
59. J. Mizusaki, K. Amano, S. Yamauchi et al., *Solid State Ionics* **1987**, *22*, 313.
60. I. Riess, M. Gödickemeier, L. J. Gauckler, *Solid State Ionics* **1996**, *90*, 91.
61. A.-M. Svensson, S. Sunde, K. Nisancioglu, *J. Electrochem. Soc.* **1997**, *144*, 2719.
62. Da. Yu. Wang, A. S. Nowick, *J. Electrochem. Soc.* **1979**, *126*, 1155.
63. F. H. van Heuveln, H. J. M. Boumeester, *J. Electrochem. Soc.* **1997**, *144*, 134.
64. B. A. Boukamp, B. A. van Hassel, I. C. Vinke et al., *Electrochim. Acta* **1993**, *38*, 1817.
65. Y. Jaing, S. Wang, Y. Zhang et al., *Solid State Ionics* **1998**, *110*, 111.
66. J. Maier, *Angew. Chem.* **1993**, *105*, 558.
67. M. Gödickemeier, K. Sasaki, L. J. Gauckler et al., *J. Electrochem. Soc.* **1997**, *144*, 1635.
68. I. Riess, Electrode processes in solid oxide fuel cells, in *Oxygen Ion and Mixed Conductors and their Technological Applications*, NATO Advanced Study Institute Series E (Eds.: H. L. Tuller, J. Schoonman, I. Riess), Kluwer Academic Publishers, Norwell, 2000, pp. 21–56, Vol. 338.
69. J. Mizusaki, K. Fueki, T. Mukaibo, *Bull. Chem. Soc. Jpn.* **1975**, *48*, 428.
70. C. Rosenkranz, J. Janek, *Solid State Ionics* **1995**, *82*, 95.
71. T. M. Gür, R. A. Huggins in *Fast Ion Transport in Solids* (Eds.: P. Vashishta, J. N. Mundy, G. K. Shenoy), Elsevier, Amsterdam, New York, 1979.
72. C. G. Vayenas, S. Bebelis, I. V. Yentekakis et al., *Catal. Today* **1992**, *11*(3) 303–442.
73. D. Tsiplakides, S. G. Neophytides, O. Enea et al., *J. Electrochem. Soc.* **1997**, *144*, 2072.
74. S. G. Neophystides, D. Tsiplakides, P. Stonehart et al., *Nature* **1994**, *370*, 45.

2.7
Polyelectrolytes in Solution and at Surfaces

Roland R. Netz
Max-Planck Institute for Colloids and Interfaces, Potsdam, Germany

David Andelman
School of Physics and Astronomy Raymond and Beverly Sackler Faculty of Exact Sciences Tel Aviv University, Ramat Aviv, Israel

Abstract

This chapter deals with charged polymers (polyelectrolytes) in solution and at surfaces. The behavior of polyelectrolytes (PEs) is markedly different from that of neutral polymers. In bulk solutions, that is, disregarding the surface effect, there are two unique features to charged polymers: first, owing to the presence of long-ranged electrostatic repulsion between charged monomers, the polymer conformations are much more extended, giving rise to a very small overlap concentration and high solution viscosity. Second, the presence of a large number of counterions increases the osmotic pressure of PE solutions, making such polymers water soluble as this is of great importance to many applications. At surfaces, the same interplay between monomer–monomer repulsion and counterion degrees of freedom leads to a number of special properties. In particular, the adsorption behavior depends on both the concentration of polymers and the added salt in the bulk. We first describe the adsorption behavior of single PE molecules and discuss the necessary conditions to obtain an adsorbed layer and characterize its width. Depending on the stiffness of the PE, the layer can be flat and compressed or coiled and extended. We then proceed and discuss the adsorption of PEs from semidilute solutions. Mean-field theory profiles of PE adsorption are calculated as a function of surface charge density (or surface potential), the amount of salt in the system, and the charge fraction on the chains. The phenomenon of charge inversion is reviewed and its relevance to the formation of multilayers is explained. The review ends with a short overview of the behavior of grafted PEs.

2.7.1
Introduction

PEs are charged macromolecules that are extensively studied not only because of their numerous industrial applications but also from a purely scientific interest [1–4]. The most important property of PEs is their water solubility giving rise to a wide range of nontoxic, environmentally friendly, and cheap formulations. On the theoretical side, PEs combine the field of statistical mechanics of charged systems with the field of polymer science and offer quite a number of surprises and challenges.

The polymers considered in this review are taken as linear and long polymer chains, containing a certain fraction of electrically charged monomers. Chemically, this can be achieved, for example, by substituting neutral monomers with acidic ones. Upon contact with water, the acidic groups dissociate into positively charged protons, which bind immediately to water molecules, and negatively charged monomers. Although this process effectively charges the polymer molecules, the counterions make the PE solution electroneutral on macroscopic length scales.

The counterions are attracted to the charged polymers via long-ranged Coulomb interactions, but this physical association typically only leads to rather loosely bound counterion clouds around the PE chains. Because PEs are present in a background of a polarizable and diffusive counterion cloud, there is a strong influence of the counterion distribution on the PE structure, as will be discussed at length in this review. Counterions contribute significantly toward bulk properties, such as the osmotic pressure, and their translational entropy is responsible for the generally good water solubility of charged polymers. In addition, the statistics of PE chain conformations are governed by intrachain Coulombic repulsion between charged monomers, resulting in more extended and swollen conformations of PEs as compared to neutral polymers.

All these factors combined are of great importance when considering PE adsorption to charged surfaces. We distinguish between physical adsorption, where chain monomers are attracted to surfaces via electrostatic or nonelectrostatic interactions, and chemical adsorption, where a part of the PE (usually the chain end) is chemically bound (grafted) to the surface. In all cases, the long-ranged repulsion of the dense layer of adsorbed PEs and the entropy associated with the counterion distribution are important factors in the theoretical description.

2.7.2
Neutral Polymers in Solution

Before reviewing the behavior of charged polymers, let us describe some of the important ideas underlying the behavior of neutral polymer chains in solution.

2.7.2.1 Flexible Chain Statistics

The chains considered in this review are either flexible or semiflexible. The statistical thermodynamics of flexible chains is well developed and the theoretical concepts can be applied with a considerable degree of confidence [5–9]. Long and flexible chains have a large number of conformations, a fact that plays a crucial role in determining their behavior in solution. When flexible chains adsorb on surfaces, they form a *diffusive* adsorption layer extending away from the surface into the solution. This is in contrast to semiflexible or rigid chains, which can form dense and compact adsorption layers.

The main parameters used to describe a flexible polymer chain are the polymerization index N, which counts the number of repeat units or effective monomers along the chain, and the Kuhn length a, being the size of one effective monomer or the distance between two neighboring effective monomers. The effective monomer size ranges from a few Å for synthetic polymers to a few nanometers for biopolymers [7]. The effective monomer size a is not to be confused with the actual size b of one chemical monomer; in general, a is greater than b owing to chain stiffening effects, as will be explained in detail later on. In contrast to other molecules or particles, a polymer chain contains not only translational and rotational degrees of freedom but also a vast number of conformational degrees of freedom. For typical polymers, different conformations are produced by torsional rotations of the polymer backbone bonds.

A simple description of flexible chain conformations is achieved with the freely jointed chain model in which a polymer consisting of $N+1$ monomers is represented by N bonds defined by bond vectors $\mathbf{r}_j$ with $j=1,\ldots N$. Each bond vector has a fixed length $|\mathbf{r}_j|=a$ corresponding to the Kuhn length, but otherwise is allowed to rotate freely. For the freely jointed chain model, the monomer size b equals the effective monomer size a, $b = a$. Fixing one of the chain ends at the origin, the position of the $(k + 1)$-th monomer is given by the vectorial sum

$$\mathbf{R}_k = \sum_{j=1}^{k} \mathbf{r}_j \quad (1)$$

Because two arbitrary bond vectors are uncorrelated in this simple model, the thermal average over the scalar product of two different bond vectors vanishes, $\langle \mathbf{r}_j \cdot \mathbf{r}_k \rangle = 0$ for $j \neq k$, while the mean-squared bond vector length is simply given by $\langle \mathbf{r}_j^2 \rangle = a^2$. It follows then that the mean-squared end-to-end radius is proportional to the number of monomers

$$\langle \mathbf{R}_N^2 \rangle = Na^2 \quad (2)$$

The same result is obtained for the mean quadratic displacement of a freely diffusing particle and eludes to the same underlying physical principle, namely, the statistics of Markov processes.

In Fig. 1(a) we show a snapshot of Monte-Carlo simulations of a freely jointed chain consisting of 100 monomers, each being represented by a sphere of diameter b (being equal here to a, the effective monomer size). The bar represents a length of $10b$, which according to Eq. (2) is the average distance between the chain ends. Indeed, the end-to-end radius gives a good idea of the typical chain radius.

The freely jointed chain model describes ideal Gaussian chains and does not account for interactions between monomers that are not necessarily close neighbors along the backbone. Including these interactions will give a different scaling behavior for long polymer chains. The end-to-end radius, $R = \sqrt{\langle R_N^2 \rangle}$, can be written more generally for $N \gg 1$ as

$$R \simeq aN^\nu \quad (3)$$

For an ideal polymer chain (no interactions between monomers), the above result implies $\nu = 1/2$. This result holds only for polymers in which the attraction between monomers (as compared with the monomer–solvent interaction) cancels the steric repulsion (which is due to the fact that the monomers cannot penetrate each other). This situation can be achieved in special solvent conditions called *theta* solvents.

More generally, polymers in solution can experience three types of solvent conditions, with theta solvent condition being intermediate between "good" and "bad" solvent conditions. The solvent quality depends mainly on the specific chemistry determining the interaction between the solvent molecules and monomers. It also can be changed by varying the temperature.

The solvent is called *good* when the monomer–solvent interaction is more favorable than the monomer–monomer one. Single polymer chains in good solvents have "swollen" spatial configurations, reflecting the effective repulsion between monomers. For good solvents, the steric repulsion dominates and the polymer coil takes a more swollen structure, characterized by an exponent $\nu \simeq 3/5$ [7]. This spatial size of a polymer coil is much smaller than the extended contour length $L = aN$ but larger than the size of an ideal chain $aN^{1/2}$. The reason for this peculiar behavior is entropy combined with the favorable interaction between monomers and solvent molecules in good solvents. As we will see later, it is the conformational freedom of polymer coils that leads to salient differences between polymer and simple liquid adsorption.

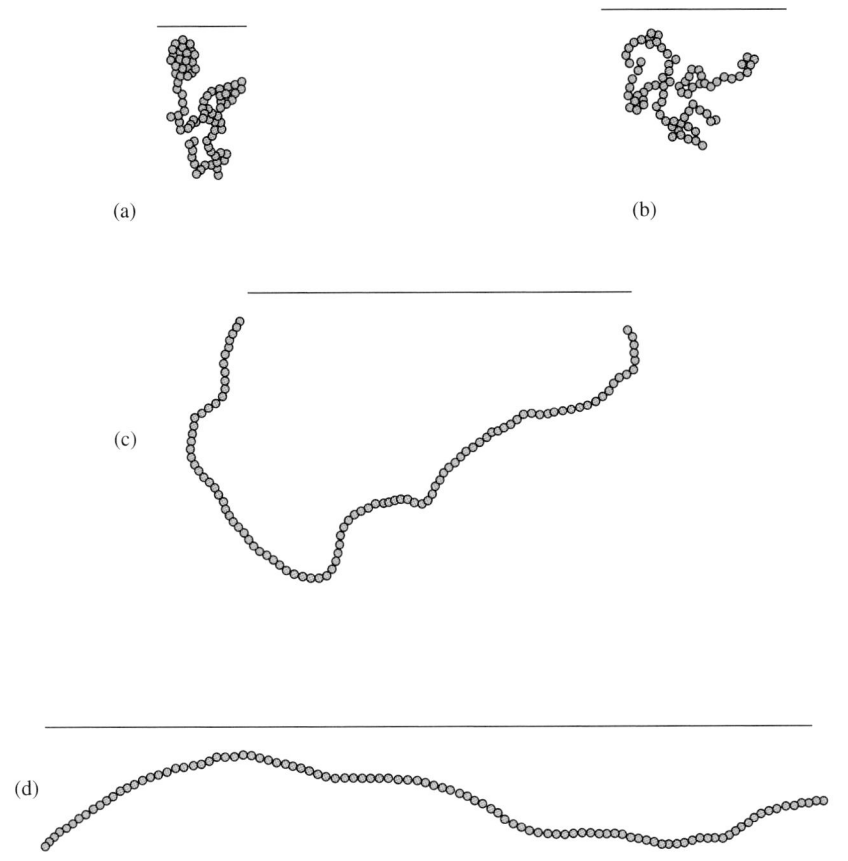

Fig. 1 Snapshots of Monte-Carlo simulations of a neutral and semiflexible chain consisting of $N = 100$ monomers of diameter b, which defines the unit of length. The theoretical end-to-end radius R is indicated by a straight bar. The persistence lengths used in the simulations are (a) $\ell_0 = 0$, corresponding to a freely jointed (flexible) chain, leading to an end-to-end radius $R/b = 10$; (b) $\ell_0/b = 2$, leading to $R/b = 19.8$; (c) $\ell_0/b = 10$, leading to $R/b = 42.4$; and (d) $\ell_0/b = 100$, leading to $R/b = 85.8$.

In the opposite case of "bad" (sometimes called *poor*) solvent conditions, the effective interaction between monomers is attractive, leading to collapse of the chains and to their precipitation from the solution (phase separation between the polymer and the solvent). It is clear that in this case, the polymer size, like any space-filling object embedded in three-dimensional space, scales as $N \sim R^3$, yielding $\nu = 1/3$.

2.7.2.2 Semiflexible Chain Statistics

Beside neglecting monomer–monomer interaction, the freely jointed chain model does not take into account the chain elasticity, which plays an important role for some polymers, and leads to more rigid structures. This stiffness can be conveniently characterized by the persistence length ℓ_0, defined as the length over which the tangent vectors at different locations on the chain are correlated. In other words, the persistence length gives an estimate for the typical radius of curvature, while taking into account thermal fluctuations. For synthetic polymers with *trans-cis* conformational freedom of the chain backbone, the stiffness is due to fixed bond angles and hindered rotations around individual backbone bonds. This effect is even more pronounced for polymers with bulky side chains, such as poly-DADMAC, because of steric constraints, and the persistence length is of the order of a few nanometers.

Biopolymers with a more complex structure on the molecular level tend to be stiffer than simple synthetic polymers. Some typical persistence lengths encountered in these systems are $\ell_0 \approx 5$ mm for tubulin, $\ell_0 \approx 20$ μm for actin, and $\ell_0 \approx 50$ nm for double-stranded DNA. Because some of these biopolymers are charged, we will discuss at length the dependence of the persistence length on the electrostatic conditions. In some cases the main contribution to the persistence length comes from the repulsion between charged monomers.

To describe the bending rigidity of neutral polymers, it is easier to use a continuum model [6], in which one neglects the discrete nature of monomers. The bending energy (rescaled by the thermal energy, $k_B T$) of a stiff or semiflexible polymer of contour length L is given by

$$\frac{\ell_0}{2} \int_0^L ds \left(\frac{d^2 \mathbf{r}(s)}{ds^2}\right)^2 \quad (4)$$

where $d^2\mathbf{r}(s)/ds^2$ is the local curvature of the polymer. We assume here that the polymer segments are nonextendable, that is, the tangent vectors are always normalized, $|d\mathbf{r}(s)/ds| = 1$. Clearly, this continuum description will only be good if the persistence length is larger than the monomer size. The mean-squared end-to-end radius of a semiflexible chain is known and reads [6]

$$R^2 = 2\ell_0 L + 2\ell_0^2 (e^{-L/\ell_0} - 1) \quad (5)$$

where the persistence length is ℓ_0 and the total contour length of a chain is L. Two limiting behaviors are obtained for R from Eq. (5): for long chains, $L \gg \ell_0$, the chains behave as flexible ones, $R^2 \simeq 2\ell_0 L$; while for rather short chains, $L \ll \ell_0$, the chains behave as rigid rods, $R \simeq L$. Comparison with the scaling of the freely jointed chain model (Eq. 2) shows that a semiflexible chain can, for $L \gg \ell_0$, be described by a freely jointed chain model with an effective Kuhn length of

$$a = 2\ell_0 \quad (6)$$

and an effective number of segments or monomers

$$N = \frac{L}{2\ell_0} \quad (7)$$

In this case the Kuhn length takes into account the chain stiffness and is independent from the monomer length. This monomer size is denoted by b whenever there is need to distinguish between the monomer size b and the persistence length ℓ_0 (or Kuhn length a). In Fig. 1 we show snapshots taken from Monte-Carlo simulations of a semiflexible chain consisting of 100 polymer beads of diameter b. The persistence length is varied from $\ell_0 = 2b$ (Fig. 1b), over $\ell_0 = 10b$ (Fig. 1c), to $\ell_0 = 100b$ (Fig. 1d). Comparison with the freely jointed chain model (having no persistence length) is given in Fig. 1(a) ($a = b$, $\ell_0 = 0$). It is seen that as the persistence length is increased, the chain structure becomes more expanded. The theoretical prediction for the average end-to-end radius R (Eq. 5) is shown as the black bar on the figure and

2.7.3
Properties of Polyelectrolytes in Solution

For PEs, electrostatic interactions provide the driving force for their salient features and have to be included in any theoretical description. The reduced electrostatic interaction between two pointlike charges can be written as $z_1 z_2 v(r)$ where

$$v(r) = \frac{e^2}{k_B T \varepsilon r} \tag{8}$$

is the Coulomb interaction between two elementary charges, z_1 and z_2 are the valences (or the reduced charges in units of the elementary charge e), and ε is the medium dielectric constant. Throughout this review, all energies are given in units of the thermal energy $k_B T$. The interaction depends only on the distance r between the charges. The total electrostatic energy of a given distribution of charges is obtained from adding up all pairwise interactions between charges according to Eq. (8). In principle, the equilibrium behavior of an ensemble of charged particles (e.g. a salt solution) follows from the partition function, that is, the weighted sum over all different microscopic configurations, which – via the Boltzmann factor – depends on the electrostatic energy of each configuration. In practice, however, this route is very complicated for several reasons:

1. The Coulomb interaction (Eq. 8) is long-ranged and couples many charged particles. Electrostatic problems are typically *many-body problems*, even for low densities.
2. Charged objects in most cases are dissolved in water. Like any material, water is polarizable and reacts to the presence of a charge with polarization charges. In addition, and this is by far a more important effect, water molecules carry a permanent dipole moment that partially orients in the vicinity of charged objects. Note that for water, $\varepsilon \approx 80$, so that electrostatic interactions and self-energies are much weaker in water than in air (where $\varepsilon \approx 1$) or some other low-dielectric solvents. Still, the electrostatic interactions are especially important in polar solvents because in these solvents, charges dissociate more easily than in unpolar solvents.
3. In biological systems and most industrial applications, the aqueous solution contains mobile salt ions. Salt ions of opposite charge are drawn to the charged object and form a loosely bound counterion cloud around it. They effectively reduce or *screen* the charge of the object. The effective (screened) electrostatic interaction between two charges $z_1 e$ and $z_2 e$ in the presence of salt ions and a polarizable solvent can be written as $z_1 z_2 v_{DH}(r)$, with the Debye–Hückel (DH) potential $v_{DH}(r)$ given on the linear-response level by

$$v_{DH}(r) = \frac{\ell_B}{r} e^{-\kappa r} \tag{9}$$

The Bjerrum length ℓ_B is defined as

$$\ell_B = \frac{e^2}{\varepsilon k_B T} \tag{10}$$

and denotes the distance at which the Coulombic interaction between two unit charges in a dielectric medium is equal to thermal energy ($k_B T$). It is a measure of the distance below which the Coulomb energy is strong enough to compete with the thermal fluctuations; in water at room temperatures, one finds $\ell_B \approx 0.7$ nm. The exponential decay is characterized by the so-called screening length κ^{-1}, which is related to the salt concentration c_{salt} by

$$\kappa^2 = 8\pi z^2 \ell_B c_{salt} \qquad (11)$$

where z denotes the valency of the screening ions ($z:z$ salt). At physiological conditions the salt concentration is $c_{salt} \approx 0.1$ M and for monovalent ions ($z = 1$) this leads to $\kappa^{-1} \approx 1$ nm.

The so-called DH interaction (Eq. 9) embodies correlation effects due to the long-ranged Coulomb interactions in a salt solution using linear-response theory. In the following we estimate the range of validity of this approximation using simple scaling arguments. The number of ions that are correlated in a salt solution with concentration c_{salt} is of the order of $n \sim \kappa^{-3} c_{salt}$, where one employs the screening length κ^{-1} as the scale over which ions are correlated. Using the definition $\kappa^2 = 8\pi z^2 \ell_B c_{salt}$, one obtains $n \sim (z^2 \ell_B c_{salt}^{1/3})^{-3/2}$. The average distance between ions is roughly $r \sim c_{salt}^{-1/3}$. The typical electrostatic interaction between two ions in the solution thus is $U \sim z^2 \ell_B / r \sim z^2 \ell_B c_{salt}^{1/3}$, and we obtain $U \sim n^{-2/3}$. Using these scaling arguments, one obtains that either (1) many ions are weakly coupled together (i.e. $n \gg 1$ and $U \ll 1$) or (2) a few ions interact strongly with each other ($n \simeq U \simeq 1$). In the first case, and in the absence of external fields, the approximations leading to the DH approximation (Eq. 9) are valid.

The DH approximation forms a convenient starting point for treating screening effects, since (owing to its linear character) the superposition principle is valid and the electrostatic free energy is given by a sum over the two-body potential (Eq.9). However, we will at various points in this review also discuss how to go beyond the DH approximation, for example, in the form of the nonlinear Poisson–Boltzmann theory (see Sect. 2.7.5) or a box model for the counterion distribution (see Sect. 2.7.6).

2.7.3.1 Isolated Polyelectrolyte Chains

We discuss now the scaling behavior of a single semiflexible PE in the bulk, including chain stiffness and electrostatic repulsion between monomers. For charged polymers, the effective persistence length is increased owing to electrostatic repulsion between monomers. This effect modifies considerably not only the PE behavior in solution but also their adsorption characteristics.

The scaling analysis is a simple extension of previous calculations for flexible (Gaussian) PEs [10–12]. The semiflexible polymer chain is characterized by a bare persistence length ℓ_0 and a linear charge density τ. Using the monomer length b and the fraction of charged monomers f as parameters, the linear charge density can be expressed as $\tau = f/b$. Note that in the limit where the persistence length is small and comparable to a monomer size, only a single length scale remains, $\ell_0 \simeq a \simeq b$.

Many interesting effects, however, are obtained in the general case treating the persistence length ℓ_0 and the monomer size b as two independent parameters. In the regime where the electrostatic energy is weak, and for long enough contour length L, where $L \gg \ell_0$, a polymer coil will be formed with a radius R unperturbed by the electrostatic repulsion between monomers. According to Eq. (5), we get $R^2 \simeq 2\ell_0 L$. To estimate when the electrostatic interaction will be sufficiently strong to swell the polymer coil, we recall that the electrostatic energy (rescaled by the thermal energy k_BT) of a homogeneously charged sphere of total charge Q and radius R is

$$W_{el} = \frac{3\ell_B Q^2}{5R} \quad (12)$$

The exact charge distribution inside the sphere only changes the prefactor of order unity and is not important for the scaling arguments. For a polymer of length L and line charge density τ, the total charge is $Q = \tau L$. The electrostatic energy of a (roughly spherical) polymer coil is then

$$W_{el} \simeq \ell_B \tau^2 L^{3/2} \ell_0^{-1/2} \quad (13)$$

The polymer length at which the electrostatic self-energy is of order k_BT, that is, $W_{el} \simeq 1$, follows as

$$L_{el} \simeq \ell_0 (\ell_B \ell_0 \tau^2)^{-2/3} \quad (14)$$

and defines the electrostatic blob size or electrostatic polymer length. We expect a locally crumpled polymer configuration if $L_{el} > \ell_0$, that is, if

$$\tau \sqrt{\ell_B \ell_0} < 1 \quad (15)$$

because the electrostatic repulsion between two segments of length ℓ_0 is smaller than the thermal energy and is not sufficient to align the two segments. This is in accord with more detailed calculations by Joanny and Barrat [11]. A recent general Gaussian variational calculation confirms this scaling result and in addition yields logarithmic corrections [12]. Conversely, for

$$\tau \sqrt{\ell_B \ell_0} > 1 \quad (16)$$

electrostatic chain–chain repulsion is already relevant on length scales comparable to the persistence length. The chain is expected to have a conformation characterized by an effective persistence length ℓ_{eff}, larger than the bare persistence length ℓ_0, that is, one expects $\ell_{eff} > \ell_0$.

This effect is clearly seen in Fig. 2, where we show snapshots of Monte-Carlo simulations of a charged chain of 100 monomers of size b each and bare persistence length $\ell_0/b = 1$ and several values of κ^{-1} and τ. The number of monomers in an electrostatic blob can be written according to Eq. (14) as $L_{el}/\ell_0 = (\tau^2 \ell_B \ell_0)^{-2/3}$ and yields for Fig. 2(a) $L_{el}/\ell_0 = 0.25$, for Fig. 2(b) $L_{el}/\ell_0 = 0.63$, for Fig. 2(c) $L_{el}/\ell_0 = 1.6$, and for Fig. 2(d) $L_{el}/\ell_0 = 4$. Accordingly, in Fig. 2(d) the electrostatic blobs consist of four monomers, and the weakly charged chain crumples at small length scales. A typical linear charge density reached with synthetic PEs such as Polystyrenesulfonate (PSS) is one charge per two carbon bonds (or, equivalently, one charge per monomer), and it corresponds to $\tau \approx 4 \text{ nm}^{-1}$. Since for these highly flexible synthetic PEs the bare persistence length is of the order of the monomer size, $\ell_0 \simeq b$, the typical charge parameter for a fully charged PE therefore is roughly $\tau^2 \ell_B \ell_0 \approx 3$ and is intermediate between Fig. 2(a and b). Smaller linear charge densities can always be obtained by replacing some of the charged monomers on the polymer backbone with neutral ones, in which case the crumpling observed in Fig. 2(d) becomes relevant. Larger bare persistence lengths can be achieved with biopolymers or synthetic PEs with a conjugated carbon backbone.

The question now arises as to what are the typical chain conformations at much larger length scales. Clearly, they will be influenced by the repulsions. Indeed, in the *persistent regime*, obtained for $\tau \sqrt{(\ell_B \ell_0)} > 1$, the polymer remains locally stiff even for contour lengths larger

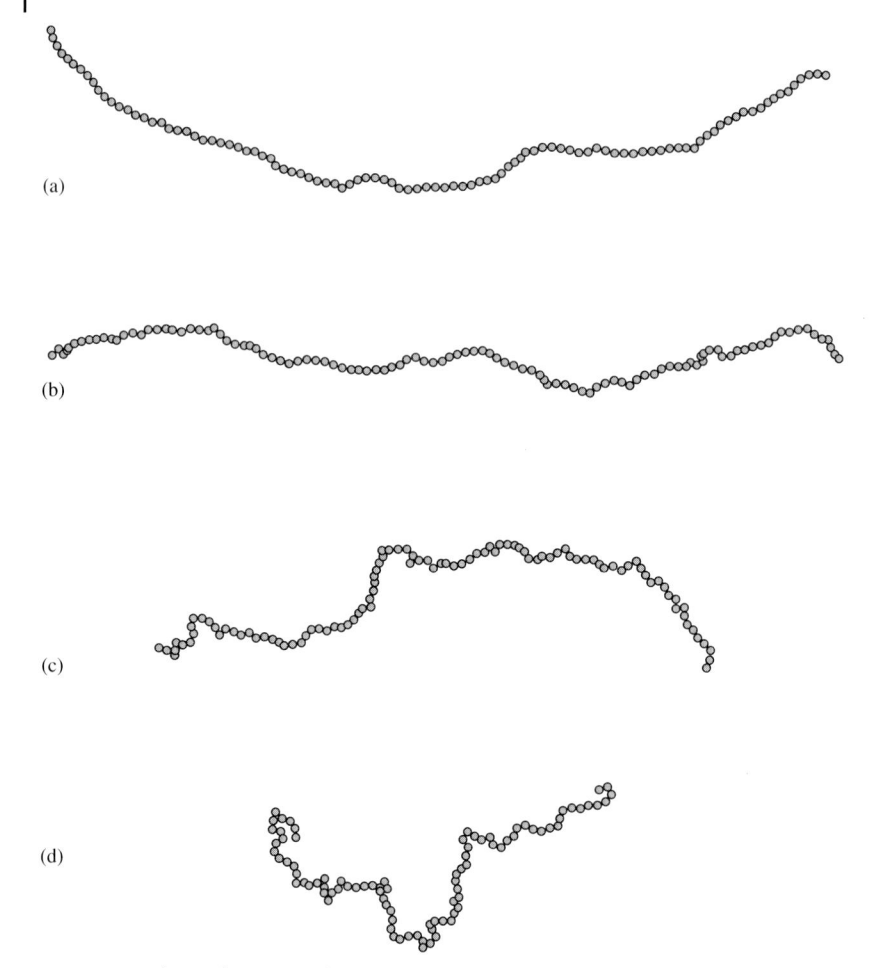

Fig. 2 Snapshots of Monte Carlo simulations of a polyelectrolyte chain of $N = 100$ monomers of size b, taken as the unit length. In all simulations the bare persistence length is fixed at $\ell_0/b = 1$, and the screening length and the charge interactions are tuned such that the electrostatic persistence length (ℓ_{OSF}) is constant and $\ell_{OSF}/b = 100$, see Eq. (18). The parameters used are (a) $\kappa^{-1}/b = \sqrt{50}$ and $\tau^2 \ell_B \ell_0 = 8$; (b) $\kappa^{-1}/b = \sqrt{200}$ and $\tau^2 \ell_B \ell_0 = 2$; (c) $\kappa^{-1}/b = \sqrt{800}$ and $\tau^2 \ell_B \ell_0 = 1/2$; and (d) $\kappa^{-1}/b = \sqrt{3200}$ and $\tau^2 \ell_B \ell_0 = 1/8$. Noticeably, the weakly charged chains crumple at small length scales and show a tendency to form electrostatic blobs.

than the bare persistence length ℓ_0, and the effective persistence length is given by

$$\ell_{\text{eff}} \simeq \ell_0 + \ell_{OSF} \tag{17}$$

The electrostatic persistence length, first derived by Odijk and independently by Skolnick and Fixman, reads [13–15]

$$\ell_{OSF} = \frac{\ell_B \tau^2}{4\kappa^2} \tag{18}$$

and is calculated from the electrostatic energy of a slightly bent polymer using the linearized DH approximation (Eq. 9).

It is valid only for polymer conformations that do not deviate too much from the rodlike reference state. The electrostatic persistence length gives a sizable contribution to the effective persistence length only for $\ell_{OSF} > \ell_0$. This is equivalent to the condition

$$\tau\sqrt{\ell_B\ell_0} > \ell_0\kappa \qquad (19)$$

The persistent regime is obtained for parameters satisfying both conditions (16) and (19). Another regime called the *Gaussian regime* is obtained in the opposite limit of $\tau\sqrt{(\ell_B\ell_0)} < \ell_0\kappa$.

The electrostatic persistence length is visualized in Fig. 3, in which we present snapshots of Monte-Carlo simulations of a charged chain consisting of 100 monomers of size b. The bare persistence length was fixed at $\ell_0 = b$, and the charge-interaction parameter was chosen to be $\tau^2\ell_B\ell_0 = 2$, close to the typical charge density in experiments on fully charged synthetic PEs. The snapshots correspond to varying screening lengths of (1) $\kappa^{-1}/b = \sqrt{2}$,

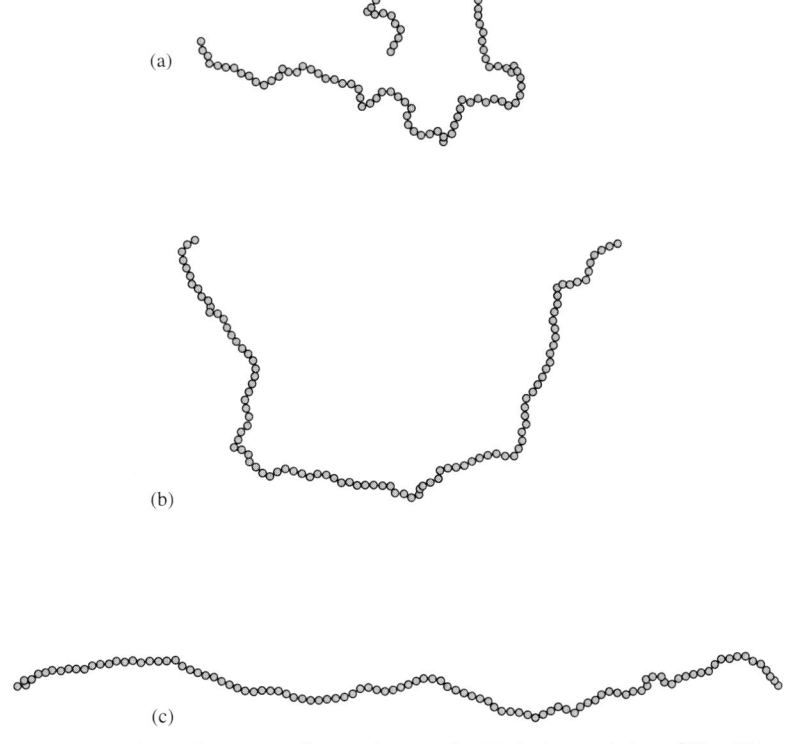

Fig. 3 Snapshots of Monte-Carlo simulations of a PE chain consisting of $N = 100$ monomers of size b. In all simulations, the bare persistence length is fixed at $\ell_0/b = 1$ and the charge-interaction parameter is chosen to be $\tau^2\ell_B\ell_0 = 2$. The snapshots correspond to varying screening lengths of (a) $\kappa^{-1}/b = \sqrt{2}$, leading to an electrostatic contribution to the persistence length of $\ell_{OSF}/b = 1$; (b) $\kappa^{-1}/b = \sqrt{18}$, leading to $\ell_{OSF}/b = 9$; and (c) $\kappa^{-1}/b = \sqrt{200}$, leading to $\ell_{OSF}/b = 100$. According to the simple scaling principle (Eq. 17), the effective persistence length in the snapshots (a–c) should be similar to the bare persistence length in Fig. 1(b–d).

leading to an electrostatic contribution to the persistence length of $\ell_{OSF} = b$ (Fig. 3a), (2) $\kappa^{-1}/b = \sqrt{18}$, or $\ell_{OSF} = 9b$ (Fig. 3b), and (3) $\kappa^{-1}/b = \sqrt{200}$, equivalent to $\ell_{OSF} = 100b$ (Fig. 3c). According to the simple scaling principle (Eq. 17), the effective persistence length in the snapshots (Fig. 3a–c) should be similar to the bare persistence length in Fig. 1(b–d), and indeed, the chain structures in Figs. 3(c) and 1(d) are very similar. Figs. 3(a) and 1(b) are clearly different, although the effective persistence length should be quite similar, mostly because of self-avoidance effects that are present in charged chains and are discussed in detail in Sect. 2.7.3.1.2.

For the case in which the polymer crumples on length scales larger than the bare bending rigidity, that is, for $L_{el} > \ell_0$ or $\tau\sqrt{(\ell_B \ell_0)} < 1$, the electrostatic repulsion between polymer segments is not strong enough to prevent crumpling on length scales comparable to ℓ_0, but can give rise to a chain stiffening on larger length scales, as explained by Khokhlov and Khachaturian [10] and confirmed by Gaussian variational methods [12]. Figure 4 schematically shows the PE structure in this regime, where the chain on small scales consists of Gaussian blobs of chain length L_{el}, within which electrostatic interactions are not important. On larger length scales, electrostatic repulsion leads to a chain stiffening, so that the PE forms a linear array of electrostatic blobs. To quantify this effect, one defines an effective line charge density of a linear array of electrostatic blobs with blob size $R_{el} \simeq \sqrt{(\ell_0 L_{el})}$,

$$\tilde{\tau} \simeq \frac{\tau L_{el}}{R_{el}} \simeq \tau \left(\frac{L_{el}}{\ell_0}\right)^{1/2} \quad (20)$$

Combining Eqs. (18) and (20) gives the effective electrostatic persistence length for a string of electrostatic blobs,

$$\ell_{KK} \simeq \frac{\ell_B^{1/3} \ell_0^{-2/3} \tau^{2/3}}{\kappa^2} \quad (21)$$

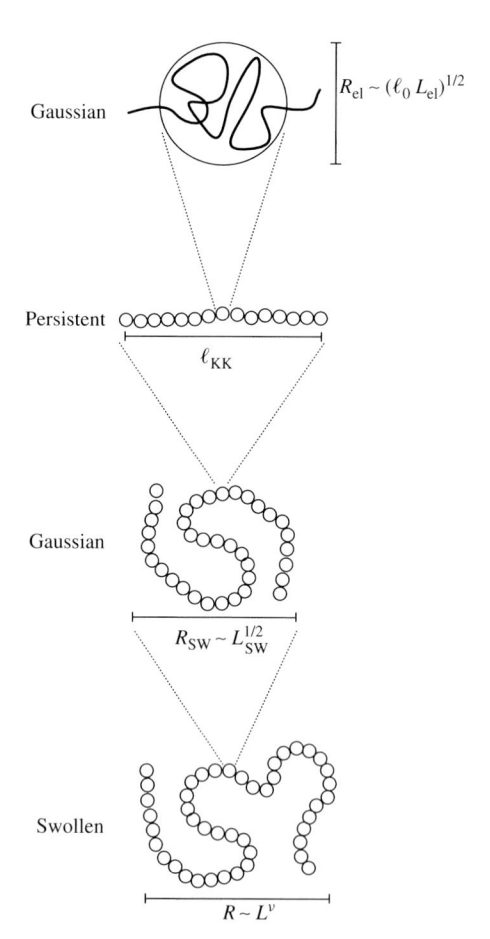

Fig. 4 Schematic view of the four scaling ranges in the Gaussian-persistent regime. On spatial scales smaller than R_{el}, the chain behavior is Gaussian; on length scales larger than R_{el} but smaller than ℓ_{KK}, the Gaussian blobs are aligned linearly. On length scales up to L_{sw}, the chain is isotropically swollen with an exponent $\nu = 1/2$, and on even larger length scales, self-avoidance effects become important and ν changes to $\nu = 3/5$.

This electrostatic stiffening is only relevant for the so-called *Gaussian-persistent regime* valid for $\ell_{KK} > R_{el}$, or equivalently

$$\tau\sqrt{\ell_B \ell_0} > (\ell_0 \kappa)^{3/2} \qquad (22)$$

When this inequality is inverted, the Gaussian-persistence regime crosses over to the Gaussian one.

The crossover boundaries (Eqs. 16, 19, 22) between the various scaling regimes are summarized in Fig. 5. We obtain three distinct regimes. In the persistent regime, for $\tau\sqrt{(\ell_B \ell_0)} > \ell_0 \kappa$ and $\tau\sqrt{(\ell_B \ell_0)} > 1$, the polymer takes on a rodlike structure with an effective persistence length larger than the bare persistence length and given by the OSF expression (Eq. 18). In the Gaussian-persistent regime, for $\tau\sqrt{(\ell_B \ell_0)} < 1$ and $\tau\sqrt{(\ell_B \ell_0)} > (\ell_0 \kappa)^{3/2}$, the polymer consists of a linear array of Gaussian electrostatic blobs, as shown in Fig. 4, with an effective persistence length ℓ_{KK} larger than the electrostatic blob size and given by Eq. (21). Finally, in the Gaussian regime, for $\tau\sqrt{(\ell_B \ell_0)} < (\ell_0 \kappa)^{3/2}$ and $\tau\sqrt{(\ell_B \ell_0)} < \ell_0 \kappa$, the electrostatic repulsion does not lead to stiffening effects at any length scale.

The persistence length ℓ_{KK} was also obtained from Monte-Carlo simulations with parameters shown in Fig. 2(d), where chain crumpling at small length scales and chain stiffening at large length scales occur simultaneously [16–20]. However, extremely long chains are needed in order to obtain reliable results for the persistence length, since the stiffening occurs only at intermediate length scales and therefore fitting of the tangent–tangent correlation function is nontrivial. Nevertheless, simulations point to a different scaling than that in Eq. (21), with a dependence on the screening length closer to a linear one, in qualitative agreement with experimental results [3]. The situation is complicated by the fact that more recent theories for the single PE chain give different results, some confirming the simple scaling results described in Eqs. (18) and (21) [12, 21, 22], some confirming Eq. (18) while criticizing Eq. (21) [11, 23, 24]. This issue is not resolved and is under intense current investigation. For multivalent counterions, fluctuation effects can even give rise to a PE collapse purely due to electrostatic interactions [25–27], which is accompanied by a negative contribution to the effective persistence length [28].

2.7.3.1.1 **Manning Condensation** A peculiar phenomenon occurs for highly charged PEs and is known as the Manning

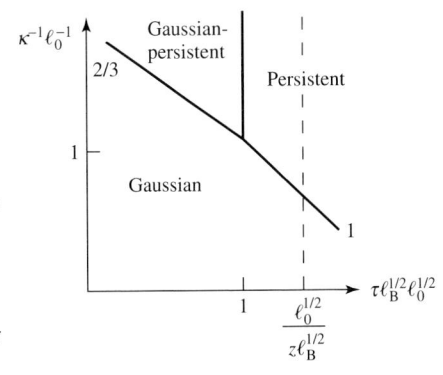

Fig. 5 Schematic phase diagram of a single semiflexible PE in bulk solution with bare persistence length ℓ_0 and line charge density τ, exhibiting various scaling regimes. High salt concentration and small τ correspond to the Gaussian regime, where the electrostatic interactions are irrelevant. In the persistent regime, the polymer persistence length is increased, and in the Gaussian-persistent regime, the polymer forms a persistent chain of Gaussian blobs as indicated in Fig. 4. The broken line indicates the Manning condensation, at which counterions condense on the polymer and reduce the effective polymer line charge density. We use a log–log plot, and the various power-law exponents for the crossover boundaries are denoted by numbers.

condensation of counterions [29, 30]. For a rigid PE represented by an infinitely long and straight cylinder with a linear charge density larger than

$$\ell_B \tau z = 1 \qquad (23)$$

where z is the counterion valency, it was shown that counterions condense on the oppositely charged cylinder even in the limit of infinite solvent dilution. Real polymers have a finite length, and are neither completely straight nor in the infinite dilution limit [31, 32]. Still, Manning condensation has an experimental significance for polymer solutions because thermodynamic quantities, such as counterion activities [33] and osmotic coefficients [34], show a pronounced signature of Manning condensation. Locally, polymer segments can be considered as straight over length scales comparable to the persistence length. The Manning condition (Eq. 23) usually denotes a region where the binding of counterions to charged chain sections begins to deplete the solution from free counterions.

Within the scaling diagram of Fig. 5, the Manning threshold (denoted by a vertical broken line) is reached typically for charge densities larger than the one needed to straighten out the chain. This holds for monovalent ions provided $\ell_0 > \ell_B$, as is almost always the case. The Manning condensation of counterions will not have a profound influence on the local chain structure since the chain is rather straight already due to monomer–monomer repulsion. A more complete description of various scaling regimes related to Manning condensation, chain collapse, and chain swelling has recently been given [35, 36].

2.7.3.1.2 Self-avoidance and Polyelectrolyte Chain Conformations
Let us now consider how the self-avoidance of PE chains comes into play, concentrating on the persistent regime defined by $\tau\sqrt{(\ell_B \ell_0)} > 1$. The end-to-end radius R of a strongly charged PE chain shows three distinct scaling ranges. For a chain length L smaller than the effective persistence length ℓ_{eff}, which according to Eq. (17) is the sum of the bare and electrostatic persistence lengths, R grows linearly with the length, $R \sim L$. Self-avoidance plays no role in this case because the chain is too short to fold back on itself.

For much longer chains, $L \gg \ell_{\mathrm{eff}}$, we envision a single polymer coil as a solution of separate polymer pieces of length ℓ_{eff}, and treat their interactions using a virial expansion. The second-virial coefficient v_2 of a rod of length ℓ_{eff} and diameter d scales as $v_2 \sim \ell_{\mathrm{eff}}^2 d$ [10]. The chain connectivity is taken into account by adding the entropic chain elasticity as a separate term. The standard Flory theory [7] for a semiflexible chain is based on writing the free energy $\mathcal{F}$ as a sum of two terms

$$\mathcal{F} \simeq \frac{R^2}{\ell_{\mathrm{eff}} L} + v_2 R^3 \left(\frac{L/\ell_{\mathrm{eff}}}{R^3}\right)^2 \qquad (24)$$

where the first term is the entropic elastic energy associated with swelling a polymer chain to a radius R and the second term is the second-virial repulsive energy proportional to the coefficient v_2 and the segment density-squared. It is integrated over the volume R^3. The optimal radius R is calculated by minimizing this free energy and gives the swollen radius

$$R \sim \left(\frac{v_2}{\ell_{\mathrm{eff}}}\right)^{1/5} L^\nu \qquad (25)$$

with $\nu = 3/5$. This swollen radius is only realized above a minimal chain length $L > L_{\mathrm{sw}} \sim \ell_{\mathrm{eff}}^7/v_2^2 \sim \ell_{\mathrm{eff}}^3/d^2$. For elongated segments with $\ell_{\mathrm{eff}} \gg d$ or, equivalently, for a highly charged PE, we obtain an intermediate range of chain lengths $\ell_{\text{eff}} < L < L_{\text{sw}}$ for which the chain is predicted to be Gaussian and the chain radius scales as

$$R \sim \ell_{\text{eff}}^{1/2} L^{1/2} \quad (26)$$

For charged chains, the effective rod diameter d is given in low salt concentrations by the screening length, that is, $d \sim \kappa^{-1}$ plus logarithmic corrections. The condition to have a Gaussian scaling range (Eq. 26) thus becomes $\ell_{\text{eff}} \gg \kappa^{-1}$. Within the persistent and the Gaussian-persistent scaling regimes depicted in Fig. 5, the effective persistence length is dominated by the electrostatic contribution and given by Eqs. (18) and (21), respectively, which in turn are always larger than the screening length κ^{-1}. It follows that a Gaussian scaling range (Eq. 26) always exists below the asymptotic swollen scaling range (Eq. 25). This situation is depicted in Fig. 4 for the Gaussian-persistent scaling regime, where the chain shows two distinct Gaussian scaling ranges at the small and large length scales. This multihierarchical scaling structure is only one of the many problems one faces when trying to understand the behavior of PE chains, be it experimentally, theoretically, or by simulations.

A different situation occurs when the polymer backbone is under bad-solvent conditions, in which case an intricate interplay between electrostatic chain swelling and short-range collapse occurs [37]. Quite recently, this interplay was theoretically shown to lead to a Rayleigh instability in the form of a necklace structure consisting of compact beads connected by thin strings [38–41]. Small-angle X-ray scattering on solvophobic PEs in a series of polar organic solvents of various solvent quality could qualitatively confirm these theoretical predictions [42].

2.7.3.2 Dilute Polyelectrolyte Solutions

It is natural to generalize the discussion of single-chain behavior to that of many PE chains at dilute concentrations. The dilute regime is defined by $c_m < c_m^*$, where c_m denotes the monomer concentration (per unit volume) and c_m^* is the concentration where individual chains start to overlap. Clearly, the overlap concentration is reached when the average bulk monomer concentration exceeds the monomer concentration inside a polymer coil. To estimate the overlap concentration c_m^*, we simply note that the average monomer concentration inside a coil with radius $R \simeq bN^\nu$ is given by

$$c_m^* \simeq \frac{N}{R^3} \simeq N^{1-3\nu} b^{-3} \quad (27)$$

For ideal chains with $\nu = 1/2$, the overlap concentration scales as $c_m^* \sim N^{-1/2}$ and thus decreases slowly as the polymerization index N increases. For rigid polymers with $\nu = 1$, the overlap concentration scales as $c_m^* \sim N^{-2}$ and decreases strongly as N increases. This means that the dilute regime for stiff PE chains corresponds to extremely low monomer concentrations. For example, taking a monomer size $b = 0.254$ nm and a polymerization index of $N = 10^4$, the overlap concentration becomes $c_m^* \approx 6 \times 10^{-7}$ nm$^{-3} \approx 10^{-3}$ mM, which is a very small concentration.

The osmotic pressure in the dilute regime in the limit $c_m \to 0$ is given by

$$\frac{\Pi}{k_B T} = \frac{f c_m}{z} + \frac{c_m}{N} \quad (28)$$

and consists of the ideal pressure of noninteracting counterions (first term) and polymer coils (second term). Note that since the second term scales as N^{-1}, it is quite small for large N and can be neglected. Hence, the main contribution to the osmotic pressure comes from the counterion entropy. This entropic term explains also why charged polymers can be dissolved in water even when their backbone is quite hydrophobic. Precipitation of the PE chains will also mean that the counterions are confined within the precipitate. The entropy loss associated with this confinement is too large and keeps the polymers dispersed in solution. In contrast, for neutral polymers there are no counterions in solution. Only the second term in the osmotic pressure exists and contributes to the low osmotic pressure of these polymer solutions. In addition, this can explain the trend toward precipitation even for very small attractive interactions between neutral polymers.

2.7.3.3 Semidilute Polyelectrolyte Solution

In the semidilute concentration regime, $c_m > c_m^*$, different polymer coils are strongly overlapping, but the polymer solution is still far from being concentrated. This means that the volume fraction of the monomers in solution is much smaller than unity, $b^3 c_m \ll 1$. In this concentration range, the statistics of counterions and polymer fluctuations are intimately connected. One example in which this feature is particularly prominent is furnished by neutron and X-ray scattering from semidilute PE solutions [43–48].

The structure factor $S(q)$ shows a pronounced peak, which results from a competition between the connectivity of polymer chains and the electrostatic repulsion between charged monomers, as will be discussed below. An important length scale, schematically indicated in Fig. 6, is the mesh size or correlation length ξ, which measures the length below which entanglement effects between different chains are unimportant. The mesh size can be viewed as the polymer (blob) scale below which single-chain statistics are valid. A semidilute solution can be thought of being composed of roughly a close-packed array of polymer blobs of size ξ.

The starting point for the present discussion is the screened interaction between two charges immersed in a semidilute PE solution containing charged polymers, their counterions, and, possibly, additional salt ions. Screening in this case is produced not only by the ions but also by the charged chain segments that can be easily polarized and shield any free charges.

Using the random-phase approximation (RPA), the effective DH interaction can be written in Fourier space as [49, 50]

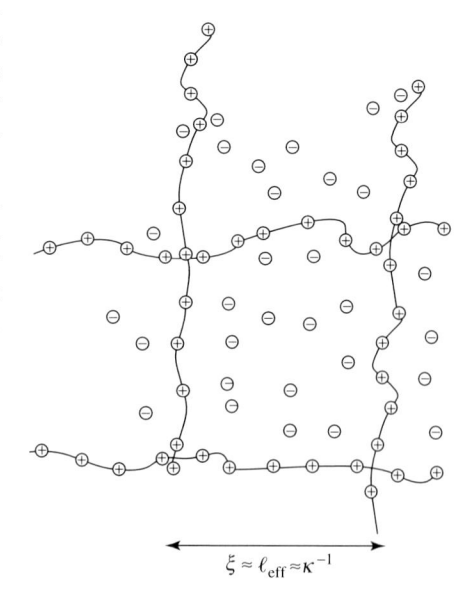

Fig. 6 Schematic view of the chain structure in the semidilute concentration range. The mesh size ξ is about equal to the effective polymer persistence length ℓ_{eff} and to the screening length κ^{-1} (if no salt is added to the system).

$\xi \approx \ell_{\text{eff}} \approx \kappa^{-1}$

$$v_{RPA}(q) = \frac{1 + v_2 c_m S_0(q)}{c_m f^2 S_0(q) + v_{DH}^{-1}(q) + v_2 c_m v_{DH}^{-1}(q) S_0(q)} \quad (29)$$

recalling that c_m is the average density of monomers in solution and f is the fraction of charged monomers on each of the PE chains. The second-virial coefficient of monomer–monomer interactions is v_2 and the single-chain form factor (discussed below) is denoted by $S_0(q)$. In the case in which no chains are present, $c_m = 0$, the RPA expression reduces to $v_{RPA}(q) = v_{DH}(q)$, the Fourier-transform of the DH potential of Eq. (9), given by

$$v_{DH}(q) = \frac{4\pi \ell_B}{q^2 + \kappa^2} \quad (30)$$

As before, κ^{-1} is the DH screening length, which is due to all mobile ions. We can write $\kappa^2 = \kappa_{salt}^2 + 4\pi z \ell_B f c_m$, where $\kappa_{salt}^2 = 8\pi z^2 \ell_B c_{salt}$ describes the screening due to added salt of valency $z:z$ and concentration c_{salt}, and the second term describes the screening due to the counterions of the PE monomers. Within the same RPA approximation, the monomer–monomer structure factor $S(q)$ of a polymer solution with monomer density c_m is given by [49, 50]

$$S^{-1}(q) = f^2 v_{DH}(q) + \frac{S_0^{-1}(q)}{c_m} + v_2 \quad (31)$$

The structure factor (or scattering function) only depends on the form factor of an isolated, noninteracting polymer chain, $S_0(q)$, the second-virial coefficient, v_2, the fraction f of charged monomers, and the interaction between monomers, which in the present case is taken to be the DH potential $v_{DH}(q)$. The structure factor of a noninteracting semiflexible polymer is characterized, in addition to the monomer length b, by its persistence length ℓ_{eff}. In general, this form factor is a complicated function that cannot be written down in closed form [51, 52]. However, one can separate between three different ranges of wave numbers q, and within each range the form factor shows a rather simple scaling behavior, namely,

$$S_0^{-1}(q) \simeq \begin{cases} N^{-1} & \text{for } q^2 < \dfrac{6}{Nb\ell_{eff}} \\ \dfrac{q^2 b \ell_{eff}}{6} & \text{for } \dfrac{6}{Nb\ell_{eff}} < q^2 < \dfrac{36}{\pi^2 \ell_{eff}^2} \\ \dfrac{qb}{\pi} & \text{for } \dfrac{36}{\pi^2 \ell_{eff}^2} < q^2 \end{cases} \quad (32)$$

For small wave numbers the polymer acts like a point scatterer, while in the intermediate wave number regime the polymer behaves like a flexible, Gaussian polymer, and for the largest wave numbers the polymer can be viewed as a stiff rod.

One of the most interesting features of semidilute PE solutions is the fact that the structure factor $S(q)$ shows a pronounced peak. For weakly charged PEs, the peak position scales as $q \sim c_m^{1/4}$ with the monomer density [45], in agreement with the aforementioned RPA results for charged polymers [49, 50]. We now discuss the scaling of the characteristic scattering peak within the present formalism. The position of the peak follows from the inverse structure factor (Eq. 31), via $\partial S^{-1}(q)/\partial q = 0$, which leads to the equation

$$q^2 + \kappa_{salt}^2 + 4\pi z \ell_B f c_m$$
$$= \left(\frac{8\pi q \ell_B f^2 c_m}{\partial S_0^{-1}(q)/\partial q} \right)^{1/2} \quad (33)$$

In principle, there are two distinct scaling behaviors possible for the peak, depending on whether the chain form factor of Eq. (32) exhibits flexible-like or rigid-like scaling. Concentrating now on the flexible case, that is, the intermediate q-range in Eq. (32), the peak of $S(q)$ scales as

$$q^* \simeq \left(\frac{24\pi \ell_B f^2 c_m}{b \ell_{\text{eff}}}\right)^{1/4} \quad (34)$$

in agreement with experimental results. A peak is only obtained if the left-hand side of Eq. (33) is dominated by the q-dependent part, that is, if $(q^*)^2 > \kappa_{\text{salt}}^2 + 4\pi z \ell_B f c_m$.

In Fig. 7(a) we show density-normalized scattering curves for a PE solution characterized by the persistence length $\ell_{\text{eff}} = 1$ nm (taken to be constant and thus independent of the monomer concentration), with monomer length $b = 0.38$ nm (as appropriate for poly-DADMAC) and charge fraction $f = 0.5$ and with no added salt. As the monomer density decreases (bottom to top in the figure), the peak moves to smaller wave numbers and sharpens, in agreement with previous implementations of the RPA. In Fig. 7(b) we show the same data in a different representation. Here we clearly demonstrate that the large-q region already is dominated by the $1/q$ behavior of the single-chain structure factor, $S_0(q)$. Since neutron-scattering data easily extend to wave numbers as high as $q \sim 5$ nm^{-1}, the stiff rodlike behavior in the high q-limit, exhibited on such a plot, will be important in interpreting and fitting experimental data even at lower q-values.

In a semidilute solution there are three different, and in principle, independent length scales: the mesh size ξ, the

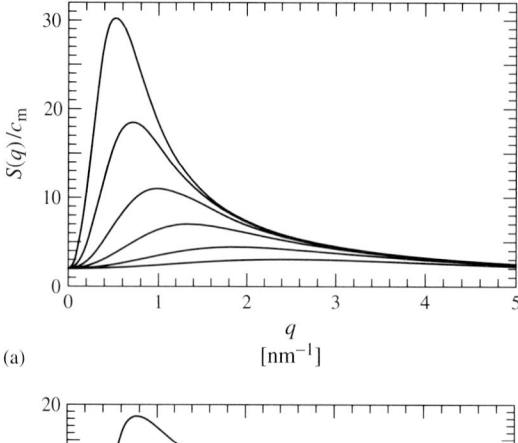

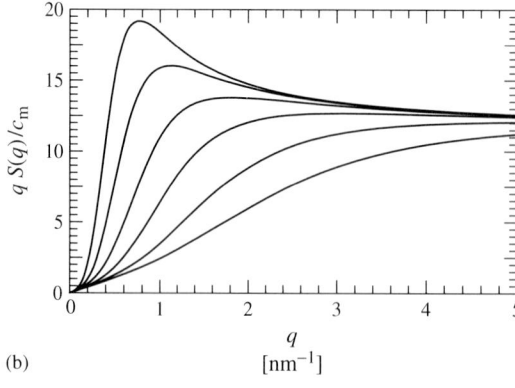

Fig. 7 (a) RPA prediction for the rescaled structure factor $S(q)/c_m$ of a semidilute PE solution with persistence length $\ell_{\text{eff}} = 1$ nm, monomer length $b = 0.38$ nm, and charge fraction $f = 0.5$ in the salt-free case. The monomer densities are (from bottom to top) $c_m = 1$ M, 0.3 M, 10 mM, 3 mM, 1 mM, and 0.3 mM. (b) For the same series of c_m values as in (a), the structure factor is multiplied by the wave number q. The semiflexibility becomes more apparent because for large q the curves tend toward a constant.

screening length κ^{-1}, and the persistence length ℓ_{eff}. In the absence of added salt, the screening length scales as

$$\kappa^{-1} \sim (z\ell_B f c_m)^{-1/2} \quad (35)$$

Assuming that the persistence length is larger or of the same order of magnitude as the mesh size, as is depicted in Fig. 6, the polymer chains can be thought of straight segments between different crossing links. Denoting the number of monomers inside a correlation blob as g, this means that $\xi \sim bg$. The average monomer concentration scales as $c_m \sim g/\xi^3$, from which we conclude that

$$\xi \sim (b c_m)^{-1/2} \quad (36)$$

Finally, the persistence length within a semidilute PE solution can be calculated by considering the electrostatic energy cost for slightly bending a charged rod. In PE solutions, it is important to include in addition to the screening by salt ions the screening due to charged chain segments. This can be calculated by using the RPA interaction (Eq. 29). Since the screening due to polymer chains is scale-dependent and increases for large separations, a q-dependent instability is encountered and leads to a persistence length [53]

$$\ell_{\text{OSF}}^{\text{sd}} \sim (b c_m)^{-1/2} \quad (37)$$

where the "sd" superscript stands for "semidilute." This result is a generalization of the OSF result for a single chain and applies to semidilute solutions. Comparing the three lengths, we see that

$$\xi \sim \ell_{\text{OSF}}^{\text{sd}} \sim \sqrt{\frac{z\ell_B f}{b}} \kappa^{-1} \quad (38)$$

Since the prefactor $\sqrt{(\ell_B f/b)}$ for synthetic fully charged polymers is roughly of order unity, one finds that for salt-free semidilute PE solutions, all three length-scales scale in the same manner with c_m, namely, as $\sim c_m^{-1/2}$, as is known also from experiments [43, 44, 54] and previous theoretical calculations [55, 56]. In simulations of many PE chains, the reduction of the chain size due to screening by PE chains was clearly seen [57–60].

2.7.4
Adsorption of a Single Polyelectrolyte Chain

After reviewing bulk properties of PE solutions, we elaborate on the adsorption diagram of a single semiflexible PE on an oppositely charged substrate. In contrast to the adsorption of neutral polymers, the resulting phase diagram shows a large region where the adsorbed polymer is flattened out on the substrate and creates a dense adsorption layer.

Experimentally, the adsorption of charged polymers on charged or neutral substrates has been characterized as a function of the polymer charge, chemical composition of the substrate, pH, and ionic strength of the solution [61, 62], as well as the substrate charge density [63–68]. Repeated adsorption of anionic and cationic PEs can lead to well-characterized multilayers on planar [69–74] and spherical substrates [75–77]. Theoretically, the adsorption of PEs on charged surfaces poses a much more complicated problem than the corresponding adsorption of neutral polymers. The adsorption process results from a subtle balance between electrostatic repulsion between charged monomers, leading to chain stiffening, and electrostatic attraction between the substrate and the polymer chain. The adsorption problem has been treated theoretically employing the uniform expansion method [78] and various continuous mean-field theories [79–83]. In all these works, the polymer density is taken to be constant in directions parallel to the surface.

The adsorption of a single semiflexible and charged chain on an oppositely charged plane [84] can be treated as a generalization of the adsorption of flexible polymers [85]. A PE characterized by the linear charge density τ is subject to an electrostatic potential created by σ, the homogeneous surface charge density (per unit area). Because this potential is attractive for an oppositely charged substrate, we consider it as the driving force for the adsorption. More complex interactions are neglected. They are due to the dielectric discontinuity at the substrate surface and are due to the impenetrability of the substrate for salt ions.

An ion in solution has a repulsive interaction from the surface when the solution dielectric constant is higher than that of the substrate. This effect can lead to desorption for highly charged PE chains. On the contrary, when the substrate is a metal, there is a possibility to induce PE adsorption on noncharged substrates or on substrates bearing charges of the same sign as the PE. See Ref. [84] for more details.

Within the linearized DH theory, the electrostatic potential of a homogeneously charged plane is

$$V_{\text{plane}}(x) = 4\pi \ell_B \sigma \kappa^{-1} e^{-\kappa x} \quad (39)$$

Assuming that the polymer is adsorbed over a layer of width δ smaller than the screening length κ^{-1}, the electrostatic attraction force per monomer unit length can be written as

$$f_{\text{att}} = -4\pi \ell_B \sigma \tau \quad (40)$$

We neglect nonlinear effects due to counterion condensation. They are described by the Gouy-Chapman (GC) theory for counterion distribution close to a charged surface. Although these effects are clearly important, it is difficult to include them systematically, and we remain at the linearized DH level.

Because of the confinement in the adsorbed layer, the polymer feels an entropic repulsion. If the layer thickness δ is much smaller than the effective persistence length of the polymer, ℓ_{eff}, as depicted in Fig. 8(a), a new length scale, the so-called deflection length λ, enters the description of the polymer statistics. The deflection length λ measures the average distance between two contact points of the polymer chain with the substrate. As shown by Odijk, the deflection length scales as $\lambda \sim \delta^{2/3} \ell_{\text{eff}}^{1/3}$ and is larger than the layer thickness δ but smaller than the persistence length ℓ_{eff} [86, 87]. The entropic repulsion follows in a simple manner from the deflection length by assuming that the polymer loses roughly a free energy of one $k_B T$ per deflection length.

On the other hand, if $\delta > \ell_{\text{eff}}$, as shown in Fig. 8(b), the polymer forms a random coil with many loops within the adsorbed layer. For a contour length smaller than $L \sim \delta^2/\ell_{\text{eff}}$, the polymer obeys Gaussian statistics and decorrelates into blobs with an entropic cost of one $k_B T$ per blob. The entropic repulsion force per monomer unit length is thus [86, 87]

$$f_{\text{rep}} \sim \begin{cases} \delta^{-5/3} \ell_{\text{eff}}^{-1/3} & \text{for } \delta \ll \ell_{\text{eff}} \\ \ell_{\text{eff}} \delta^{-3} & \text{for } \delta \gg \ell_{\text{eff}} \end{cases} \quad (41)$$

where we neglected a logarithmic correction factor that is not important for the scaling arguments. As shown in the preceding section, the effective persistence length ℓ_{eff} depends on the screening length and the line charge density; in

Fig. 8 (a) Schematic picture of the adsorbed polymer layer when the effective persistence length is larger than the layer thickness, $\ell_{\text{eff}} > \delta$. The distance between two contacts of the polymer with the substrate, the so-called deflection length, scales as $\lambda \sim \delta^{2/3}\ell_{\text{eff}}^{1/3}$. (b) Adsorbed layer for the case when the persistence length is smaller than the layer thickness, $\ell_{\text{eff}} < \delta$. In this case the polymer forms a random coil with many loops and a description in terms of a flexible polymer model becomes appropriate.

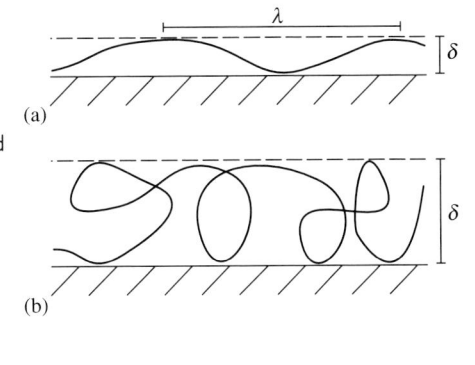

essence, one has to keep in mind that ℓ_{eff} is larger than ℓ_0 for a wide range of parameters because of electrostatic stiffening effects.

The situation is complicated by the fact that the electrostatic contribution to the persistence length is scale-dependent and decreases as the chain is bent at length scales smaller than the screening length. This leads to modifications of the entropic confinement force (Eq. 41) if the deflection length is smaller than the screening length. As can be checked explicitly, all results reported here are not changed by these modifications.

The equilibrium layer thickness follows from equating the attractive and repulsive forces (Eqs. 40 and 41). For rather stiff polymers and small layer thickness, $\delta < \kappa^{-1} < \ell_{\text{eff}}$, we obtain

$$\delta \sim (\ell_B \sigma \tau \ell_{\text{eff}}^{1/3})^{-3/5} \qquad (42)$$

For a layer thickness corresponding to the screening length, $\delta \approx \kappa^{-1}$, scaling arguments predict a rather abrupt desorption transition [84]. This is in accord with previous transfer-matrix calculations for a semiflexible polymer bound by short-ranged (square-well) potentials [88–91]. Setting $\delta \sim \kappa^{-1}$ in Eq. (42), we obtain an expression for the adsorption threshold (for $\kappa\ell_{\text{eff}} > 1$)

$$\sigma^* \sim \frac{\kappa^{5/3}}{\tau \ell_B \ell_{\text{eff}}^{1/3}} \qquad (43)$$

For $\sigma > \sigma^*$, the polymer is adsorbed and localized over a layer with a width smaller than the screening length (and with the condition $\ell_{\text{eff}} > \kappa^{-1}$ also satisfying $\delta < \ell_{\text{eff}}$). As σ is decreased, the polymer abruptly desorbs at the threshold $\sigma = \sigma^*$. In the Gaussian regime, the effective persistence length ℓ_{eff} is given by the bare persistence length ℓ_0 and the desorption threshold is obtained by replacing ℓ_{eff} by ℓ_0 in Eq. (43), that is,

$$\sigma^* \sim \frac{\kappa^{5/3}}{\tau \ell_B \ell_0^{1/3}} \qquad (44)$$

In the persistent regime, we have $\ell_{\text{eff}} \sim \ell_{\text{OSF}}$ with ℓ_{OSF} given by Eq. (18). The adsorption threshold follows from Eq. (43) as

$$\sigma^* \sim \frac{\kappa^{7/3}}{\tau^{5/3}\ell_B^{4/3}} \qquad (45)$$

Finally, in the Gaussian-persistent regime, we have an effective line charge density from Eq. (20) and a modified persistence length (Eq. 21). For the adsorption threshold, we obtain from Eq. (43)

$$\sigma^* \sim \frac{\kappa^{7/3}\ell_0^{5/9}}{\tau^{5/9}\ell_B^{7/9}} \qquad (46)$$

Let us now consider the opposite limit, $\ell_{\text{eff}} < \kappa^{-1}$. From Eq. (42), we see that the layer thickness δ is of the same order as ℓ_{eff} for $\ell_B \sigma \tau \ell_{\text{eff}}^2 \sim 1$, at which point the condition $\delta \ll \ell_{\text{eff}}$ used in deriving Eq. (42) breaks down. If the layer thickness is larger than the persistence length but smaller than the screening length, $\ell_{\text{eff}} < \delta < \kappa^{-1}$, the prediction for δ obtained from balancing Eqs. (40) and (41) becomes

$$\delta \sim \left(\frac{\ell_{\text{eff}}}{\ell_B \sigma \tau}\right)^{1/3} \qquad (47)$$

From this expression we see that δ has the same size as the screening length κ^{-1} for

$$\sigma^* \sim \frac{\ell_{\text{eff}} \kappa^3}{\tau \ell_B} \qquad (48)$$

This in fact denotes the location of a continuous adsorption transition at which the layer grows to infinity. The scaling results for the adsorption behavior of a flexible polymer (Eqs. 47–48) are in agreement with previous results [78].

In Fig. 9 we show the desorption transitions and the line at which the adsorbed layer crosses over from being flat, $\delta < \ell_{\text{eff}}$, to being crumpled or coiled, $\delta > \ell_{\text{eff}}$. The underlying PE behavior in the bulk, as shown in Fig. 5, is denoted by broken lines. We obtain two different phase diagrams, depending on the value of the parameter

$$\Sigma = \sigma \ell_0^{3/2} \ell_B^{1/2} \qquad (49)$$

For strongly charged surfaces, $\Sigma > 1$, we obtain the phase diagram as in Fig. 9(a), and for weakly charged surfaces, $\Sigma < 1$, as in Fig. 9(b). We see that strongly charged PEs, obeying $\tau \sqrt{(\ell_0 \ell_B)} > 1$, always adsorb in flat layers. The scaling of the desorption transitions is in general agreement with recent computer simulations of charged PEs [92].

2.7.4.1 Adsorption on Curved Substrates

Adsorption of PEs on curved substrates is of importance because PEs are

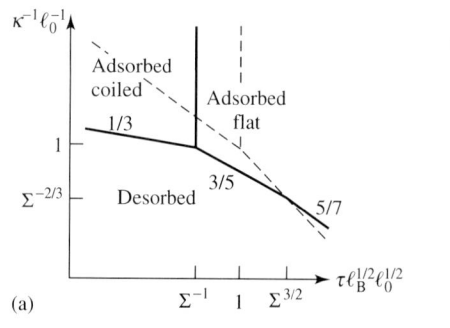

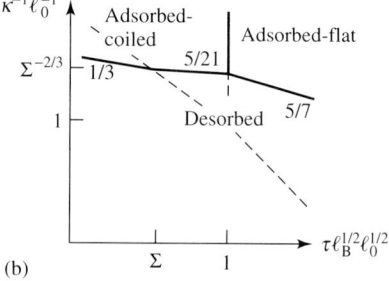

Fig. 9 Adsorption scaling diagram shown on a log–log plot for (a) strongly charged surfaces, $\Sigma = \sigma \ell_0^{3/2} \ell_B^{1/2} > 1$, and for (b) weakly charged surfaces, $\Sigma < 1$. We find a desorbed regime, an adsorbed phase in which the polymer is flat and dense, and an adsorbed phase in which the polymer shows loops. It is seen that a fully charged PE is expected to adsorb in a flat layer, whereas charge-diluted PEs can form coiled layers with loops and dangling ends. The broken lines denote the scaling boundaries of PE chains in the bulk as shown in Fig. 5. The numbers on the lines indicate the power-law exponents of the crossover boundaries between the regimes.

Fig. 10 Numerically determined adsorption diagram for a charged semiflexible polymer of length $L = 50$ nm, linear charge density $\tau = 6$ nm^{-1}, persistence length $\ell_0 = 30$ nm, interacting with an oppositely charged sphere of radius $R_p = 5$ nm. Shown is the main transition from the unwrapped configuration (at the bottom) to the wrapped configuration (at the top) as a function of sphere charge Z and inverse screening length κ. Wrapping is favored at intermediate salt concentrations. The parameters are chosen for the problem of DNA-histone complexation. (Adapted from Ref. [99].)

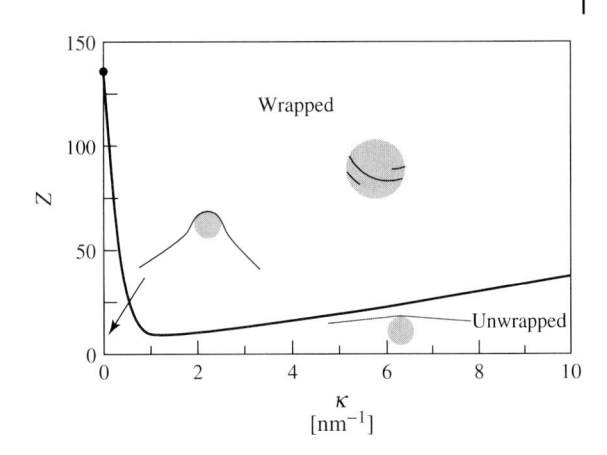

widely used to stabilize colloidal suspensions [93] and to fabricate hollow polymeric shells [75–77]. When the curvature of the small colloidal particles is large enough, it can lead to a much more pronounced effect for PE adsorption as compared with neutral polymer. This is mainly due to the fact that the electrostatic energy of the adsorbed PE layer depends sensitively on curvature [94–98]. Bending a charged polymer around a small sphere costs a large amount of electrostatic energy, which will disfavor adsorption of long, strongly charged PE at very low salt concentrations.

In Fig. 10 we show the adsorption phase diagram of a single stiff PE of finite length that interacts with an oppositely charged sphere of charge Z (in units of e). The specific parameters were chosen as appropriate for the complexation of DNA (a negatively charged, relatively stiff biopolymer) with positively charged histone proteins, corresponding to a DNA length of $L = 50$ nm, a chain persistence length of $\ell_0 = 30$ nm, and a sphere radius of $R_p = 5$ nm. The phase diagram was obtained by minimization of the total energy including bending energy of the DNA, electrostatic attraction between the sphere and the DNA, and electrostatic repulsion between the DNA segments with respect to the chain configuration [99]. Configurational fluctuations away from this ground state are unimportant for such stiff polymers.

We show in Fig. 10 the main transition between an unwrapped state, at low sphere charge Z, and the wrapped state, at large sphere charge Z. It is seen that at values of the sphere charge between $Z = 10$ and $Z = 130$, the wrapping only occurs at intermediate values of the inverse screening length $\kappa \sim c_{\text{salt}}^{1/2}$. At low salt concentrations (lower left corner in the phase diagram), the self-repulsion between DNA segments prevents wrapping, while at large salt concentrations (lower right corner in the diagram), the electrostatic attraction is not strong enough to overcome the mechanical bending energy of the DNA molecule. These results are in good agreement with experiments on DNA/histone complexes [100]. Interestingly, the optimal salt concentration, where a minimal sphere charge is needed to wrap the DNA, occurs at physiological salt concentration, for $\kappa^{-1} \approx 1$ nm. For colloidal particles of larger size and for flexible synthetic polymers, configurational fluctuations become important. They have been treated using a mean-field description in terms of the average monomer density profile around the sphere [101, 102].

2.7.5
Adsorption from Semidilute Solutions

So far we have been reviewing the behavior of single PE chains close to a charged wall (or surface). This will now be extended to include adsorption of PEs from bulk (semidilute) solutions having a bulk concentration c_m^b. As before, the chains are assumed to have a fraction f of charged monomers, each carrying a charge e, resulting in a linear charge density $\tau = f/b$ on the chain. The solution can also contain salt (small ions) of concentration c_{salt}, which is directly related to the DH screening length, κ^{-1}. For clarity purposes, the salt is assumed to be monovalent ($z = 1$) throughout Sect. 2.7.5.

We will consider adsorption only onto a single flat and charged surface. Clearly, the most important quantity is the profile of the polymer concentration $c_m(x)$ as function of x, the distance from the wall. Another useful quantity is the polymer surface excess per unit area, defined as

$$\Gamma = \int_0^\infty [c_m(x) - c_m^b] \, dx \quad (50)$$

Related to the surface excess Γ is the amount of charges (in units of e) carried by the adsorbing PE chains, $f\Gamma$. In some cases the polymer carries a higher charge (per unit area) than the charged surface itself, $f\Gamma > \sigma$, and the surface charge is overcompensated by the PE, as we will see later. This does not violate charge neutrality in the system because of the presence of counterions in solution.

In many experiments, the total amount of polymer surface excess Γ is measured as a function of the bulk polymer concentration, pH, and/or ionic strength of the bulk solution [103–110]. For reviews see, for example, Refs. [61, 62, 111, 112]. More recently, spectroscopy [105] and ellipsometry [109] have been used to measure the width of the adsorbed PE layer. Other techniques such as neutron-scattering can be employed to measure the entire profile $c_m(x)$ of the adsorbed layer [113, 114].

In spite of the difficulties in treating theoretically PEs in solution because of the delicate interplay between the chain connectivity and the long range nature of electrostatic interactions [1, 9, 115, 116], several simple approaches treating adsorption exist.

One approach is a discrete *multi-Stern layer* model [117–121], where the system is placed on a lattice whose sites can be occupied by a monomer, a solvent molecule, or a small ion. The electrostatic potential is determined self-consistently (mean-field approximation) together with concentration profiles of the polymer and small ions.

In another approach, the electrostatic potential and the PE concentration are treated as continuous functions [78, 80–82, 122–125]. These quantities are obtained from two coupled differential equations derived from the total free energy of the system. We will review the main results of the latter approach, presenting numerical solutions and scaling arguments of the mean-field profiles.

2.7.5.1 Mean-field Theory and its Profile Equations

The charge density on the polymer chains is assumed to be continuous and uniformly distributed along the chains.

Further treatments of the polymer charge distribution (annealed and quenched models) can be found in Refs. [81, 82, 123].

Within mean-field approximation, the free energy of the system can be expressed in terms of the local electrostatic potential $\psi(\mathbf{r})$, the local monomer concentration $c_m(\mathbf{r})$, and the local concentration of positive and negative ions $c^{\pm}(\mathbf{r})$. The mean-field approximation means that the influence of the charged surface and the interchain interactions can be expressed in terms of an external potential that will determine the local concentration of the monomers, $c_m(\mathbf{r})$. This external potential depends both on the electrostatic potential and on the excluded volume interactions between the monomers and the solvent molecules. The excess free energy with respect to the bulk can then be calculated using another important approximation, the ground state dominance. This approximation is used often for neutral polymers [9] and is valid for very long polymer chains, $N \gg 1$. It is then convenient to introduce the polymer order parameter $\phi(\mathbf{r})$, where $c_m(\mathbf{r}) = |\phi(\mathbf{r})|^2$, and to express the adsorption free energy $\mathcal{F}$ in terms of ϕ and ψ (and in units of $k_B T$) [80–82, 122–124]

$$\mathcal{F} = \int d\mathbf{r}\{F_{\text{pol}}(\mathbf{r}) + F_{\text{ions}}(\mathbf{r}) + F_{\text{el}}(\mathbf{r})\} \quad (51)$$

The polymer contribution is

$$F_{\text{pol}}(\mathbf{r}) = \frac{a^2}{6}|\nabla\phi|^2 + \frac{1}{2}v_2(\phi^4 - \phi_b^4)$$
$$- \mu_p(\phi^2 - \phi_b^2) \quad (52)$$

where the first term is the polymer elastic energy. Throughout this section we restrict ourselves to flexible chains and treat the Kuhn length a and the effective monomer size b as the same parameter. The second term is the excluded volume contribution where the second-virial coefficient v_2 is of order a^3. The last term couples the system to a polymer reservoir via a chemical potential μ_p, and $\phi_b = \sqrt{c_m^b}$ is related to the bulk monomer concentration, c_m^b.

The entropic contribution of the small (monovalent) ions is

$$F_{\text{ions}}(\mathbf{r}) = \sum_{i=\pm}[c^i \ln c^i - c^i - c_{\text{salt}} \ln c_{\text{salt}}$$
$$+ c_{\text{salt}}] - \mu^i(c^i - c_{\text{salt}}) \quad (53)$$

where $c^i(\mathbf{r})$ and μ^i are, respectively, the local concentration and the chemical potential of the $i = \pm$ ions, while c_{salt} is the bulk concentration of salt.

Finally, the electrostatic contributions (per $k_B T$) are

$$F_{\text{el}}(\mathbf{r})$$
$$= \frac{\left[fe\phi^2\psi + ec^+\psi - ec^-\psi - \frac{\varepsilon}{8\pi}|\nabla\psi|^2\right]}{k_B T}$$
$$\quad (54)$$

The first three terms are the electrostatic energies of the monomers, the positive ions, and the negative ions, respectively; f is the fractional charge carried by one monomer. The last term is the self-energy of the electric field where ε is the dielectric constant of the solution. Note that the electrostatic contribution (Eq. 54) is equivalent to the well-known result: $(\varepsilon/8\pi k_B T)\int d\mathbf{r}|\nabla\psi|^2$ plus surface terms. This can be seen by substituting the Poisson–Boltzmann equation (as obtained below) into Eq. (54) and then integrating by parts.

Minimization of the free energy (Eqs. 51–54) with respect to $c^{\pm}$, ϕ, and ψ yields a Boltzmann distribution for the density of the small ions, $c^{\pm}(\mathbf{r}) = c_{\text{salt}} \exp(\mp e\psi/k_B T)$, and two coupled differential equations for ϕ and ψ:

$$\nabla^2 \psi(\mathbf{r}) = \frac{8\pi e}{\varepsilon} c_{\text{salt}} \sinh\left(\frac{e\psi}{k_B T}\right)$$
$$- \frac{4\pi e}{\varepsilon}(f\phi^2 - f\phi_b^2 e^{e\psi/k_B T}) \tag{55}$$

$$\frac{a^2}{6}\nabla^2 \phi(\mathbf{r}) = v_2(\phi^3 - \phi_b^2 \phi) + \frac{f\phi e\psi}{k_B T} \tag{56}$$

Equation (55) is a generalized Poisson–Boltzmann equation including the free ions and the charged polymers. The first term represents the salt contribution and the second term is due to the charged monomers and their counterions. Equation (56) is a generalization of the self-consistent field equation of neutral polymers [9]. In the bulk, the above equations are satisfied by setting $\psi \to 0$ and $\phi \to \phi_b$.

2.7.5.2 Numerical Profiles: Constant ψ_s

When the surface is taken as ideal, that is, flat and homogeneous, the physical quantities depend only on the distance x from the surface. The surface imposes boundary conditions on the polymer order parameter $\phi(x)$ and electrostatic potential $\psi(x)$. In thermodynamic equilibrium, all charge carriers in solution should exactly balance the surface charges (charge neutrality). The Poisson–Boltzmann Equation (55), the self-consistent field Equation (56), and the boundary conditions uniquely determine the polymer concentration profile and the electrostatic potential. In most cases, these two coupled nonlinear equations can only be solved numerically.

We present now numerical profiles obtained for surfaces with a constant potential ψ_s:

$$\psi|_{x=0} = \psi_s \tag{57}$$

The boundary conditions for $\phi(x)$ depend on the nature of the short-range nonelectrostatic interactions of the monomers and the surface. For simplicity, we take a nonadsorbing surface and require that the monomer concentration will vanish there:

$$\phi|_{x=0} = 0 \tag{58}$$

We note that the boundary conditions chosen in Eqs. (57) to (58) model the particular situation of electrostatic attraction in competition with a short-range (steric) repulsion of nonelectrostatic origin. Possible variations of these boundary conditions include surfaces with a constant surface charge (discussed later) and surfaces with a nonelectrostatic short-range attractive (or repulsive) interaction with the polymer [83, 127]. Far from the surface ($x \to \infty$), both ψ and ϕ reach their bulk values and their derivatives vanish: $\psi'|_{x\to\infty} = 0$ and $\phi'|_{x\to\infty} = 0$.

The numerical solutions of the mean-field Eqs. (55) and (56) together with the boundary conditions discussed above are presented in Fig. 11, for several different physical parameters.

The polymer is positively charged and is attracted to the nonadsorbing surface held at a constant negative potential. The aqueous solution contains a small amount of monovalent salt ($c_{\text{salt}} = 0.1$ mM). The reduced concentration profile $c_m(x)/\phi_b^2$ is plotted as a function of the distance from the surface x. Different curves correspond to different values of the reduced surface potential $y_s \equiv e\psi_s/k_B T$, the charge fraction f, and the effective monomer size a.

Although the spatial variation of the profiles differs in detail, they all have a single peak that can be characterized by its height and width. This observation

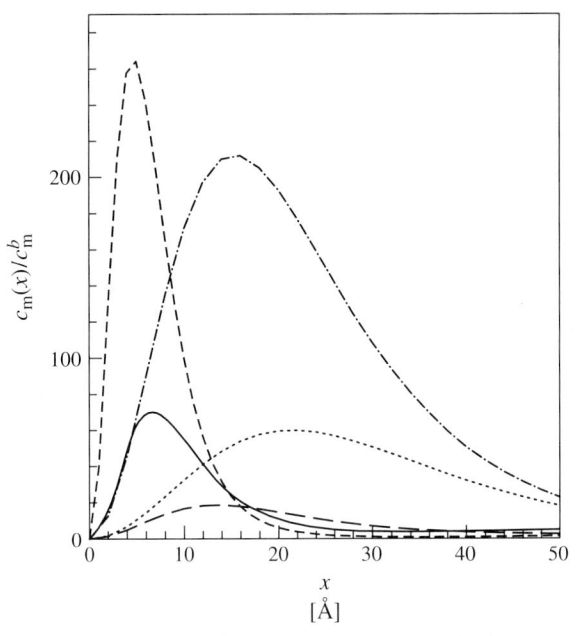

Fig. 11 Adsorption profiles obtained by numerical solutions of Eqs. (55) and (56) for several sets of physical parameters in the low-salt limit. The polymer concentration scaled by its bulk value $c_m^b = \phi_b^2$ is plotted as a function of the distance from the surface. The different curves correspond to: $f = 1$, $a = 5$ Å and $y_s = e\psi_s/k_B T = -0.5$ (solid curve); $f = 0.1$, $a = 5$ Å and $y_s = -0.5$ (dots); $f = 1$, $a = 5$ Å and $y_s = -1.0$ (short dashes); $f = 1$, $a = 10$ Å and $y_s = -0.5$ (long dashes); and $f = 0.1$, $a = 5$ Å and $y_s = 1.0$ (dot–dash line). For all cases, $\phi_b^2 = 10^{-6}$ Å^{-3}, $v_2 = 50$ Å^3, $\varepsilon = 80$, $T = 300$ K, and $c_{\text{salt}} = 0.1$ mM. (Adapted from Ref. [124].)

serves as a motivation to using scaling arguments.

2.7.5.3 Scaling Results

The numerical profiles of the previous section indicate that it may be possible to obtain simple analytical results for the PE adsorption by assuming that the adsorption is characterized by one dominant length scale D. Hence, we write the polymer order parameter profile in the form

$$\phi(x) = \sqrt{c_M} h(x/D) \quad (59)$$

where $h(x)$ is a dimensionless function normalized to unity at its maximum and c_M sets the scale of polymer adsorption, such that $\phi(D) = \sqrt{c_M}$. The free energy can now be expressed in terms of D and c_M, while the exact form of $h(x)$ affects only the numerical prefactors.

In principle, the adsorption length D depends also on the ionic strength through κ^{-1}. As discussed below, the scaling assumption (Eq. 59) is only valid as long as κ^{-1} and D are not of the same order of magnitude. Otherwise, h should be a function of both κx

and x/D. We concentrate now on two limiting regimes where Eq. (59) can be justified: (1) the low-salt regime $D \ll \kappa^{-1}$ and (2) the high-salt regime $D \gg \kappa^{-1}$. We first discuss the case of constant surface potential, which can be directly compared to the numerical profiles. Then we note the differences with the constant surface charge boundary condition in which the interesting phenomenon of charge overcompensation is discussed in detail.

2.7.5.3.1 Low-salt Regime $D \ll \kappa^{-1}$ and ψ_s = constant

In the low-salt regime the effect of the small ions can be neglected and the free energy (per unit surface area) (Eqs. 51–54) is approximated by (see also Refs. [80, 124])

$$F \simeq \frac{a^2}{6D} c_M - f|y_s|c_M D$$
$$+ 4\pi l_B f^2 c_M^2 D^3 + \frac{1}{2} v_2 c_M^2 D \quad (60)$$

In the above equation and in what follows, we neglect additional prefactors of order unity in front of the various terms that arise from inserting the scaling profile (Eq.59) into the free energy. The first term of Eq. (60) is the elastic energy characterizing the response of the polymer to concentration inhomogeneities. The second term accounts for the electrostatic attraction of the polymers to the charged surface. The third term represents the Coulomb repulsion between adsorbed monomers. The last term represents the excluded volume repulsion between adsorbed monomers, where we assume that the monomer concentration near the surface is much larger than the bulk concentration $c_M \gg \phi_b^2$.

In the low-salt regime and for highly charged PEs, the electrostatic interactions are much stronger than the excluded volume ones.

Neglecting the latter interactions and minimizing the free energy with respect to D and c_M gives

$$D^2 \simeq \frac{a^2}{f|y_s|} \sim \frac{1}{f|\psi_s|} \quad (61)$$

and

$$c_M \simeq \frac{|y_s|^2}{4\pi l_B a^2} \sim |\psi_s|^2 \quad (62)$$

recalling that $y_s = e\psi_s/k_B T$. As discussed above, these expressions are valid as long as (1) $D \ll \kappa^{-1}$ and (2) the excluded volume term in Eq. (60) is negligible. Condition (1) translates into $c_{\text{salt}} \ll f|y_s|/(8\pi l_B a^2)$. For $|y_s| \simeq 1$, $a = 5$ Å and $l_B = 7$ Å this limits the salt concentration to $c_{\text{salt}}/f \ll 0.4$ M. Condition (2) on the magnitude of the excluded volume term can be shown to be equivalent to $f \gg v_2|y_s|/l_B a^2$. These requirements are consistent with the data presented in Fig. 11.

We recall that the profiles presented in Fig. 11 were obtained from the numerical solution of Eqs. (55) and (56), including the effect of small ions and excluded volume. The scaling relations are verified by plotting in Fig. 12 the same sets of data as in Fig. 11, using rescaled variables as defined in Eqs. (61) and (62). That is, the rescaled electrostatic potential $\psi(x)/\psi_s$ and polymer concentration $c_m(x)/c_M \sim c_m(x)a^2/|y_s|^2$ are plotted as functions of the rescaled distance $x/D \sim xf^{1/2}|y_s|^{1/2}/a$. The different curves roughly collapse on the same curve.

In many experiments the total amount of adsorbed polymer per unit area Γ is measured as function of the physical characteristics of the system such as the charge fraction f, the pH of the solution, or the salt concentration c_{salt} [103–110].

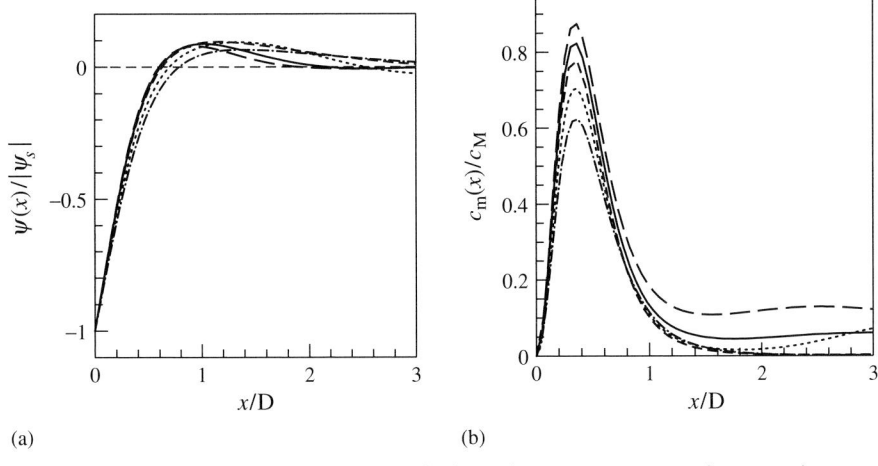

Fig. 12 Scaling behavior of PE adsorption in the low-salt regime (Eqs. 61 and 62). (a) The rescaled electrostatic potential $\psi(x)/|\psi_s|$ as a function of the rescaled distance x/D. (b) The rescaled polymer concentration $c_m(x)/c_M$ as a function of the same rescaled distance. The profiles are taken from Fig. 11 (with the same notation). (Adapted from Ref. [124].)

This quantity can be easily obtained from our scaling expressions yielding

$$\Gamma = \int_0^\infty [c_m(x) - \phi_b^2]\,dx$$

$$\simeq Dc_M \simeq \frac{|y_s|^{3/2}}{l_B a f^{1/2}} \sim \frac{|\psi_s|^{3/2}}{f^{1/2}} \quad (63)$$

The adsorbed amount $\Gamma(f)$ in the low-salt regime is plotted in the inset of Fig. 13(a). As a consequence of Eq. (63), Γ decreases with increasing charge fraction f. Similar behavior was also reported in experiments [106]. This effect is at first glance quite puzzling because as the polymer charge increases, the chains are subject to a stronger attraction to the surface. On the other hand, the monomer–monomer repulsion is stronger and, indeed, in this regime, the monomer–monomer Coulomb repulsion scales as $(fc_M)^2$ and dominates over the adsorption energy that scales as fc_M.

2.7.5.3.2 High-salt Regime $D \gg \kappa^{-1}$ and ψ_s = constant

Let us now consider the opposite case of a high ionic strength solution. Here, D is much larger than κ^{-1}, and the electrostatic interactions are short ranged with a cutoff κ^{-1}. The free energy of the adsorbing PE layer (per unit surface area) then reads

$$F \simeq \frac{a^2}{6D}c_M - f|y_s|c_M\kappa^{-1}$$
$$+ 4\pi l_B f^2 \kappa^{-2} c_M^2 D + \frac{1}{2}v_2 c_M^2 D \quad (64)$$

The electrostatic cutoff enters in two places. In the second term, only the first layer of width κ^{-1} interacts electrostatically with the surface. In the third term, each charged layer situated at point x interacts only with layers at x' for which $|x - x'| < \kappa^{-1}$. This term can be also viewed as an additional electrostatic excluded volume with $v_{el} \sim l_B(f/\kappa)^2$.

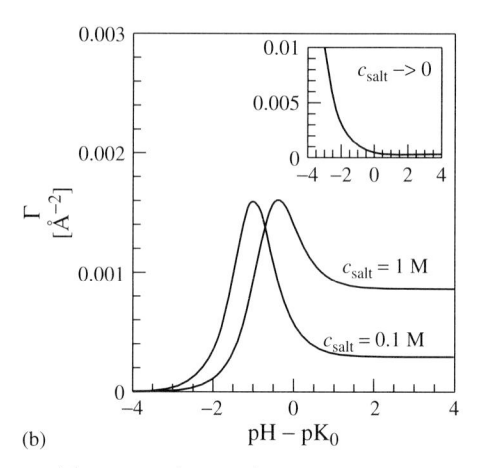

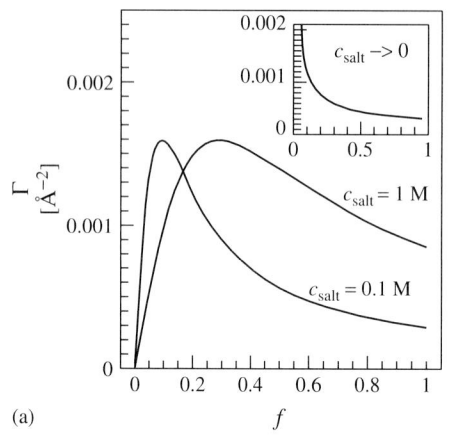

Fig. 13 Typical adsorbed amount Γ as a function of (a) the charge fraction f and (b) the pH $-$ pK$_0$ of the solution for different salt concentrations (Eq. 67). The insets correspond to the low-salt regime (Eq. 63). The parameters used for ε, T, and v_2 are the same as in Fig. 11, while $y_s = e\psi_s/k_BT = -0.5$ and $a = 5$Å. The bulk concentration $c_m^b = \phi_b^2$ is assumed to be much smaller than c_M. (Adapted from Ref. [124].)

Minimization of the free energy gives

$$D \simeq \frac{\kappa a^2}{f|y_s|} \sim \frac{c_{\text{salt}}^{1/2}}{f|\psi_s|} \quad (65)$$

and

$$c_M \sim \frac{f^2|y_s|^2/(\kappa a)^2}{f^2/c_{\text{salt}} + \alpha v_2} \quad (66)$$

yielding

$$\Gamma \sim \frac{f|y_s|c_{\text{salt}}^{-1/2}}{f^2/c_{\text{salt}} + \alpha v_2} \sim \frac{f|\psi_s|}{v_{\text{el}} + \alpha v_2} c_{\text{salt}}^{-1/2} \quad (67)$$

where α is a numerical constant of order unity that depends on the profile details.

The adsorption behavior is depicted in Figs. 13 and 14. Our scaling results are in agreement with numerical solutions of discrete lattice models (the multi-Stern layer theory) [61, 62, 111, 117–120]. In Fig. 13, Γ is plotted as function of f (Fig. 13a) and the pH (Fig. 13b) for different salt concentrations. The behavior as seen in Fig. 13(b) represents annealed PEs where the nominal charge fraction is given by the pH of the solution through the expression

$$f = \frac{10^{\text{pH}-\text{pK}_0}}{1 + 10^{\text{pH}-\text{pK}_0}} \quad (68)$$

where pK$_0 = -\log_{10} K_0$ and K_0 is the apparent dissociation constant. We note that this relation is only strictly valid for infinitely dilute monomers and that distinct deviations from it are observed for PEs because of the electrostatic repulsion between neighboring dissociating sites [126]. Still, the results in Fig. 13(b) capture the main qualitative trends of pH-dependent PE adsorption.

Another interesting observation that can be deduced from Eq. (67) is that Γ is only a function of $fc_{\text{salt}}^{-1/2}$. Indeed, as can be seen in Fig. 13, c_{salt} only affects the position of the peak and not its height.

The effect of salt concentration is shown in Fig. 14, where Γ is plotted as function of the salt concentration c_{salt} for two charge fractions $f = 0.01$ and 0.25. The curves on the right-hand side of the graph are

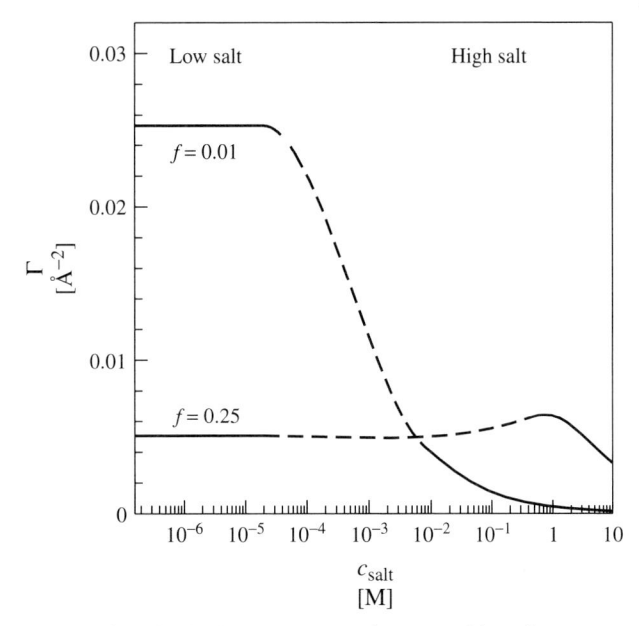

Fig. 14 The adsorbed amount Γ as a function of the salt concentration c_{salt} (Eq. 67) for $f = 0.01$ and 0.25. The solid curves on the right-hand side correspond to the scaling relations in the high-salt regime (Eq. 67). The horizontal lines on the left-hand side mark the low salt values (Eq. 67). The dashed lines serve as guides to the eye. The parameters used are $\varepsilon = 80$, $T = 300\,K$, $v_2 = 50\,\text{Å}^3$, $a = 5\,\text{Å}$, $y_s = -2.0$. (Adapted from Ref. [124].)

calculated from the high-salt expression for Γ (Eq. 67). The horizontal lines on the left-hand side of the graph indicate the low-salt values of Γ (Eq. 63). The dashed lines in the intermediate salt regime serve only as guides to the eye since our scaling approach is not valid when D and κ^{-1} are of the same order.

Emphasis should be drawn to the distinction between weakly and strongly charged PEs. For weak PEs, the adsorbed amount Γ is a monotonously decreasing function of the salt concentration c_{salt} in the whole range of salt concentrations. The reason being that the monomer–monomer Coulomb repulsion, proportional to f^2, is weaker than the monomer–surface interaction, which is linear in f.

For strongly charged PEs, on the other hand, the balance between these two electrostatic terms depends on the amount of salt. At low salt concentrations, the dominant interaction is the monomer–surface Coulomb repulsion. Consequently, addition of salt screens this interaction and increases the adsorbed amount. When the salt concentration is high enough, this Coulomb repulsion is screened out and the effect of salt is to weaken the surface attraction. At this point the adsorbed amount starts to decrease. As a result, the behavior over the whole concentration range is nonmonotonic with a maximum at some optimal value c^*_{salt}, as seen in Fig. 14.

From this analysis and from Figs. 13, 14 and Eq. (67), it is now natural to divide the high-salt regime into two subregimes according to the PE charge. At low charge fractions (subregime HS I), $f \ll f^* = (c_{salt}v_2)^{1/2}$, the excluded volume term dominates the denominator of Eq. (67) and

$$\Gamma \sim f|\psi_s|c_{salt}^{-1/2} \qquad (69)$$

whereas at high f (subregime HS II), $f \gg f^*$, the monomer–monomer electrostatic repulsion dominates and Γ decreases with f and increases with c_{salt}:

$$\Gamma \sim c_{salt}^{1/2}|\psi_s|f^{-1} \qquad (70)$$

The various regimes with their crossover lines are shown schematically in Fig. 15. Keeping the charge fraction f constant while changing the amount of salt corresponds to a vertical scan through the diagram. For weak PEs, this cut goes through the left-hand side of the diagram starting from the low-salt regime and, upon addition of salt, into the HS I regime. Such a path describes the monotonous behavior inferred from Fig. 14 for the weak PE ($f = 0.01$). For strong PEs, the cut goes through the right-hand side of the diagram, starting from the low-salt regime, passing through the HS II regime, and ending in the HS I. The passage through the HS II regime is responsible for the nonmonotonous behavior inferred from Fig. 14 for the strong PE ($f = 0.25$).

Similarly, Fig. 13(a,b) correspond to horizontal scans through the top half of the diagram. As long as the system is in the HS I regime, the adsorbed amount increases when the polymer charge fraction

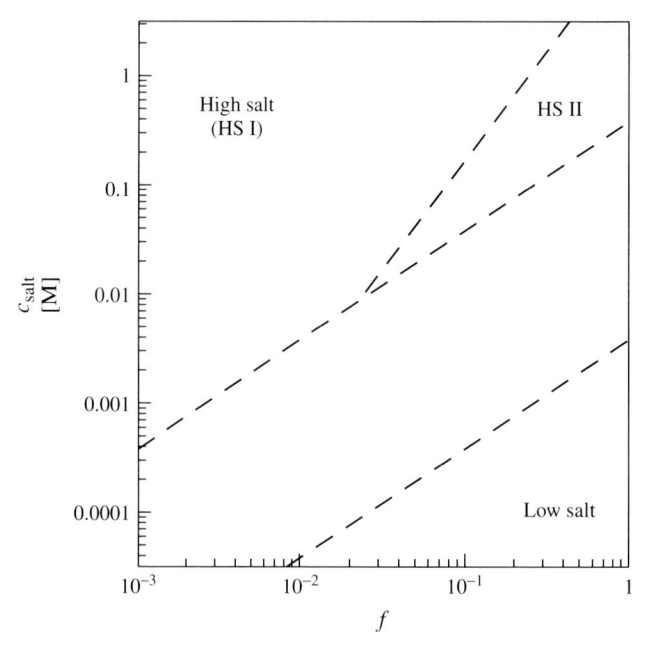

Fig. 15 Schematic diagram of the different adsorption regimes as function of the charge fraction f and the salt concentration c_{salt}. Three regimes can be distinguished: (a) the low-salt regime $D \ll \kappa^{-1}$; (b) the high-salt regime (HS I) $D \gg \kappa^{-1}$ for weak PEs $f \ll f^* = (c_{salt}v_2)^{1/2}$; and (c) the high-salt regime (HS II) $D \gg \kappa^{-1}$ for strong PEs $f \gg f^*$.

increases. As the polymer charge further increases, the system enters the HS II regime and the adsorbed amount decreases. Thus, the nonmonotonous behavior of Fig. 13. We finally note that the single-chain desorption transition for a flexible chain, which was discussed in Sect. 2.7.4, is also valid for adsorption from solutions and will lead to a desorption transition at very high salt concentrations in Fig. 15 (which is not shown for clarity).

2.7.5.4 Overcompensation of Surface Charges: Constant σ

We turn now to a different electrostatic boundary condition of constant surface charge density and look at the interesting phenomenon of charge compensation by the PE chains in relation to experiments for PE adsorption on flat surfaces, as well as on charged colloidal particles [72, 73, 75–77]. What was observed in experiments is that PEs adsorbing on an oppositely charged surface can overcompensate the original surface charge. Because PEs create a thin layer close to the surface, they can act as an effective absorbing surface to a second layer of PEs having an opposite charge compared to the first layer. Repeating the adsorption of alternating positively and negatively charged PEs, it is possible to create a multilayer structure of PEs at the surface. Although many experiments and potential applications for PE multilayers exist, the theory of PE overcompensation is only starting to be developed [83, 84, 123–125, 127, 128].

The scaling laws presented for constant ψ_s can be used also for the case of constant surface charge. A surface held at a constant potential ψ_s will induce a surface charge density σ. The two quantities are related by $d\psi/dx = 4\pi\sigma e/\varepsilon$ at $x = 0$. We will now consider separately the two limits: low salt $D \ll \kappa^{-1}$ and high salt $D \gg \kappa^{-1}$.

2.7.5.4.1 Low Salt Limit: $D \ll \kappa^{-1}$

Assuming that there is only one length scale characterizing the potential behavior in the vicinity of the surface, as demonstrated in Fig. 12(a), we find that the surface potential ψ_s and the surface charge σ are related by $\psi_s \sim \sigma eD$. In the low salt limit we find from Eq. (61)

$$D \sim (f\sigma l_B)^{-1/3} \quad (71)$$

in agreement with Eq. (47).

Let us define two related concepts via the effective surface charge density defined as $\Delta\sigma = f\Gamma - \sigma$, which is the sum of the adsorbed polymer charge density and the charge density of the bare substrate. For $\Delta\sigma = 0$, the adsorbed polymer charge exactly *compensates* the substrate charge. If $\Delta\sigma$ is positive, the PE *overcompensates* the substrate charge and more polymer adsorbs than is needed to exactly cancel the substrate charge. If $\Delta\sigma$ is positive and reaches the value $\Delta\sigma = \sigma$, it means that the PE charge is $f\Gamma = 2\sigma$ and leads to a *charge inversion* of the substrate charge. The effective surface charge consisting of the substrate charge and the PE layer has a charge density that is exactly opposite to the original substrate charge density σ.

Do we obtain overcompensation or charge inversion in the low salt limit within mean-field theory? Using scaling arguments, this is not clear since we find that $\Delta\sigma \sim f\Gamma \sim \sigma$. That is, each of the two terms in $\Delta\sigma$ scales linearly with σ, and the occurrence of overcompensation or charge inversion will depend on numerical prefactors determining the relative sign of the two opposing terms. However, if we look at the numerical solution for the mean-field electrostatic potential (Fig. 12), we see indeed that all plotted profiles have a maximum of $\psi(x)$ as function of x. An extremum in ψ means a zero local electric field. Or equivalently, using Gauss law, this means that the integrated charge density from the wall to this special extremum point (including surface charges) is exactly zero. At this point the charges in solution exactly compensate the surface charges.

2.7.5.4.2 High Salt Limit: $D \gg \kappa^{-1}$

When we include salt in the solution and look at the high salt limit, the situation is more complex. The only length characterizing the exponential decay of ψ close to the surface is the DH screening length. Hence, using $d\psi/dx|_s \sim \sigma e$ yields $\psi_s \sim \sigma e \kappa^{-1}$, and therefore from Eq. (65),

$$D \sim \frac{\kappa^2 a^2}{f \sigma l_B} \sim \kappa^2 f^{-1} \sigma^{-1} \quad (72)$$

The estimation of the PE layer charge can be obtained by using the expression for D and c_M in this high salt limit (Eqs. 63–65), yielding

$$f\Gamma \simeq \frac{\beta\sigma\,(8\pi l_B c_{\text{salt}}\kappa^{-2})}{1 + \alpha v_2 c_{\text{salt}} f^{-2}} = \frac{\beta\sigma}{1 + v_2/v_{\text{el}}} \quad (73)$$

where $v_{\text{el}} = f^2/\alpha c_{\text{salt}}$ is the electrostatic contribution to the second-virial coefficient v_2 and $\alpha > 0$ and $\beta > 1$ are positive numerical factors.

We see that $\Delta\sigma = f\Gamma - \sigma$ is a decreasing function of v_2. Charge overcompensation can occur when v_2 is smaller than v_{el} (up to a prefactor of order unity). When v_2 can be neglected in the vicinity of the surface, or when $v_2 = 0$ (theta solvents), there is always charge overcompensation, $\Delta\sigma = (\beta - 1)\sigma > 0$. This is the case of strongly charged PEs. Similar conclusions have been mentioned in Refs. [83, 127] where v_2 was taken as zero but the surface has a nonelectrostatic short range interaction with the PE. By tuning the relative strength of the surface charge density σ and the nonelectrostatic interaction, it is also possible to cause a charge overcompensation and even an exact charge inversion in a special case.

Finally, we note that the dependence of the charge parameter $\Delta\sigma$ on the amount of salt, c_{salt}, is different for constant surface charge and constant surface potential cases. While for the former, $\Delta\sigma$ is nonmonotonous and has a maximum (as mentioned above) as function of the salt concentration, in the latter case, $\Delta\sigma/\sigma$ is a monotonic decreasing function of c_{salt} (Eq. 73). This can be explained by the extra powers of c_{salt} in the latter case coming from the relation $\psi_s \sim \sigma c_{\text{salt}}^{-1/2} \sim \sigma \kappa^{-1}$.

Let us remark that in other theories the overcharging is due to lateral correlations between adsorbed PEs, which in conjunction with screening by salt ions leads to strongly overcharged surfaces [84, 128].

2.7.5.5 Final Remarks on Adsorption from Semidilute Solutions

The results presented earlier (Sect. 2.7.5) for adsorption from solutions have been derived using mean-field theory. Hence, lateral fluctuations in the polymer and ionic concentrations are neglected. In addition, we neglect the delicate influence of the charges on the PE persistence length and any deviations from ground state dominance [9]. The region of validity of the theory is for long and weakly charged polymer chains in contact with a moderately charged surface. The PE solution is placed in contact with a single and ideal surface (infinite, flat, and homogeneous). The problem reduces then to an effective one-dimensional problem depending only on the distance from the charged surface. We take very simple boundary conditions for the surface assuming that the polymer concentration is zero on the surface and keeping the surface in constant potential or constant surface charge conditions.

We find numerical solutions for the PE profile equations in various cases. These numerical solutions agree well with simple scaling assumptions describing the adsorption of PEs. Scaling expressions for the amount of adsorbed polymer Γ and the width D of the adsorbed layer, as a function of the fractional charge f and the salt concentration c_{salt}, are obtained for two cases: constant ψ_s and constant σ.

For constant ψ_s and in the low-salt regime, a $f^{-1/2}$ dependence of Γ is found. It is supported by our numerical solutions of the profile Eqs. (55) and (56) and is in agreement with experiment [106]. This behavior is due to strong Coulomb repulsion between adsorbed monomers in the absence of salt. As f decreases, the adsorbed amount increases until the electrostatic attraction becomes weaker than the excluded volume repulsion, at which point, Γ starts to decrease rapidly.

At high salt concentrations it is not possible to neglect the excluded volume interaction of the monomers since the electrostatic interactions are screened by the salt. We obtain two limiting behaviors: (1) For weakly charged PEs, $f \ll f^* = (c_{salt}v_2)^{1/2}$, the adsorbed amount increases with the fractional charge and decreases with the salt concentration, $\Gamma \sim f/\sqrt{(c_{salt})}$, owing to the monomer–surface electrostatic attraction. (2) For strong PEs, $f \gg f^*$, the adsorbed amount decreases with the fractional charge and increases with the salt concentration, $\Gamma \sim \sqrt{(c_{salt})}/f$, owing to the dominance of monomer–monomer electrostatic repulsion. Between these two regimes, we find that the adsorbed amount reaches a maximum in agreement with experiments [107, 110].

The scaling arguments are then repeated for constant σ boundary conditions. It is found that the PE can possibly cause charge overcompensation and even inversion of the nominal substrate charge, leading the way to multilayer formation of positively and negatively charged PEs. The scaling approach can serve as a starting point for further investigations. The analytical and approximated expressions are valid only in specific limits. Special attention should be directed to the crossover regime where D and κ^{-1} are of comparable size.

The problem of charge inversion is not well understood at present. Alternative approaches rely on lateral correlations between semiflexible adsorbed PE chains, which also can lead to strong overcompensation of surface charges [84, 128].

2.7.6
Polyelectrolyte Brushes

Charged polymers that are densely end-grafted to a surface are called *polyelectrolyte brushes* or *charged brushes*. They have been the focus of numerous theoretical [129–138] and experimental [139–142] studies. In addition to the basic interest, charged brushes are considered for their applications as efficient means for preventing colloids in polar media (such as aqueous solutions) from flocculating and precipitating out of solution [93]. This stabilization arises from steric (entropic) and electrostatic (energetic) repulsion. A strongly charged brush is able to trap its own counterions and generates a layer of locally enhanced salt concentration [131]. It is thus less sensitive to the salinity of the surrounding aqueous medium than a stabilization mechanism based on pure electrostatics (i.e. without polymers).

Neutral brushes have been extensively studied theoretically in the past using scaling theories [143, 144], strong-stretching theories [145–149], self-consistent field theories [150, 151], and computer simulations [152–155]. Little is known from experiments on the scaling behavior of PE brushes as compared to uncharged polymer brushes. The thickness of the brush layer has been calculated from neutron-scattering experiments on end-grafted polymers [139] and charged diblock copolymers at the air–water interface [141, 142].

Theoretical work on PE brushes was initiated by the works of Miklavic and Marcelja [129] and Misra and coworkers. [130]. In 1991, Pincus [131] and Borisov, Birshtein and Zhulina [132] presented scaling theories for charged brushes in the so-called osmotic regime, where the brush height results from the balance between the chain elasticity (which tends to decrease the brush height) and the repulsive osmotic counterion pressure (which tends to increase the brush height). In later studies, these works have been generalized to the poor solvents [133, 134] and to the regime where excluded volume effects become important, that is, the so-called quasi-neutral or Alexander regime [137].

In what follows we assume that the charged brush is characterized by two length scales: the average vertical extension of polymer chains from the wall L and the typical extent of the counterion cloud, denoted by H. We neglect the presence of additional salt, which has been discussed extensively in the original literature, and only consider screening effects due to the counterions of the charged brush. Two different scenarios emerge, as is schematically presented in Fig. 16. The counterions can either extend outside the brush, $H \gg L$, as shown in Fig. 16(a), or be confined inside the brush, $H \approx L$, as shown in Fig. 16(b). As we show now, case (b) is indicative of strongly charged brushes, while case (a) is typical for weakly charged brushes.

The free energy per unit area (and in units of $k_B T$) contains several contributions. We denote the grafting density of PEs by ρ, the counterion valency by z, recalling that N is the polymerization index of grafted chains and f their charge fraction. The osmotic free energy, F_{os}, associated with the ideal entropy cost of confining the counterions to a layer of thickness H is given by

$$F_{os} \simeq \frac{Nf\rho}{z} \ln\left(\frac{Nf\rho}{zH}\right) \quad (74)$$

F_{v_2} is the second-virial contribution to the free energy, arising from steric repulsion between the monomers (contributions due to counterions are neglected). Throughout this review, the polymers are assumed to be in a good solvent (positive second-virial coefficient $v_2 > 0$). The contribution thus reads

$$F_{v_2} \simeq \frac{1}{2} L v_2 \left(\frac{N\rho}{L}\right)^2 \quad (75)$$

Finally, a direct electrostatic contribution F_{el} occurs if the PE brush is not locally electro-neutral throughout the system, as for example is depicted in Fig. 16(a). This energy is given by

$$F_{el} = \frac{2\pi \ell_B (Nf\rho)^2}{3} \frac{(L-H)^2}{H} \quad (76)$$

This situation arises in the limit of low charge, when the counterion density profile extends beyond the brush layer, that is, $H > L$.

Fig. 16 Schematic PE brush structure. In (a), we show the weak-charge limit where the counterion cloud has a thickness H larger than the thickness of the brush layer, L. In (b), we show the opposite case of the strong-charge limit, where all counterions are contained inside the brush and a single length scale $L \approx H$ exists.

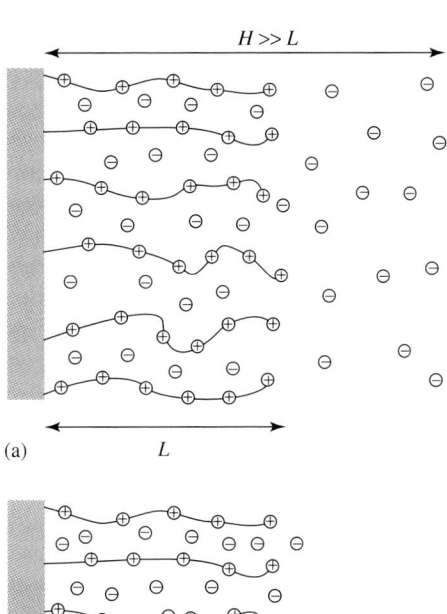

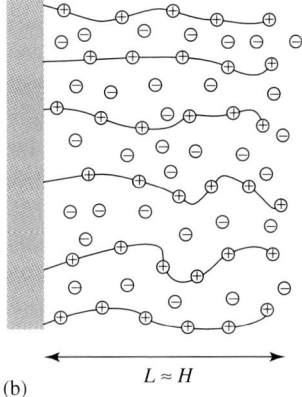

The last contribution is the stretching energy of the chains, which is

$$F_{st} = \frac{3L^2}{2Na^2}\rho \qquad (77)$$

Here, a is the monomer size or Kuhn length of the polymer, implying that we neglect any chain stiffness for the brush problem. The different free energy contributions lead, upon minimization with respect to the two length scales H and L, to different behaviors. Let us first consider the weak charging limit, that is, the situation in which the counterions leave the brush, $H > L$.

In this case, minimization of $F_{os} + F_{el}$ with respect to the counterion height H leads to

$$H \sim \frac{1}{z\ell_B N f \rho} \qquad (78)$$

which is the Gouy–Chapman length for z-valent counterions at a surface of surface charge density $\sigma = Nf\rho$. Balancing now the polymer stretching energy F_{st} and the electrostatic energy F_{el}, one obtains the so-called Pincus brush

$$L \simeq N^3 \rho a^2 \ell_B f^2 \qquad (79)$$

In the limit of $H \approx L$, the PE brush can be considered as neutral and the electrostatic energy vanishes. There are two ways of balancing the remaining free energy contributions. The first is obtained by comparing the osmotic energy of counterion confinement, F_{os}, with the polymer stretching term, F_{st}, leading to the height

$$L \sim \frac{Naf^{1/2}}{z^{1/2}} \quad (80)$$

constituting the so-called osmotic-brush regime. Finally, comparing the second-virial free energy, F_{v_2}, with the polymer stretching energy, F_{st}, one obtains

$$L \sim Na(v_2\rho/a)^{1/3} \quad (81)$$

and the PE brush is found to have the same scaling behavior as the neutral brush [143, 144]. Comparing the heights of all three regimes, we arrive at the phase diagram shown in Fig. 17. The three scaling regimes meet at the characteristic charge fraction

$$f^* \sim \left(\frac{zv_2}{N^2 a^2 \ell_B}\right)^{1/3} \quad (82)$$

and the characteristic grafting density

$$\rho^* \sim \frac{1}{N\ell_B^{1/2} v_2^{1/2}} \quad (83)$$

For large values of the charge fraction f and the grafting density ρ, it has been found numerically that the brush height does not follow any of the scaling laws discussed here [156]. This has been recently rationalized in terms of another scaling regime, the collapsed regime. In this regime, one finds that correlation and fluctuation effects, which are neglected in the discussion in this section, lead to a net attraction between charged monomers and counterions [157].

2.7.7
Conclusion

In this chapter we have reviewed the behavior of charged polymers (PEs) in solution and at interfaces, concentrating on aspects that are different from the corresponding behavior of neutral polymers.

Because charged biopolymers and isolated PE chains tend to be quite stiff due to electrostatic monomer–monomer

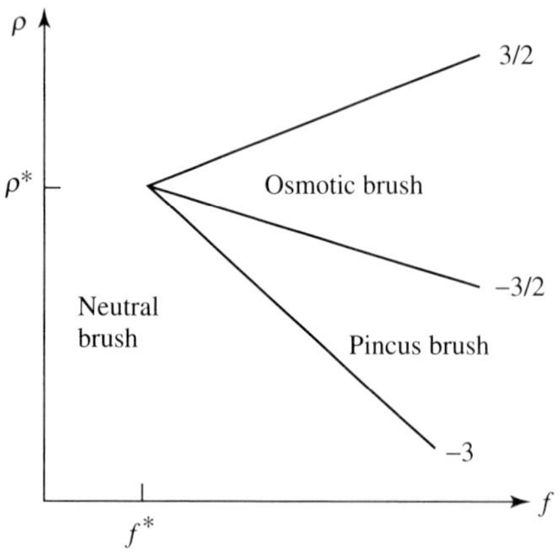

Fig. 17 Scaling diagram for PE brushes on a log–log plot as a function of the grafting density ρ and the fraction of charged monomers f. Featured are the Pincus-brush regime, where the counterion layer thickness is much larger than the brush thickness, the osmotic-brush regime, where all counterions are inside the brush and the brush height is determined by an equilibrium between the counterion osmotic pressure and the PE stretching energy, and the neutral-brush regime, where charge effects are not important and the brush height results from a balance of PE stretching energy and second-virial repulsion. The power-law exponents of the various lines are denoted by numbers.

repulsions, their chain statistics is related to that of semiflexible polymers. Neutral and charged semiflexible polymers are controlled by their bending rigidity, which is usually expressed in terms of a persistence length (see Sect. 2.7.2.2). For PEs, the electrostatic interaction considerably influences this persistence length.

In solution, we have considered the scaling behavior of a single PE (Sect. 2.7.3.1). The importance of the electrostatic persistence length was stressed. The Manning condensation of counterions leads to a reduction of the effective linear charge density (Sect. 2.7.3.1.1). Excluded volume effects are typically less important than for neutral polymers (Sect. 2.7.3.1.2). Dilute PE solutions are typically dominated by the behavior of the counterions. So is the large osmotic pressure of dilute PE solutions due to the entropic contribution of the counterions (Sect. 2.7.3.2). Semidilute PE solutions can be described by the RPA, which in particular yields the characteristic peak of the structure factor.

At surfaces, we discussed in detail the adsorption of single PEs (Sect. 2.7.4), the adsorption from semidilute solutions (Sect. 2.7.5), and the behavior of end-grafted PE chains (Sect. 2.7.6). We tried to express the PE behavior in terms of a few physical parameters such as the chain characteristics (persistence length), ionic strength of the solutions, and surface characteristics. The shape and size of the adsorbing layer is, in many instances, governed by a delicate balance of competing mechanisms of electrostatic and nonelectrostatic origin. In some cases, it is found that the adsorbing PE layer is flat and compressed, while in other cases, it is coiled and extended. Yet, in other situations, the PEs will not adsorb at all and will be depleted from the surface. We also briefly review the phenomenon of charge overcompensation and inversion, when the adsorbed PE layer effectively inverses the sign of the surface charge leading the way to formation of PE multilayers.

Important topics that we have left out are the dynamics of PE solutions, which is reviewed in Ref. [4], and the behavior of PEs under bad-solvent conditions [39–41]. In the future we expect that studies of PEs in solutions and at surfaces will be directed more toward biological systems. We mentioned in this review the complexation of DNA and histones (Sect. 2.7.4.1). This is only one of many examples of interest in which charged biopolymers, receptors, proteins, and DNA molecules interact with each other or with other cellular components. The challenge for future fundamental research will be to try to understand the role of electrostatic interactions combined with specific biological (lock–key) mechanisms and to infer on biological functionality of such interactions.

Acknowledgment

It is a pleasure to thank our collaborators I. Borukhov, J.F. Joanny, K. Kunze, L. Leibler, R. Lipowsky, H. Orland, M. Schick, and C. Seidel with whom we have been working on PE problems. We benefitted from discussions with G. Ariel, Y. Burak, and M. Ullner. One of us (DA) would like to acknowledge partial support from the Israel Science Foundation funded by the Israel Academy of Sciences and Humanities – Centers of Excellence Program and the Israel–US Binational Science Foundation (BSF) under grant no. 98-00429.

References

1. F. Oosawa, *Polyelectrolytes*, Marcel Dekker, New York, 1971.
2. H. Dautzenberg, W. Jaeger, B. P. J. Kötz et al., *Polyelectrolytes: Formation, characterization and application*, Hanser Publishers, Munich, 1994.
3. S. Förster, M. Schmidt, *Adv. Polym. Sci.* **1995**, *120*, 50.
4. J.-L. Barrat, J.-F. Joanny, *Adv. Chem. Phys.* **1996**, *94*, 1.
5. E. Eisenriegler, *Polymers Near Surfaces*, World Scientific, Singapore, 1993.
6. A. Yu. Grosberg, A. R. Khokhlov, *Statistical Physics of Macromolecules*, AIP Press, New York, 1994.
7. P. J. Flory, *Principles of Polymer Chemistry*, Cornell University, Ithaca, 1953.
8. H. Yamakawa, *Modern Theory of Polymer Solutions*, Harper and Row, New York, 1971.
9. P. G. de Gennes, *Scaling Concepts in Polymer Physics*, Cornell University, Ithaca, 1979.
10. A. R. Khokhlov, K. A. Khachaturian, *Polymer* **1982**, *23*, 1742.
11. J.-L. Barrat, J.-F. Joanny, *Europhys. Lett.* **1993**, *24*, 333.
12. R. R. Netz, H. Orland, *Eur. Phys. J. B* **1999**, *8*, 81.
13. T. Odijk, *J. Polym. Sci., Polym. Phys. Ed.* **1977**, *15*, 477.
14. T. Odijk, *Polymer* **1978**, *19*, 989.
15. J. Skolnick, M. Fixman, *Macromolecules* **1977**, *10*, 944.
16. G. A. Christos, S. L. Carnie, *J. Chem. Phys.* **1990**, *92*, 7661.
17. C. Seidel, H. Schlacken, I. Müller, *Macromol. Theory Simul.* **1994**, *3*, 333.
18. M. Ullner, B. Jönsson, C. Peterson et al., *J. Chem. Phys.* **1997**, *107*, 1279.
19. U. Micka, K. Kremer, *Phys. Rev. E* **1996**, *54*, 2653.
20. U. Micka, K. Kremer, *Europhys. Lett.* **1997**, *38*, 279.
21. H. Li, T. Witten, *Macromolecules* **1995**, *28*, 5921.
22. B.-Y. Ha, D. Thirumalai, *J. Chem. Phys.* **1999**, *110*, 7533.
23. B.-Y. Ha, D. Thirumalai, *Macromolecules* **1995**, *28*, 577.
24. T. B. Liverpool, M. Stapper, *Europhys. Lett.* **1997**, *40*, 485.
25. R. G. Winkler, M. Gold, P. Reineker, *Phys. Rev. Lett.* **1998**, *80*, 3731.
26. N. V. Brilliantov, D. V. Kuznetsov, R. Klein, *Phys. Rev. Lett.* **1998**, *81*, 1433.
27. M. O. Khan, B. Jönsson, *Biopolymers* **1999**, *49*, 121.
28. T. T. Nguyen, I. Rouzina, B. I. Shklovskii, *Phys. Rev. E* **1999**, *60*, 7032.
29. G. S. Manning, *J. Chem. Phys.* **1969**, *51*, 924.
30. G. S. Manning, *J. Chem. Phys.* **1969**, *51*, 934.
31. G. S. Manning, U. Mohanty, *Physica A* **1997**, *247*, 196.
32. M. Deserno, C. Holm, S. May, *Macromolecules* **2000**, *33*, 199.
33. C. Wandrey, D. Hunkeler, U. Wendler et al., *Macromolecules* **2000**, *33*, 7136.
34. J. Blaul, M. Wittemann, M. Ballauff et al., *J. Phys. Chem. B* **2000**, *104*, 7077.
35. H. Schiessel, P. Pincus, *Macromolecules* **1998**, *31*, 7953.
36. H. Schiessel, *Macromolecules* **1999**, *32*, 5673.
37. A. R. Khokhlov, *J. Phys. A* **1980**, *13*, 979.
38. Y. Kantor, M. Kardar, *Phys. Rev. E* **1995**, *51*, 1299.
39. A. V. Dobrynin, M. Rubinstein, S. P. Obukhov, *Macromolecules* **1996**, *29*, 2974.
40. A. L. Lyulin, B. Dünweg, O. V. Borisov et al., *Macromolecules* **1999**, *32*, 3264.
41. U. Micka, C. Holm, K. Kremer, *Langmuir* **1999**, *15*, 4033.
42. T. A. Waigh, R. Ober, C. E. Williams et al., *Macromolecules* **2001**, *34*, 1973.
43. M. Nierlich, F. Boue, A. Lapp et al., *Coll. Polym. Sci.* **1985**, *263*, 955.
44. M. Nierlich, F. Boue, A. Lapp et al., *J. Phys. (France)* **1985**, *46*, 649.
45. A. Moussaid, F. Schosseler, J. P. Munch et al., *J. Phys. II (France)* **1993**, *3*, 573.
46. W. Essafi, F. Lafuma, C. E. Williams, *J. Phys. II (France)* **1995**, *5*, 1269.
47. K. Nishida, K. Kaji, T. Kanaya, *Macromolecules* **1995**, *28*, 2472.
48. W. Essafi, F. Lafuma, C. E. Williams, *Eur. Phys. J. B* **1999**, *9*, 261.
49. V. Y. Borue, I. Y. Erukhimovich, *Macromolecules* **1988**, *21*, 3240.
50. J. F. Joanny, L. Leibler, *J. Phys. (France)* **1990**, *51*, 545.
51. J. des Cloizeaux, *Macromolecules* **1973**, *6*, 403.
52. T. Yoshizaki, H. Yamakawa, *Macromolecules* **1980**, *13*, 1518.
53. R. R. Netz, to be published.
54. M. N. Spiteri, F. Boue, A. Lapp et al., *Phys. Rev. Lett.* **1996**, *77*, 5218.

55. T. A. Witten, P. Pincus, *Europhys. Lett.* **1987**, *3*, 315.
56. J.-L. Barrat, J.-F. Joanny, *J. Phys. II (France)* **1994**, *4*, 1089.
57. M. J. Stevens, K. Kremer, *J. Chem. Phys.* **1995**, *103*, 1669.
58. M. J. Stevens, K. Kremer, *J. Phys. II (France)* **1996**, *6*, 1607.
59. H. Schäfer, C. Seidel, *Macromolecules* **1997**, *30*, 6658.
60. M. J. Stevens, S. J. Plimpton, *Eur. Phys. J. B* **1998**, *2*, 341.
61. M. A. Cohen Stuart, *J. Phys. (France)* **1988**, *49*, 1001.
62. M. A. Cohen Stuart, G. J. Fleer, J. Lyklema, et al., *Adv. Colloid Interface Sci.* **1991**, *34*, 477.
63. Y. Fang, J. Yang, *J. Phys. Chem. B* **1997**, *101*, 441.
64. J. O. Rädler, I. Koltover, T. Salditt et al., *Science* **1997**, *275*, 810.
65. T. Salditt, I. Koltover, J. O. Rädler et al., *Phys. Rev. Lett.* **1997**, *79*, 2582.
66. B. Maier, J. O. Rädler, *Phys. Rev. Lett.* **1999**, *82*, 1911.
67. H. von Berlepsch, C. Burger, H. Dautzenberg, *Phys. Rev. E* **1998**, *58*, 7549.
68. K. de Meijere, G. Brezesinski, H. Möhwald, *Macromolecules* **1997**, *30*, 2337.
69. G. Decher, J. D. Hong, J. Schmitt, *Thin Solid Films* **1992**, *210/211*, 831.
70. R. von Klitzing, H. Möhwald, *Langmuir* **1995**, *11*, 3554.
71. R. von Klitzing, H. Möhwald, *Macromolecules* **1996**, *29*, 6901.
72. G. Decher, *Science* **1997**, *277*, 1232.
73. M. Lösche, J. Schmitt, G. Decher et al., *Macromolecules* **1998**, *31*, 8893.
74. F. Caruso, K. Niikura, D. N. Furlong et al., *Langmuir* **1997**, *13*, 3422.
75. E. Donath, G. B. Sukhorukov, F. Caruso et al., *Angew. Chem., Int. Ed.* **1998**, *110*, 2323.
76. G. B. Sukhorukov, E. Donath, S. A. Davis et al., *Polym. Adv. Technol.* **1998**, *9*, 759.
77. F. Caruso, R. A. Caruso, H. Möhwald, *Science* **1998**, *282*, 1111.
78. M. Muthukumar, *J. Chem. Phys.* **1987**, *86*, 7230.
79. X. Chatellier, J.-F. Joanny, *J. Phys. II (France)* **1996**, *6*, 1669.
80. R. Varoqui, *J. Phys. (France) II* **1993**, *3*, 1097.
81. I. Borukhov, D. Andelman, H. Orland, *Europhys. Lett.* **1995**, *32*, 499.
82. I. Borukhov, D. Andelman, H. Orland in *Short and Long Chains at Interfaces* (Eds.: J. Daillant, P. Guenoun, C. Marques et al.), Edition Frontieres, Gif-sur-Yvette, 1995, pp. 13–20.
83. J. F. Joanny, *Eur. Phys. J. B* **1999**, *9*, 117.
84. R. R. Netz, J. F. Joanny, *Macromolecules* **1999**, *32*, 9013.
85. O. V. Borisov, E. B. Zhulina, T. M. Birshtein, *J. Phys. II (France)* **1994**, *4*, 913.
86. T. Odijk, *Macromolecules* **1983**, *16*, 1340.
87. T. Odijk, *Macromolecules* **1984**, *17*, 502.
88. A. C. Maggs, D. A. Huse, S. Leibler, *Europhys. Lett.* **1989**, *8*, 615.
89. G. Gompper, T. W. Burkhardt, *Phys. Rev. A* **1989**, *40*, 6124.
90. G. Gompper, U. Seifert, *J. Phys. A* **1990**, *23*, L1161.
91. R. Bundschuh, M. Lässig, R. Lipowsky, *Eur. Phys. J. E* **2000**, *3*, 295.
92. V. Yamakov, A. Milchev, O. Borisov et al., *J. Phys.: Condens. Matter* **1999**, *11*, 9907.
93. D. H. Napper, *Polymeric Stabilization of Colloidal Dispersions*, Academic Press, New York, 1983.
94. T. Wallin, P. Linse, *Langmuir* **1996**, *12*, 305.
95. T. Wallin, P. Linse, *J. Phys. Chem.* **1996**, *100*, 17 873.
96. T. Wallin, P. Linse, *J. Phys. Chem. B* **1997**, *101*, 5506.
97. E. M. Mateescu, C. Jeppesen, P. Pincus, *Europhys. Lett.* **1999**, *46*, 493.
98. R. R. Netz, J. F. Joanny, *Macromolecules* **1999**, *32*, 9026.
99. K. K. Kunze, R. R. Netz, *Phys. Rev. Lett.* **2000**, *85*, 4389.
100. T. D. Yager, C. T. McMurray, K. E. van Holde, *Biochemistry* **1989**, *28*, 2271.
101. F. von Goeler, M. Muthukumar, *J. Chem. Phys.* **1994**, *100*, 7796.
102. E. Gurovitch, P. Sens, *Phys. Rev. Lett.* **1999**, *82*, 339.
103. P. Peyser, R. Ullman, *J. Polym. Sci. A* **1965**, *3*, 3165.
104. M. Kawaguchi, H. Kawaguchi, A. Takahashi, *J. Colloid Interface Sci.* **1988**, *124*, 57.
105. J. Meadows, P. A. Williams, M. J. Garvey et al., *J. Colloid Interface Sci.* **1989**, *132*, 319.
106. R. Denoyel, G. Durand, F. Lafuma et al., *J. Colloid Interface Sci.* **1990**, *139*, 281.
107. J. Blaakmeer, M. R. Böhmer, M. A. Cohen Stuart et al., *Macromolecules* **1990**, *23*, 2301.

108. H. G. A. van de Steeg, A. de Keizer, M. A. Cohen Stuart et al., *Colloid Surf., A* **1993**, *70*, 91.
109. V. Shubin, P. Linse, *J. Phys. Chem.* **1995**, *99*, 1285.
110. N. G. Hoogeveen, Ph.D. Thesis, Wageningen Agricultural University, The Netherlands, unpublished, 1996.
111. G. J. Fleer, M. A. Cohen Stuart, J. M. H. M. Scheutjens et al., *Polymers at Interfaces*, Chapman & Hall, London, 1993, Chap. 11.
112. C. A. Haynes, W. Norde, *Colloid Surf., B* **1994**, *2*, 517.
113. P. Auroy, L. Auvray, L. Léger, *Macromolecules* **1991**, *24*, 2523.
114. O. Guiselin, L. T. Lee, B. Farnoux et al., *J. Chem. Phys.* **1991**, *95*, 4632.
115. T. Odijk, *Macromolecules* **1979**, *12*, 688.
116. A. V. Dobrynin, R. H. Colby, M. Rubinstein, *Macromolecules* **1995**, *28*, 1859.
117. H. A. van der Schee, J. Lyklema, *J. Phys. Chem.* **1984**, *88*, 6661.
118. J. Papenhuijzen, H. A. van der Schee, G. J. Fleer, *J. Colloid Interface Sci.* **1985**, *104*, 540.
119. O. A. Evers, G. J. Fleer, J. M. H. M. Scheutjens et al., *J. Colloid Interface Sci.* **1985**, *111*, 446.
120. H. G. M. van de Steeg, M. A. Cohen Stuart, A. de Keizer et al., *Langmuir* **1992**, *8*, 8.
121. P. Linse, *Macromolecules* **1996**, *29*, 326.
122. R. Varoqui, A. Johner, A. Elaissari, *J. Chem. Phys.* **1991**, *94*, 6873.
123. I. Borukhov, D. Andelman, H. Orland, *Eur. Phys. J. B* **1998**, *5*, 869.
124. I. Borukhov, D. Andelman, H. Orland, *Macromolecules* **1998**, *31*, 1665.
125. I. Borukhov, D. Andelman, H. Orland, *J. Phys. Chem. B* **1999**, *24*, 5057.
126. C. Tanford, J. C. Kirkwood, *Am. J. Chem. Soc.* **1957**, *79*, 5333.
127. D. Andelman, J. F. Joanny, *C. R. Acad. Sci. (Paris)* **2000**, *1*, 1153.
128. T. T. Nguyen, A. Y. Grosberg, B. I. Shklovskii *J. Chem. Phys.* **2000**, *113*, 1110.
129. S. J. Miklavic, S. Marcelja, *J. Phys. Chem.* **1988**, *92*, 6718.
130. S. Misra, S. Varanasi, P. P. Varanasi, *Macromolecules* **1989**, *22*, 5173.
131. P. Pincus, *Macromolecules* **1991**, *24*, 2912.
132. O. V. Borisov, T. M. Birstein, E. B. Zhulina, *J. Phys. II (France)* **1991**, *1*, 521.
133. R. S. Ross, P. Pincus, *Macromolecules* **1992**, *25*, 2177.
134. E. B. Zhulina, T. M. Birstein, O. V. Borisov, *J. Phys. II (France)* **1992**, *2*, 63.
135. J. Wittmer, J.-F. Joanny, *Macromolecules* **1993**, *26*, 2691.
136. R. Israels, F. A. M. Leermakers, G. J. Fleer et al., *Macromolecules* **1994**, *27*, 3249.
137. O. V. Borisov, E. B. Zhulina, T. M. Birstein, *Macromolecules* **1994**, *27*, 4795.
138. E. B. Zhulina, O. V. Borisov, *J. Chem. Phys.* **1997**, *107*, 5952.
139. Y. Mir, P. Auvroy, L. Auvray, *Phys. Rev. Lett.* **1995**, *75*, 2863.
140. P. Guenoun, A. Schlachli, D. Sentenac, J. M. Mays et al., *Phys. Rev. Lett.* **1995**, *74*, 3628.
141. H. Ahrens, S. Förster, C. A. Helm, *Macromolecules* **1997**, *30*, 8447.
142. H. Ahrens, S. Förster, C. A. Helm, *Phys. Rev. Lett.* **1998**, *81*, 4172.
143. S. Alexander, *J. Phys. (France)* **1977**, *38*, 983.
144. P. G. de Gennes, *Macromolecules* **1980**, *13*, 1069.
145. A. N. Semenov, *Sov. Phys. JETP* **1985**, *61*, 733.
146. S. T. Milner, T. A. Witten, M. E. Cates, *Europhys. Lett.* **1988**, *5*, 413.
147. S. T. Milner, T. A. Witten, M. E. Cates, *Macromolecules* **1988**, *21*, 2610.
148. S. T. Milner, *Science* **1991**, *251*, 905.
149. A. M. Skvortsov, I. V. Pavlushkov, A. A. Gorbunov et al., *Polym. Sci.* **1988**, *30*, 1706.
150. R. R. Netz, M. Schick, *Europhys. Lett.* **1997**, *38*, 37.
151. R. R. Netz, M. Schick, *Macromolecules* **1998**, *31*, 5105.
152. M. Murat, G. S. Grest, *Macromolecules* **1989**, *22*, 4054.
153. A. Chakrabarti, R. Toral, *Macromolecules* **1990**, *23*, 2016.
154. P. Y. Lai, K. Binder, *J. Chem. Phys.* **1991**, *95*, 9288.
155. C. Seidel, R. R. Netz, *Macromolecules* **2000**, *33*, 634.
156. F. S. Csajka, C. Seidel, *Macromolecules* **2000**, *33*, 2728.
157. F. S. Csajka, R. R. Netz, C. Seidel et al., *Eur. Phys. J. E* **2001**, *4*, 505.

3
Specific Adsorption

3.1	Introduction	327
	Boris B. Damaskin and Oleg A. Petrii	327
	References	346
3.2	**State of Art: Present Knowledge and Understanding**	349
	G. Horányi	349
3.2.1	Historical Background	349
3.2.2	The Structure of the Double Layer	350
	Recapitulation of fundamental notions	350
	The double layer in the case of specific adsorption	350
3.2.3	Thermodynamics and Models	351
3.2.3.1	The Extent of Adsorption	351
3.2.3.2	Adsorption Isotherm	354
3.2.3.3	Indication and Evaluation of Specific Adsorption	355
3.2.4	Specific Adsorption at Solid Electrodes	357
3.2.4.1	The End of the "Mercury Era"	357
3.2.4.2	Underpotential Deposition: A Limiting Case of Specific Adsorption	359
3.2.4.2.1	A Widely Used and Misinterpreted Misconception: The Electrosorption Valency	359
3.2.4.2.2	Interrelation of Anion Specific Adsorption and UPD of Metal Ions	360
3.2.4.3	The Generalization of the Thermodynamics of Monolayer Adsorption in Electrochemical Systems	361
3.2.4.4	Anion Adsorption on Well-defined Crystal Surfaces	363
3.2.5	Methods of Investigation and Illustrative Examples	366
3.2.5.1	General Remarks	366
3.2.5.2	In Situ Techniques	366
3.2.5.2.1	Optical Techniques	366
	UV-visible Techniques	366
	Ellipsometry [112, 113]	366
	Specular reflectance spectroscopy [114]	367

	Vibrational Spectroscopy (Infrared and Raman Scattering Spectroscopies)	367
	Infrared (IR) spectroscopy [115, 116]	367
	Raman Spectroscopy	367
	Second harmonic generation anisotropy from single-crystalline electrode surfaces [128]	367
	Sum and difference frequency generation at electrode surfaces [130]	368
	X-ray Methods	368
	Surface X-ray scattering (SXS)	368
3.2.5.2.2	Radiotracer Methods	368
3.2.5.2.3	Electrochemical Quartz Crystal Microbalance (EQCM) and Nanobalance (EQCN) Techniques	370
3.2.5.2.4	Scanning Tunneling Microscopy (STM) and Atomic Force Microscopy (AFM)	371
3.2.5.3	Ex Situ Methods	371
3.2.5.3.1	Electrochemical Double-layer Modeling Under Ultrahigh Vacuum Conditions	372
3.2.5.4	Illustrative Examples	372
3.2.6	Trends and Perspectives	375
	Acknowledgment	377
	References	377
3.3	**Phase Transitions in Two-dimensional Adlayers at Electrode Surfaces: Thermodynamics, Kinetics, and Structural Aspects**	**383**
	Thomas Wandlowski	383
3.3.1	Introduction	383
3.3.2	Thermodynamic Aspects	384
3.3.2.1	Phase Transitions and Order Parameter	384
3.3.2.2	Adsorption Isotherms and Lateral Interactions	385
3.3.3	Kinetics of 2D Phase Formation and Dissolution	388
3.3.3.1	Diffusion, Adsorption, Autocatalysis	388
	Mass transport controlled by diffusion	388
	Adsorption control	389
	Autocatalytically controlled surface reactions	389
3.3.3.2	Nucleation and Growth Mechanisms	390
3.3.3.2.1	Nucleation	390
	Homogeneous nucleation – concept of the critical cluster	390
	Heterogeneous nucleation on active sites	392
	Atomistic approach – small cluster model of nucleation	393
3.3.3.2.2	Growth	393
3.3.3.2.3	Coupling of Nucleation and Growth	395
	Mononucleation	395
	Polynucleation	396
3.3.3.3	Dissolution of 2D Condensed Monolayers	400

3.3.4	Examples	404
3.3.4.1	Phase Transitions in Anionic Adlayers	404
3.3.4.1.1	Introduction	404
3.3.4.1.2	Phase Formation in Halide Adlayers	405
	In Situ Results on Adlayer Structures	405
	Adlayer Phases and Phase Transitions on fcc (111) Surfaces	406
	Adlayer Phases and Phase Transitions on fcc (100) Surfaces	409
	Ag(100)/Br	409
	Au(100)/Br	411
3.3.4.1.3	Phase Transitions in Adlayers of Oxoanions	413
3.3.4.2	Phase Transitions in UPD – Adlayers	415
3.3.4.2.1	Introduction	415
3.3.4.2.2	Underpotential Deposition of Copper Ions on Au(hkl)	418
	Cu–UPD on Au(111) in sulfuric acid – structural aspects	418
	Current transients and adsorption kinetics	419
	Influence of anions	423
	Role of substrate crystallography on Cu–UPD	424
3.3.4.2.3	Underpotential Deposition of Copper Ions on Pt(hkl)	425
3.3.4.2.4	Underpotential Deposition of Lead on Ag(hkl)	430
3.3.4.3	Phase Transitions in Organic Monolayers	434
3.3.4.3.1	General Aspects	434
3.3.4.3.2	Thermodynamic Aspects	436
3.3.4.3.3	Kinetic Aspects	439
3.3.4.3.4	Case Study I – Coumarin at the Mercury–Aqueous Electrolyte Interface	439
3.3.4.3.5	Case Study II – Uracil on Au(hkl)	443
3.3.4.3.6	Case Study III – 2,2$'$-Bipyridine (2,2$'$-BP) on Au(hkl)	447
3.3.5	Conclusion and Outlook	454
	Acknowledgment	455
	References	456

3.1
Introduction

Boris B. Damaskin and Oleg A. Petrii
Moscow State University, Moscow, Russia

The adsorption of the ith solution component is defined as a phenomenon that manifests itself in the difference of this component concentration in the vicinity of the interface (c_i) and in the solution bulk ($c_i^{(b)}$). If the concentration of the ith component becomes higher than $c_i^{(b)}$ when the distance x decreases (as we approach the interface), its adsorption is positive; in the opposite case, it is negative (curves 1 and 2, respectively, in Fig. 1).

The quantity characterizing the reversible adsorption of the ith component is its surface excess Γ_i defined by the Gibbs adsorption isotherm [1] at constant temperature ($T = $ const) and pressure ($p = $ const):

$$d\gamma = -\sum_i \Gamma_i \, d\mu_i \quad (1)$$

The quantity γ is the work consumed in increasing the interface by unit surface area at $T = $ const and $p = $ const under reversible conditions, and μ_i is the chemical potential of the ith component.

In order to clarify the physical meaning of Γ_i, let us consider a binary system consisting of adsorbate molecules A and solvent molecules S, for which Eq. (1) takes the form

$$d\gamma = -\Gamma_A \, d\mu_A - \Gamma_S \, d\mu_S \quad (2)$$

However, the values μ_A and μ_S are not independent, they are interrelated by Gibbs-Duhem equation [2, 3]:

$$N_A \, d\mu_A + N_S \, d\mu_S = 0 \quad (3)$$

where N_A and $N_S = 1 - N_A$ are the mole fractions of A and S. It follows from (2) and (3) that

$$d\gamma = -\left(\Gamma_A - \frac{\Gamma_S N_A}{N_S}\right) d\mu_A \quad (4)$$

The combination in parentheses defines the so-called relative surface excess:

$$\Gamma_A^{(S)} = \Gamma_A - \frac{\Gamma_S N_A}{N_S} \quad (5)$$

which can be determined from experimental data. Particularly, for the system under consideration

$$\Gamma_A^{(S)} = -\left(\frac{\partial \gamma}{\partial \mu_A}\right)_{T,p} \quad (6)$$

In order to determine separately the values of Γ_A and Γ_S, the second equation interrelating these values should be derived. For this purpose, Γ_i is defined as the

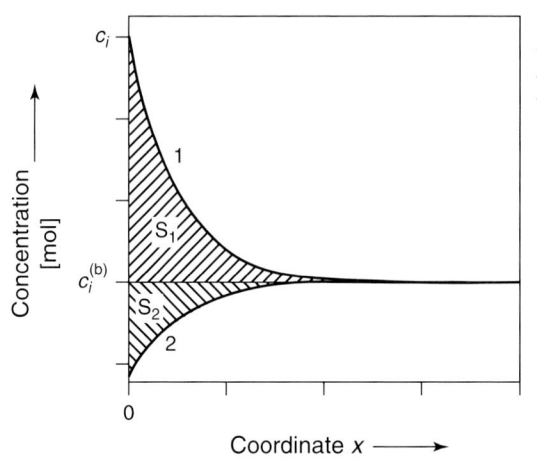

Fig. 1 Schematic dependence of concentration of the 1 and 2 components on the distance from the electrode surface.

difference between the amounts of component i in two equal solution cylinders with unit cross section, the first cylinder being normal and adjacent to the interface, and the second one being completely in the solution bulk. This definition remains ambiguous if we do not clarify the meaning of "equal." If both cylinders contain equal numbers of solvent moles, then $\Gamma_S = 0$. As follows from Eq. (5), the surface excess of A defined by imposing this condition represents $\Gamma_A^{(S)}$ (relative surface excess). If both cylinders contain equal numbers of moles of A and S, the equation

$$\Gamma_A^{(N)} + \Gamma_S^{(N)} = 0 \quad (7)$$

combined with Eq. (5) allows us to calculate another type of surface excess values:

$$\Gamma_A^{(N)} = \Gamma_A^{(S)}(1 - N_A)$$
$$\Gamma_S^{(N)} = -\Gamma_A^{(S)}(1 - N_A) \quad (8)$$

Next, for two cylinders of equal masses, the Eq. (5) should be supplemented by equation

$$M_A \Gamma_A^{(M)} + M_S \Gamma_S^{(M)} = 0 \quad (9)$$

where M_A and M_S are the molar masses of components. A simultaneous solution of Eqs. (5) and (9) results in the following expressions:

$$\Gamma_A^{(M)} = \Gamma_A^{(S)} \frac{M_S N_S}{M_A N_A + M_S N_S}$$
$$\Gamma_S^{(M)} = -\Gamma_A^{(S)} \frac{M_A N_S}{M_A N_A + M_S N_S} \quad (10)$$

Finally, if both cylinders are of equal volumes, and the molar volumes of the components V_A and V_S do not change in the course of adsorption, Eq. (5) should be supplemented by the equation

$$V_A \Gamma_A^{(V)} + V_S \Gamma_S^{(V)} = 0 \quad (11)$$

The corresponding simultaneous solution of Eqs. (5) and (11) gives the surface excesses

$$\Gamma_A^{(V)} = \Gamma_A^{(S)} \frac{V_S N_S}{V_A N_A + V_S N_S}$$
$$\Gamma_S^{(V)} = -\Gamma_A^{(S)} \frac{V_A N_S}{V_A N_A + V_S N_S} \quad (12)$$

As follows from Eqs. (8), (10), and (12), for adsorption from diluted solutions ($N_A \ll 1$), $\Gamma_A^{(S)} \approx \Gamma_A^{(N)} \approx \Gamma_A^{(M)} \approx \Gamma_A^{(V)}$. Under these conditions, the physical meaning of the surface excess Γ_i can be represented by the areas S_1 and S_2 enclosed by the curve corresponding to c_i versus x dependence in the vicinity of interface and the line $c_i^{(b)} = \text{const}$ (see shaded areas in Fig. 1). For positive adsorption, $\Gamma_i > 0$ whereas, for the negative adsorption, $\Gamma_i < 0$. In agreement with this interpretation of surface excess, its dimensionality is mol m^{-2} or mol cm^{-2}, which coincides with the dimensionality of the surface concentration. The latter can be defined as the number of moles of the ith component that find themselves in a monomolecular layer of substance closest to the surface. In contrast to the surface excess, the surface concentration is either positive or zero. For calculating the surface concentration from experimental Γ_i values, it is essential to know how c_i changes with x. The problem may be substantially simplified, if the surface excess of the ith component is located inside an adsorption monolayer, and its bulk concentration is sufficiently low ($N_i \ll 1$). Under these conditions, typical for adsorption of many organic substances on electrodes, the surface concentration practically coincides with the surface excess.

Originally, the phenomenon of specific adsorption was considered in connection with studies on mercury electrodes, the latter being ideally polarizable in a wide potential range. It should be mentioned that for ideally polarizable electrode, the total quantity of electric charge delivered to its surface is expended in the change of the *free* electrode charge, that is, the charge defined by excess or lack of electrons. Quite a number of s,p-metals (Bi, Ga, Cd, Pb, Sn, Sb, In, Tl) can also exhibit similar behavior in a certain potential interval, and are sometimes integrated into the group of mercury-like metals. Although any metal of this group undoubtedly exhibits certain specific properties, the difference appears to be not so dramatic as that which we found when going to typical d-metals (platinum group metals are the most thoroughly studied representatives).

In aqueous solutions, platinum-group metals adsorb hydrogen and oxygen and can be classified as the so-called "perfectly polarizable electrodes" in terms of Planck [4]. For perfectly polarizable electrode, one portion of electric charge delivered to its surface is expended in the change of electrode free charge, when another portion is expended in the oxidation of hydrogen adatoms or in the generation of oxygen adatoms. The complete quantity of electric charge delivered to the surface is said to be expended in the change of *total* electrode charge. It means that the state of electrode is uniquely determined by its initial state and the change of its total charge.

In the text that follows, we start with the adsorption phenomena on ideally polarizable electrodes, and then pass to perfectly polarizable electrodes.

Adsorption of various ions and molecules on electrodes is favored by their interaction with the electrode surface, and also by the squeezing of adsorbate species out of the solution bulk. The latter phenomenon is induced by the difference in free solvation energies ΔG between the surface layer and in the bulk of the solution. For positive adsorption, $\Delta G < 0$, which, in its turn, can be caused by a decrease in enthalpy ($\Delta H < 0$) and by an increase of entropy ($\Delta S > 0$) as the adsorbate species pass from the solution bulk to the surface. In principle, electrostatic forces can solely induce the interaction of ions with the electrode surface: the coulombic attraction and repulsion should result in positive and negative adsorption, respectively. In addition, at a direct contact of an ion with the metal surface (in the absence of separating solvent molecules), the ion–metal interaction can include other contributions: image forces (interaction with the charge of the opposite sign induced in the metal phase) and a covalent bonding, which is sometimes considered in the terms of donor–acceptor interaction (which can result in a partial charge transfer between adsorbed ions and metal).

If the ion–metal interaction is completely electrostatic and the squeezing out is absent, the electrolyte is surface-inactive. In the opposite case, specific adsorption takes place, and the corresponding ion is surface-active. Figure 2 shows the electrocapillary curves (dependences of γ on electrode potential E) for a mercury electrode in 0.01, 0.1, and 1 M aqueous solutions of a surface-inactive electrolyte (curves 1, 2, and 3) and the corresponding γ versus E-curves in solutions containing surface-active anions (1′, 2′ and 3′).

Gouy was the first to mention the specific behavior of anions on mercury [5]; however, the term "specific adsorption" was introduced by Stern [6] who made an attempt to treat specific adsorption quantitatively. Generally, any ion can be specifically adsorbed under definite conditions, and only under certain circumstances can specific adsorption be ignored in the first approximation.

It is conventionally supposed that the cations like Li^+, Na^+, and K^+, and anions like F^-, having small sizes and surrounded by dense solvation shells (also complex anions like HF_2^-, PF_6^-, BF_4^-), are not adsorbed specifically. Also, it is assumed that specific adsorption of anions is more pronounced than that of cations. These suppositions stemmed from the studies on a mercury electrode in aqueous solutions and cannot be generalized *a priori* to other systems.

Historically, the term "specific adsorption" goes back to the studies of adsorption

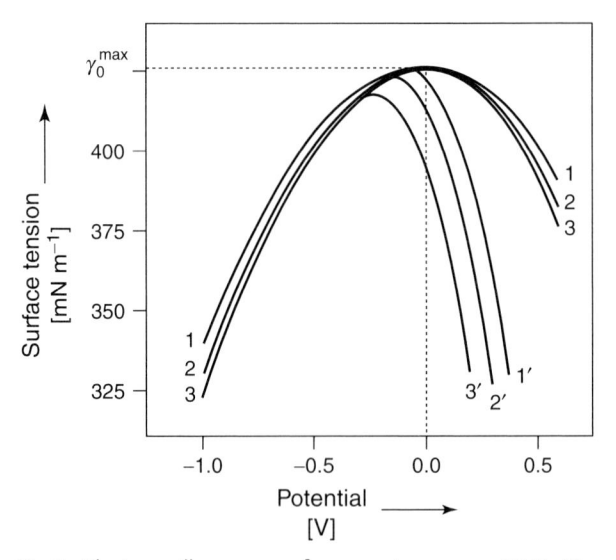

Fig. 2 Electrocapillary curves of mercury in aqueous KF (1–3) and KI (1′–3′) solutions, salt concentrations are 0.01 M (1,1′); 0.1 M (2,2′); 1 M (3,3′).

of inorganic and organic ions, but it is worthwhile to apply it also to adsorption of neutral organic molecules. The reason lies not only in the squeezing out from solution, which occurs in both cases, but also in the existence of specific interactions with the metal. That is, the molecules of aromatic compounds are bound to the electrode via the so-called π-electron interaction [7–10]; functionally substituted molecules (thiolates, amines, and so forth) assume certain interfacial orientations owing to specific interaction of substituting groups with the metal, particularly mercury [11].

According to the Lippman equation [12]

$$\sigma_M = -\left(\frac{\partial \gamma}{\partial E}\right)_{T,p,\mu} \quad (13)$$

the charge density on the metal surface σ_M is determined by the slope of the electrocapillary curve. The maximum of γ, E-curve corresponds to $\sigma_M = 0$; hence, the position of this maximum determines the potential of zero charge (pzc). In surface-inactive electrolyte solutions, the value of the pzc does not depend on concentration (Fig. 2). On the other hand, with an increase in the concentration of surface-active anion, the pzc shifts towards negative potentials, because of the increasing specific adsorption of anions.

Indeed, the condition of electrical neutrality in the 1,1-electrolyte solution is

$$\sigma_M = F\Gamma_- - F\Gamma_+ \quad (14)$$

where F stands for Faraday constant, and the subscripts correspond to the charges of the ions. Hence, at the pzc, $\Gamma_- = \Gamma_+$. However, the specifically adsorbed anions are located closer to the electrode surface as compared with surface-inactive cations. Thus, a potential drop arises near the uncharged electrode/solution interface, which is equal to the pzc shift and increases with Γ_-. The specific adsorption of cations shifts the pzc towards positive values. For simultaneous specific adsorption of anions and cations, the magnitude and the sign of the pzc shift can change with relative ionic concentrations.

The examples of pzc shifts induced by specific adsorption of anions on mercury in aqueous solutions are presented in Table 1. As demonstrated by Frumkin [13], the specific features of metal–ion interaction can be elucidated by comparing the pzc shifts and the changes of surface potential at the solution/air interface. The adsorption of ion at the latter interface results only from squeezing-out effect. For example, the change of surface potential when going from 0.01 M KCl to 2 M KI solution reaches only 52 mV [14], while the pzc shift for Hg/1 M KI (in comparison with pzc in surface-inactive electrolyte solution) is about 350 mV.

The pzc shift resulting from the adsorption of polar organic molecules is determined by the orientation of their dipoles on the electrode surface.

Tab. 1 The pzc and the pzc shift values for a mercury electrode in aqueous unimolar solutions as given by Frumkin [13]

Electrolyte	pzc (MCE) [V]	pzc shifts [V]
KOH	−0.47	0.00
K_2CO_3	−0.48	−0.01
K_2SO_4	−0.48	−0.01
K_2HPO_4	−0.52	−0.05
$NaNO_2$	−0.54	−0.07
KCl	−0.56	−0.09
KNO_3	−0.56	−0.09
KBr	−0.65	−0.18
NaCNS	−0.72	−0.25
KI	−0.82	−0.35
K_2S	−0.98	−0.51

The specific adsorption reduces the work of surface formation because of the difference in the energetic states of solvent molecules on the surface and in the bulk of the solution, and also because of the lateral interactions in the adsorbed layer. For neutral organic molecules, as a rule, the decrease in γ is observed only in the vicinity of the pzc, because at high $|\sigma_M|$ the organic molecules are squeezed out by the solvent of a higher permittivity.

The electrocapillary theory developed by Frumkin [15, 16] makes it possible to determine not only the electrode charge and pzc but also the relative surface excesses of various ions by treating the γ, E-curves. Namely, in a binary 1,1-electrolyte solution, when we use a reference electrode reversible with respect to anion (see above in Electrode Potentials),

$$d\gamma = -\sigma_M \, dE_- - 2RT\Gamma_+ \, d\ln a_\pm \quad (15)$$

whereas, in the same solution with the reference electrode reversible in respect to cation,

$$d\gamma = -\sigma_M \, dE_+ - 2RT\Gamma_- \, d\ln a_\pm \quad (16)$$

In the formulae given above, R is the universal gas constant, and $a_\pm$ is the mean activity of the electrolyte solution. By combining Eqs. (15) and (16) and introducing the notation $dE_0 = d(E_+ + E_-)/2$, one obtains

$$d\gamma = -\sigma_M \, dE_0 - RT(\Gamma_+ + \Gamma_-) \, d\ln a_\pm \quad (17)$$

On the combined potential, scale introduced above the pzc value does not depend on the concentration of the surface-inactive electrolyte. Hence, it is convenient to refer the potential to this zero point (see Fig. 2).

The properties of the total differential of a function of two variables, allow us to obtain the following expression from (17):

$$\left(\frac{\partial E_0}{\partial \ln a_\pm}\right)_{\sigma_M} = -RT\left[\frac{\partial(\Gamma_+ + \Gamma_-)}{\partial \sigma_M}\right]_{a_\pm} \quad (18)$$

At the pzc, according to Eq. (14), $\Gamma_- = \Gamma_+$, and from Eq. (18) one obtains

$$\left(\frac{\partial E_0}{\partial \ln a_\pm}\right)_{\sigma_M=0} = -2RT\left(\frac{\partial \Gamma_-}{\partial \sigma_M}\right)_{a_\pm} \quad (19)$$

Equation (19) relates the dependence of the pzc on the concentration of the electrolyte solution in the presence of specific adsorption ($\Gamma_- \neq 0$ when $\sigma_M = 0$) and its variation with the electrode charge. The dependence of the pzc of a mercury electrode on the logarithm of KI concentration was used for the first time for studying the iodide specific adsorption in [17] and later was named the Esin-Markov effect. As follows from the model theories of the electric double layer (see Sect. 3.2), the limiting slope of the aforementioned dependence should tend to the value $-RT/\lambda F$, where the coefficient $\lambda (0 < \lambda \leq 1)$ characterizes the discrete nature of the charge of specifically adsorbed anions.

The combination of Eqs. (13), (14), and (17) make it possible to calculate the dependencies of Γ_- and Γ_+ on E_0 from the experimental γ, E-curves. Figure 3 shows the corresponding dependencies of $F\Gamma_-$ and $F\Gamma_+$ on E_0 for 0.1 M solutions of surface-inactive and surface-active electrolytes. The positive adsorption of cations at $E_0 < 0$ and of anions at $E_0 > 0$ in a former solution are induced only by the coulombic attraction of ions to the oppositely charged electrode surface. In contrast, the negative adsorption of cations at $E_0 > 0$ and of anions at $E_0 < 0$ is caused by the repulsion from the surface of the same sign. In 0.1 M KF solution at $E_0 = 0$,

$\Gamma_- = \Gamma_+ \approx 0$. However, in concentrated (>1 M) solutions of surface-inactive electrolytes at $E_0 = 0$, the value of γ tends to increase with concentration. In accordance with Eq. (17), this corresponds to the negative adsorption: $\Gamma_- = \Gamma_+ < 0$. The latter is caused by hydration of ions, when a layer of solvent molecules separates ions from the electrode surface. By designating the thickness of this layer as δ and assuming that the concentration of ions near the uncharged surface sharply increases from zero to $c^{(b)}$ at $x = \delta$, we obtain $\Gamma_- = \Gamma_+ = -\delta c^{(b)}$. Experimental results confirm that, at $E_0 = 0$, the negative adsorption of a salt varies proportionally as the salt concentration in the bulk. For Hg/water interface, the slope of this line corresponds approximately to 0.3–0.4 nm, namely, a monolayer of water molecules.

As seen from Fig. 3, the specific adsorption of anions substantially affects not only Γ_- versus E_0 but also Γ_+ versus E_0 dependencies. The coulombic repulsion of anions from the negatively charged electrode surface can fully eliminate the manifestations of their specific adsorption only at sufficiently high negative E_0. In accordance with Eq. (14), the positive values of Γ_+ at $\sigma_M > 0$ correspond to $F\Gamma_- > \sigma_M$, which is known as a phenomenon of the recharging of the surface by specifically adsorbed ions. The positive values of Γ_+ often pass through a minimum (as shown in Fig. 3), and the growth of Γ_+ with the positive potential at $\sigma_M > 0$ points to the enhancement of the surface recharging. Under these conditions, the discreteness coefficient of the charge of specifically adsorbed anions is $\lambda < 1$. Specific adsorption of ions induces more complicated potential-distance dependence than compared with the corresponding decay in the absence of specific adsorption. For example, when the surface is recharged by the specifically adsorbed anions, the potential versus distance dependence should pass through a minimum. When both anions and cations are specifically adsorbed, the recharging can take place at any electrode charge. However, at $\sigma_M > 0$, the anions are located closer to the surface than cations; and, when at $\sigma_M < 0$, the opposite situation takes place.

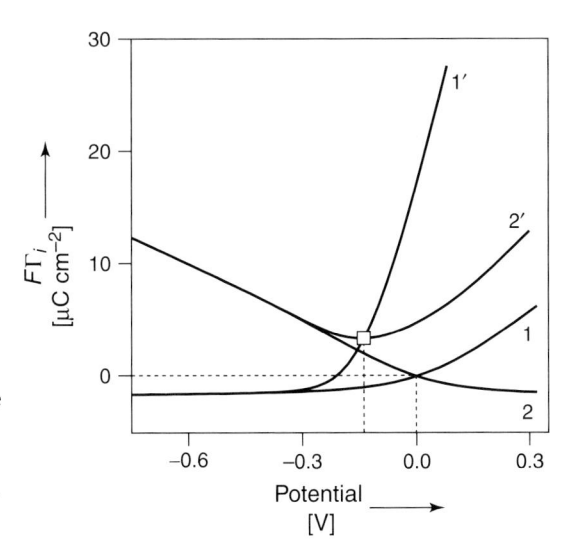

Fig. 3 Components of electrode charge associated with the surface excesses of anions (1,1′) and cations (2,2′), as function of the potential of a mercury electrode in 0.1 M aqueous solutions of KF (1,2); KI (1′ 2′).

Certain model assumptions are necessary in order to reveal the surface concentration of specifically adsorbed ions in the total surface excess Γ_i. Usually, the ionic component of the electrical double layer (EDL) is assumed to consist of the dense part and the diffuse layer separated by the so-called outer Helmholtz plane. Only specifically adsorbing ions can penetrate into the dense layer close to the surface (e.g. iodide ions), with their electric centers located on the inner Helmholtz plane. The charge density of these specifically adsorbed ions σ_1 is determined by their surface concentration $\Gamma_i^{(1)}$. Namely, for single-charged anions:

$$\sigma_1 = -F\Gamma_-^{(1)} \quad (20)$$

On the other hand, the surface excesses of all ions in the diffuse layer $\Gamma_i^{(2)}$ are defined by their coulombic interaction with the net charge $(\sigma_M + \sigma_1)$. According to the classical theory of the diffuse layer [18–20], the behavior of 1,1-electrolyte solution with concentration c is described by:

$$F\Gamma_+^{(2)} = A\sqrt{c}\left[\exp\left(-\frac{F\varphi_2}{2RT}\right) - 1\right] \quad (21)$$

$$F\Gamma_-^{(2)} = A\sqrt{c}\left[\exp\left(\frac{F\varphi_2}{2RT}\right) - 1\right] \quad (22)$$

where $A = (2RT\varepsilon_0\varepsilon)^{1/2}$, ε is the permittivity of solvent, $\varepsilon_0 = 8.854 \times 10^{-12}$ F m^{-1}, and φ_2 is the potential of the outer Helmholtz plane referred to the solution bulk and related to the value of $(\sigma_M + \sigma_1)$ by the formula

$$\varphi_2 = \frac{2RT}{F}\sinh^{-1}\left(\frac{\sigma_M + \sigma_1}{2A\sqrt{c}}\right) \quad (23)$$

If we suppose that cations in solutions of lithium, sodium, and potassium salts are surface-inactive, that is, at any given E_0, their surface excesses are located in the diffuse layer, then the value of Γ_+ determined from the experimental data should be equal to $\Gamma_+^{(2)}$. Hence, we can calculate the value of φ_2 from Eq. (21) and, then, find the contribution of anions residing in the diffuse part, $\Gamma_-^{(2)}$ from Eq. (22). Finally, we determine the specific adsorption of anions, $\Gamma_-^{(1)}$ from the experimental value of $\Gamma_- = \Gamma_-^{(1)} + \Gamma_-^{(2)}$.

This technique of calculating the surface concentration of anions was pioneered by Grahame and Soderberg [21] using the experimental dependencies of the differential capacitance C on the potential, where C is given by

$$C = \left(\frac{\partial \sigma_M}{\partial E}\right)_{a_\pm} = -\left(\frac{\partial^2 \gamma}{\partial E^2}\right)_{a_\pm} \quad (24)$$

In contrast to γ, the value of C can be measured not only for liquid electrodes but also for solid ones. Usually, the specific adsorption of ions increases the capacitance of the EDL, whereas the adsorption of organic molecules decreases it. In the framework of a simplified model of the double layer as a capacitor, this corresponds to varying the distance between the capacitor plates and decreasing the permittivity. In dilute solutions of a surface-inactive 1,1-electrolyte, a capacitance minimum appears in the vicinity of pzc (see curve 1 in Fig. 4). The position of this minimum is determined by the potential $E_0 = 0$. In solutions with relatively high (≥ 0.1 M) concentrations of surface-inactive electrolytes, this minimum of C, E_0-curve disappears, but the pzc position still corresponds to $E_0 = 0$ (curve 2 in Fig. 4). Numerical integration of these C, E_0-curves

$$\sigma_M = \int_0^{E_0} C\,dE_0 \quad (25)$$

results in the corresponding σ_M, E_0-curves (see curves 1 and 2 in Fig. 5).

Fig. 4 Dependencies of the differential capacitance of a mercury electrode on its potential in aqueous solutions of: 0.003 M KF (1); 0.1 M KF (2); 0.1 M KI (3).

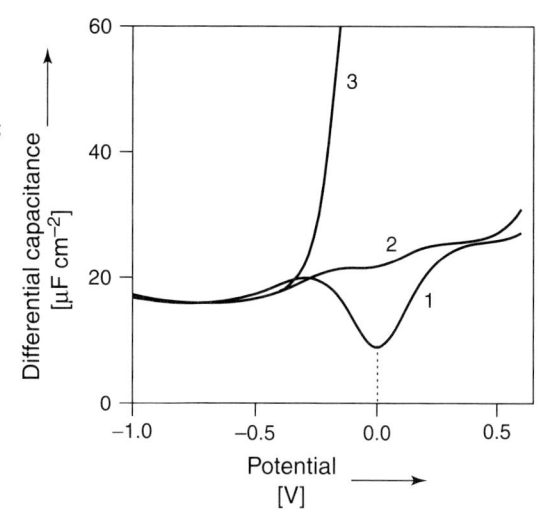

Fig. 5 Dependencies of the charge of a mercury electrode on its potential in aqueous solutions of: 0.003 M KF (1); 0.1 M KF (2); 0.1 M KI (3).

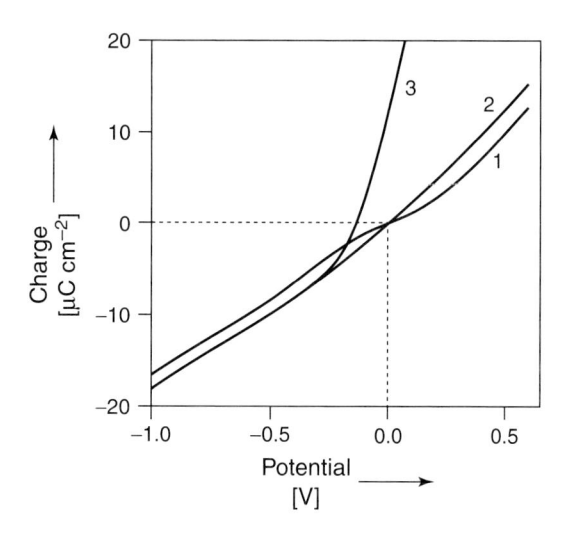

For nonsymmetrical surface-inactive electrolytes, the minimum of C shifts from the pzc. If, in the absolute magnitude the anion charge is higher than the cation charge, the minimum shifts from pzc to more negative values, and vice versa. Specific adsorption of ions also induces a shift of the capacitance minimum from the pzc. Moreover, when the specific adsorption is sufficiently pronounced, no minimum appears in the C, E_0-curves, even at extreme dilution (limited by the potentialities of capacitance measurements). In this case, the pzc should be calculated by means of back integration of C, E_0-curves. The idea of this method is as follows.

At a certain sufficiently negative potential E^*, the specific adsorption of anions disappears. Under these circumstances, the C, E_0-, σ_M, E_0-, and γ, E_0-curves measured, on the one hand, in solutions of surface-inactive electrolytes and, on the other hand, in those of surface-active electrolytes of equal concentrations with the same cation merged together (see curves 2 and 3 in Figs. 4 and 5, and also see curves 1 and 1', 2 and 2', 3 and 3' in Fig. 2). This coincidence provides a way of calculating the σ_M, E_0- and γ, E_0-curves from the curves of differential capacitance for solutions containing specifically adsorbed anions

$$\sigma_M = \sigma_M^* + \int_{E^*}^{E_0} C\, dE_0;$$

$$\gamma = \gamma_0^* - \int_{E^*}^{E_0} \sigma_M\, dE_0 \quad (26)$$

In this equation, from the data for a solution of the surface-inactive electrolyte,

$$\gamma_0^* = \gamma_0^{max} - \int_0^{E_0} \sigma_M\, dE_0 \quad (27)$$

where γ_0^{max} is the arbitrary constant. The choice of its value does not affect the values of Γ_+ and Γ_-, because, according to the Eq. (17), these values are determined not by the absolute γ values but by the changes of the surface tension with $a_\pm$.

Another way of estimating the surface concentration of the specifically adsorbed ions was proposed in Refs. [22, 23], and is based upon using binary mixtures of electrolytes of a constant total concentration, like

$$xc\, CA_* + (1-x)c\, CA \text{ or}$$
$$xc\, C_*A + (1-x)c\, CA \quad (28)$$

in which x is the mole fraction of the surface-active ions (A_*^- or C_*^+) in the mixture. It is assumed that ions C^+ and A^- are not specifically adsorbed, and their surface excesses are located in the diffuse layer. Let us designate the surface-active ion by a subscript 1, the surface-inactive ion of the same sign — by 2, and their common counterion of a constant concentration — by 3. Bearing in mind that, to a first approximation, the activity coefficient in solutions of a constant ionic strength does not change with the ratio c_1/c_2, we can write

$$d\ln a_3 \approx d\ln c_3 = 0;$$

$$d\ln a_1 \approx d\ln(xc_3) = d\ln x = \frac{dx}{x} \quad (29)$$

$$d\ln a_2 \approx d\ln[(1-x)c_3]$$
$$= d\ln(1-x) = -\frac{dx}{1-x} \quad (30)$$

Hence, the electrocapillarity equation for the systems like (28) takes the form

$$d\gamma = -\sigma_M\, dE - RT\left[\Gamma_1^{(1)} d\ln x + \Gamma_1^{(2)} \frac{dx}{x}\right.$$
$$\left. - \Gamma_2^{(2)} \frac{dx}{1-x}\right] \quad (31)$$

The surface excesses of ions 1 and 2 in the diffuse layer are caused only by the coulombic interaction of these ions with the net charge ($\sigma_M + \sigma_1$), hence, $\Gamma_1^{(2)}/\Gamma_2^{(2)} = x/(1-x)$. Under these condition, the sum of two last terms inside the brackets of Eq. (31) is equal to zero, so that

$$d\gamma = -\sigma_M\, dE - RT\Gamma_1^{(1)} d\ln x \quad (32)$$

Hence, for the systems of the type (28),

$$\Gamma_1^{(1)} = -\frac{1}{RT}\left(\frac{\partial \gamma}{\partial \ln x}\right)_E \quad (33)$$

The γ versus $\ln x$ dependence at $E =$ const, which is required for using formula (33), can be obtained either from experimental electrocapillary curves, or on the basis of a series of C, E- or σ_M, E-curves for systems like (28) with different x.

In principle, σ_M,E-curves can be obtained by various electrochemical techniques. For example, according to Ref. [24], we can use the chronocoulometry technique, in which current transients are measured following a potential step ΔE, and the variation of charge $\Delta\sigma_M$ are obtained by integrating these transients. This technique has made good use of pyridine adsorption studies on the single crystalline silver and gold [25–30].

It should be noted that deviations from Eqs. (32) and (33) can be induced by the noncoincidence of the outer Helmholtz planes for ions 1 and 2 of a system like 28 [31].

When the species i are adsorbed reversibly, their electrochemical potentials in the adsorption layer $\bar{\mu}_i^{(ads)}$ and in the solution bulk $\bar{\mu}_i^{(b)}$ should be equal: $\bar{\mu}_i^{(ads)} = \bar{\mu}_i^{(b)}$. This simple relationship does not allow one to link unambiguously the surface and bulk concentrations of ith species, because the activity coefficients in the bulk of the solution and especially in the surface layer are concentration-dependent, and, in general, these dependencies are ambiguously determined. Therefore, the dependence of the surface concentration of adsorbate or the surface coverage θ on the bulk concentration $c_i^{(b)}$ is usually described by an equation providing the best approximation of experimental data with a number of fitting parameters being as low as possible. The adsorption of neutral organic molecules is most frequently described by the Frumkin isotherm [32]:

$$\beta c_i^{(b)} = \frac{\theta}{1-\theta}\exp(-2a\theta) \quad (34)$$

where β is the constant of adsorption equilibrium, and a is the so-called attraction constant. The value of $\ln\beta$ characterizes the free adsorption energy ΔG_A^0:

$$-\Delta G_A^0 = \text{const} + RT\ln\beta \quad (35)$$

where the value of const depends on the choice of the standard state. The attraction constant reflects the set of lateral interactions between the molecules of adsorbate A and solvent S in the surface layer:

$$a = \frac{-\Delta G_{AA} + \Delta G_{AS} - \Delta G_{SS}}{RT} \quad (36)$$

The ΔG_A^0, ΔG_{AA}, ΔG_{AS} and ΔG_{SS} values, and, correspondingly, $\ln\beta$ and a values depend on the electric state of the surface, that is, on the electrode potential or charge. However, these dependences can be obtained in an explicit form only on the basis of some model of the surface layer consisting of the mixture of organic and water molecules. Unfortunately, so far, the models of various authors [33–41] have not resulted in any unified approach.

Equation (34) was deduced by Frumkin [32] as a general case of Langmuir isotherm, which corresponds to $a = 0$. A statistical derivation of Eq. (34) was carried out by Fowler and Guggenheim [42]; however, the applicability of Frumkin isotherm to the experimental data appears to be substantially wider than it follows immediately from the derivation conditions.

Figure 6 shows the dependencies of θ versus $\ln(\beta c_i^{(b)})$ calculated by Eq. (34) at various values of a. When $a = 2$, at $\theta = 0.5$, $\partial\theta/\partial\ln(\beta c_i^{(b)}) = \infty$. When $a > 2$, a steplike transition from low θ values to $\theta \approx 1$ appears on the isotherm, which corresponds to 2D-condensation of adsorbate molecules (see Sect. 3.3.2.1).

When the ions of the ith species are adsorbed in place of neutral molecules, their concentration at the outer Helmholtz plane $c_i^{(2)}$ should be substituted for the

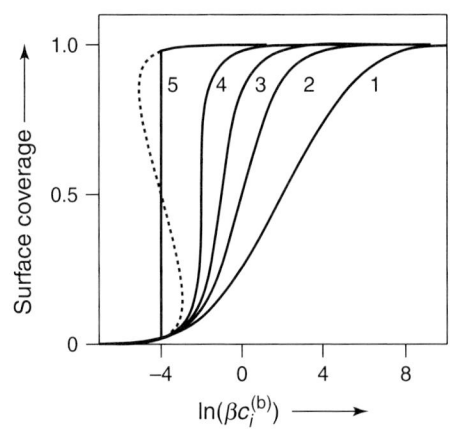

Fig. 6 Dependencies of the surface coverage on $\ln(\beta c_i^{(b)})$ calculated from Eq. (34) for various values of a: -2(1); 0(2); 1(3); 2(4); 4(5).

bulk concentration $c_i^{(b)}$ [43]. According to Boltzmann formula,

$$c_i^{(2)} = c_i^{(b)} \exp\left(\frac{-z_i F \varphi_2}{RT}\right) \quad (37)$$

According to Eq. (23), the value of φ_2, is a function of the charge of specifically adsorbed ions $\sigma_1 = z_i F \Gamma_{max} \theta$, where Γ_{max} is the limiting value of their surface concentration corresponding to $\theta = 1$. Thus, the adsorption isotherm equation for specifically adsorbed ions appears to be very complex.

At low surface coverage, when $\ln(1 - \theta) \approx -\theta$, the Frumkin isotherm (34) is rearranged into a virial isotherm:

$$\ln \beta_v + \ln c_i^{(b)} = \ln \Gamma_i^{(1)} + 2B_v \Gamma_i^{(1)} \quad (38)$$

where

$$\ln \beta_v = \ln \beta + \ln \Gamma_{max}; \quad \Gamma_i^{(1)} = \theta \Gamma_{max}$$

$$B_v = \frac{1 - 2a}{2\Gamma_{max}} \quad (39)$$

As mentioned in the preceding text, free energy of adsorption of ions or molecules, and, correspondingly, the value of $\ln \beta$ (see Eq. 35) depends on one of electrical variables: electrode potential E, electrode charge σ_M, and the potential drop in the inner part of the EDL. The model applied for describing the EDL determines the choice of the independent variable. Namely, in the framework of Frumkin model [33] for adsorption of neutral organic molecules, $\ln \beta$ is a square function of E_0; in terms of Vorotyntsev's model [44] for ionic adsorption, $\ln \beta$ is also a function of E_0, but substantially more complicated [45]; in Grahame-Parsons model [46, 47] for ionic adsorption, the value of $\ln \beta$ is considered as a function of σ_M; in contrast, in Kolotyrkin–Alekseev–Popov model [48], put forward later in Refs. [49, 50], and also in a similar Nikitas' model [40, 41], $\ln \beta$ is a linear or square function of the potential drop in the inner layer: $\Delta \varphi = E_0 - \varphi_2$.

As demonstrated in Refs. [51, 52], in the case of reversible specific adsorption of Cl$^-$, Br$^-$, and I$^-$ anions on various mercury-like electrodes from different solvents, the values of $\ln \beta$ at $\sigma_M = 0$ and $\Delta E_{\sigma_M=0}$ at $c_i^{(b)} = $ const can be correlated with a linear combination of donor (DN) and acceptor (AN) numbers of solvents:

$$y = k_0 - k_1 \text{DN} - k_2 \text{AN} \quad (40)$$

The quantity of y in Eq. (40), which is equal to $\ln \beta$ at $\sigma_M = 0$ [51] or to $\Delta E_{\sigma_M=0}$

at $c_i^{(b)}$ = const [52], is the measure of specific adsorption energy of the ith ion, and k_0, k_1, and k_2 are constants independent of the nature of solvent for the given metal and ion. In its physical meaning, the value of k_0 should represent the metal-anion bond energy in the absence of solvent.

In the physical sense, Eq. (40) is based on the supposition that the greater the specific adsorption energy of anions should be, the smaller their solvation energy (presented by the value of AN) and the weaker the bond of solvent molecules with the uncharged electrode surface (presented by the value of DN). This is induced by the donor–acceptor nature of ion–dipole interactions, and also by the interaction of metal with anions and solvent molecules in the course of their competitive adsorption on the uncharged electrode surface.

In Ref. [52], the Eq. (40) was compared with the experimental values of $y = \Delta E_{\sigma_M=0}$ at $c_i^{(b)}$ = 0.1 M, which resulted from the specific adsorption of Cl$^-$, Br$^-$, and I$^-$ anions on the electrodes made of mercury, gallium, and eutectic alloys In-Ga (16.4 at % In) and Tl-Ga (0.02 at % Tl) in water, acetonitrile, dimethylformamide, N-methylformamide, and dimethyl sulfoxide. The correlation analysis demonstrated that the bond energy in the absence of solvent (k_0) increases in the sequence I$^-$ < Br$^-$ < Cl$^-$, in agreement with the row for dissociation energy of gaseous monohalogenides of mercury and gallium. At the same time, for various donor–acceptor properties of solvent, the energy of specific adsorption on the metal/solution interface at $\sigma_M = 0$ can either increase in a sequence Cl$^-$ < Br$^-$ < I$^-$ (this "usual" sequence was observed in aqueous solutions on mercury and mercury-like metals), or decrease in the same sequence in parallel with k_0 value (namely, for gallium in aprotic solvents), or, finally, remain independent of the anion nature (e.g. on In-Ga/acetonitrile interface).

Let us consider the specific adsorption phenomena on the d-metals.

As was mentioned in the preceding text, the main difference between "perfectly polarizable" and "ideally polarizable" electrodes consists in the fact that the potential of the former is determined by the presence of a certain redox system,

$$\text{Ox} + n\bar{e} \Longleftrightarrow \text{Red} \quad (41)$$

and the state of the system is unambiguously determined by its initial state and a quantity of charge passed across the interface. For platinum metals in aqueous solutions of surface-inactive electrolytes, Ox$_1$ = H$^+$, Red$_1$ = H$_{\text{ads}}$ ($n = 1$); Ox$_2$ = O$_{\text{ads}}$ + 2H$^+$, Red$_2$ = H$_2$O ($n = 2$). A consistent thermodynamic theory of perfectly polarizable electrodes was developed in [53–60] on the basis of the initial theory of amalgam [13, 61, 62] and platinum hydrogen [63] electrodes. In this case, Gibbs equation at T, p = const takes the following form:

$$d\gamma = -\Gamma_{\text{Ox}} d\mu_{\text{Ox}} - \Gamma_{\text{Red}} d\mu_{\text{Red}} - \sum_j \Gamma_j d\mu_j \quad (42)$$

in which μ_{Ox}, μ_{Red}, Γ_{Ox}, Γ_{Red} are the chemical potentials and surface excesses of Ox and Red components, μ_j, Γ_j are the chemical potentials and surface excesses of other solution components.

It follows from Eq. (42) that, when μ_j = const, the function γ is a 3D surface, its intersects with the planes μ_{Ox} = const and μ_{Red} = const being the characteristic electrocapillary curves described by Lippman equations:

$$\left(\frac{\partial \gamma}{\partial E}\right)_{T,p,\mu_j,\mu_{Ox}} = +F\Gamma_{Red} \quad (43)$$

$$\left(\frac{\partial \gamma}{\partial E}\right)_{T,p,\mu_j,\mu_{Red}} = -F\Gamma_{Ox} \quad (44)$$

The quantities of $F\Gamma_{Ox}$ and $F\Gamma_{Red}$ are called the total charges of the surface at the corresponding choice of values remaining constant, because these values characterize the quantities of electricity required for increasing the surface by a unit area without introducing any changes in the composition of both contacting phases.

It is evident that for redox-systems more complex than (41), Gibbs adsorption equation should also become more complicated. The number of characteristic Lippman Equations should be equal to the number of independent variables in the Nernst Equation for the corresponding redox system.

In order to divide the values of $F\Gamma_{Ox}$ and $F\Gamma_{Red}$ into bound charges (A_{Ox} and A_{Red}) and free charge σ_M, certain assumptions going beyond the frames of thermodynamics are necessary.

In a certain case of a platinum hydrogen electrode, on which the equilibrium $H_{ads} \Leftrightarrow H^+ + \bar{e}$ is established, at $\mu_j = $ const,

$$d\gamma = -\Gamma_H \, d\mu_H - \Gamma_{H^+} \, d\mu_{H^+} \quad (45)$$

where

$$F\Gamma_H = -Q_1 = A_H - \sigma_M \quad (46)$$
$$F\Gamma_{H^+} = Q_2 = \sigma_M + A_{H^+} \quad (47)$$

that is, total charges $F\Gamma_H$ and $F\Gamma_{H^+}$ are the algebraic sum of bound (A_H, A_{H^+}) and free (σ_M) charges. At an excess of surface-inactive electrolyte, A_{H^+} can be

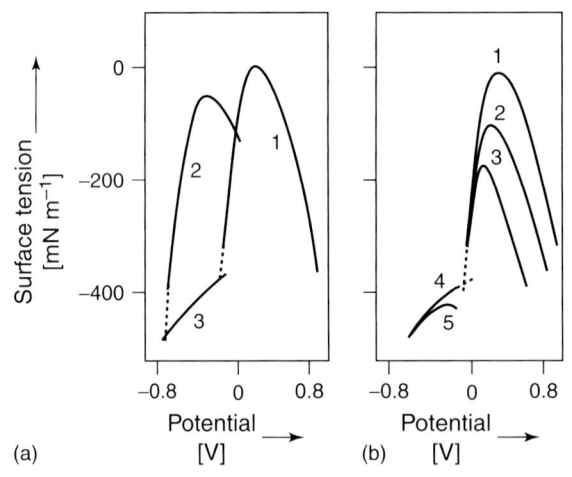

Fig. 7 Electrocapillary curves of the first (1a, 2a, 1b, 2b, 3b) and second (3a, 4b, 5b) (at $E = 0$ (RHE)) kind for a platinized platinum electrode in the following solutions:
0.005 M H_2SO_4 + 0.5 M Na_2SO_4 (1a, 1b);
0.01 M NaOH + 0.5 M Na_2SO_4 (2a); 0.01 M HCl + 1 M KCl (2b); 0.01 M HBr + 1 M NaBr (3b); 0.5 M Na_2SO_4 (3a, 4b); 0.1 M KI (5b) [64].

equated to zero, and $F\Gamma_{H^+} = \sigma_M$, that is, under certain conditions, $F\Gamma_{H^+}$ can be considered as a free surface charge.

The electrocapillary curves corresponding to Eq. (43) are called the electrocapillary curves of the first kind, and those corresponding to the Eq. (44) are called the electrocapillary curves of the second kind [60, 64]. Figure 7 shows the examples of curves of these kinds for Pt electrode. This figure demonstrates also the effects the solution pH and the anion adsorption on electrocapillary curves. It should be stressed that pH-dependent adsorption behavior of Pt hydrogen electrode is a typical feature of hydrogen-adsorbing metals, it is governed by Eq. (45), for the latter no assumptions are made on the specific adsorption of OH^- ions.

Therefore, for Pt hydrogen electrode, two pzc values should be introduced, which correspond to $\Gamma_H = 0$ and $\Gamma_{H^+} = 0$ (and to the maxima of the electrocapillary curves of the 1st and 2nd kinds), respectively.

When the concentration of a specifically adsorbed ion i is sufficiently low as compared with the supporting-electrolyte concentration, so that it can be changed without changing the chemical potential of supporting ions, the Gibbs Equation takes the following form:

$$d\gamma = -\Gamma_H \, d\mu_H - \Gamma_i \, d\mu_i \quad (48)$$

where μ_i is a chemical potential of adsorbed ion, and Γ_i – its surface excess. From this equation, the quantitative relationships for determination of Γ_i versus electrode potential dependencies by using the electrode charging curve and the shifts of electrode potential with the concentration of adsorbed ions (so-called adsorption shift of potential) [65]. The measurements of this sort gave the results that agree with direct data on ionic adsorption obtained by radiotracer techniques [66]. Equation (48) allows us to describe the effect of specifically adsorbed ions on the hydrogen adsorption. Inasmuch as ionic adsorption affects the hydrogen and oxygen adsorption, an attempt should be made in using these effects for determination of ionic adsorption on platinum group metals [67].

In contrast to mercury-like metals, the adsorption of ions on platinum metals appears to be relatively slow [68], the adsorption rate v_{ads} exponentially decreasing with the surface coverage θ in accordance with Roginskii-Zel'dovich equation

$$v_{ads} = K_{ads} \, c_i^{(b)} \exp(-\alpha f \theta) \quad (49)$$

where K_{ads} and α are constants ($0 < \alpha < 1$) and f is the inhomogeneity factor [67]. For the middle coverage of platinum metals by ions, a logarithmic dependence of θ on adsorption time t_{ads} is observed:

$$\theta = \text{const}_1 + \frac{1}{\alpha f} \ln t_{ads} \quad (50)$$

These data confirm that the adsorption of ions on platinum metals follows the general tendencies of adsorption on the energetically inhomogeneous surfaces. This conclusion agrees also with the results of analyzing the isotherms found for ionic adsorption on platinum [68], which follow the logarithmic law (Temkin isotherm):

$$\theta = \text{const}_2 + \frac{1}{f} \ln c^{(b)} \quad (51)$$

and also with the data on the isotherms of displacement of one ion by another. The latter proved to be helpful when comparing the surface activities of ions. Particularly, it was found (Fig. 8) that Cs^+ ions displace Na^+ ions adsorbed on platinum at substantially lower bulk concentration than that corresponding to the opposite

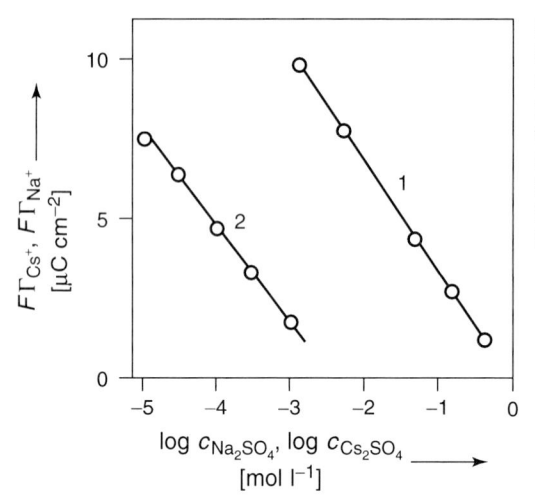

Fig. 8 Isotherms of displacement of Cs^+ cations by Na^+ (1), and of Na^+ cations by Cs^+ (2) on platinized platinum electrodes at $E = 0$ (RHE) in the following solutions: 0.0005 M H_2SO_4 + 0.001 M Cs_2SO_4 + x M Na_2SO_4 (1); 0.0005 M H_2SO_4 + 0.001 M Na_2SO_4 + x M Cs_2SO_4 (2) [69].

situation (displacement of adsorbed Cs^+ by Na^+), while the adsorption values of Cs^+ and Na^+ differ only slightly [69].

It should be stressed that the adsorption of isotherms accompanied by charge transfer differ from those measured in the absence of the latter even on uniform surfaces. In Refs. [70–74], the isotherm of adsorption with charge transfer are considered, and the field effect is introduced by two different ways: via the Nernst Equation and by considering the displacement of water molecules whose adsorption changes with potential. This is why the author refers to this as combined isotherm.

In as much as the specific adsorption is frequently associated with the donor–acceptor ion–metal interaction, the state of specifically adsorbed ions requires special attention. It is supposed that specific adsorption can result either in partial or in complete charge transfer between adsorbed species and metal. Although the idea of partial charge transfer was put forward by Lorenz [75, 76] in connection with ionic adsorption on mercury, the most impressive confirmations were obtained for d-metals.

A complete charge transfer makes it possible to introduce the term "adatom," whereas the phenomenon of adatom formation, which occurs before the thermodynamic potential of the corresponding system is reached, was called "underpotential deposition" (upd) (see Chap. 6).

A typical example of adatom formation known for a long time is the adsorption of hydrogen ions on platinum, which is accompanied by charge transfer and consequent transition to the atomic state. Oxygen adsorption can also be considered to be reversible in a certain potential region preceding the oxygen evolution.

On the basis of the generalized concept of the electrode charge, a consistent phenomenological treatment of charge transfer during the chemisorption was developed [57, 77]. Let us assume that the process of type (41) but involving only one electron occurs through an intermediate adsorption state A:

$$Ox + n_1 \bar{e} \Longleftrightarrow A \quad (52)$$

$$A + n_2 \bar{e} \Longleftrightarrow Red \quad (53)$$

It is evident that $n_1 + n_2 = 1$. The quantities n_1 and n_2 are called the formal (thermodynamic, macroscopic) charge-transfer coefficients. In the case of adsorption of Ox, the process may stop at stage (52), and, in the case of adsorption of Red, it may stop at stage (53). It can be shown [57] that the formal charge-transfer coefficients are determined by the relationships:

$$n_1 = -\frac{1}{F}\left(\frac{\partial Q_1}{\partial \Gamma_\Sigma}\right)_E = \left(\frac{\partial \Gamma_{Red}}{\partial \Gamma_\Sigma}\right)_E \quad (54)$$

$$n_2 = \frac{1}{F}\left(\frac{\partial Q_2}{\partial \Gamma_\Sigma}\right)_E = \left(\frac{\partial \Gamma_{Ox}}{\partial \Gamma_\Sigma}\right)_E \quad (55)$$

where $\Gamma_\Sigma = \Gamma_{Ox} + \Gamma_{Red}$.

The values n_1 and n_2 designate the formal coefficients of charge transfer, which characterize the overall effect of introducing the species into the surface layer – both the true (microscopic) charge transfer from species to the electrode or vise versa, and any charge redistribution in the EDL. Thus, n_i differs from zero when the charge of an adsorbed species remains unchanged, however, its adsorption changes the double-layer capacitance at a constant potential. This is why, it is hardly appropriate to consider n_j as the "electrosorption valence" [78–81]. If n electrons take part in redox process, we should use (n_1/n) and (n_2/n) instead of n_1 and n_2.

The formal charge-transfer coefficient can be experimentally determined; whereas, the determination of the true transfer coefficient based on electrochemical data requires making assumptions outside the scope of thermodynamics [82, 83]. In principle, the latter value can be determined from nonelectrochemical, particularly spectroscopic techniques sensitive directly to the electronic states of species and/or the metal.

Charge transfer can be evidenced by the enhanced adsorption of ions that proceeds simultaneously with the less-pronounced adsorption of oppositely charged ions, the latter compensating the superequivalent adsorption [66]. Namely, the adsorption of Na^+ on platinized platinum in the region of positive charges increases in the sequence $Na_2SO_4 < NaI < NaCl < NaBr$ (Fig. 9), whereas the adsorption of anions grows in a row $SO_4^{2-} < Cl^- <$

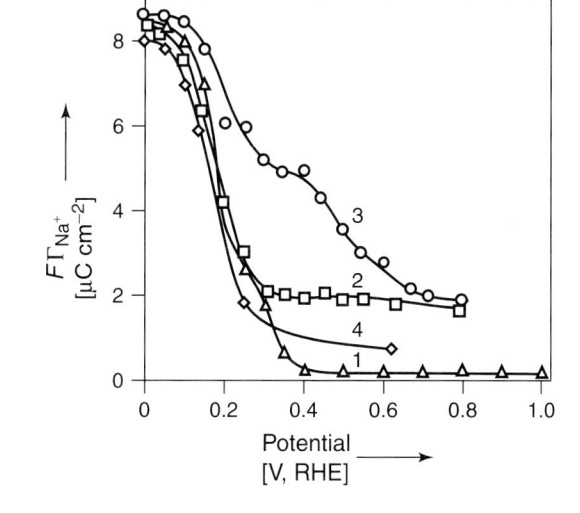

Fig. 9 Dependencies of Na^+ adsorption on the potential of a platinized platinum electrode in the following solutions: 0.0005 M H_2SO_4 + 0.0015 M Na_2SO_4 (1); 0.001 M HCl + 0.003 M NaCl (2); 0.001 M HBr + 0.003 M NaBr (3); 0.0005 M H_2SO_4 + 0.003 M NaI [66].

$Br^- < I^-$. Thus, the sequences for specific adsorbability and superequivalent adsorption disagree. This phenomenon can be explained under a supposition that the charge-transfer degree increases with the adsorbability. The approximately constant or even slightly decreasing adsorption of Na^+ observed with the growth of potential indicates that the charge-transfer degree increases as the potential shifts towards more positive values.

The charge transfer, which accompanies the adsorption processes, can affect the latter because the ionic charges are spread over the surface in a peculiar manner. Consequently, the Esin-Marov effect can be substantially weakened. A direct determination of Esin-Markov coefficient on platinum group metals [54] demonstrates no effect altogether, which can be explained by a pronounced charge transfer from adsorbed anions. According to the thermodynamic theory of the platinum hydrogen electrode, there are two Esin-Markov coefficients corresponding to the constant values of free and total charges respectively. When comparing the data for Pt and Hg electrodes, one should use the former coefficient.

Charge transfer manifests itself also by a slow establishment of adsorption equilibria mentioned in the preceding text, a slow exchange between adsorbed ions and solution, temperature effects, and the adsorption isotherms of strongly surface-active ions. The latter isotherms demonstrate a slope close to that known for the adsorption of neutral and low-polar species, namely, H_{ads}, for which the Temkin isotherm is applicable [67].

For charge transfer from adsorbed ions to the metal surface, a problem arises as to whether this charge should be attributed to the "electrode" part of the double-layer or to its ionic part. Depending on the solution of this problem, the conclusion on the pzfc (potential of zero free charge) shift induced by specific adsorption can change. As demonstrated in Ref. [84], the pzfc of a platinum electrode can shift substantially towards more positive and more negative potential values when I^- and Tl^+, respectively, are adsorbed. Thus, the direction of the pzfc shift is opposite as compared with that induced by the adsorption of the same ions on mercury-like metals. Hence, under the conditions of charge transfer, we can discuss only conventional zero free charge potentials, because, actually, the nature of the electrode surface changes.

Complete charge transfer (upd) is possible when the support-adsorbate interaction is stronger than the lateral interaction in the adlayer. Usually, upd induces monolayer coverage; however, for certain systems the formation of second and even third layer turns out to be possible, and these layers do not resemble the corresponding bulk deposits.

Consideration of perfectly polarizable electrodes presented in the preceding text is related to equilibrium conditions ($E =$ const). When dealing with nonpotentiostatic modes (linear, stepwise, sinusoidal, etc.), one always observes not only double-layer capacity but also pseudocapacity induced by the changes of surface coverage with time in the course of reactions (52, 53). Conway, Gileadi and coworkers intensively studied this problem Refs. [85, 86].

An application of the thermodynamic theory of Pt hydrogen electrode to the reversible adsorption of organic substances leads to a conclusion [87] that the maximum adsorption of neutral molecules should be observed in the potential region, where the adsorption of both hydrogen and oxygen is low (so-called double-layer region). However, this conclusion was never checked, because for the majority of adsorbates, the destruction of organic molecules in the course of adsorption was found [88], which was accompanied by their hydridization, dehydridization, association of adsorption products, and so forth. Hence, their adsorption appears to be irreversible, and the composition of adsorbed species differs from that in the solution bulk. In this case, $\bar{\mu}_i^{(ads)} \neq \bar{\mu}_i^{(b)}$; no exchange between adsorbed layer and solution occurs, the species cannot be desorbed in pure solvent, and no unambiguous relationship exists between the bulk and surface concentrations of adsorbates. Certain inorganic ions (like I^- in acid media) are also adsorbed irreversibly on platinum.

In parallel with complications induced by irreversible adsorption, new approaches to the experimental studies of surface excesses are developed. Namely, the adsorbate's state and coverage usually remain unchanged when the potential is switched off and the electrode withdrawn from solution. Moreover, some adsorbed layers remain stable even under UHV conditions, and a number of highly informative ex situ spectroscopic techniques were successfully applied to the studies of adsorption on d-metals [89–91].

Modeling of electrochemical interfaces under UHV conditions supplements the aforementioned approaches. For example, it exploits the controlled modification of the metal/vacuum interface by the various components typical of electrochemical interfaces (ions, solvent molecules) and registration of infrared reflection-absorption spectra along with work function measurements [92]. A gradual increase in the surface coverage by solvent molecules, which changes from a submonolayer up to multilayers at various doses of adsorbed ions, makes it possible to clarify the roles the ion solvation and the metallic surface solvation play in the interfacial structure, and, particularly, understand the mechanism of solvophilicity of metals.

In the potential region where formation of chemisorbed oxygen layers and phase oxides takes place, the adsorption phenomena become irreversible and extremely complicated, because almost all solution components can take part in electrode reactions and undergo electrochemical transformations. The adsorption peculiarities under these conditions differ strongly from those known for d-metals in the absence of oxygen adsorption. For example, the adsorbability sequence for halide ions on Pt at $E > 1.5$ V (RHE) transforms into $I^- < Br^- < Cl^- < F^-$ [93]. It is assumed that anions penetrate into the oxide layer with parallel dehydration, and their susceptibility to injection correlates with the reciprocal crystallographic radia. The adsorption versus potential dependence appears to be nonmonotonous and demonstrates a number of maxima; moreover, in a number of cases, the maxima of anion adsorption correspond to the maxima of cation adsorption [94–98]. It was also found that organic substances could adsorb at high anodic potentials on platinum group metals [99]. The mechanisms of these phenomena still remain unclear.

In principle, the theory of reversible electrodes can serve as the basis for developing the thermodynamic approach to the surface electrochemistry of oxide materials with metallic conductivity (iridium, ruthenium, tin dioxides, and so on) [100]. In addition, the properties of these interfaces can be considered within the framework of three models: the classical model of bound sites [101], which can be modified by considering several types of surface groups

(MUSIC model) [102, 103]; various modifications of Grahame-Parsons model [104]; and the model of nonlocal electrostatics [105]. On oxide surfaces, functional groups are present, which can be ionized reversibly upon their contact with the electrolyte solution:

$$MOH_2^+ \rightleftharpoons MOH + H^+ \quad (56)$$

$$MOH \rightleftharpoons MO^- + H^+ \quad (57)$$

$$MOH_2^+ \ldots A^- \rightleftharpoons MOH + H^+ + A^- \quad (58)$$

$$MOH + C^+ \rightleftharpoons MO^- \ldots C^+ + H^+ \quad (59)$$

where C^+ and A^- are cations and anions of the electrolyte. In this case, the specific adsorption of certain components can be considered as an increase in the binding degree of this component with the surface functional groups [106].

For oxide/solution interfaces, a pronounced correlation was revealed between the type of organic compound and their adsorption [107]. Such selectivity was noted for adsorption on metals and can point to the chemisorption nature of the metal–adsorbate interaction [108].

A modern stage in the development of notions of specific adsorption is masked by a changeover to describing this phenomenon on a microscopical level, using computational approaches to simulation of adsorbed layers. We can expect a rapid progress in these directions, because the information on the microscopic structure of adlayers becomes progressively more available because of studies on single-crystal electrodes and the results of physical surface science techniques.

There is a number of novel problems in the field of specific adsorption induced by fast progress of material science. Particularly, the role of nanostructure and nanoheterogeneity should be stressed because rather unexpected features of adlayers result from comparable sizes of adsorbate molecule and surface fragment. One of the aspects of this problem is the effect of long-living metastable surface defectiveness that is responsible for "adsorption traps," the surface centers with anomalously high adsorption ability.

Specific adsorption is rated among the most important phenomena in surface science. Its unique importance for electrochemistry is dictated by the role of specific adsorption of reactants and products, and also of inert solution components in the kinetics of electrode reactions. Adsorption can change not only reaction rates but also the mechanisms.

References

1. J. W. Gibbs, *The Collected Works of J. W. Gibbs*, Longmans, New York, 1931, p. 219, Vol. 1.
2. E. A. Guggenheim, N. K. Adam, *Proc. R. Soc. London* **1933**, *A139*, 218–236.
3. A. W. Adamson, *Physical Chemistry of Surfaces*, John Wiley & Sons, New York, 1976, p. 68.
4. M. Planck, *Ann. Phys.* **1891**, *44*, 385–428.
5. G. Gouy, *Compt. Rend.* **1908**, *146*, 1374–1376.
6. O. Stern, *Z. Elektrochem.* **1924**, *30*, 508–516.
7. M. A. Gerovich, *Doklady AN SSSR* **1954**, *96*, 543–546.
8. E. Blomgen, J. O'M. Bockris, *J. Phys. Chem.* **1959**, *63*, 1475–1484.
9. M. A. Gerovich, *Doklady Akad. Nauk SSSR* **1955**, *105*, 1278–1281.
10. E. Blomgen, J. O'M. Bockris, C. Jesch, *J. Phys. Chem.* **1961**, *65*, 2000–2010.
11. H. O. Finklea in *Electroanalytical Chemistry* (Eds.: A. J. Bard, I. Rubinstein), Marcel Dekker, New York, 1996, pp. 109–335.
12. G. Lippmann, *Ann. Chim. Phys.* **1875**, *5*, 494–549.
13. A. N. Frumkin, *Elektrokapillyarnye yavleniya i elektrodnye potentsialy (Electrocapillary Phenomena and Electrode Potentials)*, Tipographiya Sapozhnikova, Odessa, 1919.

14. A. Frumkin, *Z. Phys. Chem.* **1924**, *109*, 34–48.
15. A. Frumkin, *Z. Phys. Chem.* **1923**, *103*, 43–70.
16. A. Frumkin, *Zh. Fiz. Khim.* **1956**, *30*, 2066–2069.
17. O. A. Esin, B. F. Markov, *Acta Physicochim. URSS* **1939**, *10*, 353–364.
18. G. Gouy, *J. Phys. Radium* **1910**, *9*, 457–466.
19. D. L. Chapman, *Philos. Mag.* **1913**, *25*, 475–483.
20. D. C. Grahame, *Chem. Rev.* **1947**, *41*, 441–501.
21. D. C. Grahame, B. A. Soderberg, *J. Chem. Phys.* **1954**, *22*, 449–460.
22. H. D. Hurwitz, *J. Electroanal. Chem.* **1965**, *10*, 35–41.
23. E. Dutkiewicz, R. Parsons, *J. Electroanal. Chem.* **1966**, *11*, 100–110.
24. J. Lipkowski, C. Nguen Van Huong, C. Hinnen et al., *J. Electroanal. Chem.* **1983**, *143*, 375–396.
25. L. Stolberg, J. Lipkowski, D. E. Irish, *J. Electroanal. Chem.* **1987**, *238*, 333–353.
26. A. Hamelin, S. Morin, J. Richter et al., *J. Electroanal. Chem.* **1989**, *272*, 241–252.
27. L. Stolberg, J. Lipkowski, D. E. Irish, *J. Electroanal. Chem.* **1990**, *296*, 171–189.
28. L. Stolberg, J. Lipkowski, D. E. Irish, *J. Electroanal. Chem.* **1992**, *322*, 357–372.
29. A. Hamelin, S. Morin, J. Richter et al., *J. Electroanal. Chem.* **1990**, *285*, 249–262.
30. A. Hamelin, S. Morin, J. Richter et al., *J. Electroanal. Chem.* **1991**, *304*, 195–209.
31. W. R. Fawcett, *J. Electroanal. Chem.* **1972**, *39*, 474–477.
32. A. Frumkin, *Z. Phys. Chem.* **1925**, *116*, 466–484.
33. A. Frumkin, *Z. Physik* **1926**, *35*, 792–802.
34. R. Parsons, *J. Electroanal. Chem.* **1964**, *7*, 136–152.
35. J. O'M. Bockris, M. A. V. Devanathan, K. Müller, *Proc. R. Soc. London* **1963**, *A274*, 55–79.
36. J. O'M. Bockris, E. Gileadi, K. Müller, *Electrochim. Acta* **1967**, *12*, 1301–1321.
37. B. Damaskin, A. Frumkin, A. Chizhov, *J. Electroanal. Chem.* **1970**, *28*, 93–104.
38. M. V. Sangaranarayanan, S. K. Rangarajan, *Can. J. Chem.* **1981**, *59*, 2072–2079.
39. R. Guidelli, G. Aloisi, *J. Electroanal. Chem.* **1992**, *329*, 39–58.
40. P. Nikitas, *J. Electroanal. Chem.* **1994**, *375*, 319–338.
41. P. Nikitas, *Electrochim. Acta* **1996**, *41*, 2159–2170.
42. R. Fowler, E. A. Guggenheim, *Statistical Mechanics*, Cambridge University Press, Cambridge, 1956.
43. R. de Levie, *J. Electrochem. Soc.* **1971**, *118*, 185C–192C.
44. M. A. Vorotyntsev, *J. Res. Inst. Catalysis, Hokkaido Univ.* **1982**, *30*, 167–177.
45. B. Damaskin, U. Palm, M. Salve, *J. Electroanal. Chem.* **1987**, *218*, 65–76.
46. D. C. Grahame, R. Parsons, *J. Am. Chem. Soc.* **1961**, *83*, 1291–1296.
47. J. M. Parry, R. Parsons, *Trans. Faraday Soc.* **1963**, *59*, 241–256.
48. Ya. M. Kolotyrkin, Yu. V. Alekseev, Yu. A. Popov, *J. Electroanal. Chem.* **1975**, *62*, 135–149.
49. B. Damaskin, S. Karpov, S. Dyatkina et al., *J. Electroanal. Chem.* **1985**, *189*, 183–194.
50. B. B. Damaskin, O. A. Baturina, *Russ. J. Electrochem.* **1999**, *35*, 1186–1193.
51. W. R. Fawcett, A. Motheo, *J. Electroanal. Chem.* **1994**, *375*, 319–338.
52. B. B. Damaskin, *Russ. J. Electrochem.* **1994**, *30*, 91–94.
53. A. Frumkin, N. Balashova, V. Kazarinov, *J. Electrochem. Soc.* **1966**, *113*, 1011–1018.
54. A. Frumkin, O. Petry, A. Kossaya et al., *J. Electroanal. Chem.* **1968**, *16*, 175–191.
55. A. Frumkin, O. Petry, *Electrochim. Acta* **1970**, *15*, 391–403.
56. A. N. Frumkin, O. A. Petrii, B. B. Damaskin, *J. Electroanal. Chem.* **1970**, *27*, 81–100.
57. B. Grafov, E. Pekar, O. Petrii, *J. Electroanal. Chem.* **1972**, *40*, 179–186.
58. A. N. Frumkin, B. B. Damaskin, O. A. Petrii, *J. Electroanal. Chem.* **1974**, *53*, 57–65.
59. A. N. Frumkin, *Potentsialy nulevogo zaryada (Potentials of Zero Charge)*, Nauka, Moscow, 1982.
60. A. N. Frumkin, O. A. Petrii, *Electrochim. Acta* **1975**, *20*, 347–359.
61. A. Frumkin, *Philos. Mag.* **1920**, *40*, 363–375.
62. A. Frumkin, *Philos. Mag.* **1920**, *40*, 375–385.
63. A. Frumkin, A. Slygin, *Acta Physicochim. URSS* **1936**, *5*, 819–840.
64. A. N. Frumkin, O. A. Petrii, *Doklady Akad. Nauk SSSR* **1971**, *196*, 1387–1390.
65. O. A. Petrii, Yu. G. Kotlov, *Elektrokhimiya* **1968**, *4*, 1256–1260.
66. V. E. Kazarinov, O. A. Petrii, V. V. Topolev et al., *Elektrokhimiya* **1971**, *7*, 1365–1368.

67. V. S. Bagotzky, Yu. B. Vassiliev, J. Weber et al., *J. Electroanal. Chem.* **1970**, *27*, 31–46.
68. N. A. Balashova, V. E. Kazarinov in *Electroanalytical Chemistry* (Ed.: A. J. Bard), Marcel Dekker, New York, 1969, pp. 135–197, Vol. 3.
69. B. B. Damaskin, O. A. Petrii, V. E. Kazarinov, *Elektrokhimiya* **1972**, *8*, 1373–1377.
70. E. Gileadi, *Collect. Czech. Chem. Commun.* **1971**, *36*, 464–475.
71. E. Gileadi, *Isr. J. Chem.* **1971**, *9*, 405–412.
72. E. Gileadi, G. E. Stoner, *J. Electrochem. Soc.* **1971**, *118*, 1316–1319.
73. E. Gileadi, *Discuss. Faraday Soc.* **1973**, *56*, 228–234.
74. E. Gileadi, *Electrochim. Acta* **1987**, *32*, 221–229.
75. W. Lorenz, *Z. Phys. Chem.* **1961**, *218*, 272–276.
76. W. Lorenz, *Z. Phys. Chem.* **1973**, *254*, 123–133.
77. A. Frumkin, B. Damaskin, O. Petrii, *Z. Phys. Chem.* **1975**, *256*, 728–736.
78. K. Vetter, J. Schultze, *Ber. Bunsen-Ges. Phys. Chem.* **1972**, *76*, 920–927.
79. J. Schultze, K. Vetter, *J. Electroanal. Chem.* **1973**, *44*, 63–81.
80. K. Vetter, J. Schultze, *Ber. Bunsen-Ges. Phys. Chem.* **1972**, *76*, 927–933.
81. K. Vetter, J. Schultze, *J. Electroanal. Chem.* **1974**, *53*, 67–76.
82. J. Schultze, F. Koppitz, *Electrochim. Acta* **1976**, *21*, 327–336.
83. F. Koppitz, J. Schultze, *Electrochim. Acta* **1976**, *21*, 337–343.
84. A. N. Frumkin, Zh. N. Malysheva, O. A. Petrii et al., *Elektrokhimiya* **1972**, *8*, 599–603.
85. B. E. Conway, E. Gileady, *Trans. Faraday Soc.* **1962**, *58*, 2493–2509.
86. B. E. Conway, E. Gileady in *Modern Aspects of Electrochemistry* (Eds.: J. O'M. Bockris, B. E. Conway), Butterworths, London, 1964, pp. 347–442, Vol. 3.
87. A. N. Frumkin, *Doklady Akad. Nauk SSSR* **1964**, *154*, 1432–1436.
88. B. B. Damaskin, O. A. Petrii, V. V. Batrakov, *Adsorption of Organic Compounds on Electrodes*, Plenum Press, New York, 1971.
89. A. Hubbard, J. Stickney, M. Soriaga et al., *J. Electroanal. Chem.* **1984**, *168*, 143–166.
90. P. Andricacos, P. Ross, *J. Electroanal. Chem.* **1984**, *167*, 301–308.
91. A. Hubbard, *Heterog. Chem. Rev.* **1994**, *1*, 3–39.
92. M. J. Weaver, I. Villegas, *Langmuir*, **1997**, *13*, 6836–6844.
93. V. E. Kazarinov, *Elektrokhimiya* **1966**, *2*, 1389–1394.
94. N. V. Pospelova, A. A. Rakov, V. I. Veselovskii, *Elektrokhimiya* **1969**, *5*, 797–803.
95. A. A. Yakovleva, R. K. Bairamov, V. I. Veselovskii, *Elektrokhimiya* **1975**, *11*, 515–518.
96. V. I. Naumov, Yu. M. Tuyrin, *Elektrokhimiya* **1973**, *9*, 1011–1016.
97. V. I. Naumov, V. E. Kazarinov, Yu. M. Tuyrin, *Elektrokhimiya* **1973**, *9*, 1412–1413.
98. N. V. Pospelova, A. A. Rakov, V. I. Veselovskii, *Elektrokhimiya* **1969**, *5*, 1318–1320.
99. L. M. Mirkind, M. I. Fioshin, *Doklady Akad. Nauk SSSR* **1964**, *154*, 1163–1166.
100. O. A. Petrii, *Electrochim. Acta* **1996**, *41*, 2307–2312.
101. D. E. Yates, S. Levine, T. W. Healy, *J. Chem. Soc., Faraday Trans. 1* **1974**, *70*, 1807–1818.
102. T. Hiemstra, W. H. Van Reimsdijk, G. H. Bolt, *J. Colloid Interface Sci.* **1989**, *133*, 91–104.
103. T. Hiemstra, W. H. Van Reimsdijk, G. H. Bolt, *J. Colloid Interface Sci.* **1989**, *133*, 105–117.
104. B. B. Damaskin, *Elektrokhimiya* **1989**, *25*, 1641–1648.
105. M. A. Vorotyntsev, A. A. Kornyshev, *Elektrostatika sred s prostranstvennoi dispersiei (in Russian) (Electrostatics of the Spatially Dispersed Media)*, Nauka, Moscow, 1993.
106. H. Tamura, T. Oda, M. Nagayama et al., *J. Electrochem. Soc.* **1989**, *136*, 2782–2786.
107. E. A. Nechaev, *Zh. Fiz. Khim.* **1978**, *52*, 1494–1496.
108. M. I. Urbakh, E. A. Nechaev, *Elektrokhimiya* **1980**, *16*, 1264–1268.

3.2
State of Art: Present Knowledge and Understanding

G. Horányi
Institute of Chemistry, Budapest, Hungary

3.2.1
Historical Background

The thermodynamic treatment of ideally polarizable interphase (the term used when no charged component is common in two adjoining phases) leading to the Gibbs adsorption equation owing to the nature of the thermodynamic approach conveys no information about the structure of the interphase and has nothing to do with the forces playing role in the formation of this structure.

The strict thermodynamic analysis of the interphase is based on data available from the bulk phases (concentration variables) and the total amount of material involved in the whole system figuring in the relations expressing the relative surface excess of suitably chosen (charged or not charged) components of the system. In addition, the Gibbs equation for a polarizable interphase contains a member related to the potential difference between one of the phases (metal) and a suitably chosen reference electrode immersed in the other phase (solution) (and attached to a piece of the same metal that forms one of the phases).

The only model used in the case of such a treatment is the assumption that the existence of charged particles (ions and electrons) in the bulk phases can be considered as a physical reality.

In the historical development of the concept of electrified interphase, the modeling preceded the strict thermodynamic description and thus for a long period, only the double-layer model served as a unique basis for the elucidation of the problems connected with electrified interphases.

In constructing models for interfacial structure, only long-range electrostatic effects were considered and aside from the magnitude of the charges on the ions (and their radii) the differences in the chemical nature of the different ions were ignored. The well-known Gouy–Chapman (GC) model is based on this approach (see Chapter 2.2, Sect. 2.2.2). However, the experimental observations furnished more and more evidences that the behavior of several systems deviates significantly from that expected on the basis of GC theory. One of the well-known examples was the significant deviation of the surface excess of anions measured at Hg electrode at positive potentials. It was reasonable to assume that these deviations should be ascribed to short-range interactions between the ions and the surface of the metal. As these types of interactions should be specific for both metal and ions, the adsorption resulted by these forces could be named as *specific adsorption*.

Although different definitions and names for this phenomenon have been proposed, the most general definition (accepted by IUPAC as well) is an operative one defining specific adsorption as the deviation from the extent of the adsorption predicted by GC theory. In contrast to this, the adsorption of ions "obeying" GC theory is termed as *nonspecific adsorption*. It is important to emphasize that the distinction between specific and nonspecific adsorption is based on an arbitrarily chosen model assumption. It should, however, be borne in mind that it is impossible, by means of thermodynamic arguments, to determine the extent of the contribution of the specific adsorption to the overall surface excess concentration and to the charge density.

This means that the concept of specific adsorption involves the use of nonthermodynamic methods and model assumptions.

For the orientation of the reader, we summarize the fundamental elements of the characteristic features of the double-layer model relating to the specific adsorption on the basis of the corresponding IUPAC recommendations [1]. For a detailed survey including historical background, see elsewhere [2–7].

3.2.2
The Structure of the Double Layer

Recapitulation of fundamental notions In the case of nonspecific adsorption it is assumed that ions retain their solvation shell, and in the position of closest approach to the interface they are separated from it by one or more solvent layers. The locus of the electrical centers of nonspecifically adsorbed ions in their position of closest approach is the *outer Helmholtz plane* (OHP).

The position of this plane for the anion and the cation of the electrolyte may differ. It may also differ for the different cations and the different anions of solution. The region in which nonspecifically adsorbed ions are accumulated and distributed by the contrasting action of the electric field and thermal motion is called the *diffuse layer*. The region between the OHP and the interface is called the *inner (compact) layer*.

In accordance with the IUPAC definitions [1] denoting by x_2, the distance of OHP from the interface, and by ϕ_2, the mean electric potential at OHP, the potential drop across the inner layer is $(\phi^M - \phi_2)$, while in the diffuse layer it is $(\phi_2 - \phi^S)$ where ϕ^M and ϕ^S are the inner potentials of the metal and electrolyte, respectively. It is often convenient to take ϕ^S as zero.

The surface free charge density in the diffuse layer is denoted by σ^d and it may be regarded as made up of the charges contributed by the surface excess or deficiency of each ionic species (Γ_i)

$$\sigma^d = \sum \sigma_i^d = \sum z_i n_i^d = \sum z_i \Gamma_i \quad (1)$$

the sum being overall ionic species in the solution.

On the basis of the formal relationship

$$\frac{d(\phi^M - \phi^S)}{d\sigma^M} = \frac{d(\phi^M - \phi_2)}{d\sigma^M} + \frac{d(\phi_2 - \phi^S)}{d\sigma^M} \quad (2)$$

the differential capacitance of the whole interphase can be conceived as composed of two components, so Eq. (2) may be written as

$$C^{-1} = (C^i)^{-1} + (C^d)^{-1} \quad (3)$$

that is, the interphase behaves as a series combination of C^i, the *inner-layer capacitance*, and C^d, the *diffuse-layer capacitance*.

The double layer in the case of specific adsorption In the case of *specifically adsorbed* ions it is assumed that they penetrate into the inner layer and may (but not necessarily) come in contact with the metal surface. They are usually assumed to form a partial or complete monolayer. The locus of the electrical centers of this layer of specifically adsorbed ions is the *inner Helmholtz plane* (IHP) assumed to be at a distance x_1 from the metal surface. In certain cases the amount of charge of the ions specifically adsorbed in the IHP (σ^i) is higher than the charge on the metal phase (with the reverse sign); thus the charge of the diffuse layer has the same sign as that on the metal (superequivalent adsorption).

If the contribution to the solution charge from ions whose centers lie in the region $x_1 < x < x_2$ is denoted by σ^i, the electroneutrality condition for the interphase will be as follows:

$$\sigma^M = -\sigma^S = -(\sigma^i + \sigma^d) \quad (4)$$

The separation of σ^S into σ^i, and σ^d cannot be made without the introduction of a model. Use of the GC theory and the assumption that one of the ionic components is not specifically adsorbed enables the other σ_j^d to be calculated and so the specifically adsorbed charge due to the other ion to be obtained:

$$\sigma_j^i = \sigma_j^S - \sigma_j^d \quad (5)$$

In view of the short-range of the forces causing specific adsorption, the quantities σ_j^i are often equivalent to less than one monolayer of ions. Consequently the real distribution is often represented in models as a monolayer of ions with their centers on the IHP.

The charge in the diffuse layer is now

$$\sigma^d = -\sigma^M - \sum \sigma_j^i \quad (6)$$

and the corresponding differential capacitance (C_s^d) of the diffuse layer is

$$C_s^d = -\frac{d\sigma^d}{d(\phi_2 - \phi^S)} \quad (7)$$

Thus Eq. (3) can be replaced by

$$C^{-1} = (C^i)^{-1} + (C^d)^{-1} \left(\frac{1 + d\sum \sigma_j^i}{d\sigma^M} \right)^{-1} \quad (8)$$

The inner-layer capacity can be further analyzed to express its dependence on specific adsorption. The potential drop across the inner layer may be assumed to depend on the charge σ^M and specifically adsorbed charges σ_j^i. From this assumption it follows generally that

$$d(\phi^M - \phi_2) = \left[\frac{\partial(\phi^M - \phi_2)}{d\sigma^M} \right]_{\sigma_j^i} d\sigma^M$$
$$+ \sum \left[\frac{\partial(\phi^M - \phi_2)}{d\sigma_j^i} \right]_{\sigma^M, \sigma_k^i} d\sigma_j^i \quad (9)$$

or

$$\frac{d(\phi^M - \phi_2)}{d\sigma^M} \equiv \frac{1}{C^i} = \left[\frac{\partial(\phi^M - \phi_2)}{\partial\sigma^M} \right]_{\sigma_j^i}$$
$$+ \sum \left[\frac{\partial(\phi^M - \phi_2)}{\partial\sigma_j^i} \right]_{\sigma^M, \sigma_k^i} \left(\frac{d\sigma_j^i}{d\sigma^M} \right) \quad (10)$$

Thus the inner-layer capacitance may be related to a series of partial capacitances that depend on σ^M:

$$C^i(\sigma^M) = \left[\frac{\partial\sigma^M}{\partial(\phi^M - \phi_2)} \right]_{\sigma_j^i} \quad (11)$$

and each σ_j^i for the specifically adsorbed ions:

$$C^i(\sigma_j^i) = \left[\frac{\partial\sigma_j^i}{\partial(\phi^M - \phi_2)} \right]_{\sigma^M, \sigma_k^i} \quad (12)$$

Here the subscript σ_k^i means that all σ_j^i are kept constant except one.

3.2.3
Thermodynamics and Models

3.2.3.1 The Extent of Adsorption

In the previous section it was emphasized that the charge distribution in the double layer cannot be described without model assumptions concerning both the specific adsorption and the electrostatic adsorption. It should, however, be taken into consideration that a strict thermodynamic treatment of the interfacial phenomena lead to the determination of the extent of adsorption in terms of relative surface excesses, quantities that are not depending on the choice of the position of the plane of the interface; however, through the Gibbs–Duhem equation they depend on which component is taken as a basis for the expression of the relative surface excesses of the other components [8]. In electrochemistry, mostly the solvent is considered as such basis.

Thus, the relative surface excess concentration (Γ'_i) of some ith component referred to the solvent (water) is given by the equation:

$$\Gamma'_i = \frac{1}{A}\left(n_i + n_w \frac{x_i}{x_w}\right) = \Gamma_i - \frac{x_i}{x_w}\Gamma_w \quad (13)$$

in the case of any arbitrarily chosen reference system, where n_i and n_w are the numbers of moles of the ith component and of the water, respectively, in the entire system, x_i and x_w are the corresponding equilibrium concentrations, Γ_i the surface excess of component i with respect to the arbitrarily chosen reference system defined by

$$\Gamma_i = \frac{1}{A}(n_i - n^r x_i) \quad (14)$$

where n^r is the total amount (in moles) of the chosen reference system. It may be seen from these simple relationships that while the expression for relative surface excess contains only easily measurable variables characterizing the system studied, the Γ_i surface excess strongly depends on an arbitrary chosen parameter, n^r.

In the case of models, an adsorption layer of some definite thickness is assumed; therefore, we can write

$$n_i = n_i^a + n_i^S$$
$$n_w = n_w^a + n_w^S \quad (15)$$

where the indices a and S refer to the adsorption and liquid phase, respectively. Since

$$\frac{x_i}{x_w} = \frac{n_i^S}{n_w^S} \quad (16)$$

we obtain the relation

$$\Gamma'_i = \frac{1}{A}\left(n_i^a - n_w^a \frac{x_i}{x_w}\right) \quad (17)$$

If a monomolecular adsorption layer is assumed, for the total surface area A we have

$$\sum A_i n_i^a = A \quad (18)$$

where A_i is the surface requirement of one mole of the ith component. From this

$$n_w^a = \frac{A - \sum_{i \neq w} A_i n_i^a}{A_w} \quad (19)$$

and hence

$$\Gamma'_i = \frac{1}{A}\left(n_i^a - \frac{A - \sum A_i n_i^a}{A_w} \frac{x_i}{x_w}\right) \quad (20)$$

In dilute solutions $x_i \ll x_w$ and $x_w \approx 1$, and therefore

$$\Gamma'_i \approx \frac{1}{A} n_i^a \quad (21)$$

That is, the thermodynamically determinable relative surface excess amount is practically the same as the total amount of substance in the monomolecular layer, and on the basis of the aforementioned text, the coverage Θ_i relating to the ith component is

$$A_i \Gamma'_i = \frac{A_i n_i^a}{A} = \Theta_i \quad (22)$$

In the sense of the outlined model, the surface is always completely covered, and thus the adsorption of some dissolved component S can in essence be regarded as an exchange process, characterized in the simplest case by the equation

$$S_S + H_2O_a \rightleftharpoons S_a + H_2O_S \quad (23)$$

In a system consisting of solvent (water) and a single solute, we have for this equilibrium

$$K = \frac{a_S^a a_w^S}{a_S^S a_w^a} = \frac{f_S^a f_w^S}{f_S^S f_w^a} \cdot \frac{x_S^a x_w^S}{x_S^S x_w^a} \quad (24)$$

The a values are the activities referring to the adsorption and solution phases, while the f values are the activity coefficients.

From the above relations it follows that

$$K \frac{x_S^S}{x_w^S} = \frac{f_S^a f_w^S}{f_w^a f_S^S} \cdot \frac{n_S^a}{n_w^a} = \frac{f_S^a f_w^S}{f_S^S f_w^a} \cdot \frac{A_w n_S^a}{A - A_S n_S^a}$$

$$= \frac{f_S^a f_w^S}{f_w^S f_w^a} \cdot \frac{A_w}{A_S} \cdot \frac{\Theta_S}{1 - \Theta_S} \quad (25)$$

In dilute solutions, f_S^S / f_w^S can be regarded as constant, and in ideal mixtures as unity. Further, $x_w \approx 1$, and thus

$$K x_S^S = \frac{f_S^a}{f_w^a} \cdot \frac{A_w}{A_S} \frac{\Theta_S}{1 - \Theta_S} \quad (26)$$

This is in effect the equation describing the adsorption isotherm (see later). If the adsorption phase is also regarded as an ideal mixture, that is, $(f_S^a / f_w^a) \approx 1$, we obtain the Langmuir isotherm.

If we consider the problem in a more generalized form, the surface excess concentrations Γ_i per definition can be given by the equation:

$$\Gamma_i = \frac{1}{A}(n_i - n_i^r) \quad (27)$$

where n_i is the amount of component i in the whole system and n_i^r is the amount of component i in the reference system. The difference between Γ_i and Γ'_i can be expressed as

$$\Delta \Gamma_i = \Gamma'_i - \Gamma_i = \frac{1}{A} \frac{x_i}{x_w} n_w^a$$

$$= -\frac{1}{A} \frac{x_i}{x_w} \left(n_w - n_w^r \right) = -\frac{x_i}{x_w} \Gamma_w \quad (28)$$

According to Eq. (28), at a constant solution composition, the smaller the absolute value of Γ_w, the smaller is the absolute value of the difference between surface excess concentrations and (relative) surface excesses. Since the extent of the reference phase can be selected arbitrarily, and the value of $\Delta \Gamma_i$ depends directly on this choice, the latter is ill defined from a thermodynamic point of view. However, by introducing some nonthermodynamic assumptions, the value of $\Delta \Gamma_i$ can be estimated for several special cases. If the above-mentioned monolayer model can be considered as a reliable description of the interphase, the value of Γ_w is small, and the concentration of the species i in the solution (x_i) is also low, therefore $\Delta \Gamma_i$ is negligible, and Eq. (22) is always true.

The thickness of the adsorption layer, the diffuse part of the double layer, strongly depends on the total ion concentration, on the ionic strength. Consequently, if the ions involved in specific adsorption are present at a low concentration and the total ionic concentration is also rather low, the monolayer model cannot be used without restrictions.

There is a very important point to be stressed. Although Γ_i values are arbitrary depending on the choice of the reference system, the charge on both sides of the double layer expressed through $\sum z_i \Gamma_i$ according to

$$\sigma^M = -\sigma^S = -\sum z_i \Gamma_i = -\sum z_i \Gamma_i' \quad (29)$$

does not depend on this choice as

$$\sum z_i \Gamma_i = \frac{1}{A}\left(\sum n_i z_i - n^r \sum x_i z_i\right)$$

$$= \frac{1}{A}\sum n_i z_i \quad (30)$$

$$\sum z_i \Gamma_i = \frac{1}{A}\left(\sum n_i z_i - \frac{n_w}{x_w}\sum x_i z_i\right)$$

$$= \frac{1}{A}\sum n_i z_i \quad (31)$$

considering the very fact that owing to the electroneutrality of the bulk solution phase, $\sum x_i z_i = 0$.

3.2.3.2 Adsorption Isotherm

Adsorption in systems of neutral molecules is frequently expressed in terms of an adsorption isotherm that gives the amount adsorbed (or surface excess) as a function of bulk activity of the same species at constant temperature. In the case of electrified interphases the electrical state of the interphase must also be kept constant for the determination of the relation between surface and bulk concentrations.

Two ways for this are followed in the practical determination of the isotherms. These are the determination at constant σ^M or at constant $(\phi^M - \phi^S)$ (or at some constant cell potential). From a thermodynamic point of view, isotherms with respect to relative surface excesses may be determined at constant charge or at any well-defined constant potential. However, the interpretation and physical meaning of the results may be significantly more difficult in the case when constant cell potential is used.

In the case of adsorbed layers, by analogy with the two-dimensional pressure of spread films, a surface pressure defined as the change of interfacial tension (γ)

Tab. 1 Adsorption isotherms[a]

Name	Isotherms
Henry	$\beta a_i = RT\theta$
Langmuir	$\beta a_i = \dfrac{\theta}{1-\theta}$
Volmer	$\beta a_i = \dfrac{RT\Gamma_i}{1-b\Gamma_i}\exp\dfrac{b\Gamma_i}{1-b\Gamma_i}$
van der Waals	$\beta a_i = \dfrac{RT\Gamma_i}{1-b\Gamma_i}\exp\dfrac{b\Gamma_i}{1-b\Gamma_i}\exp(-2r\theta)$
Virial	$\beta a_i = \Gamma_i \exp(-2r\theta)$
Frumkin	$\beta a_i = \dfrac{\theta}{1-\theta}\exp(-2r\theta)$
Modified H.F.L. (Helfand, Frisch, and Lebowitz, modified by Parsons)	$\beta a_i = \dfrac{\theta}{1-\theta}\dfrac{b}{0.907}\exp\left[\dfrac{1+\theta(1-\theta)}{(1-\theta)^2}\right]\exp(-2r\theta)$

[a] a_i = activity of the ions; r = interaction parameter determinable from experiment in isotherms 1–7; b = van der Waals constant; and θ = fraction of the surface covered by ions.

caused by the addition of a given species to a base solution is introduced. At constant cell potential this is

$$\pi = \gamma_{base} - \gamma \qquad (32)$$

where γ_{base} is the interfacial tension of the base solution.

The analogue quantity at constant charge is

$$\Phi = \xi_{base} - \xi \qquad (33)$$

where

$$\xi = \gamma + \sigma E \qquad (34)$$

In both cases, all other variables should be kept constant. However, in reality the interfacial concentration of other species may be changed by the introduction of one of the components.

Various isotherms were proposed on the basis of different physical models. Three approaches should be distinguished [1].

1. The solution phase is considered as a continuum and the behavior of the adsorbed species is similar to that of the adsorbate at the vapor–condensed phase interface.
2. The adsorption process is a replacement reaction in which the adsorbing species replaces another species, normally the solvent (see previous section).
3. The interphase is regarded as a two-dimensional solution.

Some of the isotherms described in the literature are shown in Table 1.

3.2.3.3 Indication and Evaluation of Specific Adsorption

In order to show how the occurrence of specific adsorption can be demonstrated through thermodynamic relationships, we consider a simple case: solution of a salt MeX containing Me^+ and X^- ions in contact with Hg.

For this case the Gibbs adsorption equation takes the form

$$-d\gamma = \sigma^M dE_+ + \Gamma'_- d\mu_{salt} \qquad (35)$$

$$-d\gamma = \sigma^M dE_- + \Gamma'_+ d\mu_{salt} \qquad (36)$$

where E_+ and E_- are the potentials with respect to a reference electrode reversible for the cation and anion, respectively, Γ'_- and Γ'_+ are the corresponding relative surface excesses (see Eq. 13) of the ions.

The relative surface excesses can be obtained from the following relationship:

$$-\left(\frac{d\gamma}{d\mu_{salt}}\right)_{E_\pm} = \Gamma_\pm \qquad (37)$$

A very important indicator of the occurrence of specific adsorption is the so-called Esin–Markov coefficient, which in a generalized form can be given by the following equation:

$$\left(\frac{\partial E_\pm}{\partial \mu_{salt}}\right)_{\sigma^M} = \frac{1}{RT}\left(\frac{\partial E_\pm}{\partial \ln a_{salt}}\right)_{\sigma^M} \qquad (38)$$

Originally the notion of Esin–Markov coefficient was introduced for the condition of $\sigma^M = 0$, that is, for the effect observed at the potential of zero charge (E_{pzc}). In the absence of specific adsorption the Esin–Markov coefficient is zero, that is, no shift of the PZC should occur when the concentration (activity) of the electrolyte is changed. The origin of the shift at $\sigma_M = 0$ in the case of specific adsorption can be understood easily. Holding the potential at E_{pzc}, if the addition of the salt is followed by the specific adsorption of the anion or the cation, this process should result in a change in the electric state of the interphase. The charge of the specifically adsorbed species should be compensated by the charge of the counterions on the solution side of the double layer. This means that at the preselected potential value, σ^M

does not remain zero following the adsorption. In order to attain again the $\sigma^M = 0$, the potential of the electrode should be shifted to a new value. In general, specific adsorption of anions is indicated by negative shifts in potential at constant charge density, while specific cationic adsorption results in the opposite shift.

The determination of the extent of specific adsorption, remaining at the above-mentioned simplest case (MeX), follows the following pattern.

In these simplest cases the evaluation of the specific adsorption is based on the assumption that one type of ion in the electrolyte is not involved in the specific adsorption. Mostly anionic specific adsorption is calculated assuming that the cation (often an alkali ion) is not specifically adsorbed. This means that the cation is only in the diffuse layer. Determining Γ'_- and Γ'_+ from Eq. (37) (by measuring the interfacial tension), it is possible to calculate, from Γ'_+ the extent of Γ^d_-, excess concentration of the anion in the diffuse layer, by the relationships of GC theory.

The amount of specifically adsorbed anion is then obtained by the difference

$$\Gamma^{sp}_- = \Gamma_- - \Gamma^d_- \tag{39}$$

(In the case of dilute solutions $\Gamma'_- \approx \Gamma_-$). The reliability of such a determination depends on the validity of GC theory. In the case of strong specific anion adsorption, this method furnishes reliable results. The errors in the determination of Γ^d_- are greatest when specific adsorption is weak and the solution is dilute.

For the interpretation of inner-layer models in the absence of specific adsorption, it is very important to find cases and systems where there is no specific anion adsorption at all.

In the case of mercury it is accepted that specific adsorption of F^- plays no role under certain conditions.

However, this conclusion for mercury electrodes cannot automatically be carried over to other metals shown, for instance, in studies of Ag electrodes [9, 10].

The "method of mixed electrolytes" first presented in two fundamental papers by Hurwitz [11] and Dutkiewicz and Parsons [12] was an important contribution to the methods of evaluation of specific adsorption. In Ref. [11], Gibbs's adsorption equation has been derived in a general form for the case of mixed solutions of strong electrolytes at constant molal ionic strength.

It was predicted that in the case of the binary systems NaF-NaX or KF-KX (X = Cl$^-$, Br$^-$, I$^-$), investigated at constant molal ionic strength at a mercury electrode, the treatment would lead to the specifically adsorbed surface excess of Cl$^-$, Br$^-$, I$^-$, provided that the specific adsorption of F$^-$ is considered negligible, which seems a reasonable assumption. Likewise, it was demonstrated that the determination of Γ_{Cs^+} for Cs$^+$ from mixtures of the type LiCl-CsCl is possible if the specific adsorption of Li$^+$ is assumed to be weak.

In Ref. [12] the specific case of KI + KF system was discussed in detail.

Their derivation, which could be an illustrative example on how to treat a problem simultaneously in terms of thermodynamics and model approach, is as follows.

The electrocapillary equation for a pure mercury electrode in contact with an aqueous solution containing KI and KF at constant temperature and pressure may be written

$$-d\gamma = \Gamma'_{K^+} d\bar{\mu}_{K^+} + \Gamma'_{I^-} d\bar{\mu}_{I^-} + \Gamma'_{F^-} d\bar{\mu}_{F^-} \tag{40}$$

where γ is the interfacial tension, $\bar{\mu}_i$ the electrochemical potential of species i, and Γ'_i is the surface excess of species i relative to the water. If we introduce the chemical potentials of the salts and eliminate the electrochemical potentials of the anions, we may write Eq. (40) in the form

$$-d\gamma = \sigma dE_+ + \Gamma'_{I^-} d\mu_{KI} + \Gamma'_{F^-} d\mu_{KF} \quad (41)$$

where σ is the charge per unit area on the mercury surface, E_+ is the potential of the mercury electrode with respect to a reference electrode reversible to the potassium ion in the working solution, and μ_j is the chemical potential of the salt j.

Making the very plausible assumption that in a mixed-salt solution of constant total concentration, we may write

$$d\mu_{KI} = RTd\ln m_{KI} = RTd\ln x \quad (42)$$

where m_{KI} is the molal concentration of KI, $x = m_{KI}/(m_{KI} + m_{KF})$, and $m_{KI} + m_{KF}$ is constant. Similarly,

$$d\mu_{KF} = RTd\ln m_{KF} = RTd\ln(1-x)$$
$$= -RT\left\{\frac{x}{1-x}\right\} d\ln x \quad (43)$$

Substituting Eqs. (42) and (43) in Eq. (41), we obtain

$$-d\gamma = \sigma dE_+$$
$$+ \left[\Gamma'_{I^-} - \left\{\frac{x}{1-x}\right\}\right.$$
$$\left. \times \Gamma'_{F^-}\right] RTd\ln x \quad (44)$$

Now the total surface excess of iodide ion can be divided into that present in the inner layer, $\Gamma^i_{I^-}$, and that present in the diffuse layer, $\Gamma^d_{I^-}$. If it is assumed that the fluoride ion is present only in the diffuse layer, then it follows that

$$\frac{\Gamma^d_{I^-}}{\Gamma_{F^-}} = \frac{x}{1-x} = \frac{\Gamma^{d'}_{I^-}}{\Gamma'_{F^-}} \quad (45)$$

because the concentration of ions of the same charge in the diffuse layer is directly proportional to their bulk concentration. Thus it may be seen from Eq. (45) that the second term in the coefficient of $d\ln x$ in Eq. (44) is simply the surface concentration of iodide in the diffuse layer, and Eq. (44) reduces to

$$-d\gamma = \sigma dE_+ + \Gamma^i_{I^-} RTd\ln x \quad (46)$$

Finally, since the concentration of K^+ is constant and the activity coefficients are constant to a good approximation,

$$dE_+ = dE_{N.C.E.} \quad (47)$$

where $E_{N.C.E.}$ is the potential of the mercury electrode with respect to the normal calomel electrode. Thus the coefficients

$$\left(\frac{\partial\gamma}{RT\partial\ln x}\right)_{E_{N.C.E.}} = -\Gamma^i_{I^-} \quad (48)$$

$$\left(\frac{\partial\xi_{N.C.E.}}{RT\partial\ln x}\right)_\sigma = -\Gamma^i_{I^-} \quad (49)$$

give directly the surface excess of iodide ion present in the inner layer, if

$$\xi_{N.C.E.} = \gamma + \sigma E_{N.C.E.} \quad (50)$$

3.2.4
Specific Adsorption at Solid Electrodes

3.2.4.1 The End of the "Mercury Era"
During the first half of the last century, both the theoretical and experimental work in connection with electrified interphases were mainly centered around the mercury–water system.

Considering the potentialities of the available experimental methods, this was the only feasible way to understand double-layer phenomena and to create a reliable theoretical basis for the interpretation of the phenomena observed.

In principle, in the case of nonspecific adsorption the nature of the electrode metal should play a secondary role; therefore, the fundamental relationships found for this type of adsorption in the case of mercury should be of general validity, independent of the nature of the metal. In reality, the experimental results obtained with mercury, the methods for the evaluation of these experimental data, and the theoretical efforts to interpret them served as very important basis for the study of any kind of electrode. Some of these, for instance, methods of Hurwitz and of Parsons and coworkers and the Esin–Markov effect, were mentioned in the previous section.

It is evident that such methods as the application of capillary electrometer or the dropping electrode cannot be used in the case of solid electrodes. On the other hand, the measurement of surface tension and surface stress of solid electrodes is not easy and in addition, the interpretation of the data is very problematic [13–24].

Beginning from the fifties of the last century, the investigation of the behavior of solid electrodes came into the foreground. The available methods, for instance, wetting (contact) angle method, the measurement of differential capacity, were far from furnishing reliable data in the case of solid electrodes.

The practical aspects of electrocatalysis (fuel cell–oriented research) required data on the specific adsorption at electrodes with high real-surface area, while for theoretical considerations, results obtained with well-defined electrode surfaces (single-crystal surfaces) were of great importance.

The direct in situ study of the specific adsorption of ions and organic species was dictated by the desire to obtain more and more direct information on the processes occurring on the surface of the electrode. This was one of the most important factors leading to the elaboration of nonelectrochemical methods for the study of specific adsorption. For instance, radiotracer method, ellipsometry, and various spectroscopic methods (such as Raman spectroscopy) should be mentioned here.

The introduction of new methods resulted in a change in the philosophy on how to separate specific and nonspecific adsorption. According to the approach presented in the previous section, the determination of specific adsorption required the deduction of the adsorption in the diffuse double layer, calculated on the basis of model assumption, from the total adsorption determined by well-defined and reliable thermodynamic methods. In the case of some new methods, the specific adsorption of a given species is studied in the presence of a great excess of a supporting electrolyte. If this excess is several orders of magnitude higher than the concentration of the species studied, there could be no doubt that in the case of a measurable adsorption of these species (more than 10^{-11} to 10^{-12} mol cm^{-2}) the value obtained should be ascribed to specific adsorption. This consequence can be clearly demonstrated in the case of radiotracer adsorption studies.

For instance, if the adsorption of ^{35}S-labeled H_2SO_4 is studied in the presence of 1 mol dm^{-3} $HClO_4$ and the concentration of labeled sulfate is 10^{-4} mol dm^{-3}, the ratio of the adsorption of HSO_4^- and ClO_4^- ions in the diffuse part of the double layer should be 10^{-4} as well. This means that the extent of the nonspecific adsorption of HSO_4^- species under such conditions should be less than 10^{-13} mol cm^{-2}.

Despite the development and application of new mostly nonelectrochemical methods, "classical" electrochemical methods remained in use for solid electrodes.

Whatever the method used should be, the study of the specific adsorption in the presence of a supporting electrolyte has the advantage that the results can be obtained directly without any calculations based on model assumptions concerning the structure of the double layer.

3.2.4.2 Underpotential Deposition: A Limiting Case of Specific Adsorption

In the literature the specific adsorption and the underpotential deposition (UPD) processes are discussed separately. For a long period the adsorption of hydrogen on Pt, Rh, Pd electrodes at more positive potential values than the hydrogen equilibrium potential was considered as some specific and interesting phenomenon characteristic for some noble metals and hydrogen.

However, beginning from the sixties it became quite evident that the adsorption of metal ions on foreign metal substrates is often accompanied by significant charge transfer commensurable with the charge involved in hydrogen adsorption at noble metal electrodes [25–28]. These phenomena were termed as UPD characterized by the general equation

$$M_S^{n+} + ne^- \rightleftharpoons M_{ad} \quad (51)$$

From thermodynamic point of view, it is not possible to decide whether we have M^{n+} and ne^- or M_{ad} at the interface. If the equilibrium characterized by Eq. (51) is considered as a reality, only two from the three reaction partners can be considered as independent components of the (thermodynamic) system. Depending on the choice of the components, the thermodynamic charge Q (the total charge) of the electrode at a given potential should be different. This total charge should be distinguished from the so-called "free charge" corresponding to the double-layer model believed to exist on the solution side and (with opposite sign) on the metal side of the double layer formed at the interface (see later).

The study of the UPD of metal ions belongs to the group of adsorption studies in the presence of a great excess of supporting electrolyte as in most cases studied the concentration of metal ions is chosen to be very low in comparison with that of the ions of the supporting electrolyte.

The most important exception to this rule is the specific adsorption, the UPD, of H^+ ions or hydrogen adsorption on some noble metal electrodes. These systems were studied from several molar acid concentrations to very high pH values. For detailed discussion of UPD see Sects. 3.2 and 3.3.

3.2.4.2.1 A Widely Used and Misinterpreted Misconception: The Electrosorption Valency
In the classical double-layer models it is assumed that specifically adsorbed ions retain their original charge during the adsorption process.

In the sixties, Lorenz and coworkers have suggested that a purely electrostatic bond between the ion and the metal cannot be considered as a physical reality and there should be a degree of charge transfer similar to the formation of a covalent bond [29–33]. The specific adsorption on the basis of this concept was formulated as follows

$$R^z = R_{ads}^{z(1-\lambda)} + \lambda z e \quad (52)$$

where R^z is the adsorbing species of charge ze and λ is a dimensionless coefficient, denoted as the partial-charge coefficient. Thus in the generally accepted model, $\lambda = 0$. If $\lambda = 1$, then this would imply an entirely covalent bond.

The Lorenz concept of partial charge transfer was subject to general criticism by Parsons, Damaskin, and other authors and these critical views are well summarized in the IUPAC recommendations [1] in connection with these matters:

"The fraction of charge shared by the adsorbed particle may be characterized by a *partial charge number*. This term, although conceptually correct when applied to the model, is not adequate to indicate the quantity accessible to experimental measurements on a thermodynamic basis.

The accessible quantity is the *formal partial charge number* defined by

$$l_B = \frac{-(\partial \sigma^M / \partial \Gamma_B)_{E,\mu_i \neq \mu_B}}{F}$$

$$= \frac{(\partial \mu_B / \partial E)_{\Gamma_B, \mu_i \neq \mu_B}}{F} \quad (53)$$

Physically, l_B measures the average number of unit charge supplied to the electrode from the external circuit when one molecule of species B is adsorbed at constant potential."

Considering that in the overall process the adsorption of B could be accompanied by the simultaneous adsorption and desorption of other species changed, for instance, oppositely as B, the interpretation of the l_B values could be very difficult (see Sect. 3.2.4.2.2)

A similar criticism refers to the concept of electrosorption valency introduced by Vetter and Schultze starting from the partial charge concept [34, 35].

They described the adsorption equilibrium of a charged species by the following equation:

$$\nu M - OH_2 + S^z_{aq} \rightleftarrows M - S^{z+\lambda}$$
$$+ \lambda e^- + \nu H_2O_{(aq)} \quad (54)$$

where ν is the stoichiometric number, S^z is the adsorbing species, λ is the partial charge coefficient, M stands for the metallic atom, and $S^{z+\lambda}$ is the specifically adsorbed substance.

For characterizing this process they introduced the concept of "electrosorption valency," which is defined by the relation

$$\gamma' = -\frac{1}{F}\left(\frac{\partial \sigma^M}{\partial \Gamma}\right)_{\Delta \phi} = \frac{1}{F}\left(\frac{\partial \mu_s}{\partial \Delta \phi}\right)_\Gamma \quad (55)$$

where μ_s is the chemical potential of the adsorbing species in solution and $\Delta \phi$ is the potential difference across the interface.

The term "electrosorption valency" was chosen because of the analogy between the value, γ' as defined above, and the charge on the ion, z. If we compare Eqs. (53) and (55), the only difference is that instead of E in the former $\Delta \phi$ can be found in the latter. Although E is a potential referred to a reference electrode while $\Delta \phi$ is considered as a potential difference across the interface, the physical content of the two equations is almost the same. Considering the conclusions drawn in connection with Eq. (53), the electrosorption valency values determined according to the preceding definition equation could be very misleading if the possible simultaneous specific adsorption of counterions is not taken into consideration. This question will be discussed in the next section.

3.2.4.2.2 Interrelation of Anion Specific Adsorption and UPD of Metal Ions

The simultaneous occurrence of the specific

adsorption of anions and the formation of adatoms on a surface could result in significant difficulties in the calculation of the charge and mass balance required for the reliable interpretation of the overall adsorption phenomena. A number of situations should be distinguished in such cases; however, we consider two limiting cases.

1. The UPD takes place at a surface (in a potential range) where the role of the specific adsorption of anions can be neglected. In such cases the mass and charge balance can be relatively easily treated.
2. The UPD takes place on a surface (in a potential range) where the specific adsorption of the anions present in the system is significant. This group of phenomena can be divided into two subgroups.

   – Adatoms displace adsorbed anions, that is, as the coverage with respect to adatom increases the adsorbed anions decrease.
   – Specific adsorption of the anions takes place as a result of interaction with adatoms. The extent of anion adsorption is proportional to the coverage with respect to the adatoms. Even if the adatom displaces the anions adsorbed on the support metal, this effect could be compensated or overcompensated by the extent of adsorption induced by the adatoms. In the latter case an increase in the overall anion adsorption can be observed and this increase can be called *induced anion adsorption*.

Radiotracer adsorption studies furnish rapid and reliable information on these phenomena, using labeled anions in low concentration in the presence of a great excess of supporting electrolyte containing the metal ions forming the adatoms present also in a low concentration [36–42].

In the last three to four years the study of the coadsorption of anions and other solution components with adatoms plays a central role in order to clarify the structure of adlayers formed on single-crystal surfaces. In most of these cases the layer of Cu adatom formed on Pt(111) [43–51] and Au(111) [50, 52–61] surfaces was studied. Several studies were devoted to UPD of Ag on Au(111) surfaces [62–66].

3.2.4.3 The Generalization of the Thermodynamics of Monolayer Adsorption in Electrochemical Systems

The thermodynamics of adsorbed monolayers was first discussed in detail by Frumkin and coworkers for the particular case of hydrogen adsorbed on a platinum metal [67]. Starting from these studies, a generalized derivation was attempted by Parsons [68] through an illustrative example considering a particular system: a gold electrode in contact with an aqueous solution containing a low concentration of thallium perchlorate (1 mM) and a much higher concentration of potassium perchlorate (0.1 M). The main stages of this derivation will be presented here.

The interphase was described at constant temperature and pressure by the Gibbs equation in the form

$$-d\gamma = \Gamma_{TA} d\mu_{TA} + \Gamma_{KA} d\mu_{KA} + \Gamma_T d\mu_T \tag{56}$$

where Γ_i is the surface excess and μ_i is the chemical potential of species i. T represents thallium, A the anion, perchlorate, and K the cation, potassium. The three components are taken as neutral species so that the electroneutrality of the interphase is maintained. The chemical potential of the gold is, of course, invariant at constant temperature and pressure.

The chemical potential of the monolayer T may be determined from the potential (E_T) of the working electrode with respect to an electrode of pure T, both in equilibrium with T ions in the solution, that is, the electromotive force (emf) of the cell T|T$^+$|T|Au is considered

$$E_T = -\frac{(\mu_T(\theta) - \mu_T^\circ)}{z_T F} \quad (57)$$

where $\mu_T(\theta)$ is the chemical potential of T at coverage θ on the gold surface and μ_T° is the standard chemical potential of pure T.

It follows directly from Eq. (56) that three surface excesses are accessible to experiment:

$$\Gamma_{TA} = -\left(\frac{\partial \gamma}{\partial \mu_{TA}}\right)_{\mu_{KA}, \mu_T} \quad (58)$$

$$\Gamma_{KA} = -\left(\frac{\partial \gamma}{\partial \mu_{KA}}\right)_{\mu_{TA}, \mu_T} \quad (59)$$

$$\Gamma_T = -\left(\frac{\partial \gamma}{\partial \mu_T}\right)_{\mu_{TA}, \mu_{KA}} \quad (60)$$

From these equations, three Lippmann equations can be derived depending on the choice of reference electrode reversible with respect to one of the ions present in the system.

1. The potential of the working electrode is measured against a reference electrode reversible with respect to A ions (E_A)

$$Q_A = z_- \nu_- F \Gamma_{TA} = \left(\frac{\partial \gamma}{\partial E_A}\right)_{\mu_{KA}, \mu_T} \quad (61)$$

2. Reference electrode reversible with respect to K ions (E_K)

$$Q_K = z_+ \nu_+ F \Gamma_{KA} = \left(\frac{\partial \gamma}{\partial E_K}\right)_{\mu_{TA}, \mu_T} \quad (62)$$

3. Reference electrode reversible with respect to T ions (E_T)

$$Q_T = z_T F \Gamma_T = \left(\frac{\partial \gamma}{\partial E_T}\right)_{\mu_{KA}, \mu_{TA}} \quad (63)$$

z_-, z_+ and z_T are the charge of ions, ν_- is the number of anions in salt TA, and ν_+ is the number of cations in salt KA. Q_A, Q_K, and Q_T are the so-called thermodynamic or total charges of the electrode. This thermodynamic charge in general differs from the "free charge" believed to exist on the two sides of the double layer. They can be considered equal to each other in the case of ideally polarizable electrodes without any charge transfer. In such situation the charge of the electrode has a single value.

However, in the case of adsorption processes connected with charge transfer (for instance, UPD), the thermodynamic charge of the electrode has more than one value as owing to the equilibrium processes (with charge transfer), the number of the degrees of freedom for the choice of thermodynamically independent components decreases, that is, it will be less than the number of chemical components distinguishable in the system.

In the case discussed above we have the following choices.

The system is built up from

1. KA and T
2. TA and T
3. KA and TA

The Q_A, Q_K, and Q_T charge values correspond to the charge to be introduced electrically together with the corresponding material to create 1 cm² of interphase of the electrode keeping constant all the parameters characterizing the system.

3.2.4.4 Anion Adsorption on Well-defined Crystal Surfaces

Fundamental questions connected with the characterization of the double layer at well-defined single-crystal faces of solid metals have been reviewed and surveyed several times by Hamelin [6, 69–72].

Since the early 1980s, cyclic voltammetry has been used to characterize single-crystal electrodes in terms of surface order, presence or absence of defects contaminations, and so on. In some cases the voltammograms have also been used for identification of adsorbates, mostly on the basis of electric charge calculation. However, since the double-layer charging and surface reactions such as UPD and anion adsorption may occur in parallel, it is difficult to break down the total voltammetric charge into all of its individual components. Therefore, interpretation from voltammetry alone, used as a tool for the adsorbed species identification, may be ambiguous. This has been, for instance, the case with the interpretation of the "unusual adsorption states" on the Pt(111) electrode [73–82].

Although at a very early stage of the studies the appearance of the unusual peak on the voltammograms of Pt(111) was explained by anion adsorption for a long period, the view that it should be ascribed to hydrogen adsorption (or incorporation of hydrogen) was widely accepted by many researchers. Recently the hydrogen adsorption model is rejected on the basis of reexamination of the problem by the most eminent proponents of the model studying the unusual desorption states on Pt(111) electrodes by CO and I displacement method. Firstly, they arrived at the conclusion that the results obtained confirm that the charge associated with the unusual states is at least formally compatible with the exchange of one electron per three surface platinum atoms.

In accordance with this view, the occurrence of the unusual peaks on the cyclic voltammograms observed in acetic acid–acetate system was ascribed to a one-electron-transfer process of acetic acid/acetate ions according to the equations [83]

$$CH_3COOH \rightleftharpoons CH_3COO_{ads} + H^+ + e^-$$
$$\text{at pH} < 2$$

$$CH_3COO^- \rightleftharpoons CH_3COO_{ads} + e^-$$
$$\text{at pH} > 5$$

There are, however, many evidences demonstrating that the existence of the unusual states cannot be easily explained with the charge transfer to or from the anions (or other species). For instance, recent spectroscopic studies do not support the occurrence of electron transfer in the case of anion adsorption [77, 79, 84–88].

A very important and confusing observation is that the charge involved in the unusual states in various solutions is nearly independent of the nature of the anion, including organic anions. This aspect of the unusual states cannot be easily reconciled with the charge transfer to or from the anion as a mechanistic origin of the pseudocapacitive current generated by the reversibly adsorbed anions on Pt(111).

Results obtained from careful examination of the pH effects on the potentiodynamic behavior of the Pt(111) electrode in acidified perchlorate solutions [82] advocate in the favor of the assumption that water and not ClO_4^-

anions are involved in the unusual adsorption states in the double-layer region of Pt(111). Presumably H_2O dissociation and OH deprotonation are responsible for the two main electrode processes in the double-layer region of Pt(111) in $NaClO_4$ solutions.

For instance, in situ Fourier transform infrared (FTIR) spectroscopy has been used by Faguy and coworkers [79] to study the potential dependent changes in anion structure and composition at the surface of Pt(111) electrodes in HSO_4^--containing solutions. From the infrared (IR) differential normalized relative reflectance data, the maximum rate of intensity changes for three IR bands can be obtained. Two modes associated with the adsorbed anion and one mode assigned to species is not adequately described as either sulfate or bisulfate; the data are more consistent with an adsorbed $H_3O^+-SO_4^{2-}$ ion pair, possibly with the three unprotonated sulfate oxygens interacting with Pt sites.

In a series of papers by Lipkowski and coworkers, specific adsorption SO_4^{2-}, Cl^-, and Br^- on Au(111) electrode from 0.1 M $HClO_4 + x$ M K_2SO_4 and 0.1 M $HClO_4 + x$ M KCl or KBr or KI solution was studied using chronocoulometry and evaluating the data on the basis of thermodynamic of ideally polarizable electrode (x varies between 5×10^{-6} and 5×10^{-3} M) [53, 56, 89–94]. A further study was devoted to OH^- adsorption [95].

We should make a distinction between these cases and the case of adsorption from a mixed electrolyte at constant ionic strength discussed in Sect. 3.2.3.3. In this latter case, the investigated anion replaces progressively the anion of the supporting electrolyte, while in the method by Lipkowski and coworkers the concentration of the anion of the supporting electrolyte (ClO_4^-) is kept constant when the concentration of the adsorbing anion is changed. It is considered that the effect of the adsorption of the anion of the supporting electrolyte on the measured surface excess of the investigated anion can be minimized under this condition. In addition, it is assumed that in the presence of an excess of supporting electrolyte the measured surface excess is essentially equal to the surface concentration.

In the case of sulfate adsorption the surface excesses determined from the charge density data were compared with the Gibbs excesses determined from radiochemical measurements using ^{35}S-labeled sulfate solutions. Very good agreement of the data was reported [89].

In order to obtain insight into the nature of the adsorbed species, Esin–Markov coefficients for SO_4^{2-} adsorption from two series of solutions were determined: (1) at constant pH and variable K_2SO_4 concentration; (2) at constant K_2SO_4 concentration and variable pH. The Esin–Markov coefficients were used to identify the nature of the adsorbed species (SO_4^{2-} or HSO_4^-). The authors arrived at the conclusion that SO_4^{2-} ion is the adsorbed species even if HSO_4^- predominates in the bulk of the solution. Similar results were reported in Ref. [96].

This conclusion, however, seems to be questionable, in the light of thermodynamic considerations and other experimental results (see later).

Results obtained for Br^-, Cl^-, and I^- adsorption were compared. The main conclusions are as follows: All the three halides form a chemisorption bond with the gold surface. The bonds are quite polar at the negatively charged surface; however, its polarity drops significantly at the positively charged surface. At low charge densities and coverages, the bond polarity is determined by the ability of free electrons to screen the dipole formed by the adsorbed anion and its image charge in the metal. At high charge densities and coverages, the chemisorption bond has a predominantly covalent character. The strength of the halide adsorption and the covalent character of the chemisorption bond increase progressively by moving from chloride to iodide.

An interesting comparative study of the adsorption of bisulfate anion on different metals was carried out by Shingaya and Ito [88]. Adsorption of sulfuric acid species on various metal(111) electrodes in a 0.5-M H_2SO_4 solution was investigated by IR reflection absorption spectroscopy (IRAS). Pt, Au, Ag, Cu, and Rh(111) were used as electrodes. According to this study, HSO_4^- is adsorbed on all of the surfaces including Pt, Rh, Au, Ag, and Cu in a 0.5-M H_2SO_4 solution. It was found that considerable amounts of sulfate in addition to bisulfate coexist on Ag(111) in a 0.5-M H_2SO_4 solution. An interesting finding that both DSO_4^- and SO_4^{2-} are coadsorbed on Au(111) in 0.5-M $D_2SO_4 + D_2O$ solution. The interconversion of HSO_4^-/H_3O^+ and H_2SO_4 that was seen on Pt(111) was not observed on the other electrodes. The frequencies of the SO_3 symmetric and S—OH stretching bands of a bisulfate adsorbed on the electrode surfaces are related to the electronic state of each M(111) surface.

A somewhat different conclusion was reported for Ag(111) by Adzic and coworkers in a study [97] in which cyclic voltammetry and in situ IR spectroelectrochemistry were employed to investigate the ionic adsorption on the Ag(111) electrode surface in sulfate-containing solutions of different pH. It has been concluded that the sulfate anion is the predominant species adsorbed at the electrode surface in acidic and neutral media. The adsorption of sulfate in alkaline solutions is inhibited by the more strongly specifically adsorbed OH^- ions. The potential dependence of surface coverage, estimated on the basis of spectroelectrochemical data, indicated that a considerable repulsion exists among the adsorbed sulfate species. The spectral data were compared to those obtained for sulfate adsorption on Pt(111) and on underpotentially deposited Ag on the Pt(111) surface.

Voltammetric study of the adsorption of 14 anions on Ag(111) was reported in Ref. [98].

The specific anion adsorption of chloride and sulfate on Cu(111) from acidic aqueous electrolytes has been studied by Broekmann and coworkers [99] using scanning tunneling microscopy (STM) and cyclic voltammetry. It was found that at positive potentials adsorbed chloride forms a well-ordered ($\sqrt{3} \times \sqrt{3}$)R30° superstructure. At negative potentials the bare copper surface can be observed. In the case of sulfate adsorption at positive potentials, high-resolution STM images revealed an additional species that was assigned to coadsorbed water molecules. Close-packed sulfate rows are separated by zigzag chains of these water molecules.

Adsorption of sulfate ions on the three basal planes of monocrystalline copper and silver electrodes in 0.1 M $HClO_4$ was studied by radiometric and electrochemical methods [100, 101].

The adsorption of OH^- ions on the Pt(111) plane has been studied by Drazic and coworkers [102] using fast cyclic voltammetry in sodium hydroxide solutions (0.03 to 1 M), under quasi-equilibrium and Tafel approximation conditions. It was shown that the OH^- ion adsorption is an electrosorption process with one electron exchanged between an OH^- ion and the platinum surface. The electrosorption process follows the Frumkin adsorption isotherm with low-intensity repulsive interactions of the adsorbed species ($f = 2-3$). The estimated values of the standard electrochemical rate constant ($k^o = 5.6 \times 10^{-4}$ cm s^{-1}) and the standard exchange current density ($j_0^0 = 5.45 \times 10^{-2}$ A cm^{-2}) indicate a rather fast electrochemical process.

The thermodynamic and inner-layer characteristics of halide ions adsorbed on bismuth single-crystal planes from solutions of methanol and propanol was studied by an Estonian group in a series of publications [103–108]. The adsorption of Cl^-, Br^-, and I^- ions from methanolic solutions on different faces of a bismuth single-crystal electrode was studied by impedance and chronocoulometric techniques. The charge of specifically adsorbed anions σ_1 was calculated using the Hurwitz–Parsons–Dutkiewicz method. Fitting the results to the simple virial isotherm, the values of the standard Gibbs energy of adsorption $-\Delta G_A^o$ and the mutual repulsion coefficient B of the adsorbed anions were estimated. The dependence of the integral capacitance of the inner layer on the electrode charge σ was calculated according to the traditional inner-layer model using the electrode charge as the independent electrical variable. It was found that the adsorption of halide ions decreases in the sequence of the Bi single-crystal planes (011) > (001) > (111).

As a continuation of this work, in accordance with the method suggested by Lipkowski and coworkers, instead of electrode charge the electrode potential was considered as independent electrical variable for the evaluation of the experimental data.

It was found that under comparable conditions the results obtained at constant electrode potential and constant charge coincide.

3.2.5
Methods of Investigation and Illustrative Examples

3.2.5.1 General Remarks

During the last two to three decades, sophisticated methods of solid-state surface science were introduced to study the adsorption phenomena at electrochemical interfaces. These new techniques of investigation enable us to interpret adsorption phenomena at molecular level.

The main characteristic of most of the new experimental methods is the simultaneous use (coupling) of electrochemical techniques with other nonelectrochemical methods. Only some of these methods can be very briefly mentioned in the following sections. More detailed description of the methods and their application can be found in monographs and textbooks [109–111].

3.2.5.2 In Situ Techniques

3.2.5.2.1 Optical Techniques

UV-visible Techniques

Ellipsometry [112, 113] Ellipsometry is one of the earliest optical technique to be applied to the study of adsorption processes [112, 113]. It involves the analysis of the phase change and the change in amplitude ratio of polarized light reflected from a surface.

A significant limitation of the method is that a model for the adsorbed layer should be made for the evaluation of the experimental data. As the theory of ellipsometry is based on the assumption that the reflection occurs from an ideally smooth surface, the interpretation of data obtained with surfaces with nonideal topography encounters difficulties.

Specular reflectance spectroscopy [114] The principle of the method is the recording of a reflection of the incident monochromatic beam by the surface of the electrode as a function of wavelength, potential, or time.

Similar to ellipsometry, a model of the interface is required to evaluate the experimental data. Several techniques were elaborated to enhance the sensitivity of reflectance measurements (for instance, multiple reflectance).

In order to avoid the problems connected with measurement of absolute reflectivities, changes in reflectivity are detected as the potential of the electrode is modulated in the case of the so-called "modulated specular reflectance spectroscopy."

Vibrational Spectroscopy (Infrared and Raman Scattering Spectroscopies) Although UV-visible techniques were proven useful, they are not suitable for identifying adsorbates or orientation of adsorbates. This type of information could be provided by the various versions of vibrational spectroscopy.

During the last decade, vibrational spectroscopy has contributed to a great extent to the advance in electrochemical surface science

Infrared (IR) spectroscopy [115, 116] For in situ application of IR spectroscopy for the study of adsorption, there were two fundamental problems to overcome:

1. Most solvents (especially water) absorb IR very strongly.
2. How to separate the signals coming from the adsorbed layer from the noises.

One of the most commonly applied IR techniques developed to overcome these problems is the external reflectance technique. In this method, the strong solvent absorption is minimized by simply pressing a reflective working electrode against the IR transparent window of the electrochemical cell. The sensitivity problem, that is, the enhancement of the signal/noise ratio in the case of external reflectance techniques is solved by various approaches. These are, for instance, electrochemically modulated infrared spectroscopy (EMIRS), in situ FTIR (which use potential modulation), and polarization modulation infrared reflection absorption spectroscopy (PM-IRAS, FTIR) [86, 117–123].

A new development is the elaboration of synchrotron far infrared spectroscopy (SFIRS) for adsorption studies [124].

Raman Spectroscopy The major problems in the case of the application of IR spectroscopy, the strong absorption by the solvent and window material, can be avoided using Raman scattering techniques, since only visible radiation is involved in this case.

Raman scattering cross sections of typical adsorbates are low, therefore two fundamental techniques were elaborated to enhance them: resonance Raman scattering (RRS) and surface-enhanced Raman scattering (SERS) [125–127].

Second harmonic generation anisotropy from single-crystalline electrode surfaces [128] Interfacial second harmonic generation (SHG) is an optical spectroscopy that is inherently surface-sensitive. All effects that modify the electron density profile of the interface affect the SHG process. Therefore, SHG depends on the exciting and SH frequency, on surface structure, and on adsorbates and adlayers. For metal electrodes there is a remarkable dependence of SHG on potential, adsorption, and charge density at the metal, and on surface structure. Using the advanced form of the SHG anisotropy, the interference SHG anisotropy, one can determine the anisotropy of the SH field itself and can pursue how the various isotropic and anisotropic sources contributing to this field evolve with potential and adsorption. This yields information on the geometric and electronic structure of the interface and adlayer to unparalleled details.

There are mutual links between SHG at electrodes and electrochemistry, and the results of classical electrochemical measurements, simply because these methods probe different physical properties of the interfacial regime. Adsorption and potential, charge density, surface structure, and the formation of (ordered) adlayers have a significant impact on both the behavior of the electric double layer and the SH response. SHG was applied to study sulfate adsorption at Au(111) [129].

Sum and difference frequency generation at electrode surfaces [130] The investigation of vibrational and electronic properties of the electrochemical interface by using nonlinear optical techniques of visible-infrared sum (SFG) and difference (DFG) frequency generation constitutes an interesting aspect of the study of adsorbed species. The vibrational behavior of H-Pt(hkl) system, both in the underpotential and overpotential regime, adsorption of CN^- on Au(hkl) electrode (using DFG), electrode surface electronic properties were studied and measurements of the PZC were carried out [131].

X-ray Methods

Surface X-ray scattering (SXS) The first SXS study of an underpotentially deposited metal monolayer was reported more than ten years ago. In a recent review [132] it is demonstrated that this method is well suited for the study of the structure of metals, halides, and metal-halide adlayers on single-crystal electrodes. As another example, the study of the distribution of water at Ag(111) surface can be mentioned [133].

In situ X-ray absorption fine structure method (XAFS) [134] gives insight into the interaction between adsorbate and substrate.

In situ X-ray surface diffraction (SRSD) method was applied for the study of UPD of copper on Au [135].

3.2.5.2.2 **Radiotracer Methods** Of all the in situ techniques used to obtain information about adsorption phenomena at electrode surfaces, the radiotracer technique was the first direct method furnishing reliable data on the extent of adsorption of a specific preselected species. Generally, in situ radiotracer methods provide direct information on adsorption|accumulation processes on a great a variety of electrodes under various experimental conditions.

The in situ experimental techniques used by various authors are described in the relevant literature (papers and reviews) in detail [100, 136–155].

The main problem of the in situ radiotracer study of electrosorption phenomena originates from the very nature of the method, because the radiation measured consists of two main parts. The first one is that coming from the solution phase or from the solution layer contacting the electrode. The second radiation component is the radiation coming from the adsorption layer of interest for the measurement of adsorption.

The elaboration of a technique for the radiotracer adsorption measurements is equal to the search of the optimal conditions for the *minimalization* of the role of the radiation coming from the solution background. The usual classification of the various methods follows this pattern, that is, how this requirement is fulfilled. According to this principle, the methods can be divided in two main groups: the radiation of solution background is governed and minimized by self-absorption of the radiation (thin-foil method) and the background radiation intensity is minimized by mechanical means (thin-gap method). The most important and widely used representative of the first approach is the so-called foil method.

In the case of the thin-foil method, the detector "sees" simultaneously both components of the radiation coming through a thin foil (metal or metal(gold)-plated plastic film) forming the bottom of the cell. The adsorbent is either the foil itself or a thin deposited layer on the bottom of the cell serving as a mechanical support and electric conductor (if the foil is metal-plated).

The situation can be visualized by Scheme 1.

The radiation measured is $I_T = I_s + I_a$ (where I_s and I_a are the intensities of the radiation coming from the solution phase and from the adsorbed layer, respectively). Using isotopes emitting soft β-radiation (radiations characterized by high mass-absorption coefficient (μ) (^{14}C, ^{35}S, and ^{36}Cl), the self-absorption of the radiation in the solution phase is so high that the thickness of the solution layer effective in the measured solution background radiation is very low. (Similar phenomena could be observed for isotopes emitting low-energy X ray.)

In cases in which the radiation coming from the solution background would be too high in comparison with that originating from the adsorbed layer (smooth surfaces, γ-radiation), some kind of mechanical means should be applied to reduce the role of the background radiation.

The mechanical control of the background radiation can be achieved by two different ways: fixed rigid reduction of solution layer thickness (classical thin-layer or gap method) and flexible mechanical or temporary reduction of the thickness of the solution layer (electrode-lowering technique). In the case of the first version, specially designed cells are used where only a very thin-solution layer flows between the electrode and the detector as shown by Scheme 2.

This method is not used very often because of technical problems. These problems are avoided by the application of a flexible version, the so-called electrode-lowering technique. In the case of this technique, the solution gap between the electrode and detector is minimized temporarily only for the time of the measurement of the adsorption. The electrode is placed in two positions as shown by Scheme 3.

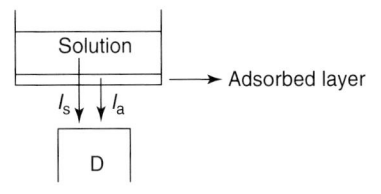

Scheme 1

Scheme 2

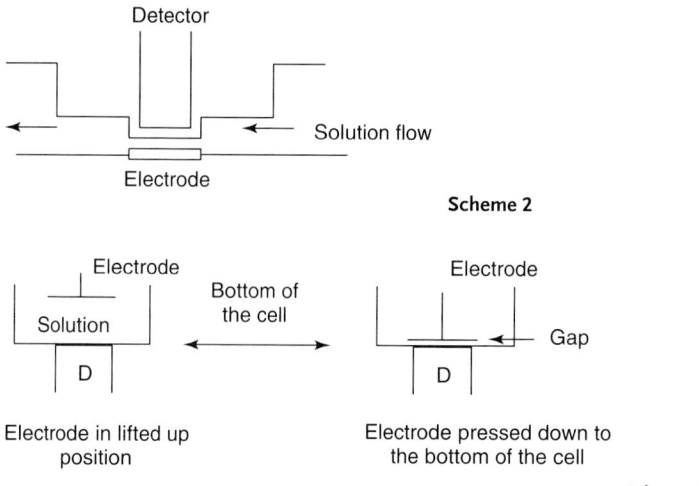

Scheme 3

Electrode in lifted up position

Electrode pressed down to the bottom of the cell

In the "lifted-up" position, the attainment of the adsorption and the electrochemical equilibrium (or steady state) proceeds without any disturbance, the detector measures only the solution background radiation. In the "pressed-down" position, the intensity measured comes from the species adsorbed at the electrode surface and from the solution layer present in the gap between the bottom of the cell and the electrode. The thickness of this gap depends on the mechanical state of the electrode surface, the stability of the bottom of the cell, and so on.

3.2.5.2.3 Electrochemical Quartz Crystal Microbalance (EQCM) and Nanobalance (EQCN) Techniques The first application of quartz crystal microbalance (QCM) in electrochemistry dates back to 1981 [156]. A review on the application of these techniques can be found in Ref. [157].

One of the main advantages of the EQCM and EQCN techniques in the study of adsorption phenomena is the possibility of obtaining simultaneous voltammetric and gravimetric response through frequency changes.

It is important to emphasize that although from the data obtained the ratio between the electric charge and the corresponding mass change can be defined for every experimental point as a function of the potential applied, one should be very cautious, at least in the case of adsorption, to draw conclusions, particularly the application of the notion of electrosorption valency should be avoided.

The mass change could be the result of several simultaneous adsorption and desorption processes involving the molecules of the solvent as well. Sometimes the frequency shifts observed could not be ascribed to specific adsorption.

Tsionsky and coworkers [158] studied the double-layer region of gold electrode in 0.1 M $HClO_4$, KNO_3, and KOH using EQCN method. The authors observed frequency shifts even though the ions of these electrolytes are not specifically adsorbed and there is no faradaic reaction that could lead to the formation of adsorbed species through charge transfer. Tsionsky and coworkers [158] developed a model in which they assumed that the liquid in a thin layer near the surface has a higher viscosity because of the high concentration of ions and the high electrical field in this region.

Some of the illustrative examples for the application of EQCN technique are collected in [157]. The electrical double-layer region of optically polished gold in 0.1-M NaF, KF, RbF, and CsF solutions in the absence and with addition of specifically adsorbing anions SO_4^{2-}, OH^-, Cl^-, and Br^- was studied, arriving at the conclusion that EQCN can be used to investigate the electrical double-layer phenomena on metal electrodes.

A combined interferometric and EQCN technique was elaborated for the study of the energetics of adsorption phenomena [159].

EQCN technique is very suitable for the study of the deposition and dissolution processes of adatoms on various electrodes [55, 159–163].

3.2.5.2.4 Scanning Tunneling Microscopy (STM) and Atomic Force Microscopy (AFM)

STM was introduced into electrochemistry during the late 1980s. The method is based on the phenomena of quantum-mechanical tunneling. It allows atomic imaging of the structural properties of both bare and adsorbate-covered surfaces. The principle of the method is as follows:

A sharp tip of a chemically inert metal is brought so close to the electrode surface that tunneling can occur and a tunneling current flows maintaining the potential difference between the tip and the electrode. Scanning the tip across the surface of the sample at constant current or height from the surface, a topographical image of the surface with atomic resolution can be obtained.

The coadsorption structure of Cu and halides on Pt(111), Pt(100), and Au(111) was investigated and interpreted in Ref. [164]. Other interesting results obtained in various systems have been reported in Refs. [48–50, 165–175].

Atomic force microscopy (AFM) similar to STM is a scanning atomic-scale probe microscopy; however, AFM does not rely on tunneling. It detects the interatomic forces between the tip and surface atoms. This method provides a very powerful complement to STM. It can be used in cases when faradaic current is flowing, that is, it offers the possibility of studying surfaces in the course of electrochemical processes. Nonconducting samples can be imaged as well.

3.2.5.3 Ex Situ Methods

In the application of in situ methods, the presence of the bulk electrolyte constitutes a significant limitation. As it was shown in the previous sections, by the use of thin-layer cells or carrying out the measurements in the presence of a thin-solution layer, the role of the bulk electrolyte phase can be reduced.

An alternative route suggested in the literature [114, 176–178] to overcome these limitations is the emersion of the electrode from the electrochemical cell and transferring it into an ultrahigh vacuum (UHV) chamber, where studies using surface-sensitive methods can be carried out.

However, this procedure involves the loss of the electrochemical environment and potential control. Auger electron spectroscopic (AES) studies [179] show that despite the evacuation required by AES, chemisorbed species remain at the surface and their identification and quantitization can be carried out in vacuum.

According to Hubbard [179], cationic species present in the emersed layer of liquid or in the electrical double layer (EDL) can also be determined using AES. However, anions X^-, present in the emersed EDL and solution, typically undergo hydrolysis to volatile HX species, which are subsequently lost to the vacuum.

In contrast to the latter statement, Kolb and coworkers reported somewhat different results claiming that under certain conditions the electrode can be removed from the electrolyte into an inert atmosphere or even into vacuum with the double layer apparently intact and the bulk electrolyte completely absent [114]. They measured the work function of a gold electrode after emersion as a function of potential in order to prove that the potential drop across the double layer remains unchanged.

The chemical composition of the emersed double layer was tested by ESCA. It was found for nonspecific adsorbing ions that the ex situ obtained data are in very good agreement with predictions from classical double-layer theory.

In the case of immersion technique the opposite way (from UHV chamber to the solution phase) is followed for the determination of the PZC of Au(111), Pt(111) electrodes [180].

3.2.5.3.1 Electrochemical Double-layer Modeling Under Ultrahigh Vacuum Conditions Simulations of the electrochemical double layer under UHV conditions allow the characterization of the chemical interaction between electrode surface and electrolyte constituents [181–186]. The electron work function measured for an adsorbate-covered metal in UHV is considered to be equivalent to the electrochemical potential for the same adsorbate–electrode pair in solution. This common potential scale represents an important link between UHV studies and electrochemistry. The potential range accessible by the investigation of substrates with different work functions can be expanded by the coadsorption of electronegative and electropositive species, such as halogens or alkali metals, which induce large work function changes. The adsorption of anions or cations from electrolytes, achieved in UHV at low temperature, is of particular interest. Solvation processes and surface reactions, which result from the electrostatic interaction between the partially ionically adsorbed species and coadsorbed H_2O, the most frequently used solvent, can be studied by applying surface-sensitive techniques. A variety of methods have been developed allowing the investigation of the surface structure and chemical composition.

The structure of sulfuric acid species adsorbed on a Pt(111) electrode has been successfully determined by using the UHV modeling [186]. The microscopic information about the structure of an adsorbed electrolyte anion and water as a solvent is indispensable to a true molecular level understanding of the electrochemical double layers and chemical reactions at the electrode surfaces.

3.2.5.4 Illustrative Examples
1. The influence of the surface structure on the anion adsorption. Potential dependence of the adsorption of sulfate ions at various crystal faces of copper as studied by radiotracer technique [101] is shown in Fig. 1.
2. Effect of the nature of the anion on the potential dependence of the adsorption. Data obtained by the method of Lipkowski and coworkers at

Au(111) electrode are summarized in Fig. 2 [93].

3. The unusual peak on Pt(111). Figure 3 shows the cyclic voltammograms of Pt(111) at various pH values [82]; A-1 and C-1 are the unusual peaks. Figure 4 shows the shift of the unusual peak following the addition of acetic acid to a HClO$_4$ solution [83].

4. The formation of a layer of adatoms. Figure 5 shows the voltammetric curve for the UPD of Ag$^+$ ions on Au(100) and the $\Gamma(E)$ relationship [187].

5. Anion specific adsorption induced by adatoms. Figure 6 demonstrates the increase in the sulfate adsorption (determined by radiotracer technique) following the formation of Cd, Cu, Hg adlayer on a polycrystalline Au support [147].

6. The application of EQCM technique coupled with voltammetry. Figure 7 shows the simultaneous recording of the cyclic voltammogram and the frequency change of the Au(111) EQCM electrode in the course of UPD of Cu [55].

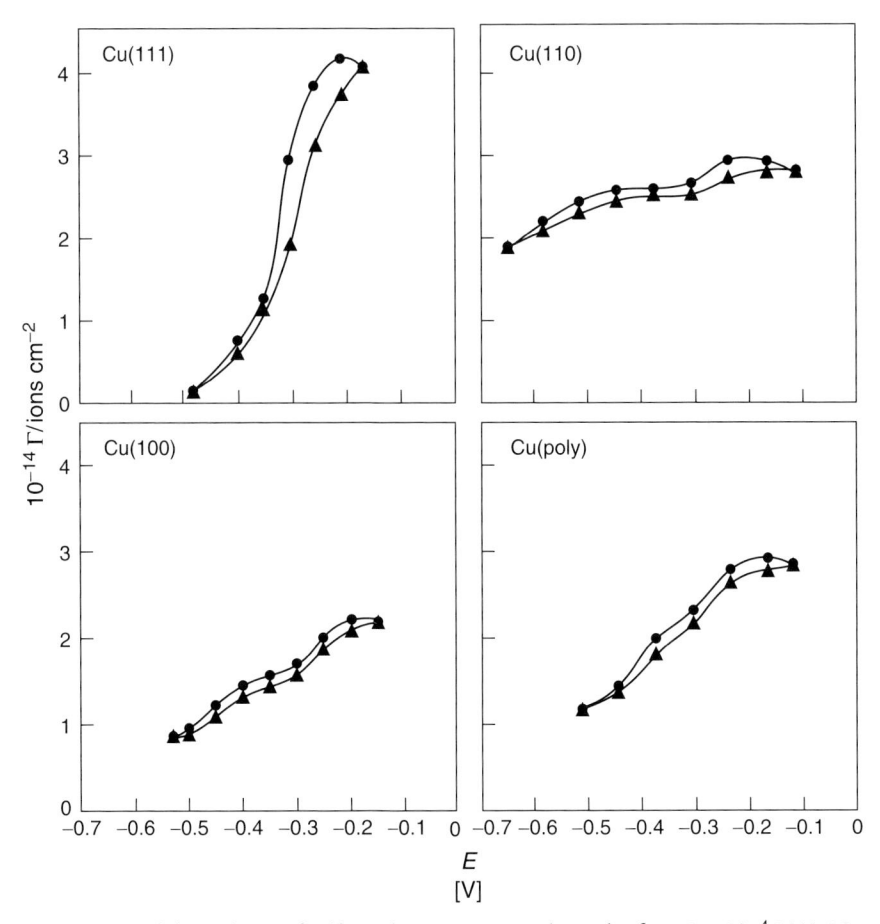

Fig. 1 Potential dependence of sulfate adsorption on Cu electrodes from 5×10^{-4} M H$_2$SO$_4$ in 0.1 M HClO$_4$; (•), positive scan; (Δ), negative scan. Potentials were measured versus Ag|AgCl|1 M Cl$^-$. (Reprinted from Ref. [101] with permission from Elsevier Science.)

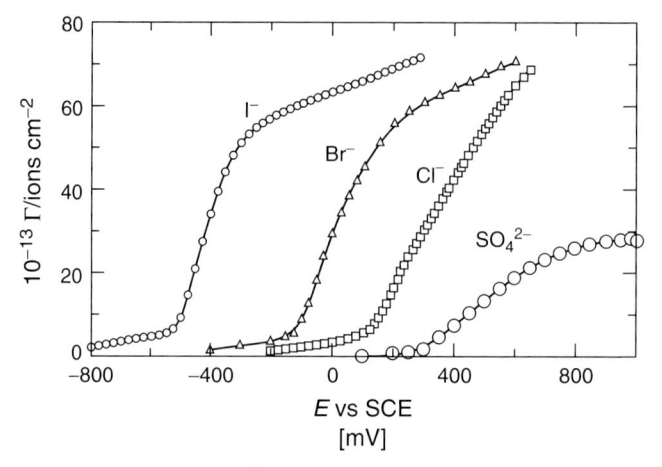

Fig. 2 Comparison of SO_4^{2-}, Cl^-, Br^-, and I^- adsorption at the Au(111) electrode surface from 0.1 M $HClO_4$ + 10^{-3} M K_2SO_4 and 0.1 M $HClO_4$ + 10^{-3} M KCl, KBr, and KI solutions. (Reprinted from Ref. [93] with permission from Elsevier Science.)

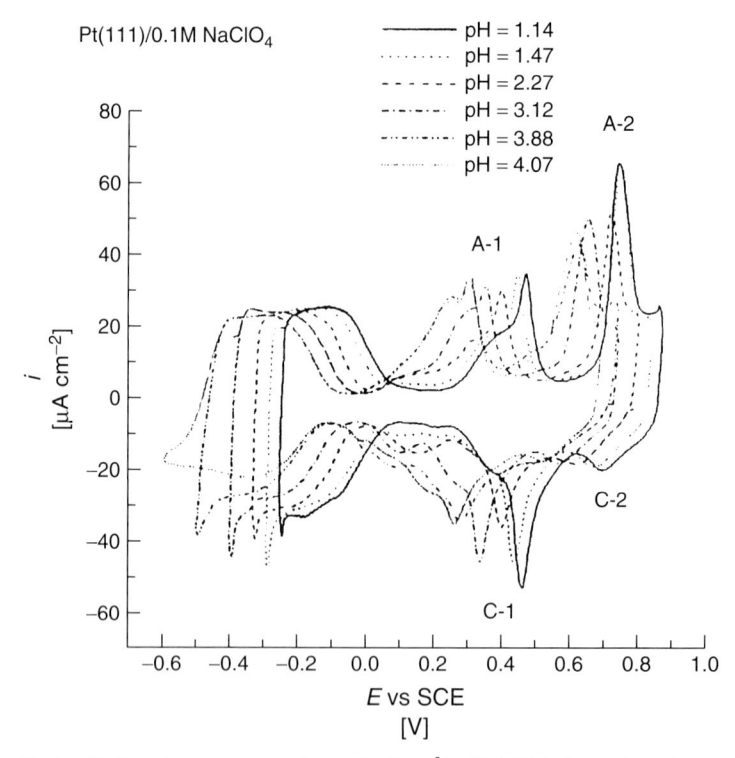

Fig. 3 Cyclic voltammograms ($v = 40$ mV s^{-1}) of Pt(111) electrode in 0.1 M $NaClO_4$ solutions of lower pH values. (Reprinted from [82] with permission from Elsevier Science.)

3.2.6
Trends and Perspectives

The classical models of the double layer were based on the primitive model in which the solvent is regarded as a dielectric continuum and the solid phase was regarded as a charged hard wall.

Recently, progress has been made toward replacing this primitive model by a microscopic or molecular model of the solvent and an electronic model of the metal [188–192]. The assumption that the solvent molecules are dipolar hard spheres and that the metal can be modeled as jellium leads to results that are nearly free of empirical parameters and that are in close agreement with experiment.

Computer simulations of the electrochemical double layer could contribute to a better understanding of the double-layer structure [193–197].

Information on the structure of water next to metal surfaces is obtained from molecular dynamics computer simulations [191]. Such simulations, although able to mimic the reality in more detail compared with simple models used in analytical theories, are still limited to the description of models. From the simulations it follows that water next to an uncharged metallic surface is perturbed to a distance of ∼1 nm. Next to the charged surface, water is reorienting and, when the external field is strong, undergoes a layering transition.

In Ref. [192] a simple model for interfacial adsorption is presented in which the three-dimensional system reduces to an equivalent two-dimensional lattice gas. This model is applied to study the UPD of copper on the (111) surface of a crystalline gold electrode in the presence of bisulfate. A cluster variation approximation

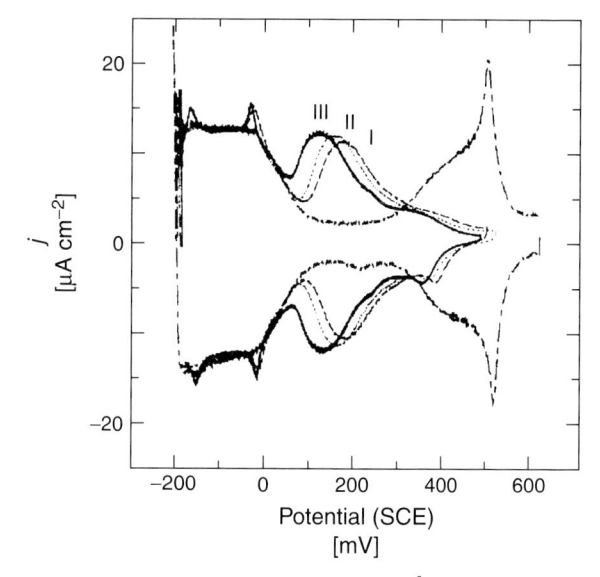

Fig. 4 Cyclic voltammograms at 20 mV s^{-1} on Pt(111) in 0.2 M HClO$_4$ at pH 0.8 with 0 mM (– · – · –), (I) 1 mM (- - - - -), (II) 3 mM (· · · · ·) and (III) 11 mM (———) CH$_3$COOH. (Reprinted from Ref. [83] with permission from Elsevier Science.)

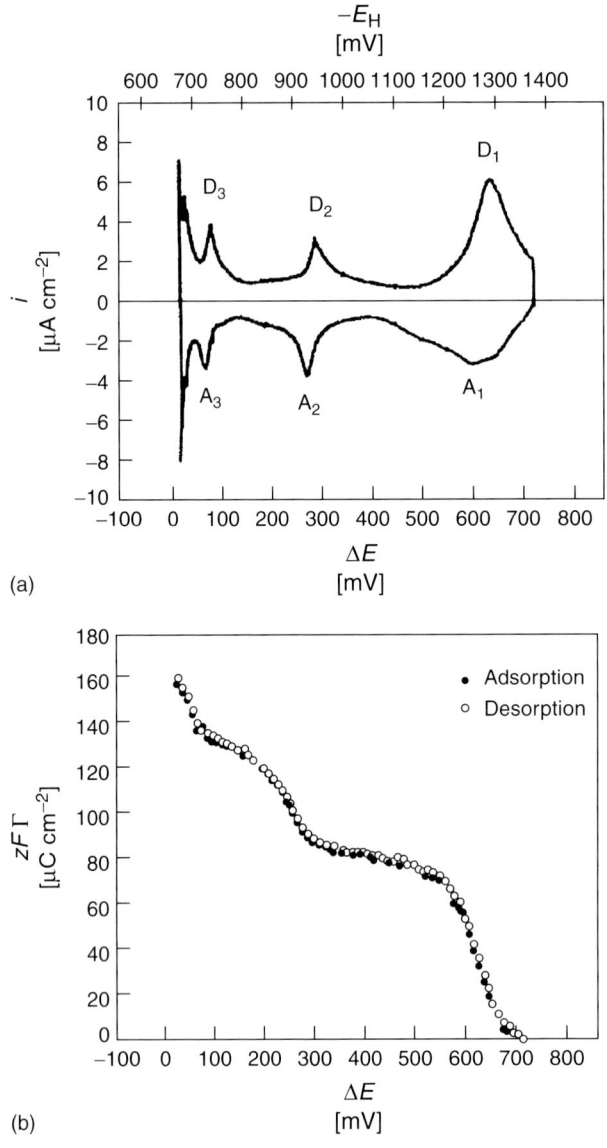

Fig. 5 (a) Cyclic voltammogram in the system Au(100)/5×10^{-3} M Ag_2SO_4 + 0.5 MH_2SO_4 at $T = 298$ K. $|dE/dt| = 7$ mV s^{-1}. A_n and D_n with $n = 1, 2, 3$ denote cathodic adsorption and anodic desorption peaks, respectively. (b) $\Gamma(E)$ isotherm of the system Au(100)/4.2×10^{-4} M Ag_2SO_4 + 0.5M H_2SO_4 at $T = 298$ K. (Reprinted from Ref. [187] with permission from Elsevier Science.)

Fig. 6 Effect of the UPD of Cd^{2+} (1), Cu^{2+} (2), and Ag^+ (3) on the potential dependence of the adsorption of HSO_4^- ions on polycrystalline gold. Curve 4 (- - - -) corresponds to the potential dependence without any addition: $c_{H_2SO_4} = 4 \times 10^{-4}$ mol dm^{-3}, $c_{Me} = 8 \times 10^{-4}$ mol dm^{-3} in 1 mol dm^{-3} $HClO_4$. (Reprinted from Ref. [147] with permission from Elsevier Science.)

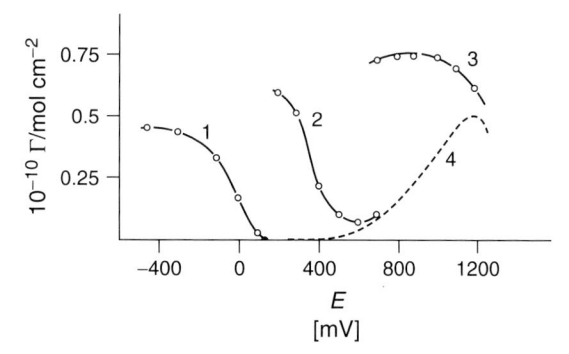

Fig. 7 Cyclic voltammogram and simultaneously recorded frequency change of the Au(111) EQCM electrode in 0.05 M H_2SO_4 (- - - -) and in 0.05 M H_2SO_4 + 5 mM $CuSO_4$. Sweep rate = 10 mV^{-1}. The potential was scanned from 0.75 V and reversed at 0.33 V. (Reprinted from Ref. [55] with permission from Elsevier Science.)

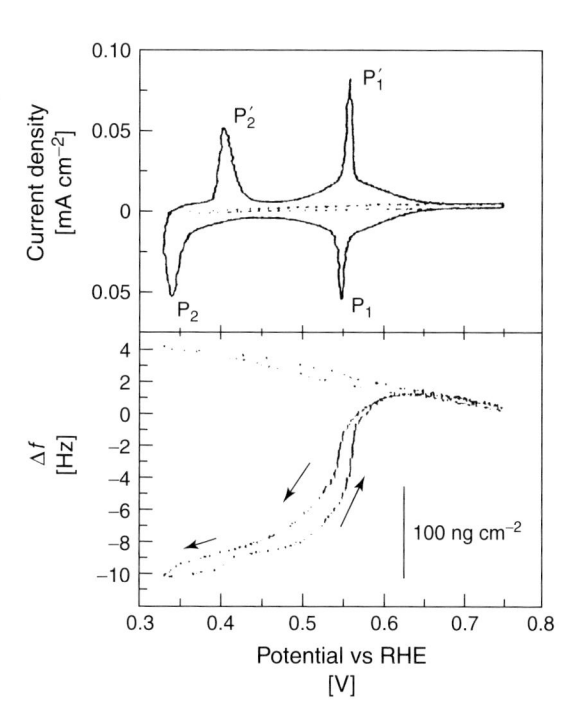

to the model system yields realistic adsorption isotherms, and the resulting model voltammogram agrees well with experiment.

Acknowledgment

Financial support from Hungarian Science Foundation (Grants T023056 and 031703) is acknowledged.

References

1. S. Trasatti, R. Parsons, *Pure Appl. Chem.* **1986**, *58*, 437–454.
2. M. A. Habib, J. O'M. Bockris, in *Comprehensive Treatise of Electrochemistry*, Double Layer (Eds.: J. O'M. Bockris, B. E. Conway, E. Yeager), Plenum Press, New York, 1980, pp. 135–219, Vol. 1.
3. A. J. Bard, L. R. Faulkner, *Electrochemical Methods, Fundamentals and Applications*,

John Wiley & Sons, New York, 1980, pp. 488–552.
4. S. Trasatti, in *Trends in Interfacial Electrochemistry* (Ed.: A. F. Silva), D. Reidel Publishing Company, Dordrecht, 1986, pp. 25–48.
5. R. Parsons, in *Trends in Interfacial Electrochemistry* (Ed.: A. F. Silva), D. Reidel Publishing Company, Dordrecht, 1986, pp. 71–81.
6. A. Hamelin, in *Trends in Interfacial Electrochemistry* (Ed.: A. F. Silva), D. Reidel Publishing Company, Dordrecht, 1986, pp. 83–102.
7. R. Parsons, in *Trends in Interfacial Electrochemistry* (Ed.: A. F. Silva), D. Reidel Publishing Company, Dordrecht, 1986, pp. 373–385.
8. D. H. Everett, *Pure Appl. Chem.* **1970**, *21*, 583–596.
9. G. Valette, R. Parsons, *J. Electroanal. Chem.* **1986**, *204*, 291–297.
10. G. Valette, R. Parsons, *J. Electroanal. Chem.* **1985**, *191*, 377–386.
11. H. D. Hurwitz, *J. Electroanal. Chem.* **1965**, *10*, 35–41.
12. E. Dutkiewicz, R. Parsons, *J. Electroanal. Chem.* **1966**, *11*, 100–110.
13. G. Láng, K. E. Heusler, *J. Electroanal. Chem.* **1995**, *391*, 169–179.
14. K. E. Heusler, G. Láng, *Electrochim. Acta* **1997**, *42*, 747–756.
15. G. Valincius, *Langmuir* **1998**, *14*, 6307–6319.
16. J. Lipkowski, W. Schmickler, D. M. Kolb et al., *J. Electroanal. Chem.* **1998**, *452*, 193–197.
17. W. Haiss, R. J. Nichols, J. K. Sass et al., *J. Electroanal. Chem.* **1998**, *452*, 199–202.
18. W. Schmicler, E. Leiva, *J. Electroanal. Chem.* **1998**, *453*, 61–68.
19. R. Guidelli, *J. Electroanal. Chem.* **1998**, *453*, 69–78.
20. G. Valincius, *J. Electroanal. Chem.* **1999**, *478*, 40–49.
21. B. M. Grafov, *Russ. J. Electrochem.* **1999**, *35*, 1029–1032.
22. G. Láng, K. E. Heusler, *J. Electroanal. Chem.* **1999**, *472*, 168–173.
23. R. Guidelli, *J. Electroanal. Chem.* **1999**, *472*, 174–177.
24. B. M. Grafov, *J. Electroanal. Chem.* **1999**, *471*, 105–108.
25. D. M. Kolb, in *Advances in Electrochemistry and Electrochemical Engineering* (Eds.: H. Gerischer, C. W. Tobias), John Wiley & Sons, New York, 1978, pp. 125–271, Vol. 11.
26. R. R. Adžić, in *Advances in Electrochemistry and Electrochemical Engineering* (Eds.: H. Gerischer, C. W. Tobias), John Wiley & Sons, New York, 1985, pp. 159–260, Vol. 13.
27. A. Aramata, in *Modern Aspects of Electrochemistry* (Eds.: J. O'M. Bockris et al.), Plenum Press, New York, 1977, pp. 181–250, Vol. 31.
28. S. Szabó, *Int. Rev. Phys. Chem.* **1991**, *10*, 207–248.
29. W. Lorenz, *Z. Phys. Chem.* **1962**, *219*, 421–423.
30. W. Lorenz, *Z. Phys. Chem. N. F.* **1967**, *54*, 191–195.
31. W. Lorenz, *Z. Phys. Chem.* **1970**, *244*, 65–84.
32. W. Lorenz, *Z. Phys. Chem.* **1971**, *248*, 161–170.
33. C. Engler, W. Lorenz, *Z. Phys. Chem.* **1978**, *259*, 1188–1190.
34. K. J. Vetter, J. W. Schultze, *Ber. Bunsen-Ges Phys. Chem.* **1972**, *76*, 927–933.
35. K. J. Vetter, J. W. Schultze, *J. Electroanal. Chem.* **1974**, *53*, 67–76.
36. A. Aramata, S. Taguchi, T. Fukuda et al., *Electrochim. Acta* **1998**, *44*, 999–1007.
37. G. Horányi, A. Aramata, *J. Electroanal. Chem.* **1997**, *437*, 259–262.
38. G. Horányi, A. Aramata, *J. Electroanal. Chem.* **1997**, *434*, 201–207.
39. G. Horányi, *ACH Models Chem.* **1997**, *134*, 33–47.
40. G. Horányi, M. Wasberg, *J. Electroanal. Chem.* **1996**, *413*, 161–164.
41. P. Zelenay, M. Gamboa-Aldeco, G. Horányi et al., *J. Electroanal. Chem.* **1993**, *357*, 307–326.
42. K. Varga, P. Zelenay, G. Horányi et al., *J. Electroanal. Chem.* **1992**, *327*, 291–306.
43. N. M. Marković, H. A. Gasteiger, P. N. Ross Jr., *Langmuir* **1995**, *11*, 4098–4108.
44. Y. Shingaya, H. Matsumoto, H. Ogasawara et al., *Surf. Sci.* **1995**, *335*, 23–31.
45. R. Gómez, H. S. Yee, G. M. Bommarito et al., *Surf. Sci.* **1995**, *335*, 101–109.
46. N. M. Marković, H. A. Gasteiger, C. A. Lucas et al., *Surf. Sci.* **1995**, *335*, 91–100.

47. L. J. Buller, E. Herrero, R. Gómez et al., *J. Chem. Soc., Faraday Trans.* **1996**, *92*, 3757–3762.
48. I. Oda, Y. Shingaya, H. Matsumoto et al., *J. Electroanal. Chem.* **1996**, *409*, 95–101.
49. J. Inukai, Y. Osawa, M. Wakisaka et al., *J. Phys. Chem. B* **1998**, *102*, 3498–3505.
50. E. Herrero, S. Glazier, L. J. Buller et al., *J. Electroanal. Chem.* **1999**, *461*, 121–130.
51. M. S. Zei, K. Wu, M. Eiswirth et al., *Electrochim. Acta* **1999**, *45*, 809–817.
52. J. Hotlos, O. M. Magnussen, R. J. Behm, *Surf. Sci.* **1995**, *335*, 129–144.
53. Z. Shi, S. Wu, J. Lipkowski, *Electrochim. Acta* **1995**, *40*, 9–15.
54. J. G. Gordon, O. R. Melroy, M. F. Toney, *Electrochim. Acta* **1995**, *40*, 3–8.
55. M. Watanabe, H. Uchida, M. Miura et al., *J. Electroanal. Chem.* **1995**, *384*, 191–195.
56. Z. Shi, S. Wu, J. Lipkowski, *J. Electroanal. Chem.* **1995**, *384*, 171–177.
57. E. D. Chabala, J. Cairns, T. Rayment, *J. Electroanal. Chem.* **1996**, *412*, 77–84.
58. W. Haiss, J. K. Sass, *J. Electroanal. Chem.* **1996**, *410*, 119–124.
59. M. Legault, L. Blum, D. A. Huckaby, *J. Electroanal. Chem.* **1996**, *409*, 79–86.
60. S. Wu, Z. Shi, J. Lipkowski et al., *J. Phys. Chem. B* **1997**, *101*, 10310–10322.
61. H. Uchida, M. Hiei, M. Watanabe, *J. Electroanal. Chem.* **1998**, *442*, 97–106.
62. P. Mrozek, Y.-E. Sung, A. Wieckowski, *Surf. Sci.* **1995**, *335*, 44–51.
63. A. J. Motheo, J. R. Santos Jr., A. Sadkowski et al., *J. Electroanal. Chem.* **1995**, *397*, 331–334.
64. P. Mrozek, Y. E. Sung, M. Han et al., *Electrochim. Acta* **1995**, *40*, 17–28.
65. E. D. Chabala, A. R. Ramadan, T. Brunt et al., *J. Electroanal. Chem.* **1996**, *412*, 67–75.
66. E. D. Chabala, T. Rayment, *J. Electroanal. Chem.* **1996**, *401*, 257–261.
67. A. N. Frumkin, O. A. Petrii, B. B. Damaskin, *J. Electroanal. Chem.* **1970**, *27*, 81–90.
68. R. Parsons, *J. Electroanal. Chem.* **1994**, *376*, 15–20.
69. A. Hamelin, in *Modern Aspects of Electrochemistry* (Eds.: B. E. Conway, R. E. White, J. O'M. Bockris), Plenum Press, New York, 1985, pp. 1–101, Vol. 16.
70. A. Hamelin, *J. Electroanal. Chem.* **1996**, *407*, 1–11.
71. A. Hamelin, A. M. Martins, *J. Electroanal. Chem.* **1996**, *407*, 13–21.
72. A. Hamelin, *J. Electroanal. Chem.* **1995**, *386*, 1–10.
73. J. Clavilier, R. Faure, G. Guinet et al., *J. Electroanal. Chem.* **1980**, *107*, 205–209.
74. F. T. Wagner, R. N. Ross Jr., *J. Electroanal. Chem.* **1983**, *150*, 141–164.
75. K. al Faaf-Golze, D. M. Kolb, D. Scherson, *J. Electroanal. Chem.* **1986**, *200*, 353–362.
76. N. Marković, M. Hanson, G. McDougall et al., *J. Electroanal. Chem.* **1986**, *214*, 555–566.
77. P. W. Faguy, N. Marković, R. R. Adžić et al., *J. Electroanal. Chem.* **1990**, *289*, 245–262.
78. E. Herrero, J. M. Feliu, A. Wieckowski et al., *Surf. Sci.* **1995**, *325*, 131–138.
79. P. W. Faguy, N. S. Marinković, R. R. Adžić, *J. Electroanal. Chem.* **1996**, *407*, 209–218.
80. A. Zolfaghari, G. Jerkiewicz, *J. Electroanal. Chem.* **1997**, *422*, 1–6
81. T. Fukuda, A. Aramata, *J. Electroanal. Chem.* **1997**, *440*, 153–161.
82. V. Lazarescu, J. Clavilier, *Electrochim. Acta* **1998**, *44*, 931–941.
83. T. Fakuda, A. Aramata, *J. Electroanal. Chem.* **1999**, *467*, 112–120
84. F. C. Nart, T. Iwasita, M. Weber, *Electrochim. Acta* **1994**, *39*, 961–968.
85. T. Iwasita, A. Rodes, E. Pastor, *J. Electroanal. Chem.* **1995**, *383*, 181–189.
86. E. Pastor, A. Rodes, T. Iwasita, *J. Electroanal. Chem.* **1996**, *404*, 61–68.
87. Y. Shingaya, K. Hirota, H. Ogasawara et al., *J. Electroanal. Chem.* **1996**, *409*, 103–108.
88. Y. Shingaya, M. Ito, *J. Electroanal. Chem.* **1999**, *467*, 299–306.
89. Z. Shi, J. Lipkowski, M. Gamboa et al., *J. Electroanal. Chem.* **1994**, *366*, 317–326.
90. Z. Shi, J. Lipkowski, S. Mirwald et al., *J. Electroanal. Chem.* **1995**, *396*, 115–125.
91. Z. Shi, J. Lipkowski, *J. Electroanal. Chem.* **1996**, *403*, 225–239.
92. Z. Shi, J. Lipkowski, S. Mirwald et al., *J. Chem. Soc., Faraday Trans.* **1996**, *92*, 3737–3746.
93. J. Lipkowski, Z. Shi, A. Chen et al., *Electrochim. Acta* **1998**, *43*, 2875–2888.
94. A. Chen, Z. Shi, D. Bizzotto et al., *J. Electroanal. Chem.* **1999**, *467*, 342–353.
95. A. Chen, J. Lipkowski, *J. Phys. Chem. B* **1999**, *103*, 682–691.
96. I. R. de Moraes, F. C. Nart, *J. Electroanal. Chem.* **1999**, *461*, 110–120.

97. N. S. Marinković, J. S. Marinković, R. R. Adžić, *J. Electroanal. Chem.* **1999**, *467*, 291–298.
98. K. J. Stevenson, X. Gao, D. W. Hatchett et al., *J. Electroanal. Chem.* **1998**, *447*, 43–51.
99. P. Broekmann, M. Wilms, M. Kruft et al., *J. Electroanal. Chem.* **1999**, *467*, 307–324.
100. S. Smoliński, P. Zelenay, J. Sobkowski, *J. Electroanal. Chem.* **1998**, *442*, 41–47.
101. S. Smoliński, J. Sobkowski, *J. Electroanal. Chem.* **1999**, *463*, 1–8.
102. D. M. Dražić, A. V. Tripković, K. D. Popović et al., *J. Electroanal. Chem.* **1999**, *466*, 155–164.
103. M. Väärtnou, P. Pärsimägi, E. Lust, *J. Electroanal. Chem.* **1995**, *385*, 115–119.
104. M. Väärtnou, P. Pärsimägi, E. Lust, *J. Electroanal. Chem.* **1996**, *407*, 227–232.
105. M. Väärtnou, E. Lust, *J. Electroanal. Chem.* **1999**, *469*, 182–188.
106. M. Väärtnou, E. Lust, *Electrochim. Acta* **1999**, *44*, 2437–2444.
107. J. Ehrlich, T. Ehrlich, A. Jänes et al., *Electrochim. Acta* **1999**, *45*, 935–943.
108. M. Väärtnou, E. Lust, *Electrochim. Acta* **2000**, *45*, 1623–1629.
109. R. Greef, R. Peat, L. M. Peter et al., *Instrumental Methods in Electrochemistry*, John Wiley & Sons, New York, 1985.
110. P. A. Christensen, A. Hamnett, *Techniques and Mechanisms in Electrochemistry*, Blackie Academic & Professional, London, 1994.
111. A. Wieckowski, (Ed.), *Interfacial Electrochemistry. Theory, Experiment, and Applications*, Marcel Dekker, New York, 1999.
112. Y. Chiu, M. A. Genshaw, *J. Phys. Chem.* **1969**, *73*, 3571–3577.
113. W.-K. Puik, M. A. Genshow, J. O'M. Bockris, *J. Phys. Chem.* **1970**, *74*, 4266–4275.
114. D. M. Kolb, in *Trends in Interfacial Electrochemistry* (Ed.: A. F. Silva), D. Reidel Publishing Company, Dordrecht, 1986, pp. 301–330.
115. A. Bewick, in *Trends in Interfacial Electrochemistry* (Ed.: A. F. Silva), D. Reidel Publishing Company, Dordrecht, 1986, pp. 331–358.
116. T. Iwasita, E. Pastor, in *Interfacial Electrochemistry, Theory, Experiment, and Applications* (Ed.: A. Wieckowski), Marcel Dekker, New York, 1999, pp. 353–372.
117. M. Weber, F. C. Nart, *Langmuir* **1996**, *12*, 1895–1900.
118. M. Weber, F. C. Nart, *Electrochim. Acta* **1996**, *41*, 653–659.
119. F. Kitamura, T. Ohsaka, K. Tokuda, *J. Electroanal. Chem.* **1996**, *412*, 183–188.
120. T. Iwasita, X. Xia, *J. Electroanal. Chem.* **1996**, *411*, 95–102.
121. J. J. Calvente, N. S. Marinković, Z. Kováčová et al., *J. Electroanal. Chem.* **1997**, *421*, 49–57.
122. W. R. Fawcett, A. A. Kloss, J. J. Calvente et al., *Electrochim. Acta* **1998**, *44*, 881–887.
123. F. Kitamura, N. Nanbu, T. Ohsaka et al., *J. Electroanal. Chem.* **1998**, *452*, 241–249.
124. C. A. Melendres, F. Hahn, *J. Electroanal. Chem.* **1999**, *463*, 258–261.
125. G. M. Brown, G. A. Hope, *J. Electroanal. Chem.* **1996**, *405*, 211–216.
126. S. Z. Zou, Y. X. Chem, B. W. Mao et al., *J. Electroanal. Chem.* **1997**, *424*, 19–24.
127. M. Bron, R. Holze, *Electrochim. Acta* **1999**, *45*, 1121–1126.
128. B. Pettinger, C. Bilger, J. Lipkowski et al., in *Interfacial Electrochemistry, Theory, Experiment, and Applications* (Ed.: A. Wieckowski), Marcel Dekker, New York, 1999, pp. 373–404.
129. S. Mirwald, B. Pettinger, J. Lipkowski, *Surf. Sci.* **1995**, *335*, 264–272.
130. A. Tadjeddine, A. le Rille, in *Interfacial Electrochemistry, Theory, Experiment, and Applications* (Ed.: A. Wieckowski), Marcel Dekker, New York, 1999, pp. 317–343.
131. A. Tadjeddine, A. le Rille, *Electrochim. Acta* **1999**, *45*, 601–609.
132. J. X. Wang, R. R. Adžić, B. M. Ocko, in *Interfacial Electrochemistry, Theory, Experiment, and Applications* (Ed.: A. Wieckowski), Marcel Dekker, New York, 1999, pp. 175–186.
133. M. F. Toney, J. N. Howard, J. Richer et al., *Surf. Sci.* **1995**, *335*, 326–332.
134. O. Endo, M. Kiguchi, T. Yokoyaama et al., *J. Electroanal. Chem.* **1999**, *473*, 19–24.
135. M. Cappadonia, K. M. Robinson, J. Schmidberger et al., *J. Electroanal. Chem.* **1997**, *436*, 73–78.
136. N. A. Balashova, V. N. Kazarinov, in *Electroanalytical Chemistry* (Ed.: A. J. Bard), Marcel Dekker, New York, 1969, pp. 135–197, Vol. 3.
137. G. Horányi, *Electrochim. Acta* **1980**, *25*, 43–57.
138. V. E. Kazarinov, V. N. Andreev, in *Comprehensive Treatise on Electrochemistry*, (Eds.: E. Yeager, J. O'M. Bockris, B. E. Conway),

Plenum Press, New York, 1990, pp. 393–443, Vol. 9.

139. A. Wieckowski, in *Modern Aspects of Electrochemistry* (Eds.: J. O'M. Bockris, B. E. Conway, R. E. White), Plenum Press, New York, 1990, pp. 65–119, Vol. 21.

140. P. Zelenay, A. Wieckowski, in *Electrochemical Interfaces: Modern Techniques for In Sity Surface Characterization* (Ed.: H. D. Abruna), VCH Publishers, New York, 1991, pp. 479–527.

141. E. K. Krauskopf, A. Wieckowski, in *Frontiers of Electrochemistry* (Eds.: P. N. Ross, J. Lipkowski), VCH Publishers, New York, 1992, pp. 119–169.

142. M. E. Gamboa-Aldeco, K. Franaszczuk, A. Wieckowski, in *The Handbook of Surface Imaging and Visualization* (Ed.: A. T. Hubbard), CRC Press, New York, 1995, p. 635.

143. G. Horányi, in *Interfacial Electrochemistry, Theory Experiment, and Applications* (Ed.: A. Wieckowski), Marcel Dekker, New York, 1999, pp. 477–491.

144. Gy. Horányi, in *A Specialist Periodical Report, Catalysis* (Ed.: J. J. Spivey), The Royal Society of Chemistry, Cambridge, 1996, p. 254–301, Vol. 12.

145. G. Horányi, *Rev. Anal. Chem.* **1995**, *14*, 1–58.

146. A. Wieckowski, P. Zelenay, K. Varga, *J. Chim. Phys.* **1991**, *88*, 1247–1270.

147. G. Horányi, E. M. Rizmayer, P. Joó, *J. Electroanal. Chem.* **1983**, *152*, 211–222.

148. Y.-E. Sung, A. Thomas, M. Gamboa-Aldeco et al., *J. Electroanal. Chem.* **1994**, *378*, 131–142.

149. D. Poškus, G. Agafonovas, *J. Electroanal. Chem.* **1995**, *393*, 105–112.

150. A. Kolics, A. E. Thomas, A. Wieckowski, *J. Chem. Soc., Faraday Trans.* **1996**, *92*, 3727–3736.

151. D. Poškus, A. Agafonovas, I. Jurgaitiene, *J. Electroanal. Chem.* **1997**, *425*, 107–115.

152. D. Poškus, *J. Electroanal. Chem.* **1998**, *442*, 5–7.

153. A. Kolics, J. C. Polkinghorne, A. Wieckowski, *Electrochim. Acta* **1998**, *43*, 2605–2618.

154. P. Waszczuk, A. Wnuk, J. Sobkowski, *Electrochim. Acta* **1999**, *44*, 1789–1795.

155. A. Wieckowski, A. Kolics, *J. Electroanal. Chem.* **1999**, *464*, 118–122.

156. T. Nomura, M. Iijima *Anal. Chim. Acta* **1981**, *131*, 97–102.

157. M. Hepel, in *Interfacial Electrochemistry, Theory, Experiment, and Applications* (Ed.: A. Wieckowski), Marcel Dekker, New York, 1999, pp. 599–630.

158. V. Tsionsky, L. Daikhin, E. Gileadi, *J. Electrochem. Soc.* **1996**, *143*, 2240–2245.

159. L. Jaeckel, G. Láng, K. E. Heusler, *Electrochim. Acta* **1994**, *39*, 1031–1038.

160. H. W. Lei, H. Uchida, M. Watanabe, *J. Electroanal. Chem.* **1996**, *413*, 131–136.

161. H. Uchida, N. Ikeda, M. Watanabe, *J. Electroanal. Chem.* **1997**, *424*, 5–12.

162. B. K. Niece, A. A. Gewirth, *J. Phys. Chem. B* **1998**, *102*, 818–823.

163. G. Gloaguen, J.-M. Léger, C. Lamy, *J. Electroanal. Chem.* **1999**, *467*, 186–192.

164. H. Matsumoto, J. Inukai, M. Ito, *J. Electroanal. Chem.* **1994**, *379*, 223–231.

165. G. Nagy, *Electrochim. Acta* **1995**, *40*, 1417–1420.

166. I. Villegas, X. Gao, M. J. Weaver, *Electrochim. Acta* **1995**, *40*, 1267–1275.

167. F. Möller, O. M. Magnussen, R. J. Behm, *Electrochim. Acta* **1995**, *40*, 1259–1265.

168. K. Ogaki, K. Itaya, *Electrochim. Acta* **1995**, *40*, 1249–1257.

169. D. Carnal, P. I. Oden, U. Müller et al., *Electrochim. Acta* **1995**, *40*, 1223–1235.

170. L. J. Wan, S. L. Yau, G. M. Swan et al., *J. Electroanal. Chem.* **1995**, *381*, 105–111.

171. J. C. Bondos, A. A. Gewirth, R. G. Nuzzo, *J. Phys. Chem.* **1996**, *100*, 8617–8620.

172. J. Inukai, S. Sugita, K. Itaya, *J. Electroanal. Chem.* **1996**, *403*, 159–168.

173. A. M. Funtikov, U. Stimming, R. Vogel, *J. Electroanal. Chem.* **1997**, *428*, 147–153.

174. W. H. Li, R. J. Nichols, *J. Electroanal. Chem.* **1998**, *456*, 135–160.

175. P. A. Christensen, A. Hamnett, in *Techniques and Mechanisms in Electrochemistry*, Blackie Academic, Chapman & Hall, London, 1994, pp. 67–89.

176. D. Hecht, J.-H. Strehblow, *J. Electroanal. Chem.* **1997**, *440*, 211–217.

177. D. Hecht, H.-H. Strehblow, *J. Electroanal. Chem.* **1997**, *436*, 109–118.

178. D. Lützenkirchen-Hecht, H.-H. Strehblow, *Electrochim. Acta* **1998**, *43*, 2957–2968.

179. A. T. Hubbard, in *Interfacial Electrochemistry, Theory, Experiment, and Applications* (Ed.: A. Wieckowski), Marcel Dekker, New York, 1999, pp. 211–229.

180. U. W. Hamm, D. Kramer, R. S. Zhai et al., *J. Electroanal. Chem.* **1996**, *414*, 85–89.

181. I. Villegas, M. J. Weaver, *J. Phys. Chem.* **1996**, *100*, 19502–19511.
182. Y. Shingaya, M. Ito, *Electrochim. Acta* **1998**, *44*, 889–895.
183. G. Pirug, H. P. Bonzei, *Surf. Sci.* **1998**, *405*, 87–103.
184. E. M. Stuve, K. Bange, J. K. Sass, in *Trends in Interfacial Electrochemistry* (Ed.: A. F. Silva), D. Reidel Publishing Company, Dordrecht, 1986, pp. 255–280.
185. G. Pirug, H. P. Bonzel, in *Interfacial Electrochemistry, Theory, Experiment, and Applications* (Ed.: A. Wieckowski), Marcel Dekker, New York, 1999, pp. 269–286.
186. Y. Shingaya, M. Ito, in *Interfacial Electrochemistry, Theory, Experiment, and Applications* (Ed.: A. Wieckowski), Marcel Dekker, New York, 1999, pp. 287–300.
187. S. Garcia, D. Salinas, C. Mayer et al., *Electrochim. Acta* **1998**, *43*, 3007–3019.
188. S. Trasatti, (Ed.), *Electrochim. Acta* **1996**, *41*(14), 2071–2338.
189. J. W. Halley, S. Walbran, D. Lee Price, in *Interfacial Electrochemistry, Theory, Experiment, and Applications* (Ed.: A. Wieckowski), Marcel Dekker, New York, 1999, pp. 1–17.
190. L. Blum, M. D. Legault, D. A. Huckaby in *Interfacial Electrochemistry, Theory, Experiment, and Applications* (Ed.: A. Wieckowski), Marcel Dekker, New York, *1999*, pp. 19–31.
191. M. L. Berkowitz, I. C. Yeh, E. Spohr, in *Interfacial Electrochemistry, Theory, Experiment, and Applications* (Ed.: A. Wieckowski), Marcel Dekker, New York, 1999, pp. 33–45.
192. G. Brown, P. A. Rikvold, S. J. Mitchell et al., in *Interfacial Electrochemistry, Theory, Experiment, and Applications* (Ed.: A. Wieckowski), Marcel Dekker, New York, 1999, pp. 47–61.
193. A. Ignaczak, J. A. N. F. Gomes, *J. Electroanal. Chem.* **1997**, *420*, 209–218.
194. M. T. M. Koper, *Electrochim. Acta* **1998**, *44*, 1207–1212.
195. W. Schmickler, (Ed.), *J. Electroanal. Chem.* **1998**, *450*(2), 157–351.
196. P. A. Bopp, A. Kohlmeyer, E. Spohr, *Electrochim. Acta* **1998**, *43*, 2911–2918.
197. E. Spohr, *Electrochim. Acta* **1999**, *44*, 1697–1705.

3.3
Phase Transitions in Two-dimensional Adlayers at Electrode Surfaces: Thermodynamics, Kinetics, and Structural Aspects

Thomas Wandlowski
Institute of Surfaces and Interfaces, Juelich, Germany

3.3.1
Introduction

Two-dimensional (2D) phase transitions on surfaces or in adlayers have received increased attention in recent years [1–4] as they are related to important aspects in surface, interfacial and materials science, and nanotechnology, such as ordered adsorption, island nucleation and growth [2, 5–7], surface reconstruction [8], and molecular electronics [9]. Kinetic phenomena such as catalytic activity and chirality of surfaces [10–12], selective recognition of molecular functions [13], or oscillating chemical reactions [14] are directly related to phase-formation processes at interfaces.

Concepts of ordering and reactivity in two dimensions can be also addressed in studies with organic, ionic, or metallic (sub)-monolayers at potentiostatically controlled electrode–electrolyte interfaces. This approach offers the advantage, in comparison to a nonelectrochemical environment, that the structural and dynamic properties of the adsorbate and the substrate can be directly tuned through the applied electrode potential. The first electrochemical studies of 2D phase transitions were mostly confined to processes occurring at ideally smooth mercury electrodes. Typical examples are the formation of compact monolayers of organic molecules or salts [15, 16] and so-called anodic films [17]. Other liquid metal electrodes employed in these studies have been gallium [18] and amalgams [19]. The use of well-defined single-crystal electrodes, such as Au(hkl), Ag(hkl), and Pt(hkl), demonstrated the important influence of the substrate material and its surface structure (crystallographic orientation, defect, and reconstruction pattern) on electrode kinetics and 2D phase transitions in electrochemically formed adlayers of organic molecules [20], ions [21], or metals [22]. Additional contributions arise from solvent molecules and the codeposition of ions of the supporting electrolyte. The combination of classical electrochemical experiments, based on measurements of current, charge density, or impedance as a function of the applied electrode potential or time, with structure-sensitive in situ techniques, such as scanning probe microscopy [23], surface X-ray scattering [24], electroreflectance [25], second harmonic generation [26] and/or vibrational spectrocopies (FTIR [27], surface-enhanced infrared reflection absorption spectroscopy (SEIRAS) [28], Raman [29], and sum frequency generation (SFG) [30]) provide a powerful strategy (1) to access macroscopic data and (2) to develop a molecular/atomic level structural and mechanistic understanding of 2D phase formation in these systems. The structure and stability of the adsorbed monolayers is affected by the symmetry of the substrate, the adsorbate–adsorbate interactions, and the corrugations in the adsorbate–substrate interaction potential.

The present chapter starts with some general remarks on the thermodynamics and kinetics of 2D phase transitions in potentiostatically generated adlayers on well-defined metal–electrolyte interfaces. Subsequently, three main groups of systems will be considered: organic and (an-)ionic monolayers and underpotential-deposited (UPD) metal films. Each subtopic is divided into an introductory overview and several case studies. The latter have been selected in order to illustrate typical approaches and the present understanding of thermodynamic, kinetic, and structural aspects of phase formation processes in 2D electrochemical systems. The review aims at covering main developments in this field, but does not attempt to be all-inclusive. We apologize for unavoidable omissions.

3.3.2
Thermodynamic Aspects

3.3.2.1 Phase Transitions and Order Parameter

Two-dimensional phase transitions occur at a clean or adsorbate-covered surface because a system in thermodynamic equilibrium seeks to minimize its free energy, F [1, 31–33]

$$F = U - TS \qquad (1)$$

One phase will supplant another at a given electrode potential E and/or temperature T because different states partition their free energy between the internal energy $U(T, E, c)$ and the entropy $S(E, T, c)$ in different ways [34–36]. The competing phases are characterized by the so-called order parameter [1]. By construction, the order parameter has a nonzero value in one phase (usually low temperature/low symmetry state) and vanishes in the other (high-temperature/high-symmetry) phase. For liquid ↔ gas or liquid ↔ solid phase transitions, the order parameter is the difference in density or the respective high-density coverage [37, 38]. Structural phase transitions, such as orientational or positional changes, may be represented by the intensity or position of a characteristic diffraction spot [39].

The behavior of the order parameter near the transition potential or temperature distinguishes two rather different transformation scenarios. In a *first-order phase transition*, the particle configuration, and consequently the order parameter, changes discontinuously. This implies that quantities such as the internal energy U or entropy S change discontinuously (but the free energy F remains continuous!), and a latent heat is involved in the transition [40, 41]. In this case, two distinct free energy curves cross one another. The system abruptly changes from one equilibrium phase to a second equilibrium phase [42, 43]. Analogous to Langmuir–Blodgett (LB) films or insoluble monolayers at the air–water interface [44], one may distinguish between gaseous, liquidlike (liquid-expanded and/or liquid-condensed), and solidlike films [2, 3, 5, 42, 43]. First-order phase transitions are characterized by (1) the phase coexistence and (2) the formation of nuclei that subsequently grow. Typical electrochemical responses comprise sharp current peaks ("spikes"), discontinuities in the charge density or capacitance versus potential curves ("capacitance pits" [45]), or the steplike blocking of charge transfer reactions [46]. Examples of first-order phase transitions in electrochemical systems are the potential-induced formation of a hydrogen-bonded uracil film on Hg [47] or Au(hkl) electrodes [20] from its disordered adsorbed monomers, the deposition of the rotated hexagonal Pb UPD monolayer on Ag(111) from citrate-containing electrolyte [48–50], or the gaseous-like ↔ c($\sqrt{2} \times 2\sqrt{2}$)R45° transition of bromide adsorbed at the aqueous electrolyte–Au(100) interface [51].

By contrast, the two competing phases become indistinguishable at the critical values of electrode potential (E_c) or temperature (T_c) for a *continuous transition* * (In modern theories, one typically distinguishes between discontinuous and continuous phase transitions. This classification is more appropriate than the classical approach of Ehrenfest, which distinguishes between first-, second-, and higher-order phase transitions [52]). The particle configuration and the order parameter change continuously, while the symmetry of the system changes discontinuously. Close to the continuous phase transition, strong fluctuations in the order parameter and in its values around its mean occur. The correlation length ξ, which is a measure of the domain size, diverges. The Landau–Lifshitz theory states that a phase transition can be continuous only if the low-temperature/low-symmetry surface space group G is a subgroup of the high-temperature/high-symmetry phase G_o, and if the respective transform function is an irreducible representation of G_o [41].

One typically finds that the order parameter of a continuous phase transition varies in the critical region as $(E - E_c)^\beta$ [53] or $(T - T_c)^\beta$ [1, 33]. The numerical value of the critical exponent β depends only on a few physical properties, such as the dimension of the local variable (order parameter) in the Hamiltonian, the symmetry of the coupling between the local variables, and the dimensionality of the system (here 2D). This property is called universality [32, 33, 54]. Systems with identical critical behavior form one universality class. Only two examples have been reported for interfacial electrochemical systems: In situ surface X-ray scattering (SXS) [53], chronocoulometry [55], and Monte Carlo (MC) simulations [56, 57] demonstrated for bromide onto a single-crystalline Ag(100)-electrode a potential-induced continuous (second-order) phase transition between a disordered, low-coverage halide adlayer at more negative potentials, and a doubly degenerate c(2 × 2) ordered phase with coverage 0.5 at more positive potentials (paragraph 3.3.4.1.2). A temperature-dependent low energy electron diffraction (LEED) study of Ag(100)/Br revealed the same type of phase transition under UHV-conditions [58]. The transition was shown to belong to the 2D-square Ising universality class, for which the exact solution has been known since the work of Onsager [59, 60]. The other example of a continuous phase transition is the temperature dependence of a 2D condensed film of 5-iodocytosine adsorbed at a mercury electrode in the neighborhood of the critical temperature, at constant adsorbate concentration and electrode potential [35] (paragraph 3.3.4.3.2).

Exact solutions, as obtained for universality classes of 2D systems such as the 2D-Ising (isomorphic to the lattice gas) or 3-state Potts model, are important but do not contain any qualitative insight into the nature of the respective continuous phase transitions. Such an insight is given by the (1) renormalization group [61], which explains how the qualitative features of the cooperative behavior arise, (2) series expansion techniques [60], (3) MC simulations [33, 38, 62] and in situ structure-sensitive experimental studies with the spatial resolution of individual atoms and/or molecules.

3.3.2.2 Adsorption Isotherms and Lateral Interactions

Adsorption phenomena including 2D phase transitions at electrochemical interfaces may be classified, according to the binding energy of the adsorbate onto the surface, in two categories. *Physisorption* corresponds to small binding energies, the substrate–adsorbate interactions are mainly due to van der Waals forces and involve almost no mixing between the orbitals of the adsorbate and the substrate [1]. Lateral interactions between the adsorbed species are dominated by dipole–dipole, dipole–induced dipole and hydrophobic interactions, π-stacking and/or hydrogen bonding [63, 64].

Chemisorption represents the formation of a surface chemical bond, which is either covalent (sharing of electrons) or ionic (electron transfer). The understanding of chemisorption phenomena is rather complex and requires knowledge on the geometrical structure of the system, adsorbate binding and charge transfer, the electronic structure of adsorbate and substrate, as well as vibrational frequencies [63]. Three dominant types of interactions may occur between chemisorbed species: dipole–dipole (direct and screened by the electrolyte), electron–electron (indirect via substrate electrons or direct at short distances), and elastic (via substrate ions) [31, 32, 65, 66].

Adsorption and phase formation is treated in three complementary regimes: the thermodynamic approach is applied to derive macroscopic properties of a system at equilibrium; principles of quantum mechanics are used to develop a microscopic understanding; and statistical mechanics establishes the connection between macroscopic and microscopic quantities, and relates the two previous approaches [63].

Intriguing aspects of both physisorption and chemisorption at electrode–electrolyte interfaces are the various phases, which may exist at the surface, and the transitions between them. The relations between coverage θ (with θ referred either to the maximum surface excess Γ_m or the maximum number of substrate sites), adsorbate concentration (activity), electrode potential, and temperature allow the construction of various types of *phase diagrams*. Graphs of $\theta = f(c)_{T,E}$ and $\theta = f(E)_{T,c}$ versus temperature as the parameter are identified as *isotherms* [1, 63].

Phase-formation phenomena at electrode–electrolyte interfaces can be conveniently treated with lattice gas concepts [38, 60, 67]. Such models consider that the entities, atoms, ions, or molecules, are fixed to particular 'cells' i, j. (M. Fisher explicitly pointed out "... that instead of imagining the particles confined to lattice sites, one may suppose that they move continuously in space divided into 'cells,' but that their interactions are determined solely by which particular cells are occupied" [52].) The configurational energy of the adlayer on a (L × L) square lattice is given, as an example, by the following Grand Canonical Hamiltonian [38, 56, 57, 63]

$$H = -\sum_{i,j} \varepsilon_{ij} s_i s_j - \bar{\mu} \sum_{ij}^{L^2} s_i \quad (2)$$

where i and j denote sites on the lattice, s_i is the occupation state of site i, which is either 0 (empty) or 1 (occupied). $\sum_{ij}$ is the sum over all pairs of sites on the lattice, and ε_{ij} is the respective pair interaction energy (three-body and higher interactions require additional terms). The geometry of the lattice may vary with the substrate surface crystallography. The surface coverage is given by [38, 63]

$$\theta = N^{-1} \sum_i s_i \quad (3)$$

with N as the total number of surface sites on the lattice. The electrochemical potential $\bar{\mu}$ is the intensive variable, which controls the coverage. In the weak solution approximation, it is related to the electrode potential E by

$$\bar{\mu} = \bar{\mu}_o + RT \ln\left(\frac{c}{c_o}\right) - F\gamma E \qquad (4)$$

with $\bar{\mu}_o$ and c_o as suitably chosen reference states, and γ as the electrosorption valency. In the limit of $\varepsilon_{ij} \to 0$ and $\theta \to 0$, Eq. (2–4) yields the Henry isotherm, and $\varepsilon_{ij} \to 0$ combined with finite values of θ results in the Langmuir isotherm [5, 60, 63].

The accurate theoretical description of the lateral interactions in the adsorbed layer requires that the complete substrate-adlayer-solution system be treated. In the simplest case, one considers only pair interactions between nearest-neighbor adsorbed species. In general, attractive (short-range) interactions will induce first-order phase transitions with typical features such as the existence of a critical temperature T_{crit}, phase coexistence at $T < T_{\mathrm{crit}}$, and island formation controlled by nucleation and growth. If the interactions are repulsive (long range), then continuous phase transitions may occur [63, 67]. In case of lateral attractive interactions, the adsorbed species occupy preferentially nearest-neighbor sites even at small concentrations Typical attractive interactions are π – stacking, hydrogen bonding, hydrophobic forces, and so on (cf. Sect. 3.3.4.3). If the minimum of the lateral interactions is located at a distance larger than the spacing between the nearest adsorption sites, the initially growing 2D phase will not match the densest packing of the monolayer. After the former phase covers the whole surface, further deposition of the adsorbate (for instance at higher electrode potentials) results in a denser film, which is likely to undergo a series of structural modifications before the maximum packing density of the first monolayer is attained. Phase diagrams of systems with lateral repulsion are more diverse (cf. Sect. 3.3.4.1). For smaller degrees of coverage, one may expect the formation of 2D lattices with large interatomic spacing, and more compact lattices may appear in a fairly complex sequence. An example is bromide on Au(100)-(1 × 1) [51]. A gaseous-like bromide adlayer undergoes a first-order phase transition to form a commensurate $c(\sqrt{2} \times 2\sqrt{2})R45°$ structure. Upon further increasing the coverage, an aligned, uniaxially incommensurate $c(\sqrt{2} \times p)R45°$ phase is formed (see section on Adlayer phases and phase transitions on fcc (100) surfaces). The commensurate/uniaxially incommensurate transition (C/UIC) has been described theoretically by Frenkel–Kontorova [68] and Frank–van der Merwe [69] on the basis of a model of harmonically bound atoms in a sinusoidal corrugation potential. The compression may be either uniform or associated with the formation of localized regions of high density (domain walls or solitons). Pokrovsky and Talapov [70] found that the incommensurability is proportional to both the reduced temperature and the (electro-)chemical potential with the same power law exponent $\beta = 0.5$. This theoretical prediction is also supported by the C/UIC–transition of bromide and chloride adlayers on Au(100) in an electrochemical environment [51].

Further compression of the uniaxial incommensurate (UI) phase may result in (1) an incommensurate phase in both directions or (2) an incommensurate rotated phase, because shear waves typically cost less energy than compression waves if the misfit is large enough [71, 72]. The deregistry transition results from the competition between the depth and width of substrate potential wells, and the dimension and magnitude of the lateral interaction energy of the adsorbed system.

The consideration of interactions between the adsorbed particles is a significant improvement of the Langmuir approach. In the simplest case, one takes into account only pair interactions between nearest neighbors, which may be attractive or repulsive, respectively. With the assumption of randomly distributed adspecies, which are not correlated, one obtains the Bragg–Williams Approximation (BWA) [60, 63, 73–76]. The interactions of the test particle are considered in the mean field of all other particles being proportional to the coverage θ multiplied by the total number of particles in the lattice (mean-field approximation or MFA). Combination with Eq. (4) leads directly to Frumkin-type equations. It involves the statistical distribution of adspecies among energetically uniform sites and neglects local fluctuations and correlations. The loop in the isotherm below the critical temperature T_{crit} is related to an artificial system forced to be homogeneous under all conditions. Especially, the coexistence of the two phases with different densities at the equilibrium phase transition under conditions of short-range attractive interactions is excluded. The models based on the BWA are more reliable if the lateral interactions can be neglected, or if the range of interactions is large [52]. The MFA breaks down severely under conditions in which phase transitions occur and ordered adlayers form [34–36, 54, 56, 60, 73].

Isotherms based on the so-called quasi-chemical approximation (QCA) are a significant improvement over the MFA [77, 78]. This gives an exact treatment of nearest neighbors, while treating the other interactions, as well as interactions of nearest neighbors with their neighbors in the MFA. Compared to the MFA (or BWA) and the QCA, a further advancement, especially at $T < T_{crit}$, is represented by the low-temperature series expansion (LTSE) [60, 73, 77–80]. This approach not only considers equilibrium adsorbate clusters up to a certain size, but also various configurations of a cluster with constant size, for example, the lattice geometry is taken explicitly into account. Exact isotherms within the lattice gas treatment may also be obtained by MC simulation [62], usually based on an MC algorithm in a Grand Canonical ensemble [38, 56, 57].

The detailed discussion and application of the aforementioned isotherms will be given in the paragraph on phase transitions in organic monolayers (Sect. 3.3.4.3).

3.3.3
Kinetics of 2D Phase Formation and Dissolution

3.3.3.1 Diffusion, Adsorption, Autocatalysis

The formation of 2D ordered, steady state adlayers at electrified interfaces is the result of several complex, nonequilibrium processes. In the simplest case, this sequence involves mass transfer of the molecules/ions from the bulk electrolyte towards the surface, adsorption and/or charge transfer, and association at the electrode surface [15, 16] (Fig. 1).

Mass transport controlled by diffusion The limiting case of nonfaradaic adlayer formation controlled exclusively by diffusion as r.d.s. was discussed by Koryta [81] and Delahay [82]. The time dependence of the surface excess Γ, in case of semiinfinite

Fig. 1 Schematic representation of the interfacial equilibria involved in the formation of ordered monolayers.

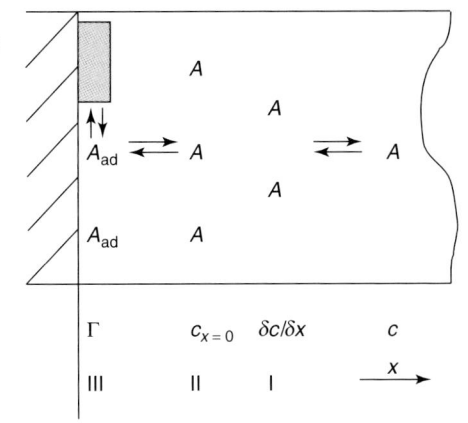

linear diffusion, is obtained from Fick's first law according to

$$\Gamma(t) = \int_0^t D\left(\frac{\partial c}{\partial x}\right)_{x=0} dt \quad (5)$$

with x as the distance from the plane of adsorption at the electrode. If the rate of adsorption is sufficiently high, so that the subsurface concentration $c_{x=0}$ is zero during the phase formation, one obtains [82, 83]

$$\theta = \frac{\Gamma}{\Gamma_m} = \frac{2c}{\Gamma_m}\sqrt{\frac{Dt}{\pi}} \quad (6)$$

Γ_m is the maximum surface excess of the condensed phase. Equation (6) is formally equivalent to the well-known Cottrell equation for faradaic processes [84]. Diffusion-controlled adsorption in the frequency domain was treated theoretically and experimentally by Melik–Gaikazyan [85, 86], Lorenz [87], and Armstrong [88].

Adsorption control The kinetics of the adsorption step in the absence of mass transport control was treated by Lorenz [87] and Delahay [89, 90] assuming either a Langmuir- or a Temkin-type formalism.

The former yields, in the limit of a negligible rate of desorption,

$$\theta = 1 - \exp(-k_{ad}c_{x=0}t) \quad (7)$$

where k_{ad} is the adsorption rate.

This theory was extended by Schuhmann to an adsorption process obeying a Frumkin isotherm [91]. The same treatment is also valid for interfacial faradaic reactions [90, 92].

Several groups derived approximate expressions for the coupling of diffusion/adsorption [85, 87, 88, 93–96] or diffusion/charge transfer-controlled processes [2, 92, 97] either in the time or in the frequency domain.

Autocatalytically controlled surface reactions Michailik and coworkers [98] proposed an autocatalytic surface process representing the time dependence of formation of a 2D condensed film of triphenylethyl phosphonium sulfate ((TPEP)$_2$SO$_4$) deposited at a mercury–electrolyte interface [99]:

$$\frac{\partial \theta}{\partial t} = k_a(1-\theta)^l \theta^p \quad (8)$$

l and p are integers ($l, p \geq 0$), with l as the reaction order with respect to the initial state and p the order with respect to the products. The analytical rate equation $\theta(t)$, derived with $l = 2$ and $p = 1$, was shown to describe quantitatively the deposition of $(TPEP)_2SO_4$ onto a mercury electrode [98].

Lorenz [100] combined diffusion, intrinsic adsorption, and surface association as separate contributions and developed the first model, which attributes a slow (dynamic) adsorption step to the 2D association of adsorbed species at an electrochemical interface.

The above-mentioned approaches suffer from a conceptual difficulty assuming that the adsorbed monomers are distributed homogeneously at the electrode. This is not unambiguously valid in case of 2D ordered ML (cf. experimental section 3.3.4). On the basis of i–t transients for 1 M pyridine in 1 M NaOH, Armstrong [101] showed that the formation of the adsorbate film proceeds according to a 2D nucleation and growth mechanism, analogous to faradaic processes such as electrocrystallization and the deposition of anodic films [17]. This approach accounts for the inhomogeneous distribution of adspecies at the electrode–electrolyte interface, and shall be discussed in the following sections in more detail.

3.3.3.2 Nucleation and Growth Mechanisms
3.3.3.2.1 Nucleation
Homogeneous nucleation – concept of the critical cluster A metal surface that is homogeneously flat, like mercury or quasi-perfect silver single crystals [2], offers no specific adsorption sites. In analogy to 3D phenomena, the nucleation of a new 2D phase may be treated on the basis of the classical nucleation theory (CNT) as formulated by Gibbs [102] and further developed by Volmer [103, 104], Farkas [105], Stranski and Kaishev [106–109], Becker and Döring [110], Zeldovich [111], and Frenkel [112]. The total Gibbs energy required to create a 2D cluster by deposition of N particles (ions, molecules) from solution is:

$$\Delta G_T(N) = \Delta G_P(N) + \Delta G_{PB}(N) \quad (9)$$

The first (favorable) term represents the free energy gain due to the formation of the new phase. The second (unfavorable) term accounts for the creation of a new phase boundary. $\Delta G_T(N)$ reaches its maximum for a critical particle number N_c. Clusters with a smaller number of particles than N_c will tend to dissolve, while larger clusters will grow further. The relation (9) is valid for any arbitrary geometrical form of the cluster [113]. The special case of a *circular patch* of radius r and height h shall be considered in more detail.

Phase formation controlled by a *charge transfer reaction* $Me^{n+} + ne^- \leftrightarrow M$ across the interface, such as 2D electrocrystallization [114, 115], UPD [2, 22, 116], and 2D anodic passivation [117, 118] gives the following expression for ΔG_T

$$\Delta G_T(r) = \frac{\pi r^2 h}{\bar{v}} n e_o \eta + 2\pi r \varepsilon \quad (10)$$

$\bar{v}$ is the specific particle volume related to the number N of deposited particles by $N = (\pi r^2 h)/\bar{v}$. $\eta = (E - E_{rev})$, with E_{rev} being the equilibrium potential, is the overvoltage and represents the driving force of the faradaic process. ε is the line tension, which has to be overcome by the expansion of the nucleus [118]. Metal deposition can occur only if η is negative, so that the Gibbs energy $\Delta G_T(r)$ of a cluster as a function of its size r first rises, reaches a maximum and then decreases (Fig. 2). The critical radius r_c and the critical free energy $\Delta G_{Tc}(r)$ are given by

$$r_c = -\frac{\varepsilon \bar{v}}{n e_o \eta} \quad (11a)$$

3.3 Phase Transitions in Two-dimensional Adlayers at Electrode Surfaces

Fig. 2 Dependence of the free energy change of formation of a disc-shaped cluster $\Delta G_T(r)$ on radius r for a charge transfer-controlled process.

and concentration according to

$$\Delta\gamma^{\#} = \int_{E_{\text{rev}}}^{E} q_f\, dE - \int_{E_{\text{rev}}}^{E} q_m\, dE \quad (13)$$

with E_{rev} being the equilibrium potential at which $\Delta\gamma^{\#} = 0$ [3, 119]. $\Delta\gamma^{\#}$ is a nonlinear function of the applied electrode potential. The free energy $\Delta G_{Tc}(r)$ and radius r_c of the critical cluster are derived as

$$\Delta G_{Tc} = -\frac{\pi\varepsilon^2}{\Delta\gamma^{\#}} \quad (14a)$$

and

$$r_c = -\frac{\varepsilon}{\Delta\gamma^{\#}} \quad (14b)$$

and

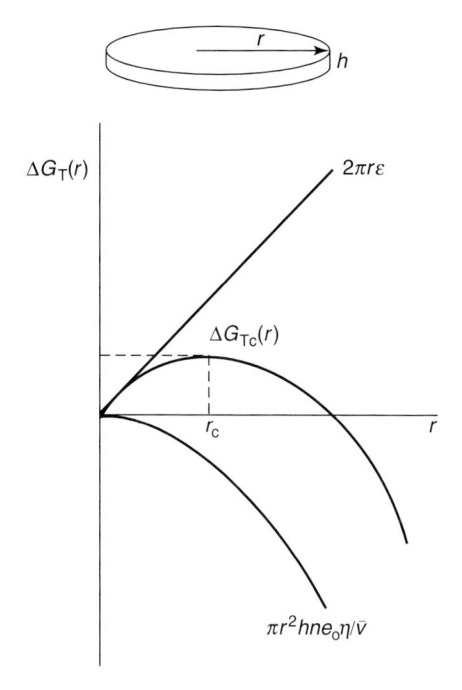

$$\Delta G_{Tc} = -\frac{\pi h \varepsilon^2 v}{n e_o \eta} \quad (11b)$$

ΔG_{Tc} is proportional to $1/\eta$.

The formation and transformation of electroinactive 2D condensed organic monolayers adsorbed at the electrode–electrolyte interface are predominantly controlled by electrocapillary forces and yield [16], instead of Eq. (11):

$$\Delta G_T(r) = \pi r^2 h \Delta\gamma^{\#} + 2\pi r\varepsilon \quad (12)$$

$\Delta\gamma^{\#}$ represents the supersaturation given by the difference in surface energy per unit area between the final (f) and the metastable (m) states [3, 119, 120]. The numerical values of the driving force are obtained by integration of the charge density curves of the two distinct phases f and m at constant pressure, temperature,

The formation of the critical nuclei occurs by thermal fluctuations and may be considered as an activation process [103]. Becker and Döring treated the homogeneous nucleation process as a series of consecutive bimolecular reactions [110]

$$M_1 + M_1 \xrightleftharpoons[\beta_2]{\alpha_1} M_2 \ldots M_i + M_1$$

$$\xrightleftharpoons[\beta_{i+1}]{\alpha_i} M_{i+1} + M_1 \longleftrightarrow \ldots \quad (15)$$

in which the growth or decay of clusters proceed by attachment or detachment of monomers M_1 to an i-mer to form an $(i+1)$ mer, respectively. The interfacial concentration of monomers is kept constant by rapid exchange with the adjacent bulk phase. The rate-determining step is assumed to be the incorporation of one additional monomer to the critical cluster M_{nc}. The solution of the set of steady state equations based on Eq. (15) then leads to the steady state nucleation rate J as:

$$J = Z\alpha^* c_{nc}$$
$$= Z\alpha^* c_1 \exp\left(\frac{-\Delta G_{Tc}}{kT}\right) \quad (16)$$

where α^* is the rate at which the monomers are added to the critical cluster of n_c atoms, ions, or molecules; c_{nc} is the equilibrium concentration of critical clusters, and Z is the nonequilibrium Zeldovich factor [111, 121]. The equilibrium concentration of critical clusters is related to the concentration of monomers, c_1, by the Boltzmann expression with the energy barrier given by the free energy of formation of the critical nucleus. The period required to reach a steady state of critical nuclei at constant supersaturation, for example, a stable distribution of clusters of the new phase controlled by "monomer" transport in the ambient phase and incorporation of building units of clusters, is identified as initial delay or induction period t' [122–127]. Toshev and coworkers reported that t' in homogeneous nucleation is usually longer than for heterogeneous nucleation [128].

Heterogeneous nucleation on active sites
Real surfaces, for instance obtained with melt-grown gold, silver, or platinum single-crystal electrodes, exhibit various sites and structural imperfections of different dimensionality, such as kinks, vacancies, monatomic steps, reconstructed surface domains, 2D islands, and holes [2, 129, 130]. Defining the number of active sites S under particular experimental conditions as N_o, one may consider the heterogeneous nucleation process as the successive incorporation of monomers M_1 into site-confined clusters $S - M_i$ according to

$$S + M_1 \underset{\beta_{s1}}{\overset{\alpha_1}{\longleftrightarrow}} S$$
$$- M_1 + M_1 \ldots S - M_{i-1}$$
$$+ M_1 \underset{\beta_{si}}{\overset{\alpha_i}{\longleftrightarrow}} S - M_i + \cdots \quad (17)$$

Instead of the steady state of (quasi) homogeneous nucleation, during which nuclei appear at constant rate, both the total number of critical nuclei and the duration of the nucleation process are confined by the maximum number of preexisting sites N_o. If there is a uniform probability with time of converting these sites into critical nuclei $S - M_{nc}$, one obtains the following first-order nucleation law as an approximate solution of the reaction sequence (17)

$$N(t) = N_o[1 - \exp(-J't)] \quad (18)$$

The quantity J' is the steady state nucleation rate per active site and may be expressed analogous to J in Eq. (16) [17, 122, 123]. Two limiting cases of the exponential law of nucleation, as given by Eq. (18), are of particular importance:

(1) $J' \gg 1$: instantaneous nucleation
$$N(t) = N_o, \quad (19)$$

Within the timescale considered, all available sites are converted to critical nuclei immediately.

(2) $J' \ll 1$: progressive nucleation
$$N(t) = J' N_o t \quad (20)$$

where the number of critical nuclei increases linearly with time and $N \ll N_o$.
The steady state rate of 2D nucleation is strongly dependent on the dimensionality

of the active sites. Staikov and coworkers reported that, at constant supersaturation, $J(\text{terrace}) < J(\text{step}) < J(\text{kink})$ [2, 130]. Active sites represent not only surface imperfections, but may also be generated by surface oxidation/reduction, reconstruction processes or ad/desorption of anions or solvent molecules [131, 132].

Fleischmann and coworkers [17, 133] and Retter [134] pointed out that the combination of active intermediates to form a critical nucleus at a specific site may lead to the power law of n–step nucleation

$$N(t) = J'' N_0 t^n \quad (21)$$

J'' represents the modified nucleation rate.

In case of homogeneous nucleation, N_0 may be attributed to the number of available lattice sites [135]. The nucleation mechanisms described earlier assume that the surface concentration of monomers remains constant throughout the transition. The formation of 2D condensed films of 5-bromocytosine [136, 137] on mercury and anodic adenine-mercury complexes [138] are examples of the so-called "truncated" nucleation. This mechanism is based on the condition that the formation of critical nuclei and their competition with growth and/or ad/desorption processes tend to decrease the available monomer concentration with time. At a certain moment, the nucleation process ceases.

Atomistic approach – small cluster model of nucleation With increasing supersaturation the number of monomers constituting the critical cluster reduces up to a few atoms or molecules, and in some particular cases of nucleation on active sites, this number was found to be close to zero [2, 139]. Macroscopic quantities, such as surface, surface free energies, and so on lose their physical meaning and the use of atomic forces of interactions becomes more reasonable. The atomistic approach for the calculation of the nucleation rate on supersaturation was first introduced by Walton [140] and then developed to a general nucleation theory by Stoyanov and Milchev [141, 142]. These authors considered the excess free energy of creation of a new "surface" (second term in Eq. (9)) by the difference between the individual binding energies of N monomers (atoms, molecules) in the bulk, $N\Psi_{1/2}$, and the dissociation energies of an N–mer cluster by including the binding energies of the monomers in the cluster as well as the interaction with the substrate, $\sum \Psi_i$. Substrate-induced strain is accounted by the introduction of a strain energy per monomer, ε_M, which yields

$$\Delta G_{\text{PB}}(N) = \left(N\Psi_{1/2} - \sum_N \Psi_i \right) + \varepsilon_M N \quad (22)$$

Applications of this atomistic approach to metal deposition on foreign substrates are summarized in [2]. Buess–Herman [143] pointed out that, in contrast to charge transfer–induced 2D adlayers, critical nuclei of surfactant molecules are quite large, often being composed of more than 100 molecules. In these cases, the CNT is still valid.

3.3.3.2.2 Growth The expansion of a supercritical nucleus through continued incorporation of monomers is called growth. If the resulting new phase is isotropic, as for instance in the case of a circular island, a single, time-independent growth rate, k_G, suffices to characterize the process. For anisotropic growth, there will be several growth rates [2]. For 2D growth, expansion of the clusters is possible only at its periphery. Referring to the elementary process involved on an ideally smooth surface, one may distinguish two main mechanisms: (1) growth controlled by the *rate of monomer incorporation* and (2) *mass transport* (surface diffusion)-controlled growth [17, 118].

Assuming that the incorporation is the rate–determining step, the number $N(t)$ of monomers belonging to an isolated circular cluster of height h obeys the equation

$$\frac{\partial N(t)}{\partial t} = 2\pi k_G r(t) h \qquad (23)$$

with

$$r(t) = \left(\frac{k_G}{\rho}\right) t \qquad (24)$$

ρ is the number of particles per unit area [17]. Normalization to the real surface area S_E and to the maximum surface excess of the condensed phase Γ_m yields the fractional coverage of this phase as

$$\frac{\partial \theta}{\partial t} = \frac{2\pi h k_G^2 M}{\rho S_E \Gamma_m} t = K_1 t \qquad (25)$$

The first in situ atomic-scale visualization of this growth mechanism was recently demonstrated for the incorporation of chloride ions at kink positions of the c(2 × 2) chloride adlayer on Cu(100) in 0.01 M HCl [144].

If the rate of advance of the growing circular center is controlled by symmetrical hemicylindrical (surface) diffusion about an axis perpendicular to the 2D nucleus one obtains

$$r(t) = A\sqrt{Dt} \qquad (26)$$

where A is a constant depending on potential, molecular weight, and density of the monomer, and D is the diffusion coefficient [145, 146]. The corresponding coverage function is given by

$$\frac{\partial \theta}{\partial t} = \frac{2\pi h A^2 D \rho}{M S_E \Gamma_m} = K_1' \qquad (27)$$

When more than one critical nucleus is formed, competition between the growing centers for the limited areas S_E takes place. Growth ceases at the point where these patches merge, domain boundaries are created, coalescence may often be ruled out [147]. Computer simulations predict for the 2D growth controlled by direct incorporation of monomers (Eden model), rather compact patches with fractal dimensions close to 2 [148]. In situ scanning tunneling microscopy (STM) experiments with hydrogen-bonded 2D condensed films of purine bases on defect-free HOPG-surfaces confirmed these findings [149]. In the diffusion-limited aggregation model, proposed by Witten and Sander, the resulting cluster has an open structure with fractal dimensions of 1.66 [150]. More complex mechanisms involve anisotropic and/or multistep growth, induction periods of growth [117, 133, 151–154], as well as defect-mediated processes [2, 17]. The case of 1D growth was treated quantitatively in the formation of surface hemimicelles of amphiphils [155]. Lorenz and coworkers pointed out that type and density of surface defects, such as monatomic steps on single-crystal surfaces, influence dramatically the steady state structures as well as the kinetics of 2D phase formation processes. They demonstrated, for instance, that barrierless nucleation and 2D growth on an equidistantly spaced stepped surface predicts rectangular i–t transients [2].

3.3.3.2.3 Coupling of Nucleation and Growth

The rate of nucleation J and the growth rate k_G are both functions of the applied supersaturation (e.g. electrode potential [2] or difference in surface energy [3], respectively). Three basic regimes may be distinguished: If the formation period of the critical cluster is much longer than the time required to cover the available electrode surface by its subsequent growth, the transformation involves just one nucleus, and is called *mononucleation*. If the rate of nucleation is much faster than the subsequent growth process, many nuclei contribute to the creation of the new phase. This regime is known as *polynucleation*. The intermediate case, which involves just a few nuclei, is named *oligonucleation*.

Mononucleation Mononucleation events at electrochemical interfaces have been observed in potential step experiments by monitoring current, charge, or capacitance as a function of time under conditions of (1) low supersaturation and (2) small, defect-free electrode surfaces. These experiments offer the unique opportunity to determine simultaneously absolute rates of the nucleation and growth processes. The formation of the critical nucleus appears stochastically in time and space (Fig. 3). The time distribution (continuous) of the birth events of critical nuclei (discrete), as indicated by the onset of the subsequent growth transient (current/charge increase, decrease of interfacial capacitance), within a sufficient long train of repetitive transient experiments under identical conditions, leads

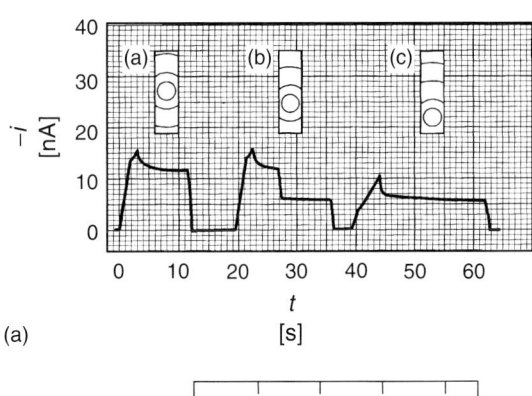

Fig. 3 (a) Current–time curves associated with the formation of single monolayers on a rectangular quasi-perfect Ag(100) electrode in 6 M AgNO$_3$ at 45 °C and an overvoltage of 2 mV, as reported by Bostanov and coworkers in Ref. [154]. The inserts show possible nucleation sites. (b) Time dependence of $-\ln(P_o)$ during the growth of a screw dislocation-free (100) face of a silver single crystal at 6.0 mV. The solid line is drawn with $JS_E = 0.0864$ nucleation per second (reprinted with permission from Ref. [159], copyright 1982 by Elsevier).

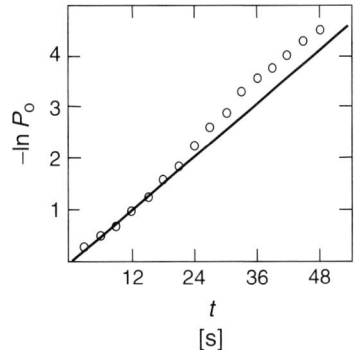

directly to the steady state nucleation rate J. The statistical analysis is based on the Poisson distribution:

$$P_N = \frac{\mu^N \exp(-\mu)}{N!} \quad (28)$$

where

$$\mu = JS_E t; \quad J \text{ in } (\text{m}^{-2}\,\text{s}^{-1}) \quad (29)$$

P_N represents the probability of having N nuclei formed at a time t, where t is a continuous variable, whereas N is discrete: $N = 0, 1, 2$, and so on [123]. Usually one evaluates the probability of observing no nucleation event, P_0, during a given time interval Δt. These data are experimentally accessible and rather unambiguous [123, 156–164]. According to Eq. (25), $\ln P_0$ should be directly proportional to t, with the proportionality constant J (Fig. 3b).

Typical examples of mononucleation are the deposition of silver on electrolytically grown Ag(100) crystals (Fig. 3) [156–159], and the formation of 2D condensed organic films, such as isoquinoline [160], thymine [163], or coumarin [162–164] on mercury electrodes. These experiments demonstrate that the rate of nucleation is more strongly dependent on the applied supersaturation than the growth rate. Extreme care has to be taken in mononucleation experiments to avoid artifacts from edges or defect contributions of the substrate or catalytic effects of impurities [161, 162].

Figure 3 illustrates, as just a small section of a train, three individual i–t transients for the deposition of a single silver monolayer on a dislocation-free Ag(100) surface having a rectangular cross section. The formation of the critical nuclei is indicated by arrows. The corresponding plot of $-\ln P_0(t)$ versus t for an overvoltage of 6 mV is shown in Fig. 3(b). The nucleation rate is strongly potential-dependent with $\ln J \sim 1/\eta$ (cf. Eqs. (11 and 16)) [156–159].

The shape of the mononucleation transient is directly related to the absolute rate of growth, since no overlapping of nuclei takes place. Figure 3(a) illustrates the isotropic growth of an expanding circular silver nucleus. Growth stops at the edges of the rectangular substrate surface. The growth rate constant was found to be of the order of 1 cm s^{-1} V^{-1} [156–159]. The carefully chosen geometry of the silver template provides, in addition, direct access to the location of the initially formed critical nucleus.

The growth of a 2D condensed organic film created by mononucleation on a hemispherical mercury electrode is often expressed in terms of a growing spherical disk by expansion of its periphery (Fig. 4a) [160, 164]:

$$i = \frac{\partial q}{\partial t} = 2\pi r q_\text{m} k_\text{G} \sin\left(\frac{k_\text{G}(t-t_\text{o})}{r}\right) \quad (30)$$

where r denotes the radius of the electrode, q_m the monolayer charge, k_G the growth rate, and t_o the onset of growth. Figure 4(b) shows a set of experimental and simulated transients for coumarin on mercury [164]. The stochastic nature of the experiment is represented by the scatter of the mononucleation transients along the time axis. The shape of the i–t traces is identical because of the use of a small, spherical (mercury) electrode.

Polynucleation At high supersaturation, many nuclei are created independently. The completion of the transition proceeds by subsequent growth of a large number of nuclei. The resulting transients, for instance triggered by a potential step, are no longer stochastic, but deterministic,

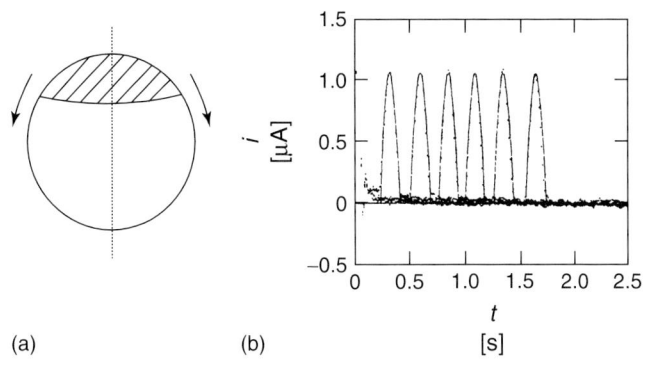

Fig. 4 (a) Scheme illustrating the growth of a single nucleus according to constant radial growth. (b) Stochastic current–time transients obtained for 0.5 M MaF + 5 mM coumarin at a hanging mercury drop electrode (HMDE) following a single potential step from $E_1 = -0.400$ V to $E_2 = -0.520$ V. The lines were drawn using Eq. (30) with $k_G/r = 17.5$ s^{-1}, $q_m = 108$ nC, and appropriate values for t_o (reprinted from Ref. [164], copyright 1995 by Bulgarian Chemical Society).

and do not depend on the shape and area of the electrode surface. In the initial stages of formation of the new phase, the individual centers can be assumed to grow rather independently of each other. The extended fractional coverage θ_x, which represents the hypothetical coverage of the unhindered growth of individual nuclei, is obtained by solving the convolution integral [17, 134, 165]

$$\frac{\partial \theta_x}{\partial t} = \int f_1(t-x) \left(\frac{\partial N}{\partial t} \right)_{t=x} dx \quad (31)$$

with $\partial N/\partial t$ given by the respective law of nucleation (cf. Eq. (18–21)) and $f_1(t-x)$ representing the growth (cf. Eq. (25, 27)). The extended coverage θ_x and the "true" coverage θ, as obtained in reality by internuclear collision and overlap of the growing centers at an advanced stage of the transition, are related within the framework of the Avrami theorem [166–171]

$$\theta = 1 - \exp(-\theta_x) \quad (32)$$

This statistical law makes it possible to correlate single cluster behavior to that of a collection of many clusters and, alternatively, to derive nucleation and growth laws from observations involving many overlapping clusters. The validity of Eq. (32) requires a random distribution of a large number of clusters, in which, each must be small with respect to the total available electrode area.

The combination of Eqs. (31 and 32) with the exponential law of nucleation (Eq. (18)), and growth of circular 2D clusters by direct incorporation of monomers (Eq. (25)) yields, as an example [172, 173]

$$\theta = 1 - \exp\left(-2K_1 N_o \left\{ t^2 - \frac{2t}{J'} + \frac{2}{J'^2}[1 - \exp(-J't)] \right\} \right) \quad (33)$$

and the limiting cases

$$\theta = 1 - \exp(-2K_1 N_o J' t^2)$$
$$\text{for } J' \longrightarrow \infty \quad (33a)$$

$$\theta = 1 - \exp\left(-\frac{2K_1 N_o J'}{3} t^3\right)$$
$$\text{for } J' \longrightarrow 0 \qquad (33b)$$

for instantaneous or progressive one-step nucleation [3, 17, 118].

Examples of the kinetic mechanisms represented by Eq. (33) and its limiting cases are the 2D formation of organic and ionic monolayers of cytosine [174], 5-bromocytosine [173], isoquinoline [175], coumarin [164], guanidinium nitrate [176], for chloride [177] and hydroxide [178] and several other so-called anodic films on mercury [179, 180], as well as the transition between the ($\sqrt{3} \times \sqrt{3}$) and the (1×1) Cu–UPD layer on Au(111) [181].

In case of surface diffusion-controlled growth (Eq. (27)), one obtains accordingly

$$\theta = 1 - \exp\left(-2K_1' N_o \times \left\{t - \frac{1}{J'}[1 - \exp(-J't)]\right\}\right) \quad (34)$$

and

$$\theta = 1 - \exp(-2K_1' N_o t)$$
$$\text{for } J' \longrightarrow \infty \qquad (34a)$$

$$\theta = 1 - \exp(-K_1' N_o J' t^2)$$
$$\text{for } J' \longrightarrow 0 \qquad (34b)$$

Phase formation kinetics based on surface diffusion-controlled growth processes have been suggested for 2D monolayers of camphor-10-sulfonate on mercury [182], physisorbed uracil films on Au(hkl) [183], as well as the commensurate/incommensurate transition c($\sqrt{2} \times 2\sqrt{2}$)R45° $\rightarrow$ c($p \times 2\sqrt{2}$)R45° of bromide on Au(100) [51].

The same strategy was applied in the derivation of rate equations for n-step nucleation according to a power law (cf. Eq. (21)) [133, 134], the combination of nucleation laws with anisotropic growth regimes [153], as well as truncated nucleation due to time-dependent concentration gradients of monomers [136]. MC simulations verified that the Avrami theorem is valid for instantaneous [184], progressive [185], and n-step nucleation according to a power law [184–187].

The above-mentioned expressions of polynucleation and growth processes, as triggered by single potential step experiments $E_1 \rightarrow E_2$, may be represented by the general Avrami equation [15, 16, 188]

$$\theta = 1 - \exp(-b_f t^m) \qquad (35)$$

b_f is a constant incorporating both rates of nucleation, J, and growth, k_w; m is related to the sum of the dimension and the time exponent in the nucleation law. The values of m provide a set of diagnostic criteria to determine possible mechanisms of the nucleation and growth processes involved (Table 1). Unfortunately, there exists no unique relation between m, b_f, and the actual kinetic mechanism on the basis of *single potential step experiments*. However, in carefully designed *secondary growth double potential step experiments* ($E_1 \rightarrow E_2 \rightarrow E_3$), one may separate nucleation and growth exploiting their different dependence on supersaturation [156, 157, 189, 190]. The rate of nucleation is a much steeper function of potential than the rate of growth. The program sequence starts with a potential pulse $E_1 \rightarrow E_2$ of large amplitude and short duration τ to create a fixed number of critical nuclei, while ensuring their insignificant growth. The second step, $E_2 \rightarrow E_3$, is chosen such that (1) further nucleation is unlikely and (2) only preformed nuclei start to grow

3.3 Phase Transitions in Two-dimensional Adlayers at Electrode Surfaces

Tab. 1 Diagnostic criteria to determine possible mechanisms of nucleation and growth based on the general Avrami equation (35)

Exponent	Nucleation	Growth	Example
1	I	SD	154
$1 < m < 2$	EL	SD	51, 154, 182, 183
2	I	DI	173, 181
	P	SD	
$2 < m < 3$	EL	DI	173, 174
3	P	DI	16, 164, 175, 176
$m > 3$	PL	DI or SD	134

Note: Abbreviations: I (P) instantaneous (progressive) nucleation; EL: exponential law of nucleation; PL: power law of nucleation; SD: surface diffusion; DI: direct incorporation.

Fig. 5 (a) Schematic charge, (b) capacitance, and (c) current transients resulting from a single potential step from E_1 to E_2. Panel (c) also shows the regime of a secondary growth double potential step experiment $E_1 \rightarrow E_2 \rightarrow E_3$.

The formation of 2D ordered monolayers at electrochemical interfaces may be triggered by an appropriate change of monomer concentration, temperature, or potential [16]. Potential step experiments are preferred for experimental simplicity. The kinetic analysis of nonfaradaic reactions is often based on capacitance or charge density transients. The surface coverage is obtained according to

$$\theta(t) = \frac{C_o - C(t)}{C_o - C_\infty} = \frac{q(t) - q_o}{q_\infty - q_o} \quad (36)$$

where C_o, C_∞ and q_o, q_∞ are the values of the differential capacitance and charge density before and after the transition (Fig. 5). In case of monolayer formation involving faradaic processes (anodic films, UPD, electrocrystallization), the current density is preferred. Numerical integration yields

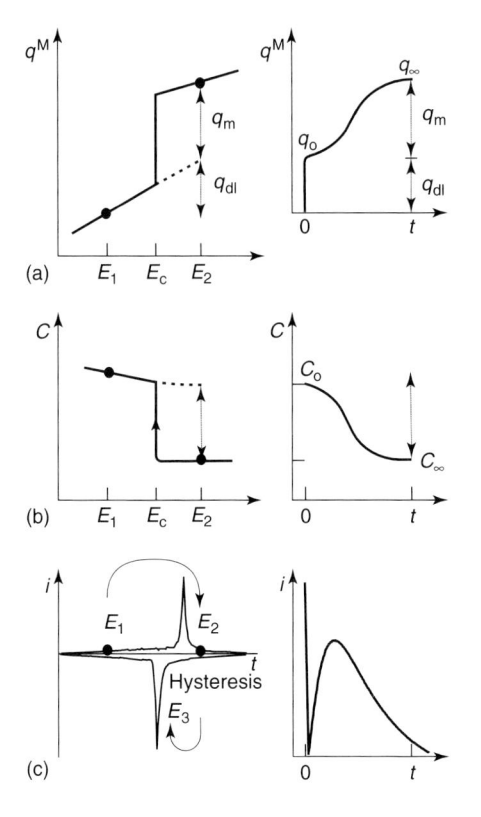

(Fig. 5). Thus, nucleation and growth are separately controlled by the parameters of the prepulse (E_2, τ) and of the secondary pulse to E_3, respectively.

$$\theta(t) = \frac{1}{q_m} \int j \, dt \quad (37)$$

q_m is the charge of a complete monolayer. Differentiation of Eq. (37) represents the strategy to transform the model equations for the time dependence of the coverage θ (cf. Eqs. (33, 34)) into the corresponding current density functions. Typical expressions are summarized in [17, 118, 146, 165, 172, 191].

Transients controlled by polynucleation and 2D growth exhibit typically sigmoidal shape of the capacitance and charge density as a function of time (nonfaradaic process) or a bell-shaped current density response (faradaic process), provided that the interference with bulk diffusion and strong coupling with ad/desorption of monomers is negligible (Fig. 5, Sect. 3.3.4). The role of bulk diffusion in 2D phase formation kinetics was analyzed by Lorenz [2], Bosco [192], and Guidelli and coworkers [193]. Rangarajan [165] and Pohlman [194] explored the consequences of the strong coupling between adsorption and nucleation/growth processes.

With the exception of the elegant mononucleation transients on silver deposition at defect–free silver electrodes [115, 156–159], the analysis of the macroscopic current, charge density, and/or capacitance transients does not provide direct access to structural information and molecular/atomistic mechanisms of 2D phase formation. Employing dynamic MC simulations of microscopic models, Rikvold and coworkers pointed out that mean-field rate equations, such as the ones based on the Avrami ansatz, are limited especially in the later stage of the overall transition [195]. For systems in which ordered phases are involved, the microscopic adlayer structure and the dynamic details of the adsorption, phase formation, and lateral diffusion processes should become important [196–198]. The combination of time-resolved dynamical MC simulations with time-resolved structure-sensitive experiments is desirable. New perspectives along these strategies developed also with the advent of time-resolved SEIRAS [199], surface X-ray scattering (SXS), [200] and in situ scanning probe techniques [201]. Some examples will be described in the application sections.

3.3.3.3 Dissolution of 2D Condensed Monolayers

The dissolution of 2D condensed monolayers represents a disorder/order transition of a high-coverage phase into a low-coverage phase. The process may be triggered by applying a potential and/or temperature perturbation [15, 16, 202]. Potential step experiments are often preferred at electrochemical interfaces because of experimental simplicity. The resulting transient is usually faster than the respective adlayer formation [3]. The overall dissolution process includes the following elementary steps (cf. Fig. 1): (1) disintegration or 2D melting of the ordered patches on the electrode surface, (2) surface diffusion and desorption of monomers, (iii) bulk diffusion into the electrolyte.

Assuming interfacial equilibrium between the condensed phase and the adsorbed noncondensed monomer species on the surface *and* within the Helmholtz layer, de Levie [83] described the dissolution transients of an ordered camphor film at a stationary mercury electrode in very dilute solution (10 µM) by semiinfinite planar diffusion according to

$$\theta_\mathrm{diss} = 1 - \theta = \frac{c - c(x=0)}{\Gamma_\mathrm{m}} \sqrt{\frac{2Dt}{\pi}} \quad (38)$$

θ as the fraction of the electrode covered by the condensed layer with Γ_m as (constant) surface excess of this phase and $c(x=0)$ as the interfacial concentration of monomers immediately adjacent to the electrode.

Under conditions of fast surface melting of the 2D condensed layer and negligible desorption of monomers into the outer Helmholtz region, Lorenz [87] and Rangarajan [165] suggested an ansatz based on the Langmuir isotherm

$$\theta_{\text{diss}} = 1 - \theta = 1 - \exp(-k_d t) \quad (39)$$

This model predicts exponential capacitance, charge density, or current versus time curves, which have been, for instance, reported for the dissolution of ordered thymine [203] and uracil [204] films on liquid electrodes. Another example, based on rather similar assumptions, is the electrodesorption of the first ML of anodically formed HgS [205].

Both models do not explain the experimentally observed dependence of the disaggregation process on (1) the dissolution overvoltage and (2) on the history of the preceding 2D condensation. Dissolution transients of ordered films formed according to polynucleation and incorporation-controlled growth indicate slower kinetics with increasing ageing time at the same potential (Fig. 6). This observation is attributed to annealing processes of adlayer defects [164, 206, 207]. Dissolution transients of ordered monolayers created from a *single nucleus* on defect-free substrates could be modeled by an inverse growth mechanism. The shape of these curves deviates significantly from that of a single exponential [163].

Faradaic admittance experiments on a mercury surface blocked by tribenzylammine [208] and in situ STM experiments with physisorbed hydrogen-bonded monolayers of guanine and adenine on HOPG [209, 210] revealed an average

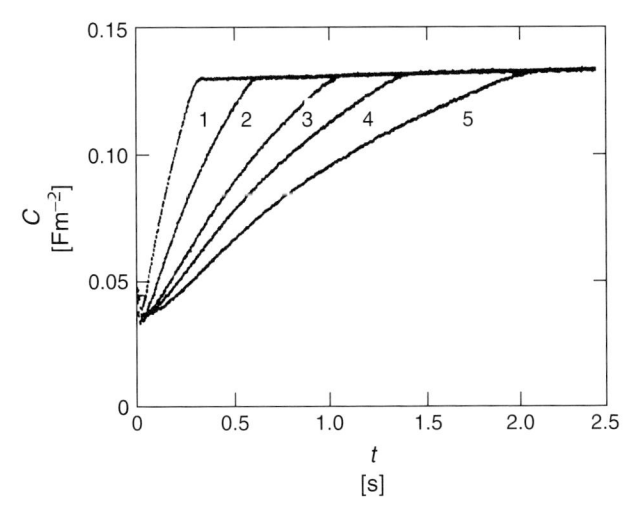

Fig. 6 Capacitance transients for 0.5 M aqueous NaF + 0.5 mM coumarin following a single potential step from $E_1 = -1.000$ V (inside the capacitance pit III) to $E_2 = -0.4265$ V at different waiting (aging) times t_o at E_1: (1) 2.5s, (2) 5.0s, (3) 10s, (4) 20s, (5) 50s. The experimental transients could be well represented by Eq. (40) in combination with Eq. (36) (reprinted from Ref. [164], copyright 1995 by Bulgarian Chemical Society).

diameter of inhomogeneously distributed 2D-ordered patches on these defect-free substrates of 1 µm to 3 µm. The domain size of organic, ionic or metal ML on real single crystals is significantly smaller, and, depending on the defect density of the substrate and the substrate–adsorbate interactions, may range between 20 nm up to 400 nm [20, 23].

Referring to the above-mentioned experimental observations, Buess–Herman and Badialli suggested a defect-mediated disordering process [3, 143, 207]: A 2D-ordered ML, formed from a large number of nuclei, is composed of polygonal patches separated by domain boundaries, which represent a lattice arrangement of line and point defects [207, 211]. These boundaries comprise sites, where the activation free energy of the phase transformation is probably lower than that required to form a hole inside the patch. In the two limiting cases, dissolution either starts from point defects or lines (Fig. 7). Point defects resulting from intersection lines, substrate defects (vacancies, atomic disorder, chemical impurities, etc.) [2], or the ad/desorption of foreign species such as solvent molecules and ions of the supporting electrolyte may trigger the creation and expansion of circular holes, which finally cause the complete dissolution of the 2D adlayer [199]. The mathematical treatment of this mechanism is isomorphic to the equations derived for polynucleation and growth (cf. Sect. 3.3.3.2), and predicts for the dissolution of 2D solidlike films sigmoidal q^M–t and C–t transients, or i–t curves exhibiting a characteristic maximum. Models based on the nucleation and spreading of holes have been applied to the reduction and dissolution of anodic films on mercury electrodes [205, 212, 213], the dielectric breakdown of bilayer membranes [214, 215], and to the transformation of the ($\sqrt{3} \times \sqrt{3}$) Cu–UPD phase on Au(111) [181]. Recently, hole nucleation and growth was also considered to describe the cathodic desorption of alkylthioles on Au(111) [216–218] and the disordering of uracil [219], cytosine [201], and uridine [220, 221] monolayers chemisorbed on Au(hkl) (paragraph 3.3.4.3.5). Experimental i–t transients have been fitted

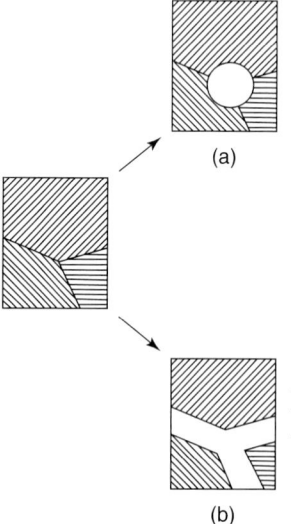

Fig. 7 Simplified scheme of disordering of a two-dimensional condensed film starting from point (a) or line defects (b). The ordered phase is represented by the hatched area.

to expressions combining instantaneous or hole nucleation according to an exponential law combined with linear or surface diffusion-controlled growth [181, 212, 213, 216–221]. Employing stepped gold electrodes with (111)-oriented terraces of different width (26, 10, 6, or 4 atoms, (110)-oriented steps) van Krieken and coworkers concluded that an increasing step density enhances the nucleation of holes within an ordered uridine layer, and simultaneously slows down their subsequent growth rate [220]. Derivations from the classical Avrami-type equations were attributed to the strong coupling of the phase dissolution with (bulk) diffusion or partial desorption [165]. Recent in situ STM experiments with physisorbed guanine ML on HOPG [210] and chemisorbed 2,2′-bipyridine on Au(111) [222] demonstrated clear evidence for hole nucleation and growth controlled disordering of a 2D condensed phase triggered by the applied electrode potential and/or temperature (cf. Sect. 3.3.4.3).

Disordering starting from line defects or grain boundaries within a 2D adlayer was first described quantitatively by Mulder, who considered the progressive shrinkage of circular patches with constant rate [211]. The model predicts a first-order melting transition [223, 224]. Taking the difference of interfacial energy $\gamma\#$ between the growing (dilute) regions, the dissolving 2D condensed phase, and the line tension ε between the two competing phases into account, Badiali and coworkers extended this approach by introducing a time-dependent propagation rate of the disordered patches [207]. The physical reason of this approach is based on the argument that dissolution leads to a progressive disappearance of line defects and consequently to a decrease of the line energy $\varepsilon \cdot l$. In parallel, the surface coverage of the disordered phase θ_{diss} increases, and the remaining adlayer fraction to be transformed reduces, which adds a positive contribution to $\Delta\gamma\#$. Assuming a squared array (average length l) of domain boundaries of the initial 2D-ordered phase leads to the following time dependence of θ_{diss}

$$\theta_{\text{diss}} = 1 - \theta = 1 - \left[\frac{\alpha}{1-(1-\alpha)\exp\left(\frac{2\alpha v_d(0)}{(\alpha-1)l}t\right)}\right]^2 \quad (40)$$

with

$$\alpha = \frac{\varepsilon}{l\Delta\gamma\#} \quad (41)$$

and $v_d(0)$ being the rate of disorder propagation at $t = 0$. This model can be adapted to any geometrical shape of domain boundaries [207]. If ε and $\gamma\#$ are known, the value of the parameter α allows the determination of L, which provides an estimation of the initial average size of the condensed phase.

Equation (40) was successfully applied to represent the disordering of physisorbed films on mercury and Au(hkl) electrodes, such as coumarin, thymine, and uracil [207, 225]. The experimental transients also support the theoretically predicted trends with ageing or healing out of domain boundaries during the polarization of the high-coverage phase (Fig. 6). In situ STM experiments of the potential-induced disordering of a cytosine ML on Au(111) indicate that dissolution initiated by point and line defects may occur even simultaneously [201]. Rather little direct experimental evidence exists at present at electrode–electrolyte interfaces for the dissolution of solidlike films by subsequent loss of long-range positional and orientation order, caused by the unbinding of dislocations and discommensurations as predicted by the KTHNY-theory [226–228]. Similarly, the interactions between adlayer and substrate surface as well as the role of substrate defects, such as vacancies, steps, and holes on the disordering of adsorbed ML at electrified interfaces have not yet been considered in detail, neither experimentally nor theoretically.

3.3.4
Examples

3.3.4.1 Phase Transitions in Anionic Adlayers

3.3.4.1.1 Introduction Anions have a strong tendency to adsorb specifically at metal surfaces, for example, to establish a direct bond with the electrode by partial loss of their hydration shell. As a consequence of the contact with the electrode, the ionic character of the anions is markedly reduced, resulting in a higher surface concentration than in case of non-specific adsorption. This effect was first observed in double-layer studies on mercury [229, 230] and later confirmed and studied in detail on single-crystal solid electrodes [231–234]. Specifically adsorbed anions can form various types of ordered structures, either more open (cf. sulfate on Au(hkl) [235, 236]) or close-packed as reported for halides on different solid electrodes [21]. Cyclic current-potential curves often reveal sharp current peaks, indicative of phase transitions within the anionic adlayers and hence of the existence of ordered phases [21, 237]. Thermodynamic data of specific anion adsorption was obtained in surface tension studies (on mercury only! [229, 238–240]), capacitance measurements [231–233], cyclic voltammetry, and chronocoulometry [234]. As an example, the following sequence of specifically adsorbed anions was found on Hg and Au(111) on the basis of the Gibbs energies of adsorption and solvation [229–232, 234, 241]:

$$F^- < ClO_4^- < SO_4^{2-} < Cl^- < Br^- < I^- \quad (42)$$

The combination of classical electrochemical measurements with ex situ transfer experiments into UHV [242], and in situ structure-sensitive studies such as electroreflectance [25], Raman and infrared (IR)-spectroscopies [29, 243], and more recently STM and SXS [39] provided detailed knowledge on energetic, electronic and structural aspects of (ordered) anion adsorption and phase formation. These experimental studies have been complemented by various theoretical approaches: (1) quantum-chemical model calculations to explore substrate–adsorbate interactions [244–246]; (2) computer simulation techniques to analyze the ion and solvent distribution near the interface [247]; (3) statistical models [67]; and (4) MC simulations [38] to describe phase transitions in anionic adlayers.

Specifically adsorbed anions can also significantly affect the stability and electrochemical reactivity of metal electrodes. Examples are the alteration of the potential (charge) distribution in the double layer known as Frumkin effect [248], the lifting of the surface reconstruction of Au(hkl) electrodes [249], the electrochemical deposition and dissolution of metals in the presence of anionic ligands [250], and the role of halides and sulfate on the oxygen reduction on Pt and Au [251].

The following chapter is focused on the structure and phase behavior of specifically adsorbed anion adlayers on solid single-crystal metal electrodes. On the

Tab. 2 Adlayer structures of halides on Au(hkl) and Ag(hkl), adapted from Ref. [21]

Electrode	Electrolyte	Structure	Coverage	Technique	Reference
Au(111)	Cl	Aligned-hex	0.508–0.5277	SXS	253, 254
	Br	Rotated-hex	0.462–0.515	SXS/STM	254
	I	$(\sqrt{3} \times \sqrt{3})R30°$	1/3	STM/LEED	255, 258
		$c(p \times \sqrt{3})$	0.33–0.41		
		rotated-hex	0.41–0.45	SXS	
Au(100)	Cl, Br	$c(\sqrt{2} \times 2\sqrt{2})R45°$	0.50	SXS, STM	51, 259, 261
		$c(\sqrt{2} \times p)R45°$	0.50–0.62		
	I	$c(p \times 2\sqrt{2})R45°$	0.46–0.49	STM	257, 263
		rotated-hex		SXS	
		$c(\sqrt{2} \times 2\sqrt{2})R45°$	0.50		
Au(110)	Br	(1×3)	2/3	STM	262
		(1×4)	3/4		
		rotated pseudohex	3/4		
		$c(2 \times p)$			
	I	$c(2 \times p)$	0.39–0.416	SXS	260
Ag(111)	Cl	Aligned-hex	0.53	STM	266
	Br	(7×7)	0.51	SXS	265
	I	$(\sqrt{3} \times \sqrt{3})R30°$	1/3	SXS	265
		$c(p \times \sqrt{3})$	0.33–0.38	STM/LEED	267
		rotated-hex	0.41		
Ag(100)	Cl, Br, I	$c(2 \times 2)$	0.50	SXS	53, 55, 268
Ag(110)	Cl, Br	$c(p \times 2)$	0.72–0.76	SXS	263

basis of selected examples with halide or oxyanion–containing adlayers, it will be demonstrated that the observed structures and 2D phase transitions can be understood as a result of competing adsorbate–adsorbate and adsorbate–substrate interactions.

3.3.4.1.2 Phase Formation in Halide Adlayers

In Situ Results on Adlayer Structures For Cl^-, Br^-, and I^- ordered structures have been observed in in situ studies on single-crystal electrodes of the face centered cubic (fcc) metals Au [51, 252–264], Ag [53, 55, 251, 265–269], Cu [270–272], Pt [23, 251, 273, 274], [275–277], Pd [278], Rh [279], and Ni [280]. A comprehensive review is given in Ref. [21]. Selected results of these studies, as obtained by in situ STM and/or SXS for ordered halide adlayers on Au(hkl) and Ag(hkl), are summarized in Table 2. Before the detailed discussion of specific examples, several general trends shall be pointed out.

Ordered halide adlayers are only observed above a critical potential E_c, which often represents rather high anion coverages. The potentials of adlayer formation and phase transitions between different ordered adlayers shift ~60 mV per decade of halide concentration. The potential ranges of the ordered phases depend strongly on E_{pzc} and the strength of metal–halide interactions. The ordered phases are stabilized by repulsive adsorbate–adsorbate interactions. Halides form predominantly close-packed hexagonal adlayers on the (111) surfaces of fcc metals reflecting isotropic adsorbate–adsorbate interactions [259, 268]. At saturation coverages the adlayer spacings are typically slightly above the halide van der Waals diameter [254]. For halide adlayers on *coinage metal surfaces* such as Au(111), Ag(111), and Cu(111) often incommensurate structures with lattice spacings, which decrease with increasing potential were found (electrocompression) [51, 254, 255, 259]. On the (111) surfaces of *transition metals* well-ordered adlayers exhibit preferentially commensurate structures indicating a stronger preference for energetically favorable adsorption sites [273–277]. Halide adlayers on fcc (100) surfaces exhibit a strong trend towards commensurate structures, with a predominant occurrence of simple c(2 × 2) adlattices. This reflects the higher substrate corrugation potential of the more open, square (100) lattice [259]. Incommensurate phases are only observed on Au(100), which represent similar trends as observed on (111) surfaces [51, 259, 268]. Only few studies of halides exist on fcc (110) surfaces, which often report pseudohexagonal adlayers, where the anions reside in the atomic rails along the (1$\bar{1}$0)-direction. An exception is Au(110)/I$^-$, where a rotated incommensurate phase was observed [257, 262, 264]. The structure and phase behavior of halide adlayers at the electrochemical interface is very similar to the adlayers formed by dissociative desorption in the gas phase in the high-coverage regime [281, 282]. Differences are found only at low coverages [283, 284]. These differences can be related to solvation effects at the electrochemical interfaces, which are most pronounced in the low-coverage regime.

The chemisorbed halide (or halogen) adlayers, in particular those on coinage metal surfaces, exhibit a structural phase behavior, which closely resembles that of classical 2D systems, such as physisorbed noble gases on graphite and metal surfaces [33, 285]. An advantage of the electrochemical environment, as compared to the solid–vacuum interface, is that the adlayers are close to equilibrium. Since coverage and adlayer structure can be directly controlled via the electrode potential, metal electrodes in halide solutions are interesting model systems for fundamental studies of phase transitions in two dimensions. Most phase transitions in ordered adlayers of halides on smooth (111) and (100) surfaces are first order, often manifested as sharp peaks in the cyclic voltammogram or double layer capacitance, and discontinuities in the adsorption isotherms. They involve phase coexistence and proceed via nucleation and growth mechanisms [51]. Phase transitions, which can be described by continuous changes of the order parameter, were reported for two electrochemical systems on (100) surfaces: the disorder-order transition for Ag(100)/Br$^-$ [53, 55, 265] and the commensurate–UI transitions for Au(100)/Br$^-$ [51, 259].

Adlayer Phases and Phase Transitions on fcc (111) Surfaces The structure of adsorbed monolayers, in particular the registry between the adlayer and the substrate, is affected by the symmetry of the substrate, the adsorbate–adsorbate interactions, and the corrugations in the adsorbate–substrate interaction potential [259]. The later is rather weak on the (111) faces of fcc metals, and of comparable magnitude to the lateral interactions between the halide adsorbates in high-density adlayers. If the average lattice constants of adlayer and substrate are different, the resulting phase is a compromise between a simple commensurate structure, which minimizes the interface energy (determined by the corrugation in the adsorption energy), and a uniformly compressed incommensurate phase, which is favored by the elastic lateral interactions in the adlayer. Frenkel and Kontorova [68] and Frank and van der Merwe [69] showed the lowest energy state corresponds to a system of commensurate regions separated by domain walls (also called solitons or misfit dislocations). When the domain walls are diffuse, such that the adsorbed species are equally spaced along the incommensurate direction, the structure is referred to as a uniformly compressed incommensurate phase. For these phases and dislocation structures, the displacement of the adsorbates from the commensurate positions can result in a long-range vertical modulation (Moire-pattern) of the adlayer, which is often observed directly by STM [252, 253, 258, 261, 267].

A whole sequence of adlayer structures is illustrated by the system Ag(111)/I$^-$, where ordered phases are observed in a wide range of potentials and surface coverages due to the strong adsorption and high polarizability of iodide [259, 267]. Starting at negative potentials in which the halide layer is disordered, a commensurate ($\sqrt{3} \times \sqrt{3}$)R30°, an UI c($p \times \sqrt{3}$), and an incommensurate rotated hexagonal structure were found with increasing potential, independently in SXS- and STM studies [259, 267] (Fig. 8). The coverage θ increases from 1/3 in the commensurate phase with iodide species residing in threefold hollow sites to 0.442 at the most positive potentials. The c($p \times \sqrt{3}$) structure is obtained by uniaxial compression of the ($\sqrt{3} \times \sqrt{3}$)R30° adlattice along the (1$\bar{1}$0) direction of the substrate.

The iodide species are arranged in a centered rectangular unit cell with sides $\sqrt{3}$ and p, where p decreases with increasing potential. The adsorbate coverage changes from about 0.355 to 0.40 in a 0.56 V wide potential range. At the most positive extreme of the c($p \times \sqrt{3}$) phase, the structure corresponds to the c(5 $\times \sqrt{3}$) high-order commensurate unit cell with a nearest-neighbor iodide spacing of 0.439 nm. Qualitatively, the SXS data show two discontinuous changes in coverage, indicating first-order phase transitions for ($\sqrt{3} \times \sqrt{3}$)R30° ↔ c($p \times \sqrt{3}$) and c($p \times \sqrt{3}$) ↔ rotated-(hex) (Fig. 8). The phase behavior of Ag(111)/I$^-$ is compared with that of iodide on an Au(111) surface, which exhibits an almost identical substrate geometry. Both incommensurate phases, the c($p \times \sqrt{3}$) and the rotated – (hex) phase, were observed by in situ STM, as well as SXS [252, 255, 258]. The commensurate ($\sqrt{3} \times \sqrt{3}$)R30° phase was only found in a limited potential range by STM and ex situ LEED studies [258], which might suggest that this structure exists only as a short-range ordered phase. The potential-dependent coverage of the incommensurate phases on Au(111) is also plotted in Fig. 8 (filled circles). For comparison, the potential scale for the data on Au(111) was shifted by −0.50 V to account for the difference in E_{pzc} of Ag(111) and Au(111) [259].

The phase behavior of iodide on Ag(111) and Au(111) is an elegant example of the stepwise loss of commensuration with increasing potential accompanied by two first-order phase transitions. This phase sequence has been predicted by Bak and coworkers [286] using Landau theory and Kadar [287] using a generalized three-state Potts model.

The orientation of the adlayers in the incommensurate hexagonal phases

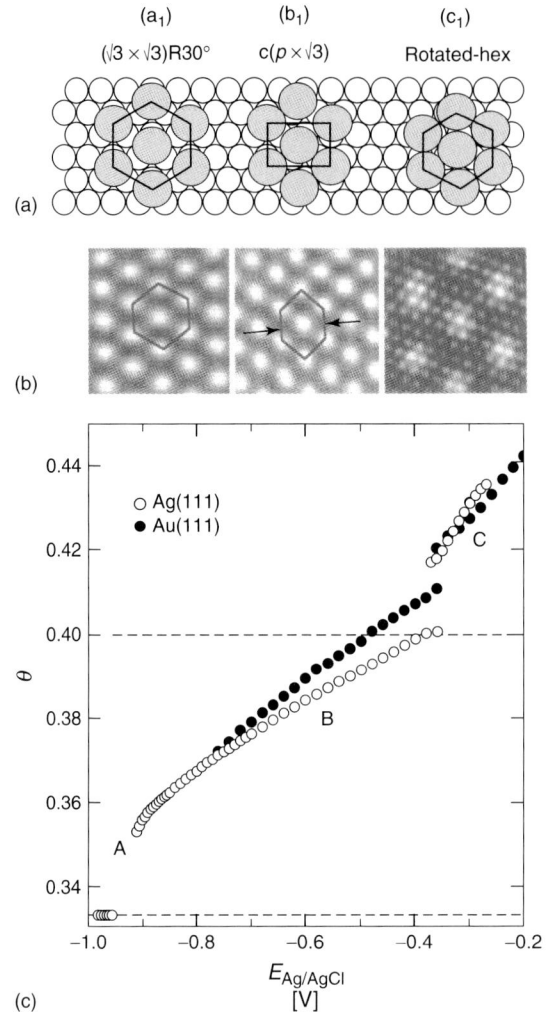

Fig. 8 (a) The structures of iodide on Ag(111) and Au(111) deduced from X-ray scattering results. They correspond to (1) the commensurate $(\sqrt{3} \times \sqrt{3})R30°$, (2) the UI $c(p \times \sqrt{3})$, and (3) the rotated hexagonal phase. (b) Corresponding in situ STM images of iodide on Au(111) (from Ref. [258]). (c) The potential-dependent coverages, θ, of iodide, determined from the in-plane diffraction, are shown versus the applied potential for Ag(111) in 0.1 M NaI (open symbols) and for Au(111) in 0.1 M KI (filled circles). The potential scale of the Au(111) data has been shifted negatively by 0.50 V to facilitate the comparison with the Ag(111) data (reprinted with permission from Ref. [265], copyright 1996 by Elsevier Science Ltd.).

exhibits a complex behavior. The adlayers are rotated relative to the $(\sqrt{3} \times \sqrt{3})R30°$ direction by an angle $\phi < 5°$, which varies continuously with adlayer density, 1° to 2.5° for Ag(111)/I$^-$ [259] and 0.5° to 4.0° for Au(111)/I$^-$ [255]. The epitaxial rotation is a well-known phenomenon in heteroepitaxial growth [285], and was first studied theoretically by Novaco and McTague [71, 72]. They showed that for an infinite incommensurate adlayer, a rotation from a high symmetry, commensurate orientation is energetically preferred if the transverse strain in the adlayer is sufficiently lower than the longitudinal strain. The total energy is minimized at a nonzero angle because of the presence of periodic lateral distortions of the adsorbate lattice (static distortion waves). The Novaco–McTague theory predicts an analytical relationship between the rotation angle ϕ and the incommensurability $\varepsilon = \sqrt{3}\theta$ in the rotated (hex) phase, and provides a reasonable description of the experimental data for iodide on Ag(111) and Au(111) [255, 259].

The continuous variation of the adlayer coverage in the UI or incommensurate phase with potential is called electrocompression [255]. Assuming an electrosorption valency $\gamma = -1$, the lateral compression κ_{2D} of the adlayer can be calculated via $\kappa_{2D} = (\gamma e_o)^{-1}(\delta A/\delta E)$, where $A = \Gamma^{-1}$ is the area per adsorbed iodide species [255, 288]. The diffraction data yield halide adlayer compressibility's in the range 1.8 to 6.9 Å^2 eV^{-1} and 4.47 to 9.40 9 Å^2 eV^{-1} for the UIC and the IC phases, respectively [21]. This is higher than the compressibility of electrochemically deposited metallic monolayers ($\kappa_{2D} = 1\text{--}2$ Å^2 eV^{-1}) [288], but significantly lower than that of physisorbed noble gas adlayers ($\kappa_{2D} = 1 - 2$ Å^2 eV^{-1}) [285]. The compressibility of metal and noble gas adlayers can be explained by the free electron model. The latter is not applicable to quantitatively describe the electrocompression of halides on Au(111). Taking into account substrate-mediated, indirect interactions, electrostatic interactions, Lennard–Jones interactions, interactions between induced dipole moments, and three-body interactions, Wang and coworkers calculated $\theta(E)$ in reasonable agreement with the experimental data [289].

Adlayer Phases and Phase Transitions on fcc (100) Surfaces With the exception of Au(100) the only stable low-order commensurate phase studied on (100) metal surfaces is the c(2 × 2) structure [53, 55, 268–270, 278]. Although the commensurate c(2 × 2) phase, where all adsorbates can occupy the preferred hollow sites, minimizes the interfacial energy between halide adlayer and metal, the resulting square adlattice is energetically unfavorable in terms of adsorbate–adsorbate interactions. An instructive example for illustrating the competing contributions of lateral interactions and corrugation potential are bromide adlayers on the structurally isomorphic Ag(100) and Au(100) surfaces.

Ag(100)/Br Recently, Wandlowski and coworkers demonstrated in a combined electrochemical and in situ SXS study the existence of a low-coverage lattice gas phase for bromide on Ag(100), and an Ising-type transition into the c(2 × 2) phase [53, 55]. Figure 9 shows the capacitance curve and the potential-dependent SXS intensity at (1/2, 1/2, 0.12), which is a first order diffraction peak of the c(2 × 2) phase, and the intensity at (1, 0, 0.12) representing the occupation of the fourfold hollow sites of the substrate. The position of the sharp peak P_2 in the capacitance curve at -0.80 V (vs. SCE) correlates well with the onset of the c(2 × 2) long-range order. For $E > -0.80$ V, the intensity at (1/2, 1/2, 0.1), which is proportional to the order parameter squared, increases according to $(E - E_c)^{2\beta}$ with the critical exponent $\beta = 0.125$ in agreement with the prediction of the 2D Ising model. The same power law was also obtained under UHV–conditions (LEED), in which the chemical potential was obtained from the coverage [58]. The intensity at (1, 0, 0.12) slowly decreases between -1.25 V and -0.60 V, for example, in the potential range of the broad capacitance peak P_1. No discontinuity is observed around the disorder/order phase transition (P_2), which indicates a continuous change in coverage, in support of a second-order phase

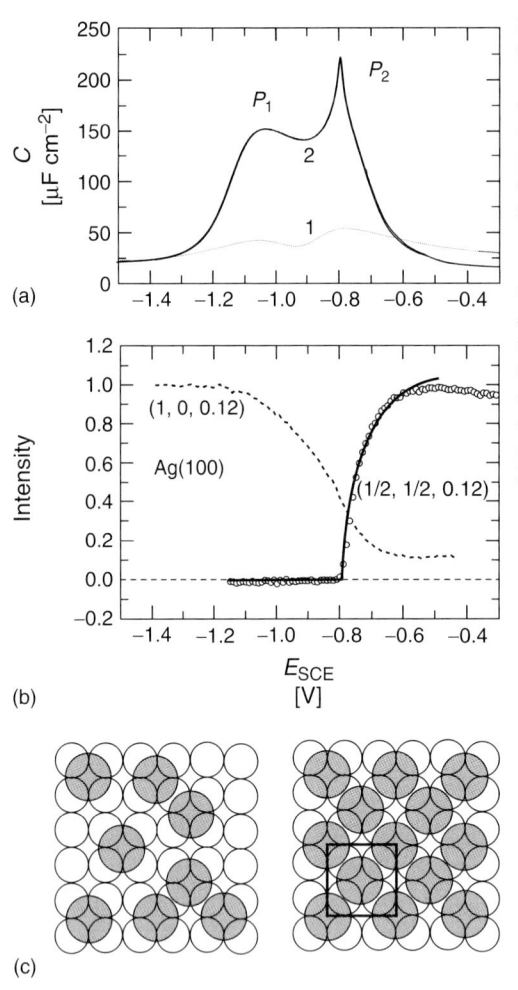

Fig. 9 (a) Capacitance versus potential curves for Ag(100) in (0.05−x) M KClO$_4$ + x M KBr; x represents the KBr concentrations: 1–0, 2–10^{-2} M. Scan rate 10 mV s^{-1}, 18 Hz and 10 mV peak-to-peak ac-amplitude. (b) Potential dependence of the normalized scattering intensity of Ag(100) in 0.04 M KClO$_4$ + 0.01 M KBr at (1/2, 1/2, 0.12) (○, reflection of the c(2 × 2) adlayer) and at (1, 0, 0.12) (dashed line, Ag(100)-(1 × 1) surface rod). (c) The real space structures of the lattice gas configuration (left) and the commensurate c(2 × 2) bromide adlayer (right) (reprinted with permission from Ref. [55], copyright 2001 by Elsevier Science Ltd.).

transition. Comparison of the bromide coverage in hollow sites, θ_{sxs}, obtained from the SXS experiments with the total coverage, θ_{tot}, measured by chronocoulometry, revealed good agreement at the highest potentials, but lower values of θ_{sxs} at intermediate coverages, which is attributed to a significant displacement of the adsorbates from hollow sites at low coverages [55] (Fig. 10a). The disorder-order transition occurs at θ_{sxs} = 0.25 and θ_{tot} = 0.35. The latter is close to the critical coverage of 0.368 obtained from simulations assuming lattice gas adsorption with a pure hard square model [290]. On the basis of an MFA, Ocko and coworkers estimated a repulsive interaction energy of 0.110 eV between bromide adsorbates on next-nearest neighbor sites [268]. Koper [56] and Mitchell and coworkers [57, 291] performed MC simulations for a lattice gas model (LGM). These authors found from fits to the Ag(100)/Br$^-$ adsorption isotherms that the lateral interactions are quite adequately described with a nearest neighbor excluded

Fig. 10 (a) Potential dependence of the normalized bromide coverage obtained chronocoulometrically (θ_{tot}, □) or by analyzing the scattering intensity in (1, 0, 0.1) position (θ_{SXS}, ○) for Ag(100) in 0.04 M KClO$_4$ + 0.01 M KBr. (b) θ_{tot} (referenced to the Ag(100)-(1 × 1) lattice) for three KBr concentrations in (0.05 − x) M KClO$_4$ on Ag(100). The solid lines represent the Monte-Carlo-simulation isotherms as calculated by Rikvold and coworkers in Ref. [291] (reprinted with permission from Ref. [55], copyright 2001 by Elsevier Science Ltd.).

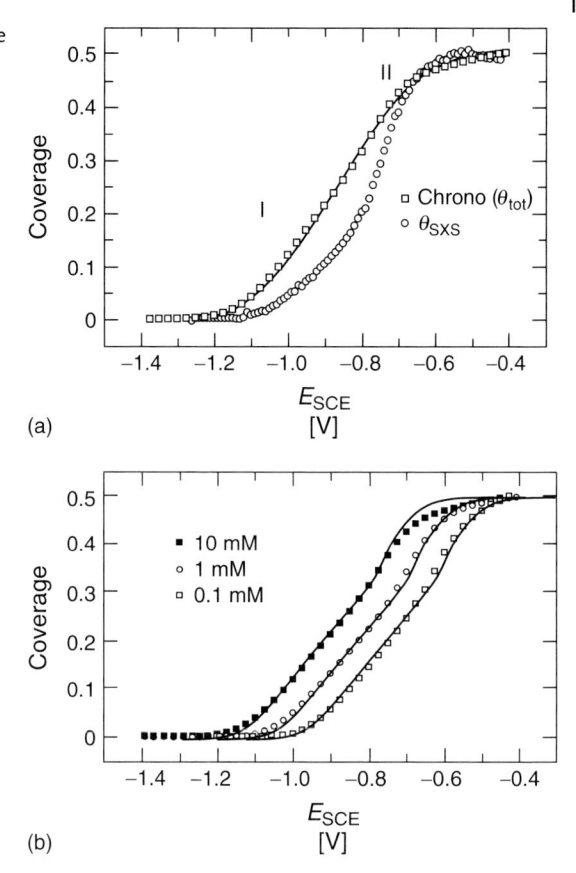

volume interaction and a (dominating) long-range dipole–dipole repulsion. The simulations quantitatively reproduced the electrochemical data and provided an excellent description of the critical properties of this continuous phase transition [57, 291] (Fig. 10b). Koper also pointed out that the QCA is a more appropriate analytical description of the experimental $\theta - E$ isotherm than the mean-field treatment (MFA or Frumkin isotherm) [56].

Au(100)/Br Figure 11 illustrates a typical current versus potential curve of bromide on Au(100) [51, 259, 261, 292]. Three characteristic current peaks are observed in the positive scan direction, and two in the negative one. Comparison with in situ SXS experiments revealed that the peak P_1 represents the transition between the hexagonal (reconstructed) and the quadratic (unreconstructed) arrangement of the gold atoms in the top layer of the substrate [51]. No other diffraction peaks were found at $E < P_2$. At the voltammetric peak P_2, a first-order phase transition between gaseous-like, randomly adsorbed bromide and a commensurate c($\sqrt{2} \times 2\sqrt{2}$)R45° halide adlayer (coverage $\theta = 0.5$), which is stable in the potential region between P_2 and P_3, takes place [51]. The rectangular unit cell contains two bromide species with nearest and next-nearest neighbor distances of

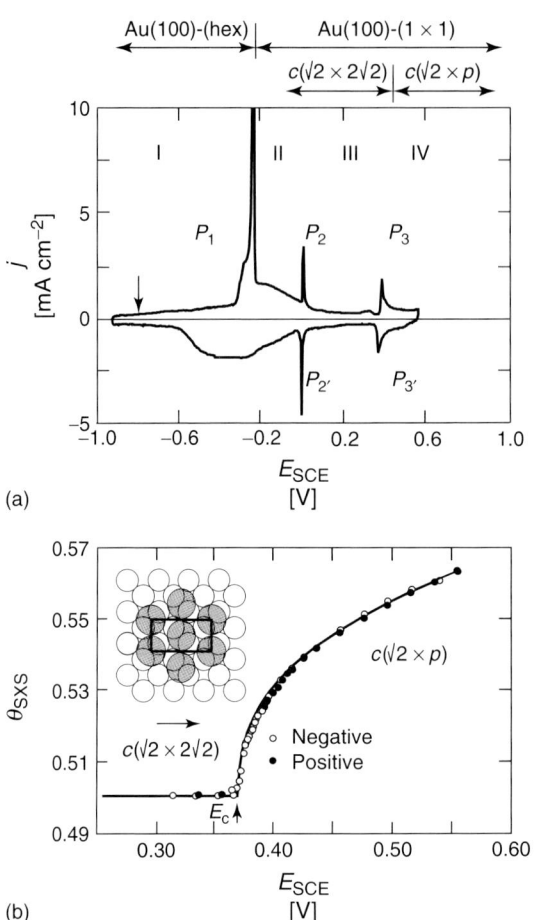

Fig. 11 (a) First current versus potential scan of a freshly flame-annealed Au(100) electrode, immersed at -0.80 V in 0.05 M NaBr, scan rate 10 mV s^{-1}. The stability regions of the various substrate surface (Au) and adlayer structures (Br) are indicated. The transitions are labeled P_1, $P_2/P_{2'}$, and $P_3/P_{3'}$, and the stability regions are named I, II, III, and IV. (b) Potential dependence of the bromide adlayer coverage as obtained by SXS for the C/UIC transition between $c(\sqrt{2} \times 2\sqrt{2})R45°$ and $c(\sqrt{2} \times p)R45°$. The open (filled) circles represent the negative (positive) going potential scan. The solid line was calculated with a power law $0.122(E - E_c)^{0.40} + 0.5$, where E_c is the concentration-dependent thermodynamically defined transition potential, as indicated in the figure (reprinted with permission from Ref. [51], copyright 1996 by American Chemical Society).

4.08 Å and 4.56 Å, respectively. The crystallographic analysis of the X-ray peak intensities suggests a registry of the $c(\sqrt{2} \times 2\sqrt{2})R45°$ adlayer with all adsorbates residing in bridge sites, rather than a mixed occupation of top and hollow sites [259]. This result is also supported by an in situ STM study of Cuesta and coworkers [261], in which no differences in the apparent height of the halide adsorbate in the $c(\sqrt{2} \times 2\sqrt{2})R45°$ unit cell were observed. The preference for the $c(\sqrt{2} \times 2\sqrt{2})R45°$ phase, with respect to the higher coordinated, and commonly on fcc(100) surfaces found $c(2 \times 2)$ phase (cf. Ag(100)/Br$^-$ [53]), suggests that the elastic interactions between the Br$^-$ adsorbates, which favor hexagonal packing, are more significant than the adsorbate–substrate interaction energy difference between the two phases. The corrugation potential for bromide on Au(100) appears to be weaker than on Ag(100). Similar trends were found in halide adlayers on Au(111) and Ag(111) [21], as well as in the corrugation potentials extracted from the calculations of Ignaczak [245].

The commensurate $c(\sqrt{2} \times 2\sqrt{2})R45°$ phase transforms at $E > 0.40$ V into a UI $c(\sqrt{2} \times p)R45°$ adlayer, where the incommensurability $\varepsilon = 2\sqrt{2}/(p-1)$ *continuously* varies from 0 (commensurate phase) to 0.12. The nearest-neighbor spacing in the $c(\sqrt{2} \times p)R45°$ structure is always the commensurate spacing $a_{Au}\sqrt{2} = 4.078$ Å, the next-nearest-neighbor spacing decreases from 4.56 Å in the commensurate phase to 4.13 Å resulting in an almost undistorted hexagonal lattice at the most positive potentials [51, 259]. As the lattice compresses the bromide species slide along the "rails" determined by the underlying gold atoms. The adlayer compression is similar to that in the rotated hexagonal phase on Au(111) [254].

The continuous change of the bromide coverage with potential upon the transition $c(\sqrt{2} \times 2\sqrt{2})R45° \rightarrow c(\sqrt{2} \times p)R45°$ can be approximated by the power law $\theta = 0.122\,(E - E_c)^\beta + 0.50$ with $\beta = 0.40$ and $E_c = 0.375$ V. The measured exponent is smaller than the theoretical value $\beta = 0.50$ predicted by Pokrovsky and Talapov for the C $\rightarrow$ UIC transition within a model of noninteracting domain walls [70]. The theoretical prediction is only supposed to apply close to the transition, where the domain walls are narrow relative to their separation. The experimental precision does not permit to quantitatively extract the exponent close to the transition, in which the theoretical prediction could be unambiguously tested. At higher incommensurabilities the bromide coverage is not only determined by the proximity to the C/UIC phase transition, but rather by the compressibility of the bromide adlayer, which may explain the deviations in the exponent. The width of the X-ray diffraction (XRD) peaks in the incommensurate direction scale quadratically with the wave vector component in this direction, and increase continuously by an order of magnitude as the C/UIC transition is approached from the UIC side.

This behavior is consistent with a cumulative disorder in the $c(\sqrt{2} \times p)R45°$ phase [259]. The results for bromide adsorption on Au(100) under electrochemical control indicate that continuous C $\rightarrow$ UIC transitions are not restricted to the vacuum environment.

The kinetics of the phase transitions of bromide adlayers on Au(100)-(1 × 1) were studied by chronocoulometric potential step experiments [51]. Current transients indicate for the disorder $\leftrightarrow$ c($\sqrt{2} \times 2\sqrt{2}$)R45° transitions in both directions an instantaneous (hole) nucleation process coupled with a Langmuir-type adsorption/desorption mechanism. Rising transients, which could be modeled with the exponential law of nucleation and 1D or surface diffusion-controlled growth, were obtained for the $c(\sqrt{2} \times 2\sqrt{2})R45° \rightarrow c(\sqrt{2} \times p)R45°$ transition when stepping into the potential regime of high incommensurability, (the uniformly compressed adlayer phase). The signal-to-noise ratio in the range of the weakly incommensurate phase was too high, and hence no direct conclusion on the "continuous" character of this phase transition around E_c (as predicted from the above-mentioned SXS results) from transient experiments was accessible. The dissolution of the UIC ($\sqrt{2} \times p$)R45° phase is characterized by monotonous i–t curves, which may be related to the considerable disorder of the $c(\sqrt{2} \times 2\sqrt{2})R45°$ when formed from the incommensurate phase [51].

3.3.4.1.3 **Phase Transitions in Adlayers of Oxoanions** The specific adsorption of oxoanions, such as sulfate/bisulfate, hydroxide, or phosphate considerably affects the shape of voltammetric and capacitance curves [21, 23, 231–233, 293]. The presence of sharp current spikes, reported, for instance, for Au(111) or Pt(111)/H$_2$SO$_4$ [237, 294, 295] and HPO$_4^{2-}$ [296, 297] was attributed to first-order phase transition involving specifically adsorbed anions. Most structural studies with oxoanions have focused on sulfate or bisulfate adsorbed on noble metal single-crystal electrodes, and in particular on the Au(111) surface [235, 298–309]. The adsorption behavior of sulfate species on Au(111) was investigated by cyclic

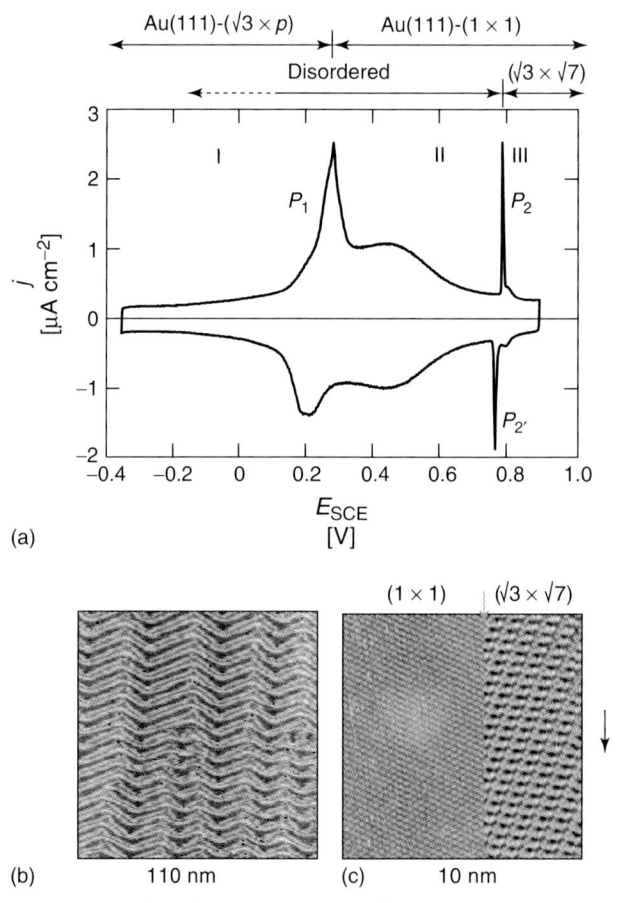

Fig. 12 (a) Cyclic voltammogram range for Au(111) in 0.05 m H$_2$SO$_4$, scan rate 10 mV s^{-1}. The stability of the various substrate (Au) and adlayer structures (HSO$_4^-$ or SO$_4^{2-}$) is indicated. (b) In situ STM image of the thermally reconstructed Au(111)-($p \times \sqrt{3}$) surface at -0.20 V. (c) First-order phase transition between randomly adsorbed sulfate species and the ordered ($\sqrt{3} \times \sqrt{7}$)19.1° phase on Au(111)-(1 × 1) during a potential scan from $E_1 = 0.750$ V to $E_2 = 0.820$ V. The scanning direction is from top to bottom. The arrow indicates the phase transition upon passing P$_2$ (reprinted from Ref. [299], copyright 1997 by VCH Verlagsgesellschaft mbH Weinheim).

voltammetry [237, 310], chronocoulometry and radiochemistry [311], microgravimetry [312], IR-spectroscopy [298, 313–315], SHG [316], and in situ STM [235, 298, 299, 317] (Fig. 12). Three potential regions can be distinguished in the steady state voltammogram for Au(111) in 0.05 M H_2SO_4, separated by well-defined current peaks. P_1 represents the lifting of the reconstruction of the gold surface [249], and implies that region I can be assigned to the reconstructed Au(111)-($p \times \sqrt{3}$) surface with minor adsorption of sulfate. Regions II and III represent the unreconstructed Au(111)-(1×1) phase. At potentials more positive than the sharp current spikes $P_2/P_{2'}$, which exhibit a small but distinct hysteresis, (bi)sulfate forms a highly ordered commensurate ($\sqrt{3} \times \sqrt{7}$)R19.1° superstructure. Chronocoulometric and radiochemical experiments [311] yielded in this potential region a maximum surface coverage of 0.2, which is significantly below the value expected for a close-packed sulfate adlayer. Infrared data revealed a C_{3v} symmetry, in which three oxygen atoms of the sulfate species interact with the substrate reflecting the symmetry match of the tetragonal anion with the trigonal gold surface [313, 314]. This assignment is in agreement with recent calculations [318]. The same ordered sulfate structure was found during in situ STM studies on (111) surfaces of Pt [300, 301], Rh [302], Cu [303–306], Ir [307], and Pd [308]. Two kinds of maxima can be seen in the STM images of the ($\sqrt{3} \times \sqrt{7}$)R19.1° structure, but only the brighter one corresponds to the adsorbed sulfate, the secondary maxima have been assigned either to coadsorbed water [302, 314] or hydronium ions [308, 309] to stabilize the oxoanion adlattice by direct hydrogen bridges.

The macroscopic kinetics of the disorder/order phase transition, giving rise to the ($\sqrt{3} \times \sqrt{7}$)R19.1° adlayer on Au(111), was studied by chronoamperometric potential step experiments [299]. The obtained experimental current transients could be represented by the exponential law of nucleation in combination with surface diffusion-controlled growth (Fig. 13). The first-order nature of this transition is supported by a recent in situ STM study [235]. Magnussen and coworkers reported on the coexistence of fluctuating ($\sqrt{3} \times \sqrt{7}$) adlayer islands and patches displaying the bare Au(111) substrate on terraces at potentials just positive of P_2. Domain boundaries were found to be mobile, and the shape of the islands was changing rapidly. In situ SEIRAS studies employing the step-scan technique might reveal further structural details of the sulfate and water species of this phase transition with a time resolution of up to 5 µs [199]. Increasing the step density of the substrate surface causes (1) the cumulative disorder of the ($\sqrt{3} \times \sqrt{7}$)R19.1° adlayer and (2) an increase of the nucleation rate [299] (Fig. 14). No current spikes P_2/P_2' were observed at a critical miscut angle of 4.7°, corresponding to a terrace width of 2.9 nm. This trend is supported by similar observations of ordered sulfate and phosphate structures on defect-rich Au(100) electrodes [309].

3.3.4.2 Phase Transitions in UPD – Adlayers

3.3.4.2.1 Introduction UPD is referred to as the deposition of a metal monolayer onto a foreign metal substrate positive of the respective bulk (Nernst) potential E_{rev}. The metal adatoms are bound more strongly to the foreign metal substrate than

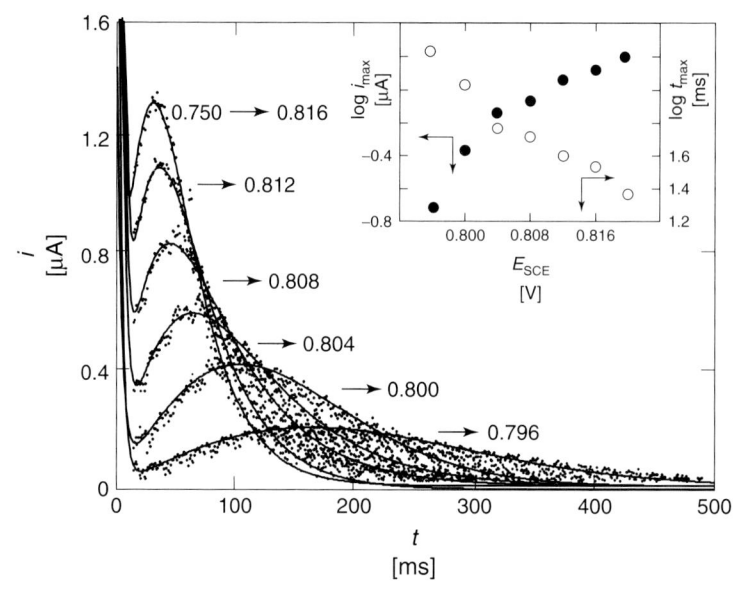

Fig. 13 Current transients $i(t)$ for Au(111), miscut $< 0.5°$, in 0.05 M H_2SO_4 obtained after a single potential step from $E_1 = 0.75$ V (region II) to various final potentials in region III. The experimental traces are given as individual data points, the solid lines represent theoretical curves calculated with the parameters of the numerical fit to a model combining (a) an adsorption process (Eq. 7) and (b) one-step nucleation according to an exponential law with surface diffusion-controlled growth (Eq. 34), (reprinted from Ref. [299]. Copyright 1997 by VCH Verlagsgesellschaft mbH Weinheim).

to a substrate of its own kind, for example, the chemical potentials of monolayer and bulk are different. In comparison to *molecular adsorption*, one observes a significant *charge transfer* (faradaic reaction !) between the deposited metal ion and the polarized electrode [116].

The UPD is often demonstrated in cyclic voltammetry. The formation (dissolution) of the first monolayer is seen by pronounced current peaks at $E > E_{rev}$, the bulk deposition occurs at $E < E_{rev}$. For many polycrystalline surfaces, it has been shown that the differences in peak potentials for the oxidative dissolution of metal monolayers and bulk correlate linearly with the difference in work function, $\Delta\Phi$, of substrate and deposit [116, 319]. This correlation is not valid at single-crystal electrodes [320–322]. New theories of UPD involving semiempirical models, DFT- and band structure calculations taking into account contributions of the electronic subsystem (related to work function differences) *and* of the ionic cores of the adsorbate, have been developed to model the trends observed on single-crystal electrodes (cf. Ref. [323]). These quantum statistical approaches are complemented by statistical mechanical models [67] and (dynamic) Monte Carlo simulations [197] on UPD systems.

The use of single-crystal electrodes in UPD–studies revealed pronounced structure specificity in metal monolayer

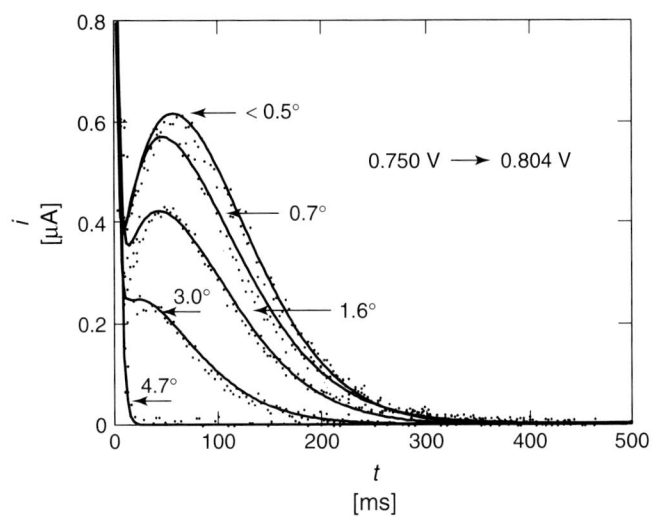

3.3 Phase Transitions in Two-dimensional Adlayers at Electrode Surfaces | 417

Fig. 14 Typical set of $i-t$ transients with $E_1 = 0.75$ V/$E_2 = 0.804$ V for stepped Au(111)_s crystals with various miscut angles (as indicated) in 0.05 M H_2SO_4. The solid lines represent calculated curves as obtained with the parameters of the numerical fit to an adsorption-nucleation model (cf. Fig. 13), (reprinted from Ref. [299], copyright 1997 by VCH Verlagsgesellschaft mbH Weinheim).

formation as was shown in the early work of Schultze [320–322], Lorenz [324], Bewick and coworkers [48–50], or Adzic [325]. Meanwhile a considerable amount of data is available, which illustrate the role of electrode material and surface crystallography, the influence of coadsorbed anions and/or solvent molecules, as well as intermetallic electronic and ionic interactions on (1) the formation/dissolution and (2) on the structural properties of these metal monolayers (cf. reviews in [2, 22, 116, 326–328]).

UPD of metals has been extensively studied with a variety of methods. Cyclic voltammetry is frequently employed to phenomenologically describe the phase behavior in UPD systems and to determine adsorption isotherms assuming full discharge of the metal ions [2, 116, 326–328]. The role of partial discharge and anion coadsorption onto the exact determination of coverage-potential isotherms was explored by the twin-electrode thin-layer technique [329] and studies with the quartz crystal microbalance (QCM) [330]. Multistep adsorption isotherm was taken as a first evidence for ordered UPD films [2, 22, 320–322]. The kinetics of adlayer deposition/dissolution, as monitored in chronoamperometric potential step experiments, was phenomenologically interpreted with homogeneous diffusion-adsorption models (cf. summary in [2]) or as first-order phase transitions controlled by nucleation and growth [48–50]. Optical properties of the metal monolayers were studied in the visible and near UV-range by electroreflectance spectroscopy [25] and second harmonic generation [331]. The use of ex situ UHV-techniques (LEED, reflection high-energy electron diffraction (RHEED), XPS, UPS,

Auger electron spectroscopy (AES)) after controlled emersion of the electrode provided important direct information on the structure, energetics, and composition of the various metal adlayers [242, 332, 333]. Major breakthroughs in the determination of adsorption sites, UPD adlayer structures, and their dynamics and reactivity properties were achieved by the use of in situ SXS techniques [24, 39, 334], as well as in situ scanning probe microscopy such as STM and atomic force microscopy (AFM) [23, 335, 336]. Very recently, several structure-sensitive in situ techniques have been applied to monitor the kinetics of UPD or dissolution. Finnefrock and coworkers [200] performed first SXS-measurements for Cu UPD on Pt(111), and Ataka and coworkers [199] demonstrated the usefulness of step-scan SEIRAS in exploring kinetic mechanisms of metal deposition processes.

The following sections are not meant to review the entire field of UPD. Emphasis will be rather given to three model systems, Cu UPD on Au(hkl), Pt(hkl) and Pb UPD on Ag(hkl) to describe, based on these examples, typical structural aspects and kinetic processes of 2D phase formation with low-dimensional metallic adlayers on foreign substrates.

3.3.4.2.2 Underpotential Deposition of Copper Ions on Au(hkl)

Cu–UPD on Au(111) in sulfuric acid – structural aspects The UPD of Cu on Au(111) in sulfuric acid electrolyte has been studied using (1) classical

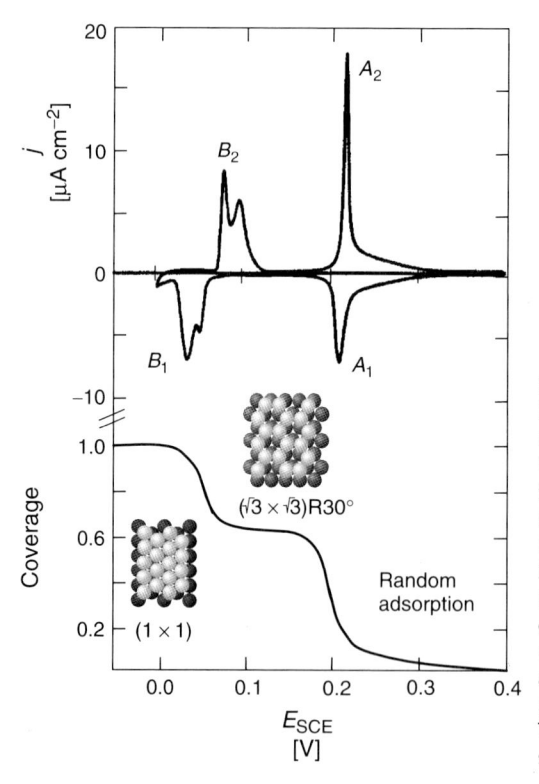

Fig. 15 Cyclic voltammogram for Cu UPD on a well-ordered Au(111) electrode in 0.1 M H_2SO_4 + 1 mM $CuSO_4$, scan rate 1 mV s^{-1} (reproduced from L. Kibler, Preparation and Characterization of Noble Metal Single-Crystal Electrodes. Copyright 2000 by International Society of Electrochemistry), and electrochemically derived Cu coverage (normalized charge referring to one complete Cu UPD ML) as a function of potential, determined by potential steps in the positive direction. The Cu adlayer structures are also shown (adapted from Ref. [360]).

electrochemical methods, such as voltammetry [320–322, 337, 338], chronocoulometry [339, 340], (2) in situ techniques such as STM [341–343], and AFM [344–346], FTIR [199, 347], X-ray absorption spectroscopy [348, 349], SXS [350, 351], and QCM [352, 353] and (3) ex situ UHV techniques (LEED, RHEED, AES) [242, 354, 355].

The cyclic voltammogram of Cu UPD on Au(111) shows two well-defined pairs of current peaks A_1/A_2 and B_1/B_2 corresponding to energetically different adsorption/desorption processes (Fig. 15) [320–322, 337]. In the first step (peak A_1), the transition between randomly adsorbed copper and (hydrogen) sulfate ions into an ordered layer of copper atoms (electrosorption valency $\gamma \sim 1.8$ [339, 340]) and coadsorbed sulfate ions takes place. The resulting $(\sqrt{3} \times \sqrt{3})R30°$ structure was first observed by ex situ LEED and RHEED experiments [354], and later confirmed by in situ SXS [350], STM [341–343], and AFM [344]. QCM [352, 353], chronocoulometric [339, 340], and FTIR-measurements [199, 347] proved the coadsorption of sulfate. The SXS study of Toney and coworkers [350] revealed that copper atoms form a commensurate honeycomb lattice (occupation of threefold hollow sites, 2/3 ML coverage), while sulfate species are adsorbed in the centers (1/3 ML coverage) above the plane of the copper atoms. Three oxygen atoms of each sulfate species are chemically bound to Cu atoms, and one points towards the electrolyte [350]. In the second step (peak B), a pseudomorphic monolayer of Cu (1×1) on Au(111) (occupation of threefold hollow sites) is formed [343]. EXAFS experiments indicate the disordered coadsorption of sulfate species on top of the copper monolayer [348, 349].

The coverage was estimated to 0.2 ML employing QCM [351, 352] and chronocoulometric [339, 340] measurements. These experimental results are corroborated by a recent theoretical work of Blum and coworkers [67, 356–358] and Rikvold and coworkers [197, 359].

Current transients and adsorption kinetics
The shape of the current peaks, the hysteresis in the peak positions between the cathodic and anodic potential sweeps (particularly for B_1 and B_2) and lattice gas simulations [197, 359] suggest that monolayer formation occurs via several first-order 2D phase transitions. Single potential step experiments revealed monotonously falling transients for peak A_1 (disordered Cu adlayer $\rightarrow$ $(\sqrt{3} \times \sqrt{3})R30°$) and rising transients for peaks A_2 $((\sqrt{3} \times \sqrt{3})R30° \rightarrow$ disordered Cu adlayer), B_1 $((\sqrt{3} \times \sqrt{3})R30° \rightarrow (1 \times 1))$, and B_2 $((1 \times 1) \rightarrow (\sqrt{3} \times \sqrt{3})R30°)$ (Fig. 16) [181]. If the potential step width is increased ("higher overpotentials"), the surface reaction becomes faster, and the current maxima are shifted to shorter times and higher current densities. At low overpotentials, the entire phase transition process can be slowed down into the timescale of seconds or minutes. The transients for peaks A_2, B_1, and B_2 could be modeled by instantaneous (hole) nucleation and 2D growth (Bewick–Fleischman–Thirsk (BFT) – theory) in parallel to a Langmuir-type adsorption/desorption process of species, with both proceeding at different electrode sites [181]. Models based on site-unspecific coupling of adsorption and nucleation [361] or solely on homogeneous adsorption processes with lateral attraction failed [362].

Additional support for the nucleation and growth controlled phase formation

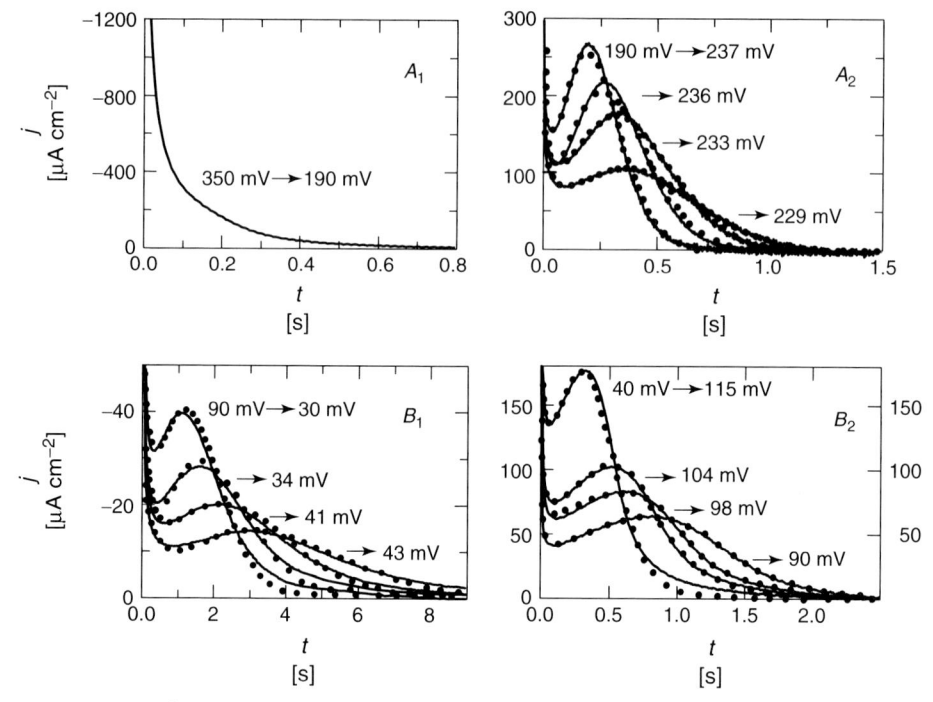

Fig. 16 Measured current transients (———) for the two-dimensional phase transitions during Cu UPD on Au(111) in 0.05 M H_2SO_4 + 1 mM $CuSO_4$ (peaks A_1, A_2, B_1, and B_2 in Fig. 15). The potential steps are indicated in the figure. The results of a least-square fit to a model combining (a) Langmuir-like adsorption according to Eq. (7) and instantaneous (hole) nucleation/2D growth represented by Eq. (33a) as two parallel processes are also shown (•). (Reprinted from Ref. [181], copyright 1994 by Elsevier Science Ltd.)

process of Cu UPD on Au(111) is given by a recent dynamic MC simulation of Rikvold and coworkers [359]. The authors employed a two-component lattice gas (Cu^{2+}, SO_4^{2-}). The microscopic dynamics includes adsorption/desorption and one-step lateral diffusion of both adsorbate species with energy barriers according to the Volmer–Erdey–Gruz theory. The simulated current profiles for the transitions $(\sqrt{3} \times \sqrt{3})R30° \rightarrow (1 \times 1)$ (B_1) and $(1 \times 1) \rightarrow (\sqrt{3} \times \sqrt{3})R30°$ (B_2) reflect one-step nucleation and growth mechanisms with the same overpotential-dependence as observed in the original experimental data. The transition of the $(\sqrt{3} \times \sqrt{3})R30°$ into the disordered, low-coverage phase (A_2) is described by a sequence of two first-order phase transitions, the first involving the desorption of 1/3 ML of Cu followed by the decay of a "long-lived" $(\sqrt{3} \times \sqrt{3})R30°$ phase of 1/3 ML Cu and 1/3 ML sulfate species [359]. Time-resolved SEIRAS experiments demonstrated that the first step may already involve the partial desorption of sulfate [199].

The reversed, qualitatively different monotonically decaying profiles of the transition between randomly adsorbed species and $(\sqrt{3} \times \sqrt{3})R30°$ (A_1) are modeled by an initial fast adsorption of Cu, which cause the collapse of the

t = 0.0 s

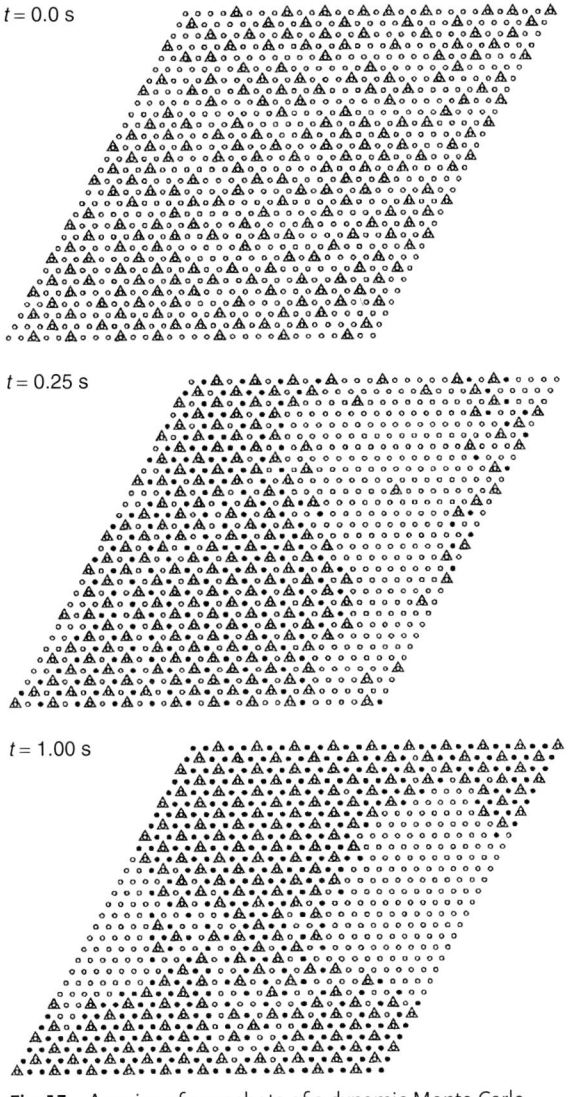

t = 0.25 s

t = 1.00 s

Fig. 17 A series of snapshots of a dynamic Monte-Carlo simulation after a negative-going potential step to 20 mV below the transition between the low coverage and mixed ($\sqrt{3} \times \sqrt{3}$) layer (peak A_1). Cu is represented by the filled circles (.), sulfate by a triangle ($\triangle$), and the unoccupied sites are open circles (.). After the step, a fraction of sulfate desorbs, but the remaining sulfate combines with the newly adsorbed copper to form a loose domain. With time, this domain fills in and grows. The MC-current response reproduces qualitatively the monotonically decreasing experimental transient in panel A_1, Fig. 16 (with permission by P. A. Rikvold, Ref. [359], copyright 1999, The Electrochemical Society).

already present sulfate into domains of a metastable ($\sqrt{3} \times \sqrt{3}$) phase with 1/3 ML Cu and sulfate species, respectively (Fig. 17). Later this fills with Cu forming the equilibrium phase with 2/3 ML Cu [359].

Time-resolved in situ STM experiments [360] performed in the so-called $x-t$ mode (one scanline is recorded as a function of time) revealed first structural details of these phase formation processes. Figure 18 indicates that

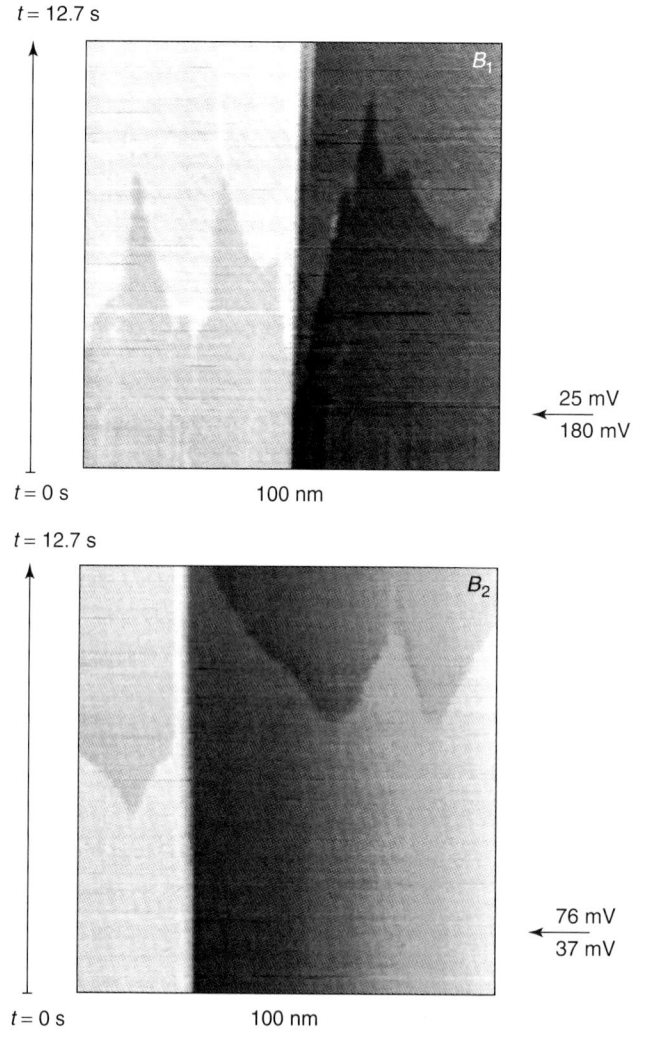

Fig. 18 (a) In situ $x-t$ STM images of the phase transition ($\sqrt{3} \times \sqrt{3}$)R30° → (1 × 1) for Cu UPD at B_1 and the reversed process at B_2 in the neighborhood of a monatomic step. The potential step regime and the corresponding time sequence are given in the figure (with permission by M. Hölzle, Ref. [360]).

the potential-induced transition ($\sqrt{3} \times \sqrt{3}$)R30° → (1 × 1) (peak B_1) starts at monatomic steps and on flat terraces. Nucleation on steps appears to be energetically preferred and somewhat faster. Growth proceeds with a constant rate. This observation is directly reflected in the corresponding current–potential and current–time curves. Measurements on well-oriented Au(111) electrodes with miscut angles ≪0.1° revealed a split peak B as a results of energetically well-separated nucleation on steps (preferentially more positive) and terraces [337, 342] (cf. Fig. 15). This separation vanishes and peak B broadens with miscut angles >1° (increasing step density). Simultaneously, the maximum of the current transients changes into monotonously falling traces indicating dominant contributions from nucleation at step edges coupled with 1D growth from these sites and/or increasing disorder of the adlayer.

STM-measurements of the reversed process (1 × 1) → ($\sqrt{3} \times \sqrt{3}$)R30° (peak B_2) clearly show that the terraces are preferential (hole) nucleation sites, and that the new phase grows linearly with time. Increasing the substrate step density modifies this mechanism; the maxima of the rising part of the corresponding current transients appear distorted. Similar trends were observed for the "desorption" process around A_2. Unfortunately, the STM and chronoamperometric experiments did not yet allow the derivation of a quantitative correlation between step density and the elementary steps of 2D phase formation of Cu–UPD.

Influence of anions Cyclic voltammetry reveals that the anions of the supporting electrolyte have a strong impact on the adlayer structure and on the deposition kinetics (cf. review in [24]). Coadsorption with Cu atoms was proven by in situ radiotracer [363], ex situ AES [364], in situ EXAFS, and SXS [348, 349, 365, 366] and chronocoulometry [339, 340, 367, 368]. In diffraction studies (LEED, RHEED [364, 369], SXS [365]) it was demonstrated that the anions are structure-determining for Cu UPD on Au(111). The adlayers at intermediate coverages are distinctly different for perchlorate, sulfate, chloride, and bromide solutions (Figs. 15, 19), and altogether do not resemble the pseudomorphic adlayer structure for Cu evaporated onto Au(111) in UHV [354]: (2.2 × 2.2) with ClO_4^-, ($\sqrt{3} \times \sqrt{3}$)R30° with SO_4^{2-}, (5 × 5) with Cl^- [364], and (4 × 4) with Br^- [365]. The strength of copper adsorption apparently decreases in the order $Cl^- > Br^- > SO_4^{2-}$ for equal concentrations of the anion in the bulk. Since Br^- is more strongly adsorbed an Au(hkl) than Cl^- indicates that the synergetic effect of the anion on the Cu adsorption does not correlate directly with the strength of anion adsorption [370].

The deposition kinetics is also strongly influenced by coadsorbed anions, as reflected in the current densities of the positive UPD peak (A_1). Cu deposition in ClO_4^- solution is slow, requiring up to 20 min to form a full monolayer. Cl^- shifts the onset of Cu deposition towards more positive values. At potentials negative of the peak at 0.265 V, a (5 × 5) structure was observed [371], which is attributed to Cl^- forming a bilayer with coadsorbed Cu (cf. discussion in [367]). Cu deposition in Br^- solution causes, around the sharp current peak at 0.260 V, the formation of a Cu-induced (4 × 4) Br^- adlayer, which fills up with increasing Cu coverage into a stochiometric CuBr pattern. At the second sharp current spike, around 0.05 V,

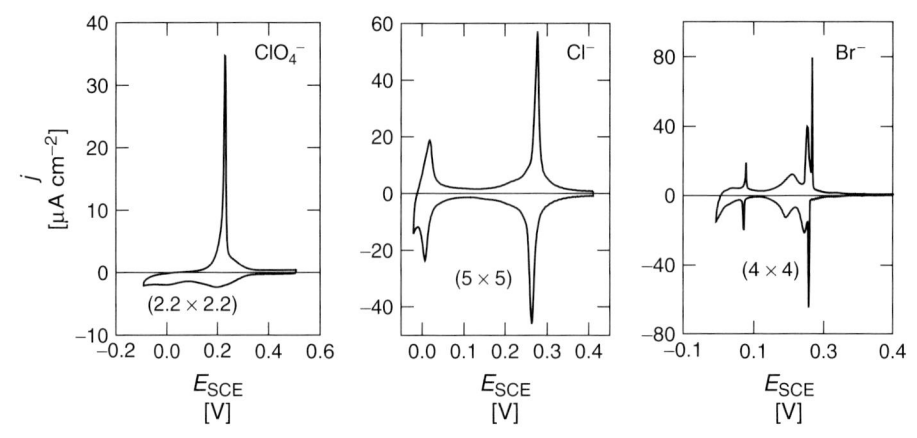

Fig. 19 Cyclic voltammograms for UPD of Cu on Au(111) in 0.1 M HClO$_4$, 0.05 M H$_2$SO$_4$, 0.1 M HClO$_4$ + 1 mM NaCl, and 0.1 M HClO$_4$ + 1 mM NaBr. The copper concentration was always 1 mM, scan rate 5 mV s^{-1}. The ordered structures in the medium coverage region are also indicated (with permission by M. Hölzle, Ref. [360]).

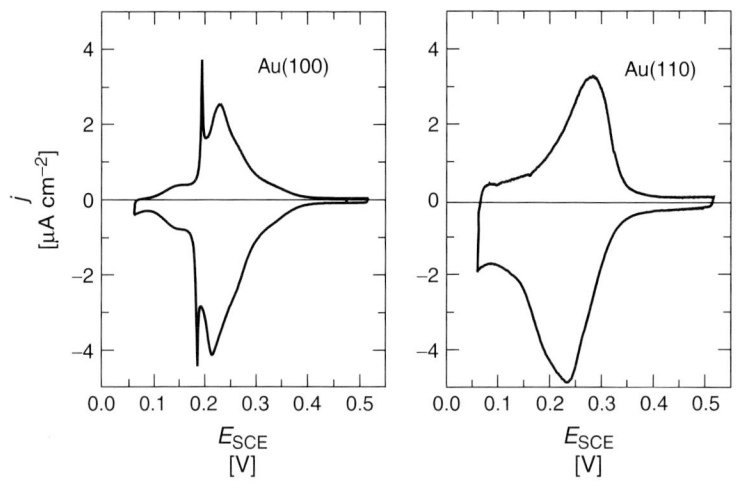

Fig. 20 Cyclic voltammograms for UPD of Cu on Au(100) and Au(110) in 0.1 M H$_2$SO$_4$ + 1 mM Cu^{2+}, scan rate 1 mV s^{-1}.

a phase transition into a (1 × 1) Cu layer with a (4 × 4) Br$^-$ structure on top of it takes place. Both transitions are quite reversible [365].

Role of substrate crystallography on Cu–UPD Figures 15 and 20 illustrate the influence of the substrate crystallographic orientation on Cu–UPD on Au(hkl) in sulfuric acid solution. The Cu ML on Au(111) is formed in two energetically distinctly different steps in 0.35 V > E > 0.05 V, with the second peak being split for quasi-perfect Au(111) surfaces into contributions from step edges and terraces (Fig. 15). The formation of the pseudomorphic Cu ML on Au(100) proceeds in two steps in 0.40 V > E > 0.17 V. The major charge contribution is consumed in the broad, more positive feature. No ordered adlayer has been reported in this potential region. At potentials negative of the sharp current spike, an ordered Cu–UPD (1 × 1) ML was found [372]. The overall deposition process appears to be strongly influenced by the presence of monatomically high gold islands, created during the lifting of the Au(100)-(hex) reconstruction [373]. On Au(110) the Cu ML is formed in a broad, single peak adsorption process [374].

3.3.4.2.3 Underpotential Deposition of Copper Ions on Pt(hkl)

The UPD of Cu on Pt(hkl) has been studied by electrochemical techniques [375–386], electroreflectance [375], radioactive labeling [387, 388], IR-spectroscopy [389, 390], electrochemical quartz crystal microbalance (EQCM) [391], in situ STM [389, 392–396], and a variety of X-ray techniques [200, 385, 397–401], as well as ex situ UHV-methods such as LEED, AES, and XPS [389, 399, 402–406].

Figure 21 shows a slow scan voltammogram of Cu–UPD on a high-quality

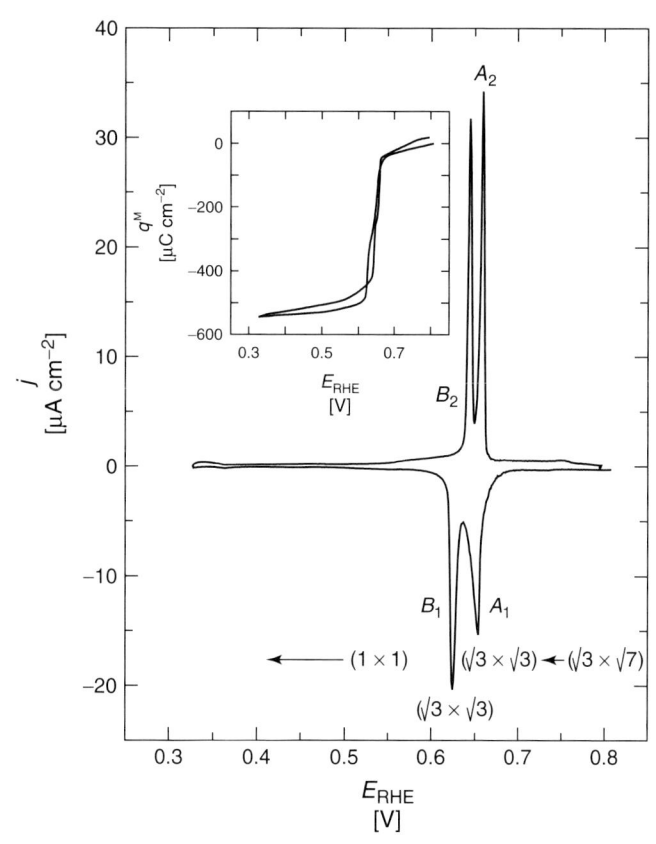

Fig. 21 UPD of Cu on a Pt(111) bead electrode in 1 mM Cu^{2+} + 0.1 M H_2SO_4, sweep rate 1 mV s^{-1}. The inset shows the integrated charge density for the positive and negative going potential scans, respectively [407].

Pt(111) electrode in 0.1 M H_2SO_4 [407]. Two pairs of narrow deposition (A_1, B_1) and dissolution (A_2, B_2) peaks, which are separated by a small, but distinct hysteresis, are developed clearly. Integrating the current between 0.70 V and 0.55 V (RHE) gives a monolayer charge of approximately 500 µC cm^{-2}. The mechanism of Cu deposition on Pt(111) in the presence of sulfuric acid is described as follows: Bisulfate ions adsorbed on the Cu-free Pt(111) surface form a ($\sqrt{3} \times \sqrt{7}$) structure [301] which gradually transforms into ($\sqrt{3} \times \sqrt{3}$)R30° [389, 390, 402]. Upon passing the first UPD peak A_1, sulfate ions and $Cu^{\delta+}$ coadsorb in a honeycomb lattice (2/3 ML of Cu, threefold hollow sites). The resulting ($\sqrt{3} \times \sqrt{3}$)R30° structure was observed by in situ STM [390, 395, 396] and SXS [408], and is in agreement with ex situ XPS and LEED investigations [389, 390, 402]. During the second UPD peak, B_2, the sulfate ions are displaced from the honeycomb center, and a (1 × 1) pseudomorphic ML is formed with the Cu residing in the threefold hollow sites of the Pt surface, on top of which the sulfate ions arrange in a ($\sqrt{3} \times \sqrt{7}$) pattern, $\theta = 0.20$ [389, 396, 397.] The reverse processes take place during the dissolution of the UPD phase upon positive potential excursion.

Systematic studies with stepped Pt(111)_s electrodes having (111) terraces and monatomic (110) [386, 407] or (100) [379, 380] steps demonstrate that the fingerprint region of Cu–UPD depends strongly on terrace size and step density. At low coverages, Cu adsorbs preferentially on steps [382, 383, 407]. Figure 22 illustrates, as an example, typical current versus potential profiles for the

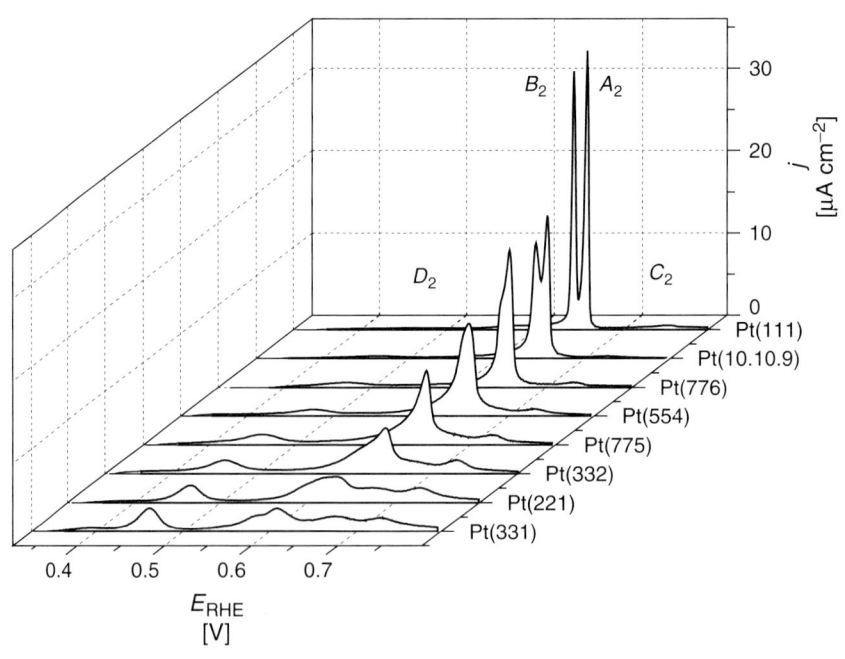

Fig. 22 Current versus potential profiles of the stripping of a Cu–UPD ML deposited on stepped Pt(n n n-2) electrodes as recorded during a positive potential scan from 0.35 V to 0.80 V, 1 mV s^{-1}, with 1 mM Cu^{2+} in 0.1 M H_2SO_4 [407].

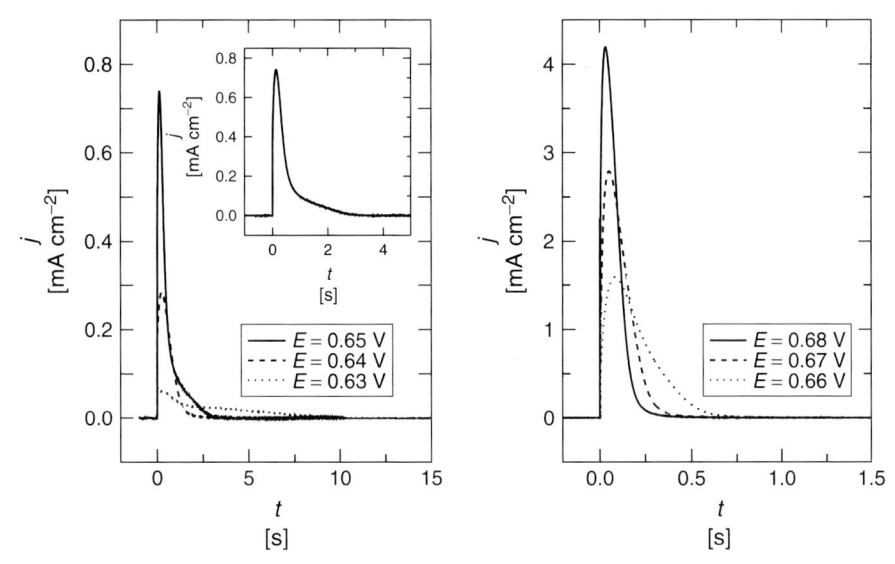

Fig. 23 Current transients of the dissolution of a Cu UPD ML on Pt(111) in 1 mM Cu^{2+} + 0.1 M H_2SO_4, obtained when stepping the potential from $E_1 = 0.50$ V to various final potentials as indicated in the figure. The transients with final potentials lower than 0.67 V could be modeled by assuming two successive hole nucleation processes according to an exponential law coupled with surface diffusion-controlled growth (cf. Eq. (30) and Eq. (33)). The inset shows, as an example, the fit (——) for the experimental transient (- - - -) $E_1 = 0.50$ V → $E_2 = 0.65$ V [407].

dissolution of the Cu–UPD phase during a slow positive potential scan (1 mV s^{-1}). The narrow peaks A_2 and B_2 representing the dissolution from terrace sites decrease and merge in a broader feature with decreasing terrace width (increasing step density). Simultaneously, two additional current peaks, D_2 and C_2, develop around 0.45 V and 0.73 V, which correspond to sulfate desorption on copper-decorated monatomic steps and, at more positive potentials than B_2 and A_2, to the Cu dissolution from previously decorated step sites [382]. The i–t transients of Cu–UPD dissolution in sulfuric acid could be modeled within the BFT-theory by two successive (Pt(111), low overvoltages) or one single (higher index, stepped electrodes) hole nucleation processes according to an exponential law coupled with surface diffusion-controlled growth (Fig. 23). Dissolution according to a 1D line-by-line controlled mechanism was recently proposed for stepped Pt-electrodes with a terrace width <6 atoms. No detailed analysis of the kinetic mechanism of the slow deposition process has been reported.

Cu–UPD on Pt(hkl) depends strongly on substrate crystallography and anions of the supporting electrolyte [378, 386, 402, 409]. On defect-free, hydrogen-cooled Pt(100)-(1 × 1) electrodes a pseudomorphic adlayer of fully discharged Cu-ions is formed (dissolved) in sulfuric acid in one single step consuming 446 µC cm^{-2} [404, 410, 411]. The stripping process is controlled by (hole) nucleation and 2D growth [407]. Cu deposition onto Pt(110)-(1 × 1) surfaces takes place in two distinct steps giving rise to a (2 × 1) and (1 × 1) overlayer periodicity [394, 403]. Cu–UPD adlayers on Pt(hkl) are stabilized by coadsorbed anions, which shift the onset of deposition towards more negative potentials according to the following sequence: $F^- < ClO_4^- < (H)SO_4^{(2)-} < Cl^- < Br^- < I^-$ [390]. Rather complex bilayer structures have been reported (Fig. 24, cf. summary in [22]). The voltammetric profiles represent two-stage processes. At potentials above the first UPD peak, chloride ions are adsorbed in a disordered phase. During the first UPD-step (disorder–order phase transition) a Cu ML is formed, which is covered by an incommensurate, close to (4×4) Cl^--adlayer. The latter transforms during the second UPD peak at negative potentials, and on top of an epitaxial (1×1) Cu adlayer, into a $(2 \times \sqrt{3})$ phase accompanied by partial desorption of Cl^- [402, 405, 412, 413]. The extend of Cu and Cl^- codeposition on Pt(111)_s-electrodes depends critically on terrace width and step density [386]. Pioneering time-resolved SXS experiments on the kinetics of phase transitions between these UPD phases were reported by Finnefrock and coworkers [200]: At the initial potential $E_1 = 0.20$ V the epitaxial $Cu(1 \times 1)$ layer is present at the surface. At the final potential $E_2 = 0.35$ V (past B2) the incommensurate CuCl structure is formed (Fig. 25a). The corresponding transient current, which is related to the dissolution of Cu and the reorganization of the Cl- adlayer, decays exponentially to zero at times well below 1 s. The X-ray response at the scattering position (0.765, 0, 0.5), which is sensitive to the 2D order of the Cu species within the UPD phase, is slower and completely different in shape (Fig. 25c). The sigmoidal SXS-transients were attributed to a 2D nucleation-and-growth process. These experimental results indicate that the processes of Cu and anion adsorption/desorption and the formation of the ordered adlayer occur at rather different timescales. This conclusion is not accessible from the analysis of just the electrochemical current response! The reversed process,

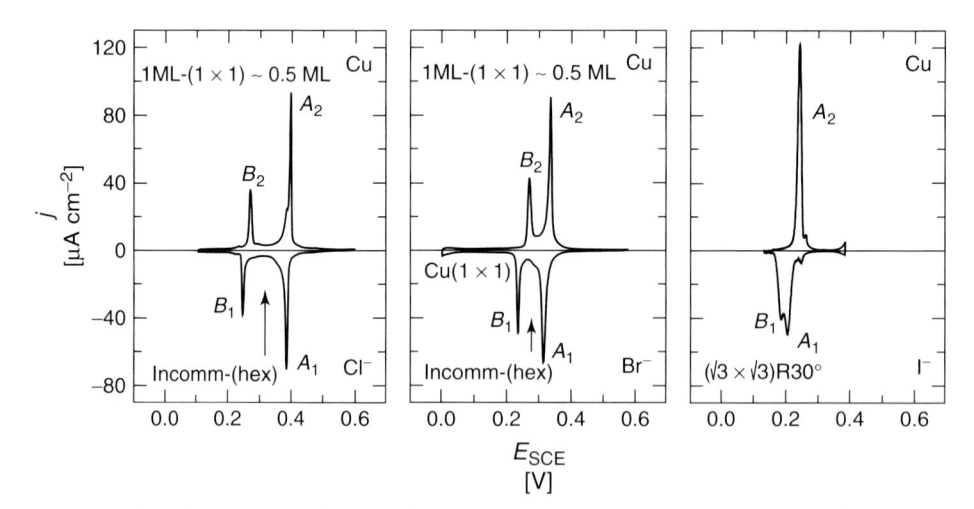

Fig. 24 Cyclic voltammograms for UPD of Cu on Pt(111) in 0.05 M H_2SO_4 + 1 mM NaCl, 0.05 M H_2SO_4 + 0.1 mM NaBr, and 0.05 M H_2SO_4 + 0.1 mM NaI (with permission by M. Hölzle, Ref. [360]). The halide adlayer structures at intermediate Cu coverages between the deposition peaks A_1 and B_1, and the Cu coverage within the respective bilayers are also indicated.

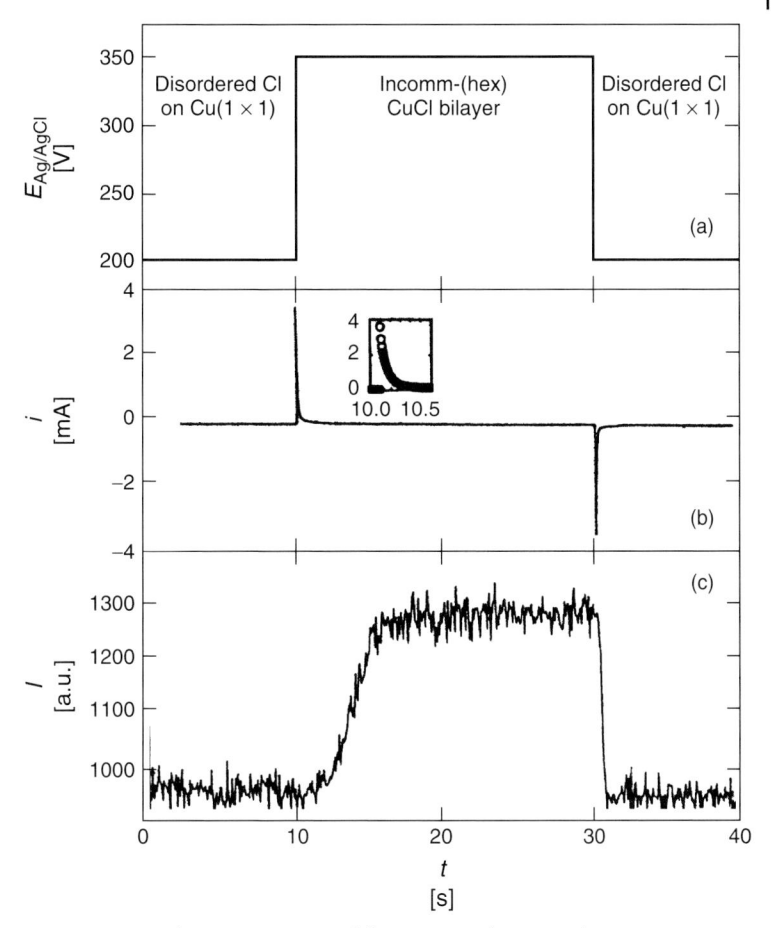

Fig. 25 Potential step experiment of the transition between the commensurate Cu(1 × 1) and the incommensurate CuCl UPD adlayer on Pt(111) in 0.1 M H_2SO_4 + 1.0 mM Cu^{2+}: (a) potential perturbation, (b) current transients (the inset shows a magnified region of the entire transient) and (c) time dependence of the scattered X-ray intensity at (0.765, 0, 0.5), which represents a characteristic diffraction rod of the incommensurate CuCl bilayer (reprinted from Ref. [200], copyright 1998 by American Physical Society).

the formation of the epitaxial (1 × 1) Cu adlayer from the incommensurate CuCl bilayer is reflected in the electrochemical and the SXS response as a single decaying function with the same time constant.

Cu–UPD on Pt(111) electrodes in the presence of bromide anions proceeds also in two steps: An incommensurate bromide adlayer transforms first into a CuBr-bilayer with Br^- arranged in an incommensurate hexagonal structure aligned along the (1,0) surface direction. At potentials negative of the second UPD peak, a pseudomorphic Cu (1 × 1) structure covered by a disordered bromide layer is formed [365, 402, 413]. The structural transitions of Cu–UPD in I^--containing electrolyte are not yet fully understood [22].

3.3.4.2.4 Underpotential Deposition of Lead on Ag(hkl)
In comparison to Cu UPD ($d = 2.560\,\text{Å}$) on Au(hkl) ($d = 2.885\,\text{Å}$) or Pt(hkl) ($d = 2.450\,\text{Å}$), lead ($d = 3.490\,\text{Å}$) assumes a larger lattice constant than the substrate material Ag(hkl) ($d = 2.885\,\text{Å}$).

The Pb–UPD on low-index silver single crystals has been studied extensively, employing electrochemical techniques [48–50, 130, 414–424] electroreflectance [420], SHG [425], STM [2, 130, 426–428], EXAFS and SXS [429–431], as well as ex situ UHV-methods such as LEED and AES [420, 432].

Figure 26 shows typical cyclic voltammograms of Pb–UPD on Ag(111) in three electrolytes containing perchlorate, acetate, or citrate ions [360, 422]. Three pairs of current peaks with a small, but distinct hysteresis between the negative and the positive potential scans developed in the presence of weakly specifically adsorbed perchlorate [48–50, 415–419, 421, 422]. The electrosorption valency of lead, γ, is 2, which indicates that the deposited metal is completely discharged and no cosorption with the anion takes place [417]. Comparison of Ag(111)-substrates prepared by chemical etching [415] with electrolytically grown quasi-perfect electrodes [419] or with thermally annealed films [130, 424] indicates that the first and third pair of peaks represent processes on defect sites, while the middle peak is attributed to terrace sites. Several authors proposed originally, on the basis of electrochemical studies, the 1D decoration of monatomic steps followed subsequently by the 2D phase formation of lead on terraces [2, 48–50, 415, 418]. The presence of more strongly adsorbing anions such as citrate, acetate [418, 421–423], or halides [433] blocks the surface defect sites. The Pb–UPD of these systems is represented by just one characteristic pair of sharp current peaks, and the onset of metal deposition shifts towards more negative potentials according to the following sequence: $\text{I}^- < \text{Br}^- < \text{Cl}^- < \text{citrate} < \text{acetate} < \text{ClO}_4^-$.

Potential step experiments of the Pb ML formation in ClO_4^- and acetate electrolyte exhibit an initial fast decay due to double layer charging followed by a plateau region and a subsequent accelerating decay to zero [130, 360, 421] (Fig. 26a). Initially, these transients were attributed to a homogeneous Frumkin-type phase formation process, involving the discharge of lead and bulk diffusion as rate-determining steps [416]. Recent electrochemical and in situ STM experiments with high-quality stepped Ag(111)-surfaces [2, 130] or thermally, in a hydrogen flame-annealed electrodes [360], indicate that the plateau region is most pronounced and does not represent a classical adsorption process. Instead, Lorenz and coworkers revised their original interpretation and described the formation transients of a Pb–UPD ML in ClO_4^- and acetate electrolyte with a model involving the lateral growth of a 2D condensed ML, which starts exclusively at monatomic steps [130, 336]. The current decay is critically dependent on step density and distribution. Formation transients of Pb UPD adlayers in solution with more strongly adsorbing anions, which block step sites (cf. citrate in [48–50, 419, 421]) exhibit characteristic bell-shaped morphology, which was quantitatively modeled by instantaneous hole nucleation in combination with a 2D growth process with constant [418, 421] or decreasing rate [360]. The critical nuclei were thought to be created upon replacement of the prior, specifically adsorbed (citrate) anions by lead species [418].

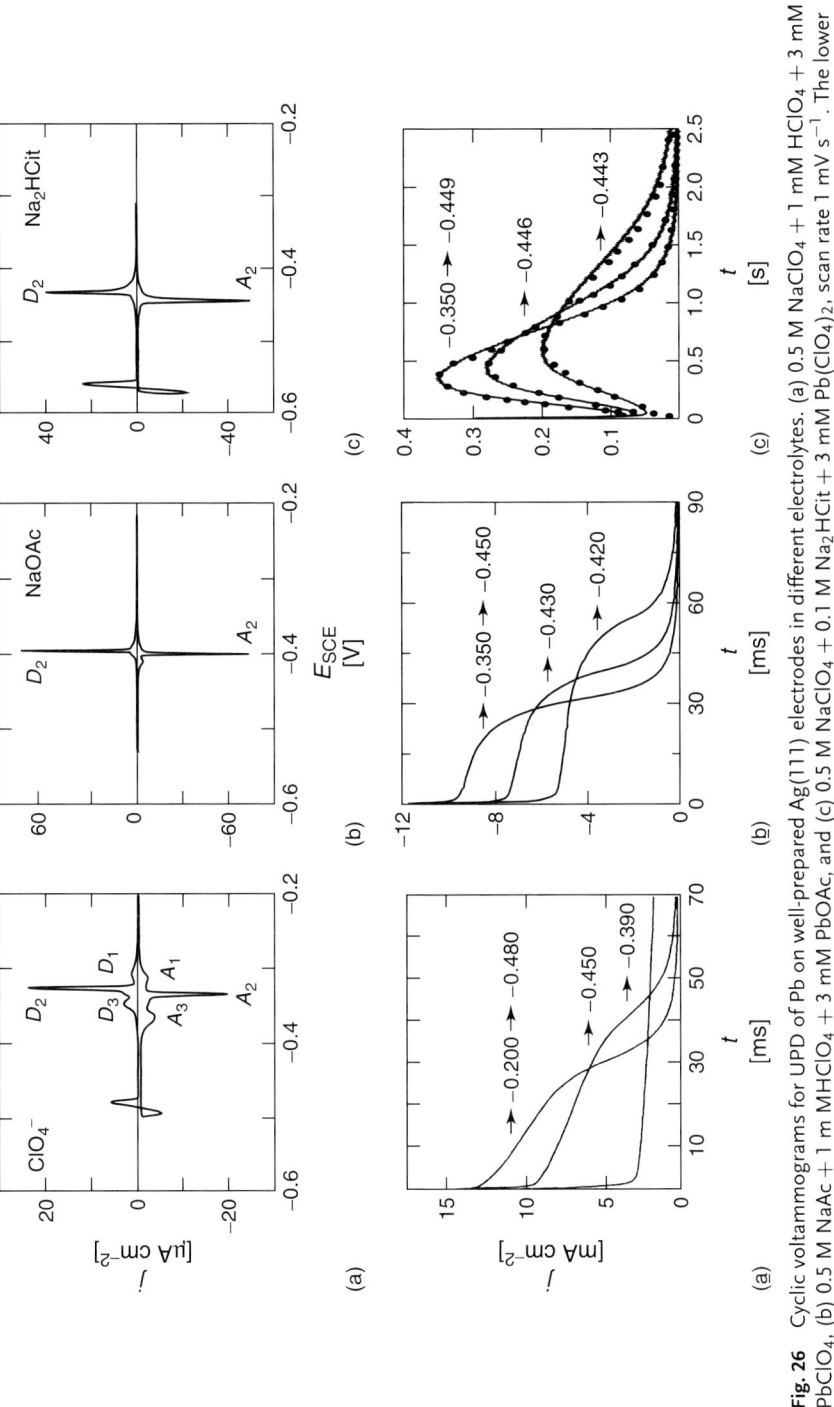

Fig. 26 Cyclic voltammograms for UPD of Pb on well-prepared Ag(111) electrodes in different electrolytes. (a) 0.5 M NaClO$_4$ + 1 mM HClO$_4$ + 3 mM PbClO$_4$, (b) 0.5 M NaAc + 1 m MHClO$_4$ + 3 mM PbOAc, and (c) 0.5 M NaClO$_4$ + 0.1 M Na$_2$HCit + 3 mM Pb(ClO$_4$)$_2$; scan rate 1 mV s^{-1}. The lower panels a, b, and c show for each electrolyte typical current transients of formation of a complete UPD Pb ML. The transients in (a) have been reproduced from Ref. [130], copyright 1997, NRC Canada. The transients of Pb UPD in citrate-containing solution (panel c) solid lines) were fitted to a model based on instantaneous hole nucleation combined with a time-dependent rate of growth (with permission by M. Hölzle, Ref. [360]).

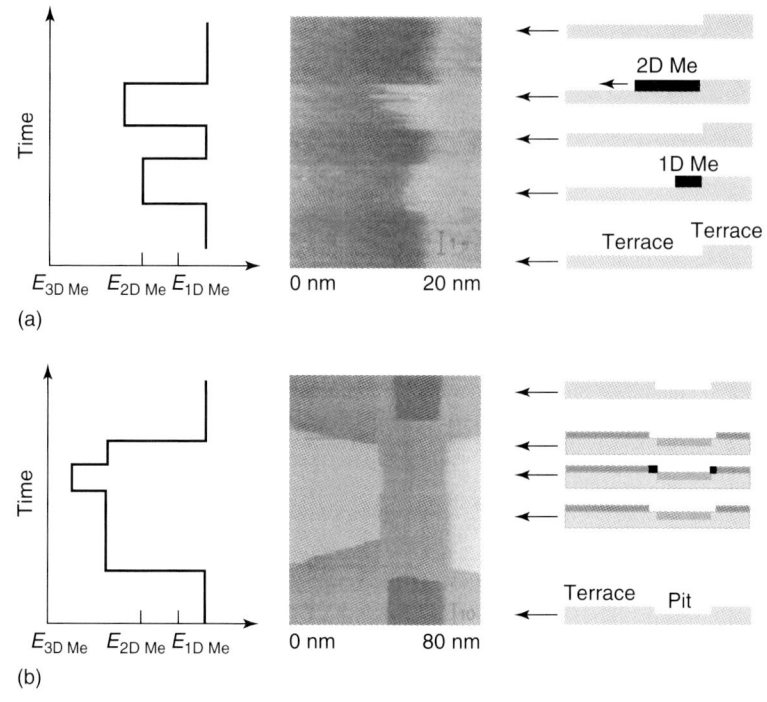

Fig. 27 (a) In situ STM line scan plot showing the step decoration by a 1D Pb phase and the initial stage of 2D phase formation on a stepped Ag(111) substrate and (b) line scans of formation of a 2D Pb UPD phase near a terrace and a monatomic deep pit for Pb UPD on Ag(111) in 4 mM Pb(ClO$_4$)$_2$ + 0.01 M HClO$_4$ (by courtesy of E. Amman from Ref. [434]).

Combined electrochemical and in situ STM experiments gave clear evidence for the stepwise formation of low-dimensional adlayer phases in the system Ag(111)/Pb^{2+}/ClO$_4^-$, H$^+$ [427, 434]. Figure 27(a) shows a series of potential-dependent line scans representing the initial stage of Pb–UPD on Ag(111). At potentials just past the first current peak in the voltammogram of Fig. 26, the decoration of monatomic steps occurs. The width of the decoration extends to about 2 nm, which is in dynamic equilibrium with an expanded 2D Pb phase at the terrace. The 1D phase either dissolves upon returning the potential towards more positive values, or grows further onto the lower terrace into a 2D-condensed lead phase at more negative potentials around the main current peak. The formation of the 2D Pb-ML represents a first-order phase transition process, as reflected by the hysteresis between the peaks A_2/D_2 in the voltammogram. The growth process is illustrated in Fig. 27(b). One may notice that a restricted region around the upper part of the step edge having a width of around 2 nm remains uncovered. Nucleation on the flat terrace (e.g. on nondefect sites) is delayed and starts only at more negative potentials corresponding to higher supersaturation. At potentials around the most negative voltammetric peak, the completion of the Pb–UPD phase on the upper terrace takes place. The dissolution of the metal adlayer proceeds in the reverse regime.

The sequence of formation/dissolution of the experimentally observed 1D and 2D Pb-phases is in agreement with theoretical predictions based on a Nernst-type equation [2]:

$$E_{Pb,nD} = E^o_{Pb,3D} + \frac{RT}{zF} \ln\left(\frac{a_{Pb2+}}{a_{Pb,nD}}\right) \quad (43)$$

with the dimensionality n, $E^o_{Pb,3D}$ the formal potential of Pb in the bulk phase, and a_{Pb2+} the Pb activity in the electrolyte. The activity of the condensed metal phases shift towards lower values with decreasing dimensionality due to stronger substrate–adsorbate interactions, and in consequence, the corresponding equilibrium potentials are shifted in the positive direction [2]. Extended polarization within the UPD-range may cause surface alloying [427].

The completed Pb UPD is metallic, and represents an incommensurate, hexagonal ML that is compressed compared with the bulk metal by 0.1–3.2%, and rotated from the substrate $(01\bar{1})$-direction by $\pm 4.5°$ [426, 427, 429–431]. The rotation of the adlayer with respect to the substrate lattice gives rise to a characteristic Moire pattern as observed in several in situ STM studies [360, 426, 427] (Fig. 28). The interaction between solvent molecules and the Pb adatoms does not influence the structure of the complete ML deposited in ClO_4^- or acetate-containing electrolyte, since the UPD phase is essentially identical to that of vapor-deposited Pb on Ag(111) at full coverage [420, 435]. The monolayer compression in the vacuum experiment (1–2%) is slightly less than for

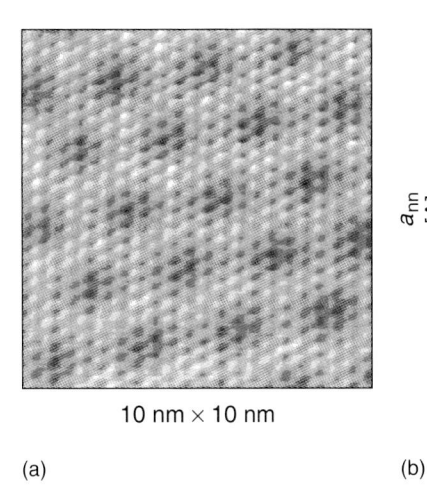

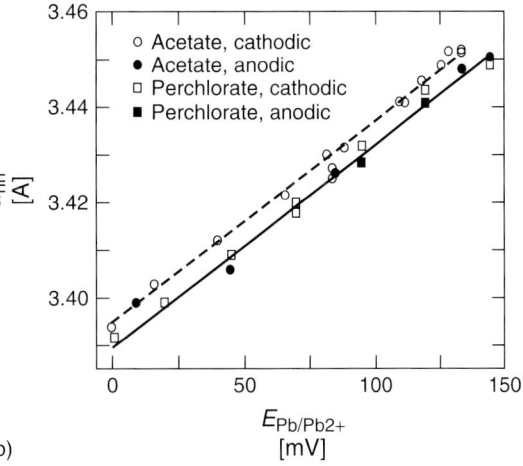

(a) 10 nm × 10 nm (b)

Fig. 28 (a) Atomic resolution STM image of a Pb UPD ML in citrate-containing electrolyte at potentials past the main deposition peak A_2 (with permission by M. Hölzle, Ref. [360]). (b) Dependence of the Pb ML near-neighbor distance a_{nn} on the electrode potential E for Pb UPD on Ag(111). The near-neighbor spacing for bulk Pb is 3.501 Å, well above the range of the plot. The solid and dashed lines are linear least-squares fits to the data for perchlorate and acetate, respectively. These lines have slopes of 0.424 ± 0.01 Å/V (perchlorate) and 0.420 ± 0.01 Å/V (acetate); their offsets are 3.390 Å (perchlorate) and 3.395 Å (acetate). (Reprinted from Ref. [431], copyright 1995 by The American Chemical Society.)

Pb UPD, probably caused by the stronger Pb–Ag bonds compared to Pb–Pb [319], although the rotational angle ϕ is about the same.

The equilibrium change of the ML Pb–Pb distance, $a_{\text{Pb-Pb}}$ with the applied electrode potential provides an estimate of the 2D compressibility of the lead ML

$$\kappa_{2D} = \frac{a_{\text{Pb-Pb}}\sqrt{3}}{ze_0}\left(\frac{\partial a_{\text{Pb-Pb}}}{\partial E}\right)_T \quad (44)$$

The experimental value $\kappa_{2D} = (1.25 \pm 0.05)$ Å^2/eV is in qualitative agreement with the results of a simplified model based on a 2D free electron gas [334, 431].

The rotational epitaxy angle estimated by Toney and coworkers from in situ SXS investigations (4.5°) is only slightly smaller than the value predicted by the model of Novaco and McTague [71, 72]. The adlayer orientation with the lowest energy is not the high symmetry direction. Instead, the adlayer prefers to rotate to nonsymmetry directions, taking advantage of the lowest energy shear wave.

A comparable discussion of Pb-UPD on the other low-index phases of silver is summarized in [2, 22].

3.3.4.3 Phase Transitions in Organic Monolayers

3.3.4.3.1 General Aspects
Organic monolayers on well-defined metal substrates may be obtained in various ways. Typical strategies are molecular beam epitaxy (MBE) [436], so-called "self-assembled monolayers" (SAM) [437] or LB films (LB) [438, 439]. SAM's are molecular assemblies formed by the spontaneous interaction of a surfactant with a solid or liquid substrate. Examples are thiols on gold or silver [437, 440, 441] as well as alkyltrichlorosilanes on oxide surfaces [442]. The order of these 2D systems is produced by a chemical reaction at the interface as the system approaches equilibrium. Unfortunately, they appear to contain a significant number of pinholes and other defects [443–445]. Constrains exist also for monolayers based on the LB technique. These studies are limited to the choice of molecules with a very low solubility in the subphase [438, 439].

Alternatively, *molecular* and *ionic* monolayers can also be obtained on conducting surfaces in an electrochemical environment [3, 15, 16, 20, 143]. This approach offers the advantage that formation and properties of a wide variety of adlayers can be controlled as required by the applied electrode (substrate) potential and subsequently characterized by structure-sensitive in situ techniques in real space and real time. For these reasons, potentiostatically or galvanostatically generated monolayers on well-defined metal electrodes have become attractive model systems and provide an important testing ground for fundamental issues in 2D physics and chemistry, such as phase transitions in adlayers and substrates surfaces [3, 7, 15, 16, 20, 143, 446]. Since the pioneering work of Frumkin [447, 448], many "equilibrium" and "dynamic" adsorption studies at metal–electrolyte interfaces have been performed with mercury [3, 15, 16, 20, 449], gallium [18], and several amalgams [19], or low-melting-point electrodes of sp-metals, such as Bi, Pb, Sn, or Zn, onto which organic molecules are usually weakly adsorbed [449, 450]. The potentiostatic formation of 2D condensed monolayers was first reported by Lorenz [451] and Vetterl [45] who found an unusual capacitance versus potential hysteresis for saturated aqueous solutions of nanoic acid and the so-called "capacitance pits" for various purine and pyrimidine bases. Since then, many other systems have been reported and quantitatively analyzed. Examples include carboxylic acids, camphor and related compounds, purine and pyrimidine and their derivatives, pyridine and bipyridines, quinolines, coumarin, thiourea, tetraalkyl-ammonium salts, and so on. The corresponding literature is reviewed in [3, 15, 16, 20, 143].

Despite the rather detailed phenomenological knowledge on the 2D phase formation in adlayers, as obtained at the mercury–electrolyte interface, further progress has been hampered by the lack of information on the structure and molecular mechanisms of film formation. New experimental and theoretical perspectives became available when single-crystal electrodes were employed as substrate materials. This step offers the advantage to combine classical electrochemical experiments with the power of structure sensitive in situ techniques, such as scanning probe microscopies (STM, AFM) [23, 452, 453], vibrational spectroscopies (IRAS [454], SEIRAS [28], SFG [30],...), electroreflectance [455], and surface plasmon microscopy [456], neutron and surface X-ray scattering [457, 458]. The formation of condensed organic films on solid electrodes was suggested by Batrakov in 1974 for the adsorption of camphor on Zn(0001) [459]. Few other examples have been reported in the literature during the last decade, mostly based on voltammetric and capacitance measurements: pyridine on Ag(210) [460], thymine on Cd(0001) [461], Au(100) [462], Au(111) [463], and Ag(hkl) [464], coumarin on Au(111) and Au(100)-(hex) [465], uridine on Au(hkl) [220, 466], uracil and camphor on Au(hkl) and Ag(hkl) [183, 193, 219, 221, 467–471]. Quantitative thermodynamic studies were carried out for pyridine [460] and several derivatives of uracil [469]. The phenomenological kinetics of film formation and dissolution, as monitored by $i-t$, q^M-t, or $C-t$ transients after single or multiple potential step protocols, have been described by models that involve (hole) nucleation and growth mechanism [183, 219, 221, 466]. The results of these "macroscopic" electrochemical experiments are still scarcely complemented by structural studies, such as IR- or Raman Spectroscopies or the use of UHV techniques after emersion of the electrode [454, 463, 472–476]. Remarkable progress in developing a "true" atomistic/molecular picture of potential-induced 2D organic phase formation on adsorbate-modified single-crystal electrodes was achieved by the application of in situ STM and AFM. Direct evidence for the formation of 2D long-range ordered structures on Au(111) has been reported for several purine and pyrimidine bases [201, 219, 454, 463, 477–480], phenol [481], pyridine [482, 483], 2,2'- and 4,4'-bipyridine [484–490], phenanthroline [491, 492], octylthiole [493], tetramethyl-thiourea [494, 495], cysteine [496], mercaptopyridine [497], and benzenethiole [498]. Tao and coworkers [484, 491] and Wandlowski and coworkers [201, 478, 479, 486, 487] have demonstrated that the dynamics of these phase formation processes can be studied successfully as a function of potential and temperature, even at a nanoscale level. Itaya and coworkers have employed iodine-modified Au(111)- and Ag(111)-electrodes to image self-organized arrays of crystal violet, methylpyridinium-phenylendivinylene and porphyrine derivatives [499–501]. The same group also reported 2D ordered structures of benzene, naphthalene, and anthracene on Pt(111), Rh(111), and Cu(111) electrodes [502, 503]. These results are complemented by a few structural studies with physisorbed films of purine bases and porphyrine derivatives at the HOPG(0001)–aqueous electrolyte interface [210, 480, 504–508]. Commensurate 2D ordered organic adlayers were also found on surfaces with quadratic symmetry, such as uracil and 2,2′-bipyridine on Au(100)-(1 × 1) [479, 487], and benzene and pyridine on Cu(100) [509].

Basic ordering principles of the above-mentioned monolayers appear to be (1) the ability to create strong and intermolecular hydrogen bonds between adjacent molecules [463, 479, 507], (2) packing constrains, molecular geometry, and dipole forces [510], (3) ion pairing [511, 512], (4) the formation of interfacial stacks due to π-electron attraction and London dispersion forces [453, 479, 484, 491], (5) hydrophobic interactions [513], as well as (6) substrate-adsorbate coordination chemistry [20, 463, 493]. Complementary in situ STM studies with uracil and 2,2′-bipyridine, potentiostatically deposited on hexagonal Au(111) or quadratic Au(100)-(1 × 1) electrodes, demonstrate that the structure of the adsorbed monolayers, and in particular the registry between the adlayer and the substrate, is affected by the symmetry of the substrate (nature of the metal, crystallographic orientation), the adsorbate–adsorbate interactions and the corrugations in the adsorbate–substrate interaction potential [478, 479, 486, 487].

The influence of the solvent on the formation and stability of 2D condensed organic phases is rather unexplored and controversial [3, 16]. Capacitance measurements of ordered 2-thiouracil films at mercury–acetonitrile interfaces showed that the addition of small amounts of water shifts the stability range and saturation capacitance of the solidlike organic phase [514]. While camphor and analogs exclude interfacial water, recent in situ SEIRAS studies on structural transitions of 4,4′-bipyridine on Au(111) demonstrate that the ordered organic adlayer is stabilized by hydrogen-bonded coadsorbed water species [476].

3.3.4.3.2 Thermodynamic Aspects The adsorption of organic molecules on ideally polarizable electrode–electrolyte interfaces is based on the Gibbs equation, which relates, at constant temperature and pressure, the specific surface work γ to the electrode potential E and the bulk activity (concentration) a_i of adsorbate i.

$$\partial \gamma = -q^M \partial E - RT \sum_i \Gamma_i \partial \ln a_i \quad (45)$$

Γ_i represents the surface excess of the component i, relative to a reference species, usually the solvent. Equation (45) is exact at liquid electrodes [515], and represents a first-order approximation for solid electrodes assuming that the elastic surface strain terms $(\partial \varepsilon / \partial E)_{T,p,a_i}$ and $(\partial \varepsilon / \partial \ln a)_{T,p,E}$ are negligible (cf. discussion in Ref. [516, 517]). The specific surface work (interfacial tension in case of liquid electrodes), charge density, and differential capacitance are related by

$$q^M = -\left(\frac{\partial \gamma}{\partial E}\right)_{a_i} \quad \text{and} \quad (46)$$

$$C = -\left(\frac{\partial^2 \gamma}{\partial E^2}\right)_{a_i} = \left(\frac{\partial q^M}{\partial E}\right)_{a_i} \quad (47)$$

The adsorption of neutral molecules at electrified interfaces is characterized by a smooth variation of γ, q^M, C and Γ_i as a function of electrode potential or temperature (dotted and dashed curve in Fig. 29) [449, 518]. The relation between surface excess Γ_i (or coverage $\theta_i = \Gamma_i / \Gamma_{im}$ with Γ_{im} as the maximum surface excess

Fig. 29 Schematic representation of the adsorption in the absence (dotted-and-dashed line) and in the presence (———) of two-dimensional phase transitions: interfacial tension γ, differential capacitance C, charge density q^M, and surface excess Γ as a function of the electrode potential E. The dashed lines represent the "adsorbate-free" base electrolyte.

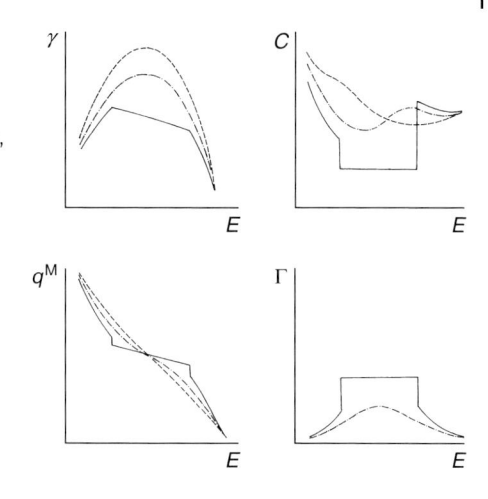

of i) and bulk activity a_i at constant temperature is expressed by adsorption isotherms, which enable the estimation of energetic parameters such as Gibbs energy of adsorption, ΔG_A, or nearest-neighbor interaction energies, ε_{AA}. 2D first-order phase transitions in the organic adlayer, such as disorder/order or reorientation processes, are represented by discontinuous changes of the interfacial properties (solid curves in Fig. 29) [3, 15, 16, 20, 143]. Often one observes a distinct hysteresis between the positive and negative going potential scans [519]. Exact values of the equilibrium transition potentials E_c, which corresponds to the coexistence of the two phases (for instance liquidlike and solidlike patches), can be obtained by double potential step experiments [143].

Historically, Frumkin-type models, which represent the Helmholtz region by a network of two or three condensers [449, 520–523] and classical thermodynamics based on a mean-field treatment [524–526], were applied first to describe 2D phase transitions in organic adlayers at metal–electrolyte interfaces as a function of concentration, potential, and temperature (Sect. 3.3.2.2). In the simplest case of one-state adsorption, the classical Frumkin isotherm

$$Bc = z = \frac{\theta}{1-\theta} \exp(-2a_F \theta) \quad (48)$$

degenerates into a vertical discontinuity if the Frumkin lateral interaction coefficient $a_F \geq 2$ [449]. The latter is related to the average nearest-neighbor interaction energy ε_{AA}^F as follows

$$\varepsilon_{AA}^F = -\frac{RT a_F}{2} \quad (49)$$

The derivation of a_F is based on the Bragg–Williams (BWA) or MFA assuming (1) the homogeneous (statistical) distribution of molecules among energetically uniform sites, and (2) neglecting local fluctuations and correlations [74–76]. Retter pointed out that a more realistic isotherm treatment of 2D condensation in organic adlayers requires the consideration of localized adsorption and short-range (nearest-neighbor interactions) [527]. This improvement was achieved by introducing LGMs with a square symmetry, such as the two-state Ising model [528–531]. Each state of the lattice is assumed to be vacant or occupied. Below a critical temperature T_c, the 2D lattice gas can split into two phases, a gaseous phase and a condensed phase. The corresponding first-order equilibrium phase transition involves a latent heat [528]. Two advanced approximations of lattice gas isotherms have been applied to describe 2D condensation in organic adlayers: The QCA considers nearest-neighbor site pairs, but treats them (still unrealistically) as independent of each other [60, 73]:

$$\ln(Bc) = \ln z$$
$$= \left[\left(\frac{(\beta - 1 + 2\theta)(1 - \theta)}{(\beta + 1 - 2\theta)\theta} \right)^2 \frac{\theta}{1 - \theta} \right] \quad (50)$$

$$\beta = \left[1 - 4\theta(1 - \theta) \times \left(1 - \exp\left(\frac{a_{Qc}}{2}\right) \right) \right]^{1/2} \quad (51)$$

The LTSE [529] considers equilibrium adsorbate clusters (up to 15 monomers) as well as various configurations of a cluster with constant size indicating that the lattice geometry is taken into account. The corresponding isotherm of the condensed phase for the square lattice is given by

$$\theta = 1 - x^8 z^{-1} - 2\left(2x^{14} - \frac{5}{2x^{16}}\right) z^{-2}$$
$$- 3\left(6x^{20} - 16x^{22} + \frac{31}{3x^{24}}\right) z^{-3} \ldots \quad (52)$$

with $z = Bc$ and $x = \exp(-a_{LTE}/4)$. Equation (52) represents a truncated series expansion, which is obtained with the assumption that contributions of clusters larger than trimers can be neglected.

Figure 30 compares the adsorption isotherms $(a_i + \ln(Bc))$ versus θ, $i = $ F, QCA, LTSE, and the temperature dependence of θ in the vicinity of the phase transition for the above three approximations and the exact LGM. According to the LGM, a first-order phase transition takes place at $(a_G + \ln(Bc)) = 0$ and at temperatures lower than T_c [60, 528, 532]. The degree of coverage of the condensed phase at the equilibrium phase transition depends on a_G, the lattice gas pair interaction coefficient, as follows:

$$\theta = 0.5 + 0.5\left[\left(1 - \sinh\left(\frac{a_G}{4}\right)\right)^{-4} \right]^{1/8} \quad (53)$$

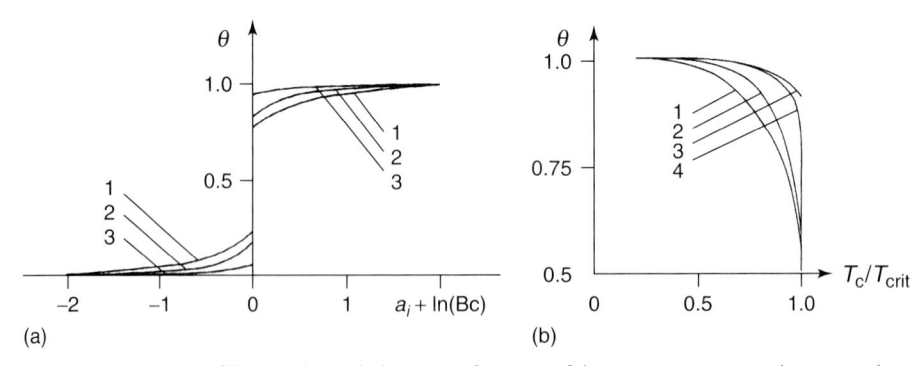

Fig. 30 (a) Coverage of the condensed phase as a function of the activity term $(a_i + \ln(Bc))$ and (b) temperature dependence (T_c/T_{crit}) of the coverage of the condensed phase, θ, for the following models: (1) BWA or MFA (Frumkin isotherm), (2) QCA, (3) LTSE, and (4) LGM. (Reprinted from Ref. [528], Copyright 1987 by Elsevier Science Ltd.)

with $a_G = -2\varepsilon_{AA}/kT$, ε_{AA} as the lateral interaction energy. The square lattice gas predicts the following critical values of θ and a_G at $T = T_c$: $\theta_c = 0.5$ and $a_{Gc} = 3.52549$ [528]. Kharkats and coworkers examined critical parameters and phase diagrams of other 2D lattice geometries [533, 534].

The degree of coverage of the condensed phase can be determined experimentally at constant potential from the saturation capacity C_s, the capacity of the noncondensed phase C_o, and the capacity of the film C according to [529, 535]

$$\theta = \frac{C_o - C}{C_o - C_s} \quad (54)$$

The potential dependence of θ at constant concentration results from combining the respective adsorption isotherms (48), (50) or (52) with the potential dependence of the adsorption coefficient B of neutral organic molecules at metal–aqueous electrolyte interfaces, which is defined as [449, 529]

$$B = B_m \exp(-\alpha(E - E_m)^2) \quad (55)$$

$$B_m = \left(\frac{1}{55.5}\right) \exp\left(\frac{\Delta G_A}{RT}\right) \quad (56)$$

$$\alpha = \frac{C_o - C_s}{2RT\Gamma_m} \quad (57)$$

with E_m, B_m, ΔG_A representing the electrode potential at maximum adsorption, the adsorption coefficient at $E = E_m$, and the Gibbs energy of adsorption.

Equations (54)–(57) and the corresponding isotherms (48), (50) or (52) predict two linear dependencies: (1) $(E_c - E_m)$ versus T_c at constant adsorbate concentration, and (2) $1/T_c$ versus $\ln i$ at $E = E_m$ with T_{crit} referring to the maximum condensation temperature T_c. Furthermore, they permit to model the capacity versus potential dependence in the region of 2D condensation [527–530]. Figure 31 illustrates experimental and theoretical $C-E$ and $\theta-T$ phase diagrams for 3 mM iodocytosine/0.5 M KCl at various temperatures [527]. The isotherm, based on the low-temperature series expansion represents the best fit of the experimental data. The analysis of this system also demonstrates that the Frumkin adsorption model (MFA) is not appropriate to model the phase diagram. These conclusions are supported by related studies with camphor-10-sulfonate [536], adenine [537], uracil [538], borneol, and 3-hydroxyadamantane [539] adsorbed on mercury electrodes as well as uracil derivatives on Ag(111) and Au(111) [469]. All studies reported nearest-neighbor interaction energies at the respective critical temperatures ranging between -4.0 and -6.0 kJ mol^{-1}.

3.3.4.3.3 Kinetic Aspects The kinetics of 2D phase formation and dissolution of organic adlayers were mostly studied by $i-t$, q^M-t or $C-t$ single or multiple potential step experiments, and analyzed on the basis of macroscopic models according to strategies described in Chapter 3.3.3. Only rather recently, modern in situ techniques such as STM [20, 201, 453, 478, 479, 484, 487, 488] and time-resolved infrared spectroscopy (SEIRAS) [475, 476] were applied to study structural aspects of these phase transitions at a molecular or atomistic level.

3.3.4.3.4 Case Study I – Coumarin at the Mercury–Aqueous Electrolyte Interface The adsorption behavior of coumarin (2H-1-benzopyran-2-one) at the mercury–electrolyte interfaces was investigated by electrocapillary, capacitance, and transient experiments [162–164, 540–543].

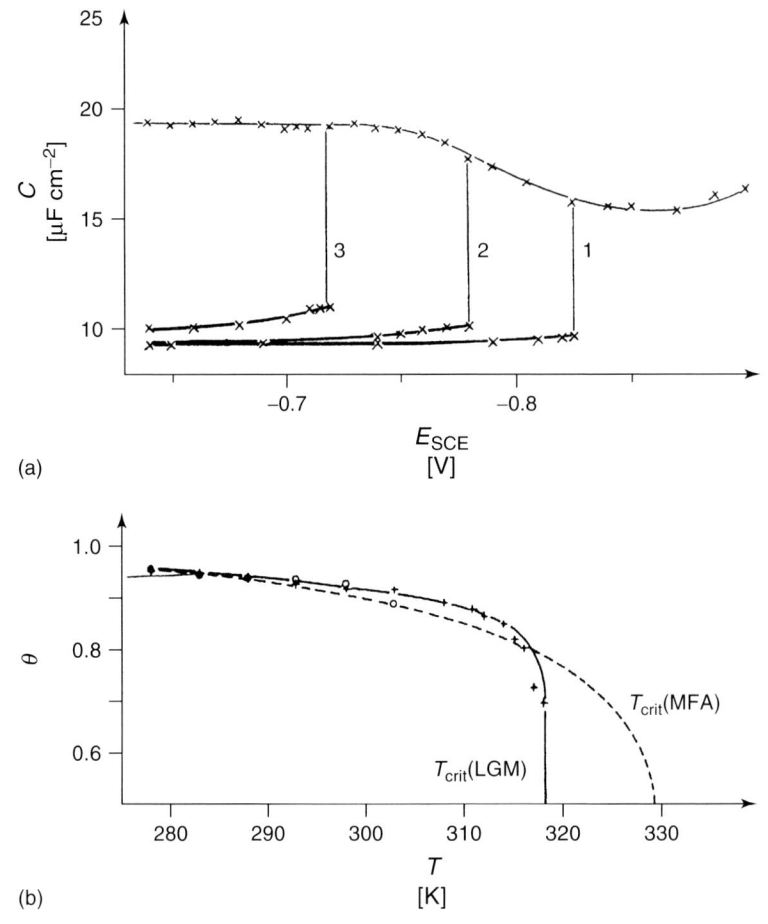

Fig. 31 (a) Capacitance-potential dependence for the system 3 mM 5-iodocytosine in 0.5 M KCl (Mc Ilvaine buffer, pH 7) at a mercury electrode at different temperatures: (1) 298 K, (2) 308 K, (3) 315 K, and (4) 319 K. The bold solid lines show the theoretical curves according to Eq. (52) calculated by nonlinear regression for the square Ising lattice (reprinted from Ref. [536], copyright 1993 by Elsevier Science Ltd.). (b) Temperature dependence of the condensation degree of coverage θ for 3 mM (o) and 15 mM (+) 5-iodocytosine in 0.5 M KCl (McIllvaine buffer, pH 7) at a mercury electrode. (- - - -) theoretical curve according to the Frumkin or MFA adsorption model; (———) theoretical curve according to the LGM (Eq. 53). (Reprinted from Ref. [527] and [529], Copyright 1984 and 1993 by Elsevier Science Ltd.)

Figure 32 shows a typical capacitance versus potential curve of 5 mM coumarin in 0.5 M NaF at a HMDE. Three different potential regions exist. Region I is interpreted as planar orientation with the coumarin molecule slightly inclined with respect to the interface and $\Gamma_{mI} = 3.1 \; 10^{-10}$ mol cm^{-2} [163]. The capacitance decreases at more negative potentials, and corresponds in region

3.3 Phase Transitions in Two-dimensional Adlayers at Electrode Surfaces

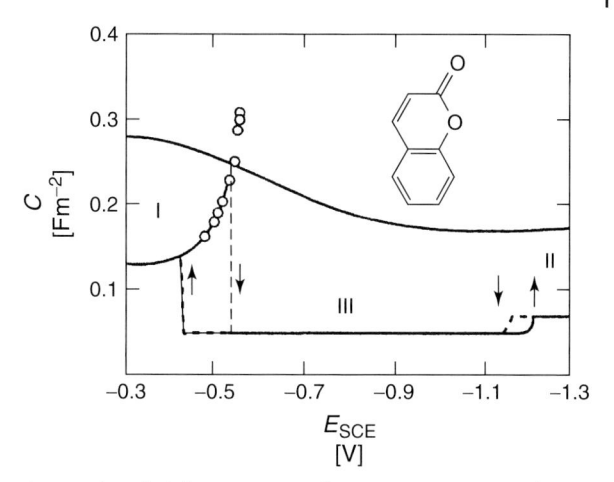

Fig. 32 Interfacial capacitance of mercury in contact with aqueous 0.5 M NaF in the absence and in the presence of 5 mM coumarin, temperature 5 °C, scan rate 5 mV s^{-1}. The scan directions are indicated by arrows in the figure. The open circles show the initial capacitance C_i observed immediately after stepping the potential from region I into region III. (Reprinted from Ref. [164], copyright 1994 by Bulgarian Chemical Society.)

II to $\Gamma_{mII} = 6.1\ 10^{-10}$ mol cm^{-2} with coumarin molecules aligned with the C(5) and C(6) carbon atoms towards the electrode surface [162]. At sufficiently high adsorbate concentrations a 2D condensed film is formed in region III, giving rise to a characteristic capacitance pit with a pronounced hysteresis at both edges [162–164]. Thomas and coworkers proposed a perpendicular orientation with coumarin carbons C(6) and C(7) pointed towards the electrode and $\Gamma_{mIII} = 8.4\ 10^{-10}$ mol cm^{-2}. The pit width decreases linearly with increasing temperature according to $(E_+ - E_-)^2$ versus T, and a critical temperature $T_{crit} = 28\ °C$ could be estimated at coumarin saturation concentration [162].

Single potential steps I → III towards the center of the capacitance pit ($E_f < -0.560$ V) yielded deterministic sigmoid-shape transients, which could be quantitatively described by progressive nucleation with 2D growth as rate limiting process, for example, $m = 3$ in the general Avrami-formalism as expressed by Eq. 35 (Fig. 33a). At smaller overvoltages with final potentials ranging between -0.530 V and -0.555 V, stochastic behavior is encountered. All capacitance transients exhibit the same values for C_i and C_f, but there is stochasticity in the onset of each transient as well as in its shape (Fig. 33b). The observed differences between the individual transients are indicative of multiple nucleation, and represent the domain of oligonucleation. By selecting E_f to be larger than -0.530 V, mononucleation events occur. Nucleation is so slow that there is negligible chance for a second nucleation during the time necessary for the first nucleus to form a complete film on the hanging mercury drop (Fig. 33c). For an isotropic

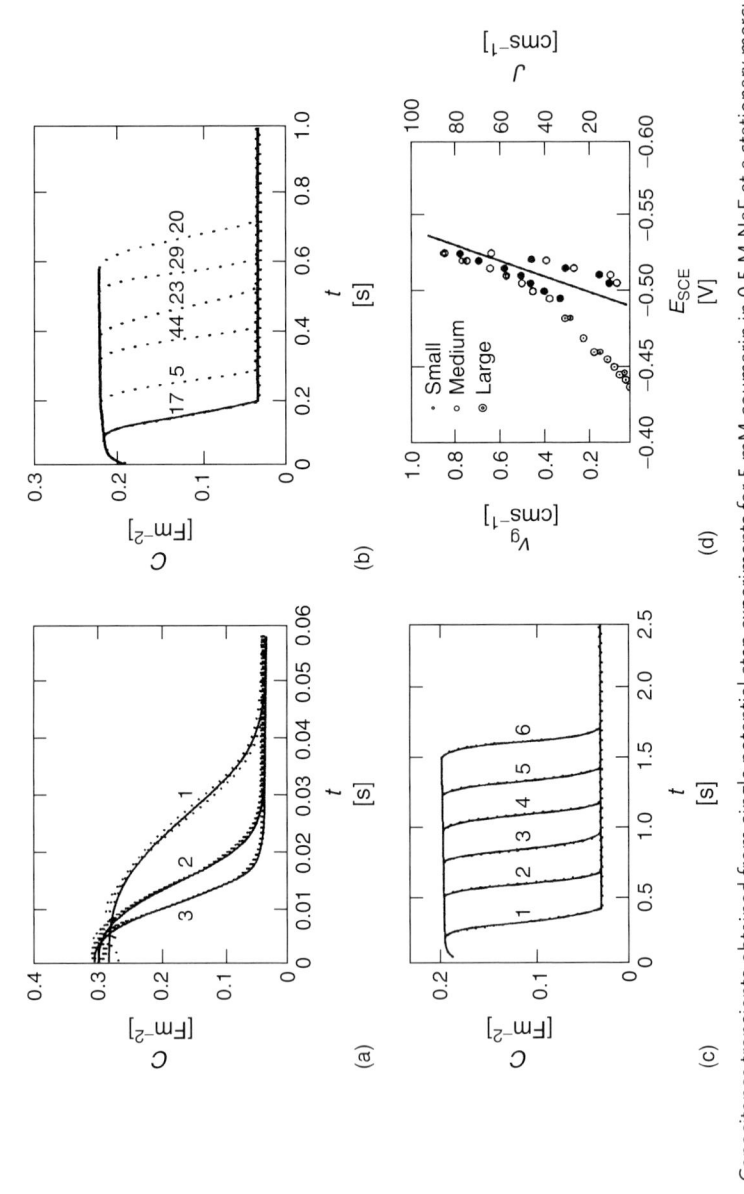

Fig. 33 Capacitance transients obtained from single potential step experiments for 5 mM coumarin in 0.5 M NaF at a stationary mercury electrode, 5 °C. (a) deterministic transients with $E_1 = -0.400$ V to $E_2 = -0.560$ V (1), -0.565 V (2), and -0.570 V (3) The solid lines were calculated by nonlinear last squares fitting to Eq. (35) with $m = 3$. (b) Transients as in (a), except that $E_2 = -0.540$ V. The transients are now stochastic. The numbers represent their sequence number in a set of 50 repeat experiments. (c) Transients as in (a), except that $E_2 = -0.520$ V. All transients reflect mononucleation and can be represented by Eq. (30). The coverage was obtained with Eq. (35). (Reprinted from Ref. [164], copyright 1994 by Bulgarian Chemical Society.)

sphere, the growth of the compact film, once it has been nucleated, is given by [544]

$$2\theta = 1 - \cos(v_g(t - t_o)/r) \qquad (58)$$

where v_g is the linear growth velocity, t_o the moment of nucleation, and r the radius of the mercury droplet, θ is obtained from Eq. (36). The potential dependence of the growth rate is plotted in Fig. 33(d) for three different drop sizes. The nucleation rate was obtained for each potential by analyzing a series of 550 mononucleation transients. Extrapolating the growth curves to t_o, and sorting in bins, provides the probability P_o that nucleation has not yet occurred in a given time interval Δt (cf. Sect. 3.3.3.2.3). Representing $-\ln P_o$ as a function of time then produces a plot from which both, the induction time and the steady state nucleation rate is obtained [163, 164, 545]. The latter is plotted in Fig. 33(d). The experiments were carefully designed to ensure that only a small fraction of the observed nucleation appears to originate at the lumen of the capillary (e.g. at defect sites!) [164].

The dissolution transients of the condensed coumarin film III depend strongly on the applied potential step regime, and show strong aging effects, which were attributed to the healing of domain boundaries. The corresponding "grain-boundary" dissolution model as developed recently by Poelman and coworkers [546] is applicable (cf. Fig. 6)

The formation of condensed coumarin monolayers was also reported on Au(111)-(1 × 1) and Au(100)-(hex) [465].

3.3.4.3.5 Case Study II – Uracil on Au(hkl)

Since the pioneering work of Vetterl [45], who reported for the first time so-called capacitance pits, thermodynamic and kinetic aspects of the phase formation of uracil and its derivatives have been studied comprehensively at the mercury–electrolyte interface (cf. reviews in [3, 15]). In 1995, Hölzle and coworkers [219] presented convincing evidence that 2D condensation of uracil is also observed on low-index gold single-crystal electrodes. On the basis of voltammetric, capacitance and transient experiments [183, 219–221, 461, 464, 462, 467, 547] and their combination with structure-sensitive methods such as in situ STM [454, 478–480], vibrational spectroscopy [454, 473, 475,], SXS [458], and UHV-based characterization strategies [463], a rather comprehensive understanding of the interfacial properties of uracil and several of its derivatives on Au(hkl), Ag(hkl), and Cd(0001) has emerged.

Figure 34 shows a typical cyclic voltammogram for 12 mM Uracil in 0.05 M H_2SO_4. The first voltammetric scan, as obtained with a flame-annealed, thermally reconstructed Au(111)-$(p \times \sqrt{3})$ electrode after immersion at -0.100 V, is indicated as dashed line. The solid line represents the steady state voltammogram, recorded after five complete potential cycles [478]. The phenomenological analysis of these electrochemical data yields the existence of four distinct interfacial regions labeled I to IV in Fig. 34 [20]: In region I the molecules are randomly (gaseous-like) adsorbed on the electrode until complete desorption occurs at sufficiently negative potentials. The formation of a 2D condensed physisorbed uracil film at more positive potentials is indicated by typical current spikes P_1/P_1' and P_2/P_2', which limit the pit region II. A marked hysteresis is observed at both pit edges when changing the direction of the potential

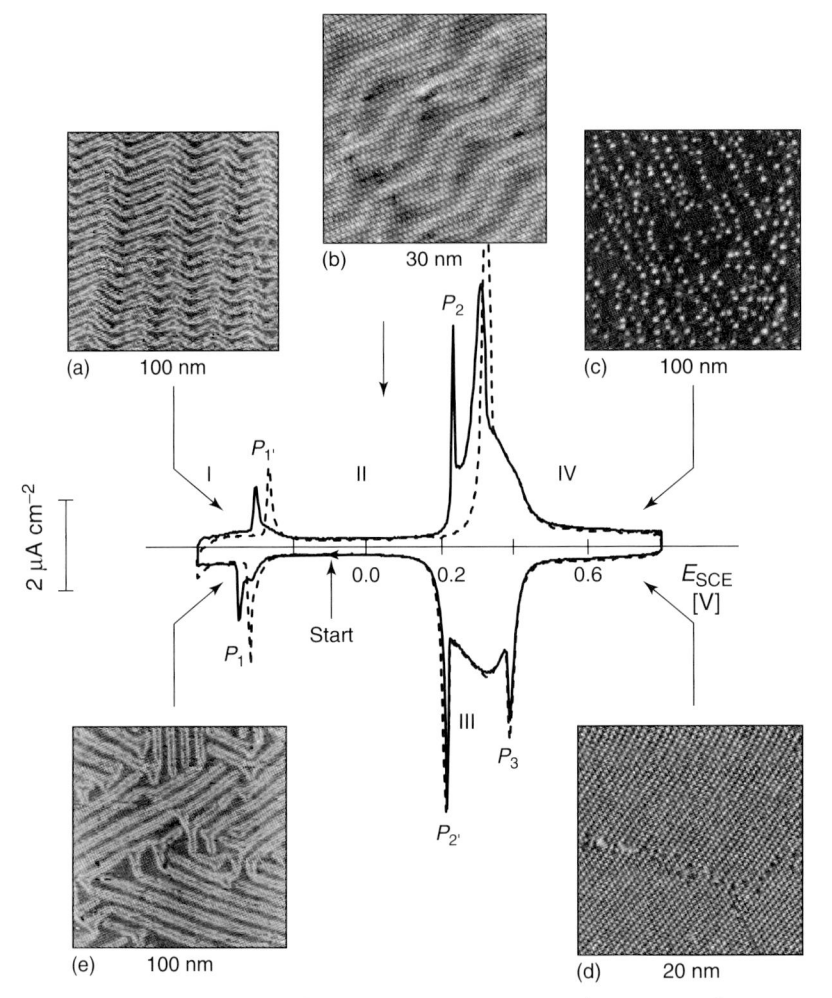

Fig. 34 Cyclic voltammograms for Au(111)/0.05 M H_2SO_4 in the presence of 12 mM uracil, scan rate 10 mV s^{-1}. The first scan as obtained with a flame-annealed, reconstructed electrode after immersion at −0.10 V is plotted as dashed line. The steady state voltammogram is shown as solid curve. The stability regions of the various adlayer phases are labelled I to IV, and they are illustrated with typical in situ STM images: (a) thermally reconstructed Au(111)-$(p \times \sqrt{3})$ surface, (b) physisorbed uracil film, (c) unreconstructed Au(111)-(1×1) surface in the presence of the chemisorbed uracil film, (d) as obtained after scanning the electrode potential from II → IV, (e) island-free area of a potential-induced reconstructed Au(111)-$(p \times \sqrt{3})$ electrode. (Reprinted from [478], copyright 1997, American Chemical Society.)

scan. The stability range of II increases with decreasing temperature. The respective critical temperatures assumes the following substrate sequence [469, 470]

$$Hg < Ag(111) < Ag(100) < Au(111)$$
$$-(p \times \sqrt{3}) < Au(100)-(1 \times 1)$$
$$\sim Au(100)-(hex) \tag{59}$$

The chronocoulometrically determined surface excess $\Gamma_{mII} = 4.2\ 10^{-10}$ mol cm^{-2} ($A = 0.40$ nm^2) and the small negative shift of E_{pzc} are consistent with a planar surface orientation of the molecules [469]. Comparative electrochemical [469], spectroscopic [475, 454], and in situ STM experiments [478, 479] on quasi-ideal Au(111) and Au(100) electrodes indeed revealed the existence of a highly ordered monolayer of uracil molecules interconnected via a network of directional hydrogen bonds, similar to that in the 3D solid. A typical high-resolution STM image and the proposed packing model with the high-order coincidence mesh are plotted in Fig. 35. One may notice that the same long-range order is found on hexagonal, densely packed (111) as well as on the more open, quadratic (100) substrate surfaces, which point to the structure-determining role of lateral molecular interactions, and to less important specific contributions of the substrate surface [478, 479]. Systematic studies with stepped electrodes Au(n, n, n-2) = Au((n(111)-(110)) having (111) terraces and monatomic (110)-oriented steps revealed that no 2D condensation occurs at miscut angles $>4°$ because of limited terrace size [475].

Single and multiple potential step experiments demonstrated that the macrokinetics of the formation of the physisorbed uracil film represents a first-order phase transition and follows the exponential law of nucleation (cf. Eq. (34)) in combination with surface diffusion-controlled growth [183]. In situ STM [20, 478, 479] and time-resolved SEIRAS studies [475] suggest that these processes are strongly related to the formation/breaking of uracil–water and water–water hydrogen bonds within the Helmholtz region.

The physisorbed uracil films II changes upon passing the sharp current peak P_2 towards positive potentials due to the break-off of the hydrogen-bonded network accompanied by a partial charge transfer process, which is strongly dependent on the solution pH and consumes up to 60 µC cm^{-2} [219, 466, 469, 475, 478].

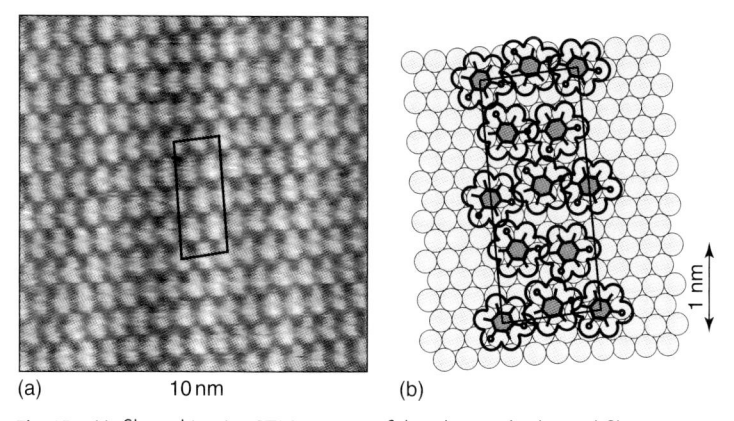

Fig. 35 Unfiltered in situ STM images of the physisorbed uracil film on Au(111)-$(p \times \sqrt{3})$/0.05 M H$_2$SO$_4$ + 3 mM uracil: (a) high-resolution image at $E = -0.05$ V, $i_T = 2$ nA, $v_T = +0.01$ V. The primitive unit cell is indicated; (b) proposed packing model. (Reprinted from [478], copyright 1997, American Chemical Society.)

At sufficiently positive potentials, in region IV, uracil forms on low- and higher-index Au(hkl) a chemisorbed, highly ordered adlayer, which (1) occurs rather independent of the adsorbate concentration, (2) is stable at temperatures as high as 100 °C, and (3) inhibits the onset of gold oxidation. This process is accompanied by the lifting of the substrate surface reconstruction, and gives rise to monatomic high gold islands as shown in Fig. 34(c). In situ STM experiments revealed on Au(111) a hexagonal densely packed arrangement of molecules giving rise to a commensurate ($\sqrt{3} \times 3$) unit cell, which contains two molecules ($\Gamma_{mIV}(111) = 7.9\ 10^{-10}$ mol cm^{-2}, $A_{IV}(111) = 0.21$ nm^2). On Au(100)-(1 × 1) electrodes, a regular sticklike pattern was obtained in region IV after equilibration. Each stick consists of four uracil molecules ($\Gamma_{mIV}(100) = 7.2\ 10^{-10}$ mol cm^{-2}, $A_{IV}(111) = 0.23$ nm^2) (Fig. 36). The steady state structures of the chemisorbed phase IV are determined by the competitive interplay of (1) short-range,

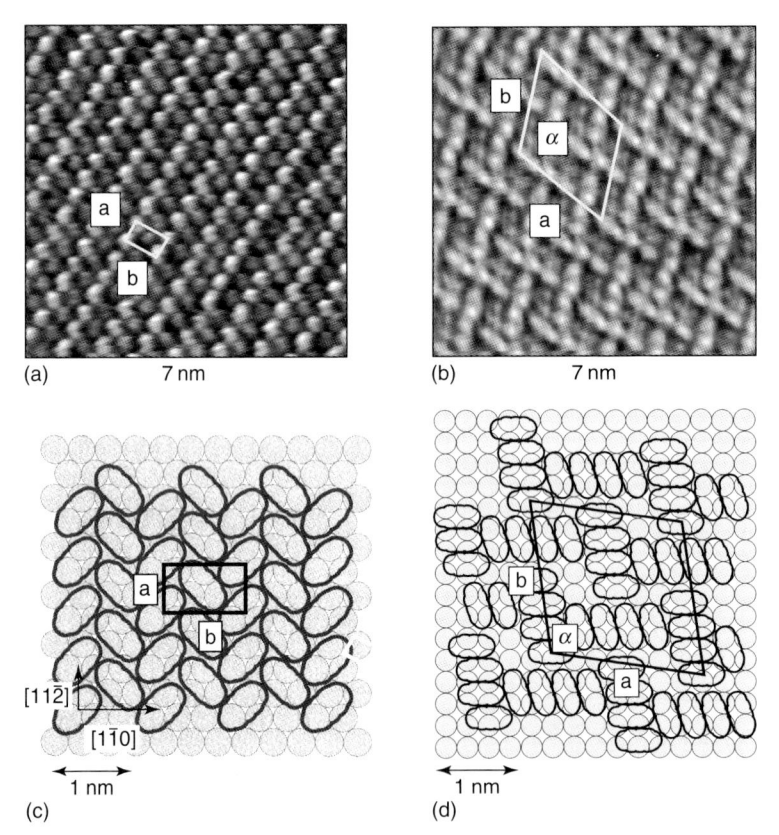

Fig. 36 High-resolution STM images and packing models of chemisorbed uracil on Au(111)-(1 × 1) (a); $E = 0.60$ V, $i_T = 2$ nA, $v_T = -0.65$ V and Au(100)-(1 × 1) (b); $E = 0.60$ V, $i_T = 1.8$ nA, $v_T = -0.52$ V in 0.05 M H$_2$SO$_4$. Characteristic dimensions of the adsorbate mesh and the respective elementary cells are indicated in the experimental images and in the proposed packing models depicted in (c) and (d), respectively.

attractive π-stacking between adjacent uracil molecules, and (2) substrate-adsorbate coordination [478, 479]. The analysis of the STM-contrast pattern, in combination with electrochemical and IR spectroscopic experiments of uracil and several of its N(1) and/or N(3)-substituted methylderivatives demonstrates that the chemisorbed phase IV corresponds to perpendicularly oriented molecules coordinated with the positively charged electrode surface via the N(3) nitrogen and both C(2)=O, C(4)=O oxygen's [469, 478, 479, 454].

The dissolution kinetics of the chemisorbed uracil phase IV on Au(111) (towards III and II) was studied by potential step measurements. The $i-t$ transients could be modeled by the exponential law [475] or instantaneous [221] hole nucleation in combination with surface diffusion-controlled growth. The corresponding two rate constants vary systematically with the defect density (1) of the chemisorbed layer (domain size, local disorder) and (2) of the substrate surface (step orientation and distribution, terrace size) [3, 221, 475]. Increasing defect density decreases the half transition time of the "dissolution" process, and the corresponding transients finally approach the instantaneous limit of hole nucleation. The STM experiments suggest that the critical hole nuclei are most probably represented by domain or point defect sites [478, 479]. Simultaneous electrochemical $i-t$ and time-resolved ATR-SEIRAS experiments in the rapid scan or the step-scan regime (Fig. 37) [475] with quasi Au(111) demonstrate that the dissolution of the chemisorbed uracil film is associated with an orientational change of uracil from perpendicular to planar. Monitoring the integrated intensity of a characteristic uracil vibration ($\nu_{C4=O}$) with a time resolution of 200 µs reveals that the spectroscopic transient response is systematically longer than the corresponding current response of the same potential step. Both approaches indicate a dissolution mechanism controlled by hole nucleation and diffusion-controlled growth, but reflect complementary properties of the phase transition [475]. Time-resolved SEIRA spectroscopy at electrochemical interfaces currently starts to develop as a powerful approach to unravel the atomistic/molecular level details of these complex interfacial processes, such as 2D phase transitions, at solid–liquid interfaces.

The brief description of uracil adlayers on Au(hkl) also demonstrates that the understanding of 2D phase transition processes on defined solid electrodes, such as single-crystalline substrates, requires the simultaneous consideration of interfacial processes associated with the adlayer as well as with the substrate surface.

3.3.4.3.6 Case Study III – 2,2′-Bipyridine (2,2′-BP) on Au(hkl) 2,2′-bipyridine is a bidentate ligand used in coordination chemistry and a typical representative of aromatic, nitrogen-containing heterocycles, which act as basic building blocks in highly specific and functionalized host lattices at defined surfaces [548, 549]. Two planar pyridine rings are connected via a C—C bond with 10% double-bond character [550]. Electrochemical studies revealed that n, n'-bipyridine [551–553] and several of its Co^{2+} and Ni^{2+} complexes [554] form 2D condensed films at the atomically smooth mercury–electrolyte interface. The adsorption of 2,2′-bipyridine on solid electrodes, such as Ag(poly), Au(hkl), and Cu(111) was studied by

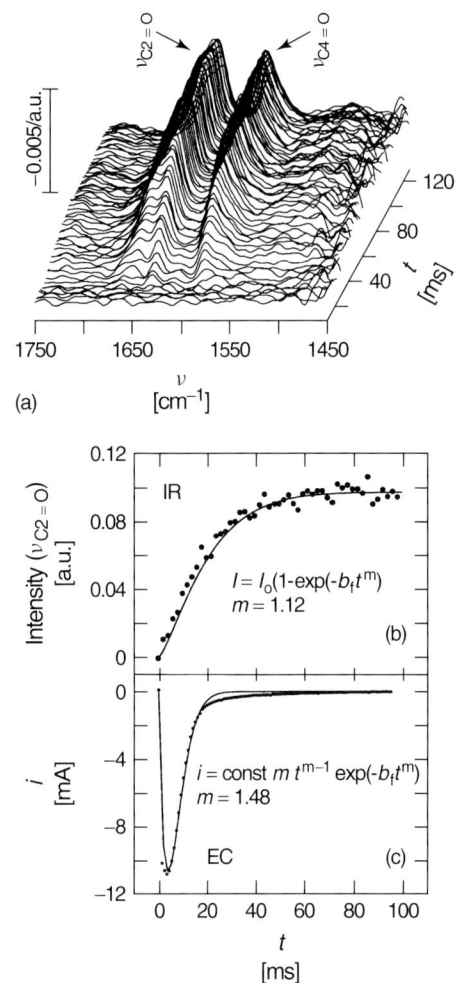

Fig. 37 Time-resolved SEIRAS–ATR experiment for the dissolution of the chemisorbed uracil monolayer on quasi-Au(111) thin-film electrodes in 0.1 M H_2SO_4 + 12 mM uracil. The potential was stepped from $E_1 = 0.60$ V ($t_1 = 30$ s) to $E_2 = 0.20$ V. (a) original set of time-resolved SEIRAS spectra (step-scan technique) with characteristic uracil vibrations in the C=O stretching region (For clarity, the "loss" spectra are plotted with an inverted sign!) (b) Time dependence of the integrated intensity of the uracil vibration $\nu_{C4=O}$ (1590 cm^{-1}). The solid lines represent fits to an Avrami-type equation (cf. Eq. (35)) with $m = 1.12$. (c) Corresponding current transient. The Avrami-exponent $m = 1.48$ points to nucleation according to an exponential law coupled with surface diffusion-controlled growth (Eq. (34) and Eq. (35), solid line). The experimental results in panels (b) and (c) indicate that the spectroscopic and the electrochemical transients probe different interfacial properties of the dissolution process and illustrate the complementary information of both approaches (reproduced from Ref. [475]).

Raman scattering [555], chronocoulometry, and SHG [556], SXS [458], FTIR [557], SEIRAS [489], and in situ STM [484, 486–489]. Lipkowski and coworkers [556] found, on the basis of electrochemical data, that 2,2′-bipyridine exhibits multistate adsorption on Au(111). Structural details of this complex phase behavior were resolved by in situ STM (Fig. 38) and SEIRAS.

2,2′-BP was dissolved in aqueous electrolyte and potentiostatically deposited onto the unreconstructed Au(111)-(1 × 1) electrode. Molecular stacks are formed via the coordination of the two ring nitrogen atoms with the positively charged gold electrode in region I (Fig. 38a, b). The distance between adjacent rows amounts to (0.96 ± 0.5) nm, the intermolecular distance is (0.38 ± 0.2) nm, and the molecules are tilted from the normal to the axis of the chain by an angle $\alpha = (28 \pm 2)°$ [484, 486, 488]. The simultaneous imaging of the hexagonal substrate surface and of the organic adlayer pattern pointed to a commensurate (4 ×

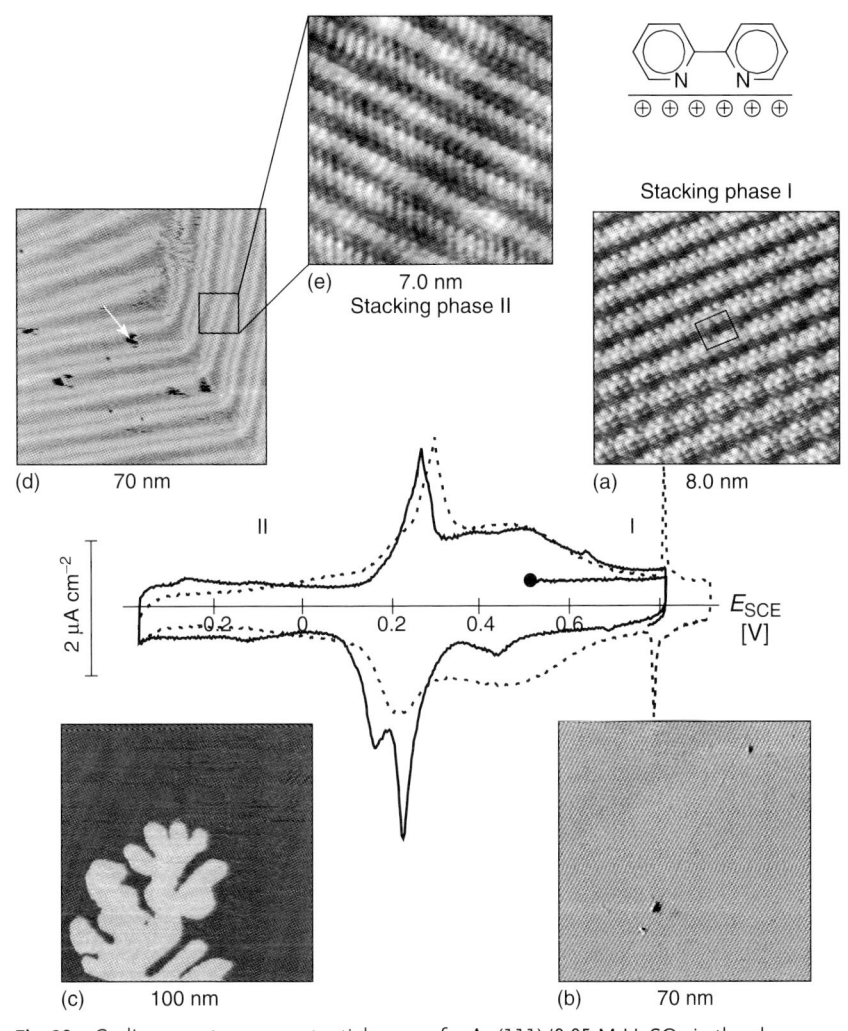

3.3 Phase Transitions in Two-dimensional Adlayers at Electrode Surfaces | 449

Fig. 38 Cyclic current versus potential curves for Au(111)/0.05 M H_2SO_4 in the absence (dotted curve) and in the presence (full line) of 3 mM 2,2′-bipyridine, scan rate 10 mV s^{-1}. (a) Unfiltered high-resolution image of the 2,2′-BP stacking structure at $E = 0.50$ V, $i_T = 2$ nA, $v_T = 10$ mV. The suggested commensurate unit cell is indicated. (b) Large domain of the anodic stacking phase I, (c) monatomic high gold "structure" decorating a step of the Au(111)-surface after a single potential scan (10 mV s^{-1}) from 0.50 V to -0.05 V (negatively charged electrode). (d, e) large scale and high-resolution image of the cathodic 2,2′-BPH$^+$-stacking layer II on Au(111)-($p \times \sqrt{3}$) at -0.20 V. (Reprinted from Ref. [486], copyright 1998 by Elsevier Science Ltd.)

$2\sqrt{3}$) unit cell, which contains three molecules, for example, $A_I \approx 0.38$ nm^2. The nitrogen atoms coordinate with the substrate surface in twofold bridge positions [486]. The structure-determining role of the substrate crystallography onto the 2,2′-BP stacking pattern is supported by experiments with the quadratic

Au(100)-(1 × 1) surface. Dretschkow and coworkers reported kinked stacking rows due to the misfit between high-order substrate coordination sites, and the potential molecular coordination sites [487].

After application of a single potential step/scan towards negative potentials, the ordered 2,2′-BP adlayer dissolves, and the entire substrate surface appears very mobile. Fractal, monatomic high gold features grow at step edges (Fig. 38c), their surface is immediately reconstructed (Fig. 38d), and soon after, a new STM-contrast pattern develops at negative charge densities in region II (Fig. 38e), which is composed of parallel molecular stacking rows with alternating molecular tilt angels of $+23°$ or $-16°$ between adjacent rows. SEIRAS-experiments indicate a torsional or inclined orientation of the two ring nitrogen atoms with respect to the electrode surface [486, 489]. Both highly ordered 2,2′-BP adlayers, in regions I and

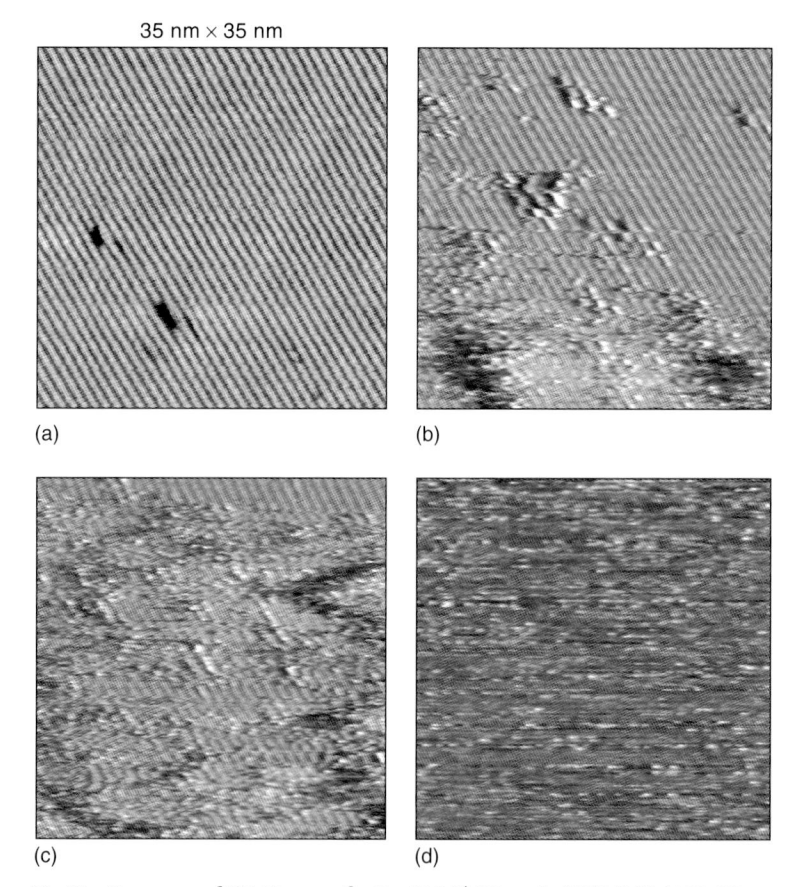

Fig. 39 Sequence of STM images for 3 mM 2,2′-BP on Au(111)/0.05 M H_2SO_4 after a single potential scan (10 mV s^{-1}) from 0.50 V to 0.24 V (potential-induced dissolution of I): (a) $E = 0.50$ V, (b) $E = 0.24$ V, $t = 3$ min, (c) $E = 0.17$ V, $t = 4.5$ min, (d) $E = 0.17$ V, $t = 8.0$ min; $i_T = 2$ nA. (Reprinted from Ref. [486], copyright 1998 by Elsevier Science Ltd.)

II, have been observed in sulfate- and in perchlorate-containing neutral or acidic electrolytes [484, 486–489].

Details of the dynamics of the rather slow potential and/or temperature-induced structural transitions could be monitored by in situ STM in real space and time [486]: The first example starts with a highly ordered 2,2′-BP stacking layer I, equilibrated at $E_i = 0.500$ V on Au(111)-(1×1) (Fig. 39a), and then the potential was scanned slowly towards negative values. The range was restricted ensuring that the gold electrode still bears a positive charge. At $E = 0.240$ V the first point dislocations occur (Fig. 39b). The active centers are formed not only on defects of the adsorbate or substrate lattices, such as point defects or domain boundaries, steps or kink positions. They are distributed evenly across entire terraces. The nucleated holes grow anisotropically with time and more negative electrode potentials until the ordered adlayer is dissolved completely. Growth proceeds preferentially along the directions of the commensurate molecular stacking rows. During this transition one can identify two phases clearly, solidlike patches exhibiting characteristic long-range order of 2,2′-BP in region I, and a fluid medium disordered on the atomic scale (Fig. 39b). The observed order/disorder transition is first order [486].

The above interpretation is supported by experiments on the temperature-induced dissolution of the ordered 2,2′-BP phase I [486]. The system was stabilized at $E_i = 0.200$ V and $T = 29.7\,°\text{C}$ (Fig. 40a), and then the temperature was slowly ramped with $0.02\,°\text{C min}^{-1}$. Firstly, point and line dislocations ("one-dimensional holes") appear (Fig. 40b). They grow anisotropically by successive stripping of molecular stacks, fast along the directions of the stacking rows, and rather slow perpendicular to it. Positional and directional fluctuations within the ordered and the disordered phases appear. Under the shown experimental conditions the ordered 2,2′-BP stacking layer dissolved completely at $33.60\,°\text{C}$ (Fig. 40d). Closer inspection of the disordered regions reveals a granular structure on a length scale exceeding the pixel resolution of the image in the $x-y$ direction. This means that a noticeable part of the modulation in the disordered area is not a time effect caused by the fast motion of adsorbed species, but rather reflects, at least in part, the static positions of particles in the disordered phase. This suggests the existence of adsorbate–substrate coordination complexes. Similar observations were reported for the 2D fluid–solid equilibrium of cesium and oxygen coadsorbed on Ru(0001) under UHV conditions [558].

In the present case, the adsorbed species "feel" the effect of the substrate corrugation potential. This situation is different from purely 2D system, like guanine or adenine on HOPG [210] or the physisorbed films of uracil or thymine on Au(hkl) [219, 478, 479], where the adsorbate–substrate interactions are very weak. There, the respective phase transitions are governed entirely by the applied electrical field and directional, lateral attractive interactions via hydrogen bonds [20, 453].

The potential and/or temperature controlled dissolution of the above order/disorder transition is completely reversible in sulfuric acid solution. A representative time sequence, after application of a potential scan from $E_i = 0.100$ V to $E_f = 0.500$ V, is plotted in Fig. 41. Within a few seconds at E_f the 2,2′-BP molecules start to order. No preferential nucleation sites, which might correlate with defect

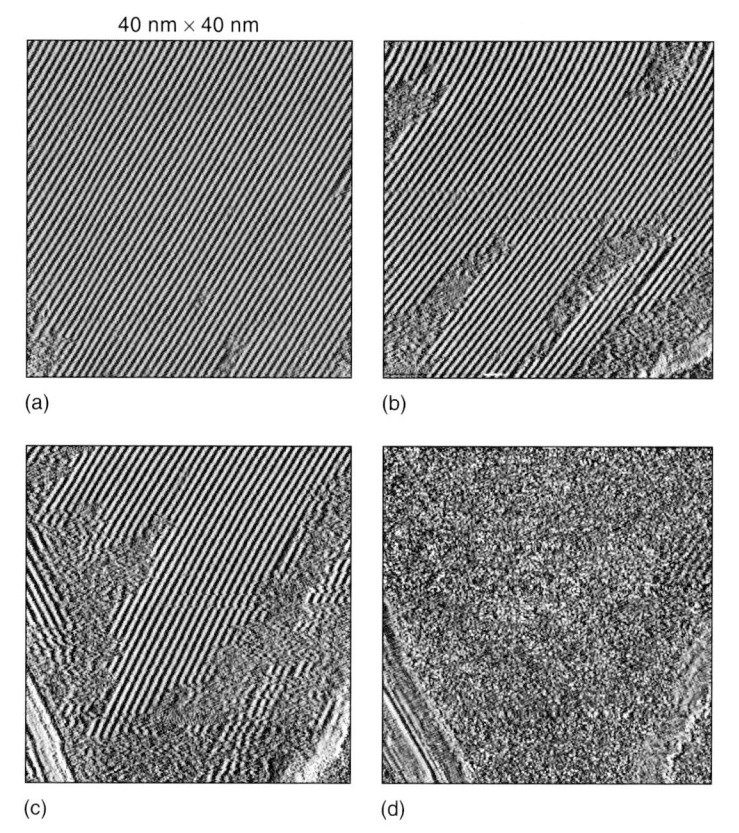

Fig. 40 Temperature-induced dissolution of the 2,2′-BP stacking phase I for 3 mM 2,2′-BP in 0.05 M H_2SO_4/Au(111). (a) The system was equilibrated at $E = 0.20$ V and 29.7 °C for 20 min. Then the temperature was ramped with 0.02 °C min^{-1}, (b) 33.14 °C, (c) 33.51 °C, (d) 33.60 °C. All images were obtained in constant current mode ($i_T = 0.25$ nA) at the same surface area. (Reprinted from Ref. [486], copyright 1998 by Elsevier Science Ltd.)

structures of the substrate surface, were found. First stacks are formed; their directions correlate but exhibit a "missing row" arrangement (Fig. 41b). With increasing observation time this metastable phase is displaced by a densely packed stacking domain (Fig. 41c) with dimensions typical for phase I. The latter grows anisotropically in two directions: (1) parallel along the main stacking direction. Individual molecules and small stacks are still in search of energetically favorable adlattice positions at the end of the growing front. (2) The second growth mechanism proceeds via incorporation of entire molecular rows perpendicular to the main stack direction. The growth and the existence of other metastable structures are strongly influenced by the nature of the supporting electrolyte anions [20, 484, 486, 488, 489]. Defects within the adlayer change their shape or heal with increasing observation time. The detailed mechanism most probably involves place exchange

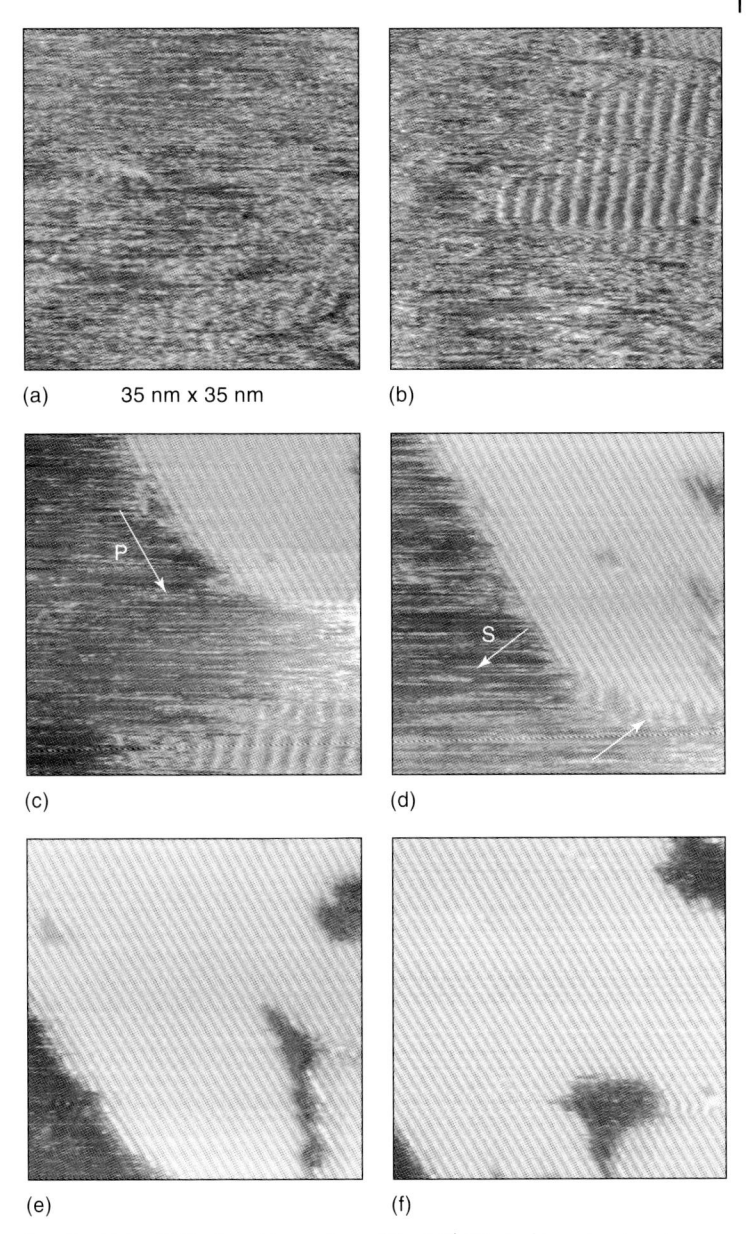

Fig. 41 Potential-induced formation of the 2,2′-BP stacking structure I on Au(111) after a potential scan from region II (10 mV s^{-1}): (a) $E_i = 0.10$ V to $E_f = 0.50$ V, (b) $E = 0.50$ V, $t = 1$ min, (c) $E = 0.50$ V, $t = 5$ min, (d) $E = 0.50$ V, $t = 8$ min, (e) $E = 0.50$ V, $t = 11$ min, (f) $E = 0.50$ V, $t = 42$ min; $i_T = 2$ nA. Solution composition 3 mM 2,2′-BP/0.05 M H$_2$SO$_4$. (Reprinted from Ref. [486], copyright 1998 by Elsevier Science Ltd.)

processes and surface diffusion of 2,2′-BP-gold complexes. A similar mechanism is proposed for the "aging" of SAMs formed by alkanthioles on gold surfaces [559].

At the end of this chapter it is remarked that the described structural details of the above-mentioned phase transitions are not accessible by macroscopic electrochemical approaches, such as measurements of the interfacial capacitance or charging current, but can be attributed to the often observed short or long-time transient responses.

3.3.5
Conclusion and Outlook

The macroscopic approach towards the "static" electrochemical interface as developed through the work of Lippmann, Gouy, Grahame, Frumkin/Damaskin, and Parsons is starting to change deeply. Over the last decade a detailed microscopic picture of the bare and adsorbate-covered electrochemical interface has emerged. These advances could be attributed to four major developments: (1) The use of well-defined single-crystal electrodes instead of polycrystalline material. (2) The combination of complementary classical electrochemical techniques, based on macroscopic measurements of current, charge density, or interfacial capacitance, with structure-sensitive in situ techniques, such as vibrational spectroscopies (FTIR, SEIRAS, surface-enhanced Raman scattering (SERS), SFG), X-ray methods (SXS, EXAFS, ...) and scanning probe microscopies (STM, AFM and related techniques). (3) The use of UHV techniques for structure-sensitive investigations after controlled emersion of the electrode and the transfer into a vacuum chamber. (4) The development of modern theoretical concepts of the electrochemical double layer, and of processes taking place there.

A main field of activities is focused on structure and reactivity in two-dimensional adlayers at electrode surfaces. Significant new insights were obtained into the specific adsorption and phase formation of anions and organic monolayers as well as into the underpotential deposition of metal ions on foreign substrates. The in situ application of structure-sensitive methods with an atomic-scale spatial resolution, and a time resolution up to a few microseconds revealed rich, potential-dependent phase behavior. Randomly disordered phases, lattice gas adsorption, commensurate and incommensurate (compressible and/or rotated) structures were observed. Attempts have been developed, often on the basis of concepts of 2D surface physics, to rationalize the observed phase changes and transitions by competing lateral adsorbate–adsorbate and adsorbate–substrate interactions.

The thermodynamic and statistical mechanical analysis of first-order and/or continuous phase transitions demonstrated clearly that models based on the MFA are often inadequate. Lattice gas models (LGM), and approximations based on the quasi-chemical treatment (QCA) or series expansion techniques (e.g. the low-temperature series expansion (SE)) are more appropriate. MC simulations have also been developed as a powerful approach to unravel the role of lateral interactions in various phase formation processes.

Impressive atomic- and molecular-scale details of steady state properties of 2D phase changes in ionic and molecular adlayers, and substrate surfaces have been obtained, mostly based on in situ STM, IR- and SXS-studies. Clear evidence was presented of the effect of crystallographic orientation, the coadsorption of ions of the supporting electrolyte and of the solvent molecules on the adlayer structures, and in a few systems on the electrode reactivity. Especially, scanning probe experiments provided important new insight into the role of defects, and their control and manipulation in 2D phase formation processes. The extension of these studies to more reactive metals, such as Fe, Ni, or Al is almost nonexistent, despite their important technological relevance.

Pioneering approaches were developed, on the basis of time-resolved in situ STM, SEIRAS, and SXS-studies, to explore structural details of dynamic processes involved in phase transitions in adlayers and substrate surfaces, and to relate these results to macrokinetic models of nucleation and growth.

In spite of this progress, major challenges exist in understanding the relationship between structure and dynamics of 2D phase changes and transitions in adlayers at electrochemical interfaces at a microscopic level. These include topics such as (1) phase changes in functionalized organic adlayers induced by controlled electrical, magnetic, or temperature perturbations; (2) the structure-determining role of interfacial water as well as dynamic relaxations; (3) the molecular-level control and modification of electrode reactivity upon phase changes in adlayer. At present, the microscopic mechanisms of reactions at adlayer-covered electrodes are only studied on a qualitative level. The "controlled" tailoring of structural properties of clean and adsorbate-covered electrodes offers fascinating opportunities to design molecular- or atomic-scale sites to trigger specific reaction pathways, as well as sensitivity and selectivity of interfacial processes. One may imagine electrochemical switches or nanoscale storage units based on phase transitions in functionalized adlayers deposited on well-structured electrodes. Fascinating perspectives towards these goals are anticipated by new methodological developments aimed to improve the time- and spatial resolution of structure-sensitive in situ techniques.

The "ultimate goal" of true insights into the physico-chemical nature of phase transitions, structural and dynamic changes in adlayers at electrodes calls for a multidisciplinary approach, combining high-level experimental techniques with advanced theoretical treatments of clean and adsorbate–modified electrified interfaces. The basic knowledge obtained in these studies comprises a challenging testing ground for the electrochemical nanostructuring of surfaces and the tailoring of surface reactivity.

Acknowledgment

The present work was supported by the Volkswagen Foundation, the Deutsche Forschungsgemeinschaft, and the Research Center Jülich. I am deeply indebted to my collaborators in the groups in Ulm, and in Jülich, namely, Th. Dretschkow, M. H. Hölzle, D. Mayer, K. Ataka, G. Nagy, and S. Pron'kin. Results of their experimental work are partially incorporated in this chapter. In addition, I would like to thank J. X. Wang, B. M. Ocko, and R. R. Adzic very much for their pleasant and fruitful cooperation during SXS-measurements at Brookhaven National Laboratory. Furthermore, I gratefully acknowledge the critical comments and interesting discussions with N. J. Tao, G. Nagy and U. Retter, and O. Magnussen for sending Ref. [21] prior to publication. Finally, I would like to thank R. de Levie, D. M. Kolb, and H. Ibach for the encouragement and the continuous support in these studies.

References

1. A. Zangwill, *Physics at Surfaces*, Cambridge University Press, Cambridge, 1988.
2. E. Budevski, G. Staikov, W. J. Lorenz, *Electrochemical Phase Formation and Growth*, VCH, Weinheim, 1996.
3. C. Buess Herman, S. Bare, M. Poelman et al., in *Interfacial Electrochemistry* (Ed.: A. Wieckowski), Marcel Dekker, New York, 1999, pp. 427–452.
4. A. Wieckowski, *Interfacial Electrochemistry*, Marcel Dekker, New York, 1999.
5. J. A. Venables, *Introduction to Surface and Thin Film Processes*, Cambridge University Press, Cambridge, 2000.
6. W. J. Lorenz, W. Plieth, *Electrochemical Nanotechnology*, Wiley/VCH, Weinheim, 1998.
7. J. Lipkowski, P. N. Ross, *Adsorption of Organic Molecules at Metal Electrodes*, VCH, New York, 1992.
8. D. M. Kolb, *Prog. Surf. Sci.* **1996**, *51*, 109–173.
9. H. S. Nalwa, *Handbook of Nanostructured Materials*, Academic Press, San Diego, 2000, Vol. 5.
10. G. A. Somorjai, *Introduction to Surface Chemistry and Catalysis*, John Wiley & Sons, New York, 1994.
11. A. Baiker, J. D. Grunwaldt, Ch. A. Müller et al., *Chimia* **1998**, *52*, 517–524.
12. A. Ahmadi, G. Attard, J. Feliu et al., *Langmuir* **1999**, *15*, 2420–2424.
13. N. J. Tao, *Phys. Rev. Lett.* **1996**, *76*, 4066–4069.
14. J. Lipkowski, P. N. Ross, *Imaging of Surfaces and Interfaces*, Wiley-VCH, New York, 1999.
15. R. de Levie, *Chem. Rev.* **1988**, *88*, 599–609.
16. C. Buess-Herman, *Prog. Surf. Sci.* **1994**, *46*, 335–375.
17. M. Fleischmann, H. Thirsk in *Advances in Electrochemistry and Electrochemical Engineering* (Ed.: P. Delahay), John Wiley & Sons, New York, 1993, pp. 123–210.
18. M. R. Moncelli, M. Innocenti, R. Guidelli, *J. Electroanal. Chem.* **1990**, *295*, 275–290.
19. M. Freymann, M. Poelmann, C. Buess-Herman, *Isr. J. Chem.* **1997**, *37*, 241–246.
20. Th. Dretschkow, Th. Wandlowski, *The Solid-Liquid Interfaces – A Surface Science Approach* (Ed.: K. Wandelt), Springer, Berlin, 2002, in press.
21. O. M. Magnussen, *Chem. Rev.* **2002**, *102*, 679–725.
22. E. Herreo, L. J. Buller, H. Abruna, *Chem. Rev.* **2001**, *101*, 1897–1930.
23. K. Itaya, *Prog. Surf. Sci.* **1998**, *58*, 121–247.
24. E. Herrero, in *Encyclopedia of Analytical Chemistry* (Ed.: R. A. Meyers), John Wiley & Sons, New York, 2002, in press.
25. D. M. Kolb, in *Spectroelectrochemistry* (Ed.: R. J. Gale), Plenum Press, New York, 1988, pp. 87–188.
26. R. Corn, in *Adsorption of Organic Molecules* (Eds.: J. Lipkowski, P. N. Ross), VCH, New York, 1992, pp. 391–408.
27. M. J. Weaver, S. Zou, in *Advances in Spectroscopy* (Eds.: R. J. H. Clark, R. E. Hesters), John Wiley & Sons, Chichester, 1998, Vol. 26, Chap. 5.
28. M. Osawa, *Bull. Chem. Soc. Jpn.* **1997**, *70*, 2861–2880.
29. Z. Q. Tian, B. Ren, in *Encyclopedia of Analytical Chemistry* (Ed.: R. A. Meyer), John Wiley & Sons, New York, 2001, pp. 9162–9201.
30. A. Tadjeddine, A. LeRille, in *Interfacial Electrochemistry* (Ed.: A. Wieckowski), Marcel Dekker, New York, 1999, pp. 317–343.
31. I. Lyuksyukov, A. G. Naumovets, V. Pokrovsky, *Two-dimensional Crystals*, Academic Press, Boston, 1992.
32. T. L. Einstein, in *Chemistry and Physics of Solid Surfaces VII* (Eds.: R. Vanselow, R. Howe), Springer, Berlin 1988, pp. 307–336.
33. B. N. J. Person, *Surf. Sci. Rep.* **1992**, *15*, 1–135.
34. U. Retter, *J. Electroanal. Chem.* **1987**, *236*, 21–30.
35. U. Retter, *J. Electroanal. Chem.* **1984**, *165*, 221–230.
36. U. Retter, *Electrochim. Acta* **1996**, *41*, 2171–2174.
37. W. Gebhardt, U. Krey, *Phasenübergänge und kritische Phänomene*, F. Vieweg & Sons, Braunschweig, 1980.
38. G. Brown, P. A. Rikvold, S. M. Mitchell et al., in *Interfacial Electrochemistry* (Ed.: A. Wieckowski), Marcel Dekker, New York, 1999, pp. 47–61.
39. J. X. Wang, R. R. Adzic, B. M. Ocko, in *Interfacial Electrochemistry* (Ed.: A. Wieckowksi), Marcel Dekker, New York, 1999, pp. 175–186.
40. P. Ehrenfest, *Leiden Comm. Suppl.* **1933**, *75b*, 1–10.

41. L. D. Landau, E. M. Lifshitz, *Statistical Physics*, Pergamon Press, New York, 1980.
42. Cl. Buess-Herman, *J. Electroanal. Chem.* **1985**, *186*, 27–39.
43. P. Nikitas, *J. Electroanal. Chem.* **1991**, *300*, 607–618.
44. G. L. Gaines, *Insoluble Monolayers at Liquid/gas Interfaces*, Wiley-Interscience, New York, 1966.
45. V. Vetterl, *Collect. Czech. Chem. Commun.* **1966**, *31*, 2105–2111.
46. J. Lipkowski, in *Modern Aspects of Electrochemistry* (Eds.: R. E. White, J. O'. M. Bockris, B. E. Conwa), Plenum Press, New York, 1992, pp. 1–99, Vol. 23.
47. Th. Wandlowski, M. Heyrovski, L. Novotney, *Electrochim. Acta* **1992**, *37*, 2663–2672.
48. A. Bewick, B. Thomas, *J. Electroanal. Chem.* **1975**, *65*, 239–244.
49. A. Bewick, B. Thomas, *J. Electroanal. Chem.* **1977**, *84*, 127–140.
50. A. Bewick, B. Thomas, *J. Electroanal. Chem.* **1977**, *85*, 329–337.
51. Th. Wandlowski, J. X. Wang, O. M. Magnussen et al., *J. Phys. Chem.* **1996**, *100*, 10 277–10 287.
52. M. Fisher, *Rep. Phys. London* **1967**, *30*, 615.
53. B. M. Ocko, J. X. Wang, T. Wandlowski, *Phys. Rev. Lett.* **1997**, *79*, 1511–1574.
54. M. Schick, *Prog. Surf. Sci.* **1981**, *11*, 245–292.
55. Th. Wandlowski, J. X. Wang, B. M. Ocko, *J. Electroanal. Chem.* **2001**, *500*, 418–434.
56. M. T. M. Koper, *J. Electroanal. Chem.* **1998**, *450*, 189–201.
57. S. J. Mitchell, G. Brown, P. A. Rikvold, *Surf. Sci.* **2001**, *471*, 125–142.
58. D. E. Taylor, E. D. Williams, R. L. Park et al., *Phys. Rev. B* **1985**, *32*, 4653–4659.
59. L. Onsager, *Phys. Rev.* **1944**, *665*, 117–125.
60. T. Hill, *Statistical Mechanics*, McGraw Hill, New York, 1956, pp. 286–353, Chap. 7.
61. K. G. Wilson, J. Kogut, *Phys. Rep.* **1974**, *12*, 75–199.
62. K. Binder, D. W. Hermann, *Monte Carlo Simulation in Statistical Physics*, Springer Series Solid State Sci., Springer, Heidelberg, 1992, Vol. 80.
63. M. C. Desjonqueres, D. Spanjaard, *Concepts in Surface Physics*, Springer, Berlin, 1996.
64. J. Israelashvilli, *Intermolecular and Surface Forces*, Academic Press, London, 1992.
65. N. D. Lang, A. R. Williams, *Phys. Rev. B* **1978**, *18*, 616–636.
66. T. L. Einstein, *CRC Crit. Rev. Solid State Mater. Sci.*, **1978**, *7*, 261–299.
67. L. Blum, D. A. Huckaby, M. Legault, *Electrochim. Acta* **1996**, *41*, 2207–2227.
68. Y. I. Frenkel, T. Kontorova, *Zh. Eksp. Teor. Fiz.* **1938**, *8*, 1340–1345.
69. F. C. Frank, J. H. van der Merwe, *Proc. R. Soc. London* **1949**, *198*, 205–216.
70. V. L. Pokrofsky, A. L. Talapov, *Phys. Rev. Lett.* **1979**, *42*, 66–67.
71. A. D. Novaco, J. P. McTague, *Phys. Rev. Lett.* **1977**, *38*, 1286–1289.
72. J. P. McTague, A. D. Novaco, *Phys. Rev. B* **1979**, *19*, 5299–5306.
73. T. Hill, *Statistical Thermodynamics*, Addison Wesley, Reading, 1962.
74. W. Bragg, E. Williams, *Proc. R. Soc. London, Ser. A* **1934**, *145*, 699–730.
75. W. Bragg, E. Williams, *Proc. R. Soc. London, Ser. A* **1935**, *151*, 540–566.
76. R. Fowler, E. Guggenheim, *Statistical Thermodynamics*; Cambridge University Press, London, 1939.
77. E. Guggenheim, *Proc. R. Soc. London, Ser. A* **1935**, *148*, 304–312.
78. H. Bethe, *Proc. R. Soc. London, Ser. A* **1935**, *150*, 552–575.
79. C. Domb, in *Phase Transitions and Critical Phenomena* (Eds.: C. Domb, M. Green), Academic Press, New York, 1974, pp. 357, Vol. 3.
80. M. Sykes, J. Essam, D. Gaumit, *J. Math. Phys.* **1983**, *6*, 283–287.
81. J. Koryta, *Collect. Czech. Chem. Commun.* **1953**, *18*, 206–213.
82. P. Delahay, I. Trachtenberg, *J. Am. Chem. Soc.* **1957**, *79*, 1362–2355.
83. R. Srinivasan, R. de Levie, *J. Electroanal. Chem.* **1986**, *205*, 303–307.
84. A. J. Bard, L. Faulkner, *Electrochemical Methods and Applications*, John Wiley & Sons, New York, 1980.
85. A. N. Frumkin, V. I. Melik-Gaikazyan, *Dokl. Akad. Nauk USSR* **1951**, *77*, 855–858.
86. V. I. Melik-Gaikazyan, *Zh. Fiz. Khim.* **1952**, *26*, 560–580, 1184–1189.
87. W. Lorenz, F. Möckel, *Z. Elektrochem.* **1956** *60*, 507–515, 939–944.
88. R. D. Armstrong, W. P. Race, H. R. Thirsk, *J. Electroanal. Chem.* **1968**, *16*, 517–529.
89. P. Delahay, D. Mohilner, *J. Am. Chem. Soc.* **1962**, *84*, 4247–4252.

90. P. Delahay, *J. Phys. Chem.* **1963**, *67*, 135–137.
91. D. Schuhmann, in *Proprietes Electriques des Interfacial Charges* (Ed.: D. Schuhmann), Masson, Paris, 1978, pp. 166–192.
92. M. Noel, K. I. Vasur, *Cyclic Voltammetry and Frontiers in Electrochemistry*, Aspect Publications, London, 1990.
93. P. Delahay, C. T. Fike, *J. Am. Chem. Soc.* **1958**, *80*, 2628–2630.
94. T. Berzins, P. Delahay, *J. Am. Chem. Soc.* **1955**, *49*, 906–909.
95. P. Delahay, I. Trachtenberg, *J. Am. Chem. Soc.* **1957**, *79*, 2355–2362.
96. L. Rampazzo, *Electrochim. Acta* **1965**, *14*, 733–741.
97. K. Jüttner, G. Staikov, W. J. Lorenz et al., *J. Electroanal. Chem.* **1977**, *80*, 67–80.
98. J. Michailik, B. B. Damaskin, *Sov. Electrochem.* **1979**, *15*, 478–481.
99. N. Emanuel, D. Knorre, *Kurs Chimitscheskoi Kinetiki*; Wyschaja schkola, Moskwa, 1962.
100. W. Lorenz, *Z. Elektrochem.* **1958**, *62*, 192–200.
101. R. D. Armstrong, *J. Electroanal. Chem.* **1969**, *20*, 168–170.
102. J. W. Gibbs, *Collected Works*, Longman's Green, London (1878), 1978.
103. M. Volmer, A. Weber, *Z. Phys. Chem.* **1926**, *119*, 277–301.
104. M. Volmer, *Kinetik der Phasenbildung*, Steinkopf, Dresden, 1939.
105. L. Farakas, *Z. Phys. Chem.* **1927**, *125*, 239.
106. I. N. Stranski, R. Kaishev, *Z. Phys. Chem.* **1934**, *B26*, 100–113.
107. I. N. Stranski, R. Kaishev, *Z. Phys. Chem.* **1934**, *B26*, 114–131.
108. I. N. Stranski, R. Kaishev, *Z. Phys. Chem.* **1934**, *B26*, 312–316.
109. I. N. Stranski, R. Kaishev, *Z. Phys. Chem.* **1934**, *A170*, 295–299.
110. R. Becker, W. Döring, *Ann. Phys.* **1935**, *24*, 719–753.
111. J. B. Zeldovich, *Acta Physicochim. (URSS)* **1943**, *18*, 1–12.
112. I. Ya. Frenkel, *J. Chem. Phys.* **1939**, *7*, 200–210; 538–546.
113. E. Budevski, G. Staikov, W. J. Lorenz, *Electrochim. Acta* **2000**, *45*, 2559–2574.
114. E. Budevski, in *Comprehensive Treatise of Electrochemistry* (Eds.: B. E. Conway, J. O. M. Bockris, E. Yeager et al.), Plenum, New York, 1983, pp. 399–450, Vol. 7.
115. E. Budevski, V. Bostanov, T. Vitanov et al., *Electrochim. Acta* **1966**, *11*, 1697–1707.
116. D. M. Kolb, in *Advances in Electrochemistry and Electrochemical Engineering* (Eds.: H. Gerischer, Ch. Tobias), John Wiley & Sons, New York, pp. 125–272, Vol. 11.
117. R. Philipp, U. Retter, *Thin Solid Films* **1992**, *207*, 42–50.
118. Southampton Electrochemistry Group, *Instrumental Methods in Electrochemistry*, Ellis Horwood, New York, 1990.
119. Cl. Buess-Herman, L. Gierst, *Colloids Surf.* **1984**, *12*, 137–150.
120. U. Retter, *J. Electroanal. Chem.* **1990**, *296*, 445–451.
121. D. Kashchiev, *Surf. Sci.* **1969**, *14*, 209–220.
122. D. Kashchiev, *Nucleation – Basic Theory and Applications*, Butterworth-Heinemann, Oxford, 2000.
123. R de Levie, in *Advances in Electrochemistry and Electrochemical Engineering* (Eds.: H. Gerischer, Ch. Tobias), John Wiley & Sons, New York, 1985, pp. 1–67, Vol. 13.
124. S. Toshev, I. Markov, *J. Cryst. Growth* **1968**, *34*, 436–445.
125. S. Toshev, I. Markov, *Ber. Bunsen-Ges. Phys. Chem.* **1969**, *73*, 184–192.
126. D. Kashchiev, *Surf. Sci.* **1969**, *18*, 389–397.
127. J. H. Sluyters, J. H. O. J. Wijenberg, W. H. Mulkder et al., *J. Electroanal. Chem.* **1989**, *261*, 263–272.
128. S. Toschev, J. Gutzow, *Phys. Status Solidi* **1967**, *21*, 683–689.
129. I. Stranski, *Z. Phys. Chem.* **1928**, *136*, 209–220.
130. G. Staikov, W. J. Lorenz, *Can. J. Chem.* **1997**, *75*, 1624–1634.
131. A. Milchev, *Electrochim. Acta* **1986**, *31*, 977–980.
132. A. Milchev, *J. Electroanal. Chem.* **1998**, *457*, 35–46.
133. P. M. Jacobs, F. C. Tompkins, in *Chemistry of the Solid State* (Ed.: W. E. Garner), Butterworths, London, 1955, pp. 181–212.
134. U. Retter, *J. Electroanal. Chem.* **1978**, *87*, 181–188.
135. W. Obretenov, I. Petrov, I. Nachev et al., *J. Electroanal. Chem.* **1980**, *109*, 195–198.
136. C. Donner, H. Baumgärtel, L. Pohlmann et al., *Ber. Bunsen-Ges. Phys. Chem.* **1996**, *100*, 403–412.
137. C. Donner, L. Pohlmann, H. Baumgärtel, *Surf. Sci.* **1996**, *345*, 363–372.

138. R. Philipp, U. Retter, *Z. Phys. Chem.* **1996**, *196*, 125–134.
139. I. Markov, *Crystal Growth for Beginners*, World Scientific, Singapore, 1995, pp. 63–146.
140. D. Walton, *J. Chem. Phys.* **1962**, *37*, 2182–2190.
141. S. Stoyanov, *Thin Solid Films* **1973**, *18*, 91–98.
142. A. Milchev, S. Stoyanov, R. Kaischev, *Thin Sold Films* **1974**, *22*, 255–265, 267–274.
143. C. Buess Herman, in *Adsorption of Organic Molecules at Metal Electrodes* (Eds.: J. Lipkowski, P. N. Ross), VCH, New York, 1992, pp. 77–118.
144. O. M. Magnussen, L. Zitzler, B. Gleich et al., *Electrochim. Acta* **2001**, *46*, 3725–3733.
145. F. C. Frank, *Proc. R. Soc. London, Ser. A* **1950**, *201*, 586–599.
146. S. Rangarajan, *Faraday Symp. Chem. Soc.* **1978**, *12*, 101–114.
147. S. Fletcher, A. Smith, *Electrochim. Acta* **1980**, *25*, 1019–1024.
148. S. H. Liu, *Solid State Phys.* **1986**, *39*, 207–211.
149. N. J. Tao, C. Z. Li, F. Cunha et al., *Scann. Microsc.* **2002**, in press.
150. T. A. Witten, L. M. Sander, *Phys. Rev. Lett.* **1981**, *47*, 1400–1403.
151. L. Pospisil, *J. Electroanal. Chem.* **1986**, *206*, 269–283.
152. L. Pospisil, *J. Phys. Chem.* **1988**, *92*, 2501–2506.
153. R. Philipp, *J. Electroanal. Chem.* **1990**, *290*, 67–78.
154. Th. Wandlowski, *J. Electroanal. Chem.* **1992**, *333*, 77–91.
155. U. Retter, *Langmuir* **2000**, *16*, 7752–7756.
156. E. Budevski, W. Bostanov, T. Vitanov et al., *Electrochim. Acta* **1966**, *11*, 1697–1707.
157. E. Budevski, W. Bostanov, T. Vitanov et al., *Phys. Status Solidi* **1966**, *13*, 577–588.
158. V. Bostanov, G. Staikov, D. K. Roe, *J. Electrochem. Soc.* **1975**, *122*, 1301–1305.
159. W. Obretenov, V. Bostanov, V. Popov, *J. Electroanal. Chem.* **1982**, *132*, 273–276.
160. L. Gierst, C. Frank, G. Quarin et al., *J. Electroanal. Chem.* **1981**, *129*, 353–363.
161. R. Sridharan, R. de Levie, *J. Electroanal. Chem.* **1987**, *218*, 287–296.
162. F. Thomas, Cl. Buess-Herman, L. Gierst, *J. Electroanal. Chem.* **1986**, *214*, 597–613.
163. R. Srinivasan, R. de Levie, *J. Phys. Chem.* **1987**, *91*, 2904–2908.
164. Th. Wandlowski, R. de Levie, *Bulg. Chem. Commun.* **1995**, *27*, 231–246.
165. E. Bosco, S. K. Rangarajan, *J. Chem. Soc., Faraday Trans.* **1981**, *177*, 483–495.
166. F. Canac, *C. R. Acad. Sci.* **1933**, *196*, 51–62.
167. A. N. Kolmogoroff, *Bull. Acad. Sci. USSR, Ser. Math. Nat.* **1937**, *3*, 355–361.
168. M. Avrami, *J. Chem. Phys.* **1939**, *7*, 1103–1111.
169. M. Avrami, *J. Chem. Phys.* **1940**, *8*, 212–224.
170. M. Avrami, *J. Chem. Phys.* **1941**, *9*, 177–184.
171. U. R. Evans, *Trans. Faraday Soc.* **1945**, *41*, 365–374.
172. M. Y. Abyaneh, M. Fleischmann, *Electrochim. Acta* **1982**, *27*, 1513–1518.
173. U. Retter, *J. Electroanal. Chem.* **1984**, *179*, 25–29.
174. U. Retter, *J. Electroanal. Chem.* **1980**, *106*, 371–375.
175. G. Quarin, Cl. Buess-Herman, L. Gierst, *J. Electroanal. Chem.* **1981**, *123*, 35–58.
176. Th. Wandlowski, G. R. Jameson, R. de Levie, *J. Phys. Chem.* **1993**, *97*, 10 119–10 126.
177. A. Bewick, M. Fleischmann, H. R. Thirsk, *Trans. Faraday Soc.* **1962**, *58*, 2200–2209.
178. R. D. Armstrong, M. Fleischmann, H. R. Thirsk, *J. Electroanal. Chem.* **1966**, *11*, 208–215.
179. J. A. Harrison, H. R. Thirsk, in *Electroanalytical Chemistry* (Ed.: A. J. Bard), Marcel Dekker, New York, 1971, pp. 67–148, Vol. 5.
180. C. Müller, J. Claret, M. Sartet, *J. Electroanal. Chem.* **1987**, *227*, 147–158.
181. M. H. Hölzle, U. Retter, D. M. Kolb, *J. Electroanal. Chem.* **1994**, *371*, 101–109.
182. R. Philipp, J. Dittrich, U. Retter et al., *J. Electroanal. Chem.* **1988**, *250*, 159–164.
183. Th. Wandlowski, Th. Dretschkow, *J. Electroanal. Chem.* **1997**, *427*, 105–112.
184. R. G. Barradas, T. J. Vandernoot, *J. Electroanal. Chem.* **1982**, *142*, 107–119.
185. R. G. Barradas, T. J. Vandernoot, *J. Electroanal. Chem.* **1984**, *176*, 151–167.
186. R. G. Barradas, M. Rennie, T. Vandernoot, *J. Electroanal. Chem.* **1983**, *144*, 455–458.
187. W. H. Mulder, J. H. O. J. Wijenberg, M. Sluyters-Rehbach et al., *J. Electroanal. Chem.* **1989**, *270*, 7–19.
188. Th. Wandlowski, *J. Electroanal. Chem.* **1992**, *333*, 77–91.

189. R. Kaishev, B. Mutafschiew, *Electrochim. Acta* **1965**, *19*, 643–652.
190. U. Retter, *J. Electroanal. Chem.* **1982**, *136*, 167–174.
191. W. Obretenov, I. Petrov, I. Nachev et al., *J. Electroanal. Chem.* **1980**, *109*, 195–198.
192. R. G. Barradas, E. Bosco, *Electrochim. Acta* **1986**, *31*, 949–963.
193. R. Guidelli, M. L. Foresti, M. Innocenti, *J. Phys. Chem.* **1996**, *100*, 47–53.
194. C. Donner, L. Pohlmann, *Langmuir* **1999**, *15*, 4898–4915.
195. R. A. Ramos, P. A. Rikvold, M. A. Novotney, *Phys. Rev. B* **1999**, *59*, 9053–9069.
196. P. A. Rikvold, A. Wieckowski, R. A. Ramos, *Mater. Res. Soc. Symp. Proc.* **1997**, *451*, 69–81.
197. P. A. Rikvold, G. G. Brown, M. A. Novotny et al., *Colloids Surf. A* **1998**, *134*, 3–14.
198. G. Brown, P. A. Rikvold, M. A. Novotney et al., *J. Electrochem. Soc.* **1999**, *146*, 1035–1040.
199. K. Ataka, G. Nishina, W. B. Cai et al., *Electrochem. Commun.* **2000**, *2*, 417–421.
200. A. C. Finnefrock, K. L. Ringland, J. D. Brock et al., *Phys. Rev. Lett.* **1998**, *81*, 3459–3462.
201. Th. Wandlowski, D. M. Lampner, S. M. Lindsay, *J. Electroanal. Chem.* **1996**, *404*, 215–226.
202. J. G. Dash, *Rev. Mod. Phys.* **1999**, *71*, 1737–1743.
203. R. Sridharan, R. de Levie, *J. Electroanal. Chem.* **1997**, *230*, 241–256.
204. Th. Wandlowski, *Habilitatition*, University Press Halle, Halle, 1990.
205. U. Retter, W. Kant; *Thin Solid Films* **1995**, *256*, 89–93, *265*, 101–106.
206. C. Franck, PhD Thesis, ULB, Brussels, 1986.
207. M. Poelman, C. Buess-Herman, J. P. Badiali, *Langmuir* **1999**, *15*, 2194–2201.
208. H. Jehring, U. Retter, E. Horn, *J. Electroanal. Chem.* **1983**, *149*, 153–166.
209. R. Srinivasan, P. Gopalan, *J. Phys. Chem.* **1993**, *97*, 8770–8775.
210. N. J. Tao, Z. Shi, *J. Phys. Chem.* **1994**, *98*, 1464–1471.
211. W. H. Mulder, *J. Electroanal. Chem.* **1994**, *366*, 287–293.
212. M. Fleischmann, H. R. Thirsk, *Electrochim. Acta* **1964**, *9*, 757–771.
213. M. Fleischmann, H. R. Thirsk, *J. Electrochem. Soc.* **1963**, *110*, 688–698.
214. I. G. Abidov, V. Arakelyan, L. V. Chernomordik et al., *Bioelectrochem. Bioenerg.* **1979**, *6*, 37–63.
215. D. Exerova, D. Kashchiev, *Contemp. Phys.* **1986**, *27*, 429–436.
216. J. J. Calvente, Z. Kovacova, M. D. Sanchez et al., *Langmuir* **1996**, *12*, 5696–5903.
217. D. F. Yang, M. Morin, *J. Electroanal. Chem.* **1997**, *429*, 1–5.
218. D. F. Yang, M. Morin, *J. Electroanal. Chem.* **1998**, *441*, 173–181.
219. M. H. Hölzle, Th. Wandlowski, D. M. Kolb, *Surf. Sci.* **1995**, *335*, 281–290.
220. M. van Krieken, C. Buess-Herman, *Electrochim. Acta* **1998**, *43*, 2831–2841.
221. S. Bare, C. Buess-Herman, *Colloids Surf., A* **1998**, *134*, 181–191.
222. Th. Dretschkow, D. Lampner, Th. Wandlowski, *J. Electroanal. Chem.* **1998**, *458*, 121–138.
223. S. T. Chui, *Phys. Rev. Lett.* **1982**, *48*, 933–935.
224. K. I. Strandburg, *Rev. Mod. Phys.* **1988**, *60*, 161–207.
225. Th. Wandlowski, 2001, unpublished
226. J. M. Kosterlitz, D. J. Thouless, *J. Phys. C* **1973**, *6*, 1181–1203.
227. B. I. Halperlin, D. R. Nelson, *Phys. Rev. Lett.* **1978**, *41*, 121–124, 519–521.
228. A. P. Young, *Phys. Rev. B* **1979**, *19*, 1855–1866.
229. D. C. Grahame, *Chem. Rev.* **1947**, *41*, 441–501.
230. J. O'. M. Bockris, B. E. Conwasy, E. Yeager, (Eds.), *Comprehensive Treatise of Electrochemistry*, Plenum Press, New York, 1980, Vol. 1.
231. A. Hamelin, T. Vitanov, E. Sevastyanonov et al., *J. Electroanal. Chem.* **1983**, *145*, 225–264.
232. A. Hamelin, in *Modern Aspects of Electrochemistry* (Ed.: J. O'. M. Bockris), Plenum Press, New York, 1987, pp. 1–101.
233. A. Hamelin, Valette, R. Parsons, *Z. Phys. Chem. N.F.* **1978**, *113*, 71–89.
234. J. Lipkowski, Z. Shi, A. Chen et al., *Electrochim. Acta* **1998**, *43*, 2875–2888.
235. O. M. Magnussen, J. Hageböck, J. Hotlos et al., *Faraday Discuss.* **1992**, *94*, 329–338.
236. M. Kleinert, A. Cuesta, L. A. Kibler et al., *Surf. Sci. Lett.* **1999**, *430*, L521–L526.
237. D. A. Scherson, D. M. Kolb, *J. Electroanal. Chem.* **1984**, *176*, 353–357.
238. G. Gouy, *Ann. Chim. Phys.* **1903**, *29*, 145–242.

239. G. Gouy, *Ann. Chim. Phys.* **1906**, *8*, 291–364.
240. G. Gouy, *Ann. Chim. Phys.* **1906**, *9*, 75–139.
241. W. Schmickler; *Chem. Rev.* **1996**, *96*, 3177–3200.
242. D. M. Kolb, *Z. Phys. Chem. N.F.* **1987**, *154*, 179–199.
243. T. Iwasita, F. C. Nart, in *Advances in Electrochemical Science and Engineering* (Eds.: H. Gerischer, C. W. Tobias), VCH, New York, 1995, pp. 126–216.
244. G. Pacchioni, P. S. Bagus, M. R. Philpott, *Z. Phys. D* **1989**, *12*, 543–546.
245. A. Ignaczak, J. A. N. F. Gomes, *J. Electroanal. Chem.* **1997**, *420*, 71–78.
246. M. T. M. Koper, R. A. van Santen, *Surf. Sci.* **1999**, *422*, 118–131.
247. E. Spohr, in *Advances in Electrochemical Sciences and Engineering* (Eds.: R. C. Alkire, D. M. Kolb), Wiley-VCH, Weinheim, 1999, pp. 1–76, Vol. 6.
248. A. N. Frumkin, *Z. Phys. Chem.* **1933**, *164A*, 121–133.
249. D. M. Kolb, *Prog. Surf. Sci.* **1996**, *51*, 109–173.
250. A. R. Despic, in *Kinetics and Mechanisms of Electrode Processes* (Eds.: B. E. Conway, J. O'. M. Bockris, E. Yeager et al.), Plenum Press, New York, 1993, pp. 451–528.
251. R. R. Adzic, J. X. Wang, *Electrochim. Acta* **2000**, *45*, 4203–4210.
252. X. Gao, M. J. Weaver, *J. Am. Chem. Soc.* **1992**, *114*, 8544–8551.
253. O. M. Magnussen, J. X. Wang, R. R. Adzic et al., *J. Phys. Chem.* **1996**, *100*, 5500–5508.
254. O. M. Magnussen, B. M. Ocko, R. R. Adzic et al., *Phys. Rev. B* **1995**, *51*, 5510–5513.
255. B. M. Ocko, G. M. Watson, J. Wang, *J. Phys. Chem.* **1994**, *98*, 897–906.
256. X. Gao, G. J. Edens, F. C. Liu et al., *J. Phys. Chem.* **1994**, *98*, 8086–8095.
257. X. Gao, G. J. Edens, M. J. Weaver, *J. Phys. Chem.* **1994**, *98*, 8074–8085.
258. T. Yamada, N. Batina, K. Itaya, *J. Phys. Chem.* **1995**, *99*, 8817–8823.
259. B. M. Ocko, O. M. Magnussen, J. X. Wang et al., *Phys. Rev. B* **1996**, *53*, R7654–R7657.
260. J. X. Wang, G. M. Watson, B. M. Ocko, *J. Phys. Chem.* **1996**, *100*, 6672–6677.
261. A. Cuesta, D. M. Kolb, *Surf. Sci.* **2000**, *465*, 310–316.
262. S. Zou, X. Gao, M. J. Weaver, *Surf. Sci.* **2000**, *452*, 44–57.
263. J. X. Wang, R. R. Adzic, T. Wandlowski et al., unpublished.
264. J. X. Wang, G. M. Watson, B. M. Ocko, unpublished.
265. B. M. Ocko, O. M. Magnussen, J. X. Wang et al., *Physica B* **1996**, *221*, 238–244.
266. G. Aloisi, A. M. Funtikov, T. Will, *J. Electroanal. Chem.* **1994**, *370*, 297–300.
267. T. Yamada, K. Ogaki, S. Okubo et al., *Surf. Sci.* **1996**, *369*, 321–335.
268. B. M. Ocko, Th. Wandlowski, *Mater. Res. Soc. Symp. Proc.* **1997**, *451*, 55–64.
269. T. Teshima, K. Ogaki, K. Itaya, *J. Phys. Chem. B* **1997**, *101*, 2046–2053.
270. P. Broekmann, M. Wilms, M. Kruft et al., *J. Electroanal. Chem.* **1999**, *467*, 307–324.
271. J. Inukai, Y. Osawa, K. Itaya, *J. Phys. Chem. B* **1998**, *102*, 1034–1040.
272. P. Broekmann, M. Anastasescu, A. Spaenig et al., *J. Electroanal. Chem.* **2001**, *500*, 241–254.
273. S. L. Yau, C. M. Vitus, B. C. Schardt, *J. Am. Chem. Soc.* **1990**, *112*, 3677–3679.
274. C. A. Lucas, N. M. Markovic, P. N. Ross, *Phys. Rev. B* **1997**, *55*, 7964–7971.
275. R. Vogel, I. Kamphausen, H. Baltruschat, *Ber. Bunsen-Ges. Phys. Chem.* **1992**, *96*, 525–530.
276. J. Inukai, Y. Osawa, M. Wakisaka et al., *J. Phys. Chem. B* **1998**, *102*, 3498–3505.
277. J. M. Orts, R. Gómez, J. M. Feliu, *J. Electroanal. Chem.* **1999**, *467*, 11–19.
278. K. Sashikata, Y. Matsui, K. Itaya et al., *J. Phys. Chem.* **1996**, *100*, 20027–20034.
279. L. J. Wan, S. L. Yau, G. M. Swain et al., *J. Electroanal. Chem.* **1995**, *381*, 105–111.
280. P. Müller, S. Ando, T. Yamada et al., *J. Electroanal. Chem.* **1999**, *467*, 282–290.
281. H. H. Farell, in *The Chemical Physics of Solid Surfaces and Heterogeneous Catalysis* (Eds.: D. A. King, D. P. Woodruff), Elsevier, New York, 1987, pp. 225–272, Vol. 36.
282. P. A. Dowden, *Crit. Rev. Solid State Mater. Sci.* **1987**, *13*, 191–210.
283. G. N. Kastanas, B. E. Koel, *Appl. Surf. Sci.* **1993**, *64*, 235–249.
284. P. J. Goddard, R. M. Lambert, *Surf. Sci.* **1977**, *67*, 180–194.
285. K. Kern, G. Comsa, in *Chemistry and Physics of Solid Surfaces* (Eds.: R. Vanselow, R. Howe), Springer, Berlin, 1988, pp. 65–108.
286. P. Bak, D. Mukamel, I. Villain et al., *Phys. Rev. B* **1979**, *19*, 1610–1613.

287. M. Kardar, A. N. Berker, *Phys. Rev. Lett.* **1982**, *48*, 1552–1555.
288. M. F. Toney, J. G. Gordon, M. G. Samant et al., *Phys. Rev. B* **1992**, *45*, 9362–9374.
289. X. Wang, R. Chen, Y. Wang et al., *J. Phys. Chem. B* **1998**, *102*, 7568–7576.
290. Z. Racz, *Phys. Rev. B* **1980**, *21*, 4012–4016.
291. S. J. Mitchell, G. Brown, P. A. Rikvold, *J. Electroanal. Chem.* **2000**, *493*, 68–74.
292. W. Haiss, PhD Thesis, Free University Berlin, Berlin, 1994.
293. F. Silva, A. Martins, in *Interfacial Electrochemistry* (Ed.: A. Wieckowski), Marcel Dekker, New York, 1999, pp. 427–461.
294. H. Angerstein-Kozlowska, B. E. Conway, A. Hamelin et al., *J. Electroanal. Chem.* **1987**, *228*, 429–453.
295. J. Clavilier, *J. Electroanal. Chem.* **1980**, *107*, 211–216.
296. J. M. Feliu, M. T. Valls, A. Aldaz et al., *J. Electroanal. Chem.* **1993**, *345*, 475–481.
297. F. Silva, M. J. Sottomayor, A. Martins, *J. Electroanal. Chem.* **1994**, *375*, 395–399.
298. G. J. Edens, X. Gao, M. J. Weaver, *J. Electroanal. Chem.* **1994**, *375*, 357–366.
299. Th. Dretschkow, Th. Wandlowski, *Ber. Bunsen-Ges. Phys. Chem.* **1997**, *101*, 749–757.
300. A. M. Funtikov, U. Linke, U. Stimming et al., *Surf. Sci.* **1995**, *324*, L343–L378.
301. A. M. Funtikov, U. Stimming, R. Vogel, *J. Electroanal. Chem.* **1997**, *428*, 147–153.
302. L. J. Wan, S. L. Yau, K. Itaya, *J. Phys. Chem.* **1995**, *99*, 9507–9513.
303. M. Wilms, P. Broekmann, M. Kruft et al., *Surf. Sci.* **1998**, *404*, 83–86.
304. M. Wilms, P. Broekmann, C. Stuhlmann et al., *Surf. Sci.* **1998**, *416*, 121–140.
305. P. Broekmann, M. Wilms, M. Kruft et al., *J. Electroanal. Chem.* **1999**, *467*, 307–324.
306. W. H. Li, R. J. Nichols, *J. Electroanal. Chem.* **1998**, *456*, 153–160.
307. L. J. Wan, M. Hara, J. Inukai et al., *J. Phys. Chem. B* **1999**, *103*, 6978–6983.
308. L. J. Wan, T. Suzuki, K. Sashikata et al., *J. Electroanal. Chem.* **2000**, *484*, 189–193.
309. A. Cuesta, M. Kleinert, D. M. Kolb, *Phys. Chem. Chem. Phys.* **2000**, *2*, 5684–5690.
310. H. Angerstein-Kozlowska, B. E. Conway, A. Hamelin et al., *Electrochim. Acta* **1986**, *31*, 1051–1061.
311. Z. Shi, J. Lipkowski, M. Gamboa et al., *J. Electroanal. Chem.* **1994**, *366*, 317–326.
312. H. Uchida, N. Ikeda, M. Watanabe, *J. Electroanal. Chem.* **1997**, *424*, 5–12.
313. Y. Shingaya, M. Ito, *Electrochim. Acta* **1998**, *44*, 745–751.
314. K. Ataka, M. Osawa, *Langmuir* **1998**, *14*, 951–959.
315. I. R. de Moraes, F. C. Nart, *J. Electroanal. Chem.* **1999**, *461*, 110–120.
316. Z. Shi, J. Lipkowski, S. Mirwald et al., *J. Electroanal. Chem.* **1995**, *396*, 115–124.
317. T. Nishizawa, T. Namata, Y. Kinoshita et al., *Surf. Sci.* **1996**, *367*, L73–L78.
318. E. M. Patrito, P. Paredes Olivera, H. Sellers, *Surf. Sci.* **1997**, *380*, 264–282.
319. D. M. Kolb, M. Przasnyski, H. Gerischer, *J. Electroanal. Chem.* **1974**, *54*, 25–38.
320. J. W. Schultze, D. Dickertmann, *Faraday Symp.* **1977**, *12*, 36.
321. J. W. Schultze, D. Dickertmann, *Faraday Symp.* **1977**, *12*, 172.
322. J. W. Schultze, D. Dickertmann, *Surf. Sci.* **1976**, *54*, 489–505.
323. E. Leiva, *Electrochim. Acta* **1996**, *41*, 2185–2206.
324. W. J. Lorenz, H. D. Hermann, N. Wütrich et al., *J. Electrochem. Soc.* **1974**, *121*, 1167–1177.
325. R. Adzic, E. Yeager, B. D. Cahan, *J. Electrochem. Soc.* **1974**, *121*, 474–484.
326. K. Jüttner, W. J. Lorenz, *Z. Phys. Chem. N.F.* **1980**, *122*, 163–185.
327. G. Kokkinidis, *J. Electroanal. Chem.* **1986**, *201*, 217–236.
328. R. Adzic, in *Advances in Electrochemistry and Electrochemical Engineering* (Eds.: H. Gerischer, C. Tobias), John Wiley & Sons, New York, 1984, pp. 159, Vol. 13.
329. K. Engelsmann, W. J. Lorenz, E. Schmidt, *J. Electroanal. Chem.* **1980**, *114*, 1–10.
330. O. Melroy, K. Kanazawa, J. J. G. Gordon et al., *Langmuir* **1986**, *2*, 697–700.
331. G. L. Richmond, M. Robinson, V. L. Shannon, *Prog. Surf. Sci.* **1988**, *28*, 1–70.
332. G. Lehmpfuhl, Y. Uchida, M. S. Zei et al., in *Imaging of Surfaces and Interfaces* (Eds.: J. Lipkowski, P. N. Ross), John Wiley & Sons, New York, 1999, pp. 57–98.
333. R. Kötz in *Advances in Electrochemical Science and Engineering* (Eds.: H. Gerischer, Ch. Tobias), VCH, Weinheim, 1990, pp. 75–109, Vol. 1.
334. M. Toney, O. R. Melroy, in *Electrochemical Interfaces – Modern Techniques for in situ*

Characterization (Ed.: H. D. Abruna), VCH, Berlin, 1991, pp. 57–132.
335. A. A. Gewirth, B. K. Niece, *Chem. Rev.* **1997**, *97*, 1129–1162.
336. G. Staikov, W. J. Lorenz, E. Budevski, in *Imaging of Surfaces and Interfaces* (Eds.: J. Lipkowski, P. N. Ross), John Wiley & Sons, New York, 1999, pp. 1–56.
337. M. H. Hölzle, V. Zwing, D. M. Kolb, *Electrochim. Acta* **1995**, *40*, 1237–1247.
338. I. H. Omar, H. J. Pauling, K. Jüttner, *J. Electrochem. Soc.* **1993**, *140*, 2187–2192.
339. Z. Shi, J. Lipkowski, *J. Electroanal. Chem.* **1994**, *364*, 289–294.
340. Z. Shi, J. Lipkowski, *J. Electroanal. Chem.* **1994**, *365*, 303–309.
341. O. M. Magnussen, J. Hotlos, R. Nichols et al., *Phys. Rev. Lett.* **1990**, *64*, 2929–2932.
342. T. Hachiya, H. Honbo, K. Itaya, *J. Electroanal. Chem.* **1991**, *315*, 275–291.
343. T. Will, M. Dieterle, D. M. Kolb, in *Nanoscale Probes of Solid/Liquid Interfaces*, NATO ASI Series E (Eds.: A. A. Gewirth, H. Siegenthaler), Kluwer, Dordrecht, 1995, pp. 137–162, Vol. 288.
344. S. Manne, P. K. Hansma, J. Massie et al., *Science* **1991**, *251*, 183–186.
345. N. Ikemiya, S. Miyaoka, S. Hara, *Surf. Sci.* **1994**, *311*, L641–L648.
346. N. Ikemiya, S. Miyaoka, S. Hara, *Surf. Sci.* **1995**, *327*, 261–273.
347. D. B. Parry, M. G. Samant, H. Seki et al., *Langmuir* **1993**, *9*, 1878–1887.
348. L. Blum, H. D. Abruna, J. White et al., *J. Chem. Phys.* **1986**, *85*, 6732–6738.
349. O. R. Melroy, M. J. Samant, G. L. Borges et al., *Langmuir* **1988**, *4*, 728–732.
350. M. F. Toney, J. N. Howard, J. Richer et al., *Phys. Rev. Lett.* **1995**, *75*, 4472–4475.
351. J. G. Gordon, O. R. Melroy, M. F. Toney, *Electrochim. Acta* **1995**, *40*, 3–8.
352. G. L. Borges, K. K. Kanazawa, J. G. Gordon et al., *J. Electroanal. Chem.* **1994**, *364*, 281–284.
353. H. Uchida, M. Hiei, M. Watanabe, *J. Electroanal. Chem.* **1998**, *452*, 97–106.
354. Y. Nakai, M. S. Zei, D. M. Kolb et al., *Ber. Bunsen-Ges. Phys. Chem.* **1984**, *88*, 340–345.
355. J. Zhang, Y. E. Sung, P. A. Rikvold et al., *J. Chem. Phys.* **1996**, *104*, 5699–5712.
356. M. Legault, L. Blum, D. A. Huckaby, *J. Electroanal. Chem.* **1996**, *409*, 79–86.
357. D. A. Huckaby, M. D. Legault, L. Blum, *J. Chem. Phys.* **1998**, *109*, 3600–3606.
358. L. Blum, M. D. Legault, D. A. Huckaby, in *Interfacial Electrochemistry* (Ed.: A. Wieckowski), Marcel Dekker, New York, 1999, pp. 19–31.
359. G. Brown, P. A. Rickvold, M. A. Novotny et al., *J. Electrochem. Soc.* **1999**, *146*, 1035–1040.
360. M. H. Hölzle, PhD Thesis, University of Ulm, Germany, 1995.
361. E. Bosco, S. K. Rangarajan, *J. Chem. Soc., Faraday Trans.* **1981**, *77*, 1673–1696.
362. K. Jüttner, G. Staikov, W. J. Lorenz et al., *J. Electroanal. Chem.* **1977**, *80*, 67–80.
363. G. Horanyi, E. M. Rizmayer, P. Jovi, *J. Electroanal. Chem.* **1983**, *152*, 211–222.
364. M. S. Zei, G. Qiao, G. Lehmpfuhl et al., *Ber. Bunsen-Ges. Phys. Chem.* **1987**, *91*, 349–353.
365. E. Herrero, S. Glazier, H. D. Abruna, *J. Phys. Chem. B* **1998**, *102*, 9825–9833.
366. S. Wu, J. Lipkowski, T. Tyliszcza et al., *Prog. Surf. Sci.* **1995**, *50*, 227–236.
367. Z. Shi, S. Wu, J. Lipkowski, *J. Electroanal. Chem.* **1995**, *284*, 171–177.
368. Z. Shi, J. Lipkowski, *J. Electroanal. Chem.* **1994**, *369*, 283–287.
369. R. Michaelis, PhD Thesis, Free University Berlin, Berlin, 1990.
370. Z. Shi, S. Wu, J. Lipkowski, *Electrochim. Acta* **1995**, *40*, 9–15.
371. N. Batina, T. Will, D. M. Kolb, *Faraday Discuss.* **1992**, *94*, 93–106.
372. F. A. Möller, O. M. Magnussen, R. J. Behm, *Phys. Rev. B* **1995**, *51*, 2484–2490.
373. R. Randler, M. Dieterle, D. M. Kolb, *Z. Phys. Chem.* **1999**, *208*, 43–56.
374. F. A. Möller, O. M. Magnussen, R. J. Behm, *Electrochim. Acta* **1995**, *40*, 1259–1265.
375. D. M. Kolb, R. Kötz, K. Yamamoto, *Surf. Sci.* **1979**, *87*, 20–30.
376. C. L. Scortichini, C. N. Reilley, *J. Electroanal. Chem.* **1982**, *139*, 233–245.
377. C. L. Scortichini, C. N. Reilley, *J. Electroanal. Chem.* **1982**, *139*, 255–260.
378. D. M. Kolb, K. A. Jaaf-Golze, M. S. Zei, *Dechema – Monographien* **1986**, *12*, 53–64.
379. Ch. Nishihara, A. Nozoye, *J. Electroanal. Chem.* **1995**, *386*, 75–82.
380. Ch. Nishihara, A. Nozoye, *J. Electroanal. Chem.* **1995**, *396*, 139–142.
381. R. Gomez, J. M. Feliu, H. D. Abruna, *Langmuir* **1994**, *10*, 4315–4323.
382. L. Buller, E. Herrero, R. Gomez et al., *J. Chem. Soc., Faraday Trans.* **1996**, *92*, 3757–3762.

383. P. Berenz, S. Tillmann, H. Massong et al., *Electrochim. Acta* **1998**, *43*, 3035–3043.
384. E. A. Meguid, P. Berenz, H. Baltruschat, *J. Electroanal. Chem.* **1999**, *467*, 50–59.
385. E. Herrero, S. Glazier, L. J. Buller et al., *J. Electroanal. Chem.* **1999**, *461*, 121–130.
386. L. Buller, E. Herrero, R. Gomez et al., *J. Phys. Chem. B* **2000**, *104*, 5932–5939.
387. K. Varga, P. Zelenay, A. Wieckowski, *J. Electroanal. Chem.* **1992**, *330*, 453–467.
388. G. Horanyi, *J. Electroanal. Chem.* **1974**, *55*, 45–51.
389. Y. Shingaya, H. Matsumoto, H. Ogasawara et al., *Surf. Sci.* **1995**, *335*, 23–31.
390. I. Oda, Y. Shingaya, H. Matsumoto et al., *J. Electroanal. Chem.* **1996**, *409*, 95–101.
391. M. Watanabe, H. Uchida, N. Ikeda, *J. Electroanal. Chem.* **1995**, *380*, 255–260.
392. H. Matsumoto, I. Oda, J. Inukai et al., *J. Electroanal. Chem.* **1993**, *356*, 275–280.
393. H. Matsumoto, I. Oda, J. Inukai et al., *J. Electroanal. Chem.* **1994**, *379*, 223–231.
394. G. Beitel, O. M. Magnussen, R. J. Behm, *Surf. Sci.* **1995**, *336*, 19–26.
395. K. Sashikata, N. Furuya, K. Itaya, *J. Electroanal. Chem.* **1991**, *316*, 361–368.
396. Z. L. Wu, Z. H. Zang, S. L. Yau, *Langmuir* **2000**, *16*, 3522–3528.
397. H. S. Yee, H. D. Abruna, *J. Phys. Chem.* **1993**, *97*, 6278–6288.
398. R. Durand, R. Faure, D. Aberdam et al., *Electrochim. Acta* **1992**, *37*, 1977–1982.
399. N. Markovic, H. A. Gasteiger, C. A. Lucas et al., *Surf. Sci.* **1995**, *335*, 91–100.
400. C. A. Lucas, N. M. Markovic, I. M. Tidswell et al., *Physica B* **1996**, *221*, 245–250.
401. E. Herrero, L. J. Buller, J. Li et al., *Electrochim. Acta* **1998**, *44*, 983–992.
402. R. Michaelis, M. S. Zei, R. S. Zhai et al., *J. Electroanal. Chem.* **1992**, *339*, 299–310.
403. R. Michaelis, D. M. Kolb *J. Electroanal. Chem.* **1992**, *328*, 341–348.
404. D. Aberdam, Y. Gauthier, R. Durand et al., *Surf. Sci.* **1994**, *306*, 114–124.
405. M. S. Zei, K. Wu, M. Eiswirthz et al., *Electrochim. Acta* **1999**, *45*, 809–817.
406. M. S. Zei, *Z. Phys. Chem.* **1999**, *208*, 77–91.
407. Th. Wandlowski, E. Herrero, J. Feliu, *J. Electroanal. Chem.*, in preparation
408. C. A. Lucas, N. M. Markovic, P. N. Ross, *Phys. Rev. B* **1997**, *56*, 3651–3654.
409. A. A. Aki, G. A. Attard, *J. Phys. Chem. B* **1997**, *101*, 4597–4606.
410. A. Bittner, J. Winterlin, G. Ertel, *Surf. Sci.* **1997**, *376*, 267–278.
411. A. A. Aki, G. A. Attard, R. Price et al., *J. Chem. Soc., Faraday Trans.* **1995**, *91*, 3585–3595.
412. N. Markovic, H. A. Gasteiger, P. N. Ross, *Langmuir* **1995**, *11*, 4098–4108.
413. I. A. Tidswell, C. A. Lucas, N. M. Markovic et al., *Phys. Rev. B* **1995**, *51*, 10205–10208.
414. D. Dickertmann, F. D. Koppitz, J. W. Schultze, *Electrochim. Acta* **1976**, *21*, 967–971.
415. W. J. Lorenz, E. Schmidt, G. Staikov et al., *Faraday Symp. Chem. Soc.* **1977**, *12*, 14–23.
416. K. Jüttner, G. G. Staikov, W. J. Lorenz et al., *J. Electroanal. Chem.* **1977**, *80*, 67–80.
417. H. Bort, K. Jüttner, W. J. Lorenz et al., *J. Electroanal. Chem.* **1978**, *90*, 413–424.
418. G. Staikov, K. Jüttner, W. J. Lorenz et al., *Electochim. Acta* **1978**, *23*, 305–313.
419. G. Staikov, K. Jüttner, W. J. Lorenz et al., *Electochim. Acta* **1978**, *23*, 319–324.
420. K. Takayanagi, D. M. Kolb, K. Kambe et al., *Surf. Sci.* **1980**, *100*, 407–422.
421. J. N. Jovicevic, V. D. Jovic, A. R. Despic, *Electrochim. Acta* **1984**, *29*, 1625–1632.
422. V. D. Jovic, J. N. Jovicevic, A. R. Despic, *Electrochim. Acta* **1985**, *30*, 1455–1464.
423. V. D. Jovic, B. M. Jovic, A. R. Despic, *J. Electroanal. Chem.* **1990**, *288*, 229–243.
424. K. J. Stevenson, D. W. Hatchett, H. S. White, *Langmuir* **1996**, *12*, 494–499.
425. D. A. Koos, V. L. Shannon, G. L. Richmond, *J. Phys. Chem.* **1990**, *94*, 2091–2098.
426. U. Müller, D. Carnal, H. Siegenthaler et al., *Phys. Rev. B* **1992**, *46*, 12899–12901.
427. D. Carnal, P. I. Oden, U. Müller et al., *Electrochim. Acta* **1995**, *40*, 1223–1235.
428. U. Schmidt, S. Vinzelberg, G. Staikov, *Surf. Sci.* **1996**, *348*, 261–279.
429. M. G. Samant, G. L. Borges, J. G. Gordon et al., *J. Am. Chem. Soc.* **1987**, *109*, 5970–5974.
430. M. G. Samant, M. F. Toney, G. L. Borges et al., *Surf. Sci.* **1988**, *193*, L29–L36.
431. M. F. Toney, J. G. Gordon, M. G. Samant et al., *J. Phys. Chem.* **1995**, *99*, 4733–4744.
432. M. E. Hansen, E. Yeager, *ACS Symp.* **1988**, *378*, 398–407.
433. L. Laguren–Davidson, F. Lu, G. Salaita et al., *Langmuir* **1988**, *4*, 224–232.
434. E. Amman, MS Thesis, University of Bern, Bern, Switzerland, 1995.

435. K. Takayanagi, *Surf. Sci.* **1981**, *104*, 527–548.
436. E. Altman, R. Colton, *Surf. Sci.* **1993**, *295*, 13–33.
437. A. Ullman, *Chem. Rev.* **1996**, *96*, 1533–1554.
438. R. E. Pagano, R. Miller, *J. Colloid. Interface Sci.* **1973**, *45*, 126–137.
439. A. Nelson, N. Auffret, *J. Electroanal. Chem.* **1988**, *244*, 99–113.
440. M. Porter, T. B. Bright, D. Allara et al., *J. Am. Chem. Soc.* **1987**, *109*, 3559–3568.
441. L. Strong, G. M. Whitsides, *Langmuir* **1988**, *4*, 546–558.
442. J. Sagiv, *J. Am. Chem. Soc.* **1980**, *102*, 92–98.
443. H. Finklea, D. A. Snider, J. Fedyk, *Langmuir* **1993**, *9*, 3360–3367.
444. P. Fenter, P. Eisenberger, K. S. Liang, *Phys. Rev. Lett.* **1993**, *70*, 2447–2450.
445. O. M. Magnussen, B. M. Ocko, M. Deutsch et al., *Nature* **1996**, *384*, 250–252.
446. A. A. Kornyshev, I. Vilfan, *Electrochim. Acta* **1995**, *40*, 109–127.
447. A. N. Frumkin, *Z. Phys. Chem.* **1925**, *116*, 466–485.
448. A. N. Frumkin, *Z. Phys.* **1926**, *35*, 792–802.
449. B. Damaskin, O. A. Petrii, V. V. Batrakov, *Adsorption of Organic Compounds on Electrodes*, Plenum Press, New York 1972.
450. R. Parsons, *Chem. Rev.* **1990**, *90*, 813–826.
451. W. Lorenz, *Z. Electrochem.* **1958**, *62*, 192–202.
452. D. A. Bonell, *Scanning Tunneling Microscopy and Spectroscopy*, VCH, New York, 1993.
453. N. J. Tao, in *Imaging of Surfaces and Interfaces* (Eds.: J. Lipkowski, P. N. Ross), John Wiley & Sons, New York, 1999, pp. 211–248.
454. W. H. Liu, W. Haiss, S. Floate et al., *Langmuir* **1999**, *15*, 4875–4883.
455. K. Takamura, F. Kusu, in *Methods od Biochemical Analysis* (Ed.: D. Glick), John Wiley & Sons, New York, 1987, pp. 155, Vol. 32.
456. H. H. Rotermund, K. Krischer, B. Pettinger, in *Imaging of Surfaces and Interfaces* (Eds.: J. Lipkowski, P. N. Ross), John Wiley & Sons, New York, 1999, pp. 139–210.
457. I. Burges, V. Zamlymny, G. Szymanski et al., *Langmuir* **2001**, *17*, 3355–3367.
458. Th. Wandlowski, B. M. Ocko, O. M. Magnussen et al., *J. Electroanal. Chem.* **1996**, *409*, 155–164.
459. V. V. Batrakov, B. B. Damaskin, Y. B. Ipatov, *Elektrokhimiya* **1974**, *10*, 216–220.
460. A. Hamelin, S. Morin, J. Richer et al., *J. Electroanal. Chem.* **1991**, *304*, 195–209.
461. A. Popov, R. Naneva, K. Dimitrov et al., *Electrochim. Acta* **1992**, *37*, 2369–2371.
462. B. Roelfs, H. Baumgärtel, *Ber. Bunsen-Ges. Phys. Chem.* **1994**, *99*, 677–681.
463. B. Roelfs, E. Bunge, C. Schröter et al., *J. Phys. Chem. B* **1997**, *101*, 754–765.
464. M. H. Hölzle, D. Krznaric, D. M. Kolb, *J. Electroanal. Chem.* **1995**, *386*, 235–239.
465. M. H. Hölzle, D. M. Kolb, *Ber. Bunsen-Ges. Phys. Chem.* **1994**, *98*, 330–335.
466. M. Scharfe, A. Hamelin, C. Buess-Herman, *Electrochim. Acta* **1995**, *40*, 61–67.
467. M. H. Hölzle, Th. Wandlowski, D. M. Kolb, *J. Electroanal. Chem.* **1995**, *394*, 271–275.
468. Th. Wandlowski, *J. Electroanal. Chem.* **1995**, *395*, 83–89.
469. Th. Wandlowski, M. H. Hölzle, *Langmuir* **1996**, *12*, 6597–6615.
470. D. Krznaric, B. Cosovic, M. H. Hölzle et al., *Ber. Bunsen-Ges. Phys. Chem.* **1996**, *100*, 1779–1790.
471. H. Striegler, M. H. Hölzle, Th. Wandlowski et al., Poster 1a-41, presented at the 47th Annual Meeting of the ISE, November 1–6, Budapest, 1996.
472. Th. Boland, B. Ratner, *Langmuir* **1994**, *10*, 3845–3852.
473. W. Haiss, B. Roelfs, S. N. Port et al., *J. Electroanal. Chem.* **1998**, *454*, 107–113.
474. M. Futamata, *J. Phys. Chem. B* **2001**, *105*, 6933–6942.
475. S. Promkin, Th. Wandlowski, unpublished.
476. Th. Wandlowski, K. Ataka, D. Mayer, *Langmuir* **2002**, la u255854.
477. N. J. Tao, J. A. DeRose, S. M. Lindsay, *J. Phys. Chem.* **1993**, *97*, 910–919.
478. Th. Dretschkow, A. S. Dakkouri, Th. Wandlowski, *Langmuir* **1997**, *13*, 2843–2856.
479. Th. Dretschkow, Th. Wandlowski, *Electrochim. Acta* **1998**, *43*, 2991–3006.
480. S. Soverby, personal communication. **1998**.
481. K. M. Richard, A. A. Gewirth, *J. Phys. Chem.* **1995**, *99*, 12288–12293.
482. G. Andreasen, M. E. Vela, R. C. Salvarezza et al., *Langmuir* **1997**, *13*, 6814–6819.
483. W. B. Cai, L. J. Wan, H. Noda et al., *Langmuir* **1998**, *14*, 6992–6998.
484. F. Cunha, N. J. Tao, *Phys. Rev. Lett.* **1995**, *75*, 2376–2379.
485. F. Cunha, N. J. Tao, X. W. Wang et al., *Langmuir* **1996**, *12*, 6410–6418.

486. Th. Dretschkow, D. Lampner, Th. Wandlowski, *J. Electroanal. Chem.* **1998**, *458*, 121–138.
487. Th. Dretschkow, Th. Wandlowski, *J. Electroanal. Chem.* **1999**, *467*, 207–216.
488. Th. Dretschkow, Th. Wandlowski, *Electrochim. Acta* **1999**, *45*, 731–740.
489. H. Noda, T. Minoha, L. J. Wan et al., *J. Electroanal. Chem.* **2000**, *481*, 62–68.
490. D. Mayer, K. Ataka, Th. Wandlowski, *J. Electroanal. Chem.* **2002**, in press.
491. F. Cunha, Q. Jing, N. J. Tao, *Surf. Sci.* **1997**, *389*, 19–28.
492. O. Dominguez, L. Echegoyen, F. Cunha et al., *Langmuir* **1998**, *14*, 821–824.
493. J. Pan, S. M. Lindsay, N. J. Tao, *Langmuir* **1993**, *9*, 1556–1560.
494. E. Bunge, R. J. Nichols, H. Baumgärtel et al., *Ber. Bunsen-Ges. Phys. Chem.* **1995**, *99*, 1243–1246.
495. E. Bunge, R. J. Nichols, B. Roelfs et al., *Langmuir* **1996**, *12*, 3060–3066.
496. A. S. Dakkouri, D. M. Kolb, R. Edelstein, Shima et al., *Langmuir* **1996**, *12*, 2849–2852.
497. L. J. Wan, H. Noda, Y. Hara et al., *J. Electroanal. Chem.* **2000**, *489*, 68–75.
498. L. J. Wan, M. Terashima, H. Noda et al., *J. Phys. Chem. B* **2000**, *104*, 3563–3569.
499. N. Batina, M. Kunikate, K. Itaya, *J. Electroanal. Chem.* **1995**, *405*, 245–250.
500. M. Kunikate, N. Batina, K. Itaya, *Langmuir* **1995**, *11*, 2337–2340.
501. K. Ogaki, N. Batina, M. Kunikate et al., *J. Phys. Chem.* **1996**, *100*, 7185–7190.
502. S. L. Yau, Y. G. Kim, K. Itaya, *J. Am. Chem. Soc.* **1996**, *118*, 7795–7803.
503. L. J. Wan, K. Itaya, *Langmuir* **1997**, *13*, 7173–7179.
504. R. Srinivasan, J. C. Murphy, R. Fainchtein et al., *J. Electroanal. Chem.* **1994**, *312*, 293–300.
505. R. Srinivasan, R. Gopalan, *J. Phys. Chem.* **1993**, *98*, 8770–8775.
506. N. J. Tao, Z. Shi, *Surf. Sci.* **1994**, *321*, L149–L156.
507. N. J. Tao, Z. Shi, *J. Phys. Chem.* **1994**, *98*, 7422–7426.
508. N. J. Tao, *Phys. Rev Lett.* **1996**, *76*, 4066–4069.
509. L. J. Wan, C. Wan, C. Bai et al., *J. Phys. Chem. B* **2001**, *105*, 8399–8402.
510. Cl. Bues-Herman, L. Gierst, N. Vanlaethem-Meuree, *J. Electroanal. Chem.* **1981**, *123*, 1–19.
511. Th. Wandlowski, G. B. Jameson, R. de Levie, *J. Phys. Chem.* **1993**, *97*, 10 119–10 126.
512. Th. Wandlowski, R. de Levie, *J. Electroanal. Chem.* **1992**, *329*, 103–127.
513. R. Guidelli, M. L. Foresti, *J. Electroanal. Chem.* **1986**, *197*, 123–141.
514. Th. Wandlowski, P. Chaiyasith, H. Baumgärtel, *J. Electroanal. Chem.* **1992**, *346*, 271–279.
515. D.M. Mohilner, in *Electroanal. Chem.* (Ed.: A. J. Bard), Marcel Dekker, New York, 1966, pp. 241–405, Vol. 1.
516. J. Lipkowski, W. Schmickler, D. M. Kolb et al., *J. Electroanal. Chem.* **1998**, *452*, 193–197.
517. S. Wu, J. Lipkowski, O. M. Magnussen et al., *J. Electroanal. Chem.* **1998**, *446*, 67–77.
518. L. Stolberg, J. Lipkowski, in *Adsorption of Organic Molecules at Metal Electrodes* (Eds.: J. Lipkowski, P. N. Ross), VCH, New York, 1992, pp. 171–238.
519. U. Retter, H. Jehring, V. Vetterl, *J. Electroanal. Chem.* **1974**, *57*, 391–397.
520. B. B. Damaskin, N. K. Akhmetov, *Sov. Elecktrochem.* **1979**, *15*, 478–481.
521. Yu. Kharkats, *J. Electroanal. Chem.* **1980**, *115*, 75–88.
522. Yu. Ya. Gurevich, Yu. I. Kharkats, *J. Electroanal. Chem.* **1978**, *86*, 245–258.
523. U. Retter, H. Lohses, *J. Electroanal. Chem.* **1982**, *134*, 243–250.
524. P. Nikitas, *J. Electroanal. Chem.* **1991**, *300*, 607–628.
525. P. Nikitas, *Electrochim. Acta* **1991**, *36*, 447–457.
526. P. Nikitas, S. Andoniou, *J. Electroanal. Chem.* **1994**, *375*, 339–356.
527. U. Retter, *J. Electroanal. Chem.* **1984**, *165*, 221–230.
528. U. Retter, *J. Electroanal. Chem.* **1987**, *236*, 21–30.
529. U. Retter, *J. Electroanal. Chem.* **1993**, *349*, 41–48.
530. U. Retter, *Electrochim. Acta* **1996**, *41*, 2171–2174.
531. R. Srinivasan, R. de Levie, S. K. Rangarajan, *Chem. Phys. Lett.* **1987**, *142*, 43–47.
532. C. Young, *Phys. Rev.* **1952**, *85*, 808–816.
533. Y. I. Kharkats, U. Retter, *J. Electroanal. Chem.* **1990**, *287*, 363–367.

534. U. Retter, *J. Electroanal. Chem.* **1992**, *329*, 81–89.
535. R. Sridharan, R. de Levie, *J. Phys. Chem.* **1982**, *86*, 4489, 4490.
536. R. Philipp, U. Retter, J. Dittrich et al., *Electrochim. Acta* **1987**, *32*, 1671–1677.
537. U. Retter, V. Vetterl, J. Jursa, *J. Electroanal. Chem.* **1989**, *274*, 1–9.
538. Th. Wandlowski, *J. Electroanal. Chem.* **1991**, *302*, 233–253.
539. S. Laushera, E. Stenina, *Sov. Electrochem.* **1992**, *28*, 78–87.
540. V. S. Griffiths, J. B. Westermore, *J. Chem. Soc.* **1963**, 4941–4945.
541. L. K. Partridge, A. C. Tansley, A. C. Porter, *Electrochim. Acta* **1967**, *12*, 1573–1580.
542. B. B. Damaskin, S. L. Dyatkina, S. I. Petrochenko, *Sov. Electrochem.* **1969**, *5*, 935–941.
543. A. A. Moussa, H. A. Ghaly, M. M. Abou-Romia, *Electrochim. Acta* **1975**, *20*, 485–497.
544. L. Gierst, C. Franck, G. Quarin et al., *J. Electroanal. Chem.* **1981**, *129*, 353–363.
545. W. Obretenov, V. Bostanov, V. Popov, *J. Electroanal. Chem.* **1982**, *132*, 273–276.
546. M. Poelman, C. Buess-Herman, J. P. Badiali, *Langmuir* **1999**, *15*, 2194–2201.
547. A. Popov, N. Dimitrov, T. Vitanov, *Electrochim. Acta* **1992**, *37*, 2373–2376.
548. P. Steel, *Coord. Chem. Rev.* **1990**, *106*, 227–265.
549. E. B. Constable, *Adv. Inorg. Chem.* **1989**, *34*, 67–138.
550. L. L. Merritt, E. Schroeder, *Acta Crystallogr.* **1956**, *9*, 801–804.
551. E. A. Mambetkaziev, A. M. Shaldybaeva, V. N. Statsyuk et al., *Sov. Electrochem.* **1975**, *11*, 1643–1645.
552. N. K. Akhmetov, R. I. Kaganovich, E. A. Mambetkaziev et al., *Sov. Electrochem.* **1977**, *13*, 248–250.
553. N. K. Akhmetov, R. I. Kaganovich, E. A. Mambetkaziev et al., *Sov. Electrochem.* **1978**, *14*, 1534–1537.
554. L. Pospisil, J. Kuta, *J. Electroanal. Chem.* **1979**, *101*, 391–398.
555. M. Kim, K. Itoh, *J. Phys. Chem.* **1987**, *91*, 126–132.
556. D. Yang, D. Bizotto, J. Lipkowski et al., *J. Phys. Chem.* **1994**, *98*, 7083–7089.
557. M. Hoon-Khosla, W. R. Fawcett, J. D. Goddard et al., *Langmuir* **2000**, *16*, 2356–2369.
558. J. Trost, J. Wintterlin, G. Ertl, *Surf. Sci.* **1995**, *329*, L583–L587.
559. G. E. Piorier, *Chem. Rev.* **1997**, *97*, 1117–1129.

4
Underpotential Deposition

4.1	Atomically Controlled Electrochemical Deposition and Dissolution of Noble Metals	471
	Shen Ye and Kohei Uosaki	471
4.1.1	Introduction	471
4.1.2	Atomically Controlled Deposition of Noble Metals	472
4.1.2.1	Platinum	475
4.1.2.2	Palladium	479
4.1.2.3	Rhodium	490
4.1.2.4	Ruthenium	494
4.1.3	Atomically Controlled Dissolution of Noble Metals	498
4.1.3.1	Palladium	499
4.1.3.2	Gold	502
4.1.4	Summary	508
	Acknowledgment	509
	References	509
4.2	Electrodeposition of Compound Semiconductors by Electrochemical Atomic Layer Epitaxy (EC-ALE)	513
	John L. Stickney, Travis L. Wade, Billy H. Flowers, Raman Vaidyanathan, and Uwe Happek	513
	Abstract	513
4.2.1	Introduction	513
4.2.1.1	Electrodeposition	514
4.2.2	Hardware	516
4.2.3	Deposition Programs	521
4.2.3.1	Cycle Steps	521
4.2.3.2	Cycles	525
4.2.4	Compound Formation	533
4.2.4.1	Toward Growing Device Structures	551
4.2.4.2	Diodes	553
	Acknowledgment	557
	References	557

4.3	Electrocatalysis on Surfaces Modified by Metal Monolayers Deposited at Underpotentials	**561**
	Radoslav Adžić	*561*
4.3.1	Introduction	561
4.3.2	Structural and Electronic Properties of Electrode Surfaces with Metal Adlayers	561
4.3.3	Electrocatalysis on Surfaces Modified with Metal Adlayers	562
4.3.3.1	Redox Reactions	562
4.3.3.2	Oxidation of Organic Molecules	564
4.3.3.2.1	Formic Acid	564
	Lead	569
	Bismuth	571
	Antimony	573
	Noble metal monolayers	574
4.3.3.2.2	Methanol	574
4.3.3.2.3	Carbon Monoxide	576
4.3.3.2.4	Other Organic Molecules	578
4.3.3.3	Electroorganic Synthesis	578
4.3.3.4	Hydrogen Evolution	581
4.3.3.5	Oxygen Reduction	584
4.3.3.5.1	Alkaline Solutions	585
4.3.3.5.2	Acid Solutions	587
4.3.3.5.3	Hydrogen Peroxide Reduction	588
4.3.3.5.4	Mechanisms of Catalytic Effects	588
4.3.3.6	Electrodeposition of Metals	592
	Acknowledgment	596
	References	596

4.1
Atomically Controlled Electrochemical Deposition and Dissolution of Noble Metals

Shen Ye and Kohei Uosaki
Hokkaido University, Sapporo, Japan

4.1.1
Introduction

Deposition and dissolution processes of noble metals are important not only for fundamental science but also for applications. As shown in Fig. 1, the noble metals are widely used in a variety of fields such as catalysis, for automobile and chemical reactions, fabrication of electronics devices and jewelry, batteries and corrosion protection [1, 2]. Deposition of a highly ordered noble metal is usually carried out by physical means such as vacuum deposition [3–6] and chemical vapor deposition (CVD) [7, 8].

Electrochemical deposition is more economical and convenient than the growth in vacuum and has been used for a long time, but the quality of an electrodeposited metal layer is generally poorer than that of a physically deposited layer. The morphology of the electrodeposited metal layer is controlled by many factors such as deposition potential, current density, temperature, concentration of metal ion/complex, pH, nature of the additives and substrates. Thus, it is essential to clarify the effects of these factors on the growth mode to obtain highly ordered metal layers.

Quantitative understanding of the electrochemical deposition and dissolution of metals became possible since Clavilier established the technique to prepare a well-defined surface of a platinum single-crystal electrode in 1980 [9]. Furthermore, the discovery of scanning tunneling microscopy (STM) [10] and atomic force microscopy (AFM) [11] soon after, made the study on the surface structure of the electrode *in situ* possible at atomic resolution. The surface structures of many noble metal single-crystal electrodes, such as Au(*hkl*) [12, 13], Pt(*hkl*) [14, 15], Pd(*hkl*) [16], Rh(*hkl*) [17], and Ir(*hkl*) [18], have been investigated by STM under electrochemical control at atomic resolution. Reality involves understanding the relationship between the electrochemical behavior and the surface structure of these electrodes.

Recently, in addition to the *in situ* STM/AFM, many other surface-analysis techniques such as surface X-ray scattering (SXS) [19, 20] and electrochemical quartz crystal microbalance (EQCM) [21, 22] have also been employed to investigate the electrochemical deposition and dissolution processes at atomic resolution. Atomically controlled electrochemical epitaxial growth and layer-by-layer dissolution

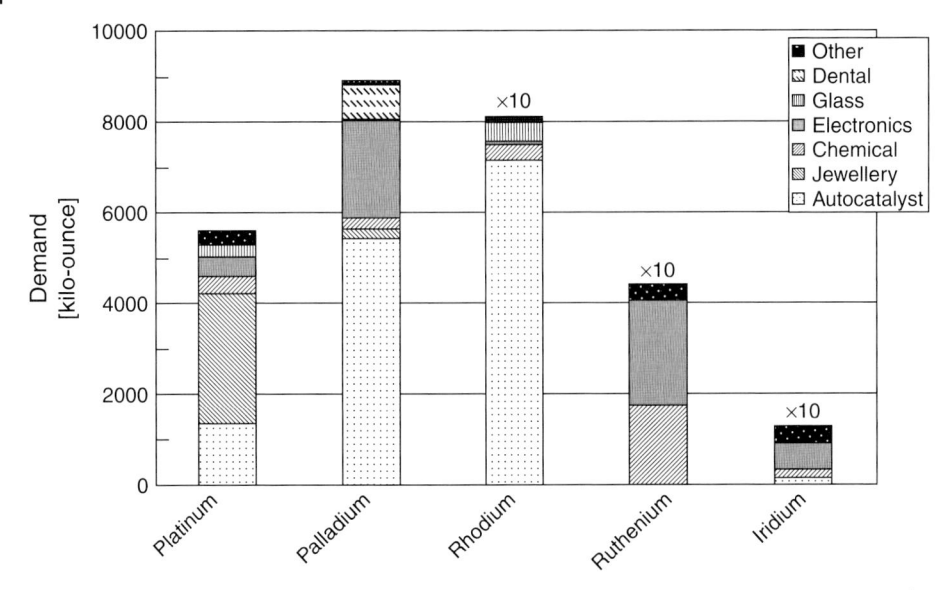

Fig. 1 Principal uses of noble metals (platinum, palladium, rhodium, ruthenium and iridium) in the world in 2000 [2]. The scales for rhodium, ruthenium and iridium are enlarged by 10 times.

of many metals have become possible now [23, 24].

In this chapter, we describe the recent *in situ* atomic resolution studies on the electrochemical deposition and dissolution of noble metals, which show extremely high catalytic activity for many chemical reactions [1, 2].

4.1.2 Atomically Controlled Deposition of Noble Metals

Depending on the binding energy and crystallographic mismatch between the deposited metal and the substrate, the growth of a metallic layer on the substrate near equilibrium condition can be classified into three mechanisms, namely, the Vomer-Weber (VW), the Stranski-Krastanov (SK), and the Frank-van der Merwe (FM) modes [25–27]. Generally, when the binding energy between the metal and the substrate is lower than that of the metal itself, a three-dimensional (3D) island growth mechanism, namely, the VW mode is favored independent of crystallographic mismatch. When the binding energy between the metal and the substrate is higher than that of the metal itself, a two-dimensional (2D) metallic monolayer can be formed in the underpotential deposition (UPD) region, and there are two possible growth mechanisms for the subsequent growth of the metallic layers depending on the crystallographic mismatch. When the crystallographic mismatch of the metal and the substrate is negligibly small, metallic layers are formed on the UPD layer on the substrate by a layer-by-layer growth, namely, the FM mode. When the crystallographic mismatch is relatively large, the UPD layer contains considerable internal strain, and therefore, the growth of the unstrained 3D metal islands on top of the strained UPD layers is energetically favored and this mechanism is called the

Tab. 1 Physical properties and electrochemical growth modes of noble metal elements on gold substrate

	Au	Pt	Pd	Rh	Ru	Os	Ir	Ag
Atomic number	79	78	46	45	44	76	77	47
Atomic weight	196.97	195.08	106.42	102.90	101.07	190.2	192.2	107.87
Lattice structure	fcc	fcc	fcc	fcc	hcp	hcp	fcc	fcc
Electronic structure	$5d^{10}6s^1$	$5d^96s^1$	$4d^{10}5s^0$	$4d^85s^1$	$4d^75s^1$	$5d^66s^2$	$5d^76s^2$	$4d^{10}5s^1$
Atomic diameter r_A [nm]	0.288	0.277	0.275	0.269	0.270	0.273	0.271	0.289
Special gravity [g cm^{-3}]	19.32	21.45	12.01	12.41	12.37	22.59	22.56	10.50
Melting point [°C]	1064	1768	1554	1963	2250	3045	2457	962
Thermal conductivity [Wcm^{-1}K^{-1}]	2.93	0.72	0.76	1.50	1.05	0.87	1.47	4.19
Resistivity [μΩ cm, 20°C]	2.2	10.6	10.8	4.6	7.3	9.5	5.1	1.62
Tensile strength [Mpa][a]	108	123	170	695	–	–	1088	147
Hardness [Vickers][a]	22	41	41	101	–	350	220	24
Surface energy v_A [Jm^{-2}] [28]	1.488	2.691	2.043	2.828	3.409	3.947	3.231	1.302
Geometry parameter r_{AB}[b]	1	0.962	0.955	0.934	0.937	0.948	0.941	1.00
Lattice mismatch Δ_{AB}[c]	0	–3.8%	–4.6%	–6.7%	–6.2%	–5.2%	–5.8%	0.17%

(continued overleaf)

Tab. 1 (continued)

	Au	Pt	Pd	Rh	Ru	Os	Ir	Ag
Surface energy mismatch Γ_{AB}[d] Electrochemical growth mode on gold substrate[e]	0 FM [29–31]	0.575 SK [32]	0.314 FM [33–36]	0.620 SK [37]	0.802 VW [38–40]	0.905	0.739	0.133 FM [41–45]

[a]Annealed;
[b]Geometric parameter $r_{AB} = r_B/r_A$ [26] where A and B denote substrate and metal, respectively. Values in the table are calculated with respect to the gold substrate;
[c]Lattice mismatches Δ_{AB} [26] are shown with respect to that of gold;
[d]The surface-energy mismatches are calculated with respect to a gold substrate following the definition given by Bauer and coworkers [26] as: $\Gamma_{AB} = 2((\nu_A - \nu_B)/(\nu_A + \nu_B))$;
[e]VW mode: Volmer-Weber mode, namely, 3D island growth; FM mode: Frank-van der Merwe mode, namely, layer-by-layer 2D mode; SK mode: Stranski-Krastanov mode, unstrained 3D islands are formed on top of strained 2D layers. See text for details.
Note: fcc: face centered cubic; hcp: hexagonal close packed.

SK mode. Thus, the metal growth mode on the substrate can be estimated qualitatively from the properties of both deposited and substrate metals.

Physical properties such as lattice and electronic structures, atomic diameters, melting point, thermal conductivity, resistivity, tensile strength, hardness as well as surface energy (v_A) [28] of the noble metal elements gold, platinum, palladium, rhodium, ruthenium, osmium, and iridium are summarized in Table 1. The same properties for silver are also shown in the same table for comparison. Following the definition given by Bauer and coworkers [26], geometry parameter (r_{AB}), lattice mismatch (Δ_{AB}) and surface-energy mismatch (Γ_{AB}) in the table are given with respect to gold, which is widely used as a substrate in studies of electrochemical epitaxial growth of noble metals. The experimentally observed electrochemical growth modes of these metals on gold are also listed in the table. These parameters play significant roles in the growth mechanism for the electrochemical deposition of these metals on various substrates. As shown in the table, the lattice mismatches (Δ_{AB}) are observed from -3.8% and -6.7% among these noble metal elements, and platinum shows the smallest mismatch (-3.8%) relative to the gold substrate. The value of Γ_{AB}, which is calculated from the surface energy of each metal with respect to a gold substrate [28], decreases in the sequence of gold (0), palladium(0.314) < platinum(0.575) < rhodium(0.620) < iridium(0.739) < ruthenium(0.802) < osmium(0.905). As proposed by Bauer and coworkers, the value of Γ_{AB} can be employed to judge the growth mode of metal deposition on various substrates [26]. Γ_{AB} values <0.5 with small lattice mismatch ($r_{AB} \leq 1.00$) are required for the 2D crystalline growth of the metal on the substrate.

If one examines the parameters shown in Table 1, one can expect that electrochemical deposition of gold and palladium on a gold electrode should be the best candidate for the layer-by-layer growth process among these noble metal elements. The observed growth modes shown in the table are in qualitative agreement with expectations based on the physical parameters.

4.1.2.1 Platinum

Platinum is the most important material in catalysis including electrocatalysis [46–48]. For example, platinum is widely used as a catalyst for automobiles [2]. Platinum and its alloys are also employed as the main catalytic materials in fuel cells, which are attracting more attention due to their high-energy conversion efficiency, flexibility, and low/clean emission than those of traditional internal-combustion engines [49, 50]. The platinized platinum electrode [51] has been used for standard reversible hydrogen electrodes (RHE) because of its extremely high reversibility for hydrogen evolution and oxidation reactions.

Ultrathin platinum layers on a chemically inactive substrate are very important systems in the field of catalysis and electrocatalysis. Many studies on epitaxial growth of platinum thin layers on various substrates have been reported in the UHV environment [4, 52–55]. The electrochemical growth of platinum layers is achieved by using a number of commercially available electroplating baths [56–58]. These studies are mainly limited to the purpose of electroplating where film thickness, purity, film strength, adhesive force to the substrate and growth rate are more important than surface atomic structure [46, 47, 51, 59–62]. The atomically controlled electrochemical deposition of platinum on a gold single-crystal substrate was realized

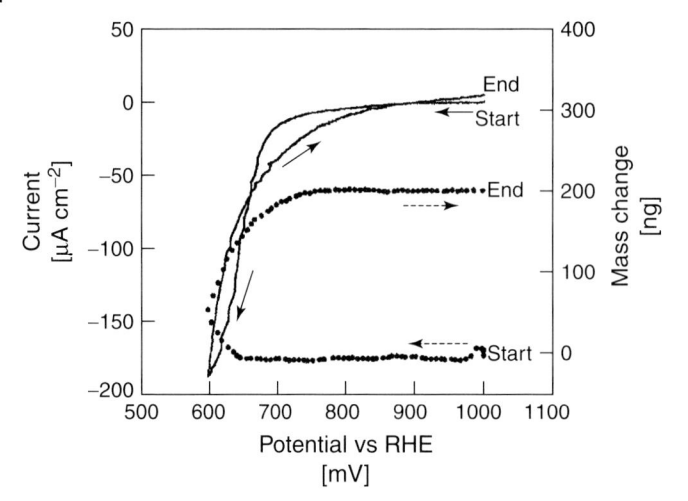

Fig. 2 Potential dependence of current (solid line) and the mass change (dotted line) at a (111)-such as gold/EQCM electrode in 50 mM HClO$_4$ + 0.6 mM H$_2$PtCl$_6$. Sweep rate: 20 mV s^{-1} [32].

recently by Uosaki and colleagues [32, 63, 64].

Figure 2 shows the potential dependence of the current and the surface mass change at an Au(111) electrode in the first potential cycle between +1.0 V and +0.6 V in 50 mM HClO$_4$ solution after 0.6 mM H$_2$PtCl$_6$ solution was added at +1.0 V (vs. RHE). Cathodic current (solid line) started to flow around +0.80 V and increased significantly as the potential became more negative than +0.70 V. The surface mass (dotted line) significantly increased when the potential became more negative than +0.65 V. Even after the sweep direction was reversed at +0.60 V, a cathodic current flowed and the surface mass continued to increase until the potential became more positive than +0.80 V. The relationship between the electric charge passed and the surface mass change showed that the surface mass increased by 48.5 g as a cathodic charge of 1 F was passed. This value is in good agreement with the calculated value for the four-electron reduction process of PtCl$_6^{2-}$:

$$PtCl_6^{2-} + 4e^- \longrightarrow Pt + 6Cl^-$$

That is, 48.8 g F^{-1}-electron, which is equal to the atomic weight of platinum (195.08), divided by 4 [32, 63, 64].

An STM image obtained at +0.95 V where neither cathodic nor anodic current flowed showed an ordered adlayer structure (40 × 40 nm^2, Fig. 3a), which is totally different from that of the Au(111) substrate. The spots with the same brightness showed a hexagonal symmetry with a nearest-neighbor distance of about 0.76 nm, while the symmetry of the adlattice was rotated by about 20 degrees from that of the Au(111) substrate. This structure can be assigned to an adlayer of the PtCl$_6^{2-}$ complex that forms the $\sqrt{7} \times \sqrt{7}R19.1°$ structure on the Au(111) surface [32, 63, 64]. The adsorption of the PtCl$_6^{2-}$ complex on the Au(111) surface has also been confirmed by the mass increase after injection of the PtCl$_6^{2-}$ complex

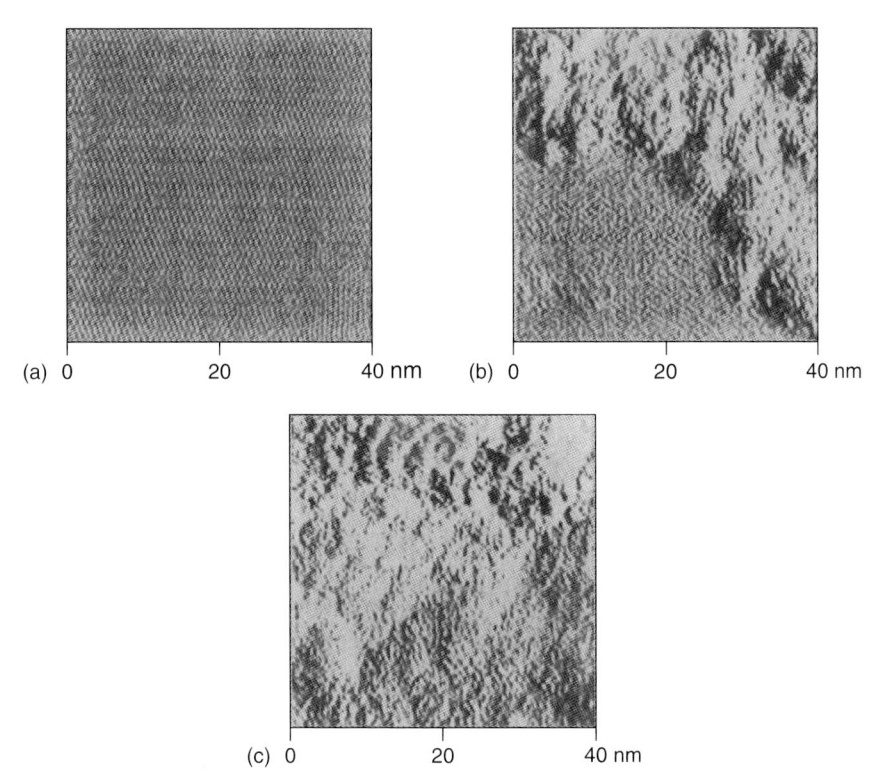

Fig. 3 STM images (40 × 40 nm^2) of platinum deposition process on an Au(111) substrate (a) at +0.95 V, (b) 10 min and (c) 30 min, respectively, after the potential was stepped from +0.95 V to +0.70 V in 50 mM H_2SO_4 + 0.05 mM H_2PtCl_6 [32].

into the blank solution at the same potential [64]. The same $\sqrt{7} \times \sqrt{7}R19.1°$ adlattice structure of the $PtCl_6^{2-}$ complex on an Au(111) electrode has been confirmed by Kolb and colleagues recently [65]. As will be described later, similar adlattice structures of metal complexes, such as $PdCl_4^{2-}$ [33], $RhCl_6^{2-}$ [37], $AuCl_4^-$ [66] and $PtCl_4^{2-}$ [67], have also been observed on the Au(III) electrode surface.

As shown in Figs. 3(b) and (c), the deposition of a platinum layer of monoatomic height was initiated as soon as the potential was stepped to +0.70 V. The growth of the first layer seemed to be essentially in a 2D mode. The deposited platinum layer grew from the top-right portion of the image (Fig. 3b) and covered the entire area of the image after prolonged deposition (Fig. 3c). The bottom-left portion of Fig. 3(b) is the Au(111) covered with a $PtCl_6^{2-}$ adlayer of $\sqrt{7} \times \sqrt{7}R19.1°$ structure (c.f. Fig. 3a). The terrace size of the deposited platinum layer seemed to be much smaller than that of the Au(111) surface before the platinum deposition. A number of small clusters were also observed on the platinum layer. These clusters are higher than the first platinum layer by a monoatomic height of the platinum layer. The third platinum layer started to grow before the growth of the second platinum layer was completed. The morphology of the subsequent platinum

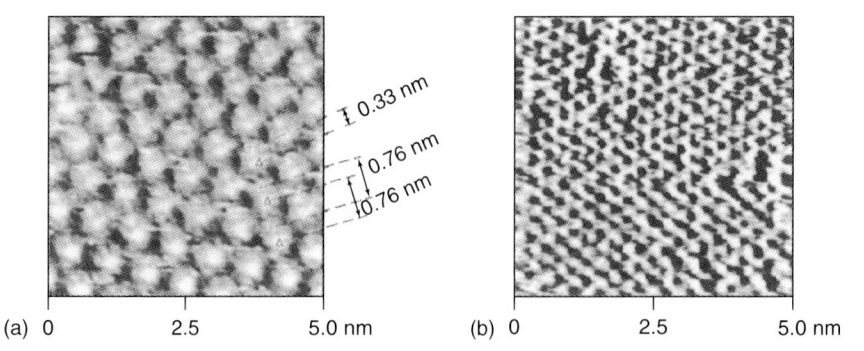

Fig. 4 High resolution STM images (5 × 5 nm^2) of (a) an Au(111) electrode obtained at +0.70 V in 50 mM HClO$_4$ + 0.6 mM H$_2$PtCl$_6$ and (b) the same electrode in 50 mM HClO$_4$ after the solution in the STM cell was replaced several times by 50 mM HClO$_4$, while the potential was held at +0.70 V after the STM image (a) was obtained. Note that the imaged portion was not the same [32].

layers became rougher than that of the first layer. This means that the growth of the platinum layer on an Au(111) electrode is not a perfect 2D growth, namely, the FM mode. However, Friedrich and colleagues observed that platinum deposition on an Au(111) surface was mainly observed at the step edges of the substrate as isolated platinum particles at the lower coverage by *ex situ* STM measurements [68]. Recent investigation on the initial stage of platinum deposition on Au(hkl) surfaces by Kolb and colleagues also showed that the nucleation of platinum occurred preferentially at the surface defects and that the mobility of platinum was apparently increased by the addition of chloride into the solution [65].

The adlayer with a structure of $\sqrt{7} \times \sqrt{7}R19.1°$ was also observed on the surface of deposited platinum, demonstrating that the PtCl$_6^{2-}$ complex adsorbed not only on the gold substrate but also on the platinum surface during the electrochemical deposition process (Fig. 4a). *In situ* STM measurement of the platinum-deposited Au(111) surface in 50 mM HClO$_4$ solution without the PtCl$_6^{2-}$ complex showed that the adlattice structure of $\sqrt{7} \times \sqrt{7}R19.1°$ disappeared but a highly ordered surface structure of a hexagonal array with a neighbor atomic distance of 0.28 nm (Fig. 4b). Although it is difficult to distinguish the distance of the unit lattice of Pt(111) (0.277 nm) and Au(111) (0.288 nm) surfaces by the present STM measurement, the image shown in Fig. 4(b) should correspond to the Pt(111) – (1 × 1) structure formed on the Au(111) substrate as both EQCM and XPS measurements confirmed the existence of platinum on the Au(111) surface [32]. Adsorption of the PtCl$_6^{2-}$ complex is considered to inhibit the 3D growth but to promote the initial 2D growth of platinum [32].

On the other hand, X-ray diffraction (XRD) measurements of the deposited platinum films (ca. 250 ML) on the Au(111) substrate demonstrated that a (111)-orientated platinum-bulk phase was definitely formed on the Au(111) electrode surface by electrochemical epitaxial growth [32].

The aforementioned experimental results suggest that the platinum layer grows on the Au(111) electrode by the

SK mechanism. This may be related to the relatively large Γ_{AB} of platinum with respect to the gold substrate (0.575) and the moderately slower surface diffusion rate of platinum atoms [69, 70].

Electrochemical characterizations of the UPD of hydrogen and copper show that the surface structure of the electrochemically deposited platinum layer on the Au(111) substrate was quite different from that of a polycrystalline platinum electrode, but more similar to that of the Pt(111) single-crystal electrode [32]. The difference in the electrochemical behavior between the platinum-deposited Au(111) and an ideal Pt(111) electrode is attributed to the relatively small (111) domain and the higher step density on the platinum-deposited surface. Recently, Watanabe and colleagues reported that the (111)-like surface structure of a platinum thin-film electrode prepared by a sputtering method can be improved by electrochemical annealing at the potential region of hydrogen adsorption in $HClO_4$ solution [71].

4.1.2.2 Palladium

As shown in Fig. 1, palladium is mainly consumed as an autocatalyst now [1, 2]. Palladium is also an important noble metal element widely applied in the development of electrocatalytic materials [46, 48, 72]. It is widely used as a promoter for electroless deposition of metals on various substrates. Palladium is also known for its extraordinary ability to absorb a large amount of hydrogen. The growth of palladium thin layers on various substrates in UHV has been investigated in detail [3–8]. Electrodeposition of palladium is widely used, and a number of commercially available electroplating baths of palladium for different purposes have been developed [56–58, 61, 72]. However, the rapid increase in the market price of palladium since 1999 may affect its applications in the industry [2].

Since Clavilier proposed a simple way to prepare a noble metal single crystal [9], many electrochemical studies have been carried out on platinum, gold, and iridium as well as palladium single-crystal electrodes. The preparation of a palladium single-crystal is, however, difficult because palladium cannot be subjected to the usual convenient flame-annealing treatment to prepare the electrode [73]. It is of great interest if one can grow a palladium epitaxial layer on other noble metal single-crystal surfaces, which are easily prepared by Clavilier's method. This provides not only a convenient method of preparing an ordered palladium surface but also a way of investigating the chemical and physical properties of a thin metallic layer, which may be different from those of the bulk material.

The investigation on the electrochemical epitaxial growth of palladium on gold and platinum single-crystal substrates was started recently. Attard and coworkers reported the irreversible deposition of submonolayer and monolayer coverage of palladium on Pt(111) electrode by the immersion technique for the first time [74]. Llorca and coworkers investigated the irreversibly adsorbed palladium on Pt(*hkl*) in acidic solution [75] and reported that electrocatalytic activity for the oxidation of formic acid on the Pt(100) electrodes was improved drastically by the palladium-adlayer modification, while that on the Pt(111) electrode was not greatly affected by the Pd-modification [76]. However, it is difficult to prepare an ultrathin film of palladium with various thickness by the immersion technique, and STM observation of the atomic structure of these surfaces is not available yet.

The electrochemical 2D growth of thin palladium layers on gold single-crystal electrodes was reported by Kolb's group for the first time [77]. They also found that the hydrogen absorption reaction at the palladium thin layers takes place at more cathodic potentials than that at a bulk-palladium electrode and does not occur at palladium films thinner than 2 ML. The observation of the electrochemical deposition of palladium at atomic resolution was not available until detailed STM and EQCM studies of the electrochemical epitaxial growth of palladium on Au(hkl) electrodes were carried out by Uosaki and colleagues [33, 34, 78–80] and later by Kolb and colleagues [35, 36].

Figure 5 shows the characteristic potential dependence of the current and the surface mass change at an Au(111) electrode in 50 mM H_2SO_4 solution containing 0.1 mM $PdCl_4^{2-}$ complex [33]. A cathodic current started to flow as soon as the potential became more negative than +0.95 V, reaching a maximum at +0.88 V and decreasing to a limiting value. When the potential scan was reversed in the positive direction, an anodic current started to flow at +0.8 V with two anodic peaks at +0.89 and +0.96 V. The mass respectively increased and decreased when the cathodic and anodic currents flowed. Quantitative analysis between the charge and the mass change show that the cathodic and anodic currents are respectively

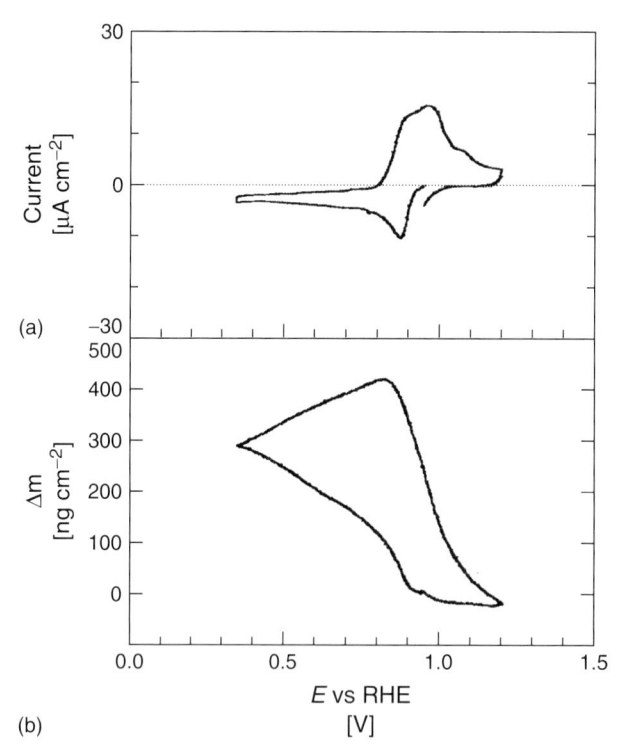

Fig. 5 Potential dependence of (a) current and (b) mass change (Δm) at an Au(111) electrode in 50 mM H_2SO_4 + 0.1 mM $PdCl_4^{2-}$. Sweep rate: 5 mV s^{-1} [33].

4.1 Atomically Controlled Electrochemical Deposition and Dissolution of Noble Metals | 481

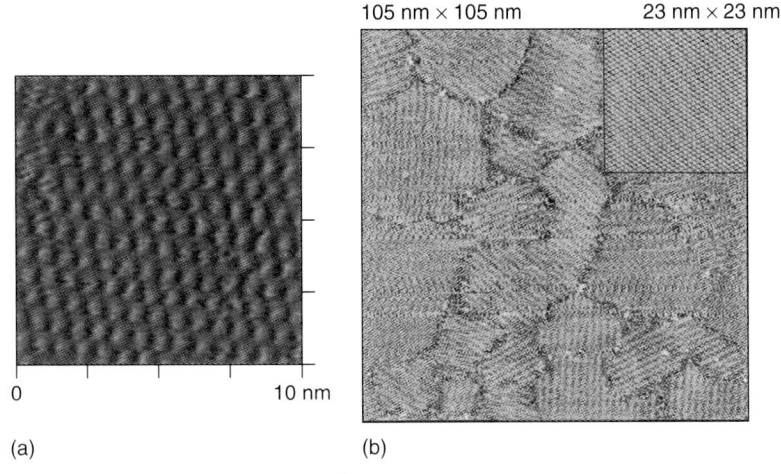

Fig. 6 (a) STM image (10 × 10 nm^2) of an Au(111) surface obtained at +0.95 V in 50 mM H$_2$SO$_4$ + 0.1 mM PdCl$_4^{2-}$ [33] and (b) STM image (105 × 105 nm^2) of an Au(111) surface at 0.63 V in +0.95 V in 0.1 M H$_2$SO$_4$ + 0.1 mM PdCl$_4^{2-}$. The inset (23 × 23 nm^2) shows the same structure at a higher resolution [35].

related to deposition and dissolution of palladium on the Au(111) surface [33]. Kolb and colleagues reported recently that the characteristic features of the current-potential curves for palladium deposition on gold single-crystal electrodes depend significantly on the surface structure and concentration of the PdCl$_4^{2-}$ complex as well as the chloride ion contained in the solution [35].

Figure 6 shows STM images of the Au(111) electrodes in dilute H$_2$SO$_4$ solution containing 0.1 mM PdCl$_4^{2-}$ at the potential where no palladium deposition takes place as reported by (a) Uosaki and colleagues [33] and (b) later by Kolb and colleagues [35]. Figure 6(a) (10 × 10 nm^2) shows an adlattice in hexagonal symmetry with a nearest-neighbor distance of 0.78 nm [33]. This structure is very similar to that of the PtCl$_6^{2-}$ adlayer on an Au(111) electrode (Fig. 3a) and is attributed to the structure of the adsorbed PdCl$_4^{2-}$ complex on the Au(111) with an adlattice structure of $\sqrt{7} \times \sqrt{7} R 19.1°$. This result is in very good agreement with the results of coverage estimated from the EQCM measurements [33]. A similar but not identical adlattice structure of the PdCl$_4^{2-}$ complex on an Au(111) substrate was observed by Kolb and colleagues as shown in Fig. 6(b) (105 × 105 nm^2) [35]. This image reveals the existence of an ordered adlayer with parallel stripes separated by about 1.8 nm and a distorted hexagonal structure (inset, 23 × 23 nm^2) incommensurate with the Au(111) substrate. The approximate next-neighbor distances were 0.81, 0.68 and 0.73 nm. Domain boundaries and small domains with three directions are clearly observed.

Figure 7 shows the surface structure on the Au(100) electrode in dilute H$_2$SO$_4$ solution containing the PdCl$_4^{2-}$ complex at the potential where no palladium deposition takes place as reported by (a) Uosaki and colleagues [34] and (b) later by Kolb and colleagues [36]. The two images show essentially the same structure of an adlattice with long and short axis lengths of

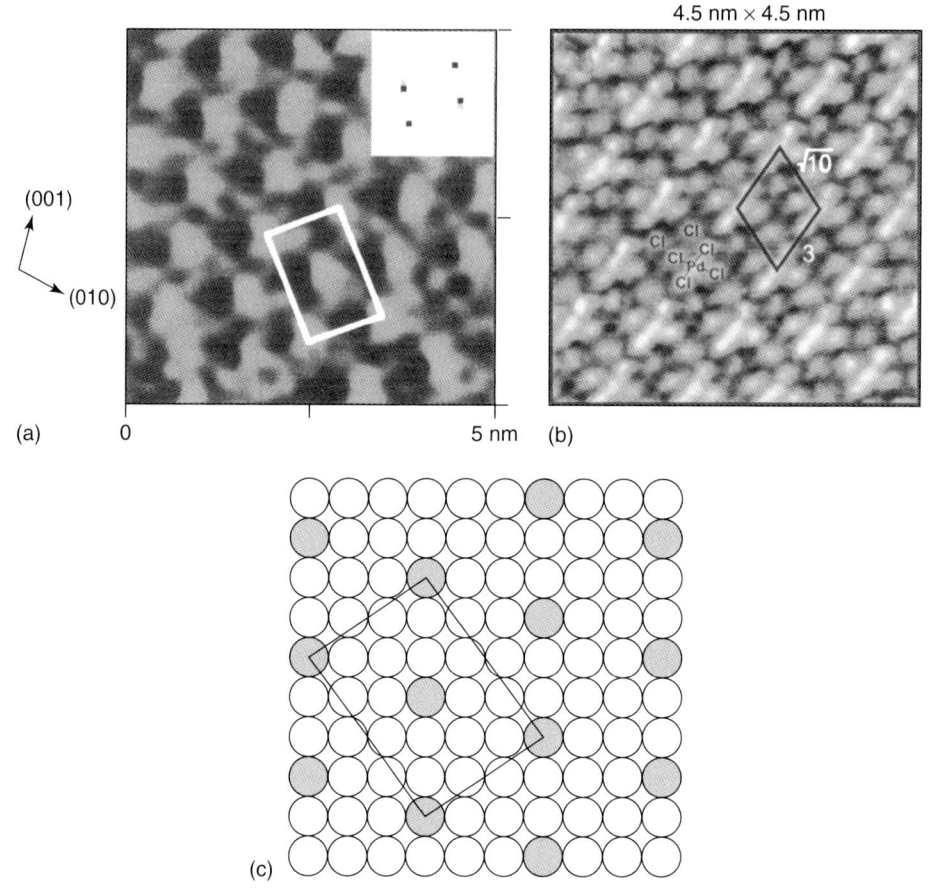

Fig. 7 (a) Lowpass filtered-STM image (5 × 5 nm^2) of an Au(100) surface at +1.05 V in 50 mM H$_2$SO$_4$ + 0.5 mM PdCl$_4^{2-}$. The inset is the Fourier transformed 2D spectrum and a unit lattice determined from the spectrum is superimposed on the image with a white box [34]; (b) STM image (4.5 × 4.5 nm^2) of an Au(100) surface at 0.61 V in 0.1 M H$_2$SO$_4$ + 0.5 mM PdCl$_4^{2-}$ + 0.6 mM HCl [36]; (c) A ball model of the PdCl$_4^{2-}$ complex adsorbed on Au(100) surface [34].

1.47 nm and 1.03 nm and can be attributed to the adlattice structure of the adsorbed PdCl$_4^{2-}$ complex on the Au(100) surface. An adlattice model is given in Fig. 7(c). The STM image of the ordered-adlayer structure on the Au(100) surface was hard to obtain when the concentration of PdCl$_4^{2-}$ species was lower than 0.05 mM, while the ordered PdCl$_4^{2-}$ adlayer was observed on the Au(111) surface under the same conditions, suggesting that the adsorption of PdCl$_4^{2-}$ on Au(100) is weaker than that on Au(111) [34].

Figure 8 shows STM images (300 × 300 nm^2) of the Au(111) electrode surface during the palladium deposition in 50 mM H$_2$SO$_4$ solution containing 0.5 mM PdCl$_4^{2-}$ [33]. Atomically flat and large terraces (Terrace-1 and Terrace-2) with monoatomic steps on the Au(111)

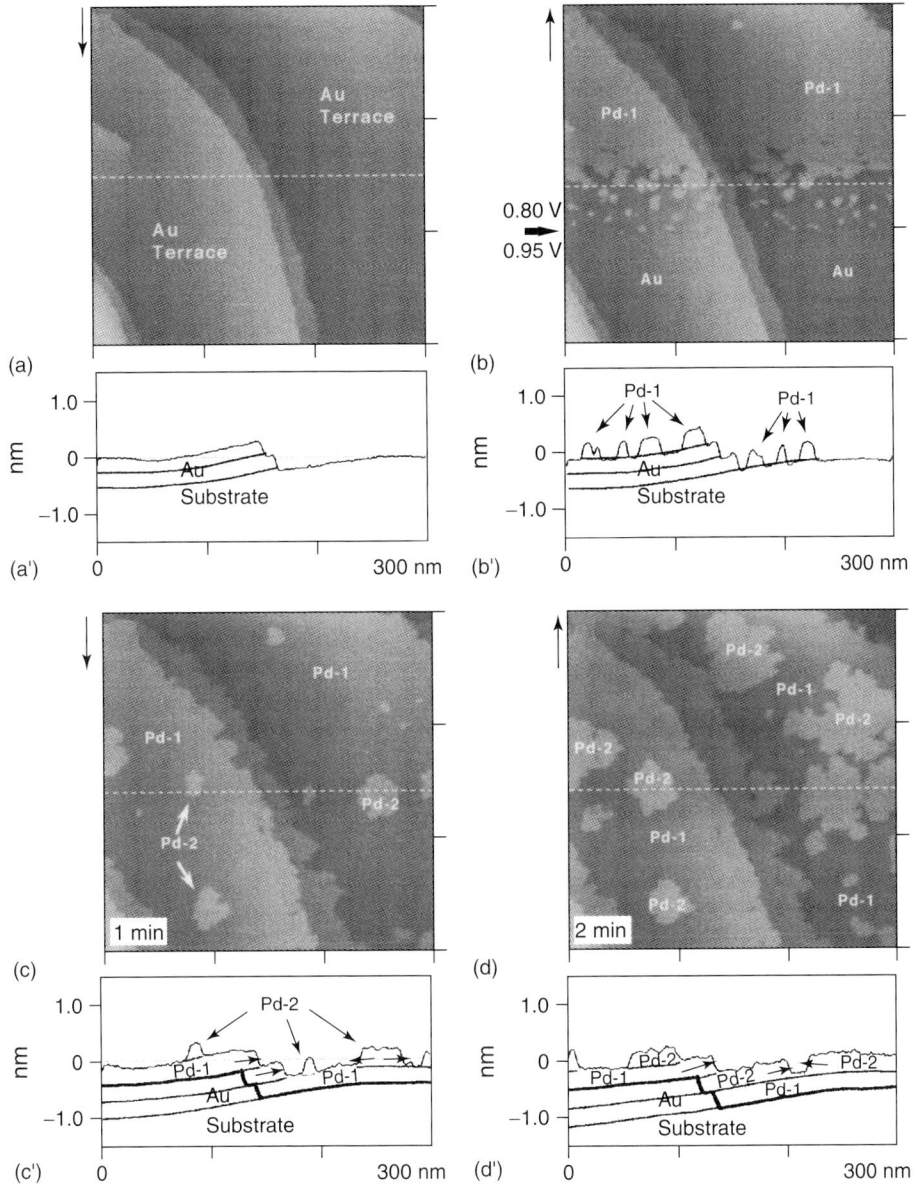

Fig. 8 (a) STM image (300 × 300 nm^2) of an Au(111) surface at +0.95 V in 50 mM H$_2$SO$_4$ + 0.1 mM PdCl$_4{}^{2-}$. Sequentially obtained STM images (300 × 300 nm^2) of the palladium deposition process on the Au(111) surface (b) 0 min, (c) 1 min, (d) 2 min, (e) 3 min, (f) 28 min, (g) 48 min and (h) 90 min after the potential was stepped from +0.95 V to +0.80 V in the same solution. A cross section along the white dotted line in each figure is shown below each image [33].

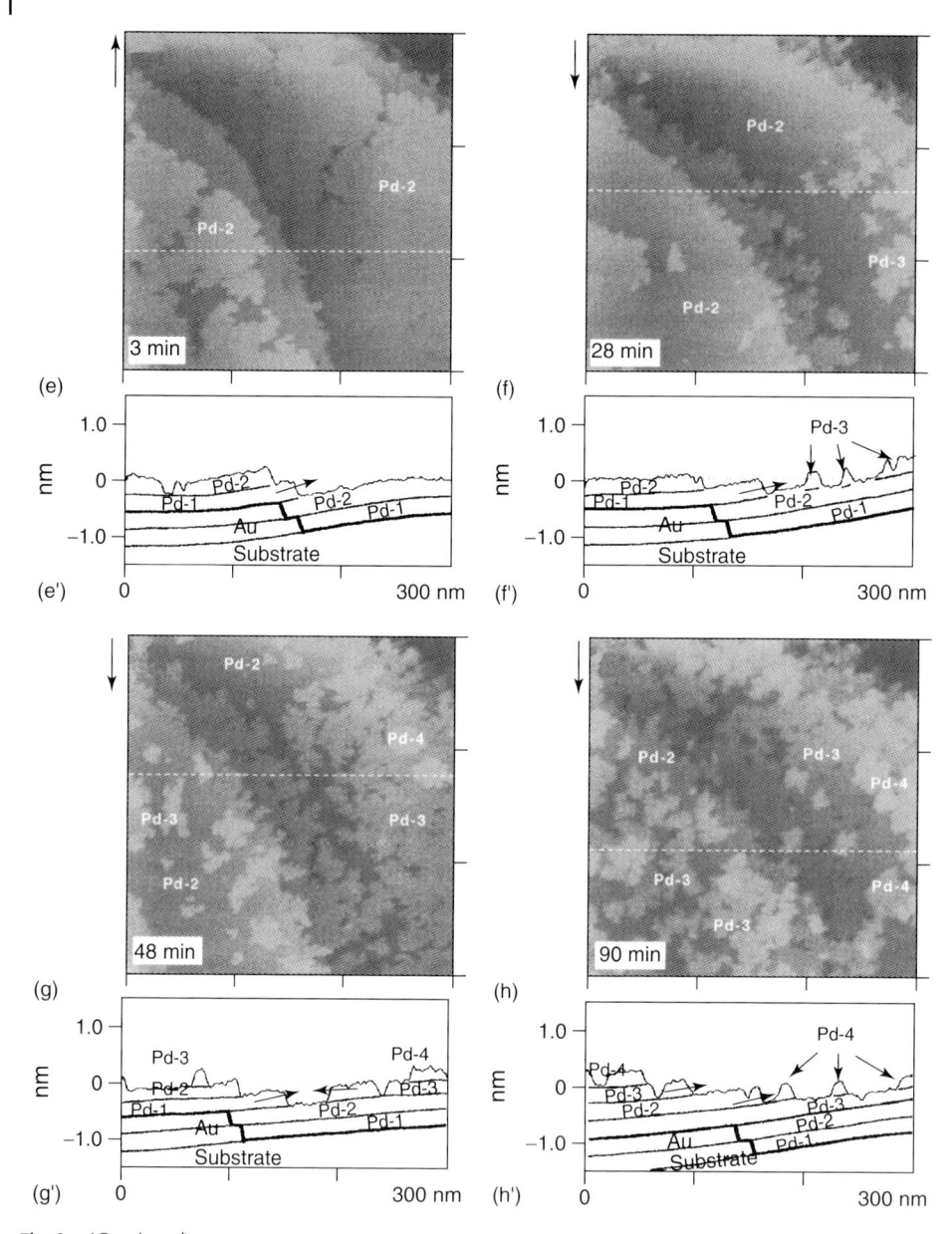

Fig. 8 (Continued)

substrate were observed at +0.95 V where no palladium deposition takes place (Fig. 8a). Figures 8(b–h) show sequentially obtained STM images of the same area at (b) 0 min, (c) 1 min, (d) 2 min, (e) 3 min, (f) 28 min, (g) 48 min and (h) 90 min after the potential of the Au(111) was stepped from +0.95 V to +0.80 V as indicated by the thick arrow in Fig. 8(b) where palladium deposition is expected to take place. The palladium nuclei were being generated on the Au(111) substrate immediately after the potential step. The cross section (Fig. 8b′) along the dotted line shows the formation of palladium islands with a monoatomic height on the large terraces of the gold substrate. In the upper part of Fig. 8(b), 2D growth of the first palladium monolayer (Pd-1) progressed not only on the large terrace but also on a narrow terrace between the step lines. This result is in contrast to that observed by Kolb and colleagues, in which the palladium layer started to preferentially grow from the step-and-edge sites [35, 77]. This discrepancy should be due to the differences in the deposition potential and solution conditions.

As shown in Figs. 8(c) and (c′), after the first palladium layer (Pd-1) was completely formed on the gold substrate, islands of the second palladium layer (Pd-2) were generated on both Terrace-1 and Terrace-2, and the first palladium layer (Pd-1) on the narrow terrace started to grow laterally further onto the Pd-1 layer on Terrace-2 from the step line. Both the Pd-2 layer on Terrace-2 and the palladium layer extended from the Pd-1 layer on the narrow terrace grew two-dimensionally and merged together, resulting in a large flat terrace of a second palladium layer on Terrace-2 (Figs. 8d, e). The cross section (Figs. 8c′,d′,e′) clearly shows the 2D growth of the palladium monoatomic layer. After the complete second palladium layer was formed on both Terrace-1 and Terrace-2, the third palladium layer (Pd-3) started to grow two-dimensionally on the terraces and from the step line (Figs. 8f,g). The growth rate of the third palladium layer was, however, slower than that of the first and the second layers. This should be due to the decrease in the local concentration of the reactant, namely, $PdCl_4^{2-}$, between the STM tip and the electrode surface, leading to a negative shift in the equilibrium potential for palladium deposition [78]. The palladium layer, which was originally first deposited on Terrace 1, was also grown laterally and finally merged with Pd-3 on Terrace 2 (Figs. 8f,g,h). The STM images in Fig. 8 demonstrate that the electrochemical deposition of palladium on the Au(111) surface is an epitaxial layer-by-layer process.

Figure 9 shows STM images (300 × 300 nm^2) of an Au(100) electrode obtained in 50 mM H_2SO_4 solution containing 0.5 mM $PdCl_4^{2-}$ [34]. A number of monoatomic-high gold islands were observed in the STM image at +1.0 V (Fig. 9a) when no palladium deposition took place. These gold islands were formed as a result of the reconstruction lifting of the Au(100) surface due to the large difference (25%) in surface densities between the reconstructed and unreconstructed Au(100) surface [81]. Figures 9(b)–(g) show STM images sequentially obtained during the palladium deposition process. As soon as the potential of the Au(100) substrate was stepped from +1.0 to +0.80 V (thick arrow in Fig. 9b), palladium nuclei were generated both on the wide terrace and on the gold islands (cross section I). The height of the palladium nuclei corresponded to that of a palladium monoatomic step (0.19 nm). Kolb and colleagues reported that the palladium nucleated at the steps and island rims on the island-covered Au(100) electrode surface and that the first layer of palladium two-dimensionally grew from these nuclei [36].

As shown in the cross section II in Fig. 9(b), the growth of the first palladium layer on the Au(100) surface

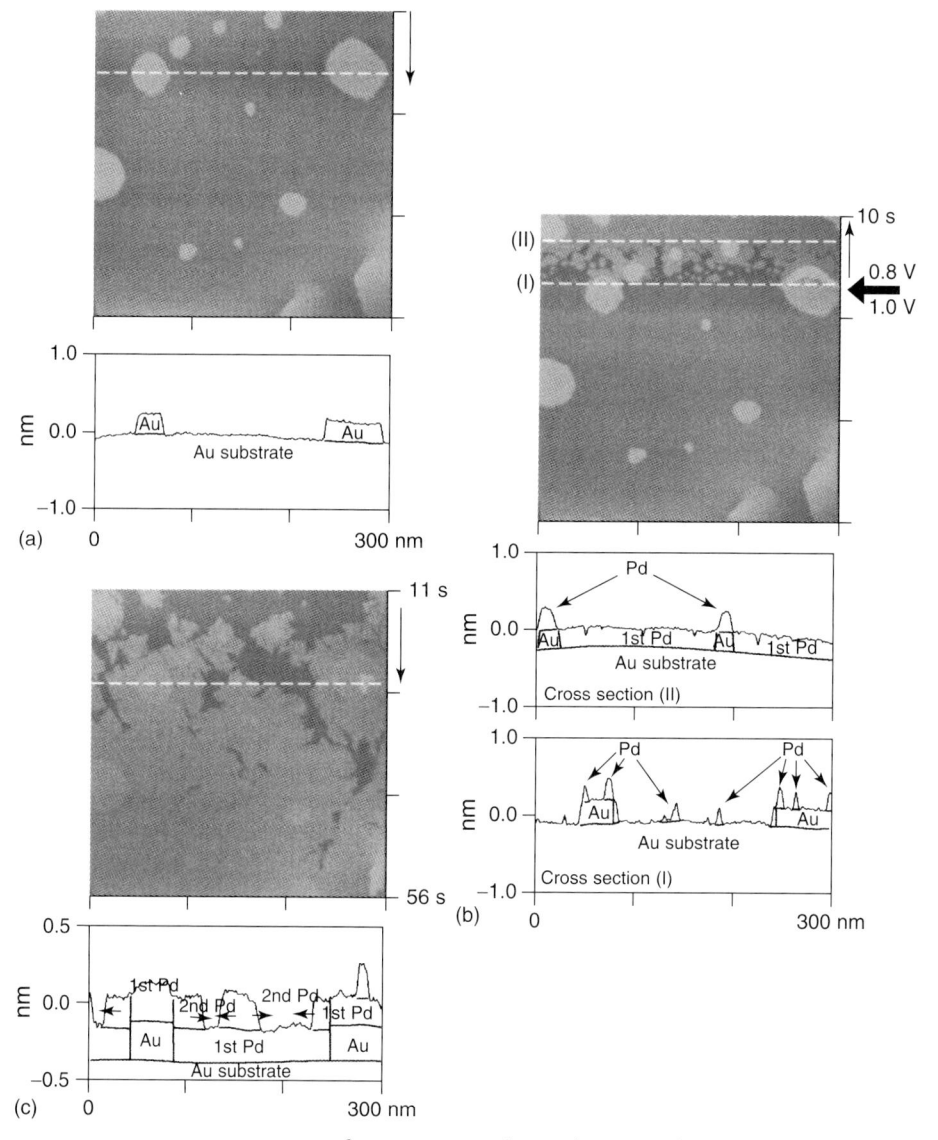

Fig. 9 (a) STM image (300 × 300 nm^2) of an Au(100) surface at +1.0 V in 50 mM H$_2$SO$_4$ + 0.5 mM PdCl$_4^{2-}$. (b)∼(g) STM image (300 × 300 nm^2) obtained sequentially during the palladium deposition process on an Au(100) surface in the same solution. Time after the potential pulse corresponding to the top and bottom of the STM image is indicated on the right side of each image. The cross section along broken line in each image is shown below [34].

was completed within 10 s after the potential step, and the second palladium layer then started to grow (Fig. 9c). As shown in the cross section in Fig. 9(c), the second layer of palladium seemed to grow laterally from the deposited palladium layer just on top of the gold islands. Growth of the second layer of

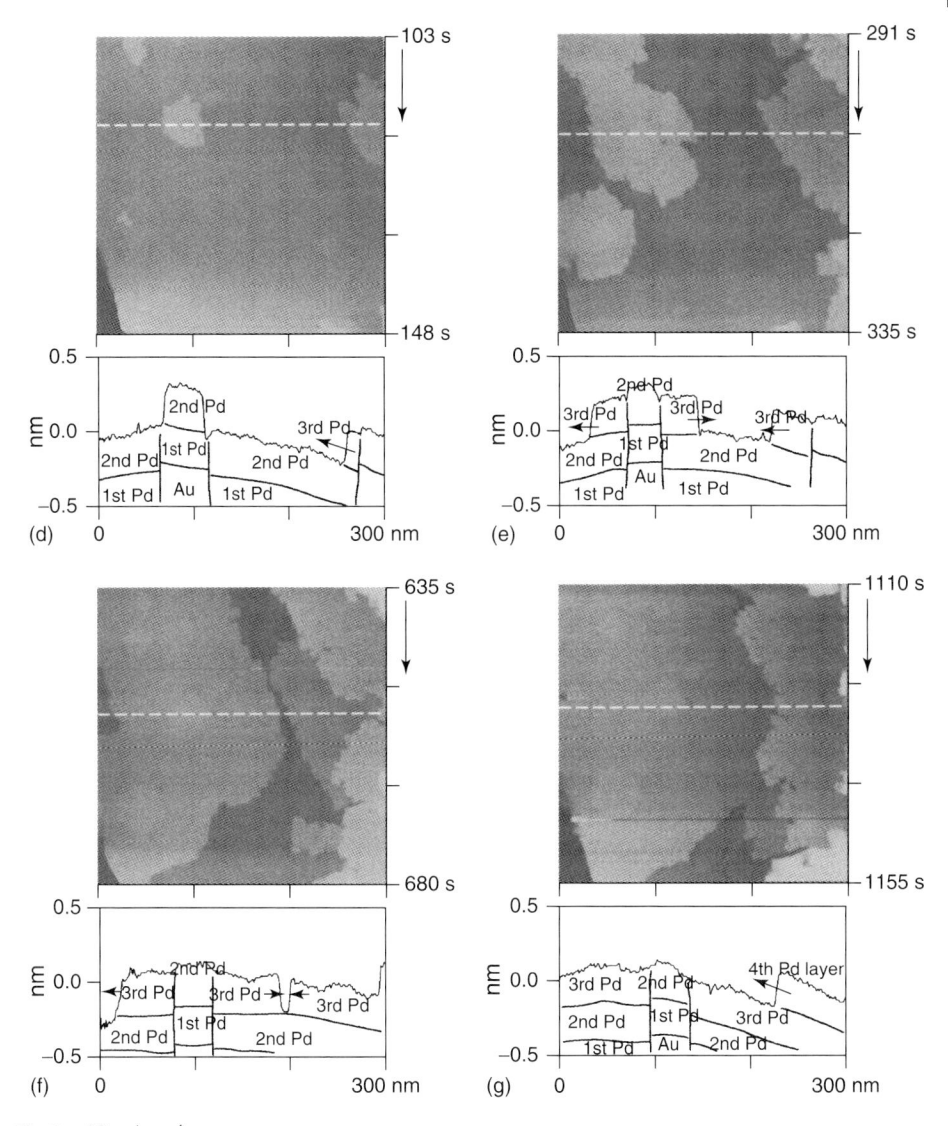

Fig. 9 (Continued)

palladium on the gold terrace was completed within about 30 s. It was noted that, although the gold terraces were covered with two layers of palladium, the gold islands were covered with only one monolayer of palladium. At this stage, the height difference between the two regions is very small but still distinguishable, reflecting the difference in atomic diameters of gold (0.288 nm) and palladium (0.275 nm).

The cross section of Fig. 9(d) (103 ∼ 148 s) clearly shows that the third layer of palladium was also initiated from the deposited palladium layer on top of the gold islands and grew laterally. As shown in the subsequent STM images in Figs. 9(e) (291 ∼ 395 s) and 9(f) (635 ∼ 680 s), each of the palladium islands grew laterally and eventually merged together. Completion of the third palladium layer took more than 1000 s, which is much longer than the time required for the growth of the first and second layers of palladium. The same phenomena were also observed in the palladium-deposition process on an Au(111) electrode and could be attributed to the concentration decrease in $PdCl_4^{2-}$ between the STM tip and the electrode surface [33, 78]. The 2D growth process of the fourth palladium layer was clearly observed on the right portion of the STM images shown in Figs. 9(f) and 9(g) (1110 ∼ 1155 sec).

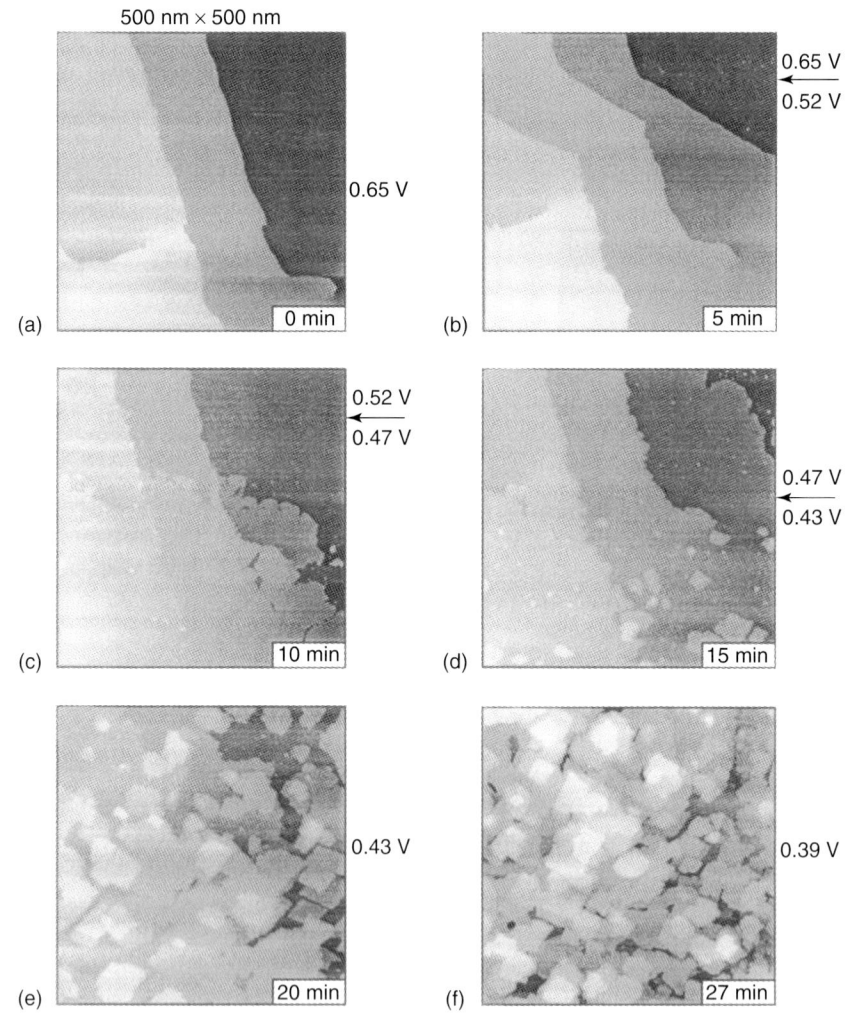

Fig. 10 Sequence of STM images for palladium deposition on island-free Au(100) in 0.1 M H_2SO_4 + 0.1 mM $PdCl_2$ + 0.6 mM HCl solution [36]. Deposition potentials as indicated in the figure. $I_T = 2$ nA.

Even after four layers of palladium were deposited on the Au(100) terrace and three layers of palladium on the gold islands, the shape of the gold islands was still distinguishable.

Recently, Kolb and colleagues compared the electrochemical deposition behavior of palladium on the island-covered and island-free unreconstructed Au(100) electrode surfaces [36]. By electrochemical annealing in 0.1 M H_2SO_4 + 0.1 mM $PdCl_2$ + 0.6 mM HCl at +0.80 V (vs. SCE; same for Kolb's work) where no palladium deposition takes place, the monoatomic-high gold islands generated by the lifting of the reconstruction disappeared instantaneously. Figure 10 shows *in situ* STM images for palladium deposition on an island-free Au(100) surface in the same solution. Palladium deposition commenced at +0.52 V, and palladium nucleated selectively at the monoatomic-high steps and grew onto the terraces (Fig. 10b). The second layer of palladium started to grow at +0.47 V by creation of small palladium islands at the steps (Fig. 10c). The fourth monolayer started to grow before the third monolayer was completed, where palladium also nucleated on the terraces (Figs. 10e–f). Square-type structures of palladium deposits were generated with the third monolayer, revealing the symmetry of the Au(100) substrate and showing a layer-by-layer epitaxial growth of palladium (Figs. 10e–f). The behavior is markedly different from those observed on the island-covered Au(100) electrode surface (Fig. 9) [34, 36]. Furthermore, when the palladium deposited on both island-covered and island-free Au(100) surfaces was stripped anodically, many small monoatomic-high islands and holes were created at those places where palladium had been deposited. These islands were much smaller than the initial gold islands. Kolb and colleagues attributed this phenomenon to the formation of a palladium-gold surface alloy on the Au(100) electrode, although such an effect was not observed on the Au(111) electrode [36]. The surface alloy was considered to be the main reason for the deviations in the electrocatalytic behavior of massive Pd(100) and thin palladium layers (<3 ML) on Au(100). However, the formation of a surface alloy during the dissolution of palladium deposited on an Au(100) electrode has not been observed by Uosaki and colleagues in their *in situ* STM [34] and SXS studies [82, 83]. This may be related to the differences between the experiments carried out in the two groups, such as electrode potential and chloride concentration as well as the *in situ* STM cell configuration.

Because STM is sensitive only to the outermost surface and the spatial resolution of STM is limited to about ±0.1 nm, the crystallographic relationship between the palladium layers and the gold substrate has remained unclear. Recently, Uosaki and colleagues carried out the *in situ* SXS measurements to elucidate the detailed structure of palladium electrochemically deposited on Au(111) and Au(100) electrode surfaces [82, 83]. The pseudomorphic layer structure was found only on the first monolayer of palladium on the Au(111) substrate, while the pseudomorphic layer-by-layer structure was observed on the Au(100) substrate up to 14 ML. The relaxation of the palladium layer on the Au(111) electrode was found between the second and fourth layer, and the palladium then started to grow from the fifth layer with (111) orientation. However, the relaxation of palladium on the Au(100) surface takes place after 14 ML and the surface became rough when its thickness was higher than 40 ML. These differences have been explained by the different adsorption sites on Au(111) (*fcc* and *hcp* sites) and Au(100) (only 4-*hollow* sites) surfaces [82, 83].

In conclusion, epitaxial layer-by-layer growth of thin palladium layers following the FM mode has been observed on both Au(111) and Au(100) electrodes. The ordered adlayer of the $PdCl_4^{2-}$ complex was observed on the surface of the palladium deposited on the Au(111) surface by Uosaki and colleague [33] and by Kolb and colleague [35]. The adlayer structure of the $PdCl_4^{2-}$ complex has not been observed clearly on the surface of the palladium deposited on the Au(100) electrode by STM so far [36, 78]. Kolb and colleague concluded that the $PdCl_4^{2-}$ complex is displaced by more strongly adsorbing chloride on the palladium overlayers on the Au(100) electrode. Uosaki and colleagues suggested that the same adlayer of the $PdCl_4^{2-}$ complex should also exist on the palladium overlays on Au(100) as that on the bare Au(100) surface and proposed that the adlayer plays a role in the 2D epitaxial growth during electrochemical deposition of palladium on the Au(100) surface similar to that observed on the Au(111) surface [34].

The electrocatalytic behavior of the thin palladium layers deposited on Au(*hkl*) surfaces for hydrogen adsorption/absorption [77, 80], oxygen reduction [79], oxide formation/reduction [80], copper UPD [33, 77], electrochemical oxidation of formic acid [84] as well as formaldehyde [80] has been investigated in detail. These electrocatalytic activities depended significantly on the surface structure and thickness of the ultrathin palladium layers [80, 84].

4.1.2.3 Rhodium

The autocatalyst is the main application for rhodium now (Fig. 1) [1, 2]. Electrodeposition of rhodium is also widely used both in corrosion prevention and in the decorative field [46, 61, 85]. Details about the progress in electroplating baths for rhodium have been summarized elsewhere [56–58, 61, 85].

The electrochemical deposition processes of rhodium at atomic resolution have been investigated first on platinum single-crystal electrodes [86–89]. From the electrochemical characterization and *in situ* infrared reflection-absorption spectroscopy (IRRAS) spectra of CO adsorbed on the rhodium deposited electrochemically on a Pt(111) electrode, Inukai and coworkers suggested that the epitaxial growth of rhodium monolayers occurs in a pseudomorphic (111) structure [86]. Figure 11 compares the voltammograms of a bulk Rh(111), an Rh monolayer/Pt(111), and Rh overlayers/Pt(111) in 0.5 M H_2SO_4 solution [89]. Although there are some small differences, the voltammogram of monolayer Rh deposited on Pt(111) is very similar to that of a massive Rh(111) electrode, some additional peaks in the hydrogen-adsorption region were observed on the thicker rhodium layers. From these types of electrochemical behavior of rhodium deposited on a Pt(111) surface, Gomez and coworkers concluded that the first rhodium layer grows in a pseudomorphic way, whereas the second and subsequent layers grow in a pseudo-layer-by-layer growth: the third layer begins to form before the second is completed (SK growth) [89]. In the case of ultrathin rhodium layers prepared by metal evaporation in UHV, the second rhodium layer begins to form prior to the completion of the first layer, although both

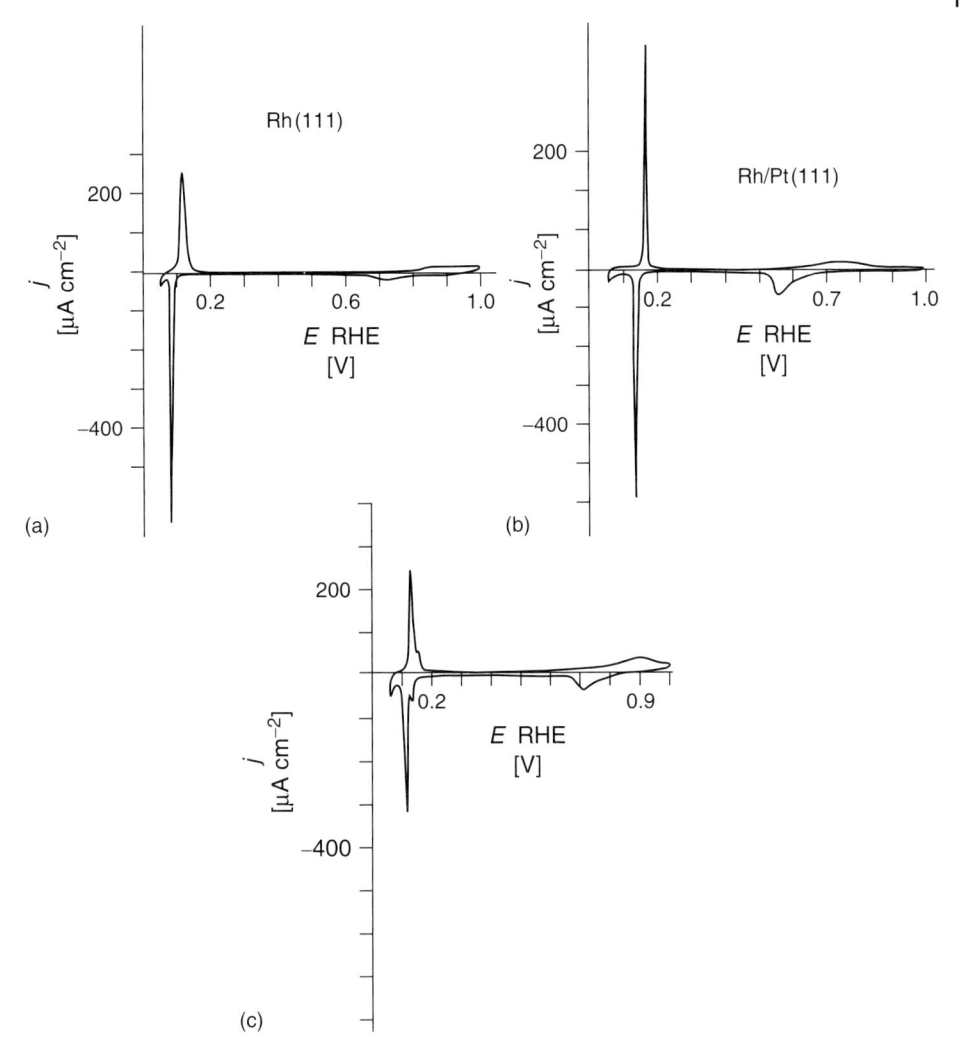

Fig. 11 CVs in 0.5 M H_2SO_4 solution for (a) a massive Rh(111) electrode, (b) Pt(111) electrode modified by one monolayer of rhodium using electroless deposition and (c) rhodium overlayer prepared by electrochemical deposition onto a monolayer Rh-modified Pt(111) electrode in 0.06 mM Rh^{3+} at +0.6 V for 620 s [89].

layers give rise to pseudomorphic (1 × 1) islands [90].

These studies are, however, characterized mainly by electrochemical measurements. The surface morphology determined by STM is still very limited, which may be related to the difficulties of keeping a platinum substrate clean during the *in situ* STM measurements.

Recently, Kolb and colleague investigated the initial stages of electrochemical deposition of rhodium on Au(111) electrodes by *in situ* STM measurements [37]. Figure 12 shows current-potential curves

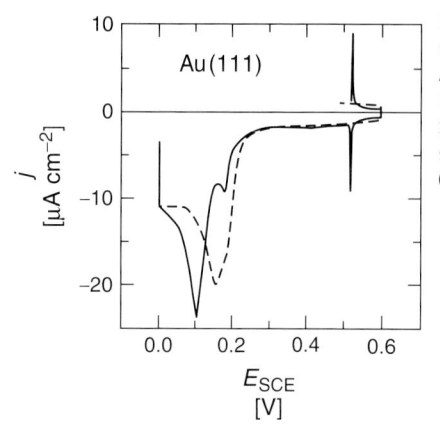

Fig. 12 CVs for a well-ordered Au(111) surface (solid line) and a 5° stepped Au(111) surface (dashed line) in 0.1 M H_2SO_4 + 0.1 mM $RhCl_3$ + 1 mM HCl. Sweep rate: 10 mV s^{-1}. The scans started at +0.5 V and finished at 0 V [37].

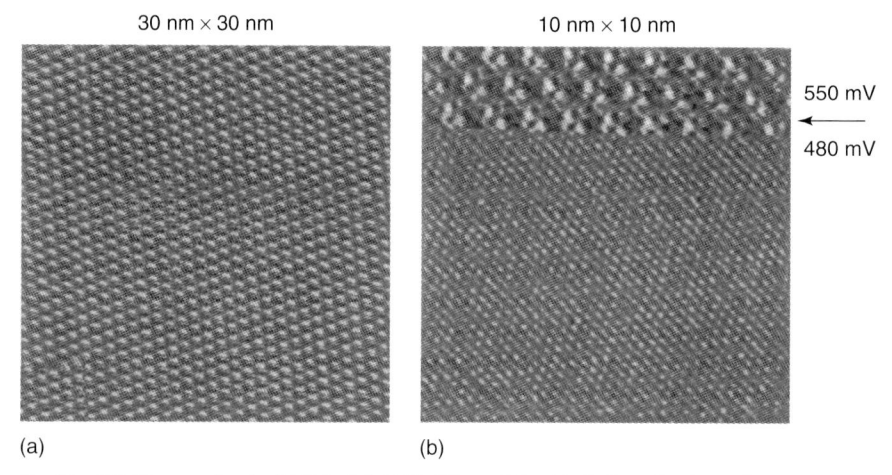

Fig. 13 STM images for Au(111) in 0.1 M H_2SO_4 + 0.1 mM $RhCl_3$ + 1 mM HCl. (a) $E = +0.65$ V, showing the ($\sqrt{13} \times \sqrt{13}$) R13.9°-superlattice pf $RhCl_6^{3-}$ and (b) the same structure at +0.55 V with the atomically resolved Au(111) surface after a potential step to +0.48 V at the moment shown by the arrow [37].

of (a) well-defined (solid line) and (b) 5° stepped Au(111) electrodes (dashed line) in 0.1 M H_2SO_4 + 0.1 mM $RhCl_3$ + 1 mM HCl. A pair of spikes was only observed on the well-defined Au(111) surface at +0.52 V (vs. SCE). The deposition of rhodium at the well-defined Au(111) surface was started around +0.20 V with a large (+0.10 V) and a small cathodic peak (+0.18 V). The peak position of the small peak coincided with the reduction peak observed at the stepped-Au(111) surface, suggesting that the rhodium deposition proceeds easier at surface defects. The rhodium deposited on a gold-electrode surface cannot be dissolved again because a passivating oxide is formed on the rhodium layer when the potential is swept to the positive region [37].

Figure 13 shows *in situ* STM images at the potential region around the current spikes. An STM image with a hexagonal structure can be observed in the potential region more positive than that of the spike (Fig. 13a). This structure is determined to be $\sqrt{13} \times \sqrt{13}R13.9°$ and is attributed to the adsorption of the $RhCl_6^{3-}$ complex. When the potential was stepped to +0.48 V, that is, negative relative to the spikes, this adlattice structure disappeared

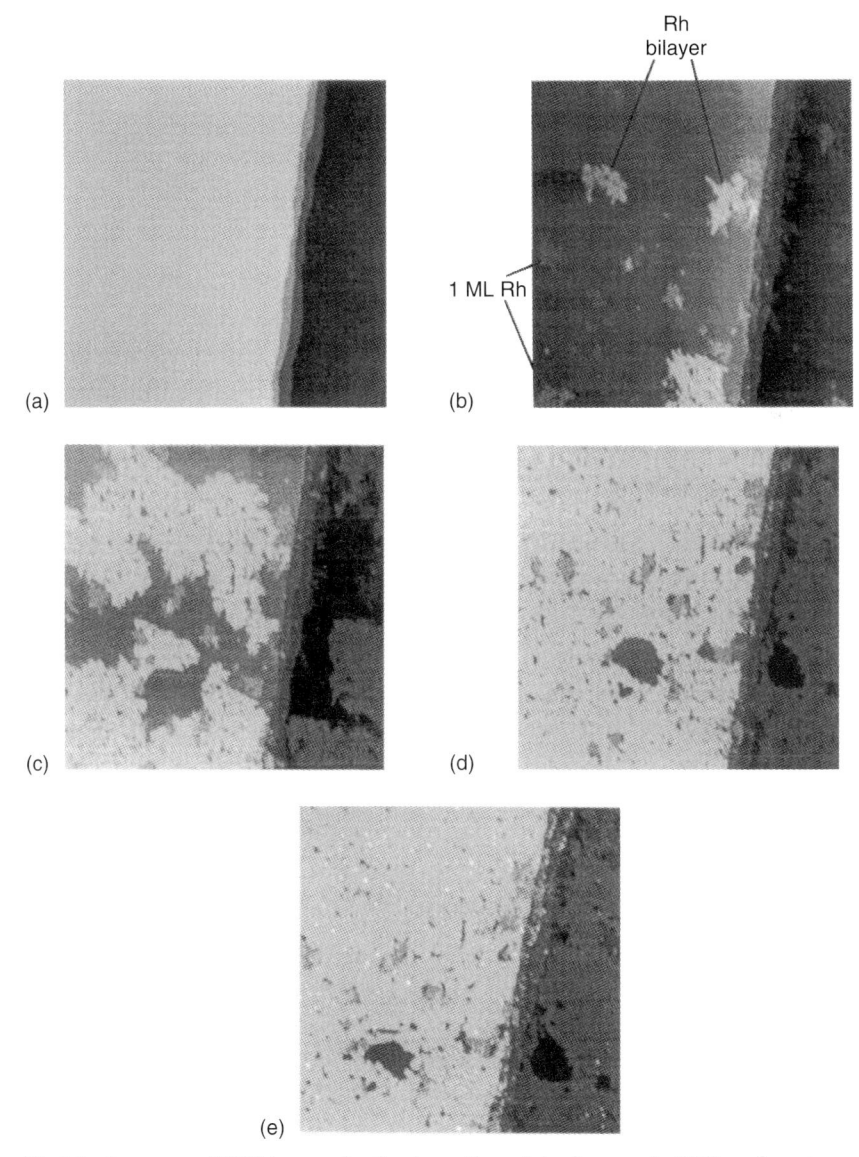

Fig. 14 Sequence of STM images for the deposition of rhodium on Au(111) surface at +0.19 V (400 × 400 nm^2) in 1 M H$_2$SO$_4$ + 0.1 mM RhCl$_3$ + 1 mM HCl. (a) Bare Au(111) surface; (b–d) growth of a rhodium bilayer on Au(111) and (e) rhodium cluster formation on the bilayer [37].

and the Au(111) substrate could be imaged atomically (Fig. 13b).

Figure 14 shows a sequence of STM images (400 × 400 nm^2) during electrochemical deposition of rhodium on an Au(111) electrode in the same solution as Fig. 13 [37]. When the potential was stepped to +0.19 V, a number of rhodium islands were formed at the terrace of the gold substrate (Fig. 14b). Most of the small rhodium islands are immediately covered by a rhodium bilayer with a total height of 0.3 nm. The adsorption of the RhCl$_6^{3-}$ complex was also observed on the surface of the rhodium bilayer. A rhodium film with a two-monolayer height covered the gold surface almost completely (Fig. 14d). The formation of the rhodium islands shows a fractal growth, indicating a low mobility of rhodium atoms on the Au(111) surface. A number of small rhodium clusters were observed on the flat rhodium bilayer (Fig. 14e). These results suggest that the rhodium deposition proceeds through the SK mode [37]. This behavior contrasts significantly with palladium on Au(111) and Au(100) electrodes where a layer-by-layer growth (Frank-van der Merwe growth mode) of palladium was observed clearly.

Electrochemical characterization also showed that the rhodium layers on Au(111) with a thickness of two monolayers or less behave similarly to a well-ordered Rh(111) single-crystal electrode, while thicker rhodium layers are similar to electrochemically disordered Rh(111), where the surface has a higher defect density [37].

4.1.2.4 Ruthenium

The principle application of ruthenium is in the electronics industry for the manufacture of resistors [2]. Ruthenium is attracting more attention because of its crucial role in electrooxidation catalysis in a direct methanol fuel cell (DMFC) [49, 50], which is a key candidate for use as a pollution-free power supply in the near future. In fact, the most efficient catalyst for the practical DMFC is the Pt/Ru based electrode, and ruthenium is known to promote the methanol oxidation on a platinum surface by providing OH$_{ad}$ species on its surface.

The electrochemical deposition of a metallic-ruthenium film is very difficult compared with that of other platinum-group elements [46, 61, 91–93]. One of the reasons may be related to the complicated electrochemistry of ruthenium deposition and the stability of the Ru-chloro complex [92]. For example, it has been reported that the RuCl$_3$ species in HClO$_4$ solution is decomposed partly into RuO^{2+}, in which Ru(IV) is present [94].

The controlled deposition of ruthenium on well-defined surfaces, such as Pt(*hkl*) [95–103] and Au(*hkl*) [38–40], has been characterized by electrochemical measurements, Fourier transform infrared reflection-absorption spectroscopy (FT-IRRAS), XPS and STM measurements. The interest in these studies is mainly concentrated on the ruthenium modification of a platinum surface because of its extreme importance in electrocatalysis. It has been demonstrated that a ruthenium-deposited Pt(111) substrate showed an extremely high activity in methanol oxidation compared to ruthenium-deposited Pt(*hkl*) electrodes with other crystallographic orientations [98, 99].

The studies of ruthenium deposition are mainly carried out by electrochemical and spontaneous (electroless) deposition procedures. Friedrich and colleagues found that the well-reproduced current-potential curves of ruthenium-modified Pt(111) electrodes were obtained when ruthenium was deposited electrochemically in the potential region between +0.8 V and +0.3 V (vs. RHE) from a freshly prepared 0.1 M H_2SO_4 containing 5 mM $RuCl_3$. The reproducibility became worse when 0.1 M $HClO_4$ was used as supporting electrolyte during the electrochemical deposition [95–97]. On the other hand, Wieckowski and colleague deposited ruthenium from 0.1 M $HClO_4$ containing

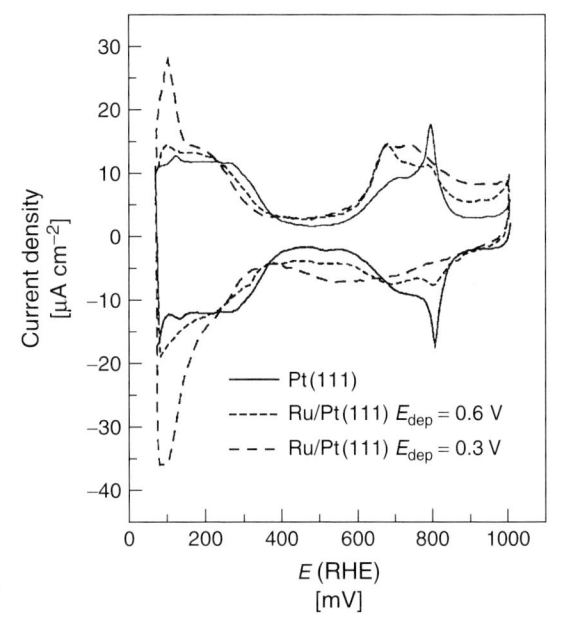

Fig. 15 CVs of Pt(111) modified by ruthenium in 0.1 M H_2SO_4 + 5 mM $RuCl_3$ at +0.6 and +0.3 V in 0.1 M $HClO_4$ in comparison to the bare Pt(111) voltammogram in 0.1 M $HClO_4$. Sweep rate: 20 mV s^{-1} [97].

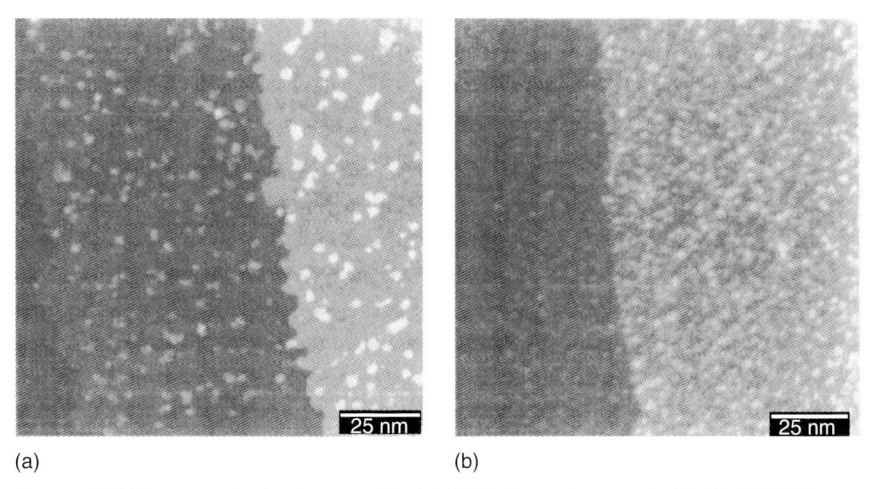

Fig. 16 STM images of ruthenium-modified Pt(111) electrodes recorded in 0.1 M $HClO_4$ at +0.5 V. Deposition was performed prior to imaging in a standard glass cell from 0.1 M H_2SO_4 + 5 mM $RuCl_3$ at +0.6 V for (a) 5 min and (b) 30 min, respectively [97].

dilute RuCl$_3$ by a spontaneous-deposition method and suggested that the RuO$_{ad}$ species was formed on the Pt(111) surface [98]. Figure 15 compared the CVs of Pt(111) before (solid line) and after (dashed and dotted lines) the electrochemical deposition of ruthenium [97]. The characteristic "butterfly" peak observed at a clean Pt(111) changed, and a new peak was observed in the hydrogen-adsorption region after ruthenium modification. Similar changes in CVs were also observed by Wieckowski and colleagues, where a spontaneous-deposition method was employed [98–100, 104]. The formation of ruthenium on the platinum surface has been confirmed by XPS, Auger, and FT-IRRAS as well as STM measurements [96–101, 104]. Because of the various oxidation states of deposited ruthenium, the coverage of ruthenium cannot be estimated reliably from the electrical charge passed. The coverage of ruthenium in these experiments was mainly determined from STM, XPS, or Auger experiments instead of electrochemical measurements.

Figure 16 shows *in situ* STM images of Pt(111) electrodes in 0.1 M HClO$_4$ at +0.5 V after electrochemical deposition of ruthenium in 0.1 M H$_2$SO$_4$ + 5 mM RuCl$_3$ at +0.6 V for (a) 5 min and (b) 30 min, respectively [96, 97]. The formation of ruthenium islands of 3 ∼ 5 nm in diameter were conformed on the terrace. It seemed to be difficult to control the size of the ruthenium islands. The coverage of ruthenium in Fig. 16(a) and (b)

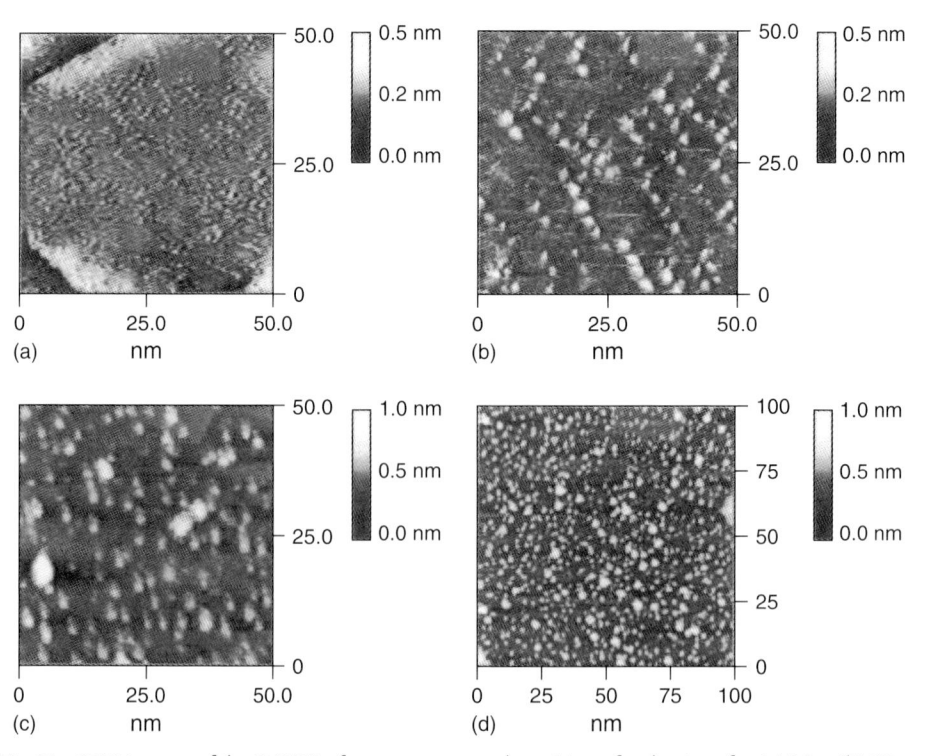

Fig. 17 STM images of the Pt(111) after spontaneous deposition of ruthenium for (a) 10 s, (b) 20 s, (c) 40 s (50 × 50 nm^2) and (d) 90 s (100 × 100 nm^2). See Table 2 for details [100].

was estimated to be 0.25 and 0.6 from the STM images and was in agreement with the calibration by XPS measurement. The density of the ruthenium islands depends on the deposition potential and deposition time [97].

Figure 17 shows STM images of a Pt(111) electrode obtained by Wieckowski and colleagues in air after immersion in 0.1 M HClO$_4$ solution containing 0.5 mM RuCl$_3$ for (a) 10 s, (b) 20 s, (c) 40 s and (d) 90 s [100]. In order to keep the Pt(111) substrate clean for longer STM-imaging times, after ruthenium deposition, the platinum surface was covered with an iodide monolayer to keep the surface free from contamination since the iodide layer has no effect in the STM topography of the Ru-modified surface. Ruthenium islands of nanometer size were observed on the Pt(111) surface and their number increased with an increase in immersion time. Table 2 summarizes the results obtained from these STM images. The islands are not uniformly distributed over the surface when the coverage is low (Fig. 17a and b). There is no selective growth of islands at the steps, indicating that the electroless deposition is not nucleated by the crystallographic defects at the surface. When the coverage is above 0.14 ML, a second layer deposit on top of the inner layer was observed, although the inner monoatomic layer has not yet been completed. The bilayer islands, corresponding to 10% of the overall islands, were observed clearly at coverage of 0.14 to 0.19 ML (Fig. 17c,d). Similar bilayer island structures of vapor-deposited ruthenium on a Pt(111) surface have been observed by Iwasita and coworkers [103]. Extending the immersion time more than 90 s does not result in higher ruthenium coverage. The limit of ruthenium coverage on the Pt(111) surface was approximately 0.2, much less than that from electrochemical deposition. When an Ru-modified electrode was polarized in hydrogen-evolution region, the amount of the ruthenium deposited also decreased, indicating that a ruthenium oxide-reduction process in that potential region may be accompanied by partial ruthenium dissolution from the surface [100].

Recently, Behm and colleagues investigated the electrodeposition of ruthenium on reconstructed [38] and unreconstructed Au(111) electrodes in detail [39]. It was found that the nucleation of ruthenium islands during the electrodeposition on a reconstructed Au(111) electrode occurs exclusively at the elbows of the Au reconstruction and that the *fcc* area of the reconstructed Au(111) surface is decorated at higher coverage [38]. However, a random nucleation process was observed on

Tab. 2 Ruthenium coverage and the ruthenium island heights as determined by STM as a function of deposition time (Fig. 18) [100]

Deposition time [s]	Ru coverage [ML]	Relative height in the island center [nm]
10	0.01	0.22
20	0.08	0.22
40	0.14	0.22 and 0.45
90	0.19	022 and 0.45
120	0.19	0.22 and 0.45

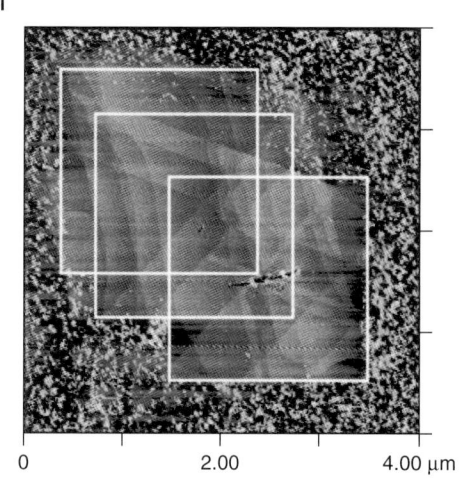

Fig. 18 *In situ* STM image (4000 × 4000 nm^2) of Au(111) surface at +0.3 V in 0.05 M H$_2$SO$_4$ + 1.6 mM RuNO(NO$_3$)$_x$(OH)$_y$ (x + y = 3). STM imagining was carried out in the 2000 nm region (shown by the white box) in the beginning when the potential was swept between +0.3 V and −0.3 V (2 mV s^{-1}) for approximately 2 hr. When the imaging region was enlarged to 4000 nm, ruthenium deposition was observed outside of the tip-scan region [40].

an unreconstructed Au(111) surface [39] as was observed on a Pt(111) electrode (cf. Figs. 16 and 17). The ruthenium is deposited in the entire Au double-layer potential region and a multilayer of ruthenium can be deposited when the potential becomes more negative than − 0.1 V (vs. Ag/AgCl).

The similar random nucleation process of ruthenium deposition on an unreconstructed Au(111) electrode was observed by Uosaki and colleague independently [40]. Figure 18 shows STM images (4000 × 4000 nm^2) at +300 mV (vs. Ag/AgCl) in 0.05 M H$_2$SO$_4$ solution containing 1.6 mM RuNO (NO$_3$)$_x$(OH)$_y$ (x + y = 3). It has been reported that the RuNO complex is more stable than that of the RuCl$_3$ species in solution [104]. Before this image was obtained, STM imaging was carried out at the narrow center portion of the image (shown by the dotted white box) for approximately 30 min, and no clear ruthenium-nucleation island was observed. When the tip scan region was enlarged, however, many ruthenium-nucleation islands were randomly formed outside of the region originally scanned, indicating a large STM-tip effect during the ruthenium deposition.

As described in the preceding text, although it became possible to grow electrochemically the ultrathin ruthenium films, which can be applied in electrocatalysis, the epitaxial growth of a metallic ruthenium film seems to be difficult at the present stage.

It should be mentioned here that the composition of the electrolyte solution used in these *in situ* STM studies of the electrochemical deposition of noble metals (platinum, palladium, rhodium and ruthenium) are quite different from those used in the real electroplating industry [56–58]. Although some experimental conditions (temperature, concentration) may be difficult for the *in situ* STM measurements, electrodeposition in a practical electrolyte bath should provide more information both in application and in fundamental fields.

4.1.3
Atomically Controlled Dissolution of Noble Metals

Knowledge about the metal-dissolution mechanism is very important for

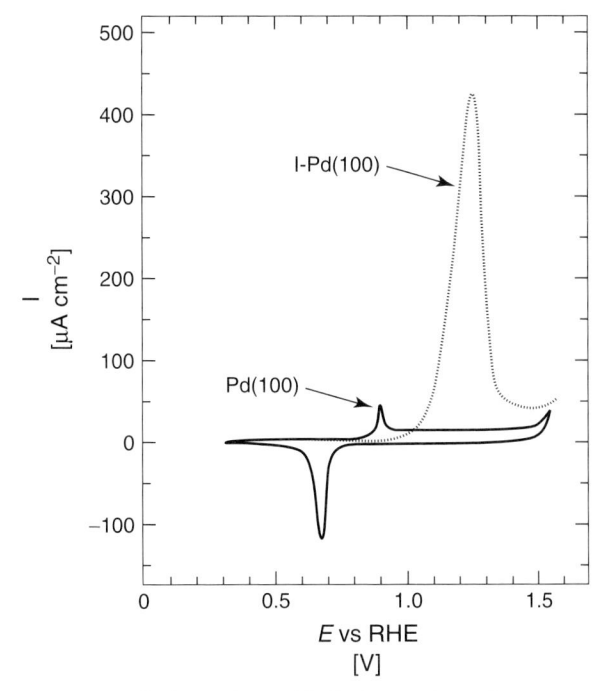

Fig. 19 CV of I-free and I-modified Pd(100) electrode in 0.05 M H_2SO_4 solution [117].

controlling the metal surface structure. Studies of anodic dissolution of metal at atomic resolution have been reviewed elsewhere [23, 24, 105]. In this section, the discussion will be concentrated on the anisotropic layer-by-layer dissolution behavior of noble metal electrodes, palladium and gold, catalyzed by an adlayer of adsorbate on the substrate. Similar anisotropic anodic dissolution processes catalyzed by the ordered adsorbed layer have also been observed on S-Ni(100), [106] I-Ni(111), [107] I-Ag(100) [108] and Cl-Cu(hkl) [109–112] as well as Cl-Au(hkl) [29–31], which will be discussed in the next section in detail.

4.1.3.1 Palladium

Soriaga and colleagues reported that the dissolution of palladium single-crystal electrodes in sulfuric solution was catalyzed by a chemisorbed iodine monolayer for the first time [113–116]. The electrochemical dissolution processes of palladium modified by iodine have been investigated in detail by a combination of ultrahigh vacuum electrochemistry (UHV-EC) and Auger as well as low-energy electron diffraction (LEED) [117, 118].

Figure 19 compares the CVs of a bare Pd(100) and an I-modified Pd(100) electrode in 0.05 M H_2SO_4 solution [117]. A small anodic peak due to the oxidation of the Pd(100) surface was observed at +0.90 V (vs. RHE) on the bare Pd(100) electrode (Reactions 1 and 2):

$$Pd + H_2O \longrightarrow Pd-OH_s + H^+ + e^- \quad (1)$$

$$Pd-OH_{(s)} \longrightarrow PdO_{(s)} + H^+ + e^- \quad (2)$$

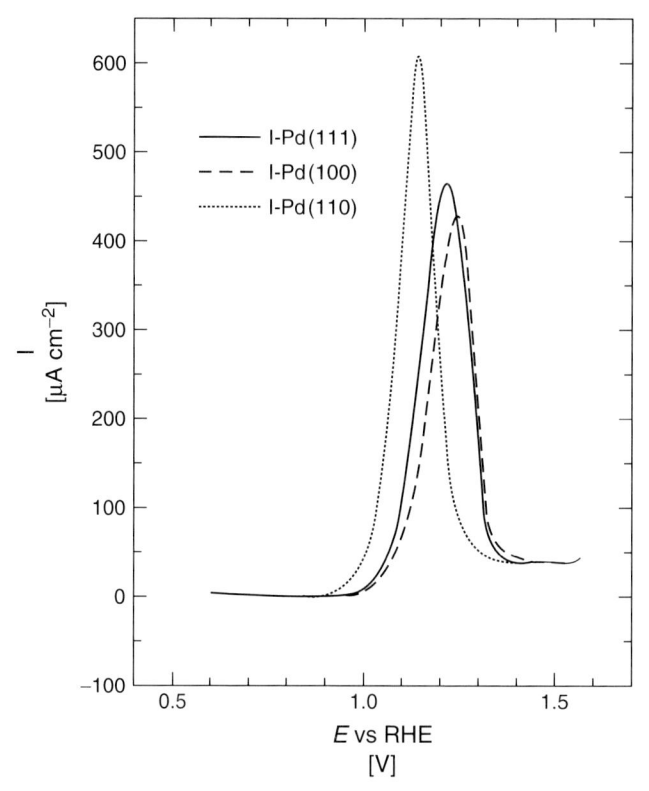

Fig. 20 CVs of I-modified Pd(111), Pd(100) and Pd(110) electrodes in 0.05 M H_2SO_4 solution [117].

This anodic peak was suppressed completely on the I-modified Pd(100) electrode, and a new anodic peak with an order-of-magnitude larger peak current than that of the I-free Pd(100) electrode was observed at +1.25 V. This new anodic peak has been attributed to the two-electron dissolution process of palladium as:

$$Pd_{(s)} \longrightarrow Pd_{(aq)}^{2+} + 2e^- \qquad (3)$$

Therefore, the anodic dissolution of palladium was enhanced significantly when the iodine adlayer was adsorbed on the palladium surface. The rate of the dissolution is directly proportional to the coverage of iodine on the surface, and the iodine coverage remains constant throughout the dissolution process [117].

Figure 20 compares the structure dependence of the I-catalyzed electrochemical dissolution of Pd(hkl) electrodes in 0.05 M H_2SO_4 solution. The dissolution rate slightly depends on the crystallographic orientation and decreases in the order of I-Pd(110) > I-Pd(111) > I-Pd(100) [117].

Figure 21 shows a sequence of *in situ* STM images of an I-modified Pd(100) electrode after the potential was stepped to +1.05 V for (a) 5, (b) 6, and (c) 7 min, where the palladium started to dissolved anodically (Fig. 19) [117]. An atomically flat Pd(100) surface with a monoatomic step was observed, and the I-catalyzed

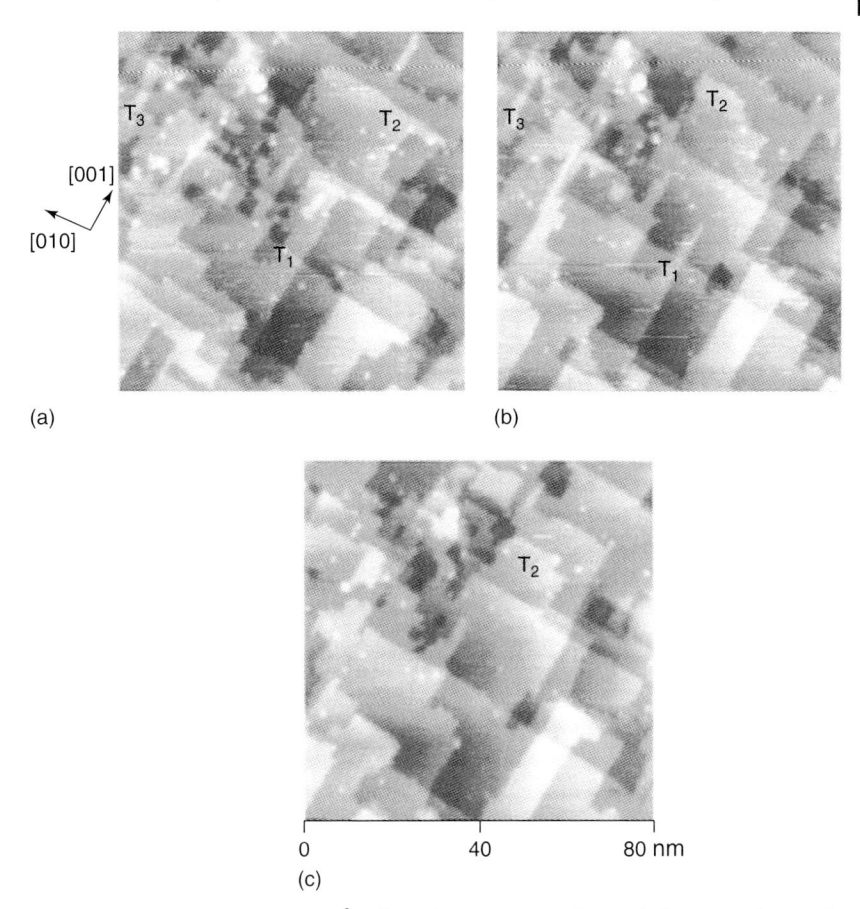

Fig. 21 STM images (80 × 80 nm^2) of Pd(100)-c(2 × 2)-I obtained after (a) 5, (b) 6 and (c) 7 min of dissolution at +1.05 V in 0.05 M H$_2$SO$_4$[117].

dissolution take places only at the step edges in a layer-by-layer mode, while etching from the pit formation was not observed at the surface. As shown in Fig. 21, terraces T$_1$, T$_2$ and T$_3$ became narrow along [001]-directed steps but only a minimal change was observed in the direction perpendicular to it, that is, an anisotropic dissolution of palladium took place. Itaya and colleagues interpret the dissolution behavior using a schematic model (Fig. 22) where iodine formed a c(2 × 2) adlattice on the Pd(100) surface [117]. The palladium atoms A and B in the corner of the top terrace can be regarded as the most reactive ones for the dissolution. Removal of atom A (dissolution along [001] direction) easily permits the iodine atom to slide down from a 4-fold site at the upper terrace onto another hollow site (H) at the lower terrace. On the other hand, iodine drop onto site H cannot be accomplished upon removal of atom B (dissolution along [010] direction) because of steric hindrance by atom A. Therefore, iodine is easier to diffuse from a high-symmetry site from the upper to a lower terrace and is considered to be

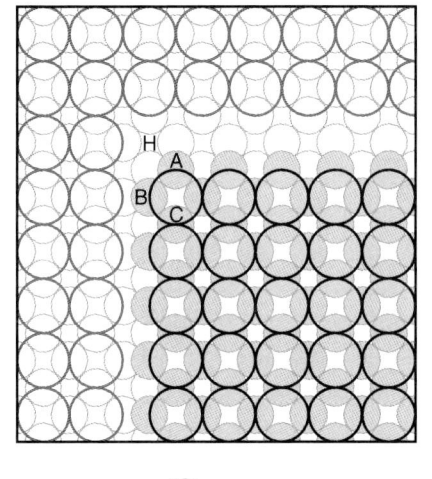

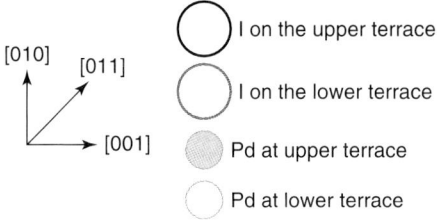

Fig. 22 Schematic model to show the anisotropic dissolution of I-Pd(100) [117].

iodine adlayers play a significant role in the iodine-catalyzed dissolution process. Therefore, the dissolution process of the I-modified Pd(111) and Pd(100) electrodes take place in the layer-by-layer sequence without disruption of the iodine-adlattice structure.

the major driving force in the anisotropic dissolution [117].

The step-selective layer-by-layer dissolution behavior was also observed on an I-Pd(111) surface. However, the anisotropic-dissolution features are not as obvious as that observed on an I-Pd(100) surface. The pit formation on the I-modified Pd(110) precludes layer-by-layer dissolution and leads to progressive disorder [117, 118].

The *ex situ* LEED [113–116] and *in situ* STM measurements with high resolution [105, 117] demonstrated that the ordered iodine adlayers were formed on the palladium surfaces as ($\sqrt{3} \times \sqrt{3}R30°$)-I-Pd(111), c(2 × 2)-I-Pd(100) and pseudohexagonal-I-Pd-(110). It is noteworthy that the same adlattice structure of iodine was observed on the palladium terrace after anodic dissolution. The ordered

4.1.3.2 Gold

Gold is known to have excellent chemical stability and is widely used in industry and as an accessory material [119, 120]. It becomes unstable and dissolves in the positive-potential region, especially in solutions containing Cl^-.

Figure 23 shows CVs of (a) Au (111) and (b) Au(100) electrodes in 0.1 M $HClO_4$ solution containing 1 mM Cl^- in the potential region between +0.2 and +1.7 V. A significant increase in the anodic current started at +1.25 V, and two anodic peaks were found at +1.40 V and +1.51 V at Au(111) electrode. When the potential became more positive than +1.55 V, the anodic current quickly decreased. The pronounced anodic current in the positive-going potential sweep was attributed to the $3e^-$ oxidative dissolution of gold (Reaction 4, forward) [29].

$$Au + 4Cl^- \longleftrightarrow AuCl_4^- + 3e^- \quad (4)$$

$$Au + H_2O \longleftrightarrow AuO + 2H^+ + 2e^- \quad (5)$$

In the more positive potential region, the dissolution reaction competed with oxide formation (Reaction 5, forward) on the gold surface and passivation was completed around +1.7 V. The cathodic peaks observed at +1.21 and +1.10 V in the negative-going potential sweep were related to the simultaneous oxide reduction

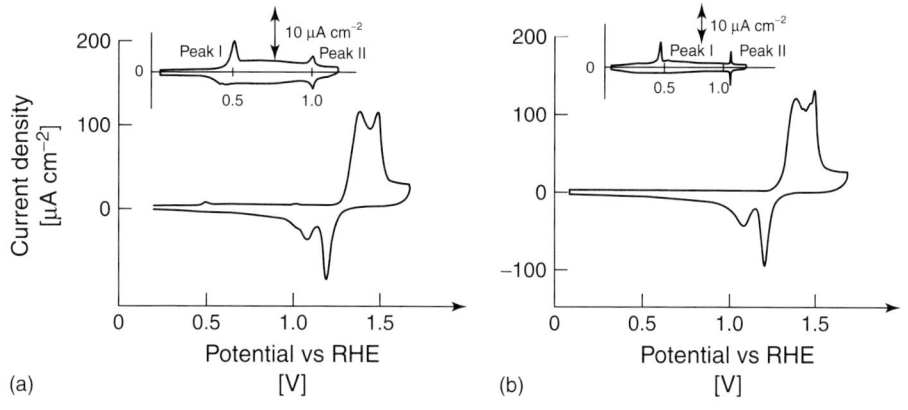

Fig. 23 CVs of (a) Au(111) and (b) Au(100) electrodes in 0.1 M HClO$_4$ solution containing 1 mM Cl$^-$ in the potential region between +0.2 V and +1.7 V. Insets show CVs observed in the potential region between +0.2 and +1.2 V. Sweep rate: 20 mV s^{-1} [31].

(Reaction 5, backward)/gold dissolution (Reaction 4, forward), and the redeposition of the dissolved gold species (Reaction 4, backward), respectively, as already confirmed by EQCM measurements [29]. Similar 3e$^-$ anodic dissolution of Au(100) electrode in acidic solution containing Cl$^-$ (Fig. 23b) is expected [31].

Enlarged CVs obtained in the potential region between +0.2 and +1.2 V in Fig. 23 show an anodic peak around 0.5 V (Peak I) and a pair of redox spikes around +1.0 V (Peak II). The Peak I can be attributed to Cl$^-$-induced lifting of the surface reconstruction on both Au(111) and Au(100) electrodes [81]. Peak II is not observed at all in the Cl$^-$-free solution and has been attributed to the formation of an ordered chloride adlayer on the Au(111) and Au(100) surfaces.

Figure 24 shows sequentially obtained *in situ* STM images (1 × 1 μm^2) of an Au(111) surface in a 0.1 M HClO$_4$ solution containing 1 mM Cl$^-$ [30, 31]. Stable large terraces and monoatomic step lines were observed at +0.8 V as was the case in Cl$^-$ free solution. The step lines, which were parallel to the [110] direction, crossed each other at 60 or 120 degrees. When the potential was slowly swept (2 mV s^{-1}) in the positive direction, drastic changes in the shape of the step lines were observed.

Figure 24 shows an STM image obtained during the potential sweep from +0.99 V (top) to +1.09 V (bottom), close to that of Peak II observed at Au(111) (Fig. 23a). Although no change was found on the terraces, several kinks were formed in the originally straight step lines. Thus, the gold surface really started to dissolve from the step sites at this potential. The outline of the step lines changed drastically on the bottom part compared to that on the top part in the STM image, demonstrating that the gold-dissolution rate increased as the electrode potential became more positive. The SXS measurement showed that the chloride coverage on the Au(111) electrode increased with a slope of about 1.5%/V after the ordered chloride layer was constructed at this potential region [121]. Thus, potential induced "compression" of the chloride adlayer and initial dissolution on the Au(111) electrode seemed to

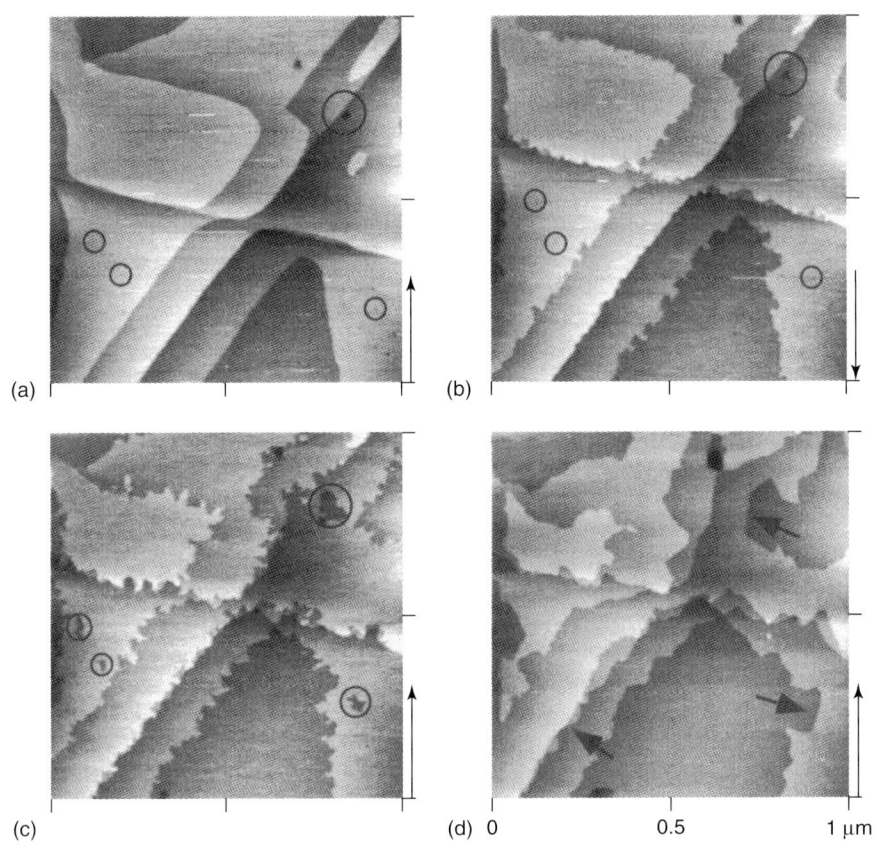

Fig. 24 A sequence of *in situ* STM images (1 × 1 μm^2) of the Au(111) surface in 0.1 M HClO$_4$ containing 1 mM Cl$^-$. (a) At +0.8 V, (b) +0.99 V ∼ +1.09 V during a potential sweep (2 mV s^{-1}) with rastering the tip downwards, (c) 0 min, (d) 6 min and (e) 40 min after the potential was swept to +1.27 V, (f) potential was stepped to +1.36 V at the moment shown by the arrow with rastering the tip downwards, (g) next scan after (f) with rastering the tip upwards, (h) 3 min later from (f) with rastering tip downwards. E_{tip} = +1.2 V. i_T = 5 nA. A compass was drawn to indicate the orientation of the surface. See the text for details [30].

take place in this potential region. The dissolution proceeded mainly at the step sites, and the originally observed straight step lines were etched like the teeth of a saw at more positive potential (+1.27 V, Fig. 24c). The dissolution also proceeded on the terrace (Fig. 24c) but only from spots where small pits or defects were found at more negative potentials (Figs. 24a,b) as indicated by the black circles. The Au(111) surface was etched mainly from its step edge in prolonged etching at +1.27 V (6 and 40 min., Figs. 24d–e). At the beginning of the dissolution, only steps along the [110] direction were observed (Figs. 24b,c). As shown by the arrows in Fig. 24d, some of the step lines were rotated about 30 degrees, and new step lines running along the [211] direction of the Au(111) substrate were observed. The step edges along the [211] direction, however, became

(e) (f) (g) (h) 0 0.5 1 μm

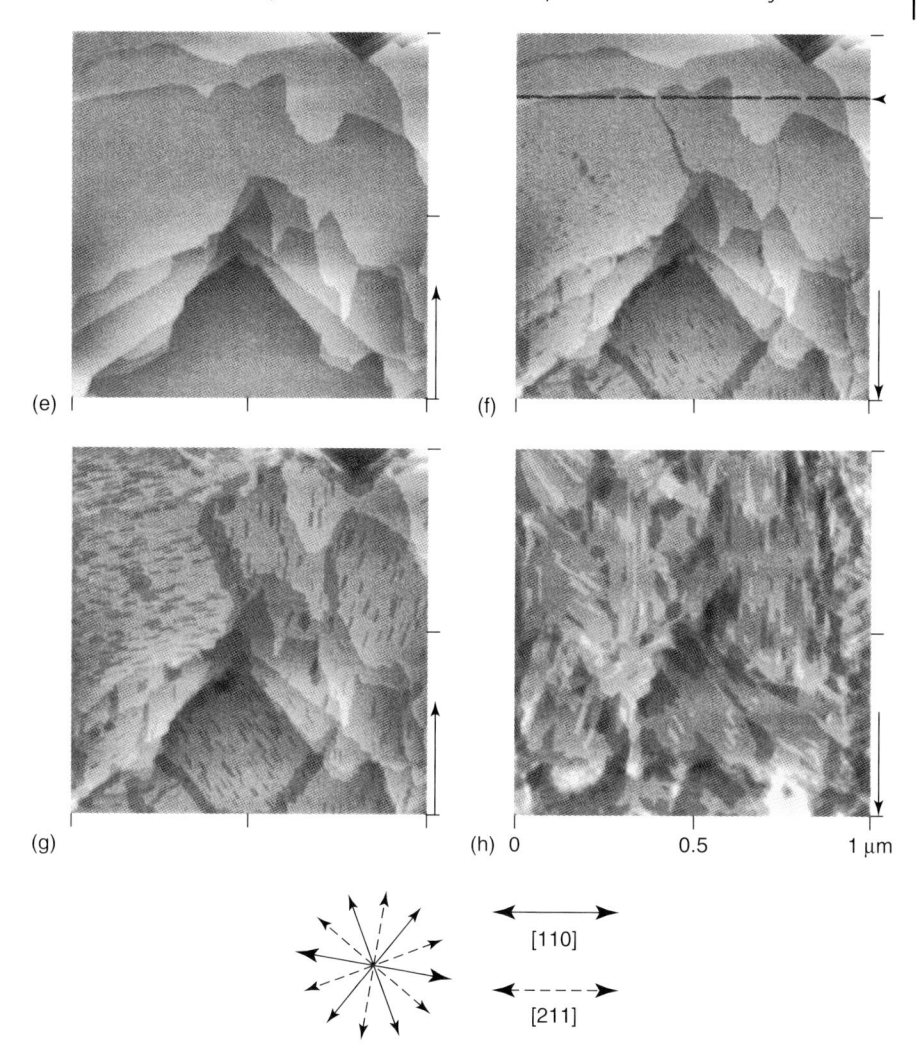

[110]
[211]

Fig. 24 (Continued)

dominant after the electrode was kept at +1.27 V for a long time (Fig. 24e).

As the potential was stepped to a more positive values (+1.36 V, Figs. 24f–h), where gold still dissolves without oxide formation, at the moment shown by an arrow in Fig. 24(f), acceleration of the dissolution rate, especially on the terraces, was observed. Many small pits were formed on the terraces. In the bottom part of the image, rectangular-shaped pits were clearly observed. In the next STM scan (Figs. 24f,g), rectangular pits were clearly observed on all terraces. It is interesting to note that the rectangular pits were parallel to each other within the same domain. Furthermore, the direction of the long axis of the rectangular pits was parallel to the [211] direction of the substrate, that is, the new step direction after dissolution (Fig. 24e). The pits grew and then merged. Figure 24(h) shows an STM image captured about 2 min after Fig. 24(g). The top layer of the Au(111) was almost etched away, and the next layer of the gold surface started to dissolve. Nearly all of the step lines were orientated along the [211] direction, and the step line along the [110] direction was absent. These results demonstrated that gold was dissolved layer by layer in an anisotropic way in which the step edges preferentially retreated along the [211] direction of the Au(111) substrate.

STM observations were also carried out in solutions of various Cl^- concentrations. It was found that the critical potential of the dissolution depended on the Cl^- concentration. The higher the Cl^- concentration, the more negative the critical potential. This potential dependence is consistent with that of Peak II. In solutions containing Cl^- at less than 0.1 mM, the dissolution rate was very low, and both the [211] and [110] step lines were observed after the electrode was kept at +1.36 V for 20 min. In a solution of higher Cl^- concentration, that is, more than 10 mM, the anisotropic characteristics were hard to observe because of the very high dissolution rate.

The anisotropic dissolution behavior was also observed on the Au(100) electrode in the similar solutions (Fig. 25). The dissolution takes place only on the step edges in the potential region around Peak II (Fig. 25a). At the beginning of the dissolution, steps only along the [110] direction were observed (Figs. 25b,c). As shown by the arrows in Fig. 25(d), many step lines rotated 45 degrees from the original [110] direction, that is, the [100] direction of the Au(100) substrate, after 10 min etching at +1.30 V. Almost all of the step lines were running along the [100] direction after being polarized at +1.30 V for longer periods (Figs. 25e,f). These results suggest that anodic dissolution of the Au(100) electrode surface is also an anisotropic layer-by-layer etching process in which the step edges preferentially retreated along the [100] direction of the Au(100) substrate [31].

As described in the preceding text, the anisotropic-dissolution behavior was observed both on Au(111) and Au(100) electrode surfaces only in acidic solution containing Cl^-. The structure of the chloride adlayer on the gold surface is considered to be one of the important origins of the anisotropic-dissolution process. Suggs and Bard observed that copper dissolved in aqueous-chloride solution from the step edges respectively retreating along the [211] and [100] directions on Cu(111) [109] and Cu(100) electrode surfaces [110]. Vogt and coworkers observed a similar anisotropic dissolution on a Cu(100) electrode surface in HCl solution and proposed that the preferential formation of the [100]-step edge resulted from its higher stability because the [100]-step edge of the Cu(100) was more stable than that of the [110]-step edges by 3 to 4 orders of magnitude [111, 112].

Figure 26 illustrates a real-space model of the ordered chloride adlayer and the underlying Au(111) at the potential of Peak II, based on the SXS measurements [121]. A chloride adlayer with a close-packed hexagonal structure is not commensurate with the gold substrate, and the interatomic distance was close to the van der Waals diameter of chloride. The highest density of the chloride adatom was found in the step lines along the [211] direction, and the adlayer was compressed more with increasing potential. As the STM results demonstrated in Fig. 24, the original step lines along the [110] direction disappeared and new [211] step lines were observed

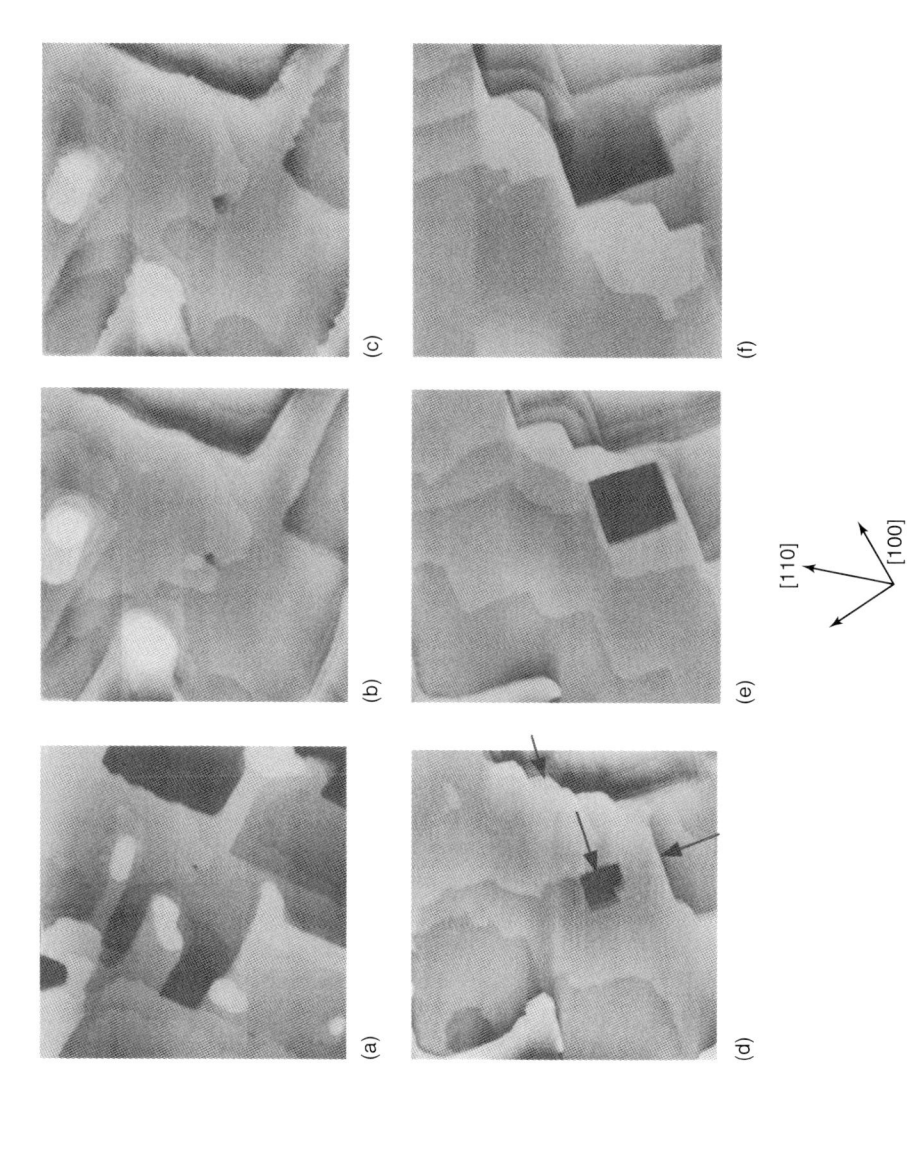

Fig. 25 A sequence of *in situ* STM images of the Au(100) surface in 0.1 M HClO$_4$ containing 1 mM Cl$^-$. (a) At +0.8 V (1 × 1 µm^2), (b) At +1.1 V, (c) 0 min, (d) 10 min, (e) 20 min, and (f) 40 min after the potential was swept to +1.30 V (0.5 × 0.5 µm^2). E_{tip} = +1.40 V. i_T = 3 nA. A compass is drawn to indicate the orientation [31].

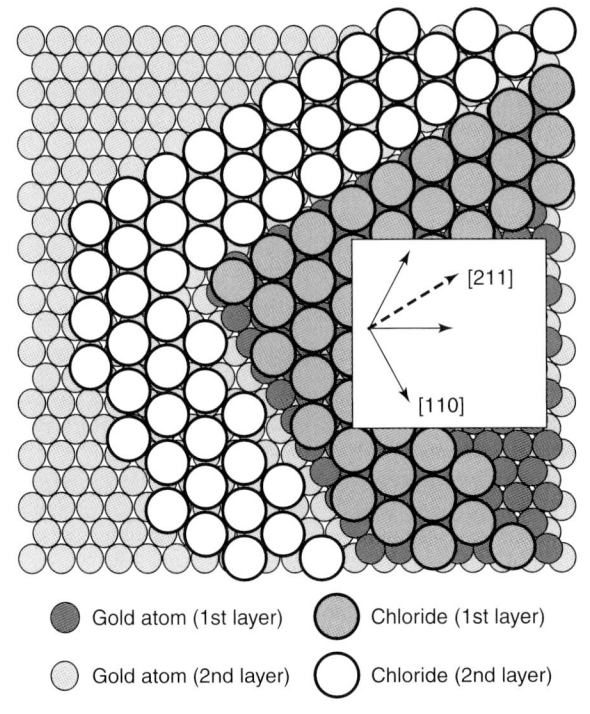

Fig. 26 A schematic model illustrating the chloride adlayer on an Au(111) surface based on the surface X-ray diffraction measurement. See the text for details [30].

on the surface after the anodic dissolution. The lateral interactions between the adsorbed chloride and gold were considered to stabilize the atom rows along the [211] direction more than those in the other directions because of its highest chloride-adatom density. In other words, a much lower etching rate on the [211] step lines was expected, and as a result, the anisotropic dissolution of Au(111) in the Cl^- containing solution was observed.

In the case of an Au(100) electrode, Wang and coworkers observed using SXS measurements that chloride formed uniaxially incommensurate $c(\sqrt{2} \times p)R45°$ adlayer structure on the Au(100) electrode surface and was compressed along the [100] direction when the potential became more positive [122]. A similar chloride adlayer structure has also been observed using in situ STM by Kolb and colleagues [123]. The chloride adlayer with a higher Cl^- adatom density on the step edges of the [100] direction will make these step edges more stable than those of [110] and other directions on the Au(100) electrode surface in the positive potential region. Because of the stability, the [110]-step lines, which were originally observed before dissolution, disappeared and only the [100]-step lines can be observed after the anodic dissolution.

4.1.4 Summary

Electrochemical deposition of noble metal elements (platinum, palladium, rhodium, and ruthenium) on substrates (mainly gold and platinum single-crystal electrode substrates) has been investigated by many modern instrumental analysis methods, such as STM, EQCM, and SXS measurements at an atomic level. As summarized in Table 1, electrochemical deposition of palladium on a gold substrate takes places in the FM mode, similar to that of silver. Platinum and rhodium grow electrochemically on a gold substrate in the SK mode. The electrochemical growth of ruthenium on a gold substrate prefers the SK mode, and a metallic-ruthenium film is still difficult to grow at the moment. These types of behavior can be understood qualitatively using the mismatch parameter of lattice and surface energy between the deposited metal and the gold substrate.

The anistropic layer-by-layer dissolution of palladium and gold are enhanced by an adlayer of iodine and chloride on the substrate surfaces.

The present information is very useful for creating electrocatalytic materials with high activity and selectivity. This is also very important for understanding the reaction mechanism of the metal epitaxy in the field of surface electrochemistry. Very rapid progress is expected in this field.

Acknowledgment

This work was partially supported by Grants-in-Aid for Scientific Research on Priority Area of "Electrochemistry of Ordered Interfaces" (No. 09237101) and for Encouragement of Young Scientists (No. 12750723, 10740314) from the Ministry of Education, Science, Sports and Culture, Japan. SY acknowledges the support from PRESTO, Japan Science and Technology Corporation (JST). SY acknowledges grants from the Iketani Science and Technology Foundation, the Japan Securities Scholarship Foundation and the Tokuyama Science and Technology Foundation.

References

1. F. R. Hartley, (Ed.), *Chemistry of the Platinum Group Metals*, Elsevier Scientific Publishers B. V., Amsterdam, 1991.
2. A. Cowley, M. Steel, *Platinum 2001*, Johnson Matthey, London, 2001.
3. G. A. Somorjai, *Surface Chemistry and Catalysis*, John Wiley & Sons, New York, 1994.
4. C. Argile, G. E. Rhead, *Surf. Sci. Rep.* **1989**, *10*, 277–356.
5. J. A. Rodriguez, *Surf. Sci. Rep.* **1996**, *24*, 223–287.
6. H. Brune, *Surf. Sci. Rep.* **1998**, *31*, 121–229.
7. T. T. Kodas, M. J. Hampden-Smith, *The Chemistry of Metal CVD*, VCH, Weinheim, 1994.
8. M. K. Hampden-Smith, T. T. Kodas, *Chem. Vapor Deposition* **1995**, *1*, 8–23.
9. J. Clavilier, R. Faure, G. Guinet et al., *J. Electroanal. Chem.* **1980**, *107*, 205.
10. G. Binnig, H. Rohrer, C. Gerber et al., *Appl. Phys. Lett.* **1982**, *40*, 178.
11. G. Binning, C. F. Quate, C. Gerber, *Phys. Rev. Lett.* **1986**, *12*, 930.
12. O. M. Magnussen, J. Hageböck, J. Hotlos et al., *Faraday Discuss.* **1992**, *94*, 329–338.
13. A. Cuesta, M. Kleinert, D. M. Kolb, *Phys. Chem. Chem. Phys. (PCCP)* **2000**, *2*, 5684–5690.
14. A. M. Funtikov, U. Stimming, R. Vogel, *Surf. Sci.* **1995**, *324*, L343–L348.
15. A. M. Funtikov, U. Stimming, R. Vogel, *J. Electroanal. Chem.* **1997**, *428*, 147–153.
16. L. J. Wan, T. Suzuki, K. Sashikata et al., *J. Electroanal. Chem.* **2000**, *484*, 189–193.
17. L. J. Wan, S. L. Yau, K. Itaya, *J. Phys. Chem. B* **1995**, *99*, 9507–9513.
18. L. J. Wan, M. Hara, J. Inukai et al., *J. Phys. Chem. B* **1999**, *103*, 6978–6983.
19. M. F. Toney, J. McBreen, *Interface* **1993**, *1*, 22–31.
20. J. X. Wang, R. R. Adzic, B. M. Ocko in *Interfacial Electrochemistry* (Ed.: A. Wieckowski), Marcel Dekker, New York, 1999, pp. 175–186.
21. D. A. Buttry, M. D. Ward, *Chem. Rev.* **1992**, *92*, 1355–1379.

22. M. Hepel in *Interfacial Electrochemistry* (Ed.: A. Wieckowski), Marcel Dekker, New York, 1999, pp. 599–630.
23. A. A. Gewirth, B. K. Niece, *Chem. Rev.* **1997**, *97*, 1129–1162.
24. K. Itaya, *Prog. Surf. Sci.* **1998**, *60*, 121–248.
25. E. Bauer, *Z. Kristallogr.* **1958**, *110*, 372–394.
26. E. Bauer, J. H. van der Merwe, *Phys. Rev. B* **1986**, *33*, 3657–3671.
27. E. Budevski, G. Staikov, W. J. Lorenz, *Electrochemical Phase Formation and Growth – An Introduction to the Initial Stages of Metal Deposition*, VCH, Weinheim, 1996.
28. L. Z. Mezey, J. Giber, *Jpn. J. Appl. Phys.* **1982**, *21*, 1569–1571.
29. S. Ye, C. Ishibashi, K. Uosaki, *J. Electrochem. Soc.* **1998**, *145*, 1614–1623.
30. S. Ye, C. Ishibashi, K. Uosaki, *Langmuir* **1999**, *15*, 807–812.
31. S. Ye, C. Ishibashi, K. Uosaki in *Scanning Probe Microscopy for Electrode Characterization and Nanometer Scale Modification*, Electrochemical Society Proceedings (Eds.: D. C. Hansen, H. S. Isaacs, K. Sieradzki), Electrochemical Society, New York, 2001, pp. 133–147, Vol. PV 2000-35.
32. K. Uosaki, S. Ye, H. Naohara et al., *J. Phys. Chem. B* **1997**, *101*, 7566–7572.
33. H. Naohara, S. Ye, K. Uosaki, *J. Phys. Chem. B* **1998**, *102*, 4366–4373.
34. H. Naohara, S. Ye, K. Uosaki, *J. Electroanal. Chem.* **1999**, *473*, 2–9.
35. L. A. Kibler, M. Kleinert, R. Randler et al., *Surf. Sci.* **1999**, *443*, 19–30.
36. L. A. Kibler, M. Kleinert, D. M. Kolb, *Surf. Sci.* **2000**, *461*, 155–167.
37. L. A. Kibler, M. Kleinert, D. M. Kolb, *J. Electroanal. Chem.* **1999**, *467*, 249–257.
38. S. Strbac, O. M. Magnussen, R. J. Behm, *Phys. Rev. Lett.* **1999**, *83*, 3246–3249.
39. S. Strbac, F. Maroun, O. M. Magnussen et al., *J. Electroanal. Chem.* **2001**, *500*, 479–490.
40. K. Uosaki, J. Ueda, S. Ye, in preparation.
41. N. Ikemiya, K. Yamada, S. Hara, *J. Vac. Sci. Technol., B* **1995**, *14*, 1369–1375.
42. N. Ikemiya, K. Yamada, S. Hara, *Surf. Sci.* **1996**, *348*, 253–260.
43. W. J. Lorenz, G. Staikov, *Surf. Sci.* **1995**, *335*, 32–43.
44. K. Ogaki, K. Itaya, *Electrochim. Acta* **1995**, *40*, 1249–1257.
45. M. J. Esplandiu, M. A. Schneeweiss, D. M. Kolb, *Phys. Chem. Chem. Phys. (PCCP)* **1999**, *1*, 4847–4854.
46. F. H. Reid, *Trans. Inst. Metal Finishing* **1970**, *48*, 115–123.
47. J. F. Llopis, F. Colom in *Encyclopedia of Electrochemistry of the Elements* (Ed.: A. J. Bard), Marcel Dekker, New York, 1976, pp. 169–219, Vol. VI.
48. S. Tanaka in *Science of Noble Metal, Application* (Ed.: S. Tanaka), Tanaka Nobel Metal, Inc., Tokyo, 1985, pp. 261–321.
49. L. Carrette, K. A. Friedrich, U. Stimming, *Chem. Phys. Chem.* **2000**, *1*, 162–193.
50. S. Wasmus, A. Kuever, *J. Electroanal. Chem.* **1999**, *461*, 14–31.
51. A. M. Feltham, M. Spiro, *Chem. Rev.* **1971**, *71*, 177–193.
52. J. W. A. Sachtler, M. A. Van Hove, J. P. Biberian et al., *Surf. Sci.* **1981**, *110*, 19–42.
53. R. F. C. Farrow, G. R. Harp, R. F. Marks et al., *J. Cryst. Growth* **1993**, *133*, 47–58.
54. T. R. Linderoth, J. J. Mortensen, K. W. Jacobsen et al., *Phys. Rev. Lett.* **1996**, *77*, 87–90.
55. M. O. Pedersen, S. Helveg, A. Ruban et al., *Surf. Sci.* **1999**, *426*, 395–409.
56. *Technical Catalog for Metal Plating*, Tanaka Noble Metal, Inc., 2000.
57. *Precious Metal Plating Process Catalog*, N. E. CHEMCAT, 2001.
58. *Technical Catalog for Metal Plating*, Japan Pure Chemical (JPC), 2001.
59. P. E. Skinner, *Platinum Met. Rev.* **1989**, *33*, 102–105.
60. A. J. Gregory, W. Levason, R. E. Noftle et al., *J. Electroanal. Chem.* **1995**, *399*, 105–113.
61. S. Tanaka in *Science of Noble Metal, Application* (Ed.: S. Tanaka), Tanaka Noble Metal, Tokyo, 1985, pp. 225–247.
62. B. Gollas, J. M. Elliott, P. N. Bartlett, *Electrochim. Acta* **2000**, *45*, 3711–3724.
63. K. Uosaki, S. Ye, Y. Oda, T. Haba et al. in *Proceedings of the Sixth International Symposium on Electrode Processes* (Eds.: A. Wieckowski, K. Itaya), Electrochemical Society, Pennington, New Jersey, 1996, pp. 168–179, Vol. PV 96-8.
64. K. Uosaki, S. Ye, Y. Oda et al., *Langmuir* **1997**, *13*, 594–596.
65. H.-F. Waibel, M. Kleinert, D. M. Kolb in *199th Meeting of the Electrochemical Society*, Washington, 2001, pp. 354.

66. W. Haiss, J. K. Sass, *J. Electroanal. Chem.* **1997**, *431*, 15–17.
67. T. Katsuzaki, K. Uosaki, unpublished.
68. K. A. Friedrich, A. Marmann, U. Stimming et al., *Fresenius' J. Anal. Chem.* **1997**, *358*, 163–165.
69. E. G. Seebauera, C. E. Allen, *Prog. Surf. Sci.* **1995**, *49*, 265–330.
70. N. Kimizuka, T. Abe, K. Itaya, *Denki Kagaku* **1993**, *61*, 796–799.
71. S.-L. Yau, T. Moriyama, H. Uchida, M. Watanabe, *Chem. Commum.* **2000**, 2279–2280.
72. J. F. Llopis, F. Colom in *Encyclopedia of Electrochemistry of Elements* (Ed.: A. J. Bard), Marcel Dekker, New York, 1976, pp. 253–275, Vol. VI.
73. A. Cuesta, L. A. Kibler, D. M. Kolb, *J. Electroanal. Chem.* **1999**, *466*, 165–168.
74. G. A. Attard, A. Bannister, *J. Electroanal. Chem.* **1991**, *300*, 467–485.
75. M. J. Llorca, J. M. Feliu, A. Aldaz et al., *J. Electroanal. Chem.* **1993**, *351*, 299–319.
76. M. J. Llorca, J. M. Feliu, A. Aldaz et al., *J. Electroanal. Chem.* **1994**, *376*, 151–160.
77. M. Baldauf, D. M. Kolb, *Electrochim. Acta* **1993**, *38*, 2145–2153.
78. H. Naohara, S. Ye, K. Uosaki, *Colloids Surf., A* **1999**, *154*, 201–208.
79. H. Naohara, S. Ye, K. Uosaki, *Electrochim. Acta* **2000**, *45*, 3305–3309.
80. H. Naohara, S. Ye, K. Uosaki, *J. Electroanal. Chem.* **2001**, *500*, 435–445.
81. A. S. Dakkouri, D. M. Kolb in *Interfacial Electrochemistry* (Ed.: A. Wieckowski), Marcel Dekker, New York, 1999, pp. 151–173.
82. M. Takahashi, Y. Hayashi, J. Mizuki et al., *Surf. Sci.* **2000**, *461*, 213–218.
83. T. Kondo, K. Tamura, M. Takahashi, J. Mizuki, K. Vosaki, *Electrochim. Acta*, in press (2002).
84. M. Baldauf, D. M. Kolb, *J. Phys. Chem.* **1996**, *100*, 11 375–11 381.
85. J. F. Llopis, I. M. Tordesillas in *Encyclopedia of Electrochemistry of the Elements* (Ed.: A. J. Bard), Marcel Dekker, New York, 1976, pp. 299–322, Vol. VI.
86. J. Inukai, M. Ito, *J. Electroanal. Chem.* **1993**, *358*, 307–315.
87. R. Gomez, A. Rodes, J. M. Perez et al., *Surf. Sci.* **1995**, *327*, 202–215.
88. R. Gomez, A. Rodes, J. M. Perez et al., *Surf. Sci.* **1995**, *344*, 85–97.
89. R. Gomez, J. M. Feliu, *Electrochim. Acta* **1998**, *44*, 1191–1205.
90. G. A. Attard, R. Price, A. Al-Akl, *Surf. Sci.* **1995**, *335*, 52–62.
91. F. H. Reid, J. C. Blake, *Trans. Inst. Metal Finishing* **1961**, *38*, 45–51.
92. J. F. Llopis, I. M. Tordesillas in *Encyclopedia of Electrochemistry of the Elements* (Ed.: A. J. Bard), Marcel Dekker, New York, 1976, pp. 277–298, Vol. VI.
93. K. R. Seddon, *Platinum Met. Rev.* **1996**, *40*, 128–134.
94. F. P. Gortsema, J. W. Cobble, *J. Am. Chem. Soc.* **1961**, *83*, 4317–4321.
95. K. A. Friedrich, K.-P. Geyzers, U. Linke et al., *J. Electroanal. Chem.* **1996**, *402*, 123–128.
96. S. Cramm, K. A. Friedrich, K.-P. Geyzers et al., *Fresenius' J. Anal. Chem.* **1997**, *358*, 189–192.
97. K. A. Friedrich, K. P. Geyzers, A. Marmann et al., *Zeitschrift fur Physikalische Chemie* **1999**, *208*, 137–150.
98. W. Chrzanowski, A. Wieckowski, *Langmuir* **1997**, *13*, 5974–5978.
99. W. Chrzanowski, A. Wieckowski in *Interfacial Electrochemistry* (Ed.: A. Wieckowski), Marcel Dekker, New York, 1999, pp. 937–954.
100. E. Herrero, J. M. Feliu, A. Wieckowski, *Langmuir* **1999**, *15*, 4944–4948.
101. H. Kim, I. R. de Moraes, G. Tremiliosi et al., *Surf. Sci.* **2001**, *474*, L203–L212.
102. W. F. Lin, M. S. Zei, M. Eiswirth et al., *J. Phys. Chem. B* **1999**, *103*, 6968–6977.
103. T. Iwasita, H. Hoster, A. John-Anacker et al., *Langmuir* **2000**, *16*, 522–529.
104. G. Tremiliosi-Filho, H. Kim, W. Chrzanowski et al., *J. Electroanal. Chem.* **1999**, *467*, 143–156.
105. K. Itaya in *Interfacial Electrochemistry* (Ed.: A. Wieckowski), Marcel Dekker, New York, 1999, pp. 187–210.
106. T. Suzuki, T. Yamada, K. Itaya, *J. Phys. Chem.* **1996**, *100*, 8954–8961.
107. P. Muller, S. Ando, T. Yamada et al., *J. Electroanal. Chem.* **1999**, *467*, 282–290.
108. T. Teshima, K. Ogaki, K. Itaya, *J. Phys. Chem.* **1997**, *101*, 2046–2053.
109. D. W. Suggs, A. J. Bard, *J. Am. Chem. Soc.* **1994**, *116*, 10 725.
110. D. W. Suggs, A. J. Bard, *J. Phys. Chem.* **1995**, *99*, 8349.
111. O. M. Magnussen, M. R. Vogt, F. A. Moller et al., *Surf. Sci.* **1996**, *367*, L33–L41.

112. M. R. Vogt, A. Lachenwitzer, O. M. Magnussen et al., *Surf. Sci.* **1998**, *399*, 49–69.
113. J. A. Schimpf, J. R. MacBride, M. P. Soriaga, *J. Phys. Chem.* **1993**, *97*, 10 518–10 520.
114. J. A. Schimpf, J. B. Abreu, M. P. Soriaga, *Langmuir* **1993**, *9*, 3331–3333.
115. A. Carrasquillo Jr., J. J. Jeng, R. J. Barriga et al., *Inorg. Chim. Acta* **1997**, *255*, 249–254.
116. M. P. Soriaga, J. A. Schimpf, A. J. Carrasquillo et al., *Surf. Sci.* **1995**, *335*, 273–280.
117. K. Sashikata, Y. Matsui, K. Itaya et al., *J. Phys. Chem. B* **1996**, *100*, 20 027–20 034.
118. M. P. Soriaga, Y. G. Kim, J. E. Soto in *Interfacial Electrochemistry* (Ed.: A. Wieckowski), Marcel Dekker, New York, 1999, pp. 249–267.
119. G. M. Schmid, M. E. Curley-Fiorino, *Gold*, Marcel Dekker, New York, 1975.
120. R. J. Puddephatt, *The Chemistry of Gold*, Elsevier Scientific Publishing Company, Amsterdam, 1978.
121. M. Magnussen, B. M. Ocko, R. R. Adzic et al., *Phys. Rev. B* **1995**, *51*, 5510–5513.
122. J. X. Wang, A. R. Adzic, M. Ocko in *Interfacial Electrochemistry* (Ed.: A. Wieckowski), Marcel Dekker, New York, 1999, pp. 175–186.
123. A. Cuesta, D. M. Kolb, *Surf. Sci.* **2000**, *465*, 310–316.

4.2
Electrodeposition of Compound Semiconductors by Electrochemical Atomic Layer Epitaxy (EC-ALE)

John L. Stickney, Travis L. Wade, Billy H. Flowers, Raman Vaidyanathan, and Uwe Happek
University of Georgia, Athens, Georgia, USA

Abstract

This chapter concerns the state of development of electrochemical atomic layer epitaxy (EC-ALE), the electrochemical analog of atomic layer epitaxy (ALE). EC-ALE is being developed as a methodology for the electrodeposition of compound semiconductors with nanoscale control. ALE is based on the formation of compounds, one monolayer (ML) at a time, using surface-limited reactions. An atomic layer of one element can be electrodeposited at a potential under that needed to deposit the element on itself, and this process is referred to as underpotential deposition (UPD). EC-ALE is the use of UPD for the surface-limited reactions in an ALE cycle.

Electrodeposition is generally performed near room temperature, avoiding problems with interdiffusion and mismatched thermal expansion coefficients. This makes EC-ALE a good candidate to form superlattices, where the compound deposited is modulated on the nanometer scale.

This chapter describes the basics of the EC-ALE cycle, the elements that have been used to form deposits, as well as the solutions, rinsing, and potential changes. It describes the hardware presently being used by this group and other research laboratories, the compounds that have been formed, and the development of deposition cycles for various compounds. It describes the status of device formation using EC-ALE, and goes over some of the problems and issues involved in developing cycles and growing films. Finally, given that EC-ALE is based on surface-limited electrochemical reactions, studies of relevant electrodeposit surface chemistry are discussed.

4.2.1 Introduction

Nanometer control in growth is a major frontier of Material Science. By constructing superlattices, nanowires, nanoclusters, and nanocrystalline materials, the electronic structure (band gap) of a semiconductor can be engineered, the wavelength of light emitted or absorbed by a compound can be adjusted over a broad range. By direct analogy with the quantum-mechanical model of a particle in a box, it is known that the smaller the box containing an electron the further apart are energy levels. This translates directly for some semiconductor structures – the smaller the thickness of the layers, or the dimensions of a particle, the larger the resulting band gap.

The primary methodologies for forming thin-film materials with atomic level control are molecular beam epitaxy (MBE) [1–6], vapor phase epitaxy (VPE) [7–9], and a number of derivative vacuum-based techniques [10]. These methods depend on controlling the flux of reactants and the temperature of the substrate and reactants. If the growth temperature in MBE and VPE is too high, especially with the formation of nanostructured materials, even 200 to 500 °C, interdiffusion of the component elements can result. In addition, different materials have different thermal expansion coefficients. If a heterojunction is formed at an elevated temperature and then cooled to room temperature, stresses and defects can develop if the coefficients do not match.

Electrodeposition is not "heat and beat," it is not a heat-driven reaction; it is generally performed near room temperature. Ideally, electrodeposition involves control of the deposition by activity of the electrons at the interface, by the potentials applied to the deposition cell.

The work described here concerns development of electrochemical methodologies to grow compound semiconductors with nanoscale or atomic layer control. That thin films of some compounds can be formed electrochemically is clear. The questions are: How much control is there over composition, structure, and morphology? What compounds can be formed, and with what quality?

The issue of epitaxy is central to the formation of high-quality electronic and optoelectronic structures. There are a number of definitions, mostly coming down to the growth of single crystals on single-crystal surfaces. In the vast majority of cases, the lattice constants of the deposit do not exactly match those of the substrate, so strain will build, and defects are generated. The question then arises: single-crystal grains of what size need to be formed on a substrate to call it epitaxy?

4.2.1.1 Electrodeposition

Electrodeposition has been used since the beginning of the century to form high-quality, mostly metallic, thin films: car bumpers, decorative plating, metal contacts, and so on. It has recently been shown that high-quality copper interconnects for ultra–large scale integration (ULSI) chips can be formed electrochemically on Si wafers, using the Dual Damascene copper-electroplating method [11, 12], and the large semiconductor companies, such as IBM, Intel, AMD, Motorola, and so on, are installing wafer-electroplating machines in their fabrication lines.

Epitaxy in electrodeposition has been studied for 40 or more years [13–17], and is similar to MBE and VPE in that the rate of deposition should be limited and have as much surface diffusion as possible. Surface diffusion is limited near room temperature, but may be enhanced due to solvation effects. Epitaxial electrodeposition is normally best carried out near equilibrium, where the deposition rate is lower and the exchange current high. The exchange current acts, in addition to surface diffusion, to move depositing atoms around. Near equilibrium, electrodeposition is dynamic, with slightly more atoms depositing than dissolving at a given time. In principle, atoms deposited in less than optimal sites redissolve, leaving only those atoms in more optimal sites. As the potential is driven from equilibrium (the formal potential), deposition rates increase, and the exchange current drops, promoting three-dimensional (3D) growth, and generally degrading epitaxy.

The quality of an elemental deposit is a function of the rate at which it deposits, the surface diffusion, its exchange current, and the substrate quality and structure. Electrodeposition of a compound thin film requires all these things, as well as stoichiometry.

Several methods and variations have been developed to electrodeposit compounds. Most of the work described in this article concerns the formation of nonoxide compounds such as II to VI and III to Vs. Oxides are probably the largest group of electrodeposited compounds (e.g.

aluminum anodization), but will not be discussed here. The electrodeposition of II to VI compounds has been extensively studied and is well reviewed in a number of articles [18–23]. The most prominent compound electrodeposition methods include codeposition, precipitation, and various two-stage techniques.

The focus of the work described here is on understanding the mechanisms of compound electrodeposition and how to control structure, morphology, and composition. The primary tool for understanding compound electrodeposition and for improving control over the process has been the methodology of EC-ALE [23–26].

Historically, EC-ALE has been developed by analogy with ALE [27–33]. ALE is a methodology used initially to improve epitaxy in the growth of thin films by MBE and VPE. The principle of ALE is to use surface-limited reactions to form each atomic layer of a deposit. If no more than an atomic layer is ever deposited, the growth will be 2D, layer-by-layer, and epitaxial. Surface-limited reactions are developed for deposition of each component element, and a cycle is formed with them. With each cycle, a compound ML is formed, and the deposit thickness is controlled by the number of cycles.

As noted, in techniques such as MBE and VPE, surface-limited reactions are generally controlled by the temperatures of the reactants and substrate. The temperature is kept high enough so that deposition over a ML sublimes, leaving only the atomic layer, forming the compound. Problems are encountered when the temperatures needed to form atomic layers of different elements are not the same, as changing the temperature between layers is difficult.

Surface-limited reactions are well known in electrochemistry, and are generally referred to as UPD [34–39]. In the deposition of one element on a second, frequently the first element will form an atomic layer at a potential under, or prior to, that needed to deposit the element on itself. One way of looking at UPD is that a surface compound, or alloy, is formed, and the shift in potential results from the favorable free energy of formation of the surface compound.

EC-ALE is the combination of UPD and ALE. Atomic layers of a compound's component elements are deposited at underpotentials in a cycle, to directly form a compound. It is generally a more complex procedure than most of the compound electrodeposition methods described earlier, requiring a cycle to form each ML of the compound. However, it is layer-by-layer growth, avoiding 3D nucleation, and offering increased degrees of freedom, atomic level control, and promoting of epitaxy.

In addition, EC-ALE offers a way of better understanding compound electrodeposition, essentially a way of breaking it down into its component pieces. It allows compound electrodeposition to be deconvolved into a series of individually controllable steps, resulting in an opportunity to learn more about the mechanisms, and gain a series of new control points for electrodeposition. The main problem with compound electrodeposition by codeposition is that the only control points are the solution composition and the deposition potential, or current density, in most cases. In an EC-ALE process, each reactant has its own solution and deposition potential, and there are generally rinse solutions as well. Each solution can be separately optimized, so that the pH, electrolyte, and additives or complexing agents are tailored to fit the precursor. On the other hand, codeposition uses only one solution, which is a compromise, required to be compatible with all reactants.

Finally, electrodeposition is orthogonal to MBE and chemical vapor deposition (CVD), as it involves growth in a condensed phase with potential control instead of temperature. This increases the variable space for producing materials, the diversity of conditions under which a compound can be formed.

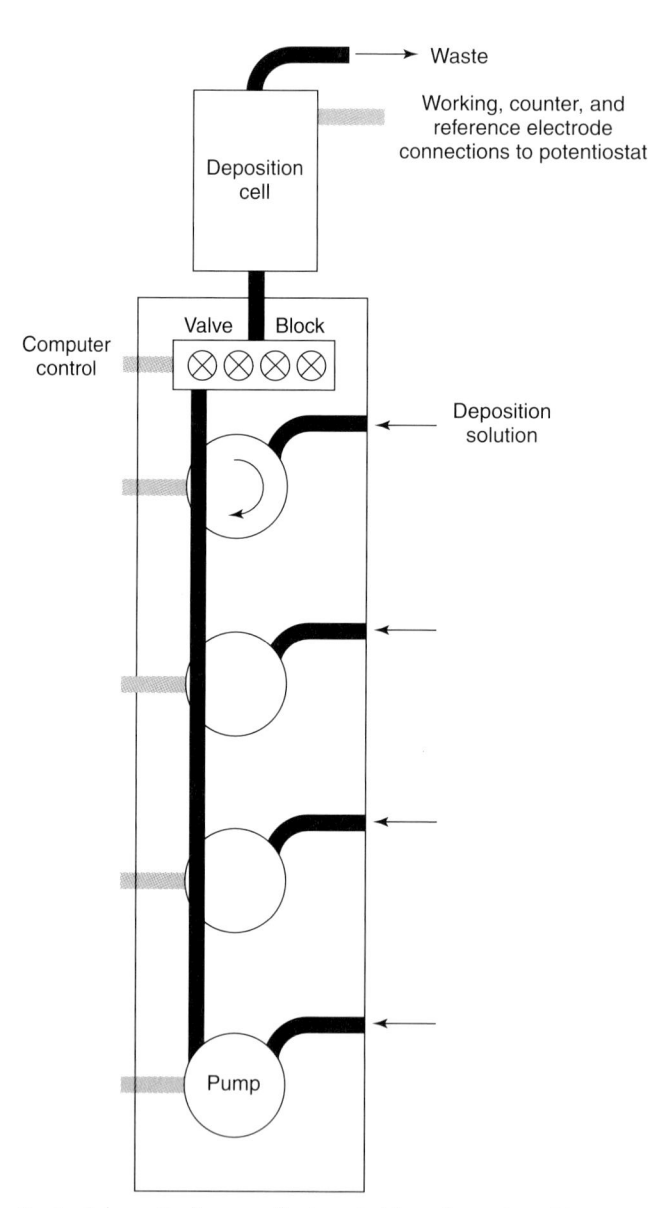

Fig. 1 Schematic diagram of automated flow electrodeposition system, for forming thin films using EC-ALE.

4.2.2 Hardware

Figure 1 is a schematic diagram of a basic electrochemical flow-deposition system used for electrodepositing thin films using EC-ALE, and Fig. 2 is a picture showing the solution reservoirs, pumps, valves, electrochemical cell, potentiostat, and computer. A number of electrochemical cell designs have been tried. A larger thin-layer electrochemical flow cell is now used (Fig. 3c) [40], with a deposition area of about 2.5 cm^2 and a cell volume of 0.1 mL, resulting in a two order of magnitude drop in solution volume, compared with the H-cell (Fig. 3b). The cell includes an indium tin oxide (ITO) auxiliary electrode, as the opposite wall of the cell from the working electrode, allowing observation of the deposit during growth, as well as providing an optimal current distribution.

Villegas and coworkers used a wall-jet configuration, with very good results, in the formation of CdTe deposits on Au [43]. The quality of their deposits appear equivalent to those recently produced by this group with the thin-layer flow cell (Fig. 3c). Figure 4 is a transmission electron micrograph(TEM) of a CdTe deposit formed using their cell and 200 EC-ALE cycles [43]. A wall-jet configuration has also been used by Foresti and coworkers, in their studies of thin-film growth using EC-ALE on Ag electrodes [44].

The pumps used in the flow system in Fig. 2 are standard peristaltic pumps (Cole

Fig. 2 Picture of one of the flow-cell systems presently used by this group.

Parmer). The main requirement for the pumps is that they are clean. If smoother pumping is required, pulse dampening or syringe pumps could be used. Foresti and coworkers have used pressurized bottles, without pumps, to deliver solution [44, 45]. In Fig. 2, there is one pump for each line, to push solution through the cell. Villegas

(a)

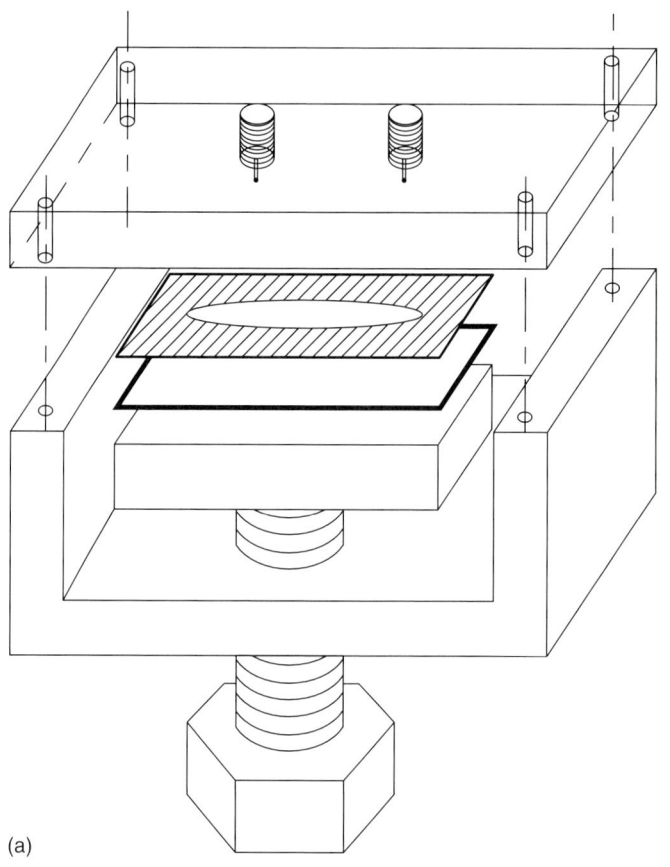

Fig. 3 Diagrams of electrochemical cells used in flow systems for thin-film deposition by EC-ALE. (a) First small thin-layer flow cell (modeled after electrochemical liquid chromatography detectors). A gasket defined the area where the deposition was performed, and solutions were pumped in and out through the top plate. (Reproduced by permission from Ref. [41].) (b) H-cell design where the samples were suspended in the solutions, and solutions were filled and drained from below. (Reproduced by permission from Ref. [42].) (c) Larger thin-layer flow cell. This is very similar to that shown in (a), except that the deposition area is larger and laminar flow is easier to develop because of the solution-inlet designs. In addition, the opposite wall of the cell is a piece of ITO, used as the auxiliary electrode. It is transparent, so the deposit can be monitored visually, and it provides excellent current distribution. The reference electrode is incorporated into the cell as well. (Adapted from Ref. [40].)

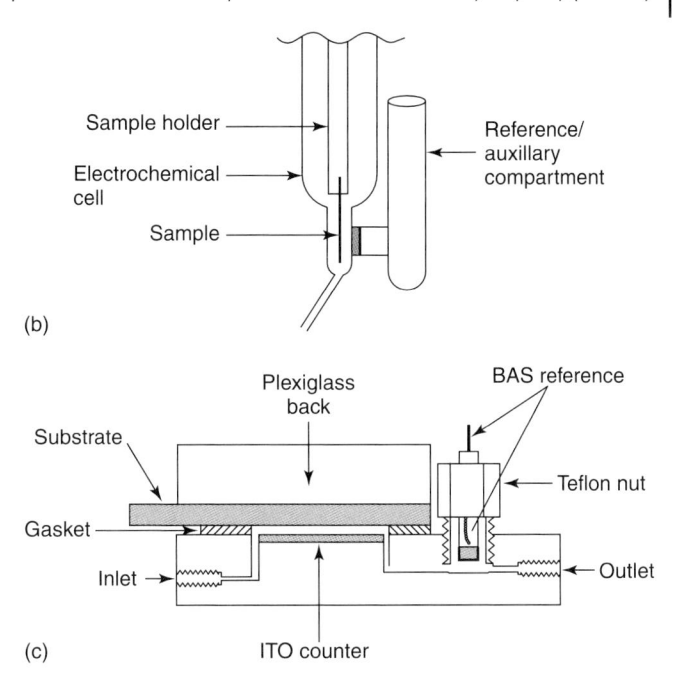

4.2 Electrodeposition of Compound Semiconductors by Electrochemical Atomic Layer Epitaxy (EC-ALE) | 519

(b) [Diagram showing Sample holder, Electrochemical cell, Sample, Reference/auxillary compartment]

(c) [Diagram showing Substrate, Gasket, Inlet, Plexiglass back, BAS reference, Teflon nut, Outlet, ITO counter]

Fig. 3 *(Continued)*

and coworkers used a single pump on the outlet to suck each solution into the cell, in an elegant simplification [43].

There are a number of vendors that sell solenoid-actuated Teflon valves, which are easily interfaced to a computer. Care must be taken to choose a design in which the internal volume at the valve outlet can be flushed easily between steps, however [41]. Rotary selection valves have been used as well, but given the number of rotations needed for a 200-cycle deposit, various failure modes revealed themselves.

As deposition of most of the relevant atomic layers involves reduction at relatively low potentials, oxygen has proven to be a major problem. It has been shown repeatedly that oxygen not rigorously excluded results in thinner deposits, if they are formed at all. For this reason, extensive sparging of the solution reservoirs is critical. Sparging alone is generally not sufficient to prevent problems with oxygen, as most tubing has some oxygen permeability. To better avoid this problem, the solution delivery tubes were threaded through larger ID tubes and into the Plexiglas box (Fig. 2) that houses the pumps and valves. The sparging N_2 was made to flow out of the solution reservoirs, through the large ID tubes (around the outside of the solution-delivery tubes) and into the box, greatly decreasing oxygen exposure. The measured oxygen content of the N_2 leaving the Plexiglas box was 10 to 30 ppm, as measured with a glove box oxygen analyzer (Illinois Instruments, model 2550).

The majority of deposits formed in this group have been on Au electrodes, as they are robust, easy to clean, have a well-characterized electrochemical behavior, and reasonable quality films can be formed by a number of methodologies. However, Au it is not well lattice-matched to

Fig. 4 Transmission electron micrograph (TEM) of a CdTe deposit formed using 200 cycle of CdTe via EC-ALE. The regular layered structure, parallel with the substrate Au lattice planes, suggest the epitaxial nature of the deposit. (Reproduced by permission from Ref. [43].)

most of the compounds being formed by EC-ALE.

Some deposits have been formed on Au single crystals. However, single crystals are too expensive to use in forming larger deposits, and have to be recycled. A number of disposable substrates have been investigated, so that the substrates do not have to be recycled, and the deposits can be kept around. The first deposits were made on cold-rolled gold foil, which proved too expensive, and polycrystalline after etching. Au vapor deposited on Si(100) at room temperature was used extensively, as the films were more reproducible and resembled Au mirrors. On the nanometer scale, however, the films consisted of 40 nm hemispheres (Fig. 5a).

Au vapor deposited on mica at 300 to 400° is known to form large (111) terraces [46–51]. Several attempts to use these films in the flow cell (Fig. 3c) resulted in delamination, due to the constant rinsing.

Presently, Au on glass is being used to form most of the deposits in the flow cells. These substrates are microscope slides, etched in HF, coated with 8 nm of Ti and then 200 nm of Au, at about 300°. The deposits do not, in general, show as many large terraces as Au on mica; however, they are orders of magnitude better than Au on Si (Fig. 5b). To improve terrace sizes, the substrates are annealed in a tube furnace at 550° for 12 h with flowing N_2. In addition, substrates are

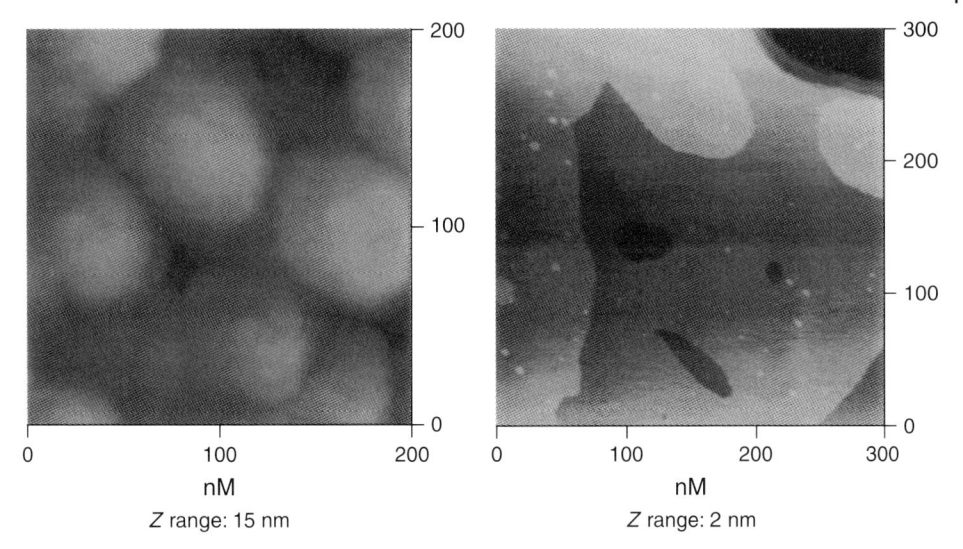

Fig. 5 (a) AFM image of Au vapor deposited on Si(100) at room temperature and (b) AFM image of Au on glass, annealed at 550 C for 12 h, in a tube furnace, in a flow of N_2.

given a brief flame anneal, in the dark with a H_2 flame to a dull orange glow, prior to use.

Some Cu substrates have been used, including Cu foils, etched foils, and vapor-deposited Cu on glass. There does not appear to be a significant difference in the quality of deposits formed on Cu versus Au, beyond that expected to result from considerations of lattice matching.

Single-crystal silver substrates have been used exclusively by Foresti and coworkers [44, 45, 52, 53]. They use macroscopic Ag single crystals, formed in-house. They have formed a number of II to IV compounds using EC-ALE, including ZnSe, CdS, and ZnS.

Semiconductors such as polycrystalline ITO on glass have been used to form deposits of ZnS [54], CdS, and CdTe, by this group, with no obvious problems. Ideally, lattice-matched semiconductor substrates could also be used to form deposit. For instance, InSb is lattice-matched with CdTe, and could be used as a substrate. The problems involve adequately preparing the substrate surfaces and understanding the electrochemistry of a compound semiconductor substrate. Work is progressing in this direction. Good quality deposits of CdSe have been formed on InP and GaAs substrates using codeposition, by Maurin and coworkers [55, 56]. They used reflection high-energy electron diffraction (RHEED) and TEM to follow the habit of CdSe growth. Their work clearly shows the applicability of high-quality commercial compoundsemiconductor wafers as substrates for compound electrodeposition.

4.2.3
Deposition Programs

4.2.3.1 Cycle Steps

As described in the introduction, UPD is the formation of an atomic layer of one element on a second element at a potential under that required to deposit the element on itself:

$$M^{2+} + 2e^- \Longleftrightarrow M_{UPD} \quad (1)$$

The formal potential is the potential at which the element begins to be stable depositing on itself:

$$M_{(s)} + M^{2+} + 2e^- \Longleftrightarrow 2M_{(s)} \quad E^{o'}_{M^{2+}/M} \quad (2)$$

The point of view in this group is that the element is more stable because a surface compound is formed, so the free energy of compound formation promotes deposition at an underpotential, accounting directly for the potential shift. This can be seen in oxidative stripping of Zn chalcogenide MLs on Au electrodes. The Zn stripping features shift to more positive potentials, as the chalcogenide layer is changed from Te to Se to S [57]. The order and magnitudes of the shifts are consistent with the differences in heats of formation for the corresponding compounds, suggesting the relationship between their electrochemical stability on the surface and the compounds formed.

Reductive UPD is the major atomic layer deposition process used in EC-ALE to form atomic layers (Eq. 1). Many metals can be obtained in a soluble oxidized form, from which atomic layers can be deposited at underpotentials. Control points are the concentration of the reactant, the potential used for the deposition, the pH, and the presence of other additives or complexing agents, which can change the activity of the reacting species.

Oxidative UPD involves the oxidation of species to form an atomic layer where the precursor contains the element in a negative oxidation state. A classic example is the formation of oxide layers on Pt and Au, where water is oxidized to form atomic layers of oxygen. Halide adsorption can be thought of similarly, where species such as Cl^-, Br^-, and I^- oxidatively adsorb on metal surfaces as halide atoms. With respect to compound formation, oxidative UPD from sulfide is a good example:

$$HS^- \Longleftrightarrow S_{(UPD)} + H^+ + 2e^- \quad (3)$$

The other chalcogenides (H_2Te and H_2Se) work as well; however, their aqueous solutions tend to be unstable, relative to sulfide. Solutions of reduced forms of the pnictides, such as H_3As, H_3Sb and H_3P, are of interest, and expected to work for oxidative UPD of atomic layers. As these species are generally gases, it is better to use a basic solution to prevent out-gassing from solution.

The same equilibrium used to described oxidative UPD (Eq. 3) can frequently be used to form atomic layers starting with more stable, oxidized forms of the elements, if a stripping step is added to the cycle. This approach has been used extensively to form atomic layers of Te [24, 41, 42, 58].

It would be nice to form CdTe using reductive UPD of both Te and Cd. If, however, acidic Te solutions are used, reductive Te UPD requires a potential near 0.0 V (Fig. 6b) in order to avoid bulk deposition. Cd UPD is optimal between -0.4 and -0.6 V. The problem is then that the Cd atomic layers strip during the Te deposition step. On the other hand, Te can be deposited at -0.5 V, where the Cd remains stable, but some bulk Te is formed along with the $Te_{(UPD)}$:

$$2HTeO_2^+ + 6H^+ + 8e^- \Longleftrightarrow$$
$$Te^o + Te_{(UPD)} + 4H_2O \quad (4)$$

The trick is that bulk Te (Te^o) can be stripped at a more negative potential in supporting electrolyte, a blank solution. In that way, only $Te_{(UPD)}$ is left, owing to

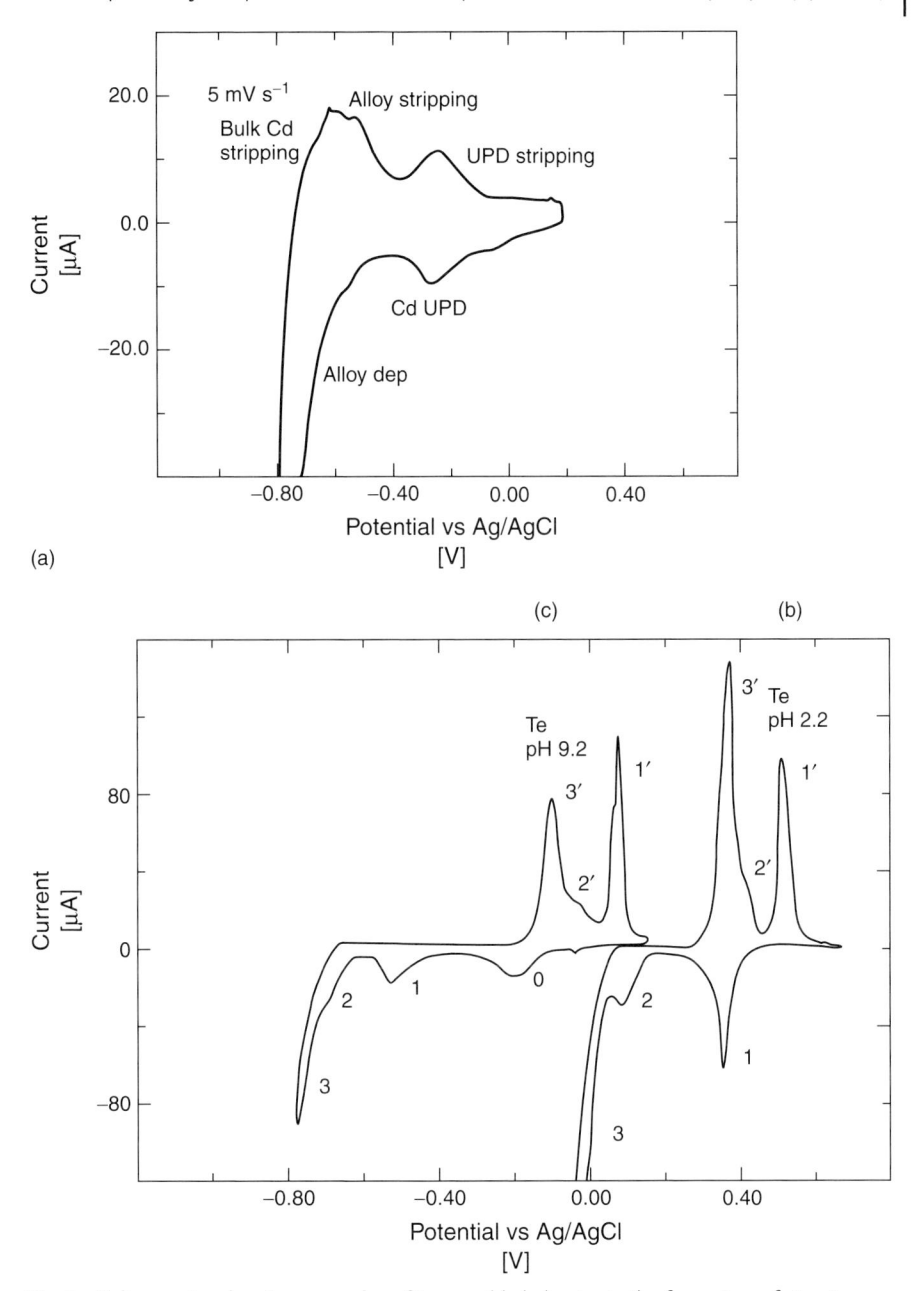

4.2 Electrodeposition of Compound Semiconductors by Electrochemical Atomic Layer Epitaxy (EC-ALE)

Fig. 6 Voltammetry showing examples of irreversible behavior in the formation of atomic layers on Au. (a) Cd, (b) Te, pH 2, (c) Te, pH 10, (d) Se, (Adapted from Ref. [59],) (e) As.

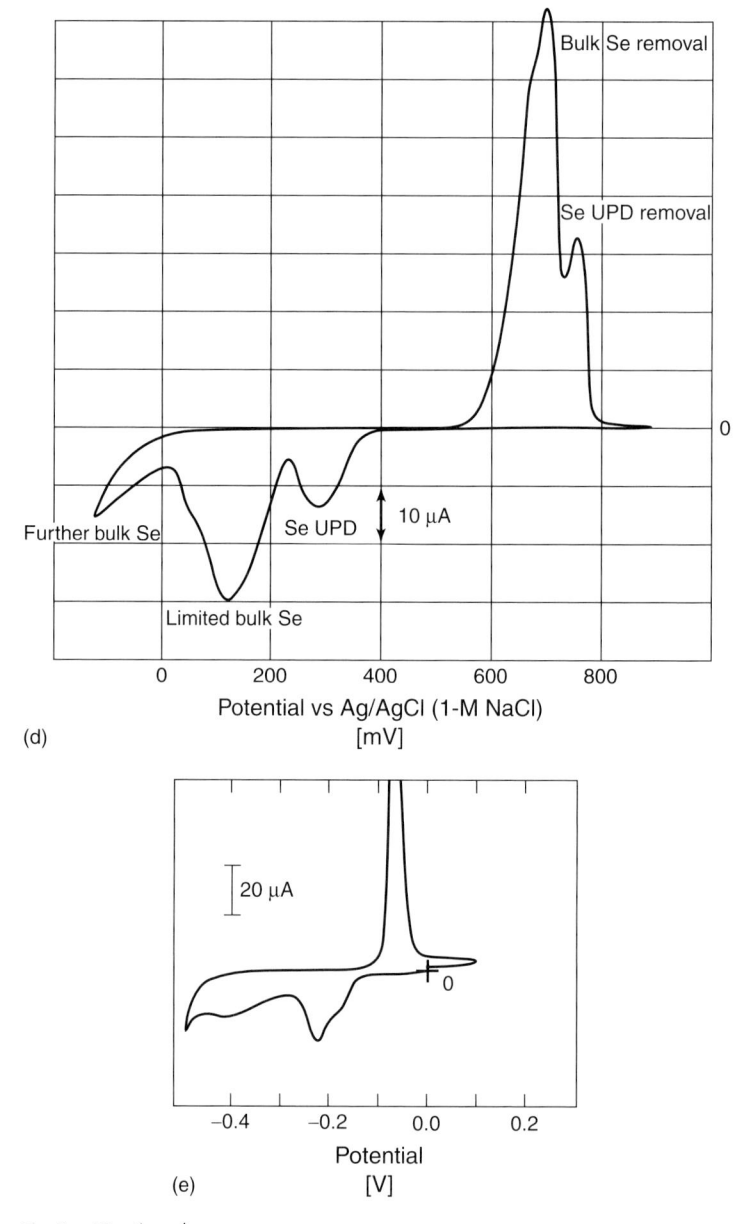

Fig. 6 (Continued)

stabilization by bonding with previously deposited Cd, forming CdTe:

$$Te^o + Te_{(UPD)} + 2H^+ + 2e^- \longleftrightarrow Te_{(UPD)} + H_2Te \quad (5)$$

Equations (4) and (5) together have essentially the same result as Eq. (3), oxidative UPD or formation of a Te atomic layer, along with conversion of some tellurite to telluride. The advantage is that F and G can be accomplished using the more stable tellurite solutions, instead of the unstable telluride.

4.2.3.2 Cycles

A cycle consists of the steps needed to form a ML of the compound. Figure 7 is a cycle diagram for CdTe and describes the deposition, stripping, and rinsing steps, as well as times and potentials.

In principle, each solution can be independent, composed of different solvents, reactants, electrolytes, buffers, and additives. In general, aqueous solutions have been used, as their electrochemical behavior is better understood, and ultrapure water can be obtained by a number of methods.

The choice of reactant has to do with whether you want to perform oxidative or reductive UPD, oxidation state, solubility, availability, purity, and price. In general, salts are used, simplifying solution preparation. Gases can be used by saturating the solution, prior to pumping it into the cell.

Reactants can also be generated electrochemically, prior to pumping the solution into the cell, or generator electrodes can be incorporated into the cell, so a precursor could be generated on the opposite wall of the cell, and allowed to diffuse across the thin solution layer to deposit. This method would have advantages in that species with limited stability could be used, as they need only last long enough to diffuse across, and only enough would be generated to form the atomic layer. This has been suggested as a way of obtaining H_3As, H_2Te, and so on as precursors for oxidative UPD.

For some elements, there are a variety of possible precursors, sulfur for instance (Fig. 8). Metal-organic precursors, such as used in metallo organic molecular beam epitaxy (MOMBE) or metallo organic vapor phase epitaxy (MOVPE), are possible precursors that can be used, if they can be made soluble in water or if a nonaqueous solvent were used. Mixed aqueous-organic solvents could improve solubility. Overall, there would be an increased probability of carbon contamination. One of the benefits

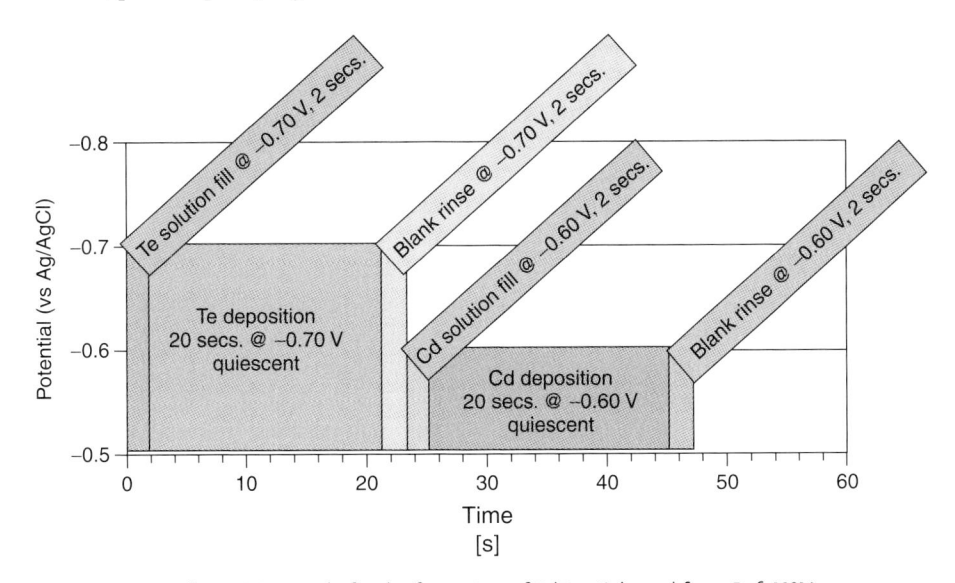

Fig. 7 Diagram of an EC-ALE cycle for the formation of CdTe. (Adapted from Ref. [60].)

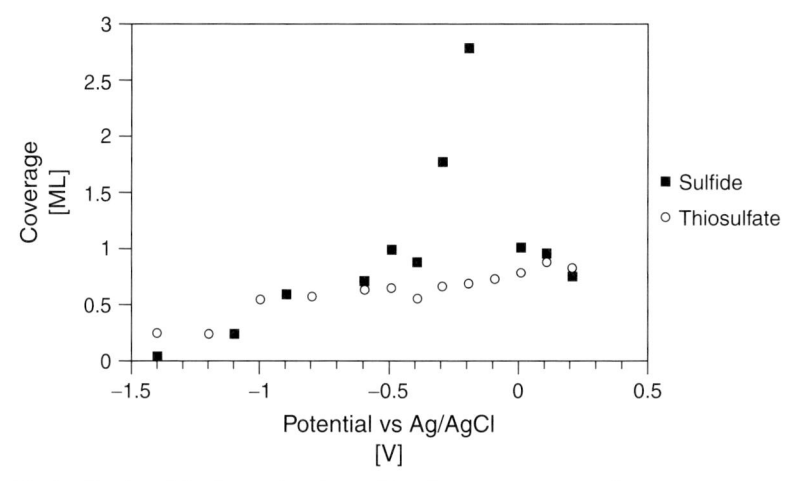

Fig. 8 Graphs of the Auger signals resulting from exposure to various sulfur-containing precursors as a function of potentials. (Adapted from Ref. [61].)

of using inorganic salts in aqueous solutions is that the number of constituents is limited, and thus the possible sources of contamination. The purities of corresponding inorganic salts are generally better than metal-organic precursors and cheaper.

The concentration of reactants can be kept very low, as only a nanomole cm^{-2} of material deposits in a given step. Use of low concentrations conserves reactants, simplifies waste handling, and minimizes the concentrations of reactant salt-born contaminants. In studies of InSb deposition, 0.02-mM solutions of the Sb precursor have been used [62].

The electrolyte is usually the major component, besides solvent, suggesting the use of the highest purity feasible. Filtering of solutions is advised to remove any particulates. The concentration determines the conductivity of the solutions, and the conductivity needed is a function of the cell design, currents expected, and range of deposition potentials to be used. At present, most studies in this group have involved 0.1- to 0.5-M electrolyte concentrations, although electrolyte concentrations of 50 mM have been tried and look promising. Currents can be kept very low, given that no more than an atomic layer is deposited at a given time. Drawbacks to high electrolyte concentrations are cost, waste, and contamination. It is anticipated that when systematic studies of trace contaminants are performed on deposits, many will be found to result from the electrolytes, pointing out the need for such studies.

Acid-based electrolytes are desirable, as the mobility of the proton allows use of more dilute solutions. However, they do promote hydrogen evolution, which can form bubbles in the cell. Good results with both sulfate and perchlorate salts have been seen. Recent results with chloride in the deposition of CdTe are encouraging as well [63].

Solution pH is an important variable, as it controls solubility, deposition potential, and precursor speciation [64]. As noted, one of the advantages of using an ALE process is that very different solutions can be used for each step in the cycle. For instance, pH 4 Cd solutions have been used with pH 10 Te solutions to form CdTe. However, some care must be taken while rinsing between solutions. On the other hand, Foresti and coworkers have used the same pH for both the metal and chalcogenide in the formation of II to VI compounds, such as CdS and CdSe on Ag single crystals [44, 45, 52, 65], in order to simplify solution exchange. To keep the Cd from precipitating in a basic solution, it was complexed with pyrophosphate.

The drawback to change the pH in each step is evident in the formation of CdTe. After Cd atomic layer formation, the Cd^{2+} solution is removed, and replaced with a basic TeO_3^{2-} solution. Cd deposition is a relatively reversible process (Fig. 6a), so there are two problems: first, the Cd^{2+} activity is lost and second, the pH is shifted from 4 to 10. First, the simple loss of Cd^{2+} activity shifts the equilibrium and can cause stripping of some of the atomic layer. In addition, shifting the pH to 10 has a similar result, as the Cd^{2+}/Cd equilibrium shifts negatively with pH. Thus maintaining the same potential and increasing the pH results in Cd formed at pH 4 stripping from the surface.

The history of electrodeposition involves control of deposit structure and morphology using additives. These can be things as simple as chloride, sulfides, and glycerol, or complex organic compounds, or traces of other metals. In EC-ALE, it is desirable to control the structure and morphology of deposits using surface-limited reactions, not additives. Some of these species may be beneficial to an EC-ALE cycle, but very little work has been done along this line. Most additive work has involved complexing agents, to shift the potential used to form an atomic layer, pH or Cl^- being good examples.

The amount of rinsing needed depends on factors such as cell design and the amount of a previous reactant that can be tolerated in a subsequent cycle step. Ideally, the cell will have good laminar flow and will rinse out easily. As noted previously, there are multiple reasons to limit the concentrations of reactants. Similar points can be made concerning buffers as they control pH. It is very important to remove buffers between steps in a cycle. In the deposition of CdTe, for instance, pH 4 Cd solutions were used along with pH 10 Te solutions. As can be seen in Figs. 6b and 6c, the potential at which bulk Te deposition occurs shifts from 0.0 V, at pH 2, to -0.7 V, at pH 10. Problems were encountered when insufficient rinsing was used between the pH 4 Cd solution and the pH 10 Te solution. Some Te was depositing from solution with a pH close to 6. Under those conditions, some bulk Te was deposited each cycle, instead of just an atomic layer, promoting 3D growth and rough deposits.

To get an idea of the amount of rinsing required, an electroactive species can be used, where the amount left in the cell after a given amount of rinsing can be determined coulometrically. Such measurements can be misleading if reactant is trapped under the gasket at the sides of the deposit (Fig. 3c). This can be avoided by using a resist to coat the substrate under the gasket, preventing electrochemical reaction of species trapped under the gasket.

Insufficient rinsing can also result in some codeposition if the previous reactant is not fully removed. The main drawback is the possibility of 3D growth, which can be hard to identify with very thin deposits. Alternatively, the rinse solution may not be important. Some high-quality CdTe films were formed in this group without using a rinse solution at all. That is, the reactant solutions were exchanged by each other, underpotential control, suggesting that some small amount of codeposition probably did occur.

The amount of electrolyte needed in a rinse solution depends on current flow during the rinse. Rinsing can be performed at open circuit in some cases, so that no electrolyte is needed. If the amount of current during the rinse under controlled potential is very low, low-electrolyte concentrations could again be used. In general, the same electrolyte concentrations have been used for rinsing and deposition solutions so far.

The starting potentials for most atomic layers, in this group, were obtained by studies of the voltammetry for an element on an Au electrode [57, 66–68], usually using a thin-layer electrochemical cell (TLEC) (Fig. 9) [69, 70]. UPD potentials on Au are not expected to be optimal for growth of a compound; however, they are generally a good start.

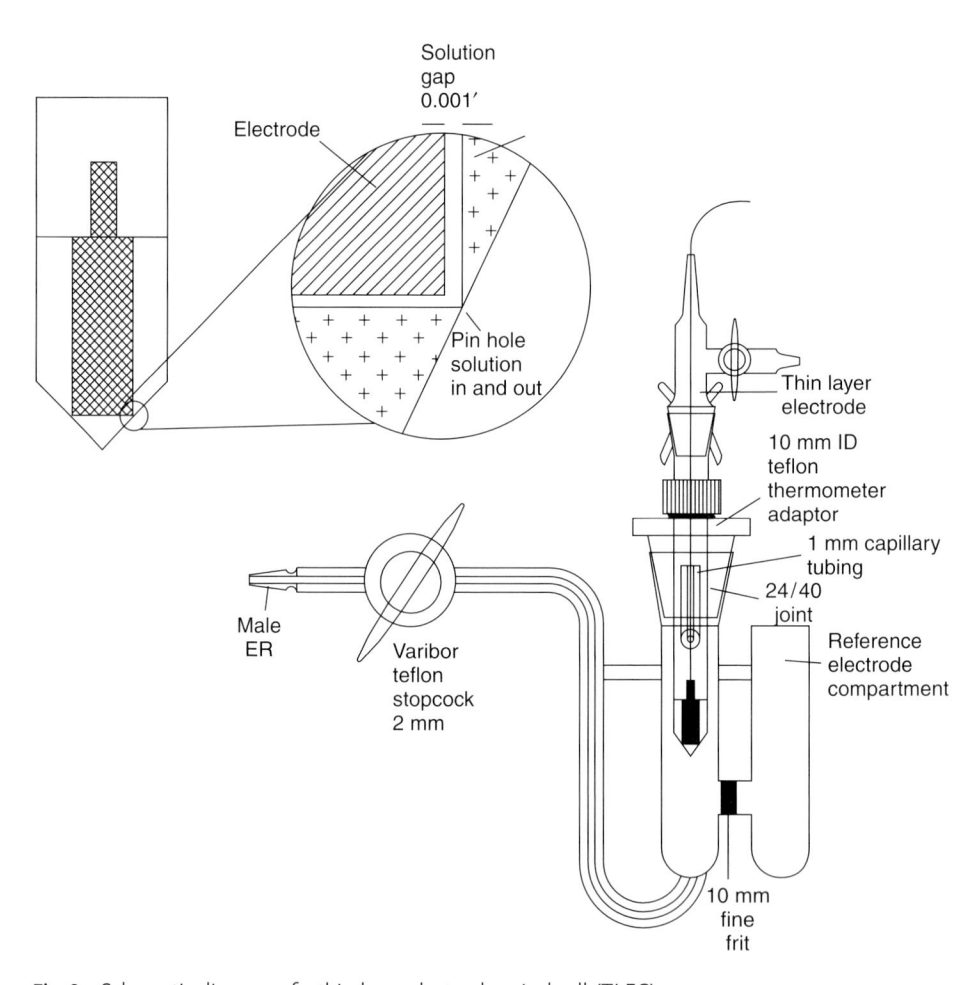

Fig. 9 Schematic diagram of a thin-layer electrochemical cell (TLEC).

The simplest model for EC-ALE is that a set of conditions is chosen for a cycle, and each cycle produces one compound ML. Ideally, the same potentials and solutions are used for each cycle of a deposition. Recently, it has become clear that this is not the optimum case for EC-ALE growth of many compounds on Au, as the initial conditions do not appear optimal from start to finish.

Graphs of the charges for Cd and Te versus the cycle # (Fig. 10) show that if a single set of potentials is used for each cycle, the first cycles are much larger than the succeeding cycles. If potentials determined from using a TLEC are used, the first cycle looks good, the charges correspond to the formation of an ML of the compound, but the charges quickly die away to nothing. On the other hand, if potentials are used that will result in something close to an ML of the compound for each cycle once steady state has been reached, the first few cycles will result in a great excess of deposition (Fig. 10).

Reasonable quality deposits have been obtained for 200-cycle runs, using these more negative "steady state" potentials from the beginning. However, that more than an ML/cycle is formed during the first few cycles is a significant problem, as it suggests bulk deposition, non–layer-by-layer, or 3D growth. That high growth rates were observed initially is understandable, in that the potentials were close to bulk growth conditions for Cd and Te.

Using these more negative potentials did result in reasonable deposits. Overall they were stoichiometric, displayed the expected X-ray diffraction (XRD) patterns without annealing, and the graphs of growth versus the number of cycles were linear (after the first few). Deposit morphologies appeared reasonable but probably not optimal, given some 3D growth during the first few cycles.

It has become clear that the potentials needed to form atomic layers shift negatively as the semiconductor films grow, especially over the first 25 cycles. The

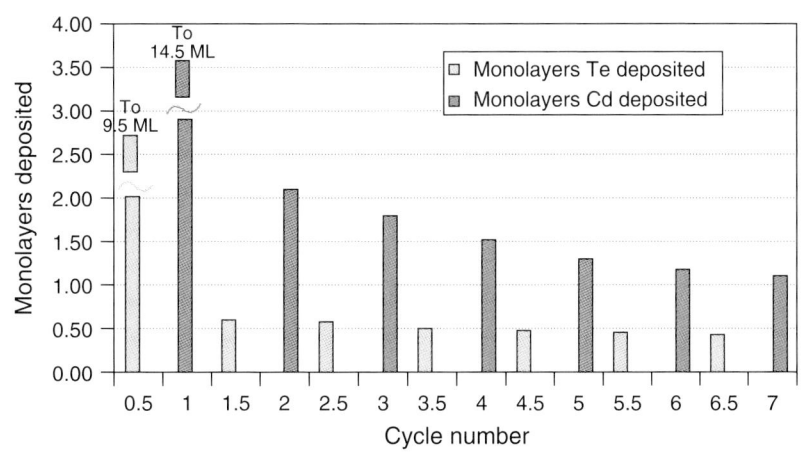

Fig. 10 Bar graphs showing the charges, in terms of monolayers, where 1 ML would be optimum, as a function of the cycle, for the first 7 cycles of CdTe deposition. The first half cycle was Te deposition, and the second Cd. A single set of potentials was used for all the cycles, and during the first cycle, the amounts deposited were excessive. (Adapted from Ref. [60].)

most logical reason for this shift is that a junction potential is developing between the Au substrate and the depositing compound semiconductor. When a program is used where the potentials are initially shifted for each cycle, steady state potentials are generally achieved after about 25 cycles, which can be maintained through the rest of the deposit without the deposited amounts changing for each cycle. Similar procedures were used by Villegas and coworkers [43] in the formation of CdTe. A reasonable potential ramp can be determined by inspection, and the potential changes programmed through the first 25 cycles or so. Some form of feedback within the cycle would be better, to regulate deposition on the fly. The most obvious feedback would be the deposition currents (or charges).

Figure 10 displays the coverages for Cd and Te deposited in each of the first few CdTe cycles, using a program in which the potentials remained constant. The coverages are the result of coulometry, two electrons for Cd and four electrons for Te. Note the excessive deposition for the first few cycles, significantly more than a ML, and the disparity between Cd and Te coverages. The relative coverages for Cd and Te should be 1:1 in Fig. 10, but seldom are in such graphs. It is not yet clear why the stoichiometry is not 1:1 from coulometry, while the stoichiometries from EPMA for the resulting deposits are 1:1. Some possible reasons for this discrepancy (Fig. 10) that have been considered include solvent or proton reduction, oxygen reduction, and charging currents. However, blank solutions provide a gauge of proton, solvent, and oxygen reduction, as well as charging currents, and they do not account directly for the asymmetry. In addition, coulometric stripping after 5 to 10 cycles of deposition agreed closely with the deposited amounts, suggesting that proton and oxygen reduction were not significant. Finally, no elemental Cd and Te were observed in XRD patterns; only diffraction peaks for CdTe and the substrate, for 200-cycle deposits, were observed.

One explanation for the observed charge disparity is that excess Cd is being deposited, and when the TeO_3^{2-} is introduced, some Te is exchanged for Cd. The TeO_3^{2-} may react chemically with the cadmium, or the current observed may be the sum of that for oxidation of part of the previously deposited excess Cd and that for reductive deposition of the Te atomic layer. In general, the currents observed during a cycle in the formation of a compound have not been easy to understand, and they are presently a major topic of study. For instance, in the formation of CdSe, a cycle similar to that for CdTe is used; however, it is the charge during Se deposition that is in excess and that for Cd, which is almost nonexistent. Again, the deposit is nearly stoichiometric, within a percent of 1:1, using EPMA. In summary, the relative charges for Cd and Te deposition are not easily interpreted in terms of deposit stoichiometry. They are, however, a key to incorporation of feedback in the EC-ALE cycle.

Ideally, each component element can be deposited at a potential optimized separately, independent of the conditions used to deposit subsequent atomic layers in the cycle. In some cases this independence is nearly realized, as in the deposition of Te atomic layers (Fig. 11) discussed previously. A broad potential region (0.7 V) is evident in Fig. 11, where the thicknesses of 200-cycle CdTe deposits do not change appreciably as the Te deposition potential is adjusted. However, Te deposition is a relatively irreversible process

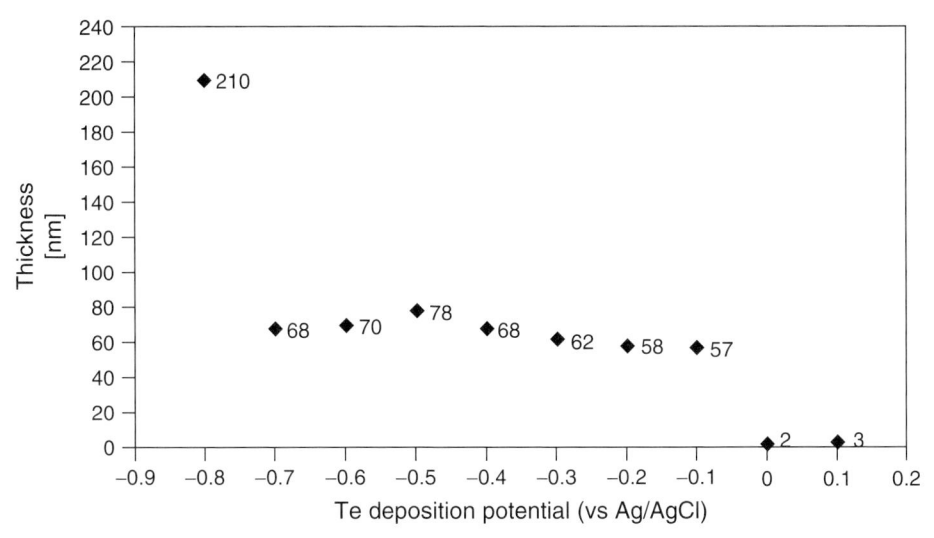

Fig. 11 Graph of CdTe deposit thickness as a function of the potential used to deposit Te. Each point represents a deposit formed with 200 cycles. (Adapted from Ref. [60].)

at pH 9 (Fig. 6c), suggesting that once it is deposited, the potential can be shifted significantly more positively (Fig. 6c) without its stripping from the surface. On the other hand, a more reversible species, such as Cd, may strip from the surface if the potential is shifted positively in a subsequent cycle step. This is exaggerated after a rinse, where the activity for Cd^{2+} has been removed.

In a system where both elements deposit reversibly, at significantly different potentials, the rinsing procedures must be carefully thought through. If, for instance, a reversible species is reductively deposited, and a more positive potential is to be used to deposit the next element, the rinse step can be performed at the more negative potential, and then shifted positively upon introduction of the second reactant, hopefully avoiding material loss during rinsing and minimizing losses during introduction. If the second element is pumped in rapidly, and with a minimum of solution, ions of the first element formed by stripping should still be close to the deposit surface, in the barrier layer, where they may redeposit as the second element deposits. On the other hand, given that the second element is reversible, the potential can be stepped positively after its introduction to the cell. Even if some bulk deposit is formed at the more negative potential, it should strip reversibly when the potential is stepped positively.

The inverse process, in which the potential is shifted negatively, is not generally a problem. The potential cannot be shifted prior to rinsing, or else bulk deposition of the first element would occur before the solution is rinsed from the cell. It is best to shift the potential after the blank rinse, just prior to introduction of the second element. In fact, the potential can be shifted after introduction of the second element, as it should not deposit at the more positive potential, but must wait until the potential is shifted negatively.

Keeping the potentials close can be beneficial, using complexing agents or the pH.

The ideal process is represented by the formation of CdS, where the Cd is first deposited from a Cd^{2+} solution by reductive UPD, and then the potential is shifted negatively to where $S_{(UPD)}$ is formed from a HS^- solution by oxidative UPD. In this way, both elements are stable on the surface during the deposition of the other.

What is the optimal time for a given step in a cycle? What is the minimum time needed to form a deposit? How much time is needed for deposition of an atomic layer and how long does it take to change solutions? Each step should be independently optimized. Pumps can easily exchange solution within a fraction of a second. The time required for deposition of an atomic layer will depend on the concentration of the solution and the kinetics for deposition. The deposition current should probably be an exponential decay, raising the question of at what point do you cut it off. This is related to fundamental questions such as: If you do not complete an ML, will defects result? There are indications that it is better to deposit less than an ML than too much, as too much will produce 3D growth. It is anticipated that an incomplete deposit will work fine, that cutting off the deposit sooner than later will probably not be a problem. The Foresti group has repeatedly developed cycles that produced only a small fraction of an ML/cycle [44, 45, 52, 53, 65]. It is probably a good idea to allow a given atomic layer deposition process to proceed as close to completion as possible, however.

With the 0.2 mM reactant solutions used in this group, deposition times as short as a couple of seconds have been used with little problem, resulting in cycle times close to 10 s. With slightly higher concentrations, and fast pumps, cycle times of a few seconds appear workable. As noted earlier, deposition times are controlled by the concentration of reactants and the kinetics of the deposition process. High concentrations result in increased atomic layer formation rates; however, they are wasteful, and may be difficult to achieve in some cases. TeO_2 solutions, for instance, are difficult to make, as are mM solutions at most pHs because of poor solubility. High concentration solutions also take longer exchange in the cell. In general, 20-s deposition times are presently used, probably much more than enough, with a couple of seconds for pumping (Fig. 7).

Of course, deposition times can be decreased by using a larger driving force, but that runs the risk of bulk deposition. It is easy to envision a cycle in which overpotentials are used, and the deposition is simply stopped after a ML of charge has passed. Such a cycle would not involve surface-limited reactions and 3D growth would be expected.

Faster pumps result in a shorter cycle. Exceeding the Reynolds number, however, results in turbulent flow. Flow patterns are frequently seen in deposits where the gasket has not been set correctly. Anything that perturbs laminar flow in the cell affects the deposit homogeneity: bubbles are a good example. If bubbles are present, even temporarily, they frequently leave their shadow in the deposit, as areas where less deposit was formed. If the gasket is not smooth, or something gets in the way of part of the inlet or outlet, the deposit quickly shows a flow pattern. Detailed studies have not yet been performed exploring the relationship between flow rates, cell design, and flow patterns, but it is clear from experience that the more laminar the flow the more homogeneous the deposit. CdTe, for example, is sensitive to flow rate, and a study of the dependence

4.2 Electrodeposition of Compound Semiconductors by Electrochemical Atomic Layer Epitaxy (EC-ALE)

on flow rate is underway. Most of the compounds, described in the next section, form homogeneously in the deposit area. CdTe, on the other hand, does not deposit well at the inlet and outlet, suggesting turbulence may affect the deposit. Slower pumps appear to help, but a detailed study is needed to determine what step in the cycle is causing the problem, and why.

Besides the flow rate, rinse times are very important. As discussed previously, if rinsing is insufficient to achieve the desired pH, deleterious effects can result. The example was given that if Te solutions are too acidic, bulk Te will form.

Generally, this group has used stop flow to provide adequate deposition times and to limit the volumes of solution needed. Villegas and coworkers have used continuous pumping, with good results [43], which may be an important alternative. Which element should be deposited first is an ongoing question. In the formation of II to VI compounds by this group, the chalcogenides have generally been deposited first. The reasons for this have more to do with the rich surface chemistry of those atomic layers than a feeling that better deposits will result. Results from two groups using Raman spectroscopy, Shannon and coworkers [71] and Weaver and coworkers [72], have suggested that higher-quality deposits of CdS are formed when the Cd is deposited first. Recent results by this group [63] indicate that no mater which is deposited first, Cd ends up being the atomic layer next to the Au substrate, suggesting it is probably better to deposit the Cd first.

4.2.4
Compound Formation

At present, the elements used in the formation of compounds by EC-ALE include the chalcogenides: S, Se, and Te; the pnictides: As and Sb; the group III metals: Ga and In; the group II metals: Zn, Cd, and Hg; as well as Cu and Co. The range of compounds accessible by EC-ALE is not clear, although the majority of work has been performed on II to VI compounds (Table 1). The III to V compounds InAs and InSb have recently been formed [40, 62]. In addition, Shannon and coworkers have begun studies of $CoSb_3$ [73] with the intent of forming thermoelectric materials.

Initially, EC-ALE was developed on the principle that reductive UPD of a metal and oxidative UPD of a main group element were required to form a working cycle. This would then limit compounds that could be formed to those containing a chalcogenide or a pnictide, as reduced forms of some of these elements were reasonably stable in aqueous solutions. Recently, it has been shown that reductive UPD of both elements is possible [40], suggesting a much larger range of compounds or possibly alloys might be formed.

Table 1 is a listing of compounds formed using EC-ALE. The first EC-ALE studies focused on CdTe [24, 25] for the historical reason that its electrodeposition had been studied the most [98–109]. The genesis for most of that work has been the desire to electrodeposit CdTe photovoltaics. The majority of the CdTe work is codeposition, after the classic paper by Kroger and coworkers [99].

The first studies of EC-ALE were stimulated by Mike Norton [66], who suggested that an electrochemical form of ALE might be possible, which prompted studies of the UPD of Cd and Te on Cu, Au, and Pt substrates, by this group. Those studies involved the use of a TLEC [69, 70] (Fig. 9), which consisted of a polycrystalline rod of the substrate metal, inserted into a vacuum-shrunk glass compartment. The

Tab. 1 EC-ALE studies

Compound	Study	Year	References
ZnTe	TLEC	1996	57
ZnSe	TLEC	1996	57
ZnSe	Flow cell deposition	1999	44
ZnS	TLEC	1996	57
ZnS	Flow cell deposition	1997	54
ZnS	TLEC	1996	57
ZnS	Flow cell deposition	1997	54
ZnS	STM studies of monolayers	1999	74
ZnS	Flow cell deposition	1999	45
ZnS	Growth of superlattice with EC-ALE	1999	75
ZnS	Size-quantized film, photoelectrochem.	2000	76
CdTe	UHV-EC of first few monolayers	1992	77
CdTe	UHV-EC of first few monolayers	1993	78
CdTe	STM	1993	79
CdTe	STM and UHV-EC	1995	80
CdTe	Flow cell	1995	41
CdTe	Review	1995	81
CdTe	Flow cells, H-cell	1998	42, 58
CdTe	UHV-EC, in situ STM	1998	82
CdTe	Wall-jet flow cell growth of thin films	1999	43
CdSe	STM	1996	83
CdSe	STM, UHV-EC	1997	84
CdSe	Flow cells, H-cell	1998	42
CdSe	SERS	1999	85
CdSe	Flow cells, large thin-layer cell	2000	86
CdS	STM	1994	87
CdS	TLEC	1994	67
CdS	Voltammetry	1994	88
CdS	STM	1996	89
CdS	STM, RRDE	1996	90
CdS	STM	1996	89
CdS	STM, surface study	1997	91
CdS	Flow cells, H-cell	1998	42
CdS	STM, voltammetry	1998	52
CdS	Raman study	1998	92
CdS	Photoelectrochemical studies of films formed by EC-ALE	1998	93
CdS	Resonance Raman study	1999	71
CdS	Photoluminescence study of a heterojunction	1999	94
CdS	Flow cells	1999	45
CdS	Growth of superlattice with EC-ALE	1999	75
HgS	Photoluminescence study of a heterojunction	1999	94

Tab. 1 (continued)

Compound	Study	Year	References
GaAs	UHV-EC	1992	68, 95
InAs	Thin films with TLEC flow system	1999	40
InSb	Thin films with TLEC flow system	2001	62
CuInSe$_2$	TLEC	1996	96
In$_2$Se$_3$	Flow cells	2001	97
CoSb	Overview of EC-ALE of CoSb	1999	73
CdS/HgS	Photoluminescence study of a heterojunction	1999	94
CdS/ZnS	Growth of superlattice with EC-ALE	1999	75
CdS/CdSe	Growth of superlattice with EC-ALE, studied by SERS	1999	85
InAs/InSb	Superlattice	2001	62

Note: STM: scanning tunneling microscopy; EC-ALE: electrochemical atomic layer epitaxy; UHV-EC: ultrahigh vacuum electrochemistry; SERS: surface-enhanced Raman scattering; RRDE: rotating ring-disk electrode.

compartment was designed to hold the electrode in the center of a roughly cylindrical glass enclosure, about 25 µm away from the glass. Two pinholes were ground into the bottom of the cell, to allow solutions to flow in and out, and for ionic conductivity. The total volume of the cell was 3.0 µL, with a 1-cm^2 electrode area.

The TLEC provided a defined environment for studies of surface-limited reactions. The large surface area to volume ratio greatly limits background reactions from traces of oxygen and other contaminants. Limited and predictable background currents facilitate coulometry and allow for more detailed coverage measurements. The reactant amounts are also limited, allowing studies with only slightly more than that needed to form atomic layers, limiting bulk-deposition and preventing bulk-stripping currents from swamping out UPD stripping. In addition, the TLEC facilitated solution exchange, which is very important for studies of EC-ALE. One solution can be flushed out and another rinsed in within a few seconds, without exposing the deposit to air. The TLEC is used to determine the potentials needed to form atomic layers of the elements on the substrate.

Following the codeposition work of Kroger and coworkers [99] and many others, acidic solutions of HTeO$_2^+$ were initially used [24, 25]. At that time, it was assumed that if Cd was deposited by reductive UPD, Te should be deposited using oxidative UPD, from a solution of H$_2$Te or a related telluride species. It was not felt that reductive UPD of both elements could be performed. Solutions of telluride proved unstable, however, oxidizing and turning purple with flecks of elemental Te, with even traces of oxygen in solution. This led to the use of the process described by Eqs. (4 and 5) [24, 25], where more than an atomic layer of Te was first deposited, from a HTeO$_2^+$ solution, and the excess Te was stripped reductively at a more negative potential using a blank solution in a separate step. The only analysis performed on those initial deposits was coulometry, as the substrates were not easily removed from the Pyrex cells (Fig. 9), and the substrate's cylindrical geometry did not lend itself to analysis by many techniques.

A flow-deposition system was subsequently developed to grow thicker films on flat disposable substrates [41]. The first edition was a small thin-layer flow cell (Fig. 3a). The majority of the system was very similar to that used today: pumps, electrochemical cell, valves, solution reservoirs, and potentiostat, all controlled by computer. Those first deposits were excessively heterogeneous, however [41]. The deposits suffered from a lack of adequate oxygen exclusion, gasket effects, and bubble problems. In addition, those deposits were grown with a single set of unoptimized deposition potentials. Scanning electron microscopy (SEM) images revealed extensive 3D growth. Interestingly, many deposits were composed of smooth bumps, suggesting they might result from layer-by-layer growth on a series of nucleation sites. The deposits were stoichiometric by EPMA, but graphs of coverage versus cycle number were not linear, as expected, but increased exponentially, clearly demonstrating increasing surface roughness or 3D growth.

There were a number of excuses made, and a number of solutions suggested for those first deposits, such as problems with valves, with gaskets, and with substrates [41]. The next set of CdTe thin-film growth studies were made using a H-cell (Fig. 3b) [42, 58]. The substrates were simply suspended in the H-cells, and the solutions were pumped in from the bottom and then drained out. This eliminated problems with gaskets and bubbles, as there were no gaskets, and the bubbles floated to the top. Substrates used in that study were Au on Si(100), which looked like a mirror but were composed of 40 nm hemispheres (Fig. 5a). The importance of oxygen exclusion was discovered while draining solutions, as some oxygen was sucked into the cell, unless an N_2 gas blanket was kept above the solution. This helped account for one of the major problems early on, reproducibility. Deposits were formed one day, but not the next, using the same conditions. Progress was exceedingly slow until the oxygen problem was better understood.

The deposit quality was greatly improved with the H-cell [42, 58], and Fig. 12 shows XRD patterns for deposits of CdTe, CdSe, and CdS made using the system. In each case, the deposits have a [110] preferred orientation, and crystallize in the zinc blende (cubic) structure. Graphs of deposit thickness versus the number of cycles were linear, a good indication of a surface-limited growth. The best deposits of CdTe were a little thin, however, corresponding to the growth of only about 0.4 ML/cycle. In hindsight, oxygen exclusion was probably still a problem, the samples were being overrinsed each cycle, and a constant set of deposition potentials were used.

As noted, the most important advance resulting from changing to the H-cells was that CdTe deposits were more reproducible. Studies of the dependence of deposit thickness on various cycle variables were possible using the H-cell design [58], including the dependency on the potentials used for Te deposition, Te stripping, and Cd deposition. Plateaus in coverage as a function of those potentials were observed, confirming the surface-limited nature of the growth process [58]. That is, the deposit thickness was zero if the

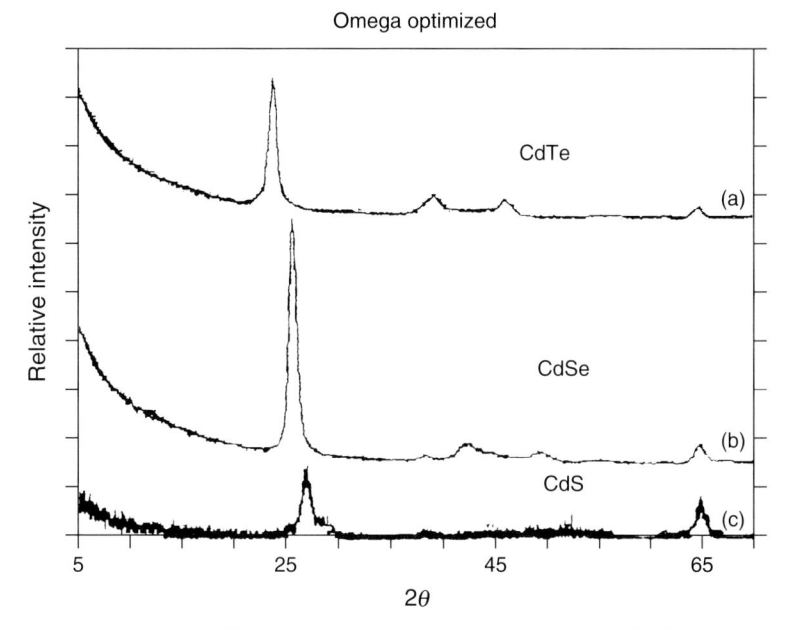

Fig. 12 XRD patterns for (a) CdTe, (b) CdSe, (c) CdS. (Reproduced with permission from Ref. [42].)

Cd deposition potential was too far positive, and if the potential was too far negative, some bulk Cd was formed for each cycle, resulting in a thick rough deposit. In between was a range of potentials where the deposit thickness was constant, close to 0.4 ML/cycle, suggesting the deposition was controlled by surface-limited reactions.

There were some major problems with the H-cells, however, such as the need for 50-gallon drums to hold the resulting chemical waste, as 100 mL of solution was used for each cycle. In addition, potential control was lost for the deposits each time the solution was drained, which can result in some deposit loss.

Subsequently, the H-cell was replaced with a larger thin-layer flow cell (Fig. 3c), with a 1×3 cm deposition area. The pumps, valves, and cell were placed inside an N_2-purged Plexiglas box, greatly improving oxygen exclusion, and an oxygen analyzer (Illinois Instruments, model 2550) was used to measure the oxygen levels in the box, 10 to 30 ppm. Use of the thin-layer flow cell dropped the solution volumes used from 100 mL/cycle to about 1 mL/cycle, and allowed potential control to be maintained throughout the deposition.

At that time, the cycle program was also changed, so that cathodic UPD was used for deposition of both elements. Use of a pH 10 tellurite solution, TeO_3^{2-}, instead of pH 2, shifted the Te UPD potential to better coincide with that for Cd UPD. A program similar to that presently used for depositing CdTe is shown in Fig. 7. A comparison of the programs in Fig. 13 clearly shows the relative simplicity of the new program and hardware, compared to the older.

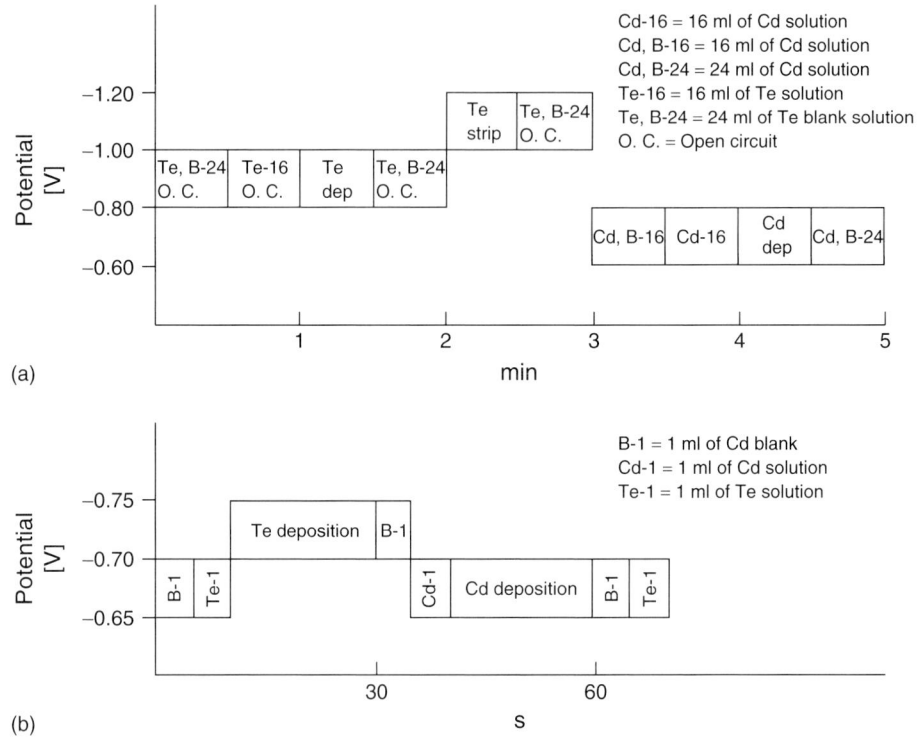

Fig. 13 Diagrams of the cycle program for use with (a) the H-cell (Fig. 3b) and (b) the large thin-layer flow cell (Fig. 3c).

As noted above, the deposits made with the H-cell design were thin, only about 0.4 ML/cycle. For a 200-cycle deposit, they appeared deep blue, the result of interference effects in the 30-nm-thick films. With the new cycle program and large thin-layer cell, the best deposits appeared gold in color, and ellipsometric measurement indicated they were very close to 1 ML/cycle. A study of the thickness as a function of the number of cycles is shown in Fig. 14, where the line is straight, indicating surface-limited control of growth. Figure 15 is a graph of the square of absorptivity X photon energy versus energy, from which a band gap of 1.55 eV has been extrapolated, equivalent to literature values for CdTe.

Figure 11 is a study of the potential dependence of deposit thickness, using 200-cycle CdTe deposits, formed with the large thin-layer flow cell, as a function of the Te deposition potential. A 0.7 V plateau in deposit thickness suggests broad flexibility in the choice of the Te deposition potential. Those results were very encouraging. At positive potentials, the coverage dropped dramatically, as no Te was deposited, so there was nothing for Cd to deposit on. At potentials below −0.7 V, some bulk Te appears to have deposited for each cycle, as the thickness greatly exceeded that expected for ML/cycle growth, and the morphology became "sandy" (Fig. 16). Sandy is a term used in this group to describe deposits where 3D growth is

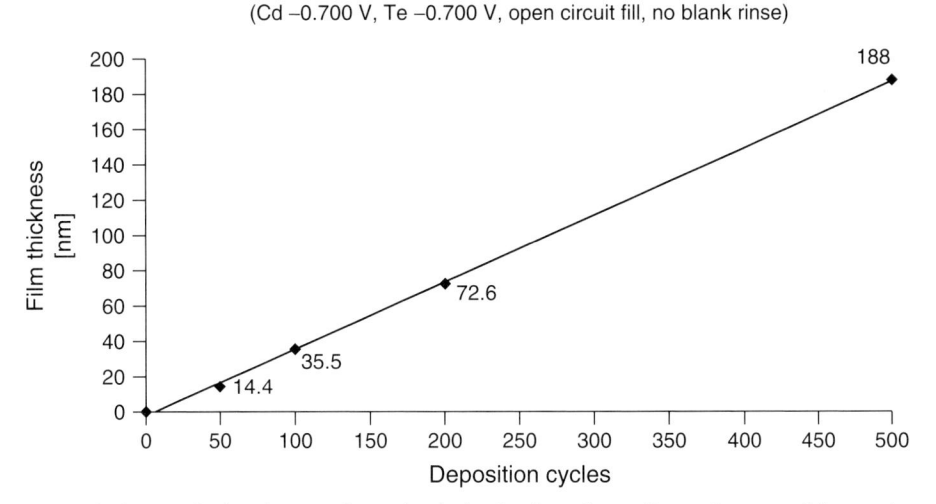

4.2 Electrodeposition of Compound Semiconductors by Electrochemical Atomic Layer Epitaxy (EC-ALE)

(Cd −0.700 V, Te −0.700 V, open circuit fill, no blank rinse)

Fig. 14 Thickness of CdTe deposits formed with the thin-layer flow cell, as a function of the number of cycles. (Adapted from Ref. [60].)

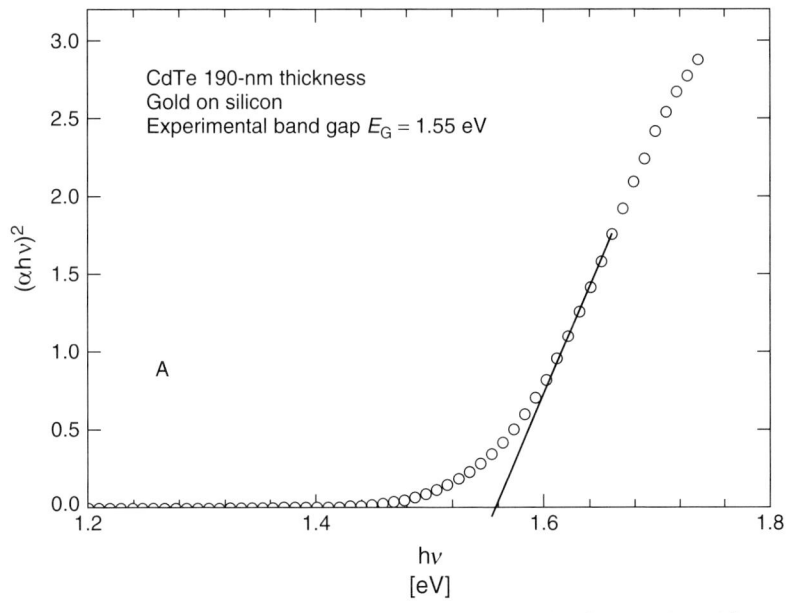

Fig. 15 Graph showing extrapolation to the band gap for CdTe films. (Adapted from Ref. [60].)

evident in an optical microscope, appearing as dark or multicolored dots, sand, when a 200-cycle deposit is examined at 1000 X with a metalographic microscope. This is a strong indicator that the growth mode is no longer layer by layer, and the deposit is roughening. Ellipsometric thickness measurements are only accurate

Fig. 16 Optical micrograph, 1000 X, taken with a metalographic microscope of CdTe deposits. The good deposit is on the left, and was produced using reasonable potentials. The bad deposit is on the right, and was produced with a program in which the potential for Te was excessively negative, leading to roughening of the deposit, and what is referred to here as sand, owing to its appearance in the microscope.

for smooth films [111], and became misleading for the films formed at such negative Te potentials. In between these potential extremes, the 0.7-V wide plateau was evident (Fig. 11), where the deposit thickness was consistent with ML/cycle growth. The presence of this plateau is consistent with surface-limited control of the deposition. Activation-limited growth would have resulted in increased deposition as the potential was decreased. Mass transfer–limited growth would have resulted in a plateau as well, but at a much higher growth rate, and would not have been expected to increase again at still more negative potentials. In addition, the deposit thickness for the plateau was not a function of the precursor concentration, consistent with surface-limited control.

Thin films of CdTe have also been produced using an automated flow-cell system by Villegas and coworkers [43]. Their films were very similar to those described above, and Fig. 4 is a TEM of one of their deposits. Their cell was a wall jet design, and they used a single pump to suck the solution from a distribution valve, through their cell. The deposit was formed in a continuous-flow mode, not the stop flow used by this group. Their design greatly simplifies the hardware and appears to work just as well as the thin-layer flow system. In addition, Villegas was the first to notice the need to adjust potentials as the deposits grow. They used a cycle incorporating the reactions shown in Eqs. (4 and 5).

Films of CdSe have been grown with the automated flow-cell systems, both using the H-cell [42] (Fig. 3b), and recently with the large thin-layer cell (Fig. 3c) [86]. Comparisons of XRD patterns have suggested that in both cases the quality of the CdSe deposits were better than equivalent CdTe films formed under similar conditions (Fig. 12b). Figure 17 shows an XRD pattern for a 200-cycle CdSe deposit,

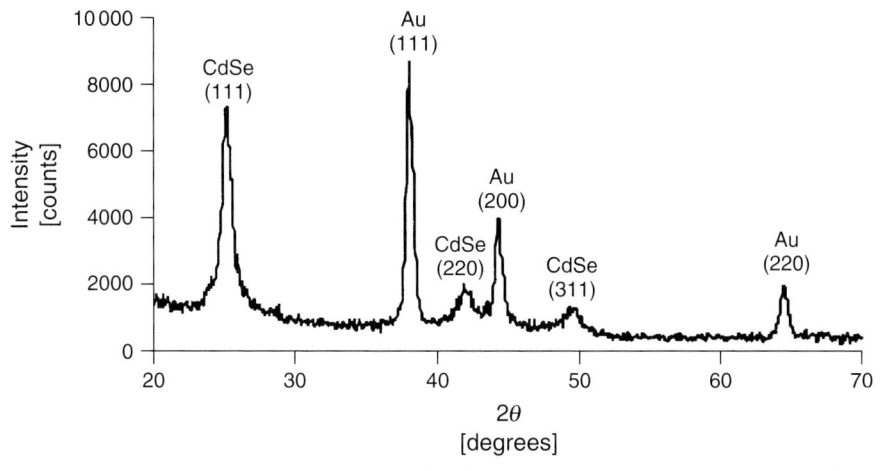

Fig. 17 XRD of CdSe formed using the large thin-layer flow cell, and 200 cycles. (Adapted from Ref. [86].)

formed using cathodic UPD for both elements. The probable reason for CdSe forming higher quality deposits than CdTe is the lattice-match with the Au substrate. That is, from atomic level studies, it is known that an ML and a half of CdTe [63, 78, 79], or CdSe [83, 84], will form a (3 × 3) surface unit cell on Au(111). The structures proposed to account for these (3 × 3) unit cells involve superposition of an ML and a half of these compounds, with the zinc blende structure, on the Au surface, where three Au lattice constants match up with two times the Cd–Cd distance in the (111) plane of CdTe or CdSe (Fig. 18). In the case of CdTe, this results in a 6% lattice mismatch, while it is less than 1% for CdSe on the same Au(111) surface. This suggests there should be fewer defects, or a larger critical thickness for CdSe deposits compared with CdTe [112].

Studies of the potential dependence of CdSe deposit thicknesses show a much shorter plateau region with CdSe than CdTe (Figs. 11 and 19). The variability in thickness evident in the plateau (Fig. 19) probably represents the standard deviation in growth and measurement at that time, rather than any significant variation in coverage as a function of potential. The plateau and trends in the graph are as expected. At potentials above −0.55 V, the deposits die off, while at potentials below −0.68 V, some bulk deposition of Se results, the growth rate exceeds an ML/cycle, and the surface roughens (sand is evident).

Typical 200-cycle CdSe deposits appear gold in color, as did the CdTe deposits. Given that the band gaps of CdSe and CdTe are both direct and about 1.6 eV, similarities in appearance are expected. Visual inspection of the 1 cm × 3 cm CdSe deposits shows them to be more homogeneous than corresponding CdTe deposits, in general.

Weaver and coworkers formed superlattices with CdSe and CdS, using EC-ALE, without an automated system [85]. They studied their relatively thin deposits by surface-enhanced Raman (SERS), examining stress build-up in the deposits.

CdS growth, by EC-ALE, has been studied by more groups than any other

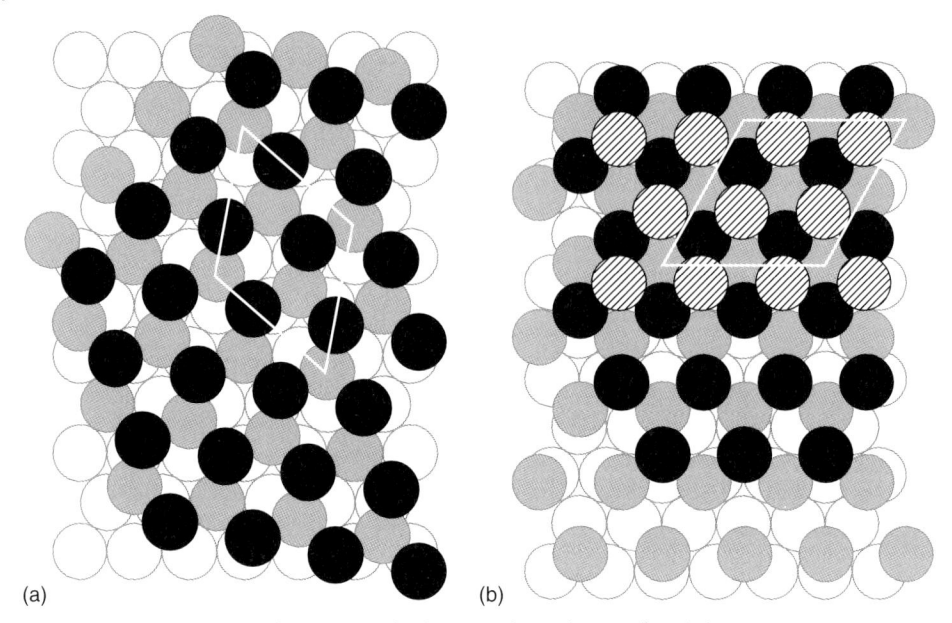

Fig. 18 Structures proposed to account for the (a) ($\sqrt{7} \times \sqrt{7}$)R19.1° and (b) (3 × 3) structures observed in the formation of monolayer of CdTe. (Adapted from Ref. [61].)

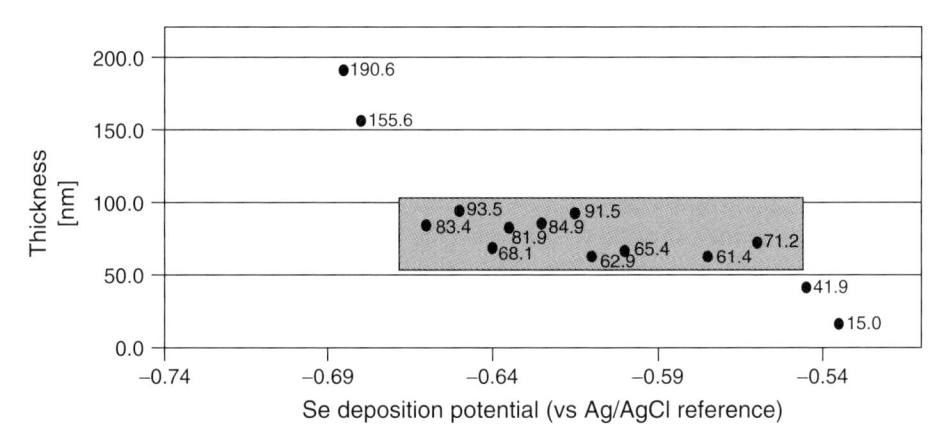

Fig. 19 Graph of the coverage of CdSe versus the potential used for Se deposition in the formation of 200-cycle deposits. (Adapted from Ref. [86].)

compound (Table 1) [42, 52, 65, 67, 75, 85, 87, 89–91, 94, 113]. Initial EC-ALE studies in this group of CdS were performed with a TLEC (Fig. 9), to determine potentials for a cycle [67]. Cd and S coverages were determined coulometrically for deposits as a function of the numbers of cycles performed. The dependence of thickness on the Cd deposition potential, for CdS deposits, revealed a plateau between -0.3 and -0.55 V, with the best deposits formed at -0.5 V, using a pH 5.9, 10 mM, $CdSO_4$

solution and a pH 11, 11 mM, Na$_2$S solution. Reductive UPD was used for the Cd atomic layers and oxidative UPD for S atomic layers.

The H-cell (Fig. 3b) was used to form CdS thin films with 200 cycles of deposition [42]. The films were a transparent yellow from visual inspection. XRD indicated the deposits were again (111) oriented, zinc blende CdS, but of a significantly lower quality than corresponding CdSe and CdTe (Fig. 12c). The films were also fairly thin, only about 0.2 ML/cycle, compared with the 0.4 ML/cycle obtained with CdTe using the H-cell, and 1 ML/cycle obtained for CdSe. EPMA suggested that the deposits were about 20% rich in S, as well. The cycle for CdS deposition had not been extensively optimized at that time. Some reasonable CdS films were formed using the large thin-layer flow cell, as well, but have not yet been well characterized. More work must be done on the cycle for CdS, and it is probable that a program in which the potentials are systematically shifted for the first 30 cycles will greatly improve the deposit quality.

An automated flow system has also been used by Foresti and coworkers to form CdS layers, with up to 150 cycles, using pH 9.2 solutions for both elements on Ag(111) electrodes [45]. In their case, the deposits appeared stoichiometric, without the excess S previously observed by this group [42]. Their cycle produced relatively thin deposits, similar to this author's, or about 1/3 ML/cycle of CdS.

EC-ALE studies of ZnTe using a TLEC were performed with up to 20 cycles of deposition [57]. Coulometry was the only analysis performed on the deposits. A plot of coverage as a function of number of cycles was linear, as expected for a surface-limited process. No thicker films have as yet been formed using the flow-deposition system. At present there is no reason to believe that the cycle developed for 20 cycles will not produce good quality deposits of any given thickness using the automated flow-cell system.

There has been a report of ZnSe thin films formed using EC-ALE in an automated deposition system, by the Foresti group [44]. Films with up to 31 cycles were formed, and produced the expected linear graph of stripping charge versus number of cycles. The slope of the graph suggested that the deposits grew at a rate of only about 0.14 ML/cycle. The reasons for this are unclear; however, most of the studies produced by the Foresti group have resulted in low coverages per cycle. It appears that they are conservative with their deposition potentials, and do not shift them as the deposits grow. As discussed earlier, as long as depositing less than an ML/cycle does not result in defects, higher-quality deposits may be a by-product of such cycles. Their report describes very careful coulometric studies of both Zn and Se coverages.

TLEC studies of the first 14 cycles of ZnSe growth have been performed by this group [57]. That study has resulted in an initial set of conditions for an EC-ALE cycle, but no thicker films have as yet been formed using an automated flow-cell deposition system, by this group.

Several groups have worked on the formation of ZnS using EC-ALE (Table 1) [23, 54, 57, 65, 75, 76, 81, 113]. The first study was again by TLEC, where it proved difficult to quantify stripping coulometry for both Zn and S separately for deposits formed with more than 5 cycles. Foresti and coworkers used a procedure in which S was reductively stripped and Zn oxidatively stripped to accurately determine both Zn and S coverages from coulometry for deposits formed with a greater number of cycles.

A ZnS cycle was automated using the H-cell design, with the idea of forming phosphor screens for field-emission displays (FED) [54]. Doping studies were performed with Ag, Cu, and Mn, and will be discussed in a subsequent section. The films showed evidence of 3D growth and roughening, and were rich in Zn (a Zn/S ratio of 1.35 from EPMA). XRD showed the deposits to be zinc blende ZnS, of variable quality, depending on the doping scheme. Deposits formed with as many as 200 cycles were produced. A plot of coverage versus cycle number was exponential; the coverage per cycle increased as more cycles were performed, consistent with some 3D growth and roughening. SEM of the deposits grown on ITO revealed small smooth bumps, suggesting that the deposits could be layer-by-layer growth on 3D nucleation sites.

From the exponential graph of coverage versus cycle number, and the presence of excess Zn, it appears that the Zn potential was pushed too far negative, causing some bulk Zn deposition and 3D growth. These ZnS deposits, formed five years ago, were deposited using the same cycle potentials throughout the deposition. Graphs of Zn/S ratio, as a function of the number of cycles, indicated that after 50 cycles the deposits were very close to stoichiometric, possibly a little S rich. Excess Zn began to show up at the 100-cycle point. It is probable that the ZnS deposits would be greatly improved, using the present hardware and shifting the potentials through the first 30 cycles.

In 1999, Foresti and coworkers published a paper describing the growth ZnS and CdS, using their automated deposition system [45]. Few details were given at that time, however. A paper looking at the photoelectrochemistry of ZnS films as a function of the layer thickness was published by Yoneyama and coworkers in 2000 [76]. Their study involved the growth of films of increasing thickness with an increasing number of EC-ALE cycles. The deposits were studied with photoelectrochemistry (photocurrent curves) and clearly showed the dependence of the band gap on the thickness of the ZnS films. The band gaps of deposits formed with more than 4 cycles, about 1.5 nm, showed the expected dependence on thickness. The thinner deposits were blueshifted from the bulk value of 3.6 eV, to about 4 eV. However, deposits formed with less than 4 cycles did not display the theoretical exponential dependence expected. This work was extended by the formation of CdS/ZnS superlattices [75], discussed in a subsequent section.

InAs was the first III to V compound grown using EC-ALE, although there was some early work concerned with the formation of GaAs [68, 95]. In those early GaAs studies, no more than a single ML was ever formed. The major problems at that time were the reactivity of Ga, and the hardware problems previously discussed: lack of oxygen exclusion and loss of potential control during rinsing.

Indium, on the other hand, is less reactive than Ga, making InAs easier to form than GaAs. Deposits of InAs of good quality have been formed using the automated thin-layer flow-deposition system (Fig. 3c) [40, 62]. Homogeneous deposits have been formed with close to 1:1 stoichiometry. EPMA suggested the deposits were somewhat rich in As, as much as 20%, but attempts to better optimize the deposition potentials did not greatly improve the EPMA results. Elemental

4.2 Electrodeposition of Compound Semiconductors by Electrochemical Atomic Layer Epitaxy (EC-ALE)

analysis, using inductively coupled plasma mass spectroscopy (ICP-MS), suggested these deposits were closer to 5% rich in As and that there may have been some In loss during EPMA analysis. Figure 20 is an XRD pattern for zinc blende InAs, formed with 500 cycles. There are no indications of As in the XRD. From Fig. 21, a plot of the $(\alpha h\nu)^2$ versus energy, the band gap was estimated to be 0.36 eV, in agreement with literature values. Band gaps for the InAs deposits appear to be sensitive functions of a number of cycle variables. Several samples resulted in band gaps closer to 0.44 eV. These blueshifts appear to result from smaller crystallites, nanoclusters, formed

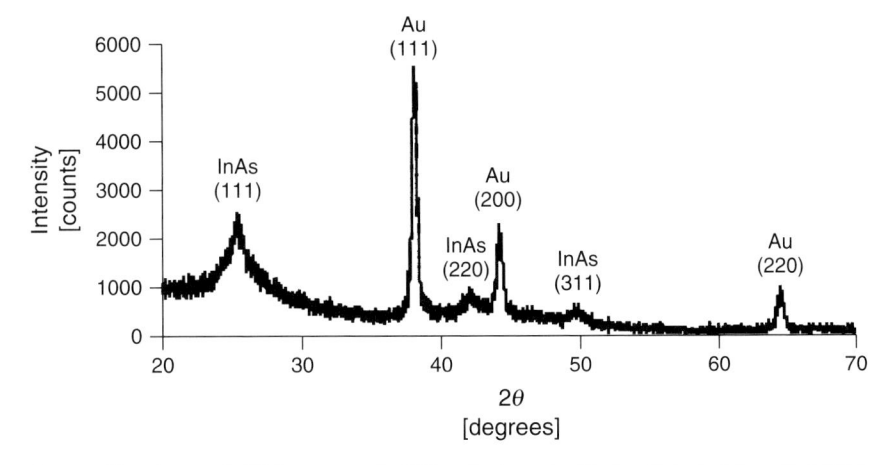

Fig. 20 Glancing-angle XRD pattern of a 270-cycle deposit of InAs. (Adapted from Ref. [62].)

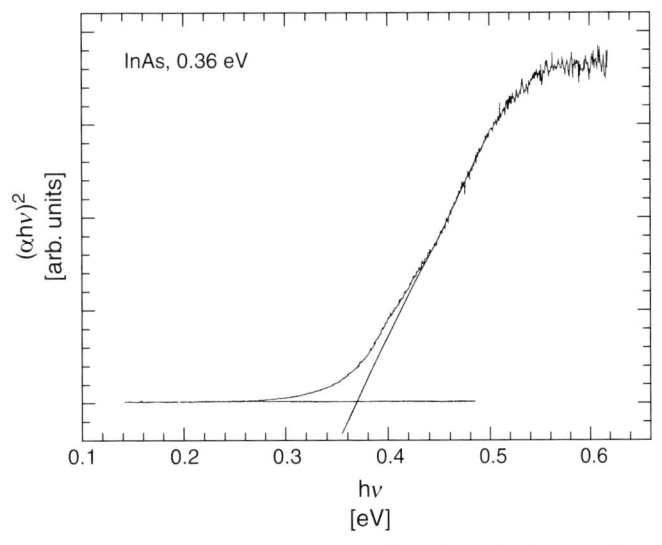

Fig. 21 Band gap measurement on InAs films formed with 200 cycles, obtained from absorption data. (Adapted from Ref. [40].)

when the deposition conditions were less than optimal.

AFM images of the InAs films show them to be relatively conformal with the Au substrate, with some texture, but little 3D growth for a 250-cycle deposit. Figure 22(a) is an image of an Au on glass substrate, annealed at 550 °C for 12 h. Figure 22(b) is of an equivalent substrate onto which 250 cycles of InAs have been grown.

The InAs program involved slowly shifting the potentials for In and As deposition negatively for the first 25 cycles, and then holding them constant. Figure 23 is a graph of cycle potentials as a function of the number of cycles. The progression is essentially exponential, asymptotically approaching the steady state values, as the junction potential is built up.

The thickness dependence of the InAs deposits as a function of the As deposition potential is shown in Fig. 24. At positive potentials, above −0.6 V, little deposit is formed, as would be expected. Below −0.6 V, a relative plateau is observed, but which gradually increases between −0.625 and −0.775 V. Below about −0.7 V, the deposits correspond to more than 1 ML/cycle, and some roughening is evident with optical microscopy. Below −0.775 V, the coverages measured with ellipsometry drop to the 1-ML/cycle level, but microscopy shows the deposits to be roughened and sandy. Coverage measurements with EPMA also indicate that these deposits are rich in As, and there is a significant increase in coverage, as expected, demonstrating that the ellipsometry readings were faulty at these negative potentials. The As potential was driven too far negative, and some bulk As was formed for each cycle. The

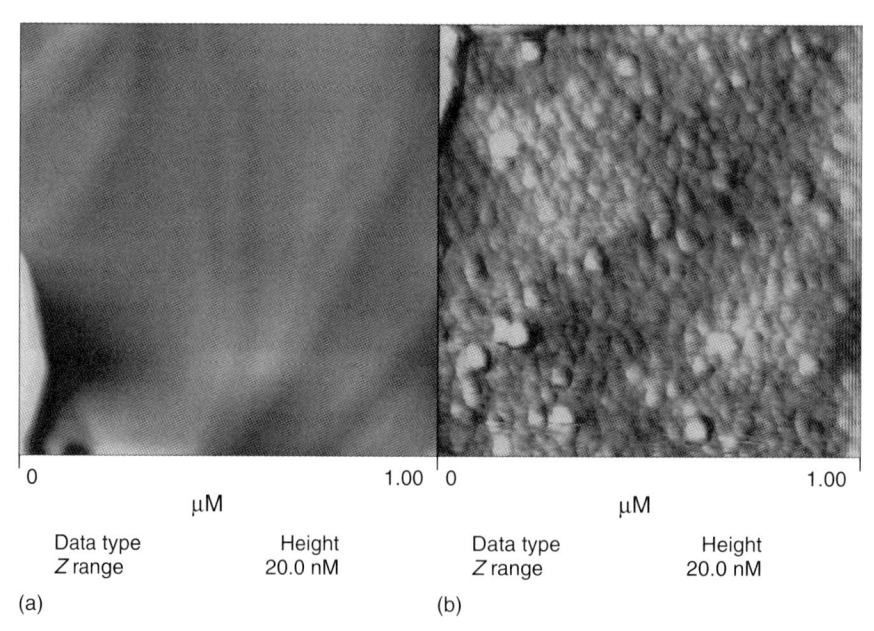

Fig. 22 AFM images of (a) the Au on glass substrate, annealed at 550 °C for 12 h and (b) equivalent substrate onto which an InAs deposit has been formed with 200 cycles of deposition.

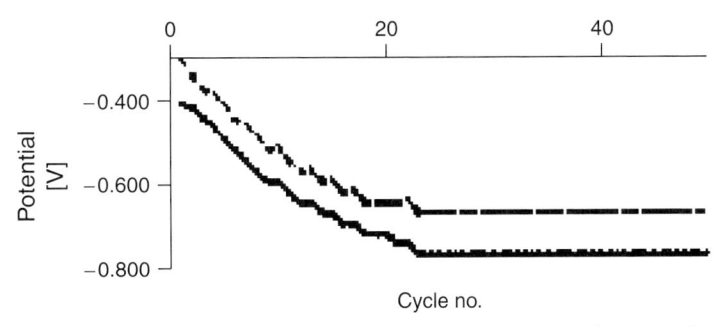

Fig. 23 A graph of the deposition potentials for In and As as a function of the cycle number in the growth of InAs. (Adapted from Ref. [62].)

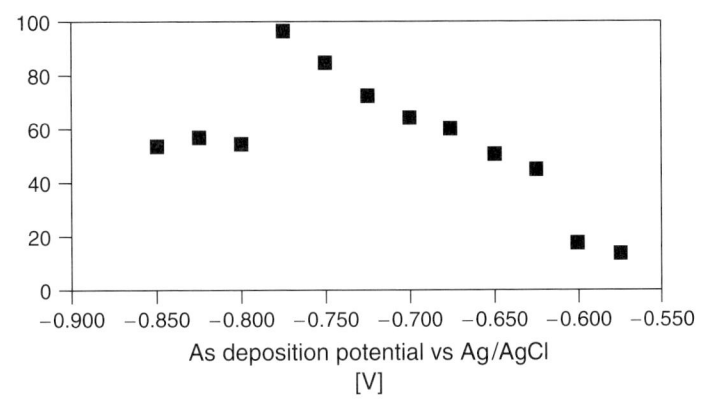

Fig. 24 A graph of deposit thickness as a function of the potential used to form As atomic layers. Each deposit was made with 200 cycles of deposition. (Adapted from Ref. [62].)

best deposits were those where the As potential was between −0.6 and −0.7 V: where the deposits did not appear rough, had close to 1:1 stoichiometry, and IR absorption indicated a band gap close to 0.36 eV.

The dependence of the InAs film thickness on the In potential used is shown in Fig. 25. Again, little deposition occurs for potentials above −0.7 V. There is a plateau between −0.725 and −0.8 V, where deposits grow at close to 1 ML/cycle, and the deposit thickness increased below −0.8 V, as some bulk In began to form for each cycle, creating a rough deposit.

Again, the thickness measurements at these negative potentials, determined with ellipsometry, appear faulty, consistent with the roughening observed. The best deposits were grown in the potential range between −0.725 and −0.8 V.

The importance of the substrate structure on the quality of the resulting deposit cannot be overemphasized. Figure 26 shows two reflection transmission IR spectra for InAs. The InAs was grown on an Au on glass substrate that had been carefully annealed. Then, half the substrate was covered, and more Au was vapor-deposited on one half at room temperature. The

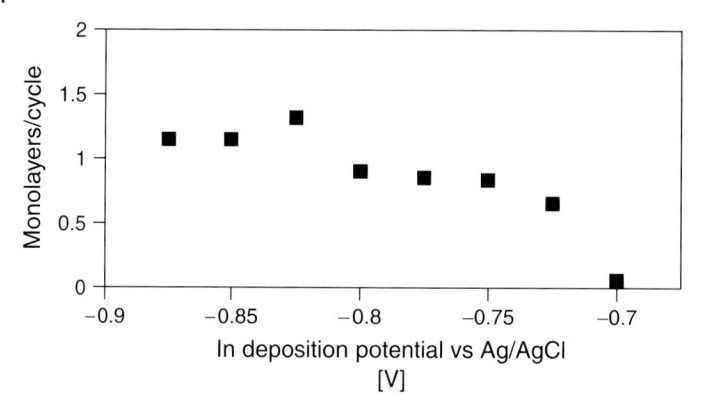

Fig. 25 Graph of the deposit thickness as a function of the In potential used. (Adapted from Ref. [62].)

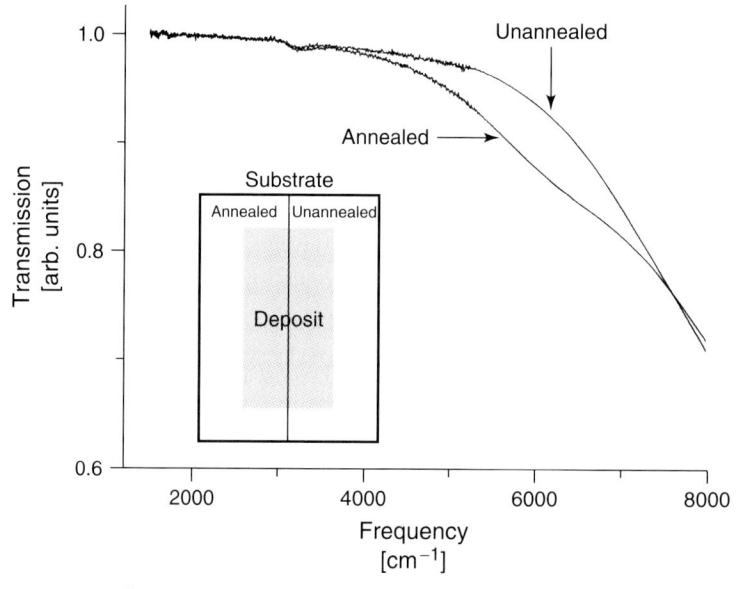

Fig. 26 IR reflection transmission spectra for InAs films formed on smooth annealed Au surface and roughened Au surface. Substrate was formed by taking a well-annealed, smooth Au on glass substrate and vapor-depositing more Au on one half, at room temperature, so that a series of 40 bumps were formed over one half of the surface. The two deposits were subsequently formed in the same electrodeposition run. (Adapted from Ref. [114].)

room-temperature Au greatly roughened half of the substrate (Fig. 27). The results are evident in Fig. 26, where the absorption of the InAs grown on the smooth side is characteristic of bulk InAs, while that on the roughened side is significantly blueshifted. The blueshift is the result of the substrate structure causing the formation of a nanocrystalline InAs deposit.

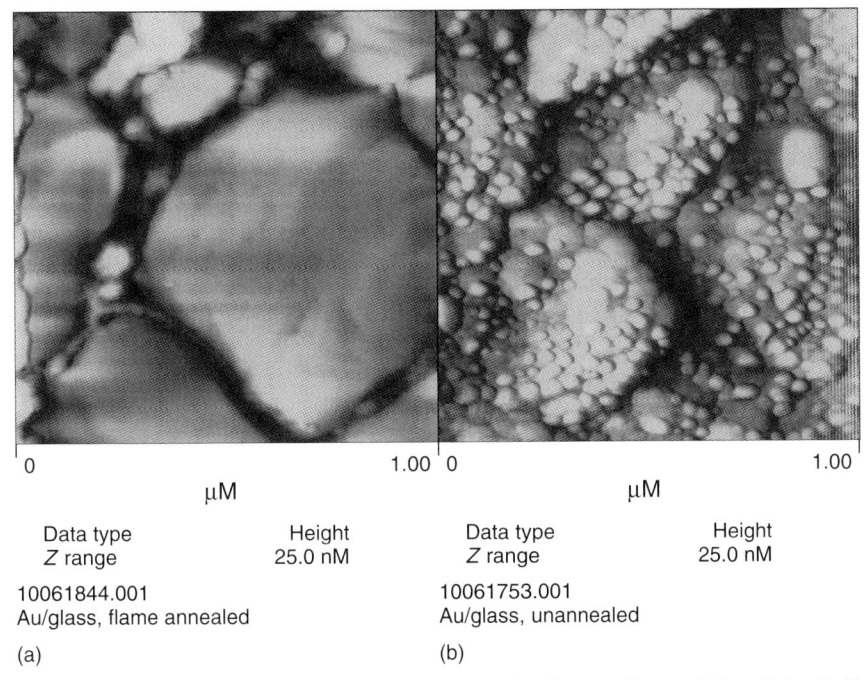

4.2 Electrodeposition of Compound Semiconductors by Electrochemical Atomic Layer Epitaxy (EC-ALE)

Data type	Height	Data type	Height
Z range	25.0 nM	Z range	25.0 nM
10061844.001		10061753.001	
Au/glass, flame annealed		Au/glass, unannealed	
(a)		(b)	

Fig. 27 AFM images of an Au on glass substrate that had been well annealed, and then half of it was covered. More Au was vapor-deposited on the other half at room temperature, forming a surface composed of 40-nm Au hemispheres. (Adapted from Ref. [114].)

Given positive results with InAs, it was felt that InSb could be grown using EC-ALE, since the voltammetry of Sb is more straightforward than that for As. That is, Sb^{+3} solutions have well-defined, relatively reversible UPD features (Fig. 28), while As^{+3} solutions show significant irreversibility (Fig. 6e). In addition, it has been shown that As is easily reduced to arsine species at modestly negative potentials [68, 95], making it difficult to keep in the deposit if other steps in the cycle are performed at overly negative potentials, one of the basic problems with GaAs formation by EC-ALE. The potentials for Ga deposition were too similar to those where As is reduced to arsine. There are reports in the literature describing the reductive conversion of Sb to stibine, H_3Sb, or related species [110, 115–119]; however, reduction of Sb, under the conditions used by this group, was very difficult. The stability of elemental Sb over a large potential range suggested that InSb formation would be tractable.

The potentials used to form InSb were applied in a roughly exponential progression, as with InAs, through the first 30 cycles. Figure 29 shows the (111) XRD peak for zinc blende InSb, from a film formed with 200 cycles on a Cu substrate. The poor quality of the XRD appears to result from the small grain size in the deposit, as seen in the AFM image in Fig. 30(b). The surface is composed of smaller crystallites than the corresponding InAs deposits (Fig. 30a). The origin of this difference appears to be the lattice constants, again. If we assume that the interface is similar to that for CdTe (zinc

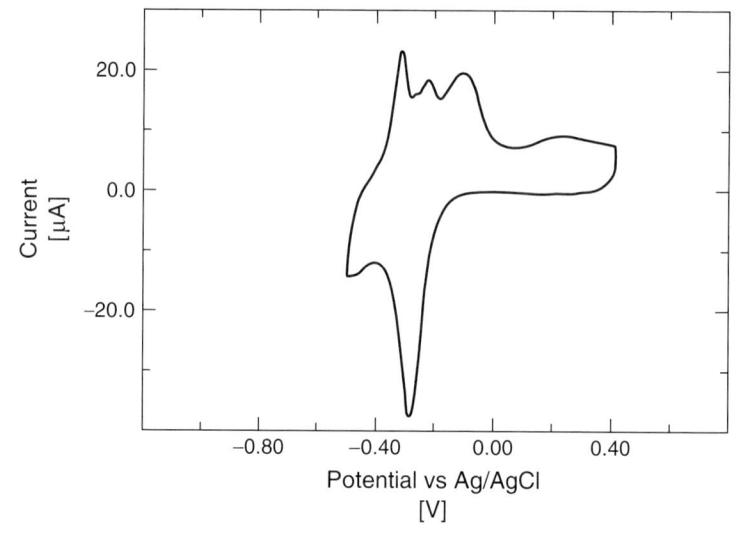

Fig. 28 Voltammetry for Sb deposition on Au, from a solution of Sb_2O_3, pH 5, 0.10 M $NaClO_4$, and approximately 50 μM Sb. (Adapted from Ref. [62].)

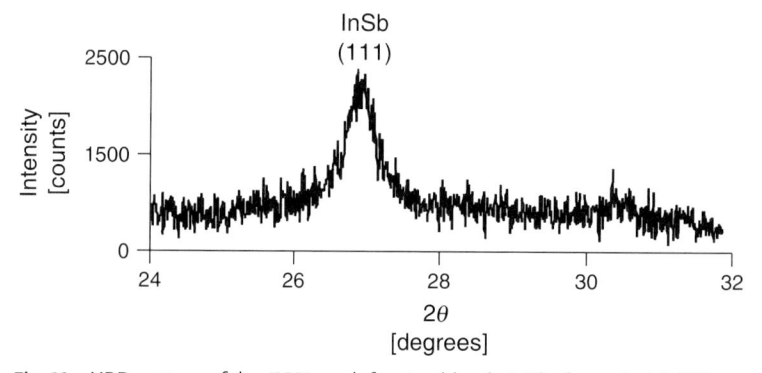

Fig. 29 XRD pattern of the (111) peak for zinc blende InSb, formed with 200 cycles on a Cu substrate. (This figure was adapted from Ref. [62].)

blende) on Au(111), then there should again be the 2:3 match between the compound and the Au surface. The lattice mismatch for InSb would be 6%, as it was for CdTe and Au, as CdTe and InSb have the same lattice constant (0.648 nm), while InAs and CdSe have the same lattice constant (0.606 nm), resulting in a much better lattice mismatch, only 1% with Au.

There have been a number of studies directed toward the electrochemical formation of $CuInSe_2$, CIS, over the last 17 years [120–132]. A number of electrodeposition methodologies have been used, including codeposition and a number of two-stage methodologies. The impetus of that work has been the excellent photovoltaic properties of CIS.

Initial TLEC studies developing an EC-ALE cycle to form CIS were published in 1996 [96]; however, a complete cycle was not accomplished at that time. Steps

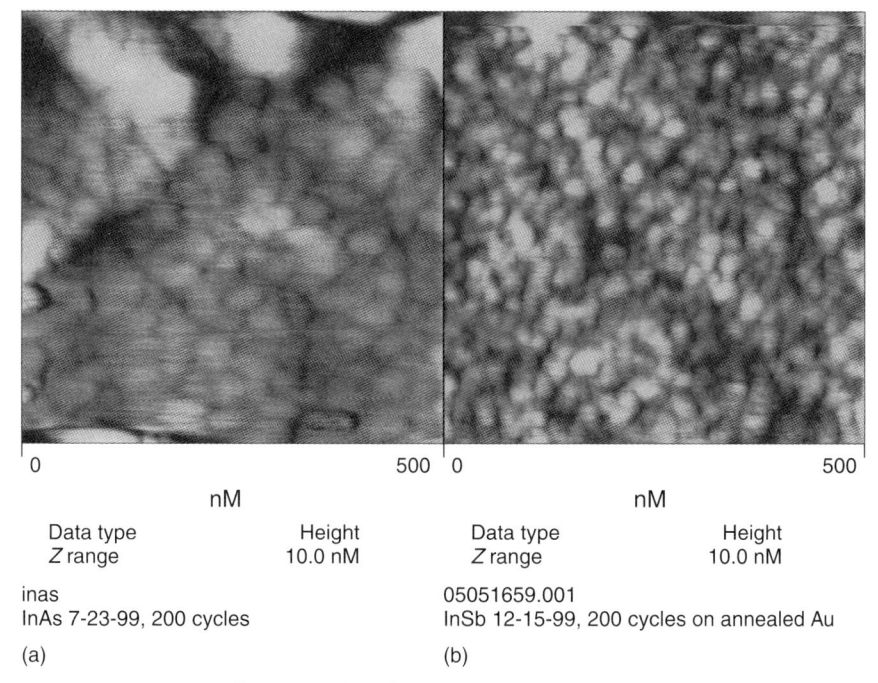

Fig. 30 AFM images of (a) 200 cycles of InAs and (b) 200 cycle of InSb, both on Au on glass substrates.

involving Se UPD, followed by Cu, Se, In, and then Se again were successful. However, keeping In on the electrode while the next atomic layer of the significantly more noble Cu was deposited, has proven problematic. Presently, these studies are being revisited using the automated flow-cell deposition system, and Cu complexation is being used to shift its deposition potential to more closely match that of In. However, problems persist in keeping the In in the deposit. To better understand the mechanism for CIS formation, studies of the EC-ALE formation of In_2Se_3 and Cu_2Se have begun.

It appears that some Cu_2Se has been formed, but not a pure phase. Work is continuing on the deposition conditions to see if different phases can be formed, and some very interesting deposit morphologies have been produced. The In_2Se_3 films are more problematic. Deposits have been made, which visually look very good, but no peaks have as yet been observed with XRD, besides those due to the Au substrate. Possible reasons for the absence of XRD would be the fact that the films are probably less than 100-nm thick, have small crystallites, and the pattern for In_2Se_3 has many peaks. Figure 31 is an AFM image of one of the deposits. EPMA indicates that the deposits are close to stoichiometric. Reflection absorption measurements give a band gap of 1.6 eV, while there are various reports of band gaps between 1.5 and 1.8 eV in the literature [97].

4.2.4.1 Toward Growing Device Structures

There are a number of ways to introduce dopants into an EC-ALE deposit. For instance, they can be introduced

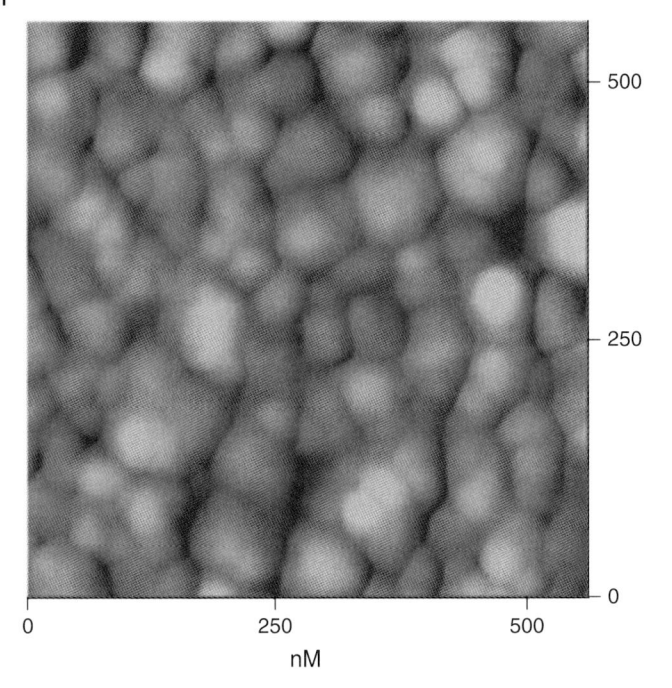

Fig. 31 An AFM image of a 200-cycle deposit of In_2Se_3. (Adapted from Ref. [97].)

homogeneously throughout the deposit or delta doped into the structure. For a relatively homogeneous distribution, low concentrations of oxidized precursors can be incorporated into the reactant solutions. By using very low concentrations, the amounts incorporated in each atomic layer will be limited. The dopant can also be incorporated in its own cycle step. Again, a low concentration would be used so that some fraction of an atomic layer is introduced for each cycle. Alternatively, a delta-doping scheme can be constructed where a fraction of an atomic layer of dopant is deposited for every set number of cycles. Any of these scenarios is a simple modification of the EC-ALE program.

Initial doping studies of ZnS were run with the idea of forming phosphor screens for flat-panel display applications [54], as noted earlier. Some of the most important commercial phosphors are based on ZnS, as a host material, with metals such as Ag, Cu, and Mn as activators. In that initial EC-ALE doping study, ZnS films were formed on Au and ITO substrates. Two methods were used, codeposition of the dopant with the Zn atomic layers and use of a separate delta-doping step every tenth cycle. No detectable dopant was found using EPMA, except for the case where Ag was codeposited with the Zn atomic layers. In that case, about 1/5 ML/cycle of Ag was deposited, probably replacing Zn. From XRD, the quality of the deposit was significantly degraded by the excess Ag incorporation. The amount of activator (Ag) was greatly in excess, judging by commercial ZnS phosphors. As noted, the Mn codeposited with Zn was not detectable with EPMA; however, cathodoluminescence of a deposit grown on ITO

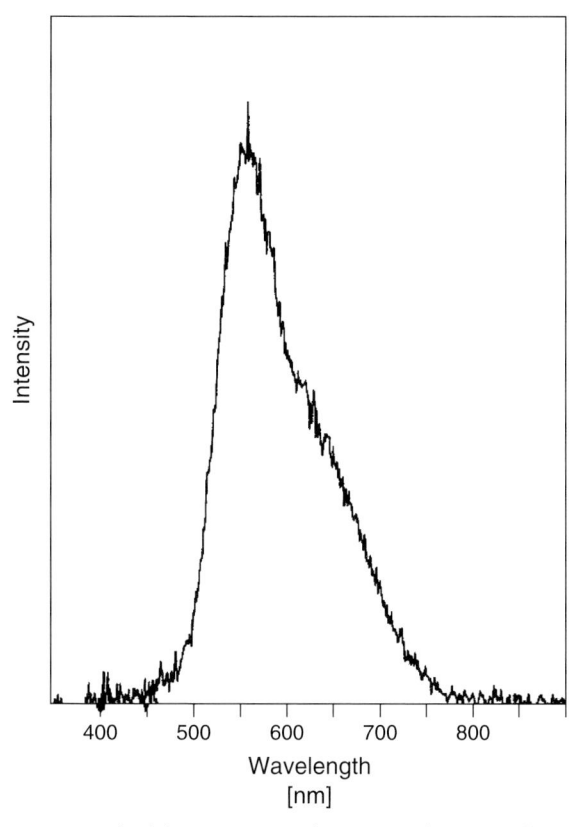

Fig. 32 Cathodoluminescence of a Mn-doped ZnS thin film, grown on ITO. (Adapted from Ref. [54].)

resulted in the emission spectrum shown in Fig. 32, which is characteristic of Mn.

4.2.4.2 Diodes

There are a number of papers in the literature concerning formation of compound semiconductor diodes by electrodeposition, the most popular structure being a CdS-CdTe-based photovoltaic. In general, CdS was deposited first on an ITO on glass substrate, followed by a layer of CdTe, usually by codeposition [133–143].

So far, there has been little work done on the formation of heterojunctions using EC-ALE. One study, however, performed by Shannon and coworkers, involved the growth of CdS on Au, using EC-ALE, which was then capped with a single ML of HgS [94]. The HgS capping layer was deposited in two ways, one using EC-ALE and the other by chemical exchange. The chemical step involved exchange of the last layer of Cd in a solution containing Hg ions. The deposits were studied with STM, electrochemistry, and photoluminescence. Photoluminescence showed excellent coupling between the electrodeposited HgS layer and the underlying CdS, while photoluminescence from the chemically formed layer was not nearly as good.

Sailor and Martin and coworkers grew an array of CdSe-CdTe nanodiodes [144] in alumite (anodized aluminum, which forms hexagonal arrays of 200-nm pours [145–149]) using a compound electrodeposition methodology called *sequential monolayer electrodeposition* (SMED) [144]. SMED involves a potential sequence in a single electrochemical bath to promote layer-by-layer growth and to minimize accumulation of elemental Se [150]. Current voltage measurements on the array revealed the rectifying behavior expected of diodes.

The first electrodeposition of a compound superlattice was by Rajeshwar and coworkers [151], where layers of CdSe and ZnSe were alternately formed using codeposition in a flow system. That study was proof of concept, but resulted in a superlattice with a period significantly greater than would be expected to display quantum confinement effects. There have since been several reports of very thin superlattices formed using EC-ALE [73, 75, 85, 113]. Surface-enhanced Raman (SERS) was used to characterize a lattice formed from alternated layers of CdS and CdSe [85]. Photoelectrochemistry was used to characterize CdS/ZnS lattices [75, 113]. These EC-ALE-formed superlattices were deposited by hand, where the cycles involved manually dipping or rinsing the substrate in a sequence of solutions.

Thicker EC-ALE-grown superlattices have recently been formed using the automated flow-deposition system [62]. A number of InAs/InSb superlattices have been formed, using a combination of cycles similar to those described in the preceding text for InAs and InSb. As superlattices are composed of alternating layers of two or more compounds, they are characterized by their period, or the repeat thickness (Fig. 33). The superlattices of InAs/InSb formed in this group were built with 10 cycles of InAs, followed by 10 cycles of InSb, making one period. A 30-cycle layer of InAs was first formed on an Au on glass substrate, before the lattices were grown, in order to help develop the junction potential. However, every time the compound was changed, twice each period, a new junction was formed. Experimentally, it appeared that the potentials needed to deposit the next

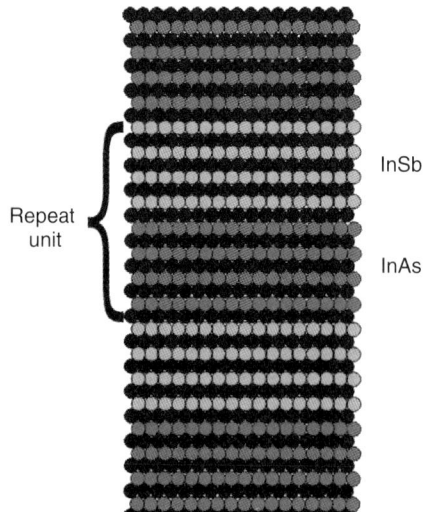

Fig. 33 Schematic diagram of a superlattice.

compound layer had to be adjusted with each cycle.

Figure 34 is a graph of the potential program used to grow the first 100 cycles of an InAs/InSb superlattice. Experience showed that the potentials had to be changed for both elements for each cycle. Those potentials and changes were determined experimentally, and were not extensively optimized. Figure 34 shows that the potentials for As and Sb were each shifted to lower potentials with successive cycles in each half period. On the other hand, the In potential was increased during the InSb half period, and decreased during the InAs half period. At present, these potential changes were arrived at by observation. It is hoped that feedback can be incorporated to account for and optimize the progressions or that the potential shifts can at least be modeled.

Given the lattice mismatch of InAs (a = 0.606 nm) with InSb (a = 0.648 nm), about 6.5%, defects are expected. At best a strained-layer superlattice would result. Figure 35 is an XRD pattern for a 41-period InAs/InSb deposit, where each period was 10 cycles of InAs followed by 10 cycles of InSb. The central [110] reflection is near 28° and is quite broad. Superlattices should display satellite peaks at angles corresponding to the period of the lattice [152]. Given that the interplaner spacings, in the [110] direction, are 0.35 nm for InAs and 0.374 nm for InSb, ideally, a period composed of 10 ML of each should be 7.24-nm thick. Using Bragg's law, with 0.1789-nm X rays, the angle between the primary peak and its satellites should be 0.71°. Figure 35 suggests the presence of two shoulders, each about 0.93° from the central peak. The quality of this preliminary XRD pattern is bad, but the symmetric shoulders are evident. On the basis of 0.93°, the period for the superlattice appears to be closer to 5.5 nm, suggesting that about 3/4 ML/cycle was actually deposited.

Elemental analysis with ICP-MS was performed on a dissolved piece of the superlattice, and indicated a deposit composition of $In_{0.49}As_{0.37}Sb_{0.14}$. It was gratifying to find the deposit close to 50% In, as expected. However, there appears to be significantly more As than Sb. The deposit was started with 30 cycles of an InAs buffer layer, so the deposit should be about 10% higher in As than Sb, but there was significantly more than 10% excess As over Sb. The absolute coverages

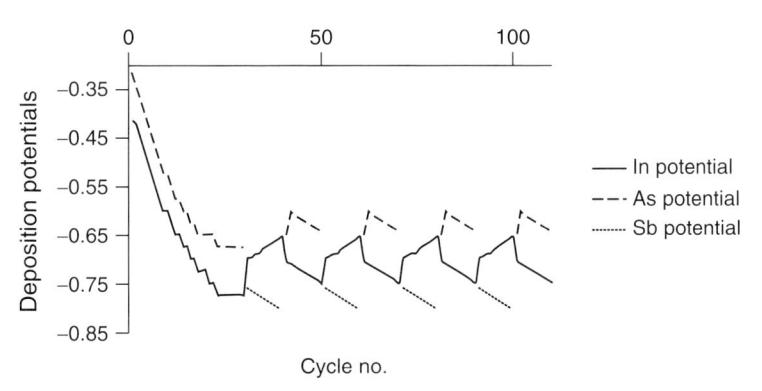

Fig. 34 Deposition potential versus cycle number for an InAs10/InSb10-41X superlattice: In (solid line), As (dot dash), Sb (dash). (Adapted from Ref. [62].)

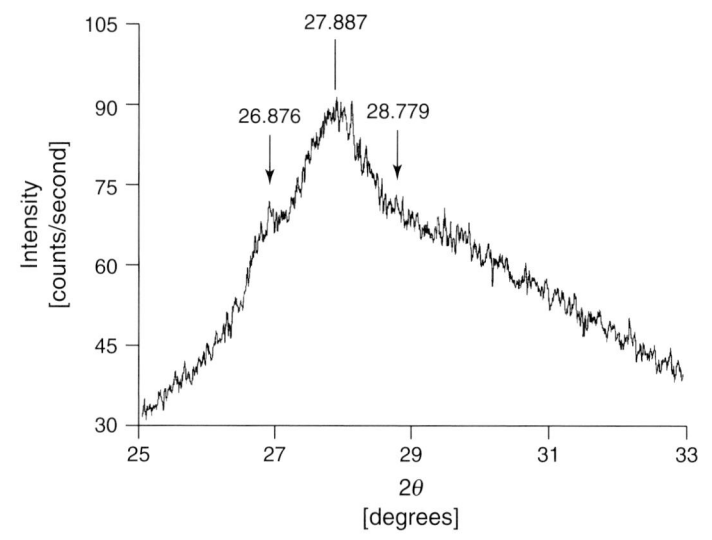

Fig. 35 Grazing incidence angle X-ray diffraction (XRD) pattern of an InAs10/InSb10-41X superlattice. The incident angle was 0.50°. (Adapted from Ref. [62].)

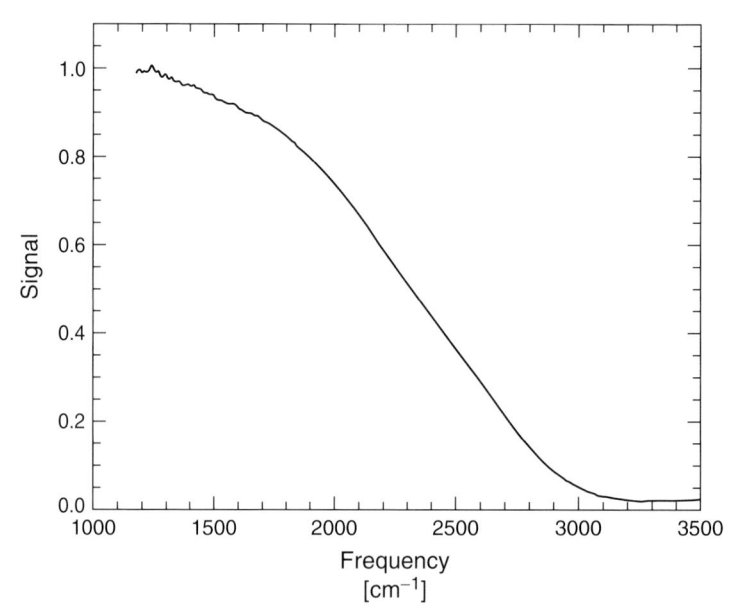

Fig. 36 Reflection IR of an InAs10/InSb10-41X superlattice. (Adapted from Ref. [62].)

from ICP-MS were about 10% less than would be expected if an ML was deposited for each cycle, suggesting somewhat more than the 3/4 ML/cycle, as indicated by the XRD satellite peaks, was deposited. The extra material is easily understood if the substrate is considered slightly rough. A roughness factor (RF) of 1.15 would rationalize the ICP-MS and XRD results, where an RF of 1.0 corresponds to an atomically flat surface. One explanation of the disparity between As and Sb is that close to a full ML of InAs was deposited for each cycle, while the coverage of InSb was closer to a 1/2 ML/cycle. This conclusion is consistent with the fact that the InSb cycle had not been well optimized.

Figure 36 is an IR reflection-transmission spectrum taken from the superlattice deposit. Very strong adsorption is evident at higher wave number, nearly 0% transmission, for the 250-nm-thick superlattice deposit. The deposit starts to absorb at very low energies, near 1000 cm^{-1}. Problems with the spectrometer's beam splitter make it hard to get a clear measure of the leading edge of the absorption, and thus a band gap measurement. However, a comparison was made between the absorptivity of the superlattice and that of a single crystal of InSb, the superlattice component with the lower band gap, around 1300 cm^{-1}. The absorptivity of the superlattice was about 20 times greater than that for single crystal InSb, indicating that the superlattice is redshifted from InSb. If the deposit was an alloy, InAs$_x$Sb$_{1-x}$, the band gap should be blueshifted from InSb, assuming it is a linear function of the mole fractions of As and Sb, and the band gaps for the two compounds, or about 0.3 eV [153]. These results strongly indicate that the deposit is not an alloy, and since the band gap is redshifted from pure InSb, it is probably a Type II superlattice [154].

Acknowledgment

Acknowledgment is made to the National Science Foundation, Divisions of Chemistry and Materials for support of this work.

References

1. A. Y. Cho, *J. Vac. Sci. Technol.* **1971**, *8*, s31.
2. A. Y. Cho, J. R. Arthur, *Prog. Solid State Chem.* **1975**, *10*, 157.
3. J. R. Arthur, *J. Appl. Phys.* **1968**, *39*, 4032.
4. J. Y. Tsao, *Materials Fundamentals of Molecular Beam Epitaxy*, Academic Press, Boston, 1993.
5. M. B. Panish, H. Temkin, *Annu. Rev. Mater. Sci.* **1989**, *19*, 209.
6. M. A. Herman, H. Sitter, *Molecular Beam Epitaxy: Fundamentals and Current Status*, Springer-Verlag, Berlin, 1989.
7. H. O. Pierson, *Handbook of Chemical Vapor Deposition*, Noyes Publications, Park Ridge, 1992.
8. F. S. Galasso, *Chemical Vapor Deposited Materials*, CRC Press, Boco Raton, 1991.
9. W. Kern in *Microelectronic Materials and Processes* (Ed.: R. A. Levy), Kluwer Academic, Dordrecht, 1989.
10. K. K. Schuegraf, *Handbook of Thin-Film Deposition Processes and Techniques*, Noyes, Park Ridge, 1988.
11. P. C. Andricacos, C. Uzoh, J. O. Dukovic et al., *IBM J. Res. Dev.* **1998**, *42*, 567.
12. P. C. Andricacos, *Interface* **1999**, *8*, 32.
13. M. Fleischmann, H. R. Thirsk, (Eds.), *Metal Deposition and Electrocrystallization*, Advances in Electrochemistry and Electrochemical Engineering, Interscience Publishers, John Wiley & Sons, New York, 1963, Vol. 3.
14. P. E. Light, D. Shanefield, *J. Appl. Phys.* **1963**, *34*, 2233.
15. K. R. Lawless, *J. Vac. Sci. Technol.* **1965**, *2*, 24.
16. H. J. Choi, R. Weil in *Electrochemical Society National Meeting* (Eds.: R. Weil, R. G. Barradas), The Electrochemical Society, 1981, p. 169, Vol. 81-6.

17. J. P. G. Farr, A. J. S. McNeil, C. A. Loong, *Surf. Technol.* **1981**, *12*, 13.
18. G. F. Fulop, R. M. Taylor, *Annu. Rev. Mater. Sci.* **1985**, *15*, 197.
19. K. Rajeshwar, *Adv. Mater.* **1992**, *4*, 23.
20. G. Hodes, *Sol. Energy Mater.* **1994**, *32*, 323.
21. R. K. Pandey, S. N. Sahu, S. Chandra, *Handbook of Semiconductor Electrodeposition*, Marcel Dekker, New York, 1996.
22. G. Hodes in *Physical Electrochemistry* (Ed.: I. Rubinstein), Marcel Dekker, New York, 1995, p. 515.
23. J. L. Stickney in *Electroanalytical Chemistry* (Eds.: A. J. Bard, I. Rubenstein), Marcel Dekker, New York, 1999, p. 75, Vol. 21.
24. B. W. Gregory, D. W. Suggs, J. L. Stickney, *J. Electrochem. Soc.* **1991**, *138*, 1279.
25. B. W. Gregory, S. J. L., *J. Electroanal. Chem.* **1991**, *300*, 543.
26. J. L. Stickney, B. W. Gregory, I. Villegas, *U.S. Patent*, University of Georgia, 1994, p. 5320736, Vol. 6.
27. T. Suntola, J. Antson, *U.S. Patent*, USA, 1977, p. 4058430.
28. C. H. L. Goodman, M. V. Pessa, *J. Appl. Phys.* **1986**, *60*, R65.
29. S. P. DenBaars, P. D. Dapkus, *J. Cryst. Growth* **1989**, *98*, 195.
30. T. F. Kuech, P. D. Dapkus, Y. Aoyagi, *Atomic Layer Growth and Processing*, Materials Research Society, Pittsburgh, 1991.
31. A. Usui, H. Watanabe, *Annu. Rev. Mater. Sci.* **1991**, *21*, 185.
32. S. Bedair, *Atomic Layer Epitaxy*, Elsevier, Amsterdam, 1993.
33. L. Niinisto, L. M., *Thin Solid Films* **1993**, *225*, 130.
34. D. M. Kolb in *Advances in Electrochemistry and Electrochemical Engineering* (Eds.: H. Gerischer, C. W. Tobias), John Wiley & Sons, New York, 1978, p. 125, Vol. 11.
35. K. Juttner, W. J. Lorenz, *Z. Phys. Chem. N. F.* **1980**, *122*, 163.
36. R. R. Adzic in *Advances in Electrochemistry and Electrochemical Engineering* (Eds.: H. Gerishcher, C. W. Tobias), Wiley-Interscience, New York, 1984, p. 159, Vol. 13.
37. A. T. Hubbard, V. K. F. Chia, D. G. Frank et al. in *New Dimensions in Chemical Analysis* (Ed.: B. L. Shapiro), Texas A & M University Press, College Station, 1985, p. 135.
38. A. A. Gewirth, B. K. Niece, *Chem. Rev.* **1997**, *97*, 1129.
39. E. Herrero, L. J. Buller, H. D. Abruna, *Chem. Rev.* **2001**, *101*, 1897.
40. T. L. Wade, L. C. Ward, C. B. Maddox et al., *Electrochem. Solid State Lett.* **1999**, *2*, 616.
41. B. M. Huang, L. P. Colletti, B. W. Gregory et al., *J. Electrochem. Soc.* **1995**, *142*, 3007.
42. L. P. Colletti, B. H. Flowers, J. L. Stickney, *J. Electrochem. Soc.* **1998**, *145*, 1442.
43. I. Villegas, P. Napolitano, *J. Electrochem. Soc.* **1999**, *146*, 117.
44. G. Pezzatini, S. Caporali, M. Innocenti et al., *J. Electroanal. Chem.* **1999**, *475*, 164.
45. M. Innocenti, G. Pezzatini, F. Forni et al. in *195th meeting of the Electrochemical Society* (Eds.: P. C. Andricacos, P. C. Searson, C. R. Simpson et al.), The Electrochemical Society, Seattle, 1999, p. 294, Vol. 99-9.
46. C. E. D. Chidsey, D. N. Loiacono, T. Sleator et al., *Surf. Sci.* **1988**, *200*, 45.
47. R. Emch, J. Nogami, M. M. Dovek et al., *J. Appl. Phys.* **1988**, *65*, 79.
48. A. L. Putnam, B. L. Blackford, M. H. Jericho et al., *Jpn. J. Appl. Phys.* **1989**, *217*, 276.
49. S. Buchholz, H. Fuchs, J. P. Rabe, *J. Vac. Sci. Technol., B* **1991**, *9*, 857.
50. J. A. DeRose, T. Thundat, L. A. Nagahara et al., *Surf. Sci.* **1991**, *256*, 102.
51. E. Holland-Moritz, J. Gordon-II, G. Borges et al., *Langmuir* **1991**, *7*, 301.
52. M. L. Foresti, G. Pezzatini, M. Cavallini et al., *J. Phys. Chem. B* **1998**, *102*, 7413.
53. M. L. Foresti, M. Innocenti, F. Forni et al., *Langmuir* **1998**, *14*, 7008.
54. L. P. Colletti, R. Slaughter, J. L. Stickney, *J. Soc. Info. Display* **1997**, *5*, 87.
55. H. Cachet, R. Cortes, M. Froment et al., *Philos. Mag. Lett.* **1999**, *79*, 837.
56. L. Beaunier, H. Cachet, M. Froment et al., *J. Electrochem. Soc.* **2000**, *147*, 1835.
57. L. P. Colletti, S. Thomas, E. M. Wilmer et al., *MRS Symp. Boston* **1996**, *451*, 235.
58. L. P. Colletti, J. L. Stickney, *J. Electrochem. Soc.* **1998**, *145*, 3594.
59. B. M. Huang, T. E. Lister, J. L. Stickney, *Surf. Sci.* **1997**, *392*, 27.
60. B. H. J. Flowers, T. L. Wade, M. Lay et al., *J. Electroanal. Chem.* **2002**, in press.
61. K. Varazo, M. D. Lay, J. L. Stickney, *Langmuir* **2002**, in prep.
62. T. L. Wade, R. Vaidyanathan, U. Happek et al., *J. Electroanal. Chem.* **2001**, *500*, 322.
63. K. Varazo, M. Lay, J. L. Stickney, *J. Electroanal. Chem.*, **2002**, in press.

64. M. J. N. Pourbaix, *Atlas of Electrochemical Equilibria in Aqueous Solutions*, Pergamon Press, Oxford, 1949.
65. M. Innocenti, G. Pezzatini, F. Forni et al., *J. Electrochem. Soc.* **2001**, *148*, c357.
66. B. W. Gregory, M. L. Norton, J. L. Stickney, *J. Electroanal. Chem.* **1990**, *293*, 85.
67. L. P. Colletti, D. Teklay, J. L. Stickney, *J. Electroanal. Chem.* **1994**, *369*, 145.
68. I. Villegas, J. L. Stickney, *J. Electrochem. Soc.* **1992**, *139*, 686.
69. A. T. Hubbard, F. C. Anson in *Electroanalytical Chemistry* (Ed.: A. J. Bard), Marcel Dekker, New York, 1970, p. 129, Vol. 4.
70. A. T. Hubbard, *Crit. Rev. Anal. Chem.* **1973**, *3*, 201.
71. B. E. Boone, A. Gichuhi, C. Shannon, *Anal. Chim. Acta 397*, **1999**, 43.
72. S. Z. Zou, M. J. Weaver, *J. Phys. Chem. B* **1999**, *103*, 2323.
73. C. Shannon, A. Gichuhi, P. A. Barnes et al. in *National Meeting of the Electrochemical Society* (Eds.: P. C. Andricacos, P. C. Searson, C. Reidsema-Simpson et al.), ECS, Seattle, 1999, p. 282, Vol. 99-9.
74. A. Gichuhi, C. Shannon, S. S. Perry, *Langmuir* **1999**, *15*, 5654.
75. H. Yoneyama, A. Obayashi, S. Nagakubo et al., *Abstracts Electrochem. Soc. Meeting* **1999**, 99-2, 2138.
76. T. Torimoto, A. Obayashi, S. Kuwabata et al., *Langmuir* **2000**, *16*, 5820.
77. D. W. Suggs, I. Villegas, B. W. Gregory et al., *J. Vac. Sci. Technol., A* **1992**, *10*, 886.
78. D. W. Suggs, J. L. Stickney, *Surf. Sci.* **1993**, *290*, 362.
79. D. W. Suggs, J. L. Stickney, *Surf. Sci.* **1993**, *290*, 375.
80. L. B. Goetting, B. M. Huang, T. E. Lister et al., *Electrochim. Acta* **1995**, *40*, 143.
81. C. K. Rhee, B. M. Huang, E. M. Wilmer et al., *Mater. Manufact. Proc.* **1995**, *10*, 283.
82. B. E. Hayden, I. S. Nandhakumar, *J. Phys. Chem. B* **1998**, *102*, 4897.
83. T. E. Lister, J. L. Stickney, *Appl. Surf. Sci.* **1996**, *107*, 153.
84. T. E. Lister, L. P. Colletti, J. L. Stickney, *Isr. J. Chem.* **1997**, *37*, 287.
85. S. Zou, M. J. Weaver, *Chem. Phys. Lett.* **1999**, *312*, 101.
86. B. H. Flowers Jr., T. L. Wade, K. Mathe et al., MRS Symposium proceedings. 2002.
87. U. Demir, C. Shannon, *Langmuir* **1994**, *10*, 2794.
88. E. S. Streltsov, I. I. Labarevich, *Dokl. Akad. Nauk Bel.* **1994**, *38*, 64.
89. U. Demir, C. Shannon, *Langmuir* **1996**, *12*, 594.
90. U. Demir, C. Shannon, *Langmuir* **1996**, *12*, 6091.
91. G. D. Aloisi, M. Cavallini, M. Innocenti et al., *J. Phys. Chem. B* **1997**, *101*, 4774.
92. A. Gichuhi, B. E. Boone, U. Demir et al., *J. Phys. Chem. B* **1998**, *102*, 6499.
93. T. Torimoto, S. Nagakubo et al., *Langmuir* **1998**, *14*, 7077.
94. A. Gichuhi, B. E. Boone, C. Shannon, *Langmuir* **1999**, *15*, 763.
95. I. Villegas, J. L. Stickney, *J. Vac. Sci. Technol., A* **1992**, *10*, 3032.
96. R. D. I. Herrick, J. L. Stickney in *New Directions in Electroanalytical Chemistry* (Eds.: J. Leddy, M. Wightman), The Electrochemical Society, Pennington, 1996, p. 186, Vol. 96-9.
97. R. Vaidyanathan, J. L. Stickney, U. Happek, *J. Electroanal. Chem.* **2002**, in press.
98. W. J. Danaher, L. E. Lyons, *Nature* **1978**, *271*, 139.
99. M. P. R. Panicker, M. Knaster, F. A. Kroger, *J. Electrochem. Soc.* **1978**, *125*, 566.
100. G. Fulop, M. Doty, P. Meyers et al., *Appl. Phys. Lett.* **1982**, *40*, 327.
101. B. M. Basol, *J. Appl. Phys.*, **1984**, *55*, 601.
102. R. N. Bhattacharya, K. Rajeshwar, *J. Electrochem. Soc.* **1984**, *131*, 2032.
103. S. S. Ou, O. M. Stafsudd, B. M. Basol, *J. Appl. Phys.* **1984**, *55*, 3769.
104. M. Takahashi, K. Uosaki, H. Kita, *J. Appl. Phys.* **1984**, *55*, 3879.
105. K. Uosaki, M. Takahashi, H. Kita, *Electrochim. Acta* **1984**, *29*, 279.
106. R. D. Engelken, T. P. V. Doren, *J. Electrochem. Soc.* **1985**, *132*, 2904.
107. G. Maurin, O. Solorza, H. Takenouti, *J. Electroanal. Chem.* **1986**, *202*, 323.
108. H. Minoura, M. Kitakata, T. Sugiura et al., *Bull. Chem. Soc. Jpn.* **1987**, *60*, 2373.
109. J. v. Windheim, A. Darkowski, M. Cocivera, *Can. J. Phys.* **1987**, *65*, 1053.
110. H. W. Salzberg, A. J. Andreatch, *J. Electrochem. Soc.* **1954**, *101*, 528.
111. H. G. Tompkins, *A User's Guide to Ellipsometry*, Academic Press, Boston, 1993.
112. J. H. van-der-Merwe, *Crit. Rev. Solid State Mater. Sci.* **1978**, *7*, 209.
113. T. Torimoto, A. Obayashi, S. Kuwabata et al., *J. Electrochem. Soc.* **2000**, in press.

114. T. L. Wade, B. H. Flowers Jr., K. Varazo et al., *Conference Proceedings of the Electrochemical Society*, Washington, 2001, 8.
115. A. Reisman, M. Berkenblit, E. C. Haas et al., *J. Electrochem. Soc.* **1954**, *101*, 387.
116. A. L. Pitman, M. Pourbaix, N. D. Zoubov, *J. Electrochem. Soc.* **1957**, *104*, 594.
117. L. Tomlinson, *Anal. Chim. Acta* **1964**, *31*, 545.
118. L. Tomlinson, *J. Electrochem. Soc.* **1964**, *111*, 592.
119. M. C. Hobson Jr., H. Leidheiser Jr., *Trans. Metallurg. Soc. AIME* **1965**, *233*, 482.
120. R. N. Bhattacharya, *J. Electrochem. Soc.* **1983**, *130*, 2040.
121. T. L. Chu, S. S. Chu, S. C. Lin et al., *J. Electrochem. Soc.* **1984**, *131*, 2182.
122. G. Hodes, T. Engelhard, C. R. Herrington et al., *Prog. Cryst. Growth Charact.* **1985**, *10*, 345.
123. R. P. Singh, S. L. Singh, S. Chandra, *J. Phys. D: Appl. Phys.* **1986**, *19*, 1299.
124. C. D. Lokhande, *J. Electrochem. Soc.* **1987**, *134*, 1728.
125. C. X. Qiu, I. Shih, *J. Appl. Phys.* **1988**, *64*, 758.
126. D. Pottier, G. Maurin, *J. Appl. Electrochem.* **1989**, *19*, 361.
127. B. M. Basol, *J. Vac. Sci. Technol., A* **1992**, *10*, 2006.
128. S. Menezes, *Appl. Phys. Lett.* **1992**, *61*, 1564.
129. J. F. Guillemoles, P. Cowache, A. Lusson et al., *J. Appl. Phys.* **1996**, *79*, 7293.
130. R. P. Raffaelle, J. G. Mantovani, R. Friedfeld, *IEEE Photovoltaic Specialists Conference*, IEEE, Anaheim, 29 September, 1997.
131. A. A. I. Al-Bassam, *Physica B* **1999**, *266*, 192.
132. M. Kemell, M. Ritala, H. Saloniemi et al., *J. Electrochem. Soc.* **2000**, *147*, 1080.
133. B. M. Basol, E. S. Tseng, R. L. Rod, *Proc. 16th IEEE Photovolt. Special. Conf. San Diego* **1982**, 805.
134. R. N. Bhattacharya, K. Rajeshwar, *J. Appl. Phys.* **1985**, *58*, 3590.
135. G. C. Morris, A. Tottszer, S. K. Das, *Mater. Forum* **1991**, *15*, 164.
136. S. K. Das, G. C. Morris, *J. Appl. Phys.* **1992**, *72*, 4940.
137. G. C. Morris, S. K. Das, P. G. Tanner, *J. Cryst. Growth* **1992**, *117*, 929.
138. T. Yoshida, *J. Electrochem. Soc.* **1992**, *139*, 2353.
139. S. K. Das, *Thin Solid Films* **1993**, *226*, 259.
140. S. K. Das, G. C. Morris, *J. Appl. Phys.*, **1993**, *73*, 782.
141. S. K. Das, G. C. Morris, *Sol. Energy Mater.* **1993**, *28*, 305.
142. S. Dennison, *J. Mater. Chem.* **1994**, *4*, 41.
143. D. Lincot, A. Kampmann, B. Mokili et al., *Appl. Phys. Lett.* **1995**, *67*, 2355.
144. J. D. Klein, R. D. Herrick, D. Palmer et al., *Chem. Mater.* **1993**, *5*, 902.
145. J. W. Diggle, T. C. Downie, C. W. Goulding, *Chem. Rev.* **1969**, *69*, 365.
146. G. E. Thompson, Y. Xu, P. Skeldon et al., *Philos. Mag. B* **1987**, *55*, 651.
147. A. Despic, V. P. Parkhutic, (Eds.), *Electrochemistry of Aluminum in Aqueous Solutions and Physics of its Anodic Oxide*, Modern Aspects of Electrochemistry, Plenum Press, New York, 1989, Vol. 20.
148. R. M. Metzger, V. V. Konovalov, M. Sun et al., *IEEE Trans. Magn.* **2000**, *36*, 30.
149. G. Zangari, D. N. Lambeth, *IEEE Trans. Magn.* **1997**, *33*, 3010.
150. A. M. Kressin, V. V. Doan, J. D. Klein et al., *Chem. Mater.* **1991**, *3*, 1015.
151. C. Wei, K. Rajeshwar, *J. Electrochem. Soc.* **1992**, *139*, L40.
152. Y. Sakuma, M. Ozeki, K. Kodama, *J. Cryst. Growth* **1991**, *115*, 324.
153. R. Pierret, *Semiconductor Device Fundamentals*, Addision-Wesley, Reading, 1996.
154. P. J. P. Tang, M. J. Pullin, S. J. Chung et al., *Semicond. Sci. Technol.* **1995**, *10*, 1177.

4.3
Electrocatalysis on Surfaces Modified by Metal Monolayers Deposited at Underpotentials

Radoslav Adžić
Brookhaven National Laboratory, Upton, New York

4.3.1
Introduction

The remarkable catalytic properties of electrode surfaces modified by monolayer amounts of metal adatoms obtained by underpotential deposition (UPD) have been the subject of a large number of studies during the last couple of decades. This interest stems from the possibility of implementing strictly surface modifications of electrocatalysts in an elegant, well-controlled way, and these bimetallic surfaces can serve as models for the design of new catalysts. In addition, some of these systems may have potential for practical applications. The UPD of metals, which in general involves the deposition of up to a monolayer of metal on a foreign substrate at potentials positive to the reversible thermodynamic potential, facilitates this type of surface modification, which can be performed repeatedly by potential control. Recent studies of these surfaces and their catalytic properties by new in situ surface structure–sensitive techniques have greatly improved the understanding of these systems.

In the mid-seventies, it was demonstrated that UPD metal adatoms can produce electrocatalytic effects on various electrochemical reactions. These include oxidation of small organic molecules, oxygen reduction, reactions of electroorganic synthesis, electrodeposition of metals, and charge transfers in redox couples. The oxidation of organic molecules that are potential fuels for fuel cells attracted special attention. In the electrodeposition of metals, in addition to causing catalytic effects, the UPD metal adlayers can cause dramatic improvement in the morphology of the deposits. Inhibition is also often observed, in particular, for H_2 evolution and hydration of hydride-forming metals by the adlayers of metals with high hydrogen-evolution overpotentials. Two reviews exist on the work prior to 1984 [1] and 1985 [2].

In this review, after a brief overview of the structural and electronic properties of metal adlayers, there are six sections describing catalytic effects on redox couples, oxidation of organic molecules, carbon monoxide, organic electrosynthesis reactions, hydrogen evolution, oxygen reduction, and metal electrodeposition. Outside the scope of this review are other UPD processes that play a role in determining the catalytic properties of electrode surfaces such as the UPD of H and OH.

4.3.2
Structural and Electronic Properties of Electrode Surfaces with Metal Adlayers

Structural and electronic properties of electrode surfaces covered by foreign metal adlayers (usually with submonolayer coverage) are expected to be considerably different from either those of the substrate or the adsorbate bulk metals. The structure of the top surface layer will be that of the adlayer, determined by the substrate surface structure, adatom size and coverage, substrate–adsorbate interaction and in some systems, by interaction with coadsorbed anions (see Section 3.2, Chapter 3 in this volume). Major advances in the understanding of metal adlayer structural behavior have been achieved by recent applications of in situ scanning probes [3] and surface X-ray scattering techniques [4].

In general, for high-coverage metal adlayer phases that form at potentials close to the reversible potentials of bulk deposition, incommensurate and hexagonal or quasi-hexagonal structures are common. The interatomic distances are potential-dependent, often having values below the distances in bulk materials. This phenomenon has been termed *electrocompression*. Commensurate adlayer structures are less common and usually form with coadsorbed anions.

The structural studies with single-crystal surfaces have clarified to a large extent the questions of electrosorption valences and partial charge of metal adatoms at large adatom coverages. Close to the reversible deposition potential, the adatoms are neutral species; otherwise, the electrocompression would not be observed. In low-coverage phases, electrosorption valences below unity may be expected, with the charge on the adatoms usually compensated by the coadsorbed anions.

The work function of the metal surface is considerably altered by the adsorption of foreign atoms. This is well established at the metal–vacuum interface. Pb adlayers on Ag cause a pronounced decrease in the work function with increasing Pb coverage [5]. A similar change can be expected at the electrode surface. The potential of zero charge (PZC) of the electrode surface, which is directly related to the work function, has been determined for the Pb adlayer on polycrystalline Ag. A value close to that was found for bulk Pb [6].

Optical properties of metal adlayers were extensively studied in the late seventies by reflectance spectroscopy, ellipsometry, surface plasmon excitation, and Mösbauer spectroscopy (See Refs. [1] and [2], and references therein). Pronounced changes of reflectivity are caused by the UPD adlayers, in particular, on substrates that exhibit large electroreflectance signals, namely, Au, Ag, and Cu. The optical properties of submonolayers and monolayers of metals were found to be markedly different from the properties of bulk phases. New insights into electronic properties of these systems have been obtained by application of synchrotron radiation–based X-ray absorption techniques (XAS). The X-ray absorption near-edge structure spectroscopy (XANES) measurements of Pb on Pt show that in the UPD region the Pb species are essentially neutral Pb atoms [7]. Figure 1 shows the effect of changes in potential on the normalized Pb XANES. The XANES indicate the presence of neutral species at all potentials except at 1.15 V where Pb is desorbed and the spectra indicate the presence of Pb^{2+}. There was no evidence of Pb interaction with oxygenated species in the UPD region.

The electronic properties of a monolayer of Pb [8] and Tl [9] on Ag(111) electrode surfaces have been calculated by using a density functional formalism. Calculations show that, as for metal surfaces in general, the excess charge in the electronic-density profile lies in front of the metal surface. The work function of the Tl monolayer on Ag(111) was found to be close to the bulk Tl value [9]. For a Pb monolayer, calculations predict almost the same interfacial capacity as for a surface of Pb(111). The latter result is in accord with the experimental data for polycrystalline Ag [6].

4.3.3
Electrocatalysis on Surfaces Modified with Metal Adlayers

4.3.3.1 Redox Reactions

The effects of the UPD adlayers on the rates of electrode reactions were first investigated for the Fe^{2+}/Fe^{3+} and Ti^{3+}/Ti^{4+} redox reactions on Au, which were found

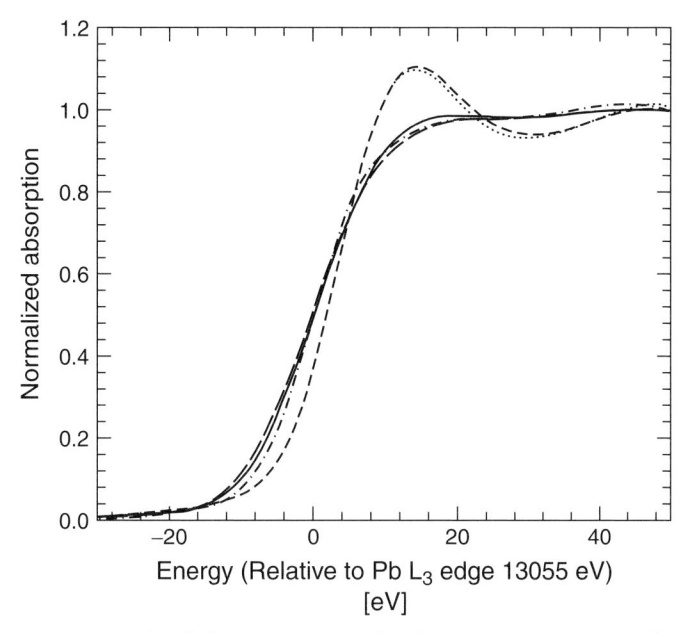

Fig. 1 Normalized Pb XANES spectra for Pb on Pt at various potentials: −0.24 V (– – –), 0.25 V (positive sweep) (———), 0.25 V (negative sweep) (·– · –·), 0.6 V (· · · ·), and 1.15 V (. . .). (Reproduced with permission from Ref. [7].)

to be slightly increased by adlayers of several metals [10]. The energy levels involved in the redox reactions are those of the adsorbates, not of the substrates, and these systems seemed ideal to provide insights into the unresolved question of the role of the nature of the metal electrode in the kinetics of the outer-sphere redox reaction. The charge transfer in these "noncatalytic" reactions takes place in the outer Helmholtz plane. Consequently, it should not be significantly affected by the metal adlayer. The data reported so far are in accord with this statement (cf. Refs. [1] and [2]), thus confirming a small role that the nature of the metal electrode has these reactions. In a recent work, Nagy and coworkers [11] took special care to remove traces of chlorides that can be coadsorbed with metal adatoms and that can increase the concentration of cations in the double layer. This effect could account for the small increase of the redox reactions by metal adlayers. Even in the absence of trace amounts of chloride, small enhancements in the Fe^{2+}/Fe^{3+} reaction rates caused by Bi on Pt and Ag on Au were found. No effect was observed for a Cu adlayer on the reaction on Au. Rhodes and coworkers [12] found an enhanced rate of a Cr^{3+}/Cr^{2+} reaction in HCl on Au by Bi, Pb, Tl, and Sn adlayers, and attributed this to the enhanced adsorption of Cl^- in the presence of adatoms. Chlorides make bridges that are necessary for this complex, inner-sphere, and redox couple reaction to occur.

Several mechanisms have been proposed to explain the small enhancement effects of metal adatoms. A change in PZC caused by metal adlayers should induce a change in the φ_2 potential and thus affect the rate of redox couples. The calculations of the φ_2 potential in 1 M H_2SO_4 solution using PZC for bulk Bi showed that the effect is too small to explain the observed rate increase [13]. The density of electronic states at the Fermi level was considered by Adžić and Despić [10] and analyzed in detail by Schmickler [14]. He suggested that the difference between the adsorbate density of states at the solvent configuration corresponding to the saddle-point of the reaction hypersurface and the density of states of the metal electrode could explain the catalytic effects. The expression for the current was derived in terms of the adsorbate density of states. Nagy and coworkers [11] reported electronic structure calculations using the extended Hückel molecular-orbital method to determine whether the magnitude of the electronic coupling in the presence of adlayers can cause catalytic effects. For the reaction at the Cu adlayer on Au, the calculation shows that the ion can approach 0.04 ± 0.02 nm closer to the Cu/Au surface than to the bare Au surface. This can increase the electronic factor by 2 to 14 times, thus increasing electronic coupling and the reaction rate, although the nature of the electronic coupling remains unchanged. Iwasita and coworkers [15] found no effect of Pb and Tl adatoms on Pt on the rate of a very fast $Ru(NH_3)_6^{2+}/Ru(NH_3)_6^{3+}$ couple. Fonseca and coworkers [16] measured the effects of Pb, Bi, and Cd on the rates of Fe^{2+}/Fe^{3+}, $Fe(CN)_6^{4-}/Fe(CN)_6^{3-}$, and $IrCl_6^{3-}/IrCl_6^{2-}$ redox reactions, and confirmed small enhancements of the rates of these redox couples.

In contrast to the simple redox reactions, the rates of the redox processes involving organic couples are considerably increased by metal adlayers on electrode surfaces. Some examples include quinhydrone, pyrocatachol/o-benzoquinonone, adrenaline/adrenalinquinone, and other quinone-type redox couples [2]. Figure 2 shows data obtained for the pyrocatechol/o-benzoquinonone system on Pt/Bi_{ad}. A considerable increase in reversibility caused by metal adlayers has been attributed by Kokkinidis [17] to the change of the reaction mechanism, from an "inner-sphere" mechanism on a Pt surface with adsorbed intermediates to an "outer-sphere" reaction on Pt/M_{ad} surfaces without adsorbed intermediates. Fonseca and coworkers [16] confirmed large effects of Pb, Bi, and Cd on the rates of several types of hydroquinone/quinone couples.

4.3.3.2 Oxidation of Organic Molecules

4.3.3.2.1 Formic Acid Catalytic effects of various magnitudes, some quite striking, are caused by monolayers of Pb, Bi, Tl, Cd, Sn, Sb, Se, and Ge on the oxidation of organic molecules that are potential fuels for fuel cells, namely, HCOOH, CH_3OH, HCHO, CH_3CHO, CO, glucose and other monosaccharides, ethylene glycol and some aliphatic alcohols on Pt, as well as on some other platinum metals (see reviews of previous work in Refs. [1, 2]). Slow rates are characteristic of the oxidation of these organic molecules even with the best catalysts because of the generation of strongly bonded intermediates – poisoning species – that reduce their activity. Upon adsorption, dehydrogenation and subsequent oxidation of adsorbed hydrogen are fast reactions that leave strongly bonded fragments on the surface. Their oxidations require very positive potentials that are impractical for fuel cell applications. The UPD metal adatoms can reduce the poisoning effects by several mechanisms (vide infra), thus causing the catalytic effects.

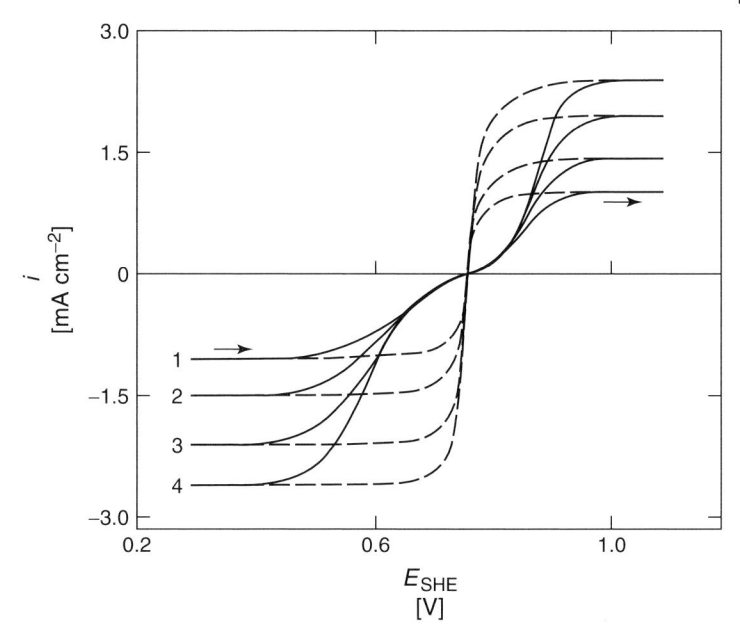

Fig. 2 Current–potential curves for a mixture of pyrocatachol (1 mM) and o-benzoquinone (1 mM) on Pt and Bi/Pt rotating disk electrodes in 0.5 M $HClO_4$ with 1 mM $Bi(ClO_4)_3$. Sweep rate 10 mV s^{-1}. Rotation frequency (1) 12.5, (2) 25, (3) 50 and (4) 75 Hz. (Reproduced with permission from Ref. [2].)

There is a consensus regarding the mechanism of the oxidation of formic acid on Pt, which involves a dual-path mechanism [18, 19] with direct oxidation to CO_2, and an indirect mechanism through a blocking intermediate:

$$\text{HCOOH} \begin{array}{c} \nearrow \text{reactive intermediate} \longrightarrow CO_2\ (E_1) \\ \\ \searrow \text{blocking intermediate} \longrightarrow CO_2\ (E_2) \end{array} \quad (1)$$

where $E_2 > E_1$. An indication for the dual-path mechanism is provided by isotope-labeling experiments of Heitbaum and coworkers [20, 21]. Strong evidence for the reactive intermediate branch comes from infrared (IR) spectroscopy data showing generation of CO_2 at potentials where CO is not oxidized [22]. On the basis of a kinetic isotope effect for the oxidation of HCOOH and DCOOH on Pt(111) and Pt(100), it was concluded that the rate-determining step is H–C bond scission [23].

The nature of the strongly bound intermediate in the oxidation of formic acid, methanol, and formaldehyde has been a long-disputed issue. CO appears to be the principal poisoning adsorbate, while other adsorbates, such as COH and CHO, are difficult to detect. These species appear less stable than CO and may be slow-reacting intermediates [24]. In the oxidation of HCOOH on Pt, COH and/or CHO were proposed as poisoning species on the basis of coulometric determination of a number of electrons per site (e.p.s.)

used for the oxidation of adsorbate [18, 19], differential electrochemical mass spectrometry (DEMS) [20], and radiotracer measurements [25]. Bewick and coworkers [26] employed electrochemically modulated IR spectroscopy (EMIRS), which clearly identified CO as the poisoning species. This was strongly supported by further work [27]. Two e.p.s. were found for several single-crystal surfaces, which also suggest that CO is a poison [24, 28]. More details on this topic can be found in reviews in Refs. [29–31].

The mechanism of the catalytic action of metal adlayers in the oxidation of organic molecules has been interpreted by using several models including a "third-body" effect, an electronic effect, and the bifunctional catalyst mechanism.

1. Third-body effect: A third-body effect, which is equivalent to an "ensemble" effect in heterogeneous catalysis, is based on the role of metal adatoms in blocking the surface sites for a side reaction that generates poisoning species, or in blocking the adsorption of the inhibiting species, which requires more than one surface site for adsorption. The reaction in the main pathway can proceed at the unoccupied sites that now form smaller ensembles. In this model, adatoms do not enhance the rate on the uncovered substrate sites. In electrocatalysis, the effect was proposed by Conway and coworkers [32] to interpret the effect of acetonitrile on formic acid oxidation on Pt. The third-body effect is operative for any adsorbate statistically distributed on the surface that is not desorbed upon adsorption of reacting species and intermediates of reactions.

2. Bifunctional mechanism: The bifunctional mechanism, often used in catalysis, has been considered in electrocatalysis by Cathro [33] and Shibata and Motto [34]. The surface is considered to have two types of sites that have distinct roles in the reaction. The substrate, such as Pt, breaks the bonds in organic molecules upon adsorption, while the adatom can adsorb oxygen-containing species that can oxidize strongly adsorbed intermediates such as CO. Oxygen adsorption on adatoms is expected to occur at potentials negative of oxygen adsorption on Pt. This facilitates CO oxidation at potentials more negative than on bare Pt, thus freeing the surface for the main reaction.

3. Electronic effects: Electronic effects can result from the modifications of the properties of uncovered near-neighbor sites and next near-neighbor sites. The adatoms modified in the interaction with substrates, may, per se, take part in reactions.

Striking catalytic effects observed in the oxidation of formic acid on Pt modified by Pb and Bi adlayers are shown in Fig. 3 [35]. Currents in anodic sweeps are enhanced by almost two orders of magnitude, which is also observed in steady state measurements [35–37]. Figure 4 shows quasi-stationary curves for the Pb/Pt system as a function of Pb coverage. (Metal adatom coverages are defined in two ways throughout the literature reviewed in this text. According to one definition, the coverage is defined as the ratio between the number of covered and total number of substrate atoms. The coverage is also defined as a number of atoms of adsorbate per one atom of substrate. The latter definition is being used more often. For large

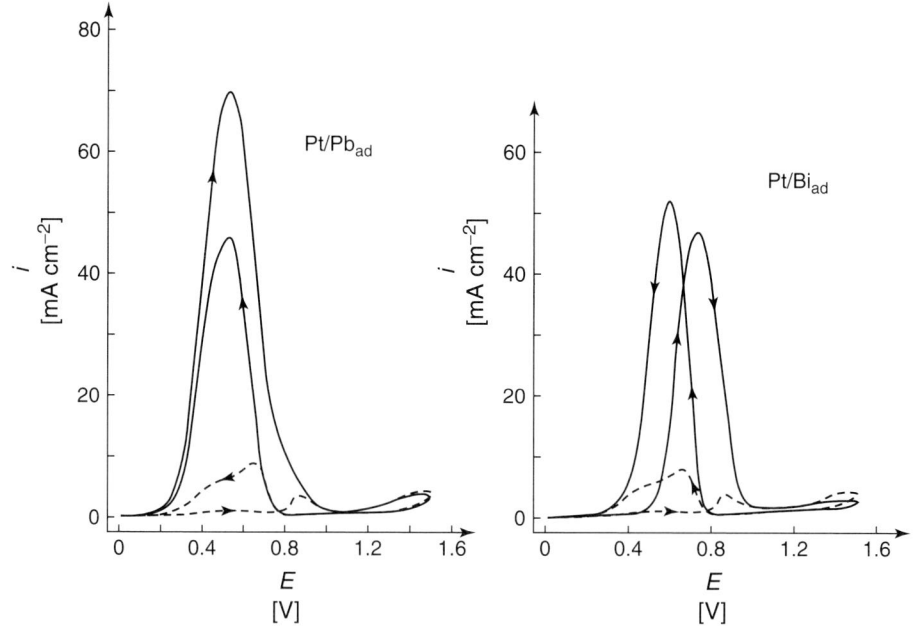

Fig. 3 Oxidation of HCOOH (0.26 M) on Pt with Pb and Bi adlayers (full lines) and on bare Pt (dashed line) in 1 M $HClO_4$. Lead concentration 1 mM; sweep rate 50 mV s^{-1}. (Reproduced with permission from Ref. [35].)

adatoms, the difference in the coverage values from the two definitions can be very large. In order to simplify comparison of results and to preserve the link to the original work when the coverages are given according to the first definition, approximate coverages calculated by using the second approach are appended in parentheses.) For the optimum coverage of $\theta_{Pb} = 0.73(0.36)$ the enhancement is about three orders of magnitude as calculated per Pb-free Pt surface. Similar effects were observed with Tl adlayers [35, 38, 39], but the effects of Cd are considerably smaller [35]. Cd was partially desorbed by HCOOH and reaction intermediates, while the Pb coverage was unaffected, as determined by reflectance spectroscopy measurements [40]. In accordance with this observation, IR measurements show no effect of Cd on CO adsorbate on Rh in HCOOH oxidation, while Pb decreased the coverage of the bridge-bonded CO [24]. Catalytic effects were observed with other platinum metals, namely, Rh, Pd, and Ir [41]. Smaller enhancements were observed than with Pt, the smallest being observed with Pd [42], which is the least poisoned Pt metal. The catalytic effects of less common UPD systems such as Se, Ge, As, and Te [43–45], Hg [38], Ag, and Cu [1] on Pt have also been reported and are smaller than those observed with Pb and Bi.

The results of the studies with single-crystal electrodes [46–51] have demonstrated the existence of a structural dependence of the magnitude of the catalytic effects. The role of the adlayer structure in determining the observed dependence has not been clearly determined so far. In most studies, the adlayer

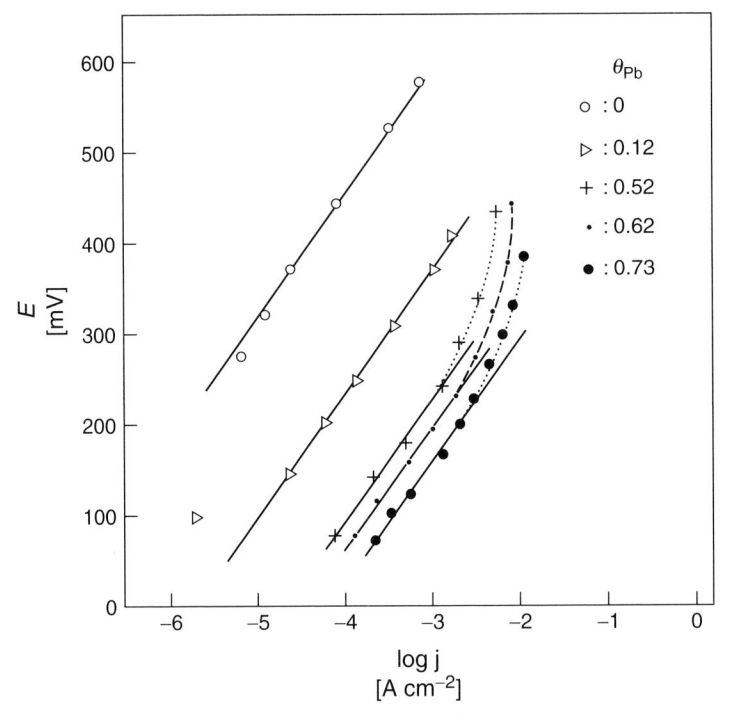

Fig. 4 Quasi-stationary current–potential plots for HCOOH (0.5 M) oxidation on sputtered Pt modified by Pb adlayer and HCOOH adsorbates in 0.1 M HClO$_4$. Current taken 2 min after stepping the potential. The current density is given with respect to the free-Pt surface (16% of the whole surface). (Reproduced with permission from Ref. [37].)

structures were derived from coverages obtained from voltammetry. It is, however, necessary to determine adlayer structures during the course of HCOOH oxidation because of the sensitivity of adatom ordering to the interaction with other adsorbates. For instance, in situ X-ray scattering techniques show that the ordered (3 × √3) Pb adlayer on Pt(111) vanishes during HCOOH oxidation, but the Pb coverage remains unaffected [52]. Further work is needed to determine whether the adlayer structure has an influence on the magnitude of the catalytic effect or whether only the adatom coverage is important.

An extension of the UPD modification of the electrocatalytic properties of electrode surfaces is a process described as an "irreversible" adsorption of metal adatoms on electrode surfaces. These adsorbates can be obtained by adsorption from cation-containing solution without the application of external potential. Janssen and Moolhyusen [53] demonstrated that immersion of clean Pt surfaces in solution containing metal salts, and subsequent emersion and removal of the excess solution, could produce adlayers of metal adatoms. Clavilier and coworkers [48, 51] used this procedure with several metal adsorbates on Pt single-crystal electrodes. These "preadsorbed" layers are quite stable. It is, however, necessary to control the electrode potential, since the adlayer coverage is potential-independent only within certain potential limits. In some cases, as for Bi on Pt, it is believed that the adsorbate changes oxidation states with potential and that therefore both species interact with a surface-forming adlayer [54]. Ex situ X-ray photoabsorption spectroscopy (XPS) data show that the charge should be attributed to OH^- adsorption rather than to a change in the valence state of adsorbed Bi [55].

There is no satisfactory description of the "irreversible" adsorption process. It was ascribed to chemical reaction with adsorbed hydrogen [53] and to a local cell mechanism through PtOH formation [56]. The interaction of cations with the surface M−OH species should also be considered as a possible mechanism, since it is known that they can strongly interact with oxidized surfaces [57, 58]. This was observed even at very positive potentials, as in the case of the oxide monolayers on Au and Pt [59]. Except for Bi, generally small catalytic effects of the preadsorbed adatoms on Pt single-crystal electrodes were reported usually at positive potentials for Sb [60], As [61], Sn [62], Se [63], and Te [64]. An analysis of metal adlayers causing the most interesting effects is presented below.

Lead Pronounced effects of Pb (Fig. 3) were interpreted in terms of a third-body effect by Adžić and coworkers [35]. Pb adatoms suppress adsorption of hydrogen and strongly bound intermediates, in particular, those interacting with two or three surface sites. The same model was assumed for Bi and Tl. Adsorbed H was considered to take part in the formation of strongly bound intermediates [18, 19]. Current responses to potential sweeps into the H adsorption region [18, 19, 65] and crystal quartz microbalance measurements [66] provide evidence of fast poison accumulation in the presence of H_{ads}. An analysis of the rates for Pb modified Pt showed that, for adatoms covering evenly half of the sites, there are no multibonded poisoning species adsorbed [41]. In agreement with the experiment, the model predicted the maximum activity at $\theta_M = 0.5(0.25)$. Pletcher and Solis [36] also found maximum activity at $\theta_M = 0.5(0.25)$. Shibata and coworkers [44] found evidence that the adatom size determines the effect that increases with the number of occupied sites. The most pronounced effect was found for Bi, the largest adatom. Pb caused the largest effects according to the data in Refs. [35, 36]. Xia and Iwasita [67] used DEMS combined with the ^{13}C labeling to make a distinction between CO_2 formed from bulk solution and from adsorbed species (Fig. 5). ^{12}C-formic acid oxidation was observed in the presence of a known Pb coverage and the adsorbate from ^{13}C-formic acid. Pb catalyzes the HCOOH oxidation from bulk solution on a bare Pt surface, and also helps in the oxidation of HCOOH adsorbates. The HCOOH oxidation could not take place for Pt completely covered by Pb and HCOOH adsorbates. The electronic interaction of Pb and Pt was assumed to take place in order to explain the catalytic effects. The maximum catalytic effect was found at $\theta_{Pb} = 0.5(0.25)$ for oxidation potentials >250 mV and $\theta_{Pb} = 0.8(0.4)$ for potentials <250 mV. At $\theta_{Pb} = 0.84(0.42)$, no adsorbate formation is possible, but the oxidation of bulk acid still takes place. The increased activity of the remaining bare Pt has been interpreted by reaction of HCOOH with H_2O modified in the interaction with Pt−Pb sites [68]. This activity is still lower than the intrinsic activity of Pt(100) determined by pulsed voltammetry [69]. Kita and coworkers [50] reported

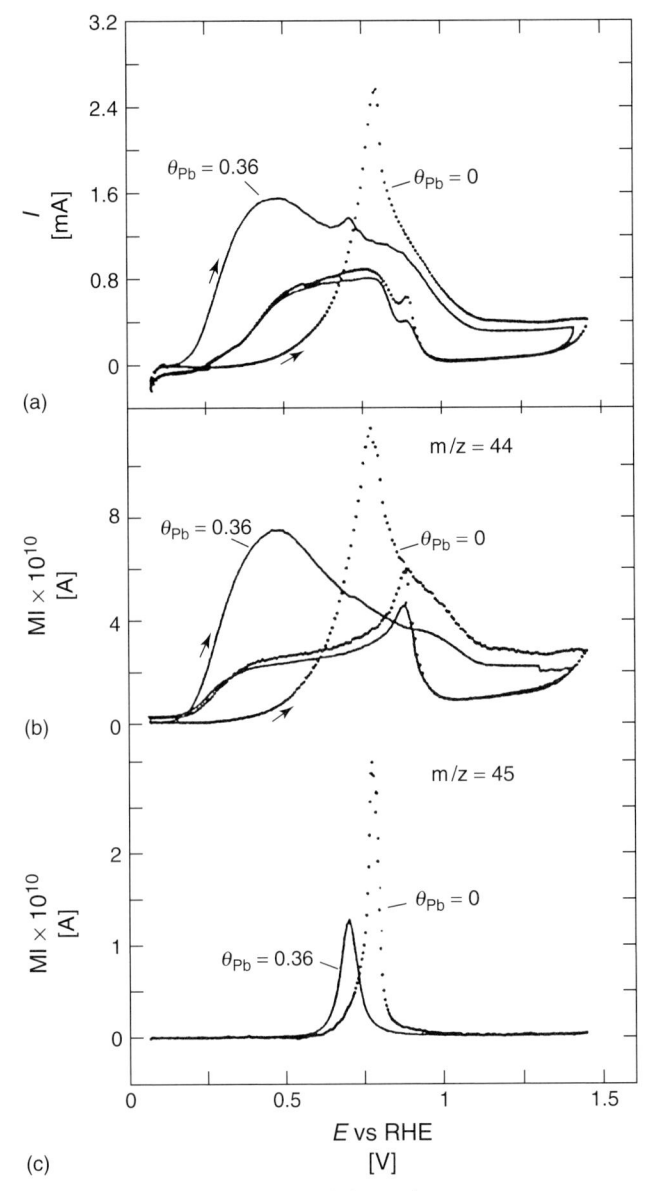

Fig. 5 Current (a) and mass signal (b,c) voltammetry curves in a 0.01 M $HClO_4$ solution containing 5 mM $H^{12}COOH$: (———) for sputtered Pt previously with modified $\theta_{Pb} = 0.36$ and $\theta_{H^{13}COOH} = 0.49$; (. . .) only $H^{13}COOH$ was adsorbed ($\theta_{HCOOH} = 0.84$); sweep rate 100 mV s^{-1}. (Reproduced with permission from Ref. [37].)

almost the same magnitude for the catalytic effects of UPD Pb and preadsorbed Pb on Pt(111). The maximum effect was observed for a Pb coverage of 0.29, which almost completely prevented the adsorption of CO. These authors also indicate that a third-body effect and a bifunctional mechanism cannot fully explain the observed effects. To summarize, for Pb, a part of the observed catalytic effect is due to a third-body effect, but there is a growing evidence for the electronic effect to be partly responsible for the catalysis.

Bismuth Large effects of Bi have been found in several studies [35, 44]. Clavilier and coworkers [48, 51] showed that Bi preadsorbed on Pt(111) enhances the kinetics of HCOOH oxidation in a large interval of coverages from 0 to 0.8(0.27) at potentials below 0.6 V. The accumulation of strongly bound intermediates is suppressed at very small Bi coverages. This suggests that the next-nearest neighbor's Pt atoms must be affected by Bi and that the third-body effect cannot account for the inhibition of the poison formation. At high Bi coverages, large currents are measured at potentials where Bi adsorbs OH [55]. Therefore, in that potential region, it could act through the bifunctional mechanism. Figure 6 shows voltammetry curves for Pt(111) with $\theta_{Bi} = 0.82(0.27)$ and the oxidation of HCOOH on that surface [48, 51]. Weaver and coworkers [70] reported an IR study of this system that shows small CO coverages on Pt(111), even in the absence of Bi. This surface is the least poisoned surface from the low-index planes. The optimal coverage of Bi was 0.2, which was interpreted in terms of an ensemble, that is, a third-body effect. An enhancement factor of approximately two orders of magnitude is cited in both works [48, 51, 70] and is similar to that observed with the UPD Bi adlayer [35].

Two roles were assumed for Bi on Pt(100) [48, 51]. First, Bi inhibits formation of poison through a third-body-type effect. The poison coverage decreases linearly and becomes zero at $\theta_{Bi} = 0.45(0.22)$. Second, the highest HCOOH oxidation rates are observed at high Bi coverage, $\sim\theta_{Bi} = 0.9(0.45)$, which indicates a direct role of Bi in the oxidation of HCOOH, that is, an electronic effect is operative in this system. The maximum rate is, however, still lower than the intrinsic rate on Pt(100). Bi forms on Pt(100) a c(2 × 2) adlayer with a coverage of $\theta_{Bi} = 0.5$, with adatoms residing at bridge positions [71, 72]. Therefore, at $\sim\theta_{Bi} = 0.45$, a small number of Pt sites apparently remains uncovered and a direct participation of Bi in the reaction appears likely. Weaver and coworkers [70] showed by IR spectroscopy a complete cessation of CO formation at $\theta_{Bi} = 0.2$, which is in good agreement with Refs. [48, 51]. The catalytic effect of Bi was ascribed to the attenuation of the degree of CO poison formation on Pt(100), which the authors interpreted in terms of a third-body effect.

Recent work involved stepped Pt surfaces with controlled adsorption of Bi on step edges [73]. For the stepped single-crystal surfaces Pt(554), Pt(332), and Pt(221), with nine-, five-, and three-atom wide terraces, respectively, the enhancement factor for HCOOH oxidation increases as the terrace width decreases. Bi appears to block adsorption of poison on reactive (110) oriented step sites and decreases the reaction ensemble size, which increases the rate on narrow (111) terraces.

Leiva and coworkers [74] reported the simulation of the catalytic effects of several adatoms by using the model in which both adatoms and poisoning species block one

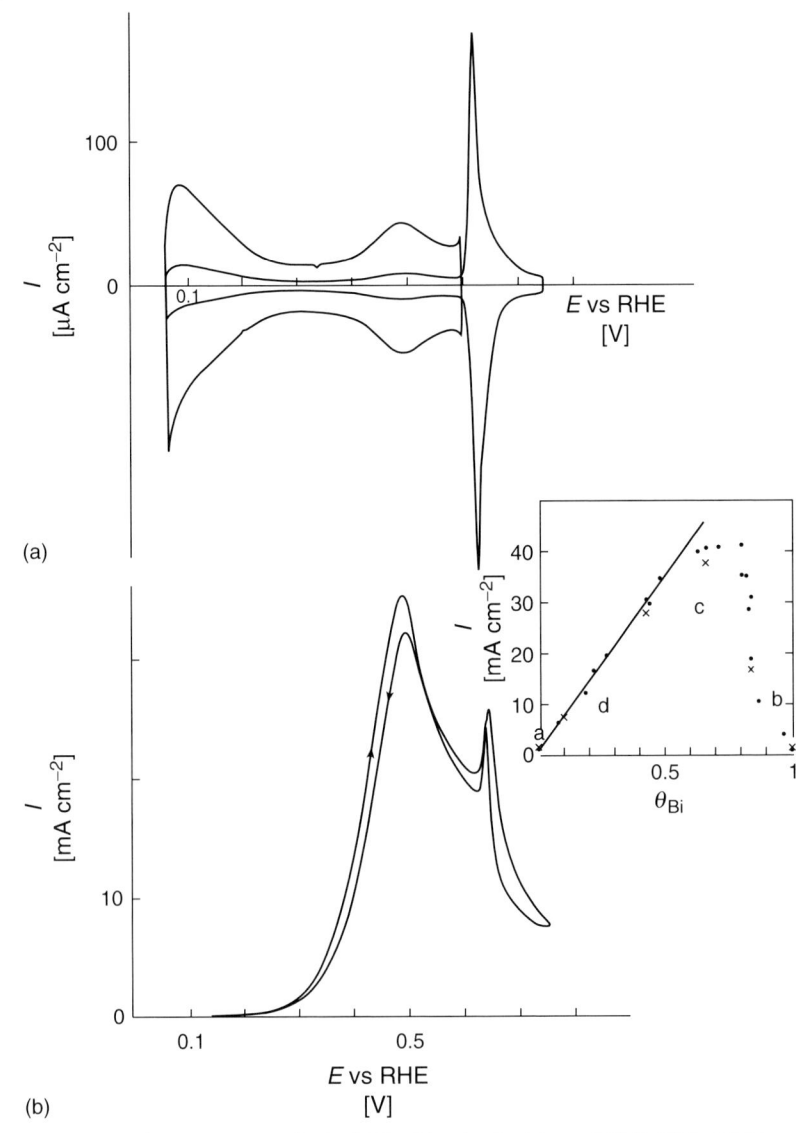

Fig. 6 (a) Voltammetry curve for Pt(111) with adsorbed $\theta_{Bi} = 0.82(0.41)$ in 0.5 M H_2SO_4. Current multiplied by 5 for short sweep. (b) First sweep for the oxidation of 0.25 M HCOOH. Insert: Peak currents at 0.5 to 0.55 V versus Bi coverage, (●) anodic and (×) cathodic sweeps. (Reproduced with permission from Refs. [48, 51].)

surface site. Eight and six near neighbors covered by adatoms protect one site from poison adsorption on square and hexagonal lattices, respectively. The adatoms are assumed to be adsorbed randomly at 50 × 50 lattices up to the coverage of θ_M, and then the poison is adsorbed up to θ_p. The oxidation currents are calculated for such a configuration for several catalytic cases. In the Pt(111)–Bi system,

Bi–Pt pairs were assumed to be responsible for the catalytic effect, without poison formation. Figure 7 shows a comparison between the experiment and the calculated curve. The agreement is surprisingly good considering the crude assumptions that adatoms cover one surface site and that they are randomly distributed. Bi is known to inhibit three H adsorption sites on Pt(111) [75]. Nevertheless, the simulation data seem to corroborate the electronic effect for the Bi–Pt system. The data for Pt(111) and Pt(100) revealed the role of substrate in determining the effect of Bi in suppressing poison adsorption (the electronic effect for Pt(111), but a third body for Pt(100)). In addition, on the same surface (Pt(100)), Bi can suppress poison formation by a third-body effect and can catalyze the oxidation of HCOOH through an electronic effect.

Antimony The effect of Sb adlayers on Pt(100) [60] were ascribed to the suppression of poison formation, and the maximum oxidation current is obtained at high coverage, $\theta_{Sb} = 0.9(0.3)$, when no poison is detected at the surface. Substantial currents are observed for this surface. The maximum activity, however, does not surpass the intrinsic activity of Pt(100). A third-body effect was found operative for this system. Kizhakevariam and Weaver [76] found that Sb inhibits the adsorption of CO on Pt (100) and Pt (111) by decreasing the twofold binding geometries deduced from the relative ν_{CO} band intensities. The optimum catalysis for

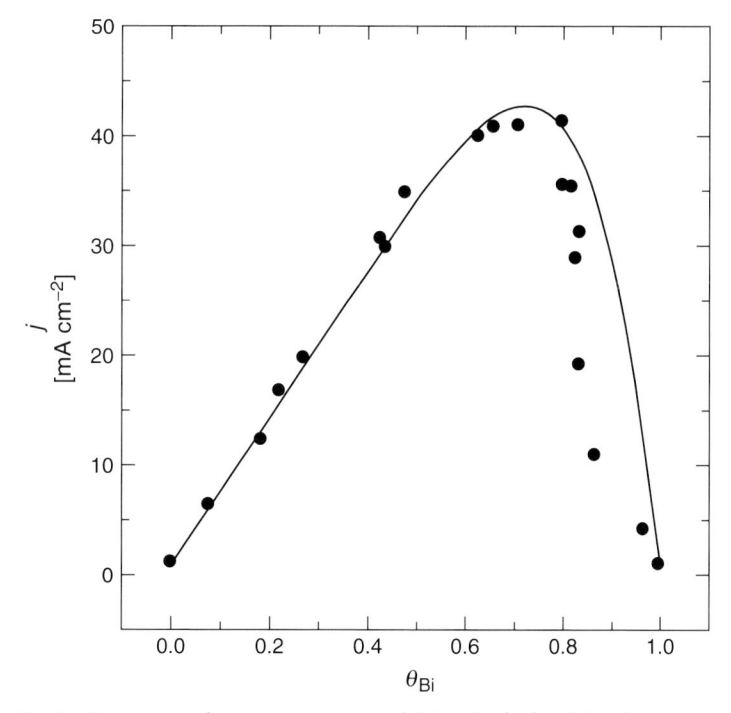

Fig. 7 Comparison between experimental (•) and calculated (——) currents versus θ_{Bi} curves for the oxidation of HCOOH 0.25 M on Bi/Pt(111) in 0.5 M H_2SO_4 (see text for details). (Reproduced with permission from Ref. [74].)

Pt(100) is observed at $\theta_{Sb} = 0.35$. Sb causes significant up-shifting of the CO band formed from HCOOH oxidation, which suggests that dissociative chemisorption is triggered at adjacent sites, electronically modified by Sb. Figure 8 shows IR spectra for Pt(100) and Pt(100) with an Sb adlayer ($\theta_{Sb} = 0.25$) and the corresponding oxidation current versus θ_{Sb} and θ_{CO} curves. The enhancement has been interpreted in-line with third-body and electronic effects [76]. Good agreement was found between the simulated effects and experimental data for the Sb–Pt(100) system in which Sb was assumed to obey the third-body effect [74].

Noble metal monolayers The systems involving a deposition of ultrathin layers of noble metals on noble metal substrates, such as Ru on Pt [77] or Pd on Pt [78] or on Au [79], exhibit enhancements of the oxidation of organic molecules. These systems are outside the scope of this review, and will be only mentioned here. In general, deposition of such layers occurs at overpotentials, and they, unlike the UPD adlayers, cannot be dissolved except at extreme positive potentials. Usually, such deposition produces 2D or 3D clusters, but pseudomorfic deposits can also be formed, as in the case of Pd on Au(111) [80, 81] and Pt on Au(111) [82]. Ru can be "spontaneously" deposited in an electroless process on Pt [83, 84]. Ru layers catalyze oxidation of methanol on Pt [83, 84], and Pd enhances the oxidation of HCOOH on Au [79] and Pt [85] single-crystal electrodes. The activity of the Pd-modified surfaces depended markedly on the Pd layer thickness and on their crystallographic orientation.

4.3.3.2.2 **Methanol** Methanol is considered the most attractive fuel for fuel cells and its reaction has been thoroughly investigated on surfaces modified by metal adatoms. The overall reaction involves the exchange of six electrons per molecule:

$$CH_3OH + H_2O \longrightarrow CO_2 + 6H^+ + 6e^- \quad (2)$$

This is a multistep reaction, which probably proceeds through a sequential dehydrogenation that can produce several intermediates, as indicated in the general scheme proposed by Bagotzky and coworkers [86]. Details of the dehydrogenation of methanol on Pt have not been established. From kinetic isotope measurements of CD_3OH and CH_3OH oxidation, it was shown that the first removal of H in the reaction

$$CH_3OH \longrightarrow CH_2OH + H^+ + e^- \quad (3)$$

is the rate-determining step [87, 88].

For methanol, more than for formic acid, evidence points out that besides CO there is at least one additional adsorbate – the COH species. Three e.p.s. values are found often for adsorbate oxidation and show that it has to contain an H atom [89]. Beden and coworkers [90, 91] reported from IR measurements the linear, bridge-bonded CO and CHO-like species or species with a carbonyl bond. By using subtractively normalized Fourier transform infrared spectroscopy (SNFTIRS), Iwasita and coworkers [92] have determined absolute absorption bands for linear-, bridge-, multibonded CO and COH adsorbates. Nichols and Bewick [93], by using a flow IR-electrochemical cell, identified COH and CH_xOH species as reactive intermediates. Recent results with DEMS indicate predominantly a mixture of CO and COH, or only CO as adsorbed intermediates [89, 94, 95]. Model calculations of methanol oxidation on Pt(111) that provide a good fit to

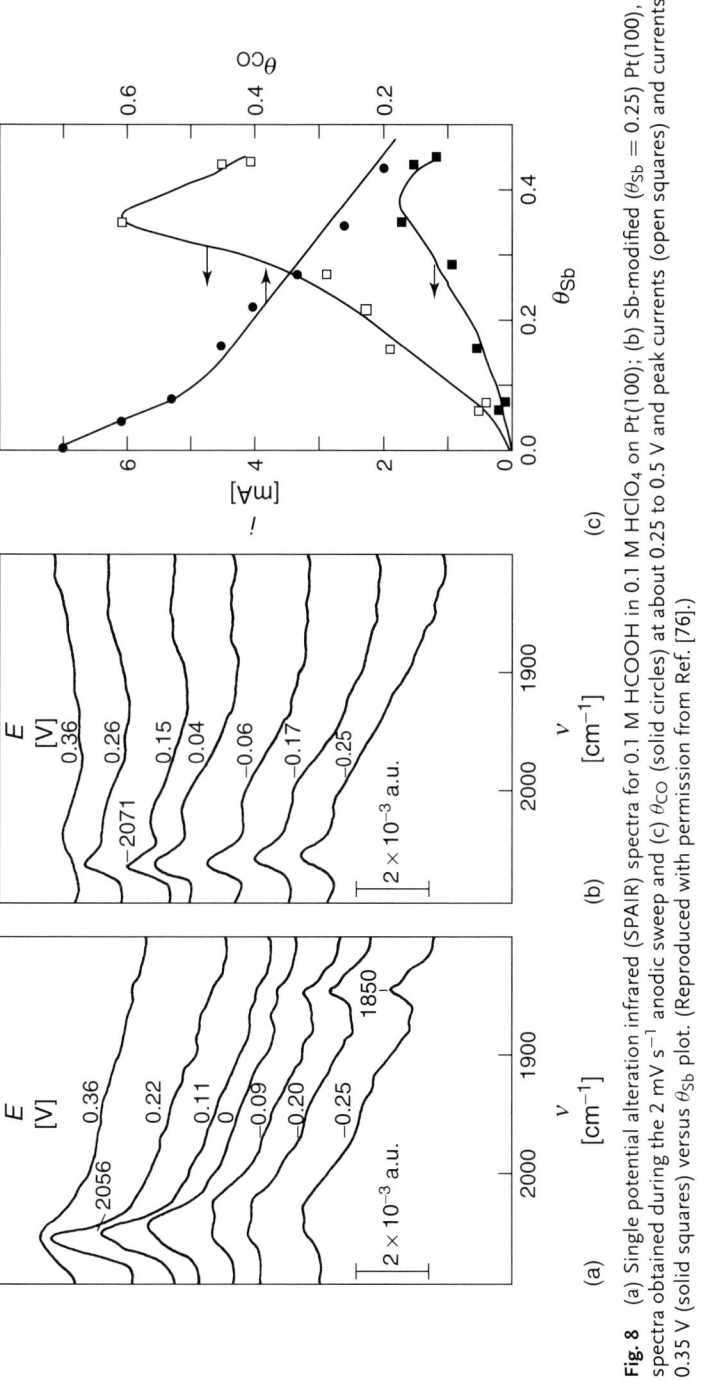

Fig. 8 (a) Single potential alteration infrared (SPAIR) spectra for 0.1 M HCOOH in 0.1 M HClO$_4$ on Pt(100); (b) Sb-modified ($\theta_{Sb} = 0.25$) Pt(100), spectra obtained during the 2 mV s^{-1} anodic sweep and (c) θ_{CO} (solid circles) at about 0.25 to 0.5 V and peak currents (open squares) and currents at 0.35 V (solid squares) versus θ_{Sb} plot. (Reproduced with permission from Ref. [76].)

the experimental data without adjustable parameters involve a reactive intermediate with the stoichiometry H : C : O [96].

Negligible catalytic [2, 97, 98] and predominantly inhibition effects [99] are observed for methanol oxidation on Pt in acid solutions. The exceptions are the effects of Sn [53, 100–102] and Ge [101, 102]. Significant catalytic effects have been observed for Pb, and Bi on Pt in alkaline solutions [103, 104]. However, carbonization of the alkaline electrolyte with CO_2 makes these systems unsuitable for practical applications. For single-crystal Pt(100) and Pt(111) surfaces, the reaction has been found to be inhibited by the Bi adlayer [49, 105].

Some agreement exists on the effects of Sn, although the results are not reproducible. The UPD of tin differs from the majority of the UPD systems [106]. Its oxidation state is extremely potential-dependent. The effects of Sn on the oxidation of CH_3OH on Pt are attractive because they occur at very low potentials. The system Pt–Sn has been investigated in three forms: Sn irreversibly adsorbed, Sn adsorbed by UPD, and ordered single-crystal Pt_3Sn alloy [107]. Vassiliev and coworkers [100] found a Temkin adsorption behavior of neutral Sn at Pt in the potential region from 0 V to 0.2. Above 0.2 V, tin undergoes a partial oxidation to Sn^+, or SnOH at ~ 0.5 V. A tenfold increase of the rate in the presence of Sn has been found and has been ascribed to the bifunctional mechanism [97]. Frelink and coworkers [108, 109] demonstrated by ellipsometry that a submonolayer of Sn oxide disappears from Pt surfaces in the presence of methanol. The reports of considerable catalytic effects of Sn [53, 101, 102] have not been reproduced in recent studies [24, 105, 110]. Inhibition was observed for the methanol oxidation on Pt(111) and Pt(100) with Sn adlayers [105], and for Pt_3Sn alloys [107]. Janssen and Moolhuysen [53] concluded that Pt is modified through a ligand, that is, through an electronic effect, while Motoo and coworkers [111] proposed that a bifunctional mechanism is operative for Sn. Bittins-Cattaneo and Iwasita [110] found that Sn(II) species are responsible for the catalytic effect, which can form a hydroxy-complex such as $Sn(OH)^+$ that can transfer O to the organic residue. The activity was ascribed to a suitably oriented adsorbed H_2O on a Pt/Sn surface. Figure 9 shows a linear sweep voltammetry and mass signal of CO_2 for Pt and Pt with adsorbed Sn. The observed enhancement of methanol oxidation is in the middle range of those reported so far. Molecular orbital studies demonstrated that Sn atoms on Pt(111) can weaken CO adsorption [112], but subsequent work showed that Sn in the Pt surface does not attract H_2O or activate the formation of the OH_{ads} species that can oxidize CO intermediates [113]. The data on Sn indicate that its state on Pt surfaces is highly dependent on surface structure and on the method of Sn deposition, which determine its catalytic properties for CH_3OH oxidation. Catalytic effects are often observed with dispersed Pt, while inhibition is usually observed with flat surfaces. While the potential for practical application of Sn-modified Pt for methanol oxidation is small, this system provides a strong support for the importance of bifunctional catalysts for methanol oxidation.

4.3.3.2.3 **Carbon Monoxide** Carbon monoxide is a model adsorbate for studying molecular adsorption in catalysis and electrocatalysis. It is a major poisoning species in the oxidation of small organic

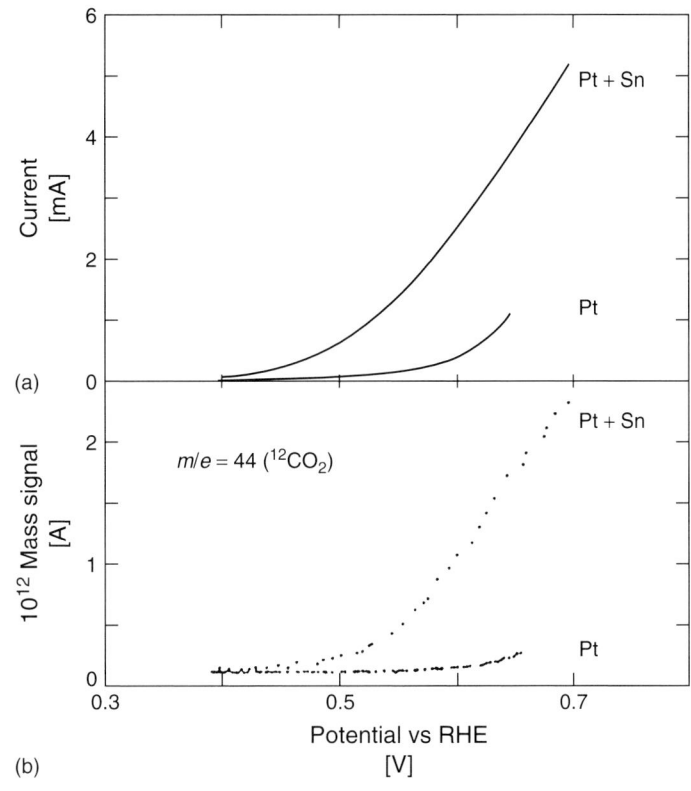

Fig. 9 Oxidation of 1 M CH$_3$OH in 0.5 H$_2$SO$_4$ on Pt and Pt with Sn adlayer ($\theta_{Sn} = 0.34(0.17)$). (a) Current and (b) mass intensity during a potential scan of 10 mV s^{-1}. (Reproduced with permission from Ref. [102].)

molecules, which is an additional motivation for studying its oxidation. The reaction on metal adlayer–covered surfaces can reveal information useful for the design of new electrocatalysts for the oxidation of methanol and reformate hydrogen. Metal adlayers affect the adsorption behavior of CO and can cause small enhancements of its oxidation. CO interacts so strongly with Pt that it can displace many metal adlayers. Lukas and coworkers [71, 72] demonstrated by SXS studies that CO causes displacement of Cu and Pb adatoms from Pt(100) and Pt(111) surfaces, while a c(2 × 2)-Bi structure on Pt(100) is unaffected by CO in solution phase. Motoo and coworkers have reported pronounced effects of Sn and Ge [101, 102], and As [114] on the oxidation of dissolved CO on Pt. Sn causes a shift of CO oxidation on Pt by 400 mV to more negative potentials. This effect was explained in terms of the bifunctional catalyst. The same mechanism was proposed for the effects of Pb, As, Sb, and Bi [115], although there is no evidence for oxygen coadsorption with Pb, while a weak evidence was reported for the other three metals.

Preadsorbed Bi adatoms were found to preferentially block CO adsorption at bridge sites, and at saturation Bi coverage, $\theta_{Bi} = 0.2$, only linear CO is observed on

Pt(111) [116]. Bi causes a decrease of CO oxidation [117] on Pt(100), which agrees with its stabilization of this adsorbate deduced from in situ IR studies [118, 119], while As has an opposite effect [119]. Both adatoms modify the CO-stripping process with a small catalytic effect observed at very positive potentials (0.7 V). The catalysis has been ascribed to the adatom-mediated oxygen transfer with a possible electronic effect for As [119]. Similar behavior is observed with Sb, which decreases the adsorption of CO on Pt(100) and Pt(111) surfaces by removing the twofold bridging geometry, as deduced from the relative ν_{CO} band intensities [76]. At the saturation Sb coverage, only linear CO is observed.

Baltruschat and coworkers [120] reported a large negative shift of CO oxidation to 0.25 V caused by Sn adsorbed at step sites on Pt (332) and Pt(755) stepped surfaces (Fig. 10). Sn adsorbs preferentially at the step sites, which can be deduced from the inhibition of H adsorption on the step sites from voltammetry curves [121]. The effect has been ascribed to a destabilization of the CO molecules due to repulsion between CO and Sn. This repulsion shifts the CO molecules into a so-called "high-coverage" state, which can be oxidized at low potentials as shown for Pt–Sn alloy surfaces [122]. (This state is not formed in oxidation of methanol on this alloy, which was offered as an explanation of the inhibition effect observed in that reaction.) The CO oxidation rates at low potentials are, however, very sluggish. The effect of Sn on the reaction on Pt(111) was found to be negligible. The activity of stepped Pt surfaces is certainly related to the observed effect of Sn on polycrystalline Pt [97], which usually has a high density of steps.

4.3.3.2.4 **Other Organic Molecules** In addition to the oxidation of formic acid and methanol, the oxidation of a considerable number of organic molecules is catalyzed by metal adlayers on Pt [1, 2]. These include formaldehyde [123], various aliphatic alcohols including ethyl, propyl, and butyl alcohols, ethylene glycol and glycerol [104], propane-diol [124–126], ascorbic acid [127], glucose and other monosaccharides [2, 128], propane [129] and lactonic acid [130]. Some of the observed effects are very large, as seen in the oxidation of HCHO on platinum metals [1, 123]. In general, the effects observed seem to be caused by reduction of adsorption of strongly bonded intermediates. Large inhibition was observed for the oxidation of hydrazine on Pt (2). The reactions of highly oxygenated molecules such as squaric, croconic, rhodizonic acids and tetrahydroxy-p-benzoquinone on Pt were also found to be affected by metal adlayers. Bi and Se were found to promote the cleavage of these molecules on Pt(111) by the third-body effect.

4.3.3.3 Electroorganic Synthesis

Electroorganic synthetic reactions on surfaces modified by metal adlayers have not received as much attention as the oxidation of small organic molecules, although it is quite conceivable that this type of surface modification can have interesting catalytic effects and can affect the selectivity of various synthetic pathways. In addition to the elimination of poisoning effects from chemisorbed organic fragments, another useful effect of metal adatoms is an increased potential window in which the electrode materials such as Pt can be used for such reactions. At potentials where the modified electrodes can be used, the electrodes of metals undergoing the UPD, such as of Pb, Bi, or Tl, are not stable.

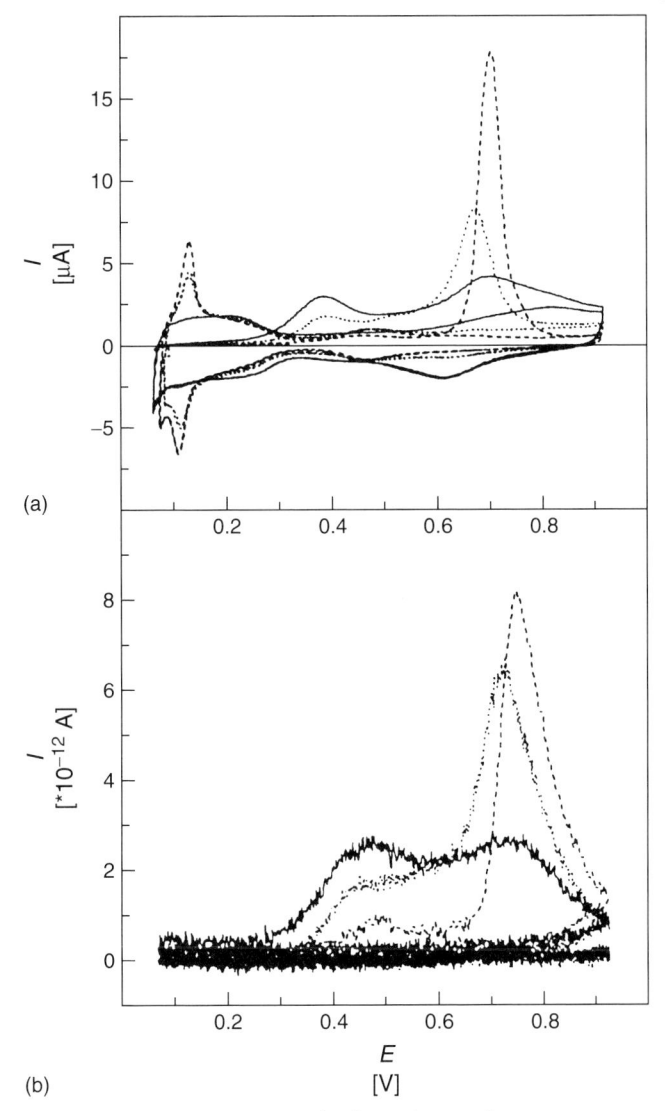

4.3 Electrocatalysis on Surfaces Modified by Metal Monolayers Deposited at Underpotentials | 579

Fig. 10 (a) Voltammetry curves for the oxidation of CO on a Sn/Pt(332) surface in 0.5 M H_2SO_4. $E_{ad} = 70$ mV, $\theta_{Sn} = 0.26$ (solid line), $\theta_{Sn} = 0.03$ (dotted line), $\theta_{Sn} = 0$ (dashed line). (b) Mass spectrometric responses for CO_2 generation. (Reproduced with permission from Ref. [120].)

Most of the work in this area was done by Kokkinidis and coworkers [2] who demonstrated catalytic effects of metal adlayers for the reduction of nitro and nitroso compounds [131–134], and for the catalytic hydrogenation of some unsaturated hydrocarbons [2]. While the reduction of the nitro group is catalyzed by metal adlayers on Pt, the reduction of nitrobenzene is partially inhibited on Ag by Pb adatoms [135].

A number of publications exist in the Russian literature about the effect of cation addition on various electroorganic synthetic reactions that are not discussed in terms of the UPD effects, although most of the systems probably involve the metal adlayer formation (see Ref. [1] and reference therein).

Electroreduction of nitro compounds is of considerable importance for electroorganic synthesis. Interesting catalytic effects were reported for the reduction of aromatic nitro compounds on Pt. Figure 11 shows that Pb, Tl, and Bi adlayers shift the half-wave potential positively by 100 to 300 mV. The catalytic effect was attributed to a change in the mechanism of the reduction of the nitro group from a catalytic hydrogenation on bare Pt to an electron-transfer mechanism on Pt/M$_{ad}$, that is, a direct electron exchange between the nitro compound and the adatom-covered electrode surface, namely,

$$(Pt): \quad R-NO_2 + 4Pt \longrightarrow R-NO_2$$
$$\xrightarrow[-H_2O]{2H_{ad}} R-N-O$$
$$\longrightarrow R-N-O$$
$$\xrightarrow[r.d.s]{mH_{ad}} \text{Products} \quad (4)$$

$$(Pt/M_{ad}): \quad R-NO_2 \xrightarrow[-H_2O]{2e^- + 2H^+} R-NO$$
$$\xrightarrow{2e^- + 2H^+} R-NHOH$$
$$\xrightarrow[-H_2O]{2e^- + 2H^+} R-NH_2 \quad (5)$$

The same change from the "catalytic" to the "electron-transfer" mechanism has been observed in benzofuroxan reduction on Pt with several metal adlayers [2].

The reduction of heterocyclic nitro compounds on platinum is also considerably catalyzed by metal adlayers. The

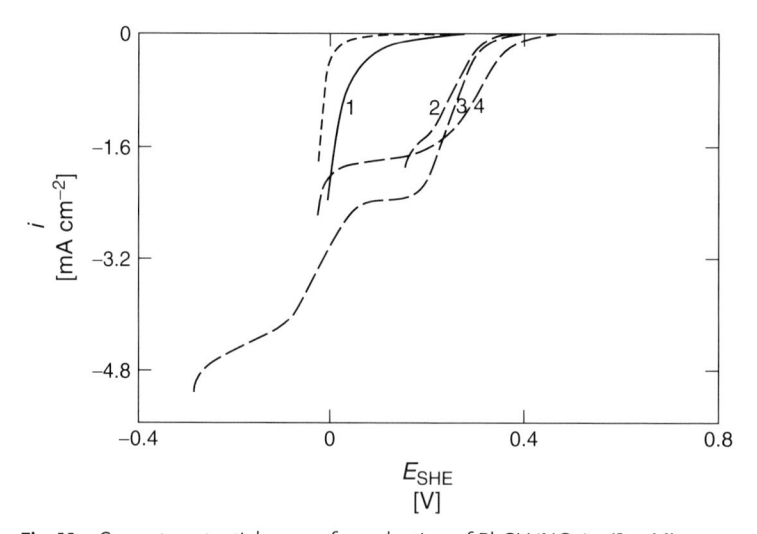

Fig. 11 Current–potential curves for reduction of PhCH(NO$_2$)$_2$ (1 mM) on a Pt rotating disk electrode in 0.5 M HClO$_4$ curve (1) and Pt with adlayers of Bi, Pb, and Tl, curves (2), (3), and (4), respectively. (- - -) Curve in base solution; concentration of cations 1 mM; sweep rate 10 mV s^{-1}; rotation frequency f = 35 Hz. (Reproduced with permission from Ref. [2].)

reaction of 3-nitro-1H-1,2,4-triazole on Pt/M_{ad} is a four-electron reduction also involving the electron-transfer mechanism, while the hydrogenation occurs on bare Pt [136]. This activity surpasses that of mercury electrodes in alkaline solutions. The reaction on Au follows an electrocatalytic and/or electron-exchange mechanism. The electrocatalytic mechanism gives a diffusion-controlled cleavage of the N–O bond, while the electron-exchange mechanism proceeds through dihyroxilamine, which can react further. On Au/M_{ad}, the reduction proceeds through dihydroxylamine [137]. Selective oxidation of lactose to lactobionic acid on Pt/Pb_{ads} with a significant enhancement of the reaction rate in comparison with the one on Pt has been reported by Druliolle and coworkers [138].

Motoo and Furuya [139, 140] demonstrated an enhancement of the ethylene reduction on Pt by Cu and Ag and inhibition by Se and Tl adlayers. Cu and Ag adatoms, which block one Pt site, desorb the ethylene molecule that blocks 2.5 Pt sites, thus leaving 1.5 sites available for H adsorption. Hydrogen adatoms at these sites facilitate a higher rate of ethylene reduction. Tl and Se adatoms cause inhibition since they block 2.5 sites as ethylene does.

4.3.3.4 Hydrogen Evolution

Hydrogen evolution has been found without exception to be inhibited by adlayers of Bi, As, Cu, and Sn on Pt [141–143], Pb, Tl, and Cd on Pt [144, 145] and Au [144, 145], and by Pb and Tl on Ag [146] electrodes. All these metals exhibit a large overpotential for hydrogen evolution. Adlayers inhibiting H_2 evolution are of interest for fundamental electrocatalysis and electrode kinetics, but they also have practical significance in promoting sorption of H into some metals, altering selectivity for some hydrogenation reactions, and in corrosion inhibition. The lattermost is possible when O_2 reduction is not enhanced by metal adlayers.

The observed inhibition on Pt has been ascribed to [144, 145]: (1) simple geometric blocking, (2) a decrease in the number of pairs of uncovered Pt atoms, which decreases the H + H recombination reaction, and (3) a change of electronic states of uncovered Pt atoms. Protopopoff and Marcus [147] have analyzed in detail the blocking effects of sulfur on H_2 evolution on Pt(110) and Pt(111) electrodes. For random adatoms distribution, the zones of deactivation overlap, and this process can be analyzed by using a simple statistical treatment [148]. For several Pt/M_{ad} systems, considerable H_2 evolution takes place when the UPD of H is completely blocked. This provides an additional proof that there is no connection between the UPD H and the intermediates in H_2 evolution.

The UPD metal adlayers have interesting effects on the reaction mechanism. Pb, Tl, and Cd at small coverages cause pronounced inhibition, but the Tafel slope for the reaction remains -30 mV, as for bare Pt. This means that the recombination of two H atoms giving H_2 ($H_{ad} + H_{ad} \to H_2$) is the rate-determining step for both surfaces. At larger Pb coverages, the Tafel slopes becomes -120 mV (Fig. 12) because there are no pairs of Pt atoms for a recombination reaction to occur, and the mechanism with the ion-plus-atom reaction ($H_{ad} + H^+ + e^- \to H_2$) rate-determining step becomes operative [144, 145]. Changes of slope from -30 to -60 and -120 mV have been reported for Pt with Ge [142, 143]. No change of Tafel slope was found for H_2 evolution on Au/Pb [144, 145] since

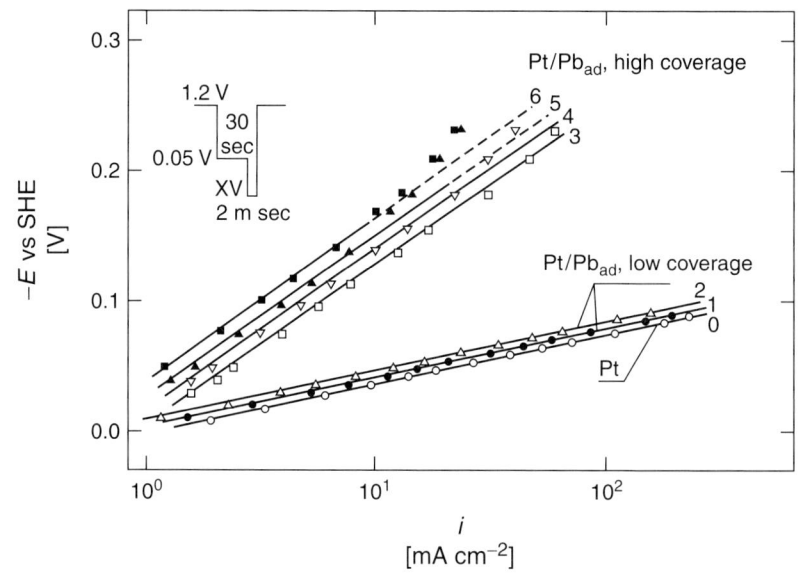

Fig. 12 Hydrogen evolution on a rotating Pt electrode in 1 M HClO$_4$ in the presence of Pb adatoms. The curves 0 to 6 correspond to θ_{Pb} = 0, 017, 0.19, 0.56, 0.67, 0.86, 0.91, respectively. Rotation rate 5000 rpm. (Reproduced with permission from Refs. [144, 145].)

on bare Au the ion-plus-atom reaction is the rate-determining step requiring one surface site.

Inhibition of H$_2$ evolution was reported for stationary single-crystal electrodes for As/Pt(111) [149], Bi/Pt(100), Sb/Pt(100), Bi/Pt(111), and Sb/Pt(111) [150, 151] surfaces. For the Bi/Pt(111) system, the polarization curves have slopes of −30 to −35 mV for θ_{Bi} < 0.04 (0.02), and for 0.14(0.07) < θ_{Bi} < 0.33(0.18), the slopes are −35 and −54 mV at small and high overpotentials, respectively [150, 151] (Fig. 13). For Pt(100), a linear decrease of the H$_2$ evolution current as a function of coverage has been observed up to θ_{Bi} = 0.35(0.18), and a steep decrease has been observed above it. For Pt(111), a large decrease of H$_2$ evolution was observed at small Bi coverages up to 0.1(0.03), and a linear decrease was seen for larger coverages. A similar observation was made for polycrystalline Pt with Pb [144, 145]. Small changes of Tafel slope for low Bi coverages were explained by a hypothetical mechanism in which a slow recombination and an ion-plus-atom reaction rate-determining steps occur in parallel. Inhibition effects of As, which is a prototype catalytic poison, have been reported for polycrystalline electrodes [152, 153] and for a Pt(111) surface [149]. Surprisingly, small inhibition was observed up to θ_{As} = 1/3(1/6) for the latter system.

The mechanism of significant H$_2$ evolution on Pt with large coverage of inhibiting metal adatoms is not clear. For Bi on Pt(100), the existence of the c(2 × 2) Bi adlayer was assumed, and it was proposed that H$_2$ evolution takes place through fourfold symmetry holes in the adlayer [150, 151]. This adlayer exists at Pt(100) [71, 72], but its structure still needs to be verified during H$_2$ evolution. In situ verification of

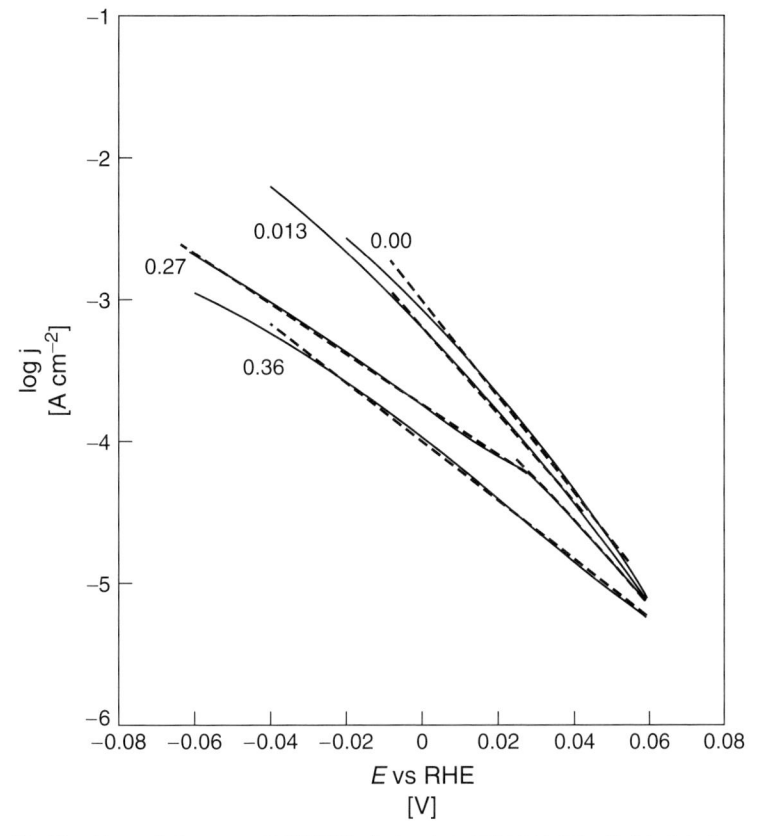

Fig. 13 H$_2$ evolution on a Bi/Pt(111) electrode in 0.5 M H$_2$SO$_4$. Tafel plots for different Bi coverages: θ_{Bi} = 0, 0.13 (0.06), 0.27(0.13), and 0.36 (0.18). (Reproduced with permission from Refs. [150, 151].)

the adlayer structure during the course of H$_2$ evolution has been obtained by SXS for Pt(111) with a Tl adlayer [154]. Figure 14 shows data for Tl adlayers on Pt(111) in solutions of two different pH values. A significant H$_2$ evolution current flows at Pt(111) despite the close-packed configuration of the Tl adlayer. The electrocompression of the Tl–Tl distance with decreasing potential is observed with this system. H$_2$ evolution has no effect on the structure of the Tl adlayer, but the Tl–Tl distance ceases to contract at its onset (Fig. 14). It is likely that in addition to defect sites, H$_2$ evolution occurs through hollow sites in the structure of the Tl adlayer. This prevents further compression with decreasing potential. These data seem to corroborate the assumption for the H$_2$ evolution through the hollow sites in the Bi adlayer on Pt(100).

Several approaches were used to quantify inhibition effects of metal adlayers. These involved calculations of currents of H$_2$ evolution based on the order-disorder theory of alloys [141, 144, 145], and simulations based on geometric [150, 151, 153] and long-range electronic effects [150, 151]. Verification of the models used in some simulations seems

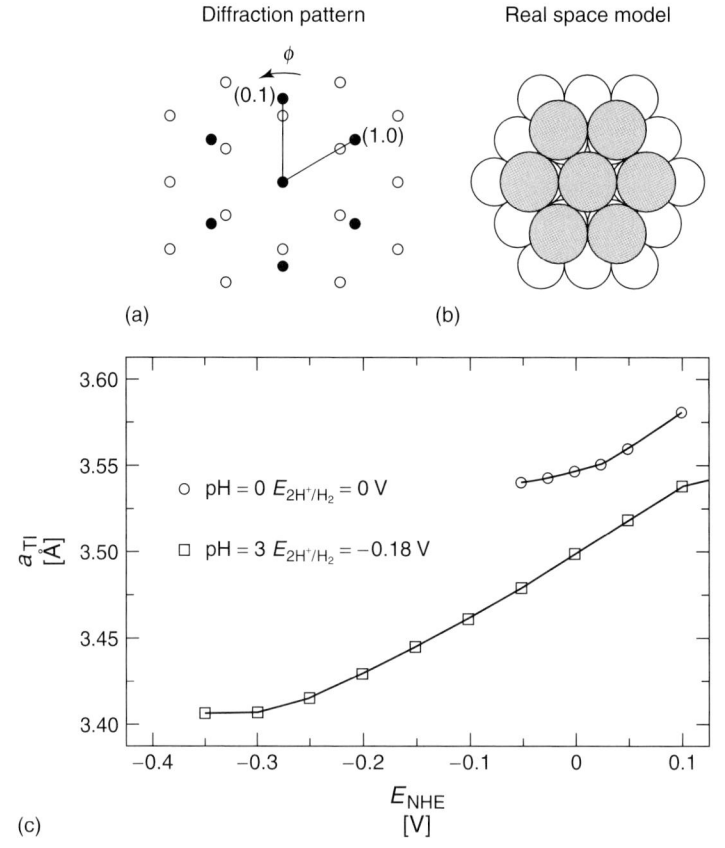

Fig. 14 (a) In-plane diffraction pattern from close-packed hexagonal Tl adlayer (open circles) on Pt(111) in 0.05 M H_2SO_4 solution with 1 mM Tl^+. Solid circles diffractions from Pt(111). (b) Real space model for Tl adlayer. (c) Tl–Tl interatomic distance determined from the rocking θ scans at diffraction positions as a function of potential. The electrocompression increases in solution with pH = 3 beyond that observed at pH = 0, but it ceases to change when H_2 evolution starts (see text). (Adapted from Ref. [154]; reproduced with permission.)

necessary in order to obtain a more complete understanding of the inhibition effects.

Related to the inhibition of H_2 evolution is the effect cations on the corrosion of iron. Dražić and Vorkapić [157] interpreted a decrease of the corrosion of Fe in sulfuric acid in the presence of Mn^{2+}, Cd^{2+}, and Zn^{2+} in terms of the inhibition of H_2 evolution by the UPD of these metals. Similar effects were observed for Pb, Tl [158, 159] and for Ge, Ga [160]. In deaerated solutions, the inhibition effect can be quite pronounced. In the presence of O_2, however, an enhanced O_2 reduction can increase corrosion rates.

4.3.3.5 Oxygen Reduction

Oxygen reduction is a multielectron reaction that involves several elementary steps in the reaction mechanisms. In aqueous solutions, oxygen reduction appears to occur by two overall pathways: a "direct" four-electron reduction, and a "peroxide" pathway that involves H_2O_2 as the intermediate [161]. For acid solutions, the reaction in a direct four-electron pathway is

$$O_2 + 4H^+ + 4e^- \longrightarrow 2H_2O$$
$$E_0 = 1.229 \text{ V vs NHE} \quad (6)$$

Two electrons are exchanged in a "peroxide" pathway:

$$O_2 + 2H^+ + 2e^- \longrightarrow H_2O_2$$
$$E_0 = 0.67 \text{ V} \quad (7)$$

Peroxide can undergo further reduction,

$$H_2O_2 + 2H^+ + 2e^- \longrightarrow 2H_2O$$
$$E_0 = 1.77 \text{ V} \quad (8)$$

or it can catalytically decompose into O_2 and H_2O. Peroxide is the final reduction product on some less active surfaces. On some surfaces, the two pathways can occur in parallel. The same pathways are operative in alkaline solutions but at different set of potentials involving intermediates and products determined by the solution pH [161].

4.3.3.5.1 Alkaline Solutions
Foreign metal adatoms cause remarkable catalytic effects on oxygen reduction on several electrode surfaces. The adatoms of Pb [155, 156, 162], Bi [163], and Tl [164, 165] cause a shift of the half-wave potential of the O_2 reduction to more positive values and an approximate doubling of the diffusion-limited current density. The latter is due to the change of the reaction mechanism from a two-electron reduction on Au into a vital four-electron reduction on Au covered with UPD metal adlayers. Catalytic effects were observed also for Pt [166], glassy carbon and graphite [2], and Ru [167] surfaces with Pb and Tl adsorbates. The latter three cases are somewhat different from the UPD adlayers on metal surfaces. For carbons, it appears that the UPD occurs through an interaction of metal with carbon surface groups. For oxidized Ru surfaces, cations interact with surface oxides.

Figure 15 shows rotating disk-ring measurements of O_2 reduction on Au and Au/Pb$_{ad}$. The upper panel shows the UPD of Pb on Au in NaOH solution. The disk current shows doubling of the diffusion-limited current, which means that four electrons are exchanged in the reaction on Pb/Au. Consequently, the ring current shows that peroxide generation is almost completely suppressed. A series reaction mechanism, involving a two-electron reduction to HO_2^-, followed by exchange of another two electrons in the reduction of HO_2^- to OH^-, has been found from the analysis of the rotating disk-ring electrode measurements, namely,

$$\text{Au:} \quad O_2 + H_2O + 2e^-$$
$$\longrightarrow HO_2^- + OH^- \quad (9)$$
$$\text{Pb/Au:} \quad O_2 + H_2O + 2e^-$$
$$\longrightarrow HO_2^- + OH^- \quad (10)$$
$$HO_2^- + H_2O + 2e^- \longrightarrow 3OH^-$$
$$(11)$$

All rate constants for this complex reaction have been determined from these measurements [163, 165]. Tl and Bi adlayers cause similar catalytic effects. Tafel plots obtained from rotating disk-ring measurements show slopes of -120 mV for Au/Pb and Au/Bi systems, while a slope of -55 mV was found for Au/Tl. The -120 mV slope, as observed for bare

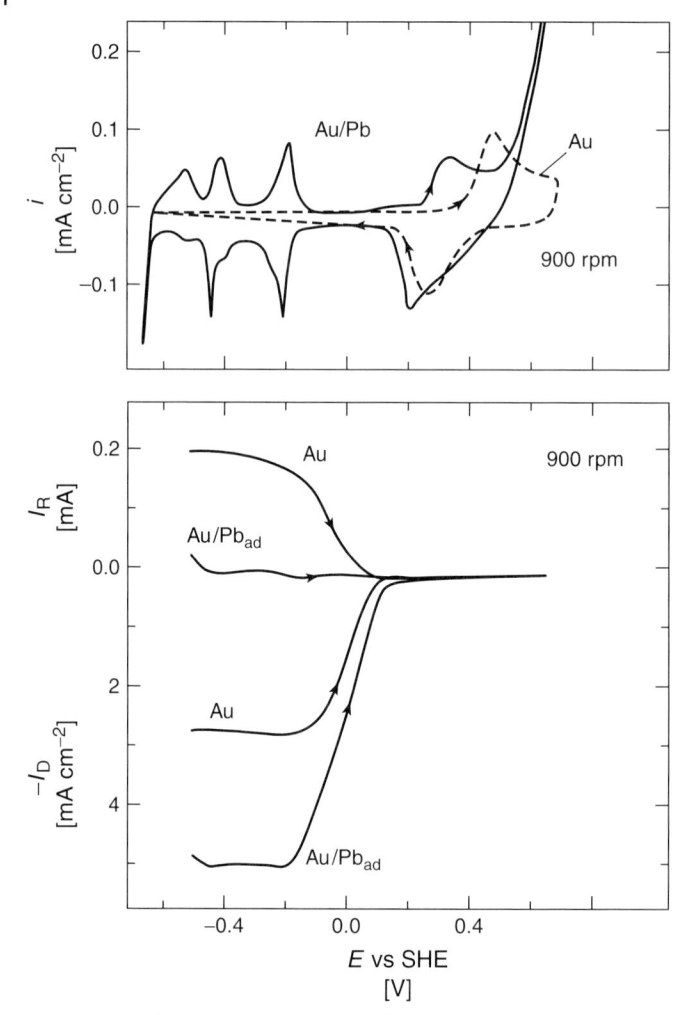

Fig. 15 O$_2$ reduction on rotating Au and Pb/Au electrodes in 1 M NaOH containing 1 mM HPbO$_2^-$. Ring potential 0.3 V, ring area 0.049 cm^2; sweep rate 50 mV s^{-1}. Upper panel: the UPD of Pb on Au. (Adapted from Refs. [155, 156]; reproduced with permission.)

Au, indicates a rate-controlling first electron exchange. The slope of −55 mV was also explained by a slow first electron exchange. A low electrosorption valency of Tl ($\gamma_{Tl} = 0.6$) had to be assumed to satisfy this slope [165].

Yeager and coworkers [162] reported interesting catalytic effects of Pb and Tl on the O$_2$ reduction on smooth and high surface area Pt in alkaline solutions. These effects shift the half-wave potential to more positive values than those on bare Pt (Fig. 16). Rotating disk-ring electrode measurements indicate a series mechanism for O$_2$ reduction, instead of a direct pathway with the O−O bond scission on bare Pt.

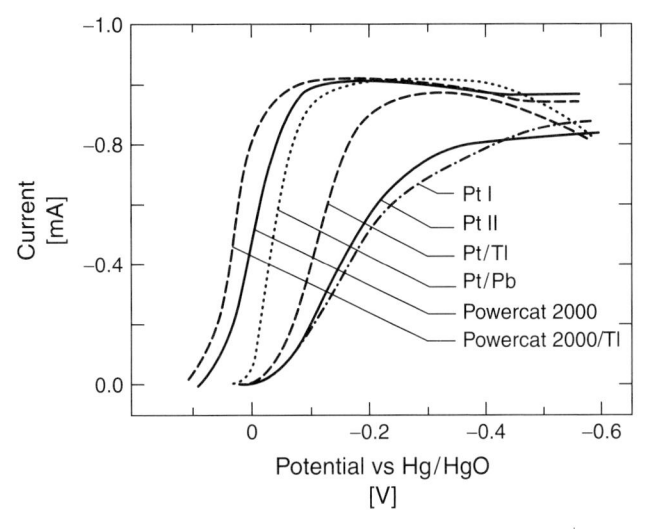

Fig. 16 O$_2$ reduction on several substrates in 1 M NaOH. Tl$^+$ concentrations: 3.2 mM for Pt and 11 mM for dispersed Pt Powercat 2000 catalyst. Pb^{2+} concentration 2 mM. Sweep rate 10 mV s^{-1}, for Powercat quasi–steady state measurements. Electrode area 0.196 cm^2. Pt I: no pretreatment, Pt II: cathodic pretreatment. (Reproduced with permission from Ref. [166].)

Although Tl and Pb adsorbates on oxidized Ru surfaces in alkaline solutions differ from the UPD adlayers, their pronounced catalytic effects are noteworthy. Oxidized Ru surfaces, which are inactive for O$_2$ reduction, support a four-electron O$_2$ reduction after surface modifications upon addition of Tl$^+$ and Pb^{2+} to the electrolyte [167]. Traces of Tl$^+$ or Pb^{2+} were sufficient to produce catalytic effects, which were assumed to be due to a pronounced modification of the electronic properties of Ru oxides in the interaction with these cations.

4.3.3.5.2 Acid Solutions

In acid solutions, metal adlayers on Au cause small enhancements of O$_2$ reduction, but except for Tl, a four-electron reduction is not observed. For Pt, a small inhibition is generally observed. A shift of 150 mV of the half-wave potentials to more positive values is caused by Bi on Au(111), while the effect of Pb is very small [168, 169]. Tl on Au(111) [170] causes a predominant four-electron reduction. O$_2$ reduction on silver is a four-electron reaction and is inhibited by Tl and Pb adlayers causing H$_2$O$_2$ generation as shown for Ag(111) and Ag(100) [168, 171]. The inhibition was dependent on the degree of coverage and the Ag crystal plane. Catalytic effects were reported for the polycrystalline Pb/Au, Bi/Au and Tl/Au, Pb/Cu, Tl/Cu [172] and Cd/Pb [173] surfaces. Inhibition effects, however, were observed for Pt with Pb, Tl, Bi [172, 174, 175], Cu [174, 176], and Ag [174] adlayers.

It is surprising that Pb and Cd at low coverages on Pt do not cause significant inhibition of a four-electron reduction in the kinetic-control region [175, 177]. This can be understood assuming a series mechanism for a four-electron reduction on Pt with adatoms. A somewhat larger inhibition was observed by Kokkinidis [172] in more concentrated solutions. These data strongly suggest that proton transfer in O_2 reduction, involving any kind of H interaction with the surface, is not likely to occur [178].

Further understanding of the catalysis of O_2 reduction by metal adlayers has been achieved by investigation of reactions on single-crystal surfaces and by structure-sensitive techniques. The catalytic effects observed with single-crystal electrodes are in agreement with those obtained with polycrystalline surfaces [1, 164]. SXS studies during the course of O_2 reduction helped to resolve the question of the coverage-dependent catalytic effect of Tl. The close-packed, rotated-hexagonal Tl phase on Au(111) was found to support predominantly two-electron reduction, while the aligned hexagonal phase in alkaline solution and patches of the (2×2) phase in acid solution, which gradually vanish during the reaction (Fig. 17), are conducive to a four-electron reduction [170].

A total inhibition of O_2 reduction on Au(111) by a Cu adlayer in the potential region where Cu forms a honeycomb $(\sqrt{3} \times \sqrt{3})R30$ structure in the absence of O_2 was determined by scanning tunnelling microscopy (STM) by Itaya and coworkers [180]. Sulfate/bisulfate is adsorbed in the center of a Cu honeycomb [181, 182], which could be the cause of the inhibition. A Cu submonolayer on Pt(111) causes inhibition by decreasing the current to approximately one half, while a Cu monolayer causes complete inhibition. The reduction of O_2 to H_2O_2 was assumed in the presence of either the simple or the honeycomb $(\sqrt{3} \times \sqrt{3})R30°$ submonolayer structure. A change from bridge to end-on adsorption of O_2 is viewed as a cause of the generation of H_2O_2 and the decrease of the current by 1/2. This conclusion should be verified by rotating disk ring electrode (RDRE) measurements. A pseudomorphic Ag monolayer and a hexagonal aligned bilayer on Pt(111) cause a complete inhibition of O_2 reduction [183]. This inhibition as a function of Ag coverage was used to determine a bridge adsorption of O_2 during reduction on Pt(111).

4.3.3.5.3 Hydrogen Peroxide Reduction

H_2O_2, which is inert on most Au surfaces, except on Au(100) and its vicinals in alkaline solutions, is particularly effectively catalyzed by several Pb, Tl, and Bi adlayers [1, 2, 165, 168]. Gewirth and coworkers [184, 185] identified by atomic force microscopy (AFM) the (2×2)-Bi structure in the potential region at which a maximum activity for H_2O_2 reduction on Au(111) is observed. The rectangular $(p \times \sqrt{3})$ Bi high-coverage adlayer on Au(111) is not catalytically active for H_2O_2 reduction (Fig. 18), which is very similar to the inactivity of a close-packed Tl adlayer on Au(111) [170]. H_2O_2 reduction is catalyzed on Au(111) with a low-coverage Tl phases, while it is almost completely suppressed by the rotated close-packed hexagonal phase [179]. For the Au(111)/Pb surface, however, the highest activity was found to occur when the surface is maximally covered by Pb islands [186].

4.3.3.5.4 Mechanisms of Catalytic Effects

Several mechanisms have been proposed to explain the catalytic effects of metal adlayers. These include (1) specific electronic properties of the metal/adlayer surface that can increase the adsorption energy of O_2 and intermediates [1], (2) redox properties of metal adlayers [164], and (3) formation of the bridge bond between O_2 and the Au

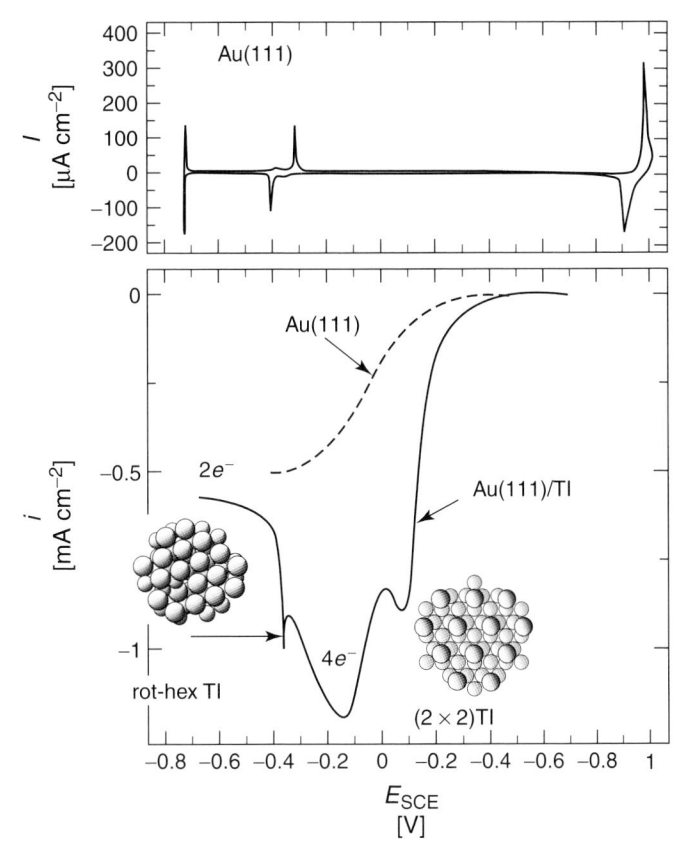

4.3 Electrocatalysis on Surfaces Modified by Metal Monolayers Deposited at Underpotentials | 589

Fig. 17 Voltammetry curve for the UPD of T on Au(111) in 0.1 M HClO$_4$ solution containing 5 mM Tl$^+$ (upper panel). Polarization curve for O$_2$ reduction on Au(111) (dashed line) and Au(111) with the (2 × 2)Tl, and rotated-hexagonal Tl adlayers. Sweep rate 10 mV s^{-1}. (Adapted from Ref. [179]; reproduced with permission.)

adatom sites, which facilitates the weakening of the O–O bond [168, 179, 184, 185]. Yeager and coworkers [166] assumed a destabilization of anodic films on Au and Pt in the interaction with metal adlayers. McIntyre and Peck [164] proposed a redox cycle in which Tl adatoms are oxidized by O$_2$ and HO$_2^-$ and are subsequently redeposited. Schriffrin [187] assumed a redox couple in which Pb^{2+} is oxidized by HO$_2^-$ to Pb^{4+}. Jüttner and coworkers [168] assumed a change from an end-on adsorption of O$_2$ on Au to a bridge adsorption on Au with metal adlayers. This favors the complete reduction of O$_2$ by weakening the O–O bond. Alvarez and Jüttner [188] reported a simulation of the catalytic effects of Pb on H$_2$O$_2$ reduction on Au(111) and Au(100) assuming essentially a lattice gas model (LGM) for the Pb adlayer at low coverages and well-defined structures at high coverages. Good agreement between the simulation and experiment is considered as a support for the bridge adsorption model. However, the assumptions regarding the Pb adlayer are not in agreement

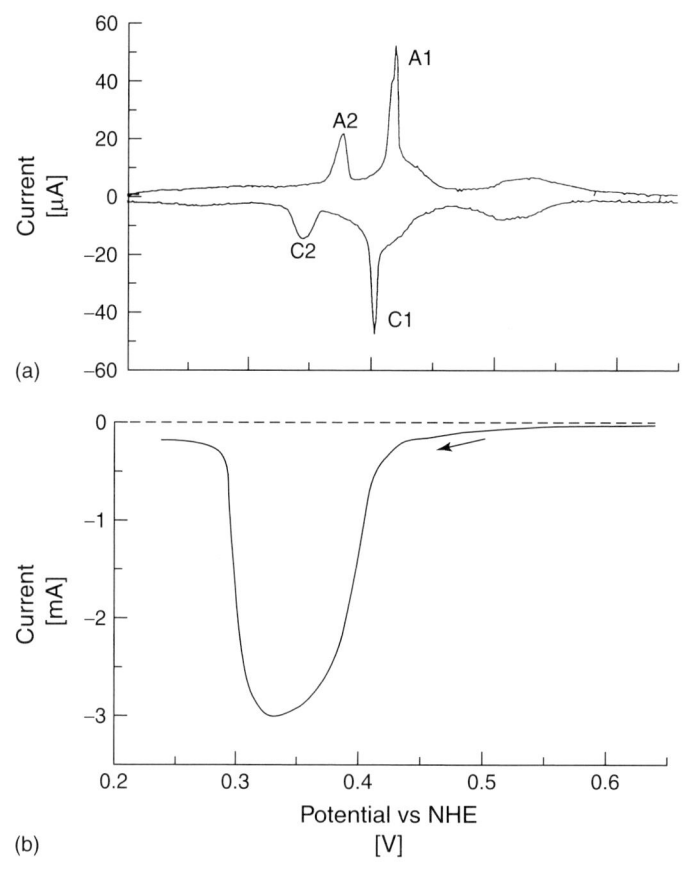

Fig. 18 (a) Voltammetry curve for the UPD of Bi on Au(111) in 0.1 M HClO$_4$ solution containing 0.5 mM Bi^{3+}. (b) Rotating disk polarization curve for H$_2$O$_2$ reduction in solution as in (a) with 10 mM H$_2$O$_2$. Rotation rate 400 rpm; sweep rate 5 mV s^{-1}. Dashed line: no Bi^{3+}. (Adapted from Ref. [185]; reproduced with permission.)

with the results of SXS studies of the Pb/Au(111) system by Toney and coworkers [189]. Additionally, the large coverage by Pb islands, not a specific structure, was found to produce a maximum activity for H$_2$O$_2$ reduction on Au(111) [190]. No ordered structure has been reported so far for Pb on Au(100). Gewirth and coworkers [184, 185] proposed a heterobimetalic bridge model for H$_2$O$_2$ interaction with the Au/Bi surface as a cause of the enhanced catalytic activity.

Adžić and Wang [191] studied the UPD of Tl on Au(111) in the presence of Br$^-$, which has three potential-dependent, commensurate Tl–Br phases that affect O$_2$ reduction in different ways. The catalytically active (3 × √3)-2TlBr phase on Au(111) was monitored by SXS during the course of oxygen reduction by measuring the structure factors for its two low-order in-plane diffractions as a function of potential. The IR voltage drop in the thin-layer cell did not allow drawing definitive conclusions regarding the stability of the active adlayer phase and the origin of the catalytic effects [192].

Most recently, Tamura and coworkers [193] have confirmed a stability of the low-coverage Bi(2 × 2) phase on Au(111) during O_2 reduction with in situ surface X-ray scattering. Under oxygen reduction conditions, both the (2 × 2)-Bi and a high-coverage ($p \times \sqrt{3}$)-2Bi adlayer structures are stable (Fig. 19). The electrochemical drop cell configuration utilized for these measurements minimizes the IR potential drop relative to that in commonly used

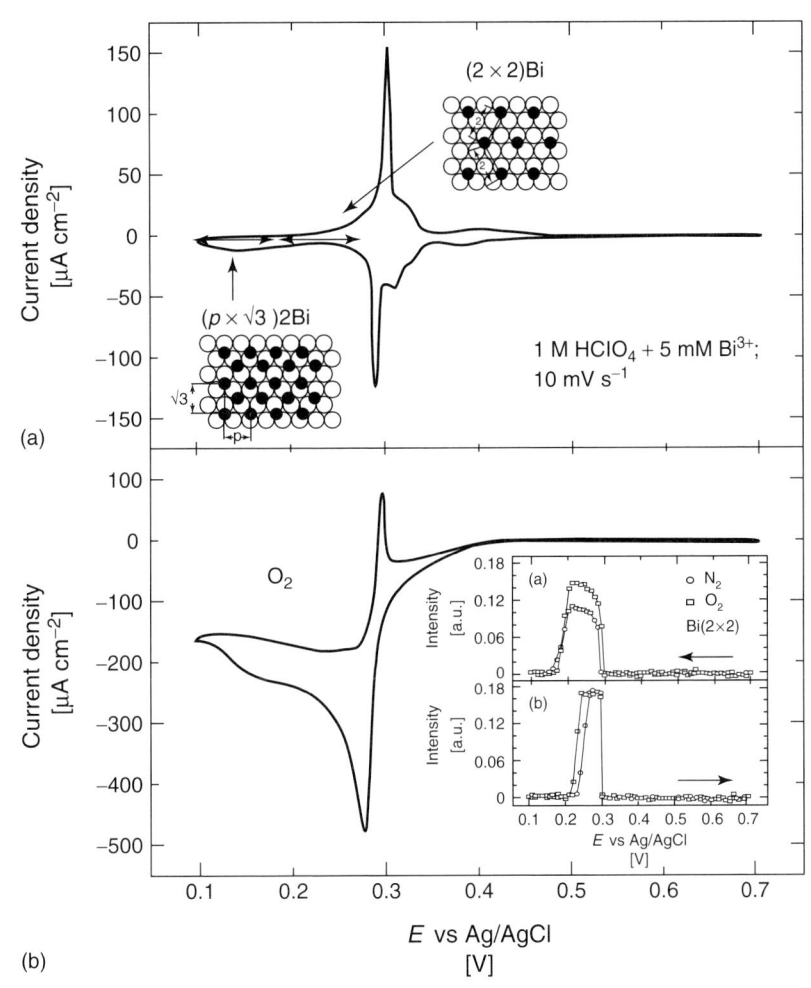

Fig. 19 Voltammetry curves for the UPD of Bi on a Au(111) electrode in 1 M $HClO_4$ with 5 mM Bi^{3+} (a) and oxygen reduction on the Au(111) surface with Bi adlayers in the same solution (b); sweep rate 10 mV s^{-1}. The potential regions of the existence of (2 × 2) and ($p \times \sqrt{3}$) Bi phases are indicated in the graph. Inset: X-ray diffraction (XRD) intensities at the (1/2, 1/2, 0.2) position (characteristic for the (2 × 2) Bi phase) as a function of potential in the presence of O_2 and N_2 obtained in a thin-layer and drop cells. $\lambda = 0.386$ Å; sweep rate 2 mV s^{-1}.

thin-layer cells. Together with rotating disk electrode measurements, the results indicate distinctly different catalytic properties for the two different ordered bismuth adlayer phases. The potential region corresponding to the (2×2)-Bi phase, the O_2 reduction is promoted to a four-electron reaction, albeit with relatively slow kinetics. The close-packed $(p \times \sqrt{3})$-2Bi phase, formed at more negative potentials, appears to have a limited number of sites available for the four-electron reduction, but the reaction kinetics on this adlayer is enhanced by the increase of overpotential. These results rule out the redox mechanism as the explanation of the catalytic effect, which leaves the electronic/adsorption effects and the bridge bond as possible origins.

4.3.3.6 Electrodeposition of Metals

Metal adatoms have profound effects on the processes of electrodeposition of metals. The effect of the addition of trace quantities of various cations to the plating solutions on the morphology of deposits has been known for quite some time from the plating patent literature. Usually, an increased brightness is observed in the presence of certain cations in plating solutions. In addition, catalytic effects are observed with some systems that shift the reactions to positive potentials. McIntyre and Peck [194] rationalized these observations in terms of the UPD adlayer effects with the example of the electrodeposition of Au from $Au(CN)_2^-$ solutions. The adatoms of Pb and Tl induce a marked positive shift of the polarization curve and extend the current density range in which bright, fine-grained deposits are formed. It has been proposed that at sufficiently high θ_M, these adatoms can act as 2D nucleation centers, with crystal growth proceeding around these centers rather than by incorporating Au atoms into the bulk lattice at the surface defects and dislocations, or by 3D growth. Davidović and Adžić [195, 196], however, have shown that Sb and Tl cause a decrease in the nucleation rate of Au deposition. The nucleation changes from instantaneous in $Au(CN)_2^-$ solutions in the absence of Sb or Tl adlayers into a progressive process in their presence (Fig. 20). For progressive nucleation, the rate can be described by the equation

$$I(t) = 2zF \, A \, N_\alpha \pi (2Dc)^{3/2} M^{1/2} \frac{t^{3/2}}{3\rho^{1/2}} \quad (12)$$

where I is the current at time t, A is the area of the electrode, N is the number of nuclei that can be obtained at a given condition, and ρ is the density of the metal. As expected for progressive nucleation, the linear I versus $t^{3/2}$ plots based on Eq. (13) have been obtained at constant E from the curves in Fig. 20(b).

Sb, a well-known "wetting" agent in gas phase metal deposition [197], causes remarkable morphological effects by producing very bright Au deposits at current densities at which rough deposits were obtained in its absence. Considerable catalytic effects of Sb adatoms were ascribed to a change in the CN^- adsorption in the presence of Sb adlayers. No Sb or Tl incorporation was found with AES, or by atomic absorption spectroscopy of the dissolved deposit.

Sieradzky and coworkers [198, 199] have recently demonstrated that the UPD of the Pb and Cu adlayers can mediate growth of ordered multilayer deposits (up to 200 layers) of Ag on Au. Atomically flat epitaxial multilayers of Ag on Au(111) have been produced in the presence of the UPD of Pb and Cu. Ag is codeposited with reversibly deposited UPD metal.

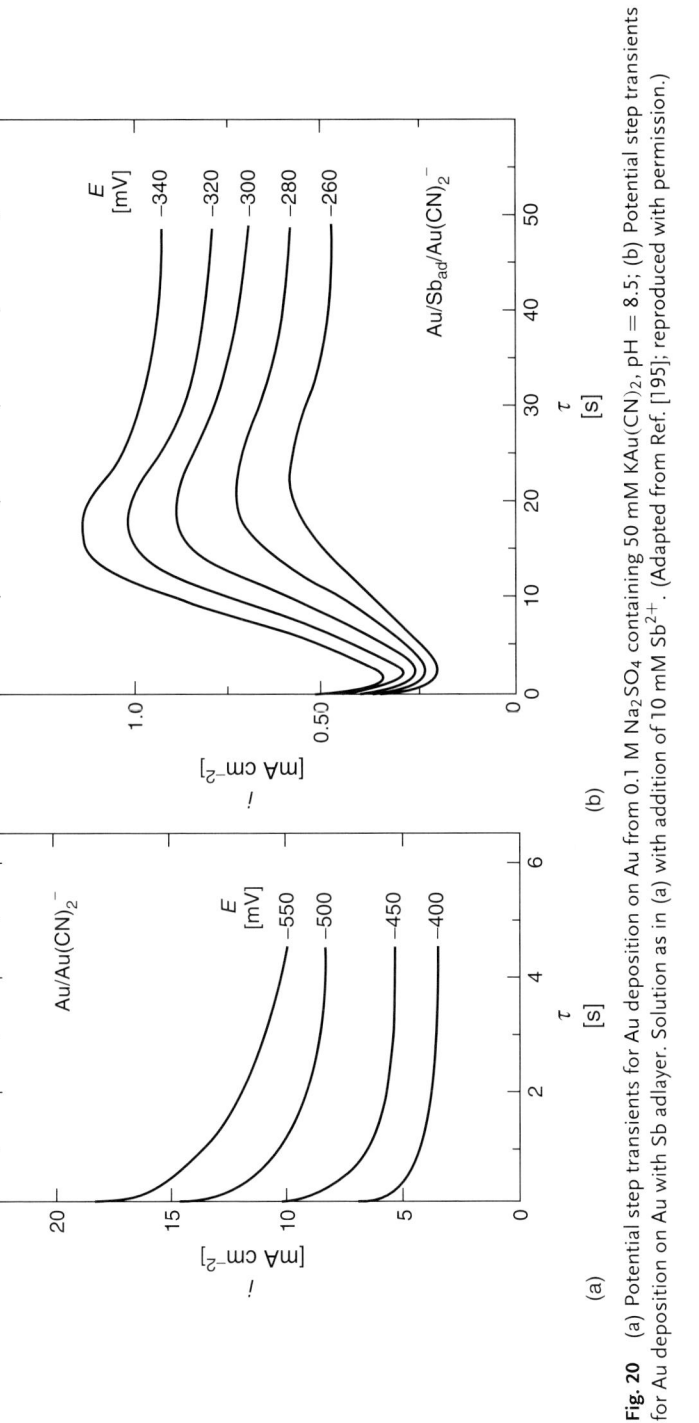

Fig. 20 (a) Potential step transients for Au deposition on Au from 0.1 M Na$_2$SO$_4$ containing 50 mM KAu(CN)$_2$, pH = 8.5; (b) Potential step transients for Au deposition on Au with Sb adlayer. Solution as in (a) with addition of 10 mM Sb^{2+}. (Adapted from Ref. [195]; reproduced with permission.)

The latter is periodically deposited and stripped from the surface, and this serves to significantly increase the density of 2D islands of Ag atoms, promoting a layer-by-layer growth mode. The UPD of the Pb monolayer covers the surface onto which the deposited Ag atoms undergo interlayer place-exchange with Pb atoms, forming 2D islands below the Pb adlayer. Upon stripping of the Pb adlayer, Ag continues to deposit with island growth. Figure 21 shows a scheme of the

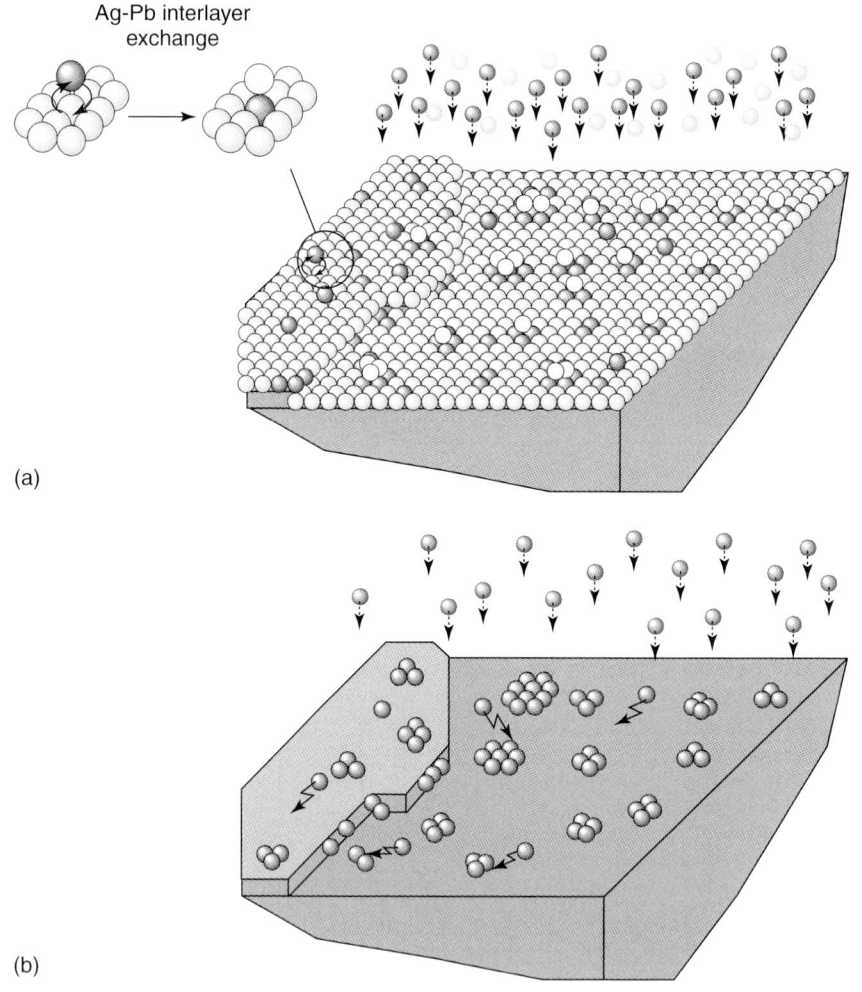

Fig. 21 Proposed mechanism of the UPD of Pb-mediated deposition of Ag. The Ag and Pb atoms are light and dark shaded spheres, respectively. Arrows indicate depositing metal cations. (a) Pb mediation: at potentials close to reversible Pb deposition, a Pb monolayer covers the surface. The deposited Ag adatoms undergo interlayer place-exchange with Pb adatoms (light green sphere) forming 2D islands below the Pb adlayer. (b) On the reverse cycle, the Pb adlayer is stripped from the surface as Ag continues to deposit resulting in island growth. (Reproduced with permission from Ref. [198].)

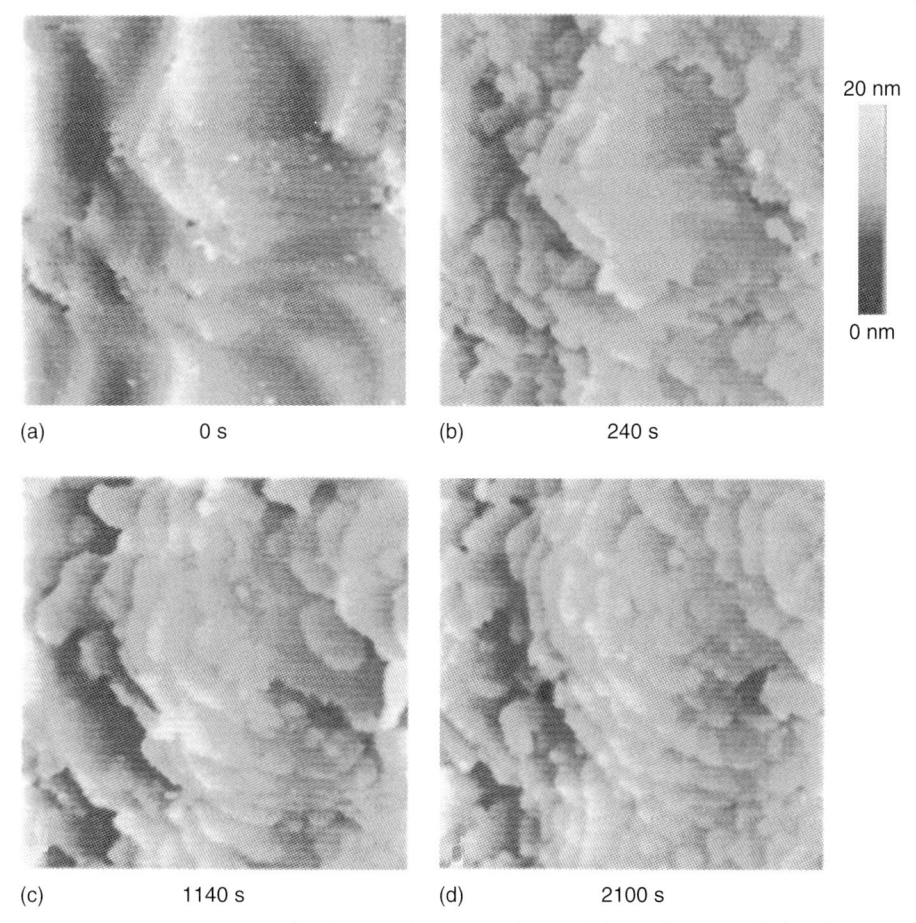

4.3 Electrocatalysis on Surfaces Modified by Metal Monolayers Deposited at Underpotentials | 595

Fig. 22 In situ STM topographs showing the time evolution of the Ag layer morphology for a deposition at 0.125 V (versus Pb/Pb^{2+}) that maintains the $\theta_{Pb} = 0.8$ in 10 mM HClO$_4$ containing 0.1 mM AgClO$_4$ and 10 mM Pb(ClO$_4$)$_2$. Scan size in each image 764 × 764 nm. (Reproduced with permission from Ref. [199].)

proposed mechanism for the Pb-mediated deposition of Ag on Au. An ordered multilayer Ag deposit can also be obtained in a constant potential electrodeposition, as discussed in Refs. [198, 199]. Figure 22 shows the Ag deposits obtained as a function of time at a potential of 0.125 V that maintains $\theta_{Pb} = 0.8$ [199].

Atomic layer control of growth is an ultimate goal of the electrodeposition of metals and nanofabrication technology, and in this context, the effects of UPD metal adlayers discussed above are very interesting. The data illustrate remarkable possibilities for further exploration of the effects of UPD adatoms in electrodeposition of metals. It would be interesting to see whether this method can produce a layer-by-layer growth in systems with larger lattice misfits between substrate and deposit and with different chemical properties.

Acknowledgment

This research was performed under the auspices of the US Department of Energy, Divisions of Chemical and Materials Sciences, Office of Basic Energy Sciences under Contract No. DE-AC02-98CH10886.

References

1. R. R. Adžić in *Advances in Electrochemistry and Electrochemical Engineering* (Ed.: H. Gerischer), John Wiley & Sons, New York, 1984, pp. 159–260, Vol. 13.
2. G. Kokkinidis, *J. Electroanal. Chem.* **1986**, *201*, 217–236.
3. A. A. Gewirth, *Chem. Rev.* **1997**, *97*, 1129–1162.
4. J. X. Wang, R. R. Adžić, B. M. Ocko in *Interfacial Electrochemistry, Theory, Experiment and Applications* (Ed.: A. Wieckowski), Marcel Dekker, New York, 1999, pp. 175–186.
5. K. Takayanagi, D. M. Kolb, K. Kambe et al., *Surf. Sci.* **1980**, *100*, 407.
6. J. T. Hupp, D. Larkin, H. Y. Liu et al., *J. Electroanal. Chem.* **1982**, *131*, 229.
7. J. McBreen, M. Sansone, *J. Electroanal. Chem.* **1994**, *373*, 227–233.
8. E. Leiva, W. Schmickler, *Electrochim. Acta* **1994**, *39*, 1015–1017.
9. E. Leiva, *Surf. Sci.* **1995**, *335*, 83–90.
10. R. R. Adžić, A. R. Despić, *J. Chem. Phys.* **1974**, *61*, 3482–3485.
11. Z. Nagy, L. A. Curtis, N. C. Hung et al., *J. Electroanal. Chem.* **1992**, *325*, 313–324.
12. A. Rhodes, J. M. Feliu, A. Aldaz et al., *J. Electroanal. Chem.* **1989**, *271*, 127–139.
13. R. R. Adžić, B. Z. Nikolić, *J. Serb. Chem. Soc.* **1992**, *57*, 887–895.
14. W. Schmickler, *J. Electroanal. Chem.* **1980**, *113*, 159.
15. T. Iwasita, W. Schmickler, J. W. Schultze, *J. Electroanal. Chem.* **1985**, *194*, 355.
16. I. Fonseca, A. C. Marin, *Port. Electrochim. Acta* **1988**, *6*, 51–68.
17. G. Kokkinidis, *J. Electroanal. Chem.* **1984**, *172*, 2665.
18. A. Cappon, R. Parsons, *J. Electroanal. Chem.* **1973**, *44*, 239.
19. A. Cappon, R. Parsons, *J. Electroanal. Chem.* **1973**, *45*, 205.
20. O. Walter, J. Willsau, J. Heitbaum, *J. Electrochem. Soc.* **1985**, *132*, 1635–1638.
21. J. Willsau, J. Heitbaum, *Electrochim. Acta* **1986**, *31*, 943.
22. S.-H. Chang, L. W. H. Leung, M. J. Weaver, *J. Phys. Chem.* **1990**, *94*, 6013.
23. A. Tripković, K. Popović, R. R. Adžić, *J. Chim. Phys.* **1991**, *88*, 1653.
24. B. Beden, J.-M. Leger, C. Lamy in *Modern Aspects of Electrochemistry* (Eds.: J. O'M. Bockris, B. C. Conway, R. E. White), Plenum Press, New York, 1992, pp. 97–264, Vol. 22.
25. V. E. Kazarinov, G. Ya. Tysyachnaya, V. N. Andreeev, *Elektrokhimiya* **1972**, *8*, 396.
26. B. Beden, A. Bewick, K. Kunimatsu et al., *J. Electroanal. Chem.* **1982**, *142*, 345.
27. D. S. Corigan, M. J. Weaver, *J. Electroanal. Chem.* **1988**, *241*, 143.
28. J. Clavilier, S. G. Sun, *J. Electroanal. Chem.* **1986**, *199*, 471–480.
29. R. Parsons, T. VanderNoot, *J. Electroanal. Chem.* **1988**, *257*, 9–45.
30. T. D. Jarvi, E. M. Stuve in *Electrocatalysis* (Eds.: J. Lipkowski, P. N. Ross), Wiley-VCH, New York, 1998, p. 75.
31. A. Hamnett in *Interfacial Electrochemistry* (Ed.: J. Wieckowski), Marcel Dekker, New York, 1999, p. 843.
32. H. Angerstein-Kozlowska, B. MacDougall, B. E. Conway, *J. Electroanal. Chem.* **1973**, *120*, 756.
33. K. J. Cathro, *J. Electroanal. Chem.* **1969**, *116*, 1608.
34. M. Shibata, S. Motoo, *J. Electroanal. Chem.* **1986**, *202*, 137.
35. R. R. Adžić, D. N. Simić, A. R. Despić et al., *J. Electroanal. Chem.* **1975**, *65*, 587–601.
36. D. Pletcher, V. Solis, *J. Electroanal. Chem.* **1982**, *131*, 309.
37. X. H. Xia, T. Iwasita, *J. Electrochem. Soc.* **1993**, *140*, 2559–2565.
38. Th. Hartung, J. Willsau, J. Heitbaum, *J. Electroanal. Chem.* **1986**, *205*, 135–149.
39. S. Ya. Vasina, R. Wetzel, L. Müller et al., *Elektrokhimiya* **1985**, *21*, 413.
40. R. R. Adžić, M. Podlavicky, *J. Phys.* **1977**, *38*, C5, 193–197.
41. R. R. Adžić, D. N. Simić, A. R. Despić et al., *J. Electroanal. Chem.* **1977**, *80*, 81–99.
42. M. D. Spasojević, R. R. Adzic, A. R. Despić, *J. Electroanal. Chem.* **1978**, *92*, 31–42.

43. S. Motoo, M. Watanabe, *J. Electroanal. Chem.* **1976**, *69*, 429.
44. M. Shibata, O. Takahashi, S. Motoo, *J. Electroanal. Chem.* **1988**, *249*, 252–264.
45. S. Motoo, M. Watanabe, *J. Electroanal. Chem.* **1979**, *98*, 203.
46. R. R. Adžić, A. V. Tripković, N. M. Marković, *J. Electroanal. Chem.* **1983**, *150*, 79–88.
47. T. Iwasita, X. Xia, E. Herrero et al., *Langmuir* **1966**, *12*, 4260–4265.
48. J. Clavilier, A. Fernandez-Vega, J. M. Feliu et al., *J. Electroanal. Chem.* **1989**, *258*, 89–100.
49. S.-C. Chang, Y. Ho, M. J. Weaver, *Surf. Sci.* **1992**, *265*, 81–94.
50. H.-W. Kei, H. Hattori, H. Kita, *Electrochim. Acta* **1996**, *41*, 1619–1628.
51. J. Clavilier, A. Fernandez-Vega, J. M. Feliu et al., *J. Electroanal. Chem.* **1989**, *261*, 113–125.
52. R. R. Adžić, J. X. Wang, unpublished.
53. M. M. P. Jansenn, J. Moolhuysen, *Electrochim. Acta* **1976**, *21*, 861–869.
54. J. Clavilier, J. M. Feliu, A. Aldaz, *J. Electroanal. Chem.* **1988**, *243*, 419.
55. U. W. Humm, D. Kramer, R. S. Zhai et al., *Electrochim. Acta* **1998**, *43*, 2969–2978.
56. J. Clavilier, J. M. Feliu, A. Aldaz, *J. Electroanal. Chem.* **1988**, *243*, 419.
57. A. Kozawa, *J. Inorg. Chem.* **1961**, *21*, 315.
58. V. E. Kazarinov, N. A. Balashova, M. I. Kulezneva, *Elektrokhimiya* **1965**, *8*, 975.
59. R. R. Adžić, N. M. Marković, *Electrochim. Acta* **1985**, *30*, 1473.
60. A. Fernandez-Vega, J. M. Feliu, A. Aldaz et al., *J. Electroanal. Chem.* **1989**, *258*, 101–113.
61. A. Fernandez-Vega, J. M. Feliu, A. Aldaz et al., *J. Electroanal. Chem.* **1991**, *305*, 229–240.
62. X. H. Xia, *Electrochim. Acta* **1999**, *45*, 1057–1066.
63. M. J. Llorca, E. Herrero, J. M. Feliu et al., *J. Electroanal. Chem.* **1994**, *373*, 217–225.
64. E. Herrero, M. J. Llorca, J. M. Feliu et al., *J. Electroanal. Chem.* **1995**, *394*, 161–167.
65. J. Clavilier, R. Parsons, R. Durand et al., *J. Electroanal. Chem.* **1981**, *124*, 321.
66. M. Zhang, C. P. Wilde, *J. Electroanal. Chem.* **1995**, *390*, 59–68.
67. X. H. Xia, T. Iwasita, *J. Electrochem. Soc.* **1993**, *143*, 2559–2565.
68. T. Iwasita, R. Dalbeck, E. Pastor et al., *Electrochim. Acta* **1994**, *39*, 1817–1823.
69. J. Clavilier, *J. Electroanal. Chem.* **1987**, *236*, 87.
70. S.-C. Chang, Y. Ho, M. J. Weaver, *Surf. Sci.* **1992**, *265*, 81–94.
71. C. A. Lukas, N. M. Marković, B. N. Grgur et al., *Surf. Sci.* **2000**, *448*, 65–76.
72. C. A. Lukas, N. M. Marković, P. N. Ross, *Surf. Sci.* **2000**, *448*, 77–86.
73. S. P. E. Smith, H. D. Abruña, *J. Electroanal. Chem.* **1999**, *467*, 43–49.
74. E. Leiva, T. Iwasita, E. Herrero et al., *Langmuir* **1997**, *13*, 6287–6293.
75. J. M. Feliu, A. Fernandez-Vega, J. M. Orts et al., *J. Chim. Phys.* **1991**, *88*, 1493–1518.
76. N. Kizhakevariam, M. J. Weaver, *Surf. Sci.* **1994**, *310*, 183–197.
77. S. Motto, M. Watanabe, *J. Electroanal. Chem.* **1979**, *98*, 203.
78. G. A. Attard, A. Banister, *J. Electroanal. Chem.* **1991**, *300*, 467.
79. M. Baldauf, D. M. Kolb, *J. Phys. Chem.* **1996**, *100*, 11 375–11 381.
80. L. A. Kilber, M. Kleinert, R. Randler et al., *Surf. Sci.* **1999**, *443*, 19–30.
81. H. Naohara, S. Ye, K. Uosaki, *J. Phys. Chem. B* **1998**, *102*, 4366.
82. K. Uosaki, S. Ye, H. Naohara et al., *J. Phys. Chem. B* **1997**, *101*, 7566.
83. E. Herrero, K. Franszczuk, A. Wieckowski, *J. Electroanal. Chem.* **1993**, *361*, 269.
84. W. Chrzarnowski, H. Kim, A. Wieckowski, *Catal. Lett.* **1998**, *50*, 69.
85. M. J. Llorca, A. J. Feliu, A. Aldaz et al., *J. Electroanal. Chem.* **1994**, *376*, 151–160.
86. Yu. B. Vassilyev, V. S. Bagotzky, O. A. Khazova, *Elektrokhimiya* **1975**, *11*, 1505.
87. K. Franaszczuk, E. Herrero, P. Zelenay et al., *J. Phys. Chem.* **1992**, *96*, 8509.
88. E. Herrero, P. Zelenay, A. Wieckowski, *J. Phys. Chem.* **1994**, *98*, 5074.
89. S. Wilhelm, T. Iwasita, W. Vielstich, *J. Electroanal. Chem.* **1987**, *238*, 383–391.
90. B. Beden, F. Hahn, S. Juanto et al., *J. Electroanal. Chem.* **1987**, *225*, 215–225.
91. S. Juanto, B. Beden, F. Hahn et al., *J. Electroanal. Chem.* **1987**, *237*, 119–129.
92. T. Iwasita, F. C. Nart, B. Lopez et al., *Electrochim. Acta* **1992**, *37*, 2361–2367.
93. R. J. Nichols, A. Bewick, *Electrochim. Acta* **1988**, *33*, 1691–1694.
94. H. Kita, H.-W. Lei, *J. Electroanal. Chem.* **1995**, *388*, 167–177.
95. T. Iwasita, X. Xia, E. Herrero et al., *Langmuir* **1996**, *12*, 4260–4265.

96. S. Srimarlu, T. D. Jarvi, E. M. Stuve, *J. Electroanal. Chem.* **1999**, *467*, 132–142.
97. Yu. B. Vassiliev, V. S. Bagotskii, N. Y. Osetrova et al., *J. Electroanal. Chem.* **1979**, *97*, 63.
98. R. R. Adžić, *Isr. J. Chem.* **1979**, *18*, 166–181.
99. M. Shibata, S. Motoo, *J. Electroanal. Chem.* **1987**, *229*, 385.
100. Yu. B. Vassiliev, V. S. Bagotskii, N. Y. Osetrova et al., *J. Electroanal. Chem.* **1979**, *97*, 63
101. S. Motoo, M. Watanabe, *J. Electroanal. Chem.* **1976**, *69*, 429.
102. M. Watanabe, S. Motoo, *J. Electroanal. Chem.* **1975**, *60*, 259–267.
103. B. Beden, F. Kadirgan, C. Lamy et al., *J. Electroanal. Chem.* **1982**, *142*, 171.
104. G. Kokkinidis, D. Jannakoudakis, *J. Electroanal. Chem.* **1983**, *153*, 185–200.
105. S. A. Campbell, R. Parsons, *J. Chem. Soc., Faraday Trans.* **1992**, *88*, 833–841.
106. E. Lamy-Pitara, L. El Ouazzani-Benhima, J. Barbier et al., *J. Electroanal. Chem.* **1994**, *372*, 233–242.
107. A. N. Haner, P. M. Ross, *J. Phys. Chem.* **1991**, *95*, 3740.
108. T. Frelink, W. Vischer, A. P. Cox et al., *Electrochim. Acta* **1995**, *40*, 1537–1543.
109. T. Frelink, W. Vischer, J. A. R. van Veen, *Surf. Sci.* **1995**, *335*, 353–360.
110. B. Bittins-Cattaneo, T. Iwasita, *J. Electroanal. Chem.* **1987**, *238*, 151.
111. M. Shibata, N. Furuya, M. Watanabe et al., *J. Electroanal. Chem.* **1989**, *263*, 97–102.
112. P. Shiller, A. B. Anderson, *Surf. Sci.* **1990**, *236*, 225–232.
113. A. Anderson, E. Grantscharova, S. Seong, *J. Electrochem. Soc.* **1996**, *143*, 2075–2082.
114. S. Motoo, M. Watanabe, *J. Electroanal. Chem.* **1980**, *111*, 261.
115. M. Watanabe, M. Shibata, S. Motoo, *J. Electroanal. Chem.* **1985**, *187*, 161.
116. S.-C. Chang, M. J. Weaver, *Surf. Sci.* **1991**, *241*, 11.
117. B. E. Hayden, A. J. Murray, R. Parsons et al., *J. Electroanal. Chem.* **1996**, *409*, 51.
118. W.-F. Lin, S. Sun, Z.-W. Tian, *J. Electroanal. Chem.* **1994**, *364*, 1–7.
119. E. Herrero, A. Rodes, J. M. Perez et al., *J. Electroanal. Chem.* **1995**, *393*, 87–96.
120. H. Massong, S. Tillman, T. Langkau et al., *Electrochim. Acta* **1998**, *44*, 1379–1388.
121. P. Berenz, S. Tillmann, H. Massong et al., *Electrochim. Acta* **1998**, *43*, 3035–3043.
122. K. Wang, H. A. Gasteiger, N. M. Marković et al., *Electrochim. Acta* **1996**, *41*, 2587–2593.
123. S. Motoo, M. Shibata, *J. Electroanal. Chem.* **1982**, *139*, 119.
124. G. Kokkinidis, D. Jannakoudakis, *J. Electroanal. Chem.* **1982**, *133*, 307.
125. R. R. Adžić, M. S. Levata, *J. Serb. Chem. Soc.* **1982**, *47*, 83.
126. F. Kadirgan, B. Beden, C. Lamy, *J. Electroanal. Chem.* **1983**, *143*, 135.
127. K. Takamura, M. Sakamoto, *J. Electroanal. Chem.* **1980**, *113*, 273.
128. G. Kokkinidis, J. M. Leger, C. Lamy, *J. Electroanal. Chem.* **1988**, *242*, 221.
129. S. B. Brummer, M. J. Turner, H. Feng, *J. Appl. Electrochem.* **1976**, *6*, 377.
130. H. Druliolle, K. B. Kokoh, B. Beden, *J. Electroanal. Chem.* **1995**, *385*, 77–83.
131. G. Kokkinidis, E. Coutouli-Argilopoulou, *Electrochim. Acta* **1985**, *30*, 1611.
132. C. Hasiotis, G. Kokkinidis, *Electrochim. Acta* **1992**, *37*, 1231.
133. G. Kokkinidis, G. Papanastatiou, *Electrochim. Acta* **1989**, *34*, 803.
134. G. Kokkinidis, A. Papoutis, G. Papanastasiou, *J. Electroanal. Chem.* **1993**, *359*, 253.
135. G. Kokkinidis, K. Jüttner, *Electrochim. Acta* **1981**, *26*, 971.
136. G. Kokkinidis, K. Hasiotis, D. Sazou, *Electrochim. Acta* **1990**, *35*, 1957–1964.
137. A. Papoutsis, G. Kokkinidis, *J. Electroanal. Chem.* **1994**, *371*, 231–239.
138. H. Druliolle, K. B. Kokoh, B. Beden, *J. Electroanal. Chem.* **1995**, *385*, 77–83.
139. N. Furuya, S. Motoo, *J. Electroanal. Chem.* **1979**, *100*, 771.
140. N. Furuya, S. Motoo, *J. Electroanal. Chem.* **1982**, *139*, 105.
141. N. Furuya, S. Motoo, *J. Electroanal. Chem.* **1976**, *72*, 165–170.
142. N. Furuya, S. Motoo, *J. Electroanal. Chem.* **1979**, *98*, 195.
143. N. Furuya, S. Motoo, *J. Electroanal. Chem.* **1979**, *99*, 19.
144. R. R. Adžić, M. D. Spasojević, A. R. Despić, *Electrochim. Acta* **1979**, *24*, 569–576.
145. R. R. Adžić, M. D. Spasojević, A. R. Despić, *Electrochim. Acta* **1979**, *24*, 577–579.
146. J. A. M. Abd. E.-Halim, K. Jüttner, W. J. Lorenz, *J. Electroanal. Chem.* **1979**, *106*, 193.
147. E. Protopopoff, P. Marcus, *J. Chim. Phys.* **1991**, *88*, 1423–1452.

148. G. A. Martin in *Metal-Support and Metal-Additive Effects in Catalysis* (Eds.: B. Imelik, C. Naccache, G. Coudurier et al.), Elsevier, Amsterdam, 1982, p. 315.
149. J. Clavilier, J. M. Feliu, A. Fernandez-Vega et al., *J. Electroanal. Chem.* **1990**, *294*, 193–208.
150. R. Gomez, A. Fernandez-Vega, J. M. Feliu et al., *J. Phys. Chem.* **1993**, *97*, 4769–4776.
151. R. Gomez, J. M. Feliu, A. Aldaz, *Electrochim. Acta* **1997**, *42*, 1675–1683.
152. N. Furuya, S. Motoo, *J. Electroanal. Chem.* **1977**, *78*, 243–256.
153. L. J. Gao, B. E. Conway, *J. Electroanal. Chem.* **1995**, *395*, 261–271.
154. R. R. Adžić, J. X. Wang, O. M. Magnussen et al., *J. Phys. Chem.* **1996**, *100*, 14 721–14 725.
155. R. R. Adzic, A. V. Tripkovic, N. M. Marković, *J. Electroanal. Chem.* **1980**, *114*, 21.
156. R. R. Adžić, A. V. Tripkovic, R. T. Anatasoski, *J. Electroanal. Chem.* **1978**, *94*, 231.
157. D. M. Dražić, L. Vorkapić, *Corros. Sci.* **1978**, *18*, 907.
158. K. Juttner, *Werkst. Korros.* **1980**, *31*, 358.
159. D. M. Dražić, V. Nakic, 29th ISE Meeting, Budapest, Aug. 25, **1978**.
160. D. M. Dražić, C. S. Hao, *Corros. Sci.* **1983**, *23*, 683.
161. E. B. Yeager, *Electrochim. Acta* **1984**, *29*, 1527.
162. R. Amadelli, J. Molla, P. Bindra, E. Yeager, *J. Electrochem. Soc.* **1981**, *128*, 2706.
163. R. R. Adžić, N. M. Marković, A. V. Tripković, *Bull. Soc. Chim. Belg.* **1980**, *45*, 399.
164. J. D. E. McIntyre, W. F. Peck Jr. in *Proceedings of the 3rd Symposium on Electrode Processes* (Eds.: S. Bruckenstein, B. Miller, J. D. E. McIntyre et al.), The Electrochemical Society, Princeton, 1980, Vol. 80–3.
165. R. Amadelli, N. M. Marković, R. R. Adžić et al., *J. Electroanal. Chem.* **1983**, *159*, 391.
166. R. Amadelli, J. Molla, E. Yeager, *J. Electroanal. Chem.* **1981**, *126*, 265.
167. N. Anastasijević, Z. M. Dimitrijević, R. R. Adžić, *Electrochim. Acta* **1992**, *37*, 457–464.
168. S. M. Sayed, K. Jüttner, *Electrochim. Acta* **1983**, *28*, 1635.
169. K. Jüttner, *Electrochim. Acta* **1984**, *29*, 1597–1604.
170. R. R. Adžić, J. Wang, B. M. Ocko, *Electrochim. Acta* **1995**, *40*, 83–89.
171. A. Zwetanova, K. Jüttner, *J. Electroanal. Chem.* **1981**, *119*, 149–164.
172. G. Kokkinidis, D. Jannakoudakis, *J. Electroanal. Chem.* **1984**, *162*, 163.
173. P. Chartier, A. Sehiili, H. Nguyen Cong, *Electrochim. Acta* **1983**, *28*, 853.
174. R. R. Adžić, A. R. Despić, Z. Phys. Chem. N.F. **1975**, *98*, 95.
175. S. A. S. Machado, A. A. Tanaka, E. R. Gonzales, *Electrochim. Acta* **1994**, *39*, 2591.
176. T. Abe, G. M. Swain, K. Sashikata et al., *J. Electroanal. Chem.* **1995**, *382*, 73.
177. N. Marković, E. Yeager, R. R. Adžić, *ESM Extended Abstracts* **1985**, *85*(1), 919.
178. R. R. Adžić in *Electrocatalysis* (Eds.: J. Lipkowski, P. Ross), Wiley-VCH Press, New York, 1998, pp. 197–242.
179. R. R. Adžić, J. X. Wang in *Oxygen Electrochemistry* (Eds.: R. R. Adžić, F. C. Anson, K. Kinoshita), The Electrochemical Society, Pennington, 1996, p. 61, Vol. 95-26.
180. T. Abe, Y. Miki, K. Itaya, *Bull. Chem. Soc. Jpn.* **1994**, *67*, 2075.
181. Z. Shi, J. Lipkowski, *J. Electroanal. Chem.* **1994**, *365*, 303.
182. M. F. Toney, J. N. Howard, J. Richer et al., *Phys. Rev. Lett.* **1995**, *75*, 4472–4475.
183. R. R. Adžić, J. X. Wang, *J. Phys. Chem.* **1998**, *102*, 8988–8993.
184. C. -H. Chen, A. A. Gewirth, *J. Am. Chem. Soc.* **1992**, *114*, 5439–5540.
185. I. Oh, M. E. Biggin, A. A. Gewirth, *Langmuir* **2000**, *16*, 1397–1406.
186. C.-H. Chen, N. Washburn, A. A. Gewirth, *J. Phys. Chem.* **1993**, *97*, 9754–9760.
187. D. E. J. Schiffrin in *Electrochemistry, Specialist Periodical Reports* (Ed.: D. Pletcher), Royal Society of Chemistry, London, 1983, p. 126, Vol. 8.
188. M. Alvarez, K. Jüttner, *Electrochim. Acta* **1988**, *33*, 33–39.
189. M. F. Toney, J. G. Gordon, M. Samant et al., *J. Phys. Chem.* **1995**, *99*, 4733–4744.
190. C.-H. Chen, N. Washburn, A. A. Gewirth, *J. Phys. Chem.* **1993**, *97*, 9754–9760.
191. R. R. Adžić, J. X. Wang, *J. Phys. Chem.* **2000**, *104*, 869–872.
192. R. R. Adzic, J. X. Wang, in preparation.
193. K. Tamura, B. Ocko, J. Wang et al., *J. Phys. Chem.* **2002**, *106*, 3896–3901.
194. J. D. E. McIntyre, W.F, Peck, *J. Electrochem. Soc.* **1976**, *123*, 1800.
195. Dj. Davidović, R. R. Adžić, *Electrochim. Acta* **1988**, *33*, 103–108.
196. Dj. Davidović, R. R. Adžić, *J. Serb. Chem. Soc.* **1988**, *53*, 499–510.

197. M. Copel, M. C. Reuter, E. Kaxiras et al., *Phys. Rev. Lett.* **1989**, *63*, 632.
198. K. Sieradzki, S. R. Branković, N. Dimitrov, *Science* **1999**, *284*, 138–141.
199. S. R. Branković, N. Dimitrov, K. Sieradzki, *Electrochem. Solid State Lett.* **1999**, *2*, 443–445.

Index

a

adlayer 404–413, 415, 417f., 423, 429f., 432–436, 439, 448, 450, 455, 476, 478f., 481f., 490, 500, 506, 508, 561f., 564, 566–569, 571, 573f., 576–585, 587ff., 592, 594
Adžić, R. R. 365, 417, 455, 561, 564, 569, 590, 592
Alexander (S.) regime 316
Alexejev, Yu. V. see also Kolotyrkin–Alexejev–Popov theory
Alleen, M. P. 138
Alvarez, M. 589
Amman, E. 432
Amokrane (S.) and Badiali, phenomenological theory 148f., 208f.
amphiphilic isotherm 178f.
Andelman, D. 282
Ångstrøm, J. A. 145
Ariel, G. 319
Armstrong, R. D. 389f.
Aspnes, D. E. 80
Ataka, A. 418, 455
atomic force microscopy (AFM) 188, 191, 199, 204f., 225–248, 371, 419, 454, 471, 521, 546, 549, 551, 588
atomic layer epitaxy (ALE) see also EC–ALE 513, 515, 526, 533
auger electron spectroscopy (AES) 188, 371f., 418, 423, 425, 430, 435, 496, 592
Avrami theorem (Avrami, M.), statistical law 397–400, 403, 448

b

Badiali, J. P. 140, 156, 402f.
Bagotskaja, I. A. 192
Bagotzky, V. S. 574
Baltruschat 578
Bard, A. J. 253, 506

Barrat, J.-L. 289
Batrakov, V. V. 435
Bauer, E. 475
Becker, R. 390f.
Behem's group 113, 497
Bellier, J. P. 204
Bérard, D. R. 149
Berkowitz, M. see Xia and Berkowitz
Bewick, A. 417, 566, 574
Bewick–Fleischmann–Thirsk (BFT) theory 419, 427
Birshtein, T. M. 316
Bittins–Cattaneo, B. 576
Bjerrum, N. 43
– length 104, 106, 287
– theory 174
Bloch electrons 72
Blum, L. 419
Bockris, J. O'M. see also double-layer model by Bockris–Devanathan–Müller 253
Bogolyubov–Born–Green–Ivon theory 186
Boltzmann, L. 42f., 70, 338, 392
– constant 99, 134, 200, 231, 261, 263, 287
– distribution 99, 141, 305
Borisov, O. V. 316
Borkowska, Z. 204
Born, M. 179, 181, 225
Born–Green–Ivon equation 182
Born–Oppenheimer approximation 84
Borukhov, I. 319
Bosco, E. 400
Bosio, L. 87
Bostanov, V. 395
boundary layer 140, 142, 155f., 162, 179
Bragg's law (Bragg, W.) 555
Bragg–Williams approximation (BWA) 388, 437f.
Buess-Herman, C. 393, 402
Burak, Y. 319
Burgess, I. 246

Butler–Volmer equation 42, 53, 271, 273, 278
Butt, H.-J. 225

c
Campbell, S. D. 241
capillary waves 118f., 121f., 156
Car–Parinello simulation method (Car, R./Parinello, M.) 84, 139, 153
catalysis 253, 279, 471, 475, 494, 564, 566, 569, 571ff., 576, 578, 588
Cathro, K. J. 566
Chao, D. Y. C. 202
Chapman, D. L. see also GCSG 42, 248
chemical polished (CP) 202f.
chemical vapor deposition (CVD) 471, 516
chronocoulometry 404, 410f., 413, 415, 419, 448
Churaev, N. V. 227
CIS formation 550f.
classical nucleation theory (CNT) 390, 393
Clavilier J. 204, 213, 471, 479, 568, 571
commensurate/uniaxially incommensurate transition (C/UIC) 387, 412f.
common diffuse layer (CDL) 63, 198
compact layer 40, 42ff., 49, 51f., 54f., 58f., 62, 64, 70, 94f., 97–104, 110f., 124, 135, 164, 170, 173f., 176, 184, 186, 198, 350
compact-layer capacitance 44, 46f., 51f., 54, 59, 61f., 64f., 69, 71, 75ff., 99f., 103, 108, 110, 124, 171
compact-layer differential capacity 78
constant face element (CFE) 201
Conway, B. E. 344, 566
Cottrell equation 389
Coulomb, Ch. 3, 90, 104, 138, 173, 183, 282f., 287f., 308f., 311, 315, 334
counterion 283, 294, 304, 316f., 355
crystallographic
 – defect 497
 – inhomogeneity effect 38, 61f., 195f., 200
 – mismatch 471f.
 – orientation 61, 65, 195, 202, 205, 214f., 218, 383, 424, 436, 454, 500
 – structure 201, 204, 208, 211
Cuesta, A. 412
Cui, Q. 186
CV see voltammetry, cyclic

d
Daikhin, L. I. 153, 156, 199
Damaskin, B. B. 206, 327, 360, 454
Davidović, Dj. 592

Debye, P. 35, 38, 43, 248
 – plasma 122
 – (screening) length 57, 59f., 64, 66f., 69, 93, 101, 104, 120, 134, 142, 153, 156f., 165, 198f., 200, 237, 241, 243f., 269
Debye–Falkenhagen time 38
Debye–Hückel (DH) theory for electrolyte solutions 82, 99, 134, 184f., 226, 287f., 290, 296f., 300, 304, 314
density profiles 90ff.
Delahay, P. 388f.
Derjaguin, B. see DLVO 227, 236
 – Derjaguin's approximation 238f.
Despić, A. R. 564
Devanathan, M. A. V. see double-layer model by Bockris–Devanathan–Müller
differential electrochemical mass spectrometry (DEMS) 566, 569, 574
diffuse layer 41, 43ff., 49, 51, 54f., 57–62, 64ff., 94f., 97–101, 104, 106f., 110, 136, 163–166, 170, 176, 180, 182, 184, 186, 189, 195ff., 200, 204, 210, 334, 336, 350f., 356f., 358
diffuse-layer capacitance 44, 46, 48, 51, 53, 59, 65ff., 101, 108, 166f., 174, 182, 189, 192, 193, 195f., 199, 210, 350
Dimitrov (D. I.) and Raev study 90
direct methanol fuel cell (DMFC) 494
discreteness-of-charge effect 97f.
discreteness-of-charge factor 97ff.
dispersion 45, 226, 229, 242
dissociation constant 238
DLVO theory see Derjaguin–Landau–Verwey–Overbeek 226–229, 239ff.
DNA 303, 319
Doering (D. L.) and Madey model 151
Döhring, W. 390f.
Döppenschmidt, A. 241
Donnan, F. G. 12, 35
 – equation 35
double-layer capacitance/capacity 42, 66, 122, 136, 143, 172, 213, 343, 406
double-layer model by Bockris, Devanathan, and Müller 83
double layer structure 151, 175, 242, 269, 359, 364, 375
double layer theory 133, 136, 144, 149, 153, 158f., 349f., 359, 372
Dražić, D. M. 365, 584
Dretschkow, Th. 450, 455
Drudelike model 80
Druliolle, H. 581
Drummond, C. L. 240, 244
Ducker, W. A. 243f.

Dutkiewicz, E. see also Hurwitz–Parsons–Dutkiewicz method   356

e
Ehrenfest, P.   385
electric double layer (EDL/edl)   7, 9, 20, 35ff., 40ff., 53, 56f., 58–62, 64, 82–84, 97f., 107, 109f., 117, 133, 139, 144, 162–165, 169f., 173, 182, 188, 198, 201, 203ff., 210, 213, 225ff., 240, 242f., 334, 338, 343, 371f.
electrocapillary equation   167f.
electrochemical atomic layer epitaxy (EC–ALE)   513, 515–523, 527, 530, 533ff., 541–544, 549–555
electrochemical cells (EC) see also solid-state electrochemical cells   253ff., 261–264, 266, 268, 271, 278ff., 499, 574
electrochemical grown (EG)   202f., 206f.
electrochemical impedance spectroscopy (EIS)   201
electrochemical polished (EP)   202f., 204–208
electrochemical quartz crystal microbalance (EQCM) method   370, 373, 425, 471, 476, 478, 480f., 503, 509, 569
electrochemical quartz crystal nanobalance electrode (EQCN) method   370f.
electrocompression   562, 584
electrode polarization   66, 164, 173, 364
electrode potential
 – absolute   18, 20
 – fixed   107
 – relative   18
electrolyte
 – metal–liquid (LE)   33f.
 – metal–solid (SE)   33ff.
electromotive force (emf)   5, 7f., 16
electron energy levels   19
electron spin resonance (ESR)   267
electron transfer (ET)   42, 53f., 84, 93, 174, 268, 275–277, 179
electronic conduction   258, 261, 265, 267
electronic spillover effect   83
electronic polarizability   143, 145
electroosmosis   56
electrophoresis   57
electroreflectance of s-polarized light   80f., 112, 404
electrostatic double-layer force   225–228, 235, 237ff., 241, 243, 248
electrostatic repulsion   57, 226, 234, 241, 244, 246, 282, 288f., 292, 303, 310, 315
ellipsometry   366f., 538f., 562
EMIRS see IRAS
EPMA analysis   530, 536, 543f., 546, 551f.

Erdey see Volmer–Erdey–Gruz theory
Esin (O. A.) and Markov coefficient/effect   97, 332, 344, 355, 358, 364
Euclid   67
Ewald summation   86, 88, 90
EXAFS experiments   419, 423, 430, 454

f
Faguy, P. W.   364
Faraday, M.   33, 172, 191, 201, 213, 261, 370, 389f., 399ff., 416
 – constant   94
Farkas, L.   390
Faulkner, L. R.   253
Fermi, E.   19, 35, 117
 – level   262, 275, 564
Fick's first law   389
field-emission displays (FED)   544
Fielden, M. L.   244
Finnefrock, A. C.   418
Fisher, M.   386
Fixman, M.   290
Fleischmann, M. see also BFT   393, 419
Flory theory (Flory, P. J.)   294
Flory–Huggins   97
Flowers, B. H.   513
Fonseca, I.   564
Foresti, M. L.   204, 521, 527, 532, 543f.
Foster, K.   150
Fourier, J.-B. J.   118f., 156, 296
 – transform (FT)   66, 154, 297, 482
 – FT-IRRAS see also IRRAS   494, 496
 – transform infrared (FTIR) spectroscopy see also IR spectroscopy, SNFTIRS   213, 364, 367, 383, 419, 448, 454
Fowler, R.   337
fractal carpet model   201
Frelink, T.   576
Frank (F. C.)–van der Merwe (FM) model   387, 407, 472, 474, 478, 490, 494, 509
Frenkel, Y. I.   390, 407
 – pairs   259, 267
Frenkel–Kontorova model   387, 407
Freundlich isotherm   178
Friedel oscillation   72
Friedrich, K. A.   478, 494
Frumkin, A.   22, 54, 56, 61, 97f., 162, 188, 191, 208, 211, 218, 331f., 337, 361, 388, 404, 430, 434, 437, 439, 440, 454
 – Frumkin's isotherm   106f., 176, 178f., 190, 337f., 354, 366, 389, 410, 437
Funke, K.   267
Furuya, N.   581

g

Gaber see Luggin–Gaber
Galvani, L. (potential)   4ff., 9, 18, 20, 35, 37, 39, 165, 186, 253, 262, 269ff., 277f.
galvanostatic–intermittent–titration–technique (GITT)   268
Gao, X.   112, 204
Gauß, C. F.   67, 199f., 284, 288f., 292f., 295, 297, 300f., 314
Gavach, C.   163
Gerber, C.   225
Gewirth, A. A.   588, 590
Gibbs, J. W.   4, 8, 12, 16, 39, 41, 179, 181f., 221, 364, 366, 390, 404, 437
– adsorption   40f., 52, 93f., 167, 177, 190, 221, 327, 339f., 349, 355f., 439
– electrocapillary equation   93, 338, 341, 349, 361, 436
– thermodynamics   39, 167
Gibbs–Duhem equation/identity   93, 327, 352
Gileadi, E.   253, 344
Girault, H. H.   155
Gomez, R.   213, 490
Grahame, D. C.   39, 43, 158f., 162, 334, 454
– ansatz   69
– Graham's curve   71
– model   44, 46, 48ff., 98
– parameterization   70
– treatment   59
Grahame–Parsons
– isotherm   106, 109
– theory/treatment   99, 103, 107, 110, 151, 338, 346
Grotthuss mechanism   258, 261
Guggenheim, E. A.   4, 337
Gouy, G.   42, 248, 330, 454
– length   123, 199
– theory   193
Gouy–Chapman (GC) model   43, 46, 49ff., 61, 70, 82, 85, 93f., 107, 121ff., 133–136, 140ff., 153, 155, 158, 163, 165, 171, 182f., 185f., 189, 196, 199, 300, 317, 349, 351, 356
Gouy–Chapman–Stern (GCS) model   69, 85
Gouy–Chapman–Stern–Grahame (GCSG) model   42, 44, 99, 189, 191, 193, 196, 204, 206
Guidelli, R.   136, 400

h

Hall coefficient   267
Halley, J. W.   148, 153
Halperlin, B. I. see also KTHNY theory
Hamaker constant   234, 239

Hamelin, A. see also Valette–Hamelin approach   113, 204, 363
Hamiltonian model   140f., 385f.
Hamm, U. W.   213
Hammett, A.   16
Hansen's method   167
Happek, U.   513
Haussdorf fractal dimension   67
Hebb (M. H.)–Wagner polarization method   267f.
Heinz, W. F.   246
Heitbaum, J.   565
Helmholtz, H. v.   42ff., 401, 437, 445
– capacity   135f., 142–145, 149
– layer   71, 164f., 189, 400
– model   41, 44, 52
– plane   54, 57, 97f., 101, 104, 111, 164, 173f., 273, 334, 337, 350f., 563
– surface   68
Helmholtz–Stern model   110
Henderson, P.   9, 85, 143, 145, 148
Henry isotherm   96, 102, 178, 354, 387
hexagonal close-packed system (hcp)   206
highest occupied molecular orbital (HOMO)   16
high salt (HS I/II) regime   312f.
Hillier, A. C.   241f.
Hölzle, M. H.   422, 424, 428, 431, 433, 443, 455
Hoh, J. H.   246
Horányi, G.   349
Houng, N. V.   204
Hubbard, A. T.   371
Huber, T.   158
Hückel molecular-orbital method see also Debye–Hückel theory   564
Hurwitz, H. D.   356, 358
Hurwitz–Parsons–Dutkiewicz method   366
hydration force   228f., 239, 248
hydrophilicity (of electrode)   205f., 210, 219, 221f.
hydrophobic anion   173
hydrophobic ion   117f., 123f.
hydrophilic ion   117f.
hypernetted chain approximation (HCA/HNC)   137f., 182, 186

i

Ibach, H.   455
ideal polarizability interval   36ff.
Ignaczak, A.   412
immersion method (IM)   213
independent diffuse layers (IDL)   61, 196ff.

inductively coupled plasma mass spectroscopy (ICP–MS)   545, 555, 557
infrared (IR) reflection absorption spectroscopy (IRAS/IRRAS) see also Raman spectroscopy   365, 367, 404, 415, 425, 435, 447, 490, 565, 571
– electrochemically modulated infrared spectroscopy (EMIRS)   367, 566
– polarization modulation infrared reflection absorption spectroscopy (PM–IRAS)   367, 435
– synchrotron far infrared spectroscopy (SFIRS)   367
inner-layer capacitance/capacity   189, 191f., 195ff., 202, 205, 210, 214, 219, 350f., 366
interface between two immiscible electrolyte solutions (ITIES)   33f., 117, 122f., 162, 163f., 168, 172, 182, 186
interface
– nonpolarizable   162ff., 167, 188
– polarizable   162f., 167, 169, 188
interfacial electrochemistry   125
ion transfer   274f., 277f.
Ishino, T.   241
Ising, anisotropic 2D model   112, 114f., 117, 136, 156, 385, 437
Ising model phase diagram   115
Itaya, K.   435
Ito, M.   365
IUPAC definitions/recommendations   349, 360
Iwasita, T.   213, 497, 564, 569, 574, 576

j

Janssen, M. M. P.   568, 576
jellium model   72, 75f., 81, 83, 87, 139, 144–149, 159
Joanny, J. F.   289, 319
Johnson, K. L. see Johnson–Kendall–Roberts theory
Johnson–Kendall–Roberts theory   230
Jüttner, K.   589

k

Kaishev, R.   390
Kanevskii, E. A.   20
Kardar, M.   407
Kékicheff, P.   244
Kendall, K. see Johnsohn–Kendall–Roberts theory
Kharkats, Yu.   439
Kharkats–Ulstrup model   180ff.
Khachaturian, K. A.   292

Khmelevaya, L. P.   206
Khokhlov, A. R.   292
Kibler, L.   418
Kirkwood, J. G.   184, 186
Kita, H.   569
Kittel, C.   19
Kizhakevariam, N.   573
Klein, J.   228
Koenig, F. O.   162, 169
Kohn, W.   72
Kohn–Sham scheme/calculation   72f., 113
Kokai, U.   471
Kokkinidis, G.   564, 579, 588
Kolb, D. M.   204, 372, 455, 478, 480f., 485, 489ff., 508
Kolotyrkin (Ya. M.), Alexejev, and Popow, model   99, 338
Kontorova, T. see Frenkel–Kontorova model
Koper, M. T. M.   410f.
Kornyshev, A. A.   33, 114, 144
Kornyshev–Vilfan theory   114, 116
Korotkov, A. P.   206
Koryta, J.   388
Kosterlitz (J. M.)–Thouless universality class   117
Kosterlitz–Thouless–Halperlin–Nelson–Young theory (KTHNY)   404
Krämer, D.   151
Kroger, F. A.   254, 533
Kröger–Vink notation   259, 261, 263
Kuhn length   283, 286, 305, 317
Kunze, K. K.   319

l

Landau, L. D. see also DLVO
Landau theory   407
Landau–Lifshitz theory see also Lifshitz   385
Lang model (Lang, M. D.)   81
Langmuir, I.   387ff., 413, 419f.
– isotherm   178, 337, 353, 354, 401
Langmuir–Blodgett (LB) film   117, 235, 384, 434
Laplace equation (Laplace, de P. S.)   179, 243
lattice gas model (LGM)   155f., 158, 410, 437f., 440, 454, 589
Lazarova, E. M.   211,
Le Chatelier's principle   143
Lecoeur, J.   204
Leibler, L.   319
Leikis, D.   192
Leiva, E.   571
Leiva–Schmickler model   202, 206, 208, 210
Lennard–Jones interaction/potential   83, 150f., 409

Levie, R. de 400, 455
Lifshitz theory (E. M. Lifshitz) 227
Lipkowski, J. 319, 364, 366, 372
Lippmann equation of electrocapillarity
  (G. Lippmann) 38, 164, 331, 340, 362
liquid electrolyte/liquid electrolytes (LE)
  172, 254–259, 261f., 265, 267f., 270ff.
liquid junction (LJ) 36
liquid-state electrochemistry (LSE) 253,
  256f., 261f., 264, 266–272, 275f., 279
Llorca, M. J. 479
local density approximation 144
Lorenz, W. J. 342, 359f., 389f., 394, 400f.,
  417, 430, 434
low-energy electron diffraction (LEED) 188,
  202, 385, 407, 409, 417, 419, 423, 425f.,
  430, 499, 502
lowest unoccupied molecular orbital
  (LUMO) 16
low-temperature series expansion (LTSE)
  388, 438
Lucas, C. A. 577
Luggin–Gaber 17
Lust, E. 68, 188, 204,

m

Madey, T. E. see Doering and Madey model
Magnussen, O. M. 113, 415, 455
Mandelbrot, B. B. 201
Manne, S. 246
Manning condensation (Manning, G. S.)
  293f.
Marcelja, S. 316
Marcus, P. 581
Markin, V. S. 162, 167
Markov, B. F. see also Esin and Markov
  effect 284
Mayer, D. 455
Mc Ilvaine buffer 440
McIntyre, J. D. E. 80, 589, 592
McTague, J. P. see Novaco–McTague theory
mean field approximation (MFA) 136,
  141, 157, 388, 410f., 437–440, 454
mean spherical approximation (MSA) 85,
  138, 142f., 147, 158, 182
measurement an analysis of surface in-
  teraction forces (MASIF) 228ff., 241
Melik–Gaikazyan, V. I. 389
metallo organic molecular beam epitaxy
  (MOMBE) 525
metallo organic vapor phase epitaxy
  (MOVPE) 525
Michailik, J. 389
microgravimetry 415

Miklavic, S. J. 316
Milchev, A. 393
Miller index (J. D. Miller) 188
Misra, S. 316
Mitchell, S. M. 410
mixed ionic electronic conductor (MIEC)
  256, 258, 260f., 263, 266–271, 273, 278
Miyatani, T. 247
Mössbauer spectroscopy 562
molecular beam epitaxy (MBE) 434, 513–516
molecular dynamics (MD), computer
  simulation method of statistical mechanics 83f., 90, 118f., 123, 125, 220
monolayer (ML) 176f., 183, 191, 208,
  241, 246, 329, 350f., 353, 361f., 383,
  384, 387ff., 391, 396, 398, 400, 402,
  409, 416f., 419, 423, 426, 433f., 436,
  445, 448, 454, 472, 479f., 485, 489f.,
  494, 497, 499, 513, 515, 522, 525, 529f.,
  532, 534, 537f., 540–544, 546f., 553,
  555, 557, 561f., 564, 569, 574, 588, 594
monomer 282–286, 288–291, 294–300,
  304ff., 308–312, 315, 317f., 384, 392ff., 401,
  438
Monte Carlo (MC), computer simulation
  method of statistical mechanics 83ff., 122,
  138, 143, 157f., 186, 284ff., 289ff., 293, 385,
  388, 398, 400, 404, 410f., 416, 420f., 454
Moolhyusen, J. 568, 576
Morse potential 150
Motoo, S. 566, 576f., 581
Moylan, T. E. 213
MPB see Poisson–Boltzmann modified theory
Müller, K. see double-layer model by Bockris–Devanathan–Müller
Mulder W. H. 403
multilayer 176, 282, 313, 315, 345
Myland, J. C. 253
MVN theory see Verwey–Nielsen model

n

Nagy, G. 455
Nagy, Z. 563f.
Nazmutdinow, R. R. 147
Nelson, D. R. see KTHNY theory
Nernst, W. 5, 8, 12, 415
– equation 340, 342, 433
– voltage 262–265
Nernst–Donnan theory 168
Netz, R. R. 282
Newton, I. 139
– Newton's equation of motion 234

Nichols, R. J.   574
Nikitas' model (P. Nikitas)   338
Nielsen, K. F. see Verwey–Nielsen theory
normal calomel electrode (NCE)   11, 357
Norton, M. L.   533
nuclear magnetic resonance (NMR)   16, 267
Novaco (A. D.)–McTague theory   408, 434

o

Ocko, B. M.   113, 410, 455
Odijk, T.   290, 300
Ohm, G. S.   37f.
open-circuit voltage   263f., 268
orientational distribution/structure   88ff.
Orland, H.   319
Ornstein–Zernicke equation/relation   137, 182
osmotic stress method   229, 239
Overbeek, J. T. G. see DLVO theory

p

Palm, U.   208
Parsons, R. see also Hurwitz–Parsons–Dutkiewicz method   136, 356, 358, 360f., 454
Parsons model   98
Parsons–Zobel (PZ), plot method   48, 51ff., 59, 61ff., 76, 124, 135, 191ff., 196, 199f., 206, 210f.
Pauli, W.   35, 76
Pecina, O.   156
Peck, W. F.   589, 592
Pereira, C. M.   158
Pethica, J. B.   235
Petrii, O. A.   327
Philpott simulation study (Philpott, M. R.)   90
piezoelectric translator   236
Pincus, P.   316ff.
Planck, M.   9, 162
Pletcher, D.   569
PM–IRAS see IRAS
Poelman, M.   443
Pohlman, L.   400
Poisson–Boltzmann equation   42, 49, 66, 69, 120, 122f., 134, 155ff., 173, 183f., 237f., 248, 288, 305f.
– modified theory (MPB)   155, 182f., 185f.
Poisson equation   87, 182f., 396
Pokrovsky, V. L.   387, 413
polarization   52, 54f., 66, 71, 89, 102, 111, 171, 179, 183, 258, 287, 329, 339, 344, 349, 362, 403, 592

polycrystalline (PC)   58, 60f., 63, 66, 146, 188, 202–221, 373, 454, 479, 520f., 533, 562, 578, 582, 587
– electrode (surface)   63, 66, 78, 190ff., 194–197f., 200f., 206, 416
polyelectrolyte (PE), polymers   282ff., 287ff., 291–300, 302ff., 307–319
Popov, Yu. A. see also Kolotyrkin–Alexejev–Popov theory   206, 219, 222
position sensitive detector (PSD)   230
potential of zero charge (pzc/PZC/p.z.c.), a fundamental parameter in electrochemistry   33, 38–42, 44, 48f., 51, 54, 58–66, 68, 70f., 76f., 95, 97, 100f., 134, 136, 143, 146f., 149–154, 156, 188, 190, 192, 195, 210f., 214, 216, 220f., 269, 331f., 334, 344, 355, 368, 372, 562ff.
potential of zero free charge (PZFC)   188
potential of zero total charge (PZTC)   188, 198, 213–217
potential window   16, 172
Potts 3-state model   385, 407
Preuss, M.   244
Prica, M.   238
Price, D.   153
Pronkin, S.   455
Protopopoff, E.   581
pseudopotential   148

q

quartz crystal microbalance (QCM) see also EQCM   417, 419
quasi-chemical approximation (QCA)   122, 388, 411, 438, 454
quasi-elastic laser scattering (QELS)   119
Quate, C. F.   225

r

Rabinovich, Y. I.   227
Rademacher, M.   247
Raev, N. D. see Dimitrov and Raev study
Raiteri, R.   241
Rajeshwar, K.   554
Raman spectroscopy see also RRS, SERS   367, 383, 404, 435, 448, 533
random-phase approximation (RPA)   296–299
Rangarajan, S.   400f.
Rayleigh instability   295
Ready, A. K. N.   253
recharging effect   95
redox potential   15ff., 278, 339f., 343
– reaction   562ff.

reference hypernetted chain (RHNC) approximation 138, 149
reflection high-energy electron diffraction (RHEED) 417, 419, 423, 521
regular 2D/3D lattice/density 105f.
resonance Raman scattering (RRS) 367
Retter, U. 393
reversible hydrogen electrodes (RHE) 13, 342f., 426, 475f., 495, 499
Reynolds number 532
Rhodes, A. 563
Riess, I. 253
Rice, O. K. 144
Rickert, H. 253
Rikvold, P. A. 400, 410, 419f.
Roberts, A. D. see Johnson–Kendall–Roberts theory
Roginskii–Zeldovich equation see Zeldovich factor 341
rotaring disk ring electrode (RDRE) measurement 588
rotating ring–disk electrode (RRDE) 534f.
Rotsch, C. 247
roughness
– effect 66
– factor (RF) 58–61, 66, 196f., 199f., 557
– function 67f., 154, 156, 199ff.
– model 199

S

Sader, J. E. 233
Sander, L. M. 394
saturated calomel electrodes (SCE) 11, 16, 204f., 210f., 213, 242, 409, 489, 492
scanning electron microscopy (SEM) 536, 544
scanning tunnelling microscopy (STM) 112f., 188, 191, 202, 204, 365, 371, 394, 401, 403ff., 407f., 411, 415, 418f., 422f., 430, 432f., 435f., 439, 443–448, 450f., 454, 471, 476–486, 488–498, 500–509, 534f., 553, 595
Schick, M. 319
Schiffrin, D. H. 155, 589
Schmalzried, H. 254
Schmickler, W. 133, 145, 148, 158, 202, 564
Schottky pairs (Schottky, W.) 259
Schuhmann, D. 389
Schultze, J. W. 360, 417
Seidel, C. 319
self-assembled monolayers (SAM) see also ML 434, 454
Senden, T. A. 244

sequential monolayer electrodeposition (SMED) 554
SERS see surface-enhanced Raman scattering
SFIRS see IRAS
Sham, L. J. see Kohn–Shame scheme
Shannon, C. 533
Shibata, M. 566, 569
Shingaya, Y. 365
Sieradzky, K. 592
Silva, A. F. 222
Skolnick, J. 290
Smoluchowski equation 56
Soderberg, B. A. 334
solid electrolyte (SE) see also electrolyte 255–258, 261ff., 266–269ff., 273, 277ff.
solid-oxide fuel cells 253
solid-state batteries 253
solid-state (electrochemical) cells see also EC 255f.
solid-state electrochemistry (SSE) 253, 256f., 260–270, 273–279
solid-state sensors 253
Solis, V. 569
Soriaga, M. P. 499
spectroscopy see also IR spectroscopy, Raman spectroscopy 80, 304, 358, 367, 445
Spohr, E. 33, 90, 153, 159
Stafiej, J. 141
Staikov, G. 393
standard hydrogen electrode (SHE) 11f., 15, 21, 203, 207, 212–217
Stern, O. see also GCS 42, 96f., 135, 304, 330
– approximation 98
– theory 97f., 142, 164
Stickney, J. L. 513
Stilinger, F. H. 186
Stockmayer model 82
stoichiometry 259f., 267, 360, 423, 529f., 536, 543, 551, 576
Stokes equation 258
Stoyanov, S. 393
Stranski–Krastanov (SK) mode 390, 472, 474f., 479, 490, 494, 509
subtractively normalized Fourier transform infrared spectroscopy (SNFTIRS) see also Fourier 574
Suggs, D. W. 506
surface charge density 92f.
surface-enhanced infrared reflection absorption spectroscopy (SEIRAS) see also IRAS 383, 400, 415, 418, 420, 435f., 439, 445, 447f., 450, 454

surface-enhanced Raman scattering (SERS) see also RRS 367, 448, 454, 534f., 541, 554
surface forces apparatus (SFA) 228ff., 239, 241
surface polarizability 144, 147f.
surface x-ray scattering (SXS/SXRS) 90, 188, 204, 220, 295f., 368, 383, 385, 400, 404f., 407–413, 418f., 423, 426, 428ff., 434f., 443, 454, 471, 489, 503, 506, 508f., 562, 568, 577, 583, 588, 590f.
Sutton, A. P. 235

t

Taft 16
Talapov, A. L. 387, 413
Tamura, K. 591
Tao, N. J. 435, 455
Teflon capillary 206, 220
Temkin, M. I. 97, 389, 576
 – isotherm 341, 344
Theopilou sum (Theopilou, A. K.) 73
thermal desorption spectroscopy (TDS) 219
thermogravimetry (TGA) 267
thin-layer electrochemical cell (TLEC) 528f., 533ff., 542f., 550
third-body effect 566, 571, 573
Thirsk, H. R. see BFT
Thomas, A. 35
Thomas, F. 441
Thomas–Fermi theory 144
Thouless, D. J. see KTHNY theory
Tildesley, D. J. 138
Toney, M. F. 90, 419, 424, 590
Torii, A. 233
Torrie, G. M. 186
Toshev, S. 392
transitorial layer 162f.
transmission electron micrograph (TEM) 517, 520f., 540
Trasatti, S. 144f., 151, 202, 204, 216, 218
triple phase boundary (TPB) 256, 269, 276
Tsionsky, V. 370
two-dimensional (2D) phase transitions 383, 385

u

Ullmer, M. 319
Ulstrup, L. see Kharkats–Ulstrup model
ultrahigh vacuum (UHV/u.h.v.) conditions 111, 114, 204, 213, 216f., 345, 371f., 385, 404, 409, 417, 419, 423, 425, 430, 435, 451, 454, 475, 479, 490, 499, 534
 – UHV-EC see also EC 534f.

ultra-large scale integration (ULSI) 514
underpotential deposition (UPD) 359–364, 375, 384, 390, 398f., 402, 415–420, 422–434, 472, 479, 490, 513, 515, 521f., 528, 532f., 535, 537, 541, 543, 549, 551, 562, 564, 567f., 571, 574, 576, 578, 580f., 584f., 587, 589f., 592, 594f.
Uosaki, K. 476, 480f., 489f., 498, 524f., 561
Urbakh, M. 156

v

Valette, G. 222
Valette–Hamelin approach 191f., 196, 199, 204
Valleau, J. P. 186
van Krieken, M. 403
van der Merwe, J. H. see Frank–van der Merwe model
van der Waals
 – diameter 406, 506
 – isotherm 354
 – attraction 226
 – force 57, 226f., 231, 239, 241, 386
vapor phase epitaxy (VPE) 513ff.
Vayenas, C. G. 279
Verwey, E. J. W. see DLVO and Verwey–Nielsen theory
Verwey–Nielsen model (MVN) 155f., 158, 163f., 186
Vetter, K. J. 360
Vetterl, V. 434, 443
Vidyanathan, R. 513
Vilfan, I. see Kornyshev–Vilfan theory
Villain, J. 116
Villegas, I. 530, 533, 540
Virial isotherm 354
Vitanov, T. 202, 206
Vogt, M. R. 506
Volkov, A. G. 162, 167
Volmer, M. 390
Volmer isotherm see also Butler–Volmer equation 354
Volmer–Erdey–Gruz theory
Volmer–Weber (VW) mode 472, 474
Volta, (potential) 6f., 18, 20, 22, 190
voltammetry 17, 112, 419, 523, 528, 534, 550, 570, 572, 576, 578f., 589ff.
 – cyclic (CV) 204, 211, 363, 365, 373–377, 404, 406, 415ff., 418f., 423f., 428, 430, 443, 444, 492, 495f., 499f., 503
Vorkapić, L. 584
Vorotynsev, M. A. 33, 204, 338

w

Wade, T. L.   513
Wagner, C. see Hepp–Wagner method
Wagner, F. T.   213, 254
Walton, D.   393
Wandlowski, Th.   383, 409, 435
Wang, J. X.   409, 508, 590
Watanabe, M.   479
Watts–Tobin model (Watts–Tobin, R. J.)   135
Weaver, M. J.   113, 533, 571, 573
Whitney, R. W.   162
Wieckowski, A.   495ff.
Williams, E. see BWA
Witten, T. A.   394

x

Xia, X. H.   213, 569
Xia and Berkowitz   151
x-ray absorption near–edge structure spectroscopy (XANTES)   562
x-ray absorption spectroscopy (XAS)   419, 562f.
x-ray diffraction (XRD)   112, 155, 229, 368, 413, 478, 508, 529f., 536f., 540f., 543ff., 549–552, 555ff., 591
x-ray photoabsorption spectroscopy (XPS)   478, 494, 496f., 569
x-ray scattering see SXS/SXRS

y

Ye, S.   471
Yeager, E.   586, 589
Yoneyama, H.   544
Young, A. P. see KTHNY theory
yttria stabilized zirconia (YSZ), oxygen ion conductor   13, 256, 263, 269, 276f., 279
Yufei, C.   174

z

Zeldovich factor (Zeldovich, J. B.)   390f.
Zhulina, E. B.   316